Applied Mathematics Reviews

Volume 1

Applied Mathematics Reviews

Volume 1

Editor

George A Anastassiou

The University of Memphis, USA

World Scientific

Singapore • New Jersey • London • Hong Kong

Published by

World Scientific Publishing Co. Pte. Ltd.

P O Box 128, Farrer Road, Singapore 912805

USA office: Suite 1B, 1060 Main Street, River Edge, NJ 07661

UK office: 57 Shelton Street, Covent Garden, London WC2H 9HE

British Library Cataloguing-in-Publication Data
A catalogue record for this book is available from the British Library.

APPLIED MATHEMATICS REVIEWS VOL. 1

ISBN 981-02-4339-1

This book is printed on acid and chlorine free paper.

Printed in Singapore by World Scientific Printers

Dedicated to my daughters,
Angela and Peggy

PREFACE

This volume contains fourteen high-quality reviews of main subareas of Applied Mathematics. These are survey articles containing also original results, written by top people of the field from around the world. Applied Mathematics plays a central, very active role in contemporary science, shaping our lifes better. Feeling that this volume is a very important research tool to the use of Applied Mathematicians, we intend to publish in the future subsequent volumes.

In the following we would like to mention the important topics that are reviewed in this volume: convergence of generalized singular integrals, computational concepts in numerical analysis, approximation in modular spaces, simultaneous approximations, minimal projections, copositive approximation, neuro-fuzzy adaptive control, sampling theory, stable financial modeling, performance measurements in finance, evaluation of statistical functionals, dynamical systems for the solution of nonlinear ill-posed problems, extrapolation in partial differential equations, and interpolation and approximation with wavelets.

This applied mathematics review volume is suitable for libraries in the mathematical sciences, engineering and economics, also good for individual researchers and graduate students (classroom and research) of the related fields.

The editor would like to thank Mrs. Kate MacDougall of Warren, RI, for the excellent, prompt typing job preparing this volume.

The Editor
George A. Anastassiou
University of Memphis, Memphis, TN
February 1, 2000

CONTENTS

CONVERGENCE OF GENERALIZED SINGULAR INTEGRALS TO THE UNIT, MULTIVARIATE CASE

George A. Anastassiou and Sorin G. Gal

Department of Mathematical Sciences, The University of Memphis, Memphis, TN 38152, USA

E-mail: anastasg@msci.memphis.edu, ganastss@memphis.edu

Department of Mathematics, University of Oradea, Str. Armatei Romane 5, 3700 Oradea, Romania

E-mail: galso@math.uoradea.ro

In this article we study the degree of L^p-approximation ($1 \le p \le +\infty$) to the unit, by multivariate variants of the Jackson-type generalizations of Picard, Gauss-Weierstrass and Poisson-Cauchy singular integrals.

These results are extensions of those in the univariate case proved by the authors in [2].

1 Introduction

Let f be a function defined on $\mathbf{R}^m$ with values in $\mathbf{R}$.

Let $x = (x_1, \dots, x_m)$, $h = (h_1, \dots, h_m) \in \mathbf{R}^m$. Let us denote

$$\Delta_h^r f(x) = \sum_{i=0}^{r} (-1)^{r-i} \binom{r}{i} f(x+ih), \quad r \in \mathbf{N}.$$

We define the rth L^p-modulus of smoothness over $\mathbf{R}^m$, $1 \le p \le +\infty$, by

$$\omega_r(f;\delta)_p = \sup\{\|\Delta_h^r f(\cdot)\|_{L^p(\mathbf{R}^m)};\ |h| \le \delta\},$$

where $|h| = (|h_1|, |h_2|, \dots, |h_m|)$, $\delta = (\delta_1, \dots, \delta_m)$, $|h| \le \delta$ means $|h_i| \le \delta_i$, $i = \overline{1,m}$ and

$$\|f\|_{L^p(\mathbf{R}^m)} = \left\{\int_{-\infty}^{+\infty} \cdots \int_{-\infty}^{+\infty} |f(x_1, \dots, x_m)|^p dx_1 \dots dx_m\right\}^{1/p}, \quad \text{if } 1 \le p < +\infty,$$

$$\|f\|_{L^\infty(\mathbf{R}^m)} = \sup\{|f(x_1, \dots, x_m)|;\ x_i \in \mathbf{R},\ i = \overline{1,m}\}, \text{ if } p = +\infty.$$

When $f \in L^p_{2\pi}(\mathbf{R}^m) = \{f : \mathbf{R}^m \to \mathbf{R};\ f$ is 2π-periodic in each variable and $\|f\|_{L^p_{2\pi}(\mathbf{R}^m)} < +\infty\}$, the rth L^p-modulus of smoothness is defined as above using the definitions

$$\|f\|_{L^p_{2\pi}(\mathbf{R}^m)} = \left\{\int_{-\pi}^{\pi} \cdots \int_{-\pi}^{\pi} |f(x_1,\ldots,x_m)|^p dx_1 \ldots dx_m\right\}^{1/p}, \text{ if } 1 \le p < +\infty,$$

$$\|f\|_{L^p_{2\pi}(\mathbf{R}^m)} = \sup\{|f(x_1,\ldots,x_m)|;\ x_i \in [-\pi,\pi],\ i = \overline{1,m}\}, \text{ if } p = +\infty.$$

Next, for $\xi = (\xi_1,\ldots,\xi_m) > 0$ (i.e. $\xi_i > 0$, $i = \overline{1,m}$), we consider the multivariate variants of the Jackson-type generalizations of Picard, Poisson-Cauchy and Gauss-Weierstrass singular integrals introduced in [1] by

$$P_{n,\xi}(f;x) = -\frac{1}{\prod_{i=1}^{m}(2\xi_i)} \sum_{k=1}^{n+1} (-1)^k \binom{n+1}{k} \int_{-\infty}^{+\infty} \cdots \int_{-\infty}^{+\infty} f(x_1 + kt_1,\ldots,x_m + kt_m) \prod_{i=1}^{m} e^{-|t_i|/\xi_i} dt_1 \ldots dt_m,$$

$$Q_{n,\xi}(f;x) = -\frac{1}{\prod_{i=1}^{m}\left[\frac{2}{\xi_i}\tan^{-1}\left(\frac{\pi}{\xi_i}\right)\right]} \sum_{k=1}^{n+1} (-1)^k \binom{n+1}{k} \int_{-\pi}^{\pi} \cdots \int_{-\pi}^{\pi} \frac{f(x_1 + kt_1,\ldots,x_m + kt_m)}{\prod_{i=1}^{m}(t_i^2 + \xi_i^2)} dt_1 \ldots dt_m,$$

and

$$W_{n,\xi}(f;x) = = -\frac{1}{\prod_{i=1}^{m}(2C(\xi_i))} \sum_{k=1}^{n+1} (-1)^k \binom{n+1}{k} \int_{-\pi}^{\pi} \cdots \int_{-\pi}^{\pi} f(x_1 + kt_1,\ldots,x_m + kt_m) \prod_{i=1}^{m} e^{-t_i^2/\xi_i^2} dt_1 \ldots dt_m,$$

respectively, where $C(\xi_i) = \int_0^{\pi} e^{-t_i^2/\xi_i^2} dt_i$, $i = \overline{1,m}$, $x = (x_1,\ldots,x_m) \in \mathbf{R}^m$.

Let us denote

$$Erf(x_i) = \frac{2}{\sqrt{\pi}} \int_0^{x_i} e^{-t_i^2} dt_i, \quad x_i \in \mathbf{R}.$$

We notice that

$$\frac{1}{2\xi_i}\int_{-\infty}^{+\infty} e^{-|t_i|/\xi_i}dt_i = 1,$$

$$\int_{-\pi}^{\pi}\frac{dt_i}{t_i^2+\xi_i^2} = \frac{2}{\xi_i}\tan^{-1}\left(\frac{\pi}{\xi_i}\right), \quad \xi_i \in \mathbf{R},\ \xi_i > 0$$

and that $\frac{2}{\pi}\tan^{-1}\frac{\pi}{\xi_i}$, $Erf\left(\frac{\pi}{\sqrt{\xi_i}}\right)$ tend to 1 as $\xi_i \to 0$.

The purpose of this article is to study the degree of L^p-approximation ($1 \le p \le +\infty$) to the unit, by the above singular integrals, extending those results of the univariate case in [2].

2 Main Results

The first main result is

Theorem 2.1. *Let $X = L^1(\mathbf{R}^m)$ (for $P_{n,\xi}$), $X = L^1_{2\pi}(\mathbf{R}^m)$ (for $W_{n,\xi}, Q_{n,\xi}$), $\xi \in \mathbf{R}^m$, $\xi > 0$, $n \in \mathbf{N}$, $f \in X$.*

Then

$$\|f - P_{n,\xi}(f)\|_X \le \left[\sum_{k=0}^{n+1}\binom{n+1}{k}k!\right]^m \omega_{n+1}(f;\xi)_X, \quad \xi > 0, \tag{1}$$

$$\|f - W_{n,\xi}(f)\|_X \le \left[\frac{\int_0^{+\infty}(u+1)^{n+1}e^{-u^2}du}{\int_0^{\pi}e^{-u^2}du}\right]^m \omega_{n+1}(f;\xi)_{L^1_{2\pi}(\mathbf{R}^m)},$$

$$0 \le \xi \le 1, \tag{2}$$

$$\|f - Q_{n,\xi}(f)\|_X \le \left[\prod_{i=1}^m K(n,\xi_i)\right]\omega_{n+1}(f;\xi)_X, \quad \xi > 0, \tag{3}$$

where

$$K(n,\xi_i) = \left[\int_0^{\pi/\xi_i}\frac{(u+1)^{n+1}}{u^2+1}du\right] / \tan^{-1}\frac{\pi}{\xi_i}, \quad i = \overline{1,m}.$$

Proof. Let $f \in X = L^1(\mathbf{R}^m)$. We have

$$f(x) - P_{n,\xi}(f;x) = \frac{1}{\prod_{i=1}^m(2\xi_i)}\int_{-\infty}^{+\infty}$$

$$\cdots \int_{-\infty}^{+\infty} (-1)^{n+1} \Delta_t^{n+1} f(x) \prod_{i=1}^{m} e^{-|t_i|/\xi_i} dt_1 \ldots dt_m, \quad (4)$$

for all $x = (x_1, \ldots, x_m)$, $t = (t_1, \ldots, t_m)$, $\xi = (\xi_1, \ldots, \xi_m) \in \mathbf{R}^m$, $\xi > 0$.

This implies (reasoning as in the univariate case in [2, proof of Theorem 2.1])

$$\|f - P_{n,\xi}(f)\|_{L^1(\mathbf{R}^m)} \leq \frac{1}{\prod_{i=1}^{m}(2\xi_i)} \int_{-\infty}^{+\infty}$$

$$\cdots \int_{-\infty}^{+\infty} \omega_{n+1}(f; \xi(|t|/\xi))_{L^1(\mathbf{R}^m)} \prod_{i=1}^{m} e^{-|t_i|/\xi_i} dt_1 \ldots dt_m,$$

where $\omega_{n+1}(f; \xi(|t|/\xi))_{L^1(\mathbf{R}^m)} = \omega_{n+1}(f; \xi_1(|t_1|/\xi_1), \ldots, \xi_m(|t_m|/\xi_m))_{L^1(\mathbf{R}^m)}$.

By denoting $\lambda = \sum_{i=1}^{m} |t_i|/\xi_i \in \mathbf{R}$, we get

$$\begin{aligned}
\omega_{n+1}(f; |t|)_{L^1(\mathbf{R}^m)} &\leq \omega_{n+1}(f; \lambda\xi)_{L^1(\mathbf{R}^m)} \\
&\leq (\lambda + 1)^{n+1} \omega_{n+1}(f; \xi)_{L^1(\mathbf{R}^m)} \\
&= \left(1 + \sum_{i=0}^{m} |t_i|/\xi_i\right)^{n+1} \omega_{n+1}(f; \xi)_{L^1(\mathbf{R}^m)} \\
&\leq \left[\prod_{i=1}^{m}(1 + |t_i|/\xi)\right]^{n+1} \omega_{n+1}(f; \xi)_{L^1(\mathbf{R}^m)}.
\end{aligned}$$

It follows that

$$\|f - P_{n,\xi}(f)\|_{L^1(\mathbf{R}^m)} \leq \frac{\omega_{n+1}(f; \xi)_{L^1(\mathbf{R}^m)}}{\prod_{i=1}^{m}(2\xi_i)} \int_{-\infty}^{+\infty}$$

$$\cdots \int_{-\infty}^{+\infty} \left(1 + \sum_{i=1}^{m} |t_i|/\xi_i\right)^{n+1} \prod_{i=1}^{m} e^{-|t_i|/\xi_i} dt_1 \ldots dt_m$$

$$\leq \frac{\omega_{n+1}(f; \xi)_{L^1(\mathbf{R}^m)}}{\prod_{i=1}^{m}(2\xi_i)} \int_{-\infty}^{+\infty}$$

$$\cdots \int_{-\infty}^{+\infty} \left[\prod_{i=1}^{m} (1 + |t_i|/\xi_i) \right]^{n+1} \prod_{i=1}^{m} e^{-|t_i|/\xi_i} dt_1 \ldots dt_m$$

$$= \text{(see [2, proof of Theorem 2.1])}$$

$$= \left[\sum_{k=0}^{n+1} \binom{n+1}{k} k! \right]^m \omega_{n+1}(f;\xi)_{L^1(\mathbf{R}^m)},$$

which proves (1).

Then, for $f \in X = L^1_{2\pi}(\mathbf{R}^m)$, we obtain

$$f(x) - W_{n,\xi}(f;x) = \frac{1}{\prod_{i=1}^{m}(2C(\xi_i))} \int_{-\pi}^{\pi}$$

$$\cdots \int_{-\pi}^{\pi} (-1)^{n+1} \Delta_t^{n+1} f(x) \prod_{i=1}^{m} e^{-t_i^2/\xi_i^2} dt_1 \ldots dt_m, \quad (5)$$

for all $x = (x_1, \ldots, x_m)$, $t = (t_1, \ldots, t_m)$, $\xi = (\xi_1, \ldots, \xi_m)$, $\xi_i > 0$, $i = \overline{1,m}$.

Reasoning exactly as in the univariate case in [2, proof of Theorem 2.1], (5) implies (for $0 < \xi_i \leq 1$, $i = \overline{1,m}$)

$$\|f - W_{n,\xi}(f)\|_X \leq \frac{\omega_{n+1}(f;\xi)_{L^1_{2\pi}(\mathbf{R}^m)}}{\prod_{i=1}^{m}(2C(\xi_i))} \int_0^{\pi}$$

$$\cdots \int_0^{\pi} \left[\prod_{i=1}^{m} (1 + t_i/\xi_i) \right]^{n+1} \prod_{i=1}^{m} e^{-t_i^2/\xi_i^2} dt_1 \ldots dt_m$$

$$\leq \left[\frac{\int_0^{+\infty} (u+1)^{n+1} e^{-u^2} du}{\int_0^{\pi} e^{-u^2} du} \right]^m \omega_{n+1}(f;\xi)_X,$$

which proves (2).

Finally, for $x, t, \xi \in \mathbf{R}^m$, $\xi > 0$, $f \in X = L^1_{2\pi}(\mathbf{R}^m)$, we get

$$f(x) - Q_{n,\xi}(f;x) = \frac{1}{\prod_{i=1}^{m} \left(\frac{2}{\xi_i} \tan^{-1} \frac{\pi}{\xi_i} \right)} \int_{-\pi}^{\pi}$$

$$\cdots \int_{-\pi}^{\pi} \frac{(-1)^{n+1}\Delta_t^{n+1} f(x)}{\prod_{i=1}^{m}(t_i^2+\xi_i^2)} dt_1 \ldots dt_m, \tag{6}$$

which implies (see the univariate case [2, proof of Theorem 2.1])

$$\|f - Q_{n,\xi}(f)\|_X \leq \left[\prod_{i=1}^{m} K(n,\xi_i)\right] \omega_{n+1}(f;\xi)_X,$$

where $K(n,\xi_i) = \left[\int_0^{\pi/\xi_i} \frac{(u+1)^{n+1}}{u^2+1} du\right] / \tan^{-1}\frac{\pi}{\xi_i}$, $i = \overline{1,m}$.

The theorem is proved. □

The second main result is

Theorem 2.2. *Let* $X = L^p(\mathbf{R}^m)$ *(for* $P_{n,\xi}$*),* $X = L^p_{2\pi}(\mathbf{R}^m)$ *(for* $W_{n,\xi}, Q_{n,\xi}$*),* $\xi \in \mathbf{R}^m$, $0 < \xi_i \leq 1$, $i = \overline{1,m}$, $n \in \mathbf{N}$, $1 < p < +\infty$, $\frac{1}{p} + \frac{1}{q} = 1$, $f \in X$.

Then

$$\|f - P_{n,\xi}(f)\|_X \leq \left(\frac{2}{q}\right)^{m/q} \|g\|^m_{L^p(\mathbf{R}_+)} \omega_{n+1}(f;\xi)_X,$$

where $g(u) = (u+1)^{n+1} e^{-u/2}$, $u \in \mathbf{R}_+$,

$$\|f - W_{n,\xi}(f)\|_X \leq \left[\left(\sqrt{\frac{\pi}{2q}}\right)^{1/q} \|h\|_{L^p(\mathbf{R}_+)} \Big/ \int_0^{\pi} e^{-u^2} du\right]^m \omega_{n+1}(f;\xi)_X,$$
$$0 < \xi \leq 1,$$

where $h(u) = (u+1)^{n+1} e^{-u^2/2}$,

$$\|f - Q_{n,\xi}(f)\|_X \leq \left[\prod_{i=1}^{m} K_p(n,\xi_i)\right] \omega_{n+1}(f;\xi)_X, \quad 0 \leq \xi \leq 1,$$

where $K_p(n,\xi_i) = \left[\int_0^{\pi/\xi_i} [(u+1)^{(n+1)p}/(u^2+1)] du / \tan^{-1}\frac{\pi}{\xi_i}\right]^{1/p}$.

Proof. Let first $f \in X = L^p(\mathbf{R}^m)$, $\frac{1}{p} + \frac{1}{q} = 1$ and denote $C_1(\xi_i) = \frac{1}{(2\xi_i)^p}\left(\frac{4\xi_i}{q}\right)^{p/q}$. By (4) and similarly with [2, proof of Theorem 2.2], we obtain

$$\|f - P_{n,\xi}(f)\|_X^p \leq \prod_{i=1}^{m} C_1(\xi_i) \int_{-\infty}^{+\infty}$$

$$\cdots\int_{-\infty}^{+\infty}\left[\omega_{n+1}(f;|t|)_X\prod_{i=1}^{m}e^{-|t_i|/(2\xi_i)}\right]^p dt_1\ldots dt_m$$

$$\leq\prod_{i=1}^{m}[2C_1(\xi_i)]\omega_{n+1}^p(f;\xi)_X\int_0^{+\infty}$$

$$\cdots\int_0^{+\infty}\left[\prod_{i=1}^{m}(1+t_i/\xi_i)\right]^{(n+1)p}\prod_{i=1}^{m}e^{-t_ip/(2\xi_i)}dt_1\ldots dt_m$$

$$=\left(\frac{2}{q}\right)^{m/q}\omega_{n+1}^p(f;\xi)_X\left[\int_0^{+\infty}(u+1)^{(n+1)p}e^{-pu/2}du\right]^m,$$

which implies the first estimate in the statement.

Now, let $f\in X=L_{2\pi}^p(\mathbf{R}^m)$, $\frac{1}{p}+\frac{1}{q}=1$. By (5) and by the proof of Theorem 2.2 in the univariate case in [2], we obtain

$$\|f-W_{n,\xi}(f)\|^p\leq\left(\prod_{i=1}^{m}[2C_2(\xi_i)]\right)\omega_{n+1}^p(f;\xi)_X$$

$$\cdot\int_0^{\pi}\cdots\int_0^{\pi}\left[\prod_{i=1}^{m}(t_i/\xi_i+1)\right]^{(n+1)p}\prod_{i=1}^{m}e^{-t_i^2p/(2\xi_i^2)}dt_1\ldots dt_m,$$

where

$$C_2(\xi_i)=\frac{1}{[2C(\xi_i)]^p}\left(\sqrt{\frac{2\pi}{q}}\xi_i\right)^{p/q}\left(Erf\left(\pi\sqrt{\frac{q}{2}}\cdot\frac{1}{\xi_i}\right)\right)^{p/q}.$$

Reasoning as in the univariate case (see [2, proof of Theorem 2.2]) we get

$$\|f-W_{n,\xi}(f)\|_X\leq\left[\left(\sqrt{\frac{\pi}{2q}}\right)^{1/q}\|h\|_{L^p(\mathbf{R}_+)}/\int_0^{\pi}e^{-u^2}du\right]^m\omega_{n+1}(f;\xi)_X,$$

where $h(u)=(u+1)^{n+1}e^{-u^2/2}$.

Finally, for $f\in X=L_{2\pi}^p(\mathbf{R}^m)$, $\frac{1}{p}+\frac{1}{q}=1$, by (6) and by the proof of Theorem 2.2 in [2]) we obtain (omitting the details)

$$\|f-Q_{n,\xi}(f)\|_X^p$$
$$\leq\left(\prod_{i=1}^{m}\left[\int_0^{\pi/\xi_i}([u+1]^{(n+1)p}/(u^2+1))du/\tan^{-1}\frac{\pi}{\xi}\right]\right)\omega_{n+1}^p(f;\xi)_X,$$

which proves the theorem. □

The last main result deals with the uniform convergence.

Theorem 2.3. *Let* $X = L^\infty(\mathbf{R}^m)$ *(for* $P_{n,\xi}$*),* $X = L^\infty_{2\pi}(\mathbf{R}^m)$ *(for* $W_{n,\xi}, Q_{n,\xi}$*),* $\xi \in \mathbf{R}^m$, $0 < \xi \le 1$, $n \in \mathbf{N}$, $f \in X$. *Then*

$$\|f - P_{n,\xi}(f)\|_X \le \left[\sum_{k=0}^{m} \binom{n+1}{k} k!\right]^m \omega_{n+1}(f;\xi)_X, \quad \xi > 0,$$

$$\|f - W_{n,\xi}(f)\|_X \le \left[\frac{\int_0^{+\infty} (u+1)^{n+1} e^{-u^2}\,du}{\int_0^{\pi} e^{-u^2}\,du}\right]^m \omega_{n+1}(f;\xi)_X, \quad 0 < \xi \le 1,$$

$$\|f - Q_{n,\xi}(f)\|_X \le \left[\prod_{i=1}^{m} K(n,\xi_i)\right] \omega_{n+1}(f;\xi)_X, \quad \xi > 0,$$

where $K(n,\xi_i)$ *is given in Theorem 2.1.*

Proof. By (4), (5), (6) and reasoning exactly as in the proof of Theorem 2.1, we easily obtain the statement of the theorem. □

Remark. Theorems 2.1-2.3 show us that while the generalized operators $P_{n,\xi}$ and $W_{n,\xi}$ give very good estimates (such that if $\xi \to 0$, i.e. $\xi_i \to 0$, $i = \overline{1,m}$, then $P_{n,\xi}(f) \to f$, $W_{n,\xi}(f) \to f$), for the generalized operator $Q_{n,\xi}(f)$ in general this does not happen, because when $\xi_i \to 0$, we have $K_p(n,\xi_i) \to +\infty$, for all $1 \le p < +\infty$.

However, under some smoothness conditions on f, (as for example if $\left|\frac{\partial^{|k|} f}{\partial x_1^{k_1} \ldots \partial x_m^{k_m}}\right| \le M$ on $\mathbf{R}^m$, for all $|k| \in \{0,1,\ldots,n+1\}$, where $|k| = k_1 + \ldots + k_m$, $k_i \in \mathbf{N} \cup \{0\}$, $i = \overline{1,m}$) and reasoning as in the univariate case [2], we easily obtain that $P_{n,\xi}(f) \to f$, as $\xi \to 0$.

References

1. Anastassiou, G.A. and Gal, S.G., Global smoothness preservation by multivariate singular integrals, *Bull. Austral. Math. Soc.*(to appear).

2. Anastassiou, G.A. and Gal, G.S., Convergence of generalized singular integrals to the unit, univariate case, submitted.

TWO CONTEMPORARY COMPUTATIONAL ASPECTS OF NUMERICAL ANALYSIS

Ioannis K. Argyros
Department of Mathematics, Cameron University, Lawton, OK 73505
E-mail: ioannisa@cameron.edu

In this study we report advances in two computational aspects of Numerical Analysis. In Part 1, we use inexact Newton-like iterates to approximate a solution of a nonlinear equation in a Banach space. Solving a nonlinear equation using Newton-like iterates at each stage is very expensive in general. That is why we consider inexact Newton-like methods, where Newton-like equations are solved only approximately and in some unspecified manner. In the second part we examine the connection between inexact Newton-like methods and their discretized versions. In particular we show that the behavior of the discretized process is asymptotically the same as that for the original iteration and consequently, the number of steps required by the two processes to converge to within a given tolerance is essentially the same. Our results compare favorably with earlier ones. Some numerical examples are also provided to show that our results apply where others fail.

1 Part 1

1.1 Introduction

We are concerned with approximating a solution x^* of the equation

$$F(x) = 0, \tag{1}$$

where F is a nonlinear operator defined on a Banach space E_1 with values in a Banach space E_2 with the properties:

(a) there exists a point $x^* \in E_1$ with $F(x^*) = 0$;

(b) F is m-times continuously Fréchet-differentiable in a neighborhood of x^* where m is an integer with $m \geq 2$;

(c) $F'(x^*)$ is nonsingular;

Note that $F'(x^*) \in L(E_1, E_2)$ the space of bounded linear operators from E_1 into E_2 denotes the first Fréchet-derivative of F evaluated at $x = x^*$, whereas $F^{(m)}(x) \in L(E_1^m, E_2)$ [9], [17] is the m-th Fréchet-derivative of operator F.

Inexact Newton method is any procedure which, given on initial guess y_0, generates a sequence $\{y_n\}$ $(n \geq 0)$ of approximation to y^* as follows:

For $n = 0$ Step 1 until convergence DO.

Find some step x_n which satisfies

$$F'(y_n)x_n = -F(y_n) + r_n, \quad (n \geq 0) \tag{2}$$

where

$$\frac{\|r_n\|}{\|F(y_n)\|} \leq \eta_n, \quad (n \geq 0). \tag{3}$$

Set

$$y_{n+1} = y_n + x_n \quad (n \geq 0). \tag{4}$$

The numbers η_n depend on y_n $(n \geq 0)$. In particular for $\eta_n = 0$ $(n \geq 0)$ we obtain Newton's method [2]–[9], [16]–[20].

In the elegant paper [13] the local behavior of inexact methods is analyzed in the special case when $E_1 = E_2 = \mathbf{R}^i$ $(i \in N)$ However, instead of condition (b) above the same assumption but on the first Fréchet-derivative is made. We observed that under our assumption all results obtained in [13] also hold under our hypotheses. Moreover, we observed that our upper error bounds on the distances involved are smaller than the corresponding ones in [10], [13]. This is due to the passing from the first to the mth Fréchet-derivative. Such an approach has produced sufficient conditions for the convergence of Newton's method to a locally unique solution of Eq. (1) different from the classical Newton–Kantorovich hypothesis [4], [9], [16], [17]. In particular we show: inexact Newton methods are locally convergent if the forcing sequence is uniformly less than one; we show how to choose a forcing sequence which preserves the rapid convergence of Newton's method. Finally, in Remarks 1 and 3 our results compare favorably with the corresponding ones in [11], [13].

1.2 Convergence Analysis

We will need a result on the mean value of mth Fréchet-differentiable operators.

Remark 1. It follows from the hypotheses that there exists constants d, α_i $(2 \leq i \leq m,\ m \geq 2$ an integer) such that

$$d \geq \|F'(y^*)^{-1}\| \quad \text{and} \quad \alpha_i \geq \|F^{(i)}(y^*)\| \quad (2 \leq i \leq m). \tag{5}$$

Consider the approximations for $p := y - y^*$, $b_1 := y^* + \theta_1 p$, $b_i := y^* + \theta_i(b_{i-1} - y^*)$, $\theta_i \in [0,1]$:

$$\begin{aligned} F'(y) - F'(y^*) &= F'(y) - F'(y^*) - F''(y^*)(y - y^*) + F''(y^*)(y - y^*) \\ &= \int_0^1 \{F''(y^* + \theta_1 p) - F''(y^*)\} d\theta_1 p + \int_0^1 F''(y^*) p d\theta_1 \end{aligned}$$

$$= \int_0^1 \int_0^1 F''(b_2)(b_1 - y^*)pd\theta_2 d\theta_1 + \int_0^1 F''(y^*)pd\theta_1$$

$$= \cdots = \int_0^1 \cdots \int_0^1 [F^{(m)}(b_{m-1}) - F^{(m)}(y^*)](b_{m-2} - y^*)$$

$$\cdots (b_1 - y^*)pd\theta_{m-1} \cdots d\theta_2 d\theta_1$$

$$+ \int_0^1 \cdots \int_0^1 F^{(m)}(y^*)(b_{m-2} - y^*) \cdots (b_1 - y^*)pd\theta_{m-1}d\theta_{m-2} \cdots d\theta_2 d\theta_1$$

$$+ \int_0^1 \cdots \int_0^1 F^{(m-1)}(y^*)(b_{m-3} - y^*) \cdots (b_1 - y^*)pd\theta_{m-2} \cdots d\theta_2 d\theta_1$$

$$+ \cdots + \int_0^1 F''(y^*)pd\theta_1; \tag{6}$$

and for $a_1 = y^* + \theta_1 p$, $a_i = y^* + \theta_i(a_{i-1} - y^*)$, $i = 2, \ldots, m-1$ as in (6) we have:

$$F(y) - F(y^*) - F'(y^*)(y - y^*)$$

$$= \int_0^1 \cdots \int_0^1 [F^{(m)}(a_{m-1}) - F^{(m)}(y^*)](a_{m-2} - y^*)$$

$$\cdots (a_1 - y^*)(1 - \theta_1)pd\theta_{m-1} \cdots d\theta_2 d\theta_1$$

$$+ \int_0^1 \cdots \int_0^1 F^{(m)}(y^*)(a_{m-2} - y^*) \cdots (a_1 - y^*)(1 - \theta_1)pd\theta_{m-1} \cdots d\theta_2 d\theta_1$$

$$+ \cdots + \int_0^1 F''(y^*)(1 - \theta_1)p^2 d\theta_1. \tag{7}$$

Since F' is continuous in a neighborhood of y^*, for given $d_0 < d$, by (6) there exists $\delta_0 > 0$ so that $\|F'(y) - F'(y^*)\| \leq d_0$, whenever $y \in \overline{U}(y^*, \delta_0) = \{y \in E_1 \mid \|y - y^*\| \leq \delta_0\}$. It follows from the Banach Lemma on invertible operators [17] that $F'(y)$ is invertible and

$$\|F'(y)^{-1}\| \leq d_1, \tag{8}$$

where, $d_1 = d(1 - dd_0)^{-1}$. Moreover, we obtain

$$\begin{aligned} \|F'(y^*)^{-1} - F'(y)^{-1}\| &= \|F'(y^*)^{-1}[F'(y) - F'(y^*)]F'(y)^{-1}\| \\ &\leq dd_1\|F'(y^*) - F'(y)\| \leq d_0 d_1 d. \end{aligned} \tag{9}$$

Furthermore, for any $\varepsilon > 0$ there exists $\delta > 0$ such that $\|F^{(m)}(y) - F^{(m)}(y^*)\| < \varepsilon$ if $\|y - y^*\| < \delta$. Hence, we have:

$$\|F(y) - F(y^*) - F'(y^*)(y - y^*)\|$$

$$\leq \left[\frac{\alpha_2}{2!} + \frac{\alpha_3}{3!}\|y - y^*\| + \cdots + \frac{\alpha_m + \varepsilon}{m!}\|y - y^*\|^{m-2}\right]\|y - y^*\|^2. \quad (10)$$

We can now provide a local convergence result of inexact Newton iterates and m-Fréchet differentiable operators.

Theorem 1. *Assume:*

(a) $\eta_n \leq \eta_{\max} < t < 1$; (11)

(b) *for* $\mu = \max\{\|F'(y^*)\|, \|F'(y^*)^{-1}\|\}$; *$t$ sufficiently small; y_0 close enough to y^* there exist ε, q small enough so that for $\|y_0 - y^*\| \leq \varepsilon$,*

$$t_0 + t_1\mu\varepsilon + \cdots + t_{m-1}\mu^{m-1}\varepsilon^{m-1} < 1, \quad (12)$$

where,

$$t_0 \geq (1 + \mu q)t, \ \ t_1 \geq 2\alpha_2(1 + \mu q)t\mu^2, \ t_2 \geq \frac{4}{3}\alpha_3(1 + \mu q)t\mu^3, \ldots, \quad (13)$$

$$t_{m-2} = (1 + \mu q)\mu^{m-1}\alpha_{m-1}\left[\frac{2}{(m-1)!} + 1\right],$$
$$t_{m-1} = (1 + \mu q)t\mu^m\left[\frac{2q}{m!} + \frac{2\alpha_m}{m!} + \frac{q + \alpha_m}{(m-1)!}\right].$$

Then, the inexact Newton sequence $\{y_n\}$ $(n \geq 0)$ generated by (4) converges to y^*. Moreover, the following error bounds hold for all $n \geq 0$

$$\|y_{n+1} - y^*\|_* \leq t_0\|y_n - y^*\|_* + t_1\|y_n - y^*\|_*^2 + \cdots + t_{m-1}\|y_n - y^*\|_*^m, \quad (14)$$

where

$$\|x\|_* = \|F'(y^*)x\|. \quad (15)$$

Proof. From the nonsingularity of operator $F'(y^*)$, it follows that

$$\frac{1}{\mu}\|x\| \leq \|x\|_* \leq \mu\|x\| \quad \text{for } x \in E_1. \quad (16)$$

By hypotheses there exist ε, q sufficiently small, so that if

$$\|x - y^*\| \leq m^2\varepsilon, \quad (17)$$

then,

$$\|F^m(x) - F^m(y^*)\| \leq q, \quad (18)$$
$$\|F'(x)^{-1} - F(y^*)^{-1}\| \leq q, \quad (19)$$

$$\begin{aligned}&\|F(x) - F(y^*) - F'(y^*)(x - y^*)\| \\ &\quad \le \left[\frac{\alpha_2}{2!} + \frac{\alpha_3}{3!}\|x - y^*\| + \cdots + \frac{\varepsilon + \alpha_m}{m!}\|x - y^*\|^{m-2}\right]\|x - y^*\|^2. \quad (20)\end{aligned}$$

Such an ε exists by the discussion in Remark 1 (see in particular (6), (9) and (11)).

We prove (14) by mathematical induction on n. We get

$$\begin{aligned}\|y_k - y^*\|_* &\le t_0\|y_{k-1} - y^*\|_* + (t_1\mu\|y_{k-1} - y^*\|)\|y_{k-1} - y^*\|_* \\ &\quad + \cdots + t_{m-1}(\mu\|y_{k-1} - y^*\|)^{m-1}\|y_{k-1} - y^*\|_* \\ &\le [t_0 + \mu t_1\|y_0 - y^*\| + \cdots + t_{m-1}\mu^{m-1}\|y_0 - y^*\|^{m-1}]\|y_{k-1} - y^*\|_* \\ &\le (t_0 + t_1\mu\varepsilon + \cdots + t_{m-1}\mu^{m-1}\varepsilon^{m-1})\|y_{k-1} - y^*\|_* \le \|y_{k-1} - y^*\|_*.\end{aligned}$$

That is,

$$\begin{aligned}\|y_k - y^*\| &\le \mu\|y_k - y^*\|_* \le \mu\|y_{k-1} - y^*\|_* \le \mu\|y_0 - y^*\|_* \\ &\le m^2\|y_0 - y^*\| \le m^2\varepsilon.\end{aligned}$$

Hence, estimates (18)–(20) hold with $x = y_k$. Moreover, x_k exists satisfying (2). Using the approximation

$$\begin{aligned}F'(y^*)(y_{k+1} - y^*) = [I &+ F'(y^*)[F'(y_k)^{-1} - F'(y^*)^{-1}] \\ &\cdot\{r_k + [F'(y_k) - F'(y^*)](y_k - y^*) \\ &- [F(y_k) - F(y^*) - F'(y^*)(y_k - y^*)]\}, \quad (21)\end{aligned}$$

taking norm, we obtain

$$\begin{aligned}\|y_{k+1} - y^*\|_* \le [1 &+ \|F'(y^*)\|\,\|F'(y_k)^{-1} - F'(y^*)^{-1}\|] \\ &\cdot[\|r_k\| + \|F'(y_k) - F'(y^*)\|\,\|y_k - y^*\| \\ &+ \|F(y_k) - F(y^*) - F'(y^*)(y_k - y^*)\|]. \quad (22)\end{aligned}$$

Using (5)–(7) for $y = y_k$, (18)–(20) we obtain

$$\begin{aligned}\|F'(y_k) - F'(y^*)\| &\le \frac{q}{(m-1)!}\|y_k - y^*\|^{m-1} + \frac{\alpha_m}{(m-1)!}\|y_k - y^*\|^{m-1} \\ &\quad + \frac{\alpha_{m-1}}{(m-2)!}\|y_k - y^*\|^{m-2} + \cdots + \alpha_2\|y - y^*\|, \quad (23)\end{aligned}$$

$$\begin{aligned}&\|F(y_k) - F(y^*) - F'(y^*)(y_k - y^*)\| \\ &\quad \le \left[\frac{\alpha_2}{2!} + \frac{\alpha_3}{3!}\|y_k - y^*\| + \cdots + \frac{q + \alpha_m}{m!}\|y_k - y^*\|^{m-2}\right]\|y_k - y^*\|^2, \quad (24)\end{aligned}$$

and

$$\begin{aligned}\|F(y_k)\| &\le \|F'(y^*)(y_k - y^*)\| + \|F(y_k) - F(y^*) - F'(y_*)(y_k - y^*)\| \\ &\le \|y_k - y^*\|_* + \left[\frac{\alpha_2}{2!} + \frac{\alpha_3}{3!}\|y_k - y^*\| \right. \\ &\quad \left. + \cdots + \frac{q+\alpha_m}{m!}\|y_k - y^*\|^{m-2}\right] \|y_k - y^*\|^2. \end{aligned} \tag{25}$$

By (23), (24), and (25) estimate (22) gives

$$\begin{aligned}\|y_{k+1} - y^*\|_* &\le (1+\mu q)\left\{\eta_{\max}\left[\|y_k - y^*\|_* + \left(\frac{\alpha_2}{2!} + \frac{\alpha_3}{3!}\|y_k - y^*\| \right.\right.\right. \\ &\quad \left.\left. + \cdots + \frac{q+\alpha_m}{m!}\|y_k - y^*\|^{m-2}\right)\|y_k - y^*\|^2\right] \\ &\quad + \frac{q}{(m-1)!}\|y_k - y^*\|^m + \frac{\alpha_m}{(m-1)!}\|y_k - y^*\|^m + \cdots + \alpha_2\|y_k - y^*\|^2 \\ &\quad \left. + \left(\frac{\alpha_2}{2!} + \frac{\alpha_3}{3!}\|y_k - y^*\| + \cdots + \frac{q+\alpha_m}{m!}\|y_k - y^*\|^{m-2}\right)\|y_k - y^*\|^2\right\} \\ &\le (1+\mu q)t\|y_k - y^*\|_* + (1+\mu q)t\|y_k - y^*\|^2\left(\frac{\alpha_2}{2!} + \alpha_2 + \frac{\alpha_2}{2!}\right) \\ &\quad + (1+\mu q)t\|y_k - y^*\|^3\left(\frac{\alpha_3}{3!} + \alpha_3 + \frac{\alpha_3}{3!}\right) \\ &\quad + \cdots + (1+\mu q)t\|y_k - y^*\|^{m-1}\left[\frac{\alpha_{m-1}}{(m-1)!} + \alpha_{m-1} + \frac{\alpha_{m-1}}{(m-1)!}\right] \\ &\quad + (1+\mu q)t\|y_k - y^*\|^m\left[\frac{q+\alpha_m}{m!} + \frac{q+\alpha_m}{(m-1)!} + \frac{q+\alpha_m}{m!}\right] \\ &\le t_0\|y_k - y^*\|_* + t_1\|y_k - y^*\|_*^2 \\ &\quad + \cdots + t_{m-2}\|y_k - y^*\|_*^{m-1} + t_{m-1}\|y_k - y^*\|_*^m \quad \text{(by (13))}.\end{aligned}$$

By the choice of ε and (12), we get $\lim_{n\to\infty} y_n = y^*$.

That completes the proof of Theorem 1. □

Remark 2. It can easily be seen from the proof of Theorem 1 that parameter t appearing in (12) and (13) can be replaced by $\eta_{\max}$. Estimate (14) is comparable to (2.1), Theorem 2.3 in [13, p. 402]. Indeed from (14) and (13) we can set:

$$t_A := t_0 + t_1\mu\|y_0 - y^*\| + t_2\mu^2\|y_0 - y^*\|^2 + \cdots + t_{m-1}\mu^{m-1}\|y_0 - y^*\|^{m-1}, \tag{26}$$

(ours), where t has been replaced by $\mu_{\max}$, and

$$t_T = (1+\mu q)[\eta_{\max}(1+\mu q) + 2\mu q] \quad \text{(theirs)}. \tag{27}$$

Estimates (14) and 2.1 in [13] become

$$\|y_{n+1} - y^*\|_* \leq t_A \|y_n - y^*\|_* \tag{28}$$

and

$$\|y_{n+1} - y^*\|_* \leq t_T \|y_n - y^*\|_*, \tag{29}$$

for all $n \geq 0$, respectively. Note that we have also assumed (without loss of generality that (18)–(20) above and (2.3)–(2.5) in [13] hold for the same ε and q. It then follows from (28) and (29) that our error bounds (14) are better than (2.1) in [13] provided that

$$t_A < t_T. \tag{30}$$

By (13), (26) and (27) it follows that (30) is true if $\|y_0 - y^*\|$ sufficiently small. Since then (30) is true if

$$t_0 < t_T, \tag{31}$$

which is true.

The following lemma is crucial.

Lemma 1. *For any $\varepsilon > 0$ there exists $\delta > 0$ such that:*

$$\begin{aligned}
&\left\{\|F'(y^*)^{-1}\|^{-1} - \left[\frac{\alpha_2}{2!} + \frac{\alpha_3}{3!}\|y - y^*\| \right.\right.\\
&\qquad \left.\left. + \cdots + \frac{\varepsilon}{m!}\|y - y^*\|^{m-2}\right]\|y - y^*\|\right\}\|y - y^*\| \\
&\leq \left\{\|F'(y^*)\| - \left[\frac{\alpha_2}{2!} + \frac{\alpha_3}{3!}\|y - y^*\| \right.\right.\\
&\qquad \left.\left. + \cdots + \frac{\varepsilon}{m!}\|y - y^*\|^{m-2}\right]\|y - y^*\|\right\}\|y - y^*\| \\
&\leq \|F(y)\| \leq \left\{\|F'(y^*)\| + \left[\frac{\alpha_2}{2!} + \frac{\alpha_3}{3!}\|y - y^*\| \right.\right.\\
&\qquad \left.\left. + \cdots + \frac{\varepsilon}{m!}\|y - y^*\|^{m-2}\right]\|y - y^*\|\right\}\|y - y^*\|,
\end{aligned} \tag{32}$$

if, $\|y - y^*\| < \delta$.

Proof. By the discussion in Remark 1, for any $\varepsilon > 0$ there exists $\delta > 0$ sufficiently small so that (10) is satisfied if $\|y - y^*\| < \delta$. Since,

$$F(y) = F'(y^*)(y - y^*) + [F(y) - F(y^*) - F'(y^*)(y - y^*)], \tag{33}$$

we get by (10) and (33),

$$\|F(y)\| \le \|F'(y^*)\| \cdot \|y - y^*\| + \left[\frac{\alpha_2}{2!} + \frac{\alpha_3}{3!}\|y - y^*\| + \cdots + \frac{\varepsilon}{m!}\|y - y^*\|^{m-2}\right] \|y - y^*\|^2,$$

and

$$\begin{aligned}
\|F(y)\| &\ge \|F'(y^*)\| \cdot \|y - y^*\| - \|F(y) - F(y^*) - F'(y^*)(y - y^*)\| \\
&\ge \left\{ \|F'(y^*)\| - \left[\frac{\alpha_2}{2!} + \frac{\alpha_3}{3!}\|y - y^*\| \right.\right. \\
&\quad \left.\left. + \cdots + \frac{\varepsilon}{m!}\|y - y^*\|^{m-2}\right] \|y - y^*\| \right\} \|y - y^*\| \\
&\ge \left\{ \|F'(y^*)^{-1}\|^{-1} - \left[\frac{\alpha_2}{2!} + \frac{\alpha_3}{3!}\|y - y^*\| \right.\right. \\
&\quad \left.\left. + \cdots + \frac{\varepsilon}{m!}\|y - y^*\|^{m-2}\right] \|y - y^*\| \right\} \|y - y^*\|,
\end{aligned}$$

if $\|y - y^*\| < \delta$.

That completes the proof of Lemma 1. □

Definition 1. Operator $F^{(m)}$ is Hölder continuous with exponent $\lambda \in (0, 1]$ at y^* if there exists $\alpha_{m+1} \ge 0$ such that:

$$\|F^{(m)}(y) - F^{(m)}(y^*)\| \le \alpha_{m+1}\|y - y^*\|^{\lambda} \tag{34}$$

for $\|y - y^*\|$ sufficiently small.

Lemma 2. *If $F^{(m)}$ is Hölder continuous with exponent $\lambda \in (0, 1]$ at y^*, then*

$$\begin{aligned}
&\|F(y) - F(y^*) - F'(y^*)(y - y^*)\| \\
&\quad \le \frac{\alpha_{m+1}}{(\lambda+1)(\lambda+2)\cdots(\lambda+m)}\|y - y^*\|^{\lambda+m} + \frac{\alpha_m}{m!}\|y - y^*\|^m \\
&\qquad + \cdots + \frac{\alpha_2}{2!}\|y - y^*\|^2,
\end{aligned} \tag{35}$$

for $\|y - y^*\|$ sufficiently small.

Proof. By (5), (7) and (34) we obtain

$$\begin{aligned}
&\|F(y) - F(y^*) - F'(y^*)(y - y^*)\| \\
&\quad \le \int_0^1 \cdots \int_0^1 \|F^{(m)}(a_{m-1}) - F^{(m)}(y^*)\| \, \|a_{m-2} - y^*\|
\end{aligned}$$

$$
\begin{aligned}
&\cdots \|a_1 - y^*\|(1-\theta_1)\|p\| d\theta_{m-1} \cdots d\theta_2 d\theta_1 \\
&+ \int_0^1 \cdots \int_0^1 \|F^{(m)}(y^*)\| \, \|a_{m-2} - y^*\| \\
&\cdots \|a_1 - y^*\|(1-\theta_1)\|p\| d\theta_{m-1} \cdots d\theta_2 d\theta_1 \\
&+ \cdots + \int_0^1 \|F''(y^*)\|(1-\theta_1)\|p\|^2 d\theta_1 \\
\leq\ & \frac{\alpha_{m+1}}{(\lambda+1)(\lambda+2)\cdots(\lambda+m)} \|y - y^*\|^{\lambda+m} \\
&+ \frac{\alpha_m}{m!}\|y - y^*\|^m + \cdots + \frac{\alpha_2}{2!}\|y - y^*\|^2,
\end{aligned}
$$

which shows (35) for $\|y - y^*\|$ sufficiently small.

To make the paper as self-contained as possible we repeat some crucial definitions which can be found in [13], [20].

Definition 2. Let $\{y_n\}$ $(n \geq 0)$ be a sequence which converges to y^*. Then,

(a) $y_n \to y^*$ superlinearly if $\|y_{n+1} - y^*\| = O(\|y_n - y^*\|)$ as $n \to \infty$;

(b) $y_n \to y^*$ with strong order at least r $(r > 1)$ if

$$\|y_{n+1} - y^*\| = \theta(\|y_n - y^*\|^r) \quad \text{as } n \to \infty;$$

(c) $y_n \to y^*$ with weak order at least r $(r > 1)$ if

$$\limsup_{n\to\infty} \|y_n - y^*\|^{1/r^n} < 1 \quad \text{as } n \to \infty.$$

Note also that given continuous nonnegative real-valued functions f_1 and f_2, we write

$$f_1(y_n) = O(f_2(y_n)) \quad \text{as } n \to \infty, \text{ if } \limsup_{n\to\infty} \frac{f_1(y_n)}{f_2(y_n)} = 0,$$

and

$$f_1(y_n) = \theta(f_2(y_n)) \quad \text{as } n \to \infty, \text{ if } \limsup_{n\to\infty} \frac{f_1(y_n)}{f_2(y_n)} < +\infty.$$

Theorem 2. *Assume inexact Newton iterates $\{y_n\}$ $(n \geq 0)$ converge to y^*. Then, $y_n \to y^*$ superlinearly if and only if*

$$\|r_n\| = O(\|F(y_n)\|) \quad \textit{as } n \to \infty.$$

Morever, if $F^{(m)}$ is Hölder continuous with exponent $\lambda \in (0,1]$ at y^*, then $y_n \to y^*$ as $n \to \infty$ with order at least $1+\lambda$ if and only if

$$\|r_n\| = \theta(\|F(y_n)\|^{1+\lambda}) \quad \textit{as } n \to \infty.$$

Proof. The proof is similar to Theorem 3.3 [13, p. 404], but there exist non-trivial modifications. At first, assume $y_n \to y^*$ as $n \to \infty$ superlinearly. By the approximation

$$\begin{aligned} r_n = {} & [F(y_n) - F(y^*) - F'(y^*)(y_n - y^*)] + [F'(y_n) - F'(y^*)](y_n - y^*) \\ & + [F'(y^*) + F'(y_n) - F'(y^*))](y_{n+1} - y^*). \end{aligned} \tag{36}$$

taking norms and using (23) and (24), we obtain

$$\begin{aligned} \|r_n\| & \le \|F(y_n) - F(y^*) - F'(y^*)(y_n - y^*)\| + \|F'(y_n) - F'(y^*)\| \, \|y_n - y^*\| \\ & \quad + [\|F'(y^*)\| + \|F'(y_n) - F'(y^*)\|]\|y_{n+1} - y^*\| \\ & \le O(\|y_n - y^*\|) + O(1)\|y_n - y^*\| \\ & \quad + [\|F'(y^*)\| + O(1)]O(\|y_n - y^*\|). \end{aligned} \tag{37}$$

Hence, by Lemma 2 we get

$$\|r_n\| = O(\|y_n - y^*\|) = O(\|F(y_n)\|) \quad \text{as } n \to \infty.$$

Conversely, assume $\|r_n\| = O(\|F(x_n)\|)$. By (21) we also get

$$\begin{aligned} \|y_{n+1} - y^*\| & \le [\|F'(y^*)^{-1}\| + \|F'(y_k)^{-1} - F'(y^*)^{-1}\|] \\ & \quad \cdot [\|r_n\| + \|F'(y_n) - F'(y^*)\| \, \|y_n - y^*\| \\ & \quad + \|F(y_n) - F(y^*) - F'(y^*)(y_n - y^*)\|] \\ & = [\|F'(y^*)^{-1}\| + O(1) \cdot [O(\|F(y_n)\|) \\ & \quad + O(1)\|y_n - y^*\| + O(\|y_n - y^*\|)]. \end{aligned}$$

Hence, by Lemma 2 we get

$$\|y_{n+1} - y^*\| = O(\|F(y_n)\|) + O(\|y_n - y^*\|) = O(\|y_n - y^*\|) \quad \text{as } n \to \infty.$$

If $F^{(m)}$ is Hölder continuous with exponent λ at x^*, then using (35) instead of (10) the proof is essentially the same.

That completes the proof of Theorem 2. □

The following result give a characterization of weak order of convergence.

Theorem 3. *Assume that inexact Newton iterates $\{y_n\}$ $(n \ge 0)$ converge to y^*. If $F^{(m)}$ is Hölder continuous with exponent $\lambda \in (0,1]$ at y^*, then $y_n \to y^*$*

as $n \to \infty$ *with weak order at least* $1+\lambda$ *if and only if* $r_n \to 0$ *as* $n \to \infty$ *with weak order at least* $1+\lambda$.

Proof. The crucial estimates (3.1)–(3.5) in Theorem 3.4 [13, p. 405] hold here also by (32), (8), (9), (34), and (35). Hence, the proof of our theorem folows.

That completes the proof of Theorem 3. □

The following result relates forcing sequences and inexact Newton iterates.

Corollary 1. *Assume inexact Newton iterates* $\{y_n\}$ $(n \geq 0)$ *converge to* y^*. *Then*

(a) $y_n \to y^*$ *superlinearly if* $\lim_{n\to\infty} \eta_n = 0$;

(b) $y_n \to y^*$ *with order at least* $1+\lambda$ *if* F'' *is Hölder continuous with exponent* $\lambda \in (0,1]$ *at* y^* *and*

$$\eta_n = \theta(\|F(y_n)\|^{\lambda}) \quad \text{as } n \to \infty;$$

(c) $y_n \to y^*$ *with weak order at least* $1+\lambda$ *if* $F^{(m)}$ *is Hölder continuous with exponent* $\lambda \in (0,1]$ *at* y^* *and* $\{\eta_n\}$ $(n \geq 0)$ *converges to* 0 *with weak order at least* $1+\lambda$.

Favorable comments for our Theorems 2 and 3 when compared to the corresponding Theorems 3.3 and 3.4 in [13] similar to the ones made in Remark 1 after Theorem 1 can now follow. Another convincing result is given by the following:

Theorem 4. *Let* $F : D \subseteq E_1 \to E_2$ *be an operator such that:*

(a) *there exists a* $y^* \in D$ *with* $F(y^*) = 0$;

(b) F *is m-Fréchet-differentiable in a neighborhood of* y^* *contained in* D;

(c) $F'(y^*)$ *is nonsingular on* D;

(d) *there exists nonnegative constants* a, α_i, $2 \leq i \leq m$, c, d, h, γ *such that:*

$$\|F^{(m)}(y) - F^{(m)}(y^*)\| \leq a\|y - y^*\|^{1+\gamma} \text{ for } \|y - y^*\| \text{ sufficiently small,} \tag{38}$$

$$\|F^{(i)}(y^*)\| \leq \alpha_i, \tag{39}$$

$$\|F'(y^*)^{-1}\| \leq d, \tag{40}$$

$$\|F'(y^*)\| \leq h, \tag{41}$$

and

$$\|r_n\| \leq c\|F(y_n)\|^2 \quad (n \geq 0). \tag{42}$$

Then, there exists $\varepsilon > 0$, such that if $y_0 \in \overline{U}(y^*, \varepsilon) = \{y \in E_1 \mid \|y - y^*\| \leq \varepsilon\}$, then inexact sequence $\{y_n\}$ $(n \geq 0)$ generated by (4): $y_n \to y^*$ quadratically in the sense that

$$\|y_{n+1} - y^*\| \leq c_0 \|y_- y^*\|^2 \quad (n \geq 0), \tag{43}$$

where

$$c_0 = d\left\{ \frac{2a\varepsilon^{m+\gamma-1}}{(3+\gamma)m!} + \frac{2\alpha_m}{m!}\varepsilon^{m-2} + \cdots + \alpha_2 \right.$$
$$\left. + 2c\left[h + \frac{a\varepsilon^{m+\gamma}}{(2+\gamma)(3+\gamma)m!} + \frac{\alpha_m \varepsilon^{m-1}}{m!} + \cdots + \frac{\alpha_2}{2!}\varepsilon\right]^2 \right\}.$$

Moreover, the following estimate holds:

$$\limsup_{n\to\infty} \frac{\|y_{n+1} - y^*\|}{\|y_n - y^*\|^2} \leq \beta_0 = d(\alpha_2 + 2ch^2). \tag{44}$$

Proof. Choose $\varepsilon > 0$ so that $F'(y)$ is nonsingular for all $y \in U(y^*, \varepsilon)$. Let $\varepsilon > 0$ be such that

$$d\left\{ \frac{2a\varepsilon^{m+\gamma-1}}{(3+\gamma)m!} + \frac{2\alpha_m \varepsilon^{m-2}}{m!} + \cdots + \alpha_2 \right.$$
$$\left. + 2c\left[h + \frac{a\varepsilon^{m+\gamma}}{(2+\gamma)(3+\gamma)m!} + \frac{\alpha_m \varepsilon^{m-1}}{m!} + \cdots + \frac{\alpha_2}{2!}\varepsilon\right]^2 \right\}\varepsilon \leq \frac{1}{2}, \tag{45}$$

and $F^{(m)}$ is Lipschitz continuous satisfying (38) in $U(y^*, \varepsilon)$. Using $\|y_0 - y^*\| \leq \varepsilon$, (45), (38)–(41) and the approximation (6) multiplied by $F'(y^*)^{-1}$ we get

$$\|F'(y^*)^{-1}[F'(y_0) - F'(y^*)]\| \leq d\left[\frac{a\|y_0 - y^*\|^{m+\gamma}}{(2+\gamma)(m-1)!} + \frac{\alpha_m \|y_0 - y^*\|^{m-1}}{(m-1)!} \right.$$
$$\left. + \frac{\alpha_{m-1}}{(m-1)!}\|y_0 - y^*\|^{m-2} + \cdots + \alpha_2 \|y_0 - y^*\| \right]$$
$$\leq d\left[\frac{a\|y_0 - y^*\|^{m+\gamma-1}}{(2+\gamma)(m-1)!} + \frac{\alpha_m \|y_0 - y^*\|^{m-2}}{(m-1)!} + \cdots + \alpha_2 \right] \|y_0 - y^*\|$$
$$\leq d\left[\frac{a\varepsilon^{m+\gamma-1}}{(2+\gamma)(m-1)!} + \frac{\alpha_m \varepsilon^{m-2}}{(m-1)!} + \cdots + \alpha_2 \right]\varepsilon \leq \frac{1}{2},$$

and the Banach Lemma on invertible operators [9], [17], we deduce $F'(y_0)$ is nonsingular and

$$\|F'(y_0)^{-1}\| \leq \frac{\|F'(y^*)^{-1}\|}{1 - \|F'(y^*)^{-1}[F'(y_0) - F'(y^*)]\|} \leq 2d. \tag{46}$$

Hence, there exists $x_0 \in E_1$ satisfying

$$\|F'(y_0)x_0 + F(y_0)\| = \|r_0\| \leq c\|F(y_0)\|^2.$$

Consequently, $y_1 = y_0 + x_0$ satisfies

$$\begin{aligned} y_1 - y^* &= y_0 - y^* - F'(y_0)^{-1}[F(y_0) - F(y^*) - r_0] \\ &= F'(y_0)^{-1}[F(y^*) - F(y_0) - F'(y_0)(y^* - y_0) + r_0] \\ &= F'(y_0)^{-1}\left\{\int_0^1 F''[y_0 + t(y^* - y_0)](1-t)(y^* - y_0)^2 dt + r_0\right\}. \end{aligned} \quad (47)$$

We also have

$$F(y_0) - F(y^*) - F'(y^*)(y_0 - y^*) = \int_0^1 F''[y^* + t(y_0 - y^*)](1-t)(y_0 - y^*)^2 dt,$$

which together with (38), (39), and (41) give

$$\begin{aligned} \|F(y_0)\| = \|F(y_0) - F(y^*)\| &\leq h\|y_0 - y^*\| + \frac{a\|y_0 - y^*\|^{m+\gamma+1}}{(2+\gamma)(3+\gamma)m!} \\ &+ \frac{\alpha_m}{m!}\|y_0 - y^*\|^m + \cdots + \frac{\alpha_2}{2!}\|y_0 - y^*\|^2, \end{aligned} \quad (48)$$

Using (38)–(42), and (46)–(48), we get

$$\begin{aligned} \|y_1 - y^*\| \leq \|F'(y_0)^{-1}\| &\left[\left\|\int_0^1 F''[y_0 + t(y^* - y_0)](1-t)(y^* - y_0)^2 dt\right\|\right. \\ &\left. + c\|F(y_0)\|^2\right] \\ \leq 2d &\left\{\frac{a\|y_0 - y^*\|^{m+\gamma+1}}{(3+\gamma)m!} + \frac{\alpha_m}{m!}\|y_0 - y^*\|^m\right. \\ &+ \cdots + \frac{\alpha_2}{2!}\|y_0 - y^*\|^2 \\ &+ c\left[h\|y_0 - y^*\| + \frac{a\|y_0 - y^*\|^{m+\gamma+1}}{(2+\gamma)(3+\gamma)m!} + \frac{\alpha_m}{m!}\|y_0 - y^*\|^m\right. \\ &\left.\left. + \cdots + \frac{\alpha_2}{2!}\|y_0 - y^*\|^2\right]^2\right\} \\ \leq d &\left\{\frac{2a\|y_0 - y^*\|^{m+\gamma-1}}{(3+\gamma)m!} + \frac{2\alpha_m}{m!}\|y_0 - y^*\|^{m-2} + \cdots + \alpha_2\right. \\ &+ 2c\left[h + \frac{a\|y_0 - y^*\|^{m+\gamma}}{(2+\gamma)(3+\gamma)m!} + \frac{\alpha_m\|y_0 - y^*\|^{m-1}}{m!}\right. \end{aligned}$$

$$+\cdots+\frac{\alpha_2}{2!}\|y_0-y^*\|\Bigg]^2\Bigg\}\|y_0-y^*\|^2$$
$$\leq c_0\|y_0-y^*\|^2. \tag{49}$$

By (45), $\|y_1-y^*\|\leq\frac{1}{2}\|y_0-y^*\|\leq\varepsilon$, so that $y_1\in U(y^*,\varepsilon)$. Hence, we showed $\|y_2-y^*\|\leq\frac{1}{2}\|y_1-y^*\|$, and that (43) holds for $n=0$.

The proof of the induction step is identical.

Finally, since $y_n\longrightarrow y^*$ as $n\longrightarrow\infty$, estimate (44) follows from (43) by letting $\varepsilon\longrightarrow 0$.

That completes the proof of Theorem 4. □

Remark 3. Let F' be Lipschitz continuous in a neighborhood of y^* contained in D with constant a_0. Then Theorem 2.2 in [10, p. 409] or Theorem 3.3 in [13, p. 404] give error estimates

$$\|y_{n+1}-y^*\|\leq d\left[a_0+2c\left(h+\frac{a_0\varepsilon}{2}\right)^2\right]\|y_n-y^*\|^2\quad(n\geq 0), \tag{50}$$

and

$$\limsup_{n\to\infty}\frac{\|y_{n+1}-y^*\|}{\|y_n-y^*\|^2}\leq\beta_1=d(a_0+2ch^2)=\beta_0+d(a_0-\alpha_2). \tag{51}$$

However, if for example $\alpha_2\leq a_0$, then our estimate (44) is an improvement over (51) (see also the example that follows).

Remark 4. As in [10], [11], [13] the results obtained here can be used for projection methods such as Arnoldi's method, the generalized minimum residual method (GMRES), the generalized conjugate residual method (GCR), and for combined inexact-Newton/finite-difference projection methods.

Remark 5. The results obtained here can also be used to solve autonomous differential equations [9], [17] of the form

$$F'(x)=P(F(x)), \tag{52}$$

where $P:E_2\longrightarrow E_1$ is a known continuously Fréchet-differentiable operator at y^*. Since $F'(y^*)=P(F(y^*))=P(0)$, $F''(y^*)=F'(y^*)P'(F(y^*))=P(0)P'(0)$, we can deal with the results obtained here without actually knowing the solution y^*.

Example 1. Let $E_1=E_2=\mathbf{R}$, and consider the functions F, P defined by

$$F(x)=e^x-\alpha,\quad P(x)=x+\alpha,\quad \alpha>0.$$

Then, functions F, P satisfy (52), $F'(y^*)=\alpha=h=d^{-1}$, $F''(y^*)=\alpha=\alpha_2$, $y^*=\ln\alpha$, and for $y\in U(y^*,\varepsilon)$, $\varepsilon>0$ we get $a_0=\alpha e^{\varepsilon}>\alpha=\alpha_2$.

Hence, our upper error bound (44) is smaller than upper bound (51).

2 Part 2

2.1 Introduction

The goal of this part is to extend the validity of the mesh independence principle for inexact Newton-like methods initiated in a previous paper [8]. Let us consider the problem of approximating a locally unique solution x^* of the equation

$$F(x) + Q(x) = 0, \tag{53}$$

where F, Q are nonlinear operators defined on some closed convex subset D of a Banach space E_1 with values in a Banach space E_2.

Let $x_0 \in D$ be fixed and define the inexact Newton-like method for all $n \geq 0$ by

$$y_n = x_n - A(x_n)^{-1}(F(x_n) + Q(x_n)) \tag{54}$$

and

$$x_{n+1} = y_n - z_n. \tag{55}$$

Here, $A(x_n)$ denotes a linear operator which is an approximation to the Fréchet derivative $F'(x_n)$ of F evaluated at $x = x_n$ $(n \geq 0)$. The points $z_n \in D$ for all $n \geq 0$, and are determined in such a way that the iteration $\{x_n\}$ $(n \geq 0)$ converges to a solution x^* of Eq. (53). The convergence of both methods have been examined extensively by others and us, [1], [2], [5], [8], [14], in connection with the mesh independence principle. Since the iterates of the inexact Newton-like method (54)–(55) (for $z_n = 0$, $n \geq 0$ or not) can rarely be computed in infinite dimensional spaces, (54)–(55) can be replaced in practice by a family of discretized equations

$$P(a) = 0 \tag{56}$$

indexed by some real number $h > 0$, where P is a nonlinear operator between finite dimensional spaces E^1 and E^2. Let the discretization on E_1 be defined by the bounded linear operators $L : E_1 \to E^1$. Consider also as in [8] the iteration $\{a_n\}$ $(n \geq 0)$ given for all $n \geq 0$ by

$$b_n = a_n - S(a_n)^{-1}P(a_n), \quad a_0 = L(x_0) \tag{57}$$

and

$$a_{n+1} = b_n - d_n. \tag{58}$$

here $S(a_n)$ denotes a linear operator which is an approximation to the Fréchet derivative $P'(a_n)$ of P evaluated at $a = a_n$ for all $n \geq 0$. The points $d_n \in E^1$ for all $n \geq 0$, and are determined in such a way that the iteration $\{a_n\}$ $(n \geq 0)$

converges to a solution a^* of Eq. (56). Note that all symbols introduced in (56)–(58) really depend on h. That is $P = P_h$, $L = L_h$, $S = S_h$, etc. But we do not use the latter to simplify the notation.

The mesh independence principle (proved in the studies listed above) asserts that the number of steps required by the two processes to converge to within a given tolerance is essentially the same. Here we show that this is true for our inexact Newton-like method (54)–(55). In contrast with [8] we consider a nonuniform discretization and use norms corresponding to the domain of definitions of the operators. Our results can be reduced to the ones obtained earlier for appropriate choices of the factors involved. We make use of the sufficient conditions for the convergence of the inexact Newton-like methods (54)–(55) that we found in [8].

The importance of the formulation of an efficient mesh size strategy based upon the mesh-independence has been extensively discussed in [1], [2], [5], [8], [14] and the references there. We finally apply our results to solve a two point boundary value problem.

2.2 Preliminaries

The norms in all spaces will be denoted by the same symbol $\|\ \|$. For any bounded linear operator from E_1 to E_2 or from E^1 to E^2, the induced norm will be used.

We find it convenient to introduce the following:

(G_1) Assume there exists a function $\bar{\alpha} : D^3 \to [0, +\infty)$ such that

$$\|A(y)^{-1}[F(y) - F(x) - A(x)(y - x) + F'(y)(z - y)]\| \leq \bar{\alpha}(x, y, z) \quad (59)$$

for all $x, y, z \in D$.

(G_2) There exist continuous, nondecreasing functions w, w_1, and w_2 such that

$$w : D \to [0, +\infty), \quad w_1, w_2 : [0, R] \to [0, +\infty)$$

with $w_1(0) = w_2(0) = 0$ and a sequence $\{z_n\}$ $(n \geq 0)$ of points with

$$\|z_i\| \leq w(z_i) \leq w_1(r) \quad \text{for all } i \geq 0 \quad (60)$$

and for all $k \in N$

$$\sum_{i=0}^{k+1} \|z_i\| \leq \sum_{i=0}^{k+1} w(z_i) \leq w_2(r) \quad (61)$$

for all $z_i \in D$ and $r \in (0, R]$. We note that (61) implies that the sequence $\{z_n\}$ $(n \geq 0)$ is null.

(G_3) There exist continuous, nondecreasing functions $w_3, w_4 : [0, R] \to [0, +\infty)$ with $w_3(0) = w_4(0) = 0$ such that

$$\overline{\alpha_i} = \bar{\alpha}(x_i, y_i, x_{i+1}) \le w_3(r) \quad \text{for all } i \ge 0 \tag{62}$$

and for all $k \in N$

$$\sum_{i=0}^{k+1} \overline{\alpha_i} \le w_4(r) \tag{63}$$

for all $x_i, y_i, x_{i+1} \in D$ and for $r \in (0, R]$.

Let us define the functions $\varphi, \varphi_1, \varphi_2 : [0, R] \to [0, +\infty)$ by

$$\begin{aligned} \varphi(r) = r - [&w_2(r) + (C(r,r,r) + C_1(r,r) \\ &+ C_2(r,r) + b_1 + b_2 + b_3)r + w_4(r)], \end{aligned} \tag{64}$$

$$\varphi_1(r) = r - \varphi(r) \tag{65}$$

and

$$\varphi_2(r) = C(0,r,r) + C_2(0,r) + C_3(0,r,r) + b_1 + b_4 + w_5(r) \quad \text{for } r \in [0, R], \tag{66}$$

where w_5 is a function with the same properties as function w_1 above.

By the hypotheses on the C and w functions above, there exist constants p_0, p_1, p_2, p_3, c_3, c_4, c_5, c_6 with $c_4 > p_3$ and an integer $h_1 \ge 0$ such that for

$$0 < p_0 \le c_4 - c_5\delta_h \le r_0(h) = r(h) \le c_3\delta_h \le p_1 \le r_0 \le R \tag{67}$$

and $r(h) = 0$ when $a_0 = a^*$ or $h = \infty$, the following are true for sufficiently small $r_0, R > 0$ and $h \ge h_1$

$$0 < \delta_2 \le \varphi_1(r(h)) \le p_3 \quad \text{and} \quad \theta(r(h)) < 1, \tag{68}$$

where

$$\theta(r) = C(R, r, r + R) + C_2(R, r) + C_3(R, r, r + R) + b_1 + b_4 \tag{69}$$

provided that $b_1 + b_4 \in [0, 1)$.

We can now show that for all $h \ge h_4$, where

$$h_4 = \max\{h_1, h_2, h_3\} \quad \text{where } c_3 > c_6, c_6 \ge c_2 \tag{70}$$

and h_2, h_3 are integers such that

$$\delta_{h_2} \le \frac{c_4 - \delta_3}{c_2 + c_5} \quad \text{and} \quad \delta_{h_3} \le \frac{\delta_2}{c_3 - c_6}$$

the following is true

$$0 < c_2\delta_h \le \varphi(r(h)) \le c_6\delta_h. \tag{71}$$

Indeed from (67) $r(h)-p_3 \le \varphi(r(h)) \le r(h)-p_2$. It is enough to show that $r(h)-p_3 > 0$ and $r(h)-p_2 \le c_6\delta_h$, which will be true if $c_4 - c_5\delta_h - p_3 \ge c_2\delta_h$ and $c_3\delta_h - p_2 \le c_6\delta_h$ respectively. The last inequalities are true by the choice of h and (70).

Similar arguments can show that for sufficiently large $h \ge h_4$ there exist $p_4, p_5, p_6, p_7, c_7, c_8, c_9$ such that for

$$0 < p_4 \le c_7 - c_8\delta_h \le r^*(h) \le c_9\delta_h \le p_5 \le r^* = \|x_0 - x^*\|, \quad p_5 \le p_1 \tag{72}$$

and $r^*(h) = 0$ when $L(x^*) = a^*$ or $h = \infty$, the following is true for a sufficiently small $r^* \le r_0$

$$0 < p_6 \le \varphi_1(r^*(h)) \le p_7 < 1. \tag{73}$$

2.3 Convergence Analysis

We can now state the following version of a semilocal Newton-Mysovskii-type theorem whose proof can essentially be found in [8, Thm. 1].

Theorem 5. *Let F, Q be operators defined on some closed convex subset D of a Banach space E_1 with values in a Banach space E_2.*

Assume:

(i) *F is Fréchet-differentiable on D whereas Q is only continuous there;*

(ii) *Let $x_0 \in D$, $A(y)$ be invertible on D and assume:*

$$(G_4) \quad \left\| \int_0^1 A(y)^{-1}(F'(x) + t(y-x)) - A(x))(y-x)dt \right\| \le [C(\|x-x_0\|, \|y-x_0\|, \|y-x\|) + b_1]\|y-x\|, \tag{74}$$

$$\|A(y)^{-1}(F'(x) - A(x))(y-x)\| \le [C_1(\|x-x_0\|, \|y-x_0\|) + b_2]\|y-x\|, \tag{75}$$

$$\|A(y)^{-1}(Q(y) - Q(x))\| \le [C_2(\|x-x\|, \|y-x_0\|) + b_3]\|y-x\| \tag{76}$$

$$\|A(y)^{-1}(A(x) - A(y))(y-x)\| \le [C_3(\|x-x_0\|, \|y-x_0\|, \|y-x\| + b_4]\|y-x\| \tag{77}$$

for all $x, y \in D$. The real functions C, C_3, C_1 and C_2 are assumed to be continuous and nondecreasing on $[0,R]^3$, $[0,R]^3$, $[0,R]^2$ and $[0,R]^2$

respectively for some $R > 0$. *Moreover* $C(0,0,0) = C_1(0,0) = C_2(0,0) = C_3(0,0,0) = 0$ *and* b_1, b_2, b_3, b_4 *are fixed nonnegative real numbers.*

(iii) *There exists a minimum nonnegative number* r_0 *such that*

$$T(r_0) \leq r_0, \ r_0 \leq R \quad \text{and} \quad U(x_0, r_0) \leq D, \tag{78}$$

where $T(r) = s_0 + \varphi_1(r)$.

(iv) *Sequences* $\{\alpha_n\}$, $\{l_n\}$, $\{\bar{\alpha}_n\}$, $\{z_n\}$ $(n \geq 0)$ *satisfy condtions* (G_2) *and* (G_3) *on* D *and* $\|z_n\| \leq l_n$ $(n \geq 0)$.

(v) *Condition* (G_1) *is satisfied on* D;

(vi) *numbers* r_0, R *satisfy inequality*

$$\theta(r_0) < 1, \tag{79}$$

where

$$\theta(r) = C(R, r, r + R) + C_2(R, r) + C_3(R, r, r + R) + b_1 + b_4. \tag{80}$$

Then

(a) *scalar sequence* $\{t_n\}$ $(n \geq 0)$ *generated by relations* (80)–(81) *in* [13] *is monotonically increasing and bounded above by its limit, which is number* r_0;

(b) *sequences* $\{x_n\}$, $\{y_n\}$ $(n \geq 0)$ *generated by relations* (54)–(55) *are well defined, remain in* $U(x_0, r_0)$ *for all* $n \geq 0$ *and converge to a solution* x^* *of the equation* $F(x) + Q(x) = 0$, *which is unique in* $U(x_0, R)$; *and*

(c) *following error estimates are true for all* $n \geq 0$:

$$\begin{gathered}\|y_n - x_n\| \leq s_n - t_n, \|x_{n+1} - y_n\| = \| - z_n\| \leq t_{n+1} - s_n,\\ \|x_n - x^*\| \leq r_0 - t_n, \ \|y_n - x^*\| \leq r_0 - s_n,\\ \|y_{n+1} - x_{n+1}\| = \|A(x_{n+1})^{-1}(F(x_{n+1} + Q(x_{n+1}))\| \leq \bar{v}_{n+1}\end{gathered}$$

and

$$\begin{aligned}\|y_n - x_n\| \leq \|x^* - x_n\| + [C(\|x_n - x_0\|, \|x^* - x_0\|, \|x^* - x_n\|) + b_1\\ + C_2(\|x_n - x_0\|, \|x^* - x_0\|)]\|x_n - x^*\|.\end{aligned}$$

We will need the following result on local convergence.

Theorem 6. *Let $F, Q : D \subset E_1 \to E_2$ be nonlinear operators as in Theorem 5, and assume:*

(i) *There exists a simple solution $x^* \in D$ of equation $F(x) + Q(x) = 0$;*

(ii) *condition (G_4) is satisfied on $U(x^*, r^*)$ (for $x_0 = x^*$) and $U(x^*, r^*) \subseteq D$;*

(iii) *there exists a sequence $\{z_n\}$ $(n \geq 0)$ of points from D satisfying*

$$\|z_n\| \leq g_n = g(z_n) \leq w_5(r) \tag{81}$$

where $g : U(x^*, r^*) \to [0, +\infty)$ is continuous; and

(iv) *the constants b_1, b_4 are such that $b_1 + b_4 \in [0, 1)$.*

Then the following are true:

(a) *For a sufficiently small $r^* \in (0, R]$*

$$0 < \varphi_2(r^*) < 1. \tag{82}$$

(b) *Sequences $\{y_n\}$, $\{x_n\}$ $(n \geq 0)$ are well defined, remain in $U(x^*, r^*)$ for all $n \geq 0$ and $\lim_{n\to\infty} x_n = \lim_{n\to\infty} y_n = x^*$. Moreover, the solution x^* of Eq. (53) is unique in $U(x^*, r^*)$.*

Furthermore, the following estimates are true for all $n \geq 0$:

$$\|x_{n+1} - x^*\| \leq \gamma_n \|x_n - x^*\| \leq \gamma \|x_n - x^*\| \tag{83}$$

and

$$\|y_n - x^*\| \leq \delta_n \|x_n - x^*\| \leq \delta \|x_n - x^*\|, \tag{84}$$

where

$$\begin{aligned} \delta_n &= C(0, \|x_n - x^*\|, \|x_n - x^*\|) + C_2(0, \|x_n - x^*\|) \\ &\quad + C_3(0, \|x_n - x^*\|, \|x_n - x^*\|) + b_1 + b_4, \end{aligned} \tag{85}$$

$$\delta = C(0, r^*, r^*) + C_2(0, r^*) + C_3(0, r^*, r^*) + b_1 + b_4, \tag{86}$$

$$\gamma_n = \delta_n + g_n \tag{87}$$

and

$$\gamma = \delta + w_5(r^*). \tag{88}$$

The points z_n $(n \geq 0)$ appearing in (61) depend on x_n $(n \geq 0)$, whereas the ones appearing in (81) depend on x_n $(n \geq 0)$ and (maybe) the point x^*. That is why we can choose the functions g, w_5 (see (81)) to be the same or different from the functions w, w_1 (see (61)) respectively.

Theorem 7. *Assume that hypotheses of Theorems 5 and 6 are true. Then point x^* is the unique solution of the equation $F(x) + Q(x) = 0$ in connected region D_0 given by*

$$D_0^* = \bigcup_{n=0}^{\infty} U(x_n, R) \cap D. \tag{89}$$

Proof. Let R be such that $\theta(R) < 1$, then from (66) and (69) we obtain $\varphi_2(R) < 1$ also. Hence we can set $r^* = R$. We observe that under the hypotheses of Theorem 1 we have $x_{n+1} \in U(x_n, R)$ since

$$\|x_{n+1} - x_n\| \leq \|x_1 - x_0\| \leq r_0 \leq R,$$

so that D^* is a connected set. We have, clearly $U(x^*, R) \cap D \subseteq D^*$. Moreover we have that $x^* \in U(x_0, R)$. Suppose now there exists $y^* \in D$ such that $F(y^*) + Q(y^*) = 0$. We choose x_0 such that $x_0 \in U(y^*, R)$, which is equivalent to $y^* \in U(x_0, R)$. Theorem 6 guarantees that the solution y^* is unique in $U(y^*, R)$. Assume that there exists another solution

$$y^{**} \in U(x_0, R) \setminus U(y^*, R)$$

with $F(y^*) + Q(y^*) = 0$. Then y^{**} must be the unique solution in $U(y^*, R)$ and the sequence of Newton-like iterates (54)–(55) starting at x_0 must converge to y^{**}—which is a contradiction to the uniqueness of the sequence of Newton-like iterates. Hence, y^* is unique in $U(x_0, R)$. The same argument can be extended over all Newton-like iterates x_n.

That completes the proof of the theorem.

2.4 The Asymptotic Mesh Independence Principle

In Section 2.2 we provided sufficient conditions under which the inexact Newton-like iteration generated by (54)–(55) converges to a locally unique solution x^* of Eq. (53). Since the formal procedure (54)–(55) can rarely be executed in infinite dimensional spaces, (53) is replaced by a family of discretized equations

$$F_h(v^h) + Q_h(v^h) = 0, \quad h \geq 0 \tag{90}$$

where $F_h : D_h \subseteq E_1^h \to E_2^h$ is a nonlinear operator defined on a convex domain D_h of a finite dimensional subspace $E_1^h \subseteq E_2$ with values in a finite-dimensional subspace $E_2^h \subseteq E_2$. That is we are interested in Galerkin methods.

We introduce the discretized-inexact Newton-like iterations of the form

$$y_h^n = x_h^n - A_h(x_h^n)^{-1}(F_h(x_h^n) + Q_h(x_h^n)) \tag{91}$$

$$x_h^{n+1} = y_h^n - z_h^n \quad n \geq 0. \tag{92}$$

We wish to choose the operators F_h in such a way that a solution x_h^* of (90) can be found for each h so that

$$\lim_{h\to\infty} x_h^* = x^*. \tag{93}$$

In practice we will solve the finite dimensional linear equation

$$A_h(x_h^n)\Delta x_h^n = -(F_h(x_h^n) + Q_h(x_n^h)), \tag{94}$$

then set

$$y_h^n = x_h^n + \Delta x_h^n, \tag{95}$$

and

$$x_h^{n+1} = x_h^n + \Delta x_h^n - z_n^h \quad (n \geq 0, h \geq 0). \tag{96}$$

We restrict ourselves to a subset $E_3 \subseteq E_1$ containing elements that have better smoothness properties than the generic elements of E_1. That is we assume that

$$\{x^*, x_n, y_n, z_n, x_n - x^*, y_n - x^*, \Delta x_n\} \subseteq E_3 \quad (n \geq 0). \tag{97}$$

Let $\pi_h : E_1 \to E_1^h$, $h \geq 0$ be a family of linear projection operators. We assume that the family $\{\pi_h\}$ $(h \geq 0)$ satisfy a stability condition of the form

$$\|\pi_h(x)\| \leq q_h\|x\|, \quad x \in E_3,\ q_h \leq q < \infty. \tag{98}$$

Since $\|\pi_h^2(x)\| \leq q_h\|\pi_h(x)\|$ and $\pi_h^2 = \pi_h$, we get $\|\pi_h(x)\| \leq q_h\|\pi_h(x)\| \Rightarrow$

$$q_h \geq 1 \quad (h \geq 0). \tag{99}$$

Moreover, we assume that

$$\|x - \pi_h(x)\| \leq \delta_h\|x\|, \quad x \in E_3\ (h \geq 0). \tag{100}$$

Using the triangle inequality we can get

$$\|\pi_h(x)\| \leq \|x - \pi_h(x)\| + \|x\| \leq \delta_h\|x\| + \|x\| = (1 + \delta_h)\|x\|.$$

Hence any best possible approximation (98) will imply that

$$q_h \leq 1 + \delta_h. \tag{101}$$

We will also assume that

$$\delta_{h+1} \le \delta_h \quad (h \ge 0) \tag{102}$$

and

$$\lim_{h\to\infty} \delta_h = 0. \tag{103}$$

Relations (99) and (101) now lead to the asymptotic stability condition

$$\lim_{h\to\infty} q_h = 1. \tag{104}$$

Thus our discretization method is characterized by the family

$$\{F_h, Q_h, A_h(\cdot), \pi_h, \delta_h\} \quad h \ge 0. \tag{105}$$

We assume that the domains D_h are such that

$$U_h(\pi_h(x^*), p_5) \subseteq D_h \subseteq D \quad h \ge 0,\ p_5 \le r^*. \tag{106}$$

Furthermore we assume that the discretization is consistent if the conditions (G_5) and (G_6) that follow are satisfied:

(G_5) If $u_j \in E_1^h$ and $u \in E_1$ are solutions of the linear equations

$$A_h(\pi_h(x))(u_h) = -(F_h(\pi_h(x)) + Q_h(\pi_h(x))), \tag{107}$$
$$A_h(x)(u) = -(F(x) + Q(x)) \tag{108}$$

where $x \in E_3 \cap D$, $\pi_h(x) \in D_h$, then

$$\|u_h - \pi_h(u)\| \le e_0\delta_h \quad (h \ge 0) \tag{109}$$

for some positive constant e_0 with $e_0 \le c_2$.

(G_6) If w_h^1, w_h^2, w_h^3, w_h^4, $w_h^5 \in E_1^h$ and $w^1, w^2, w^3, w^4, w^5 \in E_1$ are (respectively) solutions of the linear equations

$$A_h(y_h^1)(w_h^1) = \int_0^1 [F_h'(x_h^1 + t(y_h^1 - x_h^1)) - A_h(x_h^1)](y_h^1 - x_h^1)dt, \tag{110}$$
$$A_h(y_h^2)(w_h^2) = (F_h'(x_h^2) - A_h(x_h^2))(y_h^2 - x_h^2), \tag{111}$$
$$A_h(y_h^3)(w_h^3) = Q_h(y_h^3) - Q_h(x_h^3),$$
$$A_h(y_h^4)(w_h^4) = (A_h(x_h^4) - A_h(y_h^4))(y_h^4 - x_h^4), \tag{112}$$
$$A_h(y_h^5)(w_h^5) = F_h(y_h^5) - F_h(x_h^5) - A_h(x_h^5)(y_h^5) - x_h^5)$$

$$+ F_h'(y_h^5)(z_h^5 - y_h^5), \tag{113}$$

$$A(y_h^1)(w^1) = \int_0^1 [F'(x_h^1 + t(y_h^1 - x_h^1)) - A(x_h^1)](y_h^1 - x_h^2)dt, \tag{114}$$

$$A(y_h^2)(w^2) = (F'(x_h^2) - A(x_h^2))(y_h^2 - x_h^2), \tag{115}$$

$$A(y_h^3)(w^3) = Q(y_h^3) - Q(x_h^3), \tag{116}$$

$$A(y_h^4)(w^4) = (A(x_h^4) - A(y_h^4))(y_h^4 - x_h^4), \tag{117}$$

$$A(y_h^5)(w^5) = F(y_h^5) - F(x_h^5) - A(x_h^5)(y_h^5 - x_h^5) + F(y_h^5)(z_h^5 - y_h^5), \tag{118}$$

where the given points y_h^1, y_h^2, y_h^3, y_h^4, x_h^1, x_h^2, x_h^3, x_h^4, $z_h^5 \in D_h$, then

$$\|w_h^1 - \pi_h(w^1)\| \le e_1\delta_h\|y_h^1 - x_h^1\|, \tag{119}$$

$$\|w_h^2 - \pi_h(w^2)\| \le e_2\delta_h\|y_h^2 - x_h^2\|, \tag{120}$$

$$\|w_h^3 - \pi_h(w^3)\| \le e_3\delta_h\|y_h^3 - x_h^3\|, \tag{121}$$

$$\|w_h^4 - \pi_h(w^4)\| \le e_4\delta_h\|y_h^4 - x_h^4\|, \tag{122}$$

and

$$\|w_h^5 - \pi_h(w^5)\| \le e_5\delta_h \tag{123}$$

for some positive constants e_1, e_2, e_3, e_4, and e_5.

We find it convenient to define the functions C^h, C_1^h, C_2^h, C_3^h, $\bar{\alpha}^h$ by

$$C^h = q_hC, C_1^h = q_hC_1, C_2^h = q_hC_2, C_3^h = q_hC_3, \bar{\alpha}^h = e_5\delta_h + q_h\bar{\alpha} \tag{124}$$

and the constants $b_1^h, b_2^h, b_3^h, b_4^h$ by

$$b_1^h = e_1\delta_h + b_1q_h, b_2^h = e_2\delta_h + b_2q_h, b_3^h = e_3\delta_h + b_3q_h \text{ and } b_4^h = e_4\delta_h + b_4q_h. \tag{125}$$

We can now give the connection between the C^h, b^h's associated with the finite-dimensional approximating nonlinear systems with the C, b's associated with the operator equation.

Lemma 3. *Let $F, Q : D \subseteq E_1 \to E_2$ satisfy hypotheses (G_4), and suppose that the discretization method (105) satisfies (98), (100), (101), (106), and (G_6). Then the following are true for all $h \ge 0$*

$$\left\| \int_0^1 A_h(y_h^1)^{-1}[F_h'(x_h^1 + t(y_h^1 - x_h^1)) - A_h(x_h^1)](y_h^1 - x_h^1)dt \right\|$$
$$\le [C^h(\|x_h^1 - \pi_h(x^*)\|, \|y_h^1 - \pi_h(x^*)\|, \|y_h^1 - x_h^1\|)$$

$$+ b_1^h]\|y_h^1 - x_h^1\|, \tag{126}$$

$$\|A_h(y_h^2)^{-1}(F_h'(x_h^2) - A_h(x_h^2))(y_h^2 - x_h^2)\|$$
$$\leq [C_1^h(\|x_h^2 - \pi_h(x^*)\|, \|y_h^2 - \pi_h(x^*)\|) + b_2^h]\|y_h^2 - x_h^2\|, \tag{127}$$

$$\|A_h(y_h^3)^{-1}(Q_h(y_h^3) - Q_h(x_h^3))\|$$
$$\leq [C_2^h(\|x_h^3 - \pi_h(x^*)\|, \|y_h^3 - \pi_h(x^*)\|) + b_3^h]\|y_h^3 - x_h^3\|, \tag{128}$$

and

$$\|A_h(y_h^4)^{-1}(A_h(x_h^4) - A_h(y_h^4))(y_h^4 - x_h^4)\|$$
$$\leq [C_3^h(\|x_h^4 - \pi_h(x^*)\|, \|y_h^4 - \pi_h(x^*)\|, \|y_h^4 - x_h^4\|) + b_4^h]\|y_h^4 - x_h^4\| \tag{129}$$

$$\|A_h(y_h^5)^{-1}[F_h(y_h^5) - F_h(x_h^5) - A_h(x_h^5)(y_h^5 - x_h^5)$$
$$+ F_h'(y_h^5)(z_h^5 - y_h^5)] \leq \bar{\alpha}^h(x_h^5, y_h^5, z_h^5). \tag{130}$$

where C^h, C_1^h, C_2^h, C_3^h, $\bar{\alpha}^h$ and b_1^h, b_2^h, b_3^h, b_4^h are given by (124) and (125) respectively.

Proof. We will only show (126), since (127)–(130) can then follow similarly. Using the triangle inequality (98) and (100), we can obtain

$$\|w_h^1\| \leq \|w_h^1 - \pi_h(w^1)\| + \|\pi_h(w^1)\| \leq e_1\delta_h\|y_h^1 - x_h^1\| + q_h\|w^1\|. \tag{131}$$

From (74) and (110) we get

$$\|w^1\| = \left\| \int_0^1 A(y_h^1)^{-1}[F'(x_h^1 + t(y_h^1 - x_h^1)) - A(x_h^1)](y_h^1 - x_h^1)dt \right\|$$
$$\leq [C(\|x_h^1 - \pi_h(x^*)\|, \|y_h^1 - \pi_h(x^*)\|, \|y_h^1 - x_h^1\|)$$
$$+ b_1]\|y_h^1 - x_h^1\|. \tag{132}$$

Estimate (126) now follows by simply combining (131) and (132) and using the definitions (124) and (125).

That completes the proof of the lemma.

Remarks 6. (1) The pair (C_1^h, b_1^h) chosen best possible governs the convergence speed of the inexact Newton-like iterations in the subspace E_1^h. Using (124) and (125) we obtain that

$$\lim_{h\to\infty} [C^h(\|x_h^1 - \pi_h(x^*)\|, \|y_h^1 - \pi_h(x^*)\|, \|y_h^1 - x_h^1\|) + b_1^h]$$
$$\leq \lim_{h\to\infty} [C^h(p_5, p_5, p_5) + b_1^h] = C(p_5, p_5, p_5) + b_1 \tag{133}$$

which means that the discrete convergence speed is asymptotically constant and just the one for the continuous case, assuming that $C^h(p_5, p_5, p_5) = k^h(p_5) = k^h$ will be replacing

$$C^h(\|x_h^1 - \pi_h(x^*)\|, \|y_h^1 - \pi_h(x^*)\|, \|y_h^1 - x_h^1\|) \quad (h \geq 0).$$

Similar remarks can be made for the rest of the C^h, $\bar{\alpha}^h$ functions

(2) We can also set $w^h = w$ and $w_k^h = w_k$, $k = 1, 2, 3, 4, 5$ for the w functions appearing in the discretization method. These assumptions are very natural. In particular if the null sequences $\{z_n\}$, $\{d_n\}$ $(n \geq 0)$ are in $U(x_0, R) \subseteq D$, set $D_h = U_h(\pi_h(x^*), p_5)$. Choose $\pi_h(x_0) \in U(x_0, \frac{R}{3})$, $p_5 \leq \frac{R}{3}$ and $r^* \leq \frac{R}{3q}$. Then we can easily see that

$$D_h \subseteq U(x_0, R) \quad (h > 0).$$

In this case w^h functions are just the w functions restricted on D_h $(h > 0)$.

Finally we can easily see from (60) and (81) that function w_2 can be identified with the function w_5. Choose for example $z_n = d_n$ for all $n \geq 0$, $h > 0$. However we do not need to use this to prove our discretization results.

We can now state the following local results.

Theorem 8. *Under the hypotheses of Theorem 5, consider a discretization family (105) that satisfies (98), (100), (101) and (106). Moreover, assume:*

(i) *operators F_h are Fréchet-differentiable, whereas the operators Q_h are continuous on D_h;*

(ii) *operators $F_h'(x_h)$ are invertible for each $x_h \in D_h$;*

(iii) *conditions (G_5) and (G_6) are satisfied; and*

(iv) *the following are true $w^h = w$, $w_k^h = w_k$ for $k = 1, 2, 3, 4, 5$ and $h \in (0, h_4]$.*

Then there exists an index $h_0 \in N$ and a number $r^*(h)$ such that for $h \geq h_0$ and $r^*(h) \in (0, p_5]$, equation (90) has a solution x_h^* satisfying inequality

$$\|x_h^* - \pi_h(x^*)\| \leq r^*(h) \leq c_3 \delta_h. \tag{134}$$

Moreover point x_h^* is the unique solution of Eq. (90) in region

$$U_h(\pi_h(x^*), p_5) \cap D_h. \tag{135}$$

Proof. The C functions are continuous, vanish at the origin, $\lim_{h\to\infty} \delta_h = 0$, $\lim_{h\to\infty} q_h = 1$ and $b_1 + b_4 \in [0, 1)$. Hence we can find an index $h_0 > 0$

and an interval $(0, p_5]$ with $p_5 \leq r^*$ such that conditions (82) are satisfied for all $h \geq h_0 \geq h_4$ and $r^*(h) \in (0, p_5]$. We set $x_h^0 = \pi_h(x^*)$ and consider the discretized inexact Newton-like iterations generated by (91)–(92). It is convenient to introduce the approximations

$$u_h^* = A_h(\pi_h(x^*))^{-1}(F_h(\pi_h(x^*)) + Q_h(\pi_h(x^*))),$$

and

$$s_0^*(h) = \|u_h^*\|.$$

Since x^* is a solution of Eq. (53), we have

$$u^* = A(x^*)^{-1}(F(x^*) + Q(x^*)) = 0,$$

and by using condition (G_5) and (109) we get in turn

$$s_0^*(h) = \|u_h^*\| = \|u_h^* - \pi_h(u^*)\| \leq e_0\delta_h \leq c_2\delta_h. \tag{136}$$

From the analysis in Remark No. 2 there exists an index h_4 such that $w^h = w$ and $w_k^h = w_k$ $k = 1, 2, 3, 4, 5$ on $D_h = U_h(\pi_h(x^*), p_1)$ for all $h_4 \geq h$. Using (64), (124) and (125) we get

$$\begin{aligned}\varphi_h(r) = \varphi(r) &+ (1 - q_h)[C(r, r, r)r + C_1(r, r)r + C_2(r, r)r + (b_1 + b_2 + b_3)r] \\ &- (e_1 + e_2 + e_3)\delta_h \leq \varphi(r).\end{aligned} \tag{137}$$

Since $\lim_{h\to\infty} q_h = 1$ and $\lim_{h\to\infty} \delta_h = 0$ there exists an index h_5 such that (under (71))

$$c_2\delta_h \leq \varphi_h(r(h)) \leq \varphi(r(h)) \leq c_6\delta_h. \tag{138}$$

Hence from (138) we get

$$s_0^*(h) \leq \varphi_h(r(h))$$

and

$$r(h) \leq c_3\delta_h \leq p_1.$$

Hence by Theorem 5 the inexact-Newton-like iteration generated by (91)–(92) converges to a solution $x_h^* \in U_h(\pi_h(x^*), r(h))$, which is the unique solution of Eq. (90) in the region

$$D_h^* = \bigcup_{k=0}^{\infty} U_h(x_h^k, p_1) \cap D.$$

In particular x_h^* is the unique solution of Eq. (90) in the region (135).

That completes the proof of the theorem.

We note that conditions $p_1 \leq r_0$ and $p_5 \leq r^*$ (see (67) and (72)) are not used in the proof. However in many applications we may want these conditions to be true. We find it convenient to introduce the approximations

$$N = N(x_0, \varepsilon) = \left] \frac{\log(\frac{\varepsilon}{\|x_0 - x^*\|})}{\log \gamma} \right[, \quad \text{and} \tag{139}$$

$$N_h = N_h(x_h^0, \varepsilon) = \left] \frac{\log(\frac{\varepsilon}{\|x_h^0 - x_h^*\|})}{\log \gamma_h} \right[, \tag{140}$$

where $\varepsilon > 0$, $x_0 \neq x^*$, $x_h^0 \neq x_h^*$, γ, γ_h are given by (88) and $]x[= \min\{i \mid i$ integer, $i \geq x\}$.

We also note that if we want to find an integer $N = N(x_0, \varepsilon)$ such that

$$\|x_k - x^*\| \leq \varepsilon \quad \text{for } n \geq N \tag{141}$$

if

$$\|x_0 - x^*\| < p_1 \tag{142}$$

then provided that the choice of N given by (139) will do. Indeed, using (83) we get

$$\|x_n - x^*\| \leq \gamma^n \|x_0 - x^*\|. \tag{143}$$

Estimate (141) will certainly be true if

$$\gamma^n \|x_0 - x^*\| \leq \varepsilon \tag{144}$$

which is true by the choice of N. Similarly

$$\|x_h^k - x_h^*\| \leq \varepsilon \quad \text{for } n \geq N_h, \tag{145}$$

if

$$\|x_h^0 - x_h^*\| < p_5. \tag{146}$$

Since N_h and N are integer valued, they will differ by at most one whenever $\|x_0^h - x_h^*\|$, γ_h are close enough to each other. This will happen if condition (133) holds as equality and

$$x_h^0 = \pi_h(x_0). \tag{147}$$

We can now formalize the asymptotic mesh independence principle.

Theorem 9. *Assume:*

(i) *hypotheses of Theorem 8 are satisfied;*

(ii) *condition (133) holds as equality.*

Then

(a) *there exists $h_5 \geq h_0 \geq h^*$ such that for each $h \geq h_5$ the sequence of discretized Newton-like iterates (91)–(92) with starting points (147) converges to x_h^*.*

(b) *Moreover, inequalities (142) and (146) are satisfied for $h \geq h_5$ and*

$$|N_h - N| \leq 1 \quad \textit{for} \quad h \geq h_5 \tag{148}$$

where N, N_h are given by (139) and 140 respectively.

Proof. Using Theorem 5 we get that $x^* \in U(x_0, p_1)$ and hence

$$\|x_0 - x^*\| \leq p_1 \leq r_0$$

which shows (142). From the triangle inequality and (147) we obtain

$$|\|\pi_h(x_0 - x^*) - \|\pi_h(x^*) - x_h^*\|| \leq \|x_h^0 - x_h^*\| \leq \|\pi_h(x_0 - x^*)\| + \|\pi_h(x^*) - x_h^*\|$$

and by Theorem 8 and (133) we conclude that

$$\lim_{h \to \infty} \|x_h^0 - x_h^*\| = \|x_0 - x^*\|.$$

Moreover from (133) and (142) it follows that

$$\lim_{h \to \infty} r(h) = p_1 \tag{149}$$

and

$$\lim_{h \to \infty} \|x_h^0 - x_h^*\| = \|x_0 - x^*\|. \tag{150}$$

Hence by (149) and (150) we conclude that (146) and (148) are satisfied for all h sufficiently large.

That completes the proof of the theorem.

Remarks 7. (3) In all our previous results we assumed that $A(y)$ is invertible for all $y \in D$. It turns out that our results hold under the weaker condition that $A(x_0)$ is invertible only. For Theorem 5 replace $A(y)^{-1}$ by $A(x_0)^{-1}$ in (74)–(77) and add the hypotheses

$$\|A(x_0)^{-1}(A(x) - A(x_0))\| \leq C_0(\|x - x_0\|) + b_0, \quad b_0 \in [0, 1) \tag{151}$$

and

$$C_0(r_0) + b_0 < 1 \tag{152}$$

for all $x \in U(x_0, r) \subseteq U(x_0, R)$, $r \in [0, R]$, where the function C_0 is continuous and nondecreasing on $[0, R]$ with $C_0(0) = 0$. Moreover define the function C_5 on $[0, R]$ by

$$C_5(r) = [1 - (C_0(r) + b_0)]^{-1}. \tag{153}$$

Then using (151), (153) and the Banach lemma on invertible operators we get that $A(x_n)$ is invertible and

$$\|A(x_n)^{-1}A(x_0)\| \leq C_5(\|x_n - x_0\|). \tag{154}$$

Furthermore multiply the "C", w_4, $\bar{\alpha}$ functions by $C_5(r)$ (or $C_5(\|x_n - x_0\|)$) and the "b" constants by $C_5(r)$ (or $C_5(\|x_n - x_0\|)$).

With the above modifications one can easily see that the conclusions of Theorem 5 can now follow.

(4) Similarly, for Theorem 6, we can argue as in Remark No. 3 but with the following modifications x_0, C_0, b_0, C_5, w_4 are x^*, C_0^*, b_0^*, C_5^* and w_5 respectively. The "C^*" functions and the point b_0^* have properties similar to the ones without the stars. We note that they can even be taken to be equal to each other, and if this is true the rest of the results in this study can follow.

(5) The results obtained in this study will also hold if the left-hand side of (74) is replaced by the conditions

$$\|A(y)^{-1}(F'(x + t(y - x)) - A(x))(y - x)\| \quad \text{for all } t \in [0, 1],$$

or

$$\int_0^1 \|A(y)^{-1}(F'(x + t(y - x)) - A(x))(y - x)\| dt$$

or

$$\|A(y)^{-1}(F'(x + t(y - x)) - A(x))\| \, \|y - x\| \quad \text{for all } t \in [0, 1],$$

or

$$\int_0^1 \|A(y)^{-1}(F'(x + t(y - x)) - A(x))\| \, \|y - x\| dt$$

or any combination of the above conditions in a non-affine form or with $A(y) = A(x_0)$ (or not). (See the proof of Theorem 1 in [8]. In particular see relation (96).) In all cases function C and the point b, appearing at the right-hand sides of (74) may become larger which will result to larger upper bounds on the distances

$$\|y_n - x_n\|, \|x_{n+1} - y_n\|, \|x_n - x^*\| \quad \text{and} \quad \|y_n - x^*\| \quad (n \geq 0).$$

Similar remarks can be made for the left-hand sides of conditions (75), (76), and (77).

(6) Our results can be further generalized if we can find a continuous, nondecreasing and vanishing at the origin function C_6 on $[0, R]^6$ and a null sequence $\{\varepsilon_n\}$ $(n \geq 0)$ such that

$$\begin{aligned}\|y_{n+1} - x_{n+1}\| \leq C_4(&\|x_n - x_0\|, \|y_n - x_0\|, \|x_{n+1} - x_0\|, \\ &\|x_{n+1} - y_n\|, \|y_n - x_n\|, \|z_n\|) = \varepsilon_{n+1}.\end{aligned}$$

It suffices to show that the iteration $\{t_n\}$ $(n \geq 0)$ is Cauchy etc. A choice for ε_n is given by

$$\varepsilon_n = \bar{v}_n \quad (n \geq 1) \quad \text{(see (33) and (43) in [8]).}$$

Many other choices are possible.

(7) Theorem 5 can be reduced to Theorem 1 in [14] (see also [18], [19]). Set $Q(x) = 0$, $A(x) = F'(x)$ $(x \in D)$, $z_n = 0$ $(n \geq 0)$, $C_1 = C_2 = 0$ and $b_2 = b_3 = 0$. Assume that condition (1.5) in [10] is satisfied, then (74) is also satisfied if we choose $2C_3 = C(\|x - x_0\|, \|y - x_0\|, \|y - x\|) = \frac{1}{2}\omega\|y - x\|$ and $b_1 = b_4 = 0$. Similarly Theorem 5 can be reduced to the corresponding theorems in [14], [18], [19]. We also note that the results in Theorem 1.5 in [15] have been obtained under the assumption that the operator F is twice Fréchet differentiable, whereas here we only assume that F is only Fréchet differentiable. Moreover the values of the crucial "c" constants appearing in (67) and (72) can be found in [8, Remark (8)].

2.5 Applications

In this section we provide an example where our results apply whereas the ones obtained in [1], [12]–[24] do not. We will use the condition

$$\|A(x_0)^{-1}(F'(x) - F'(y))\| \leq c\|x - y\|^p \tag{155}$$

for some fixed $p \in [0, 1]$ and all $x, y \in U(x_0, R) \subseteq D$. For $p = 1$ condition (155) is called Lipschitz and has been used in [1], [12]–[24], whereas for $p \in [0, 1)$, (155) is called Hölder. In particular the results obtained in [15] cannot be used under Hölder continuity assumptions. However ours can. To show that, let us choose $A(x) = F'(x)$, $Q(x) = 0$ for all $x \in D$, $z_n = 0$ $(n \geq 0)$, $C_1 = C_2 = 0$, $w_2 = w_4 = 0$, $b_0 = b_1 = b_2 = b_3 = b_4 = 0$, and the functions C_3, C and C_0 such that

$$\begin{aligned}\frac{1}{1+p} C_3(\|x - x_0\|, \|y - x_0\|, \|y - x\|) &= C(\|x - x_0\|, \|y - x_0\|, \|y - x\|) \\ &= \frac{c}{(1+p)} \frac{\|x - y\|^p}{(1 - c\|x - x_0\|^p)}\end{aligned}$$

and

$$C_0(\|x-x_0\|) = c\|x-x_0\|^p \quad \text{(see also Remark No. 7).}$$

It can then easily be seen that conditions (78) and (79) become

$$s_0 + \frac{c}{(1+p)(1-cr^p)} r^{1+p} \leq r \tag{156}$$

and

$$\frac{c(2+p)(r_0+R)^p}{(1+p)(1-cr^p)} < 1, \tag{157}$$

respectively.

Example 2. Consider the two-point boundary value problem

$$x'' + x^{1+p} = 0, \quad p \in [0,1] \tag{158}$$
$$x(0) = x(1) = 0. \tag{159}$$

We divide the interval $[0,1]$ into n subintervals and we set $h = \frac{1}{n}$. Let $\{v_k\}$ $k = 0,1,2,\ldots,n$ be the points of subdivision with

$$p = v_0 < v_1 < \cdots < v_n = 1.$$

A standard approximation for the second derivative is given by

$$x_i'' = \frac{x_{i-1} - 2x_i + x_{i+1}}{h^2}, \quad x_i = x(v_i), \quad i = 1,2,\ldots,n-1.$$

Take $x_0 = x_n = 0$ and define operator $\tilde{F} : R^{n-1} \to R^{n-1}$ by

$$\tilde{F}(x) = H(x) - h^2\varphi(x) \tag{160}$$

$$H = \begin{bmatrix} 2 & -1 & & \\ -1 & 2 & -1 & 0 \\ \ddots & \ddots & \ddots & \\ 0 & -1 & 2 & -1 \\ & & -1 & 2 \end{bmatrix},$$

$$\varphi = \begin{bmatrix} x_1^{1+p} \\ x_2^{1+p} \\ \vdots \\ x_{n-1}^{1+p} \end{bmatrix},$$

and

$$x = \begin{bmatrix} x_1 \\ x_2 \\ \vdots \\ x_{n-1} \end{bmatrix}.$$

Then

$$\tilde{F}'(x) = H + h^2(p+1) \begin{bmatrix} x_1^p & & & 0 \\ & x_2^p & & \\ & & \ddots & \\ 0 & & & x_{n-1}^p \end{bmatrix}.$$

Newton's method cannot be applied to the equation

$$\tilde{F}(x) = 0. \tag{161}$$

We may not be able to evaluate the second Fréchet-derivative since it would involve the evaluation of quantities of the form x_i^{-p} and they may not exist.

Let $x \in R^{n-1}$, $H \in R^{n-1} \times R^{n-1}$ and define the norms of x and H by

$$\|x\| = \max_{1 \le j \le n-1} |x_j|$$

$$\|H\| = \max_{1 \le j \le n-1} \sum_{k=1}^{n-1} |h_{jk}|.$$

For all $x, z \in R^{n-1}$ for which $|x_i| > 0$, $|z_i| > 0$, $i = 1, 2, \ldots, n-1$ we obtain, for $p = \frac{1}{2}$ say,

$$\begin{aligned} \|\tilde{F}'(x) - \tilde{F}'(z)\| &= \left\| \operatorname{diag}\left\{ \left(1 + \frac{1}{2}\right) h^2 \left(x_j^{1/2} - z_j^{1/2} \right) \right\} \right\| \\ &= \frac{3}{2} h^2 \max_{1 \le j \le n-1} |x_j^{1/2} - z_j^{1/2}| \le \frac{3}{2} h^2 \left[\max_{1 \le j \le n-1} |x_j - z_j| \right]^{1/2} \\ &= \frac{3}{2} h^2 \|x - z\|^{1/2}. \end{aligned}$$

Hence, the results in [1], [8]–[24] cannot be applied here.

Newton's method can be seen as a system of linear equations of the form

$$\tilde{F}'(z_m)(z_m - z_{m+1}) = \tilde{F}(z_m) \quad m \ge 0 \quad z_0 \in R^{n-1}. \tag{162}$$

We choose $n = 10$ which gives nine equations. Since a solution would vanish at the endpoints and be positive in the interior a reasonable choice of initial approximation seems to be $130 \times \sin \pi x$. This gives us the following vector:

$$z_0 = \begin{bmatrix} 4.01524E+01 \\ 7.63785E+01 \\ 1.05135E+02 \\ 1.23611E+02 \\ 1.29999E+02 \\ 1.23675E+02 \\ 1.05257E+02 \\ 7.65462E+01 \\ 4.03495E+01 \end{bmatrix}.$$

After 4 iterations we get a vector

$$z_4 = \begin{bmatrix} 3.35740E+01 \\ 6.52027E+01 \\ 9.15664E+01 \\ 1.09168E+02 \\ 1.15363E+02 \\ 1.09168E+02 \\ 9.15664E+01 \\ 6.52027E+01 \\ 3.35740E+01 \end{bmatrix}.$$

We choose z_4 as our x_0. Our computations lead to the following results:

$$\begin{aligned} s_0 &= \|F'(x_0)^{-1}F(x_0)\| = 9.15311\mathrm{E}-05, \\ \beta &= \|(F'(x_0)^{-1}\| = 2.55882E+01, \\ c &= \frac{3}{2}h^2\beta = 0.015\beta = .383823, \end{aligned}$$

and

$$p = \frac{1}{2},$$

solving inequalities (156) and (157) we get

$$r_0 = 9.18\mathrm{E}-05 > \bar{r}_0 \quad \text{and} \quad R = 2.425626633.$$

It can easily be seen that with the above values all hypotheses of Theorem 5 are satisfied. Hence Newton's iteration (162) is well defined, remains in $U(x_0, r_0)$ and converges to a solution x^* of Eq. (161) which is unique in $U(x_0, R)$.

References

1. Allgower, E.L., Böhmer, K., Potra, F.A., and Rheinboldt, W.C. A mesh-independence principle for operator equations and their descretizations, *SIAM J. Numer. Anal.* **23**, 1(1986), 160–169.

2. Argyros, I.K. A mesh independence principle for operator equations and their discretizations under mild differentiability conditions, *Computing* **45** (1990), 265–268.

3. Argyros, I.K. On the convergence of some projection methods with perturbation, *J. Comput. Appl. Math.* **36** (1991), 255–258.

4. Argyros, I.K. Comparing the radii of some balls appearing in connection to three local convergence theorems for Newton's method, *Southwest Journal of Pure Appl. Math.* **1** (1998), 31–42.

5. Argyros, I.K. On the discretization of Newton-like methods, *Internat. J. Computer. Math.* vol. 52, (1994), 161–170.

6. Argyros, I.K. A unified approach for constructing fast two-step Newton-like methods, *Monatshefte für Mathematik* **119** (1995), 1–22.

7. Argyros, I.K. On the method of tangent hyperbolas, *J. Approx. Th. Appl.* **12**, 1, (1996), 1–23.

8. Argyros, I.K. A convergence theorem for Newton-like methods under generalized Chen–Yamamoto type assumptions, *Appl. Math. Comp.* **61**, 1, (1994), 25–37.

9. Argyros, I.K. and Szidarovszky, F. *The Theory and Applications of Iteration Methods*, C.R.C. Press, Inc., Boca Raton, Florida, U.S.A. (1993).

10. Brown, P.N. A local convergence theory for combined inexact-Newton/finite-difference projection methods, *SIAM J. Numer. Anal.* **24**, 2 (1987), 407–434.

11. Brown, P.N. and Saad, Y. Convergence theory of nonlinear Newton–Krylov algorithms, *SIAM J. Optimiz.* **4**, 2 (1994), 297–230.

12. Chen, X. and Yamamoto, T. Convergence domains of certain iterative methods for solving nonlinear equations, *Numer. Funct. Anal. Optimiz.* **10** (1 and 2), (1989), 37–48.

13. Dembo, R.S., Eisenstat, S.C., and Steihaug, T. Inexact Newton methods, *SIAM J. Numer. Anal.* **19**, 2 (1982), 400–408.

14. Deuflhard, P. and Heindl, G., Affine invariant convergence theorems for Newton's method and extensions to related methods, *SIAM J. Numer. Anal.* Vol. 16, No. 1 (1979), 1–10.

15. Deuflhard, P. and Potra, F.A. Asymptotic mesh independence of Newton-Galerkin methods and a refined Mysovskii theorem, *SIAM J. Numer. Anal.* Vol. 29, No. 5 (1992), 1395–1412.

16. Gutierez, J.M. A new semilocal convergence theorem for Newton's method, *J. Comput. Appl. Math.* **79** (1997), 131–145.

17. Kantorovich, L.V. and Akilov, E.P. *Functional Analysis in Normed Spaces.* Pergamon Press, New York (1964).

18. Mysovskii, I. On the convergence of L.V. Kantorovich's method of solution of functional equations and its applications. *Dokl. Akad. Nauk SSSR*, **70** (1950), 565–568. (In Russian.)

19. Mysovskii, I. On the convergence of Newton's method. *Trudy Mat. Inst. Steklov* **28** (1949), 145–147. (In Russian.)

20. Potra, F.A. and Ptak, V. Sharp error bounds for Newton's process, *Numer. Math.* **34** (1980), 63–72.

21. Potra, F.A. On an iterative algorithm of order 1.839 ... for solving nonlinear operator equations, *Numer. Funct. Anal. Optimiz.* **7** (1) (1984-85), 75–106.

22. Potra, F.A. On Q-order and R-order of convergence, *SIAM J. Optimiz. Th. Appl.* **63**, 3 (1989), 415–431.

23. Vainikko, G.M. Galerkin's perturbation method and the general theory of approximate methods for nonlinear equations, *USSR Comp. Math. Math. Phys.* **4** (1967), 723–751.

24. Zabrejko, P.P. and Nguen, D.F. The majorant method in the theory of Newton-Kantorovich approximations and the Ptak error estimates, *Numer. Funct. Anal. Optimiz.* **9** (5 and 6), (1987), 671–684.

LINEAR INTEGRAL OPERATORS WITH HOMOGENEOUS KERNEL: APPROXIMATION PROPERTIES IN MODULAR SPACES. APPLICATIONS TO MELLIN-TYPE CONVOLUTION OPERATORS AND TO SOME CLASSES OF FRACTIONAL OPERATORS

Carlo Bardaro and Ilaria Mantellini

Department of Mathematics University of Perugia,
Via Vanvitelli 1, 06123 Perugia, Italy
E-mail: bardaro@dipmat.unipg.it
Contact author: C. Bardaro

Dedicated to the memory of Professor Calogero Vinti,
unforgettable Teacher and Man

Here we state some modular approximation theorems for a class of linear integral operators with kernels satisfying some general homogeneity assumptions, acting on functions defined over locally compact topological groups. Moreover we study the rates of modular approximation in certain generalized Lipschitz classes, defined by means of a modular functional. Applications to Mellin convolution operators, moment type operators and Erdélyi-Kober fractional operators are given.

1 Introduction

Linear integral operators with homogeneous kernel were considered by several authors in connection with functional inequalities (see e.g. [11], [20], [26], [27], [7], [28]), and Fractional Calculus (see e.g., [19], [24], [32], [21], [7], [28]).

Also some particular cases were of interest in Approximation Theory and Calculus of Variation. For example a deep attention was dedicated to the so-called "moment type operators". From the classical work of Professor Calogero Vinti ([33], [34]), many results were obtained in connection with pointwise approximation and convergence properties with respect to variation, lenght and area. We quote here the papers [1], [36], [17], [18], [2], [21], [6], [8]. For an interesting survey on these results we refer to [35]: this is the last paper of Calogero Vinti, written during his illness, and it represents a clear proof of his great will-power and his affection to the mathematical research, which he preserved to the end. Recently in [9] general linear integral operators with homogeneous kernel are studied in connection with modular approximation theory in Musielak-Orlicz spaces.

Here we consider integral operators of the form

$$(*) \qquad (Tf)(s) = \int_G K(s,t)\, f(t)\, dt,$$

where G is a locally compact (Hausdorff) topological group, K is a kernel satisfying a general condition of homogeneity (which seems to be suitable in topological groups, and which was introduced in [7]), and f belongs to the domain of T. We study approximation properties of families of operators like $(*)$ in abstract modular spaces. These general functional spaces were extensively studied by J. Musielak (see [29]) and include the classical Orlicz spaces and the Musielak-Orlicz spaces. In Section 5 we study in particular the order of modular approximation in modular Lipschitz classes $\mathrm{lip}_\tau(\rho)$, where τ is a "comparison" function and ρ is a modular functional. These classes were introduced in [10] and represent modular versions of the classical Zygmund classes in L^p-spaces (see [16]).

In Section 6 we consider, as a particular case, a family of Mellin-type convolution operators. These operators are recently considered in a series of papers by P.L Butzer and S. Jansche ([12], [13], [14], [15]), in which a very complete theory of Mellin transform and its connection with Fourier analysis and integral operators is developed.

In Section 7 we apply the previous theory to the moment operators, thus obtaining new results about the order of approximation in modular Lipschitz classes. Finally in Section 8 we take into consideration a class of fractional integral operators introduced by Erdélyi and Kober (see [19], [24], [32]).

2 Preliminaries

Let G be a locally compact topological group, provided with its Haar measure, denoted by μ. Here, for a sake of simplicity, we suppose G abelian. Let θ be the neutral element of G and by $\mathcal{U}$ we denote a base of measurable neighbourhoods of $\theta \in G$. Let us denote by $L^0(G)$ the space of all the measurable functions $f : G \to \mathbb{R}$, finite a.e. in G.

Let $\rho : L^0(G) \to \widetilde{\mathbb{R}}_0^+ = [0, +\infty]$ be a measurable modular functional, that is ρsatisfies the following assumptions:

(i) $\rho(f) = 0$ if and only if $f = 0$ a.e.

(ii) $\rho(-f) = \rho(f)$,

(iii) $\rho(\alpha f + \beta g) \le \rho(f) + \rho(g)$, for every $f, g \in L^0(G)$, $\alpha, \beta \ge 0$, $\alpha + \beta = 1$,

(iv) $\rho(F(t, \cdot))$ is a measurable function of $t \in G$ for any globally measurable function $F : G \times G \rightarrow \mathbb{R}_0^+$.

By means of the functional ρ we introduce the vector subspace of $L^0(G)$, denoted by $L^\rho(G)$, defined by:

$$L^\rho(G) = \{f \in L^0(G) : \lim_{\lambda \to 0} \rho(\lambda f) = 0\}.$$

The subspace $L^\rho(G)$ is *the modular space generated by* ρ. A general theory of modular spaces can be found in [29] (for special cases see also [25]).

The following notions on measurable modulars will be used (see [29], [3], [4]):

(a) ρ is *quasi-convex* if there is a constant $M \geq 1$ such that:

$$\rho\left[\int_G p(t)\ h(t, \cdot)\ d\mu(t)\right] \leq M \int_G p(t)\ \rho(Mh(t, \cdot))\ d\mu(t),$$

for every $p \in L^1(G)$, $p \geq 0$, with $\|p\|_1 = 1$ and for any globally measurable $h : G \times G \rightarrow \mathbb{R}_0^+$.

(b) ρ is *monotone* if $f, g \in L^0(G)$, $|f| \leq |g|$ implies $\rho(f) \leq \rho(g)$.

(c) ρ is *finite* if $\chi_A \in L^\rho(G)$ whenever A is a measurable subset of G such that $\mu(A) < +\infty$.

(d) ρ is *absolutely finite* if it is finite and for every $\varepsilon > 0$, $\lambda > 0$, there is a $\delta > 0$ such that $\rho(\lambda \chi_B) < \varepsilon$, for any measurable subset $B \subset G$ with $\mu(B) < \delta$.

(e) ρ is *absolutely continuous* if there exists an $\alpha > 0$ such that for every $f \in L^0(G)$, with $\rho(f) < +\infty$, the following two conditions are satisfied:

(e.1) for every $\varepsilon > 0$, there is a measurable subset $A \subset G$ such that $\mu(A) < +\infty$ and $\rho(\alpha f \chi_{G \setminus A}) < \varepsilon$;

(e.2) for every $\varepsilon > 0$ there is a $\delta > 0$ such that $\rho(\alpha f \chi_B) < \varepsilon$, for any measurable subset $B \subset G$ with $\mu(B) < \delta$.

(f) ρ is τ*-bounded* if there are a constant $C \geq 1$ and a nonnegative measurable function $h : G \rightarrow \mathbb{R}_0^+$ such that $h(t) \rightarrow 0$, as $t \rightarrow \theta$, such that:

$$\rho(f(t + \cdot)) \leq \rho(Cf) + h(t),$$

for every $t \in G$, $f \in L^0(G)$ with $\rho(f) < +\infty$.

If the function h is a bounded function we will say that ρ is *strongly* τ*-bounded* and in this case we put $h_0 =: \sup_{t \in G} h(t)$.

Example 1. A classical example of modular space is the Orlicz space generated by a φ-function φ (see [29], [25]). The corresponding modular is defined by:

$$\rho(f) \equiv I_\varphi(f) = \int_G \varphi(|f(t)|)\, d\mu(t), \quad f \in L^0(G).$$

The modular I_φ satisfies all the condtitions (a) - (f).

Example 2. More generally any Musielak - Orlicz space, that is a Orlicz space generated by a φ-function depending on a parameter, is a modular space with modular (see [29], [25]):

$$\rho(f) \equiv \widetilde{I}_\varphi(f) = \int_G \varphi(t, |f(t)|)\, d\mu(t), \quad f \in L^0(G).$$

If $\varphi(\cdot, u)$ is locally integrable on G and τ-bounded (as defined in [29]), then the corresponding modular $\widetilde{I}_\varphi$ satisfies all the conditions (a) - (f).

Many other examples of modular spaces can be found in [29], [25]. Moreover in [3] the absolute continuity and absolute finiteness of some modular functionals connected with summability processes are introduced and studied.

In particular L^p-spaces are modular spaces satisfying all the conditions (a) - (f).

Let now $\eta : G \to \mathbb{R}^+$ and $K : G \times G \to \mathbb{R}_0^+$ be measurable functions. We say that K is *η-subhomogeneous* if there exist two positive constants $r \leq 1$, $R \geq 1$ such that:

$$r\eta(t+v)\; K(s,t) \leq \eta(t)\; K(s+v, t+v) \leq R\eta(t+v)\; K(s,t),$$

for every $s, t, v \in G$. We denote by $\mathcal{K}_\eta$ the class of all the $\eta-$subhomogeneous functions K.

If $r = R = 1$, we will say that K is *η-homogeneous*. In this case we have:

$$\eta(t)\; K(s+v, t+v) = \eta(t+v)\; K(s,t),$$

for every $s, t, v \in G$. We will denote by $\widetilde{\mathcal{K}}_\eta$ the class of all η-homogeneous functions K, (see [9]).

Example 3. Let $G = (\mathbb{R}^+, \cdot)$ the multiplicative group of the positive real numbers. If $K : \mathbb{R}^+ \times \mathbb{R}^+ \to \mathbb{R}_0^+$ is homogeneous of degree $\alpha \in \mathbb{R}$, then K is η-homogeneous with $\eta(t) = t^\alpha$, $t \in \mathbb{R}^+$.

Example 4. Let $G = (\mathbb{R}, +)$ be the additive line group.
If $K(s,t) = H(s-t)$, for $H \in L^1(\mathbb{R})$, then K is homogeneous of degree 0, and this implies that K is η-homogeneous with $\eta(t) = 1$, for every $t \in G$.

For $K \in \mathcal{K}_\eta$ we define the operator:

$$(Tf)(s) = \int_G K(s,t)\; f(t)\; d\mu(t), \tag{1}$$

for $f \in \mathrm{Dom}T$, i.e. for every $f \in L^0(G)$ for which $(Tf)(s)$ is well defined as an Haar integral for a.e. $s \in G$.

For $K \in \mathcal{K}_\eta$ we will use the following notations:

$$\begin{aligned} A_K &= \int_G (\eta(z))^{-1}\; K(\theta, z)\; d\mu(z) \\ \Omega_K &= \max\{|1 - rA_K|, |1 - RA_K|\} \\ A_K(U) &= \int_{G\setminus U} (\eta(z))^{-1}\; K(\theta, z)\; d\mu(z), \end{aligned}$$

for any $U \in \mathcal{U}$, and where r, R are the constants from the $\eta-$subhomogeneity of K.

If $\{K_w\}_w$ is a family of functions $K_w \in \mathcal{K}_\eta$ we will put $A_{K_w} = A_w$, $\Omega_{K_w} = \Omega_w$ and $A_{K_w}(U) = A_w(U)$, for $U \in \mathcal{U}$.

3 Some Inequalities Related to Tf

The following result gives an estimation for the operator Tf, in the modular space $L^\rho(G)$.

Theorem 1. *Let ρ be a measurable , monotone, quasi-convex and τ-bounded modular on $L^0(G)$. Let $K \in \mathcal{K}_\eta$ be such that $0 < A_K < +\infty$ and*

$$S =: \int_G (\eta(z))^{-1}\; K(\theta, z)\; h(z)\; d\mu(z) < +\infty, \tag{2}$$

where h is the function from τ-boundedness assumption on ρ.

Then for every $f \in L^\rho(G) \cap DomT$, we have:

$$\rho(Tf) \le M\rho(MCRA_K\eta f) + MA_K^{-1}S, \tag{3}$$

where M and C are the constants from (a) and (f) respectively.

Proof. Let $f \in L^\rho(G) \cap \mathrm{Dom}T$. Without loss of generality we can assume that $\rho(MCRA_K\eta f) < +\infty$. From η-subhomogeneity of K we have:

$$|(Tf)(s)| \le \int_G K(s,t)\; |f(t)|\; d\mu(t)$$

$$= \int_G K(s, s+z)\ |f(s+z)|\ d\mu(z)$$
$$\leq \int_G (\eta(z))^{-1}\ K(\theta, z)\ \eta(z+s)\ R|f(z+s)|\ d\mu(z).$$

Let us put $g = \eta f$. By monotonicity and quasi-convexity of ρ we have:

$$\rho(Tf) \leq \frac{M}{A_K} \int_G (\eta(z))^{-1}\ K(\theta, z)\ \rho[MRA_K g(z + \cdot)]\ d\mu(z).$$

From τ-boundedness of ρ we have now:

$$\begin{aligned}\rho(Tf) &\leq \frac{M}{A_K} \int_G (\eta(z))^{-1}\ K(\theta, z)\ \rho(MCRA_K g)\ d\mu(z) \\ &+ \frac{M}{A_K} S \\ &= M\rho(MRCA_K g) + MSA_K^{-1}\end{aligned}$$

and so the assertion. □

Remark 1. Note that in the proof of Theorem 1 we used only the second inequality of the η-subhomogeneity condition.
Theorem 1 implies that $Tf \in L^\rho(G)$ whenever $g = \eta f \in L^\rho(G)$. In general it is not possible to replace $g \in L^\rho(G)$ with $f \in L^\rho(G)$. There are given some examples of this in [9], in classical Orlicz spaces. However if $\eta \in L^\infty(G)$ then it is easy to show that $Tf \in L^\rho(G)$ whenever $f \in L^\rho(G)$. For $G = (\mathbb{R}^+, \cdot)$, or $G = (\mathbb{R}, +)$, this happens for example for homogeneous kernels of degree zero, as convolution operators.

Let ρ be a modular on $L^0(G)$. The map $\omega_\rho : L^0(G) \times \mathcal{U} \rightarrow \widetilde{\mathbb{R}}_0^+$ defined by:

$$\omega_\rho(f, U) = \sup_{t \in U} \rho[f(t + \cdot) - f(\cdot)]$$

is called the *ρ-modulus of continuity* (see [30], [4]).

The following theorem gives an estimation of the ρ-distance between Tf and $g = \eta\ f$.

Theorem 2. *Under the assumptions and notations of Theorem 1, we have:*

$$\begin{aligned}\rho[\frac{\lambda}{2}(Tf - g)] &\leq M\{\omega_\rho(2\lambda MRA_K g^+, U) + \omega_\rho(2\lambda MRA_K g^-, U)\} \\ &+ 2\frac{MA_K(U)}{A_K}\{\rho(4\lambda MRA_K Cg) + \rho[4\lambda MRA_K g]\} \\ &+ 2\frac{MS}{A_K} + 2\rho[2\lambda\Omega_K\ g] \qquad (4)\end{aligned}$$

for every $U \in \mathcal{U}$, $\lambda > 0$, $f \in L^0(G)$, *and* g^+, g^- *are the positive and negative parts of* g.

Proof. We split the proof in two steps. At first let us consider $f \geq 0$. Thus we have also $g \geq 0$. Let $\lambda > 0$ be fixed and let us suppose that the right hand side of (4) is finite. We have:

$$(Tf)(s) - g(s) = \int_G K(s,t)f(t)d\mu(t) - g(s),$$

for any $s \in G$. Now if $s \in G$ is such that $(Tf)(s) - g(s)$ is nonnegative, by similar reasonings as in Theorem 1, we obtain:

$$\begin{aligned}|(Tf)(s) - g(s)| \leq & \int_G K(\theta, z)\ (\eta(z))^{-1}\ R|g(z+s) - g(s)|\ d\mu(z) \\ & + |RA_K - 1|\ g(s),\end{aligned}$$

while if $s \in G$ is such that $(Tf)(s) - f(s)$ is negative, in analogous way we have:

$$\begin{aligned}|(Tf)(s) - g(s)| \leq & \int_G K(\theta, z)\ (\eta(z))^{-1}\ r\ |g(z+s) - g(s)|\ d\mu(z) \\ & + |rA_K - 1|\ g(s),\end{aligned}$$

and thus, taking into account that $r \leq R$, we finally have the following estimation:

$$\begin{aligned}|(Tf)(s) - g(s)| \leq & \int_G K(\theta, z)\ (\eta(z))^{-1}\ R|g(z+s) - g(s)|\ d\mu(z) \\ & + \Omega_K\ g(s).\end{aligned}$$

From the properties of the modular ρ we deduce, for $\lambda > 0$:

$$\begin{aligned}\rho[\lambda(Tf - g)] \leq & \rho\left[2\lambda \int_G K(\theta, z)\ (\eta(z))^{-1} R|g(z+\cdot) - g(\cdot)|\ d\mu(z)\right] \\ & + \rho[2\lambda\Omega_K g] =: J + \rho[2\lambda\Omega_K g].\end{aligned}$$

Now we evaluate J. Let $U \in \mathcal{U}$ be fixed. By monotonicity and quasi-convexity of ρ we obtain:

$$\begin{aligned}J &\leq \frac{M}{A_K} \int_G K(\theta, z)\ (\eta(z))^{-1} \rho[2MRA_K\lambda|g(z+\cdot) - g(\cdot)|]\ d\mu(z) \\ &= \frac{M}{A_K}\left\{\int_U + \int_{G\setminus U}\right\} K(\theta, z)(\eta(z))^{-1}\rho[2MRA_K\lambda|g(z+\cdot) - g(\cdot)|]\ d\mu(z) \\ &= J_1 + J_2.\end{aligned}$$

Now, $J_1 \leq M\omega_\rho(2MRA_K\lambda g, U)$. Moreover:

$$\begin{aligned}
J_2 &= \frac{M}{A_K}\int_{G\setminus U} K(\theta,z)\ (\eta(z))^{-1}\ \rho[2MA_K\lambda R|g(z+\cdot)-g(\cdot)|]\ d\mu(z)\\
&\leq \frac{M}{A_K}\int_{G\setminus U} K(\theta,z)\ (\eta(z))^{-1}\ \rho[4MRA_K\lambda g(z+\cdot)]\ d\mu(z)\\
&+ \frac{M}{A_K}\int_{G\setminus U} K(\theta,z)\ (\eta(z))^{-1}\ \rho[4MRA_K\lambda g]\ d\mu(z)\\
&= J_2' + J_2''.
\end{aligned}$$

From τ-boundedness of ρ we have:

$$\begin{aligned}
J_2' &\leq \frac{M}{A_K}\int_{G\setminus U} K(\theta,z)\ (\eta(z))^{-1}\ \rho[4MRCA_K\lambda g]\ d\mu(z) + \frac{MS}{A_K}\\
&\leq \frac{MA_K(U)}{A_K}\rho[4MRCA_K\lambda g] + \frac{MS}{A_K},
\end{aligned}$$

while $J_2'' \leq \dfrac{MA_K(U)}{A_K}\ \rho[4MRA_K\lambda g]$. Thus for nonnegative f we obtain the inequality:

$$\begin{aligned}
\rho[\lambda(Tf-g)] &\leq M\omega_\rho(2\lambda MRA_K g, U)\\
&+ \frac{MA_K(U)}{A_K}\{\rho(4\lambda MRA_K Cg) + \rho[4\lambda MRA_K g]\}\\
&+ \frac{MS}{A_K} + \rho[2\lambda\Omega_K\ g]
\end{aligned}$$

The second step is now the general case. Let $f \in L^\rho(G) \cap \mathrm{Dom}T$ be an arbitrary function. Let f^+, f^- be the positive and negative parts of f. So ηf^+, ηf^- are the positive and negative parts of g. Taking now into account that for any function f, it results $f^+ \leq |f|$, and $f^- \leq |f|$, we have $f^+, f^- \in L^\rho(G) \cap \mathrm{Dom}T$ and moreover $\rho[\lambda f^\pm] \leq \rho[\lambda f]$. Writing now, for $\lambda > 0$:

$$\rho[\frac{\lambda}{2}(Tf-g)] \leq \rho[\lambda(Tf^+ - g^+)] + \rho[\lambda(Tf^- - g^-)],$$

the inequality (4) follows, by applying the final inequality in the first step to the functions f^+ and f^-, and taking into account of the monotonicity of the modular ρ. □

Remark 2. The previous estimation extend in various directions corresponding results for Orlicz or Musielak-Orlicz spaces (see e.g. [9]).

Remark 3. From (4) we deduce that if $g = \eta f \in L^\rho(G)$ then $Tf - g \in L^\rho(G)$. Examples exist for which $f \in L^\rho(G)$ but $Tf - g \notin L^\rho(G)$.

Suppose now $\eta \in L^\infty(G)$ and $\|\eta\|_\infty > 0$. Then for $K \in \mathcal{K}_\eta$, we have:

$$\alpha(s) =: \int_G K(s,t)\, d\mu(t) \leq A_K \|\eta\|_\infty.$$

We can assume $\|\eta\|_\infty = 1$. Indeed putting $\eta^*(t) = \eta(t)/\|\eta\|_\infty$, we easily have $\mathcal{K}_{\eta^*} = \mathcal{K}_\eta$.

We have the following

Theorem 3. *Let $\eta \in L^\infty(G)$ and $\|\eta\|_\infty = 1$, and let $K \in \mathcal{K}_\eta$. Under the assumptions of Theorem 2, we have for any $\lambda > 0$, $f \in L^\rho(G) \cap DomT$, $U \in \mathcal{U}$:*

$$\begin{aligned}\rho[\lambda(Tf - f)] \leq\ & M\omega_\rho(2\lambda MRA_K f, U)\\ &+ \frac{MA_K(U)}{A_K}\{\rho(4\lambda MRA_K Cf) + \rho[4\lambda RMA_K f]\}\\ &+ \frac{MS}{A_K} + \rho[2\lambda(\alpha(\cdot) - 1)\ f(\cdot)] \end{aligned} \tag{5}$$

Proof. The proof is now more direct and we will use only the second inequality in the definition of η-subhomogeneity of K. As in Theorem 2 we can assume that the right hand side of (5) is finite. Since:

$$|(Tf)(s) - f(s)| \leq \int_G K(s,t)|f(t) - f(s)|\, d\mu(t) + |\alpha(s) - 1||f(s)|,$$

for any $s \in G$, we obtain:

$$\rho[\lambda(Tf - f)] \leq J + \rho[2\lambda(\alpha(\cdot) - 1)f(\cdot)]$$

where

$$J =: \rho\left[2\lambda \int_G K(\cdot,t)|f(t) - f(\cdot)|\, d\mu(t)\right].$$

But for $s \in G$ we have:

$$\begin{aligned}&\int_G K(s,t)\ |f(t) - f(s)|\ d\mu(t)\\ &\leq \int_G K(\theta,z)\ (\eta(z))^{-1}\ \eta(z+s)R|f(z+s) - f(s)|d\mu(z)\\ &\leq \int_G K(\theta,z)\ (\eta(z))^{-1}\ R|f(z+s) - f(s)|d\mu(z).\end{aligned}$$

Therefore

$$\begin{aligned}
J &\leq \frac{M}{A_K} \int_G K(\theta, z)\ (\eta(z))^{-1}\ \rho[2\lambda MRA_K|f(z+\cdot) - f(\cdot)|]d\mu(z) \\
&\leq \frac{M}{A_K} \left\{ \int_U + \int_{G\setminus U} \right\} K(\theta, z)\ (\eta(z))^{-1} \rho[2\lambda MRA_K|f(z+\cdot) - f(\cdot)|]d\mu(z) \\
&\leq M\omega_\rho[2\lambda MRA_K f, U] \\
&\quad + \frac{M}{A_K} \int_{G\setminus U} K(\theta, z)\ (\eta(z))^{-1}\ \rho[2\lambda MRA_K|f(z+\cdot) - f(\cdot)|]d\mu(z) \\
&= M\omega_\rho[2\lambda MRA_K f] + J'.
\end{aligned}$$

Now:

$$\begin{aligned}
J' &\leq \frac{M}{A_K} \int_{G\setminus U} K(\theta, z)\ (\eta(z))^{-1}\ \rho[4\lambda MRA_K f(z+\cdot)]\ d\mu(z) \\
&\quad + \frac{M}{A_K} \int_{G\setminus U} K(\theta, z)\ (\eta(z))^{-1}\ \rho[4\lambda MRA_K f]\ d\mu(z) \\
&\leq \frac{M}{A_K} \int_{G\setminus U} K(\theta, z)\ (\eta(z))^{-1}\ \rho[4\lambda CMRA_K f]\ d\mu(z) \\
&\quad + \frac{MS}{A_K} + \frac{MA_K(U)}{A_K} \rho[4\lambda MRA_K f]
\end{aligned}$$

and so the assertion follows. □

4 Convergence Theorems

For a given measurable function $\eta : G \rightarrow \mathbb{R}^+$, we take a family $\mathbb{K} = (K_w)_{w>0}$ of functions in $\mathcal{K}_\eta$. We shall need the constants A_w, Ω_w, $A_w(U)$ introduced is Section 2.

Let ρ be a τ-bounded modular on $L^0(G)$. If h denotes the function in the definition of $\tau-$boundedness of ρ, we will put:

$$S_w = \int_G (\eta(z))^{-1}\ K_w(\theta, z)\ h(z)\ d\mu(z).$$

We will say that $\mathbb{K}$ is *singular* if:

(K.1) $\sup_{w>0} A_w \equiv A < +\infty$, $\lim_{w\rightarrow+\infty} \Omega_w = 0$.

(K.2) For every $U \in \mathcal{U}$ we have:

$$\lim_{w\rightarrow+\infty} A_{K_w(U)} = 0.$$

Finally for a singular family $\mathbb{K} = (K_w)_{w>0}$ we consider the family of operators:

$$(T_w f)(s) = \int_G K_w(s,t)\ f(t)\ d\mu(t),$$

for any $f \in \boldsymbol{X} =: \bigcap_{w>0} \mathrm{Dom} T_w$.

In this section we will obtain modular convergence theorems for the family $\{T_w\}$.

The following result is an important tool (see [30], [4]).

Proposition 1. *Let ρ be a monotone, absolutely finite, absolutely continuous and τ-bounded modular on $L^0(G)$. Then for every $f \in L^\rho(G)$ there is $\overline{\lambda} > 0$ such that for every $\varepsilon > 0$ there is $U \in \mathcal{U}$ such that:*

$$\omega_\rho(\lambda f, U) < \varepsilon, \quad \textit{for}\ \ 0 < \lambda < \overline{\lambda}.$$

By using the previous result we can prove the following theorem.

Theorem 4. *Let $\eta : G \rightarrow \mathbb{R}^+$ be a measurable function and let ρ be a measurable, monotone, absolutely finite, absolutely continuous, quasi-convex and τ-bounded modular on $L^0(G)$. Let $\mathbb{K} = (K_w)_{w>0} \subset \mathcal{K}_\eta$ be a singular kernel. Suppose that:*

$$\lim_{w\rightarrow+\infty} S_w = 0. \tag{6}$$

Then for every $f \in \boldsymbol{X}$ such that $g = \eta f \in L^\rho(G)$ there is $\lambda > 0$ such that:

$$\lim_{w\rightarrow+\infty} \rho[\lambda(T_w f - g)] = 0. \tag{7}$$

Proof. Let $f \in \boldsymbol{X}$ and $g = \eta f \in L^\rho(G)$ and let $\overline{\lambda} > 0$ be such that for $\varepsilon > 0$

$$\omega_\rho(\lambda g^\pm, U) < \varepsilon, \quad \text{for any } 0 < \lambda < \overline{\lambda}$$

for a suitable $U \in \mathcal{U}$. Let $\lambda > 0$ so small such that:

$$4\lambda MCRA < \overline{\lambda} \ \text{ and } \ \rho[4\lambda(AR+1)g] < +\infty.$$

From (4) applied to T_w, taking into account of the definition of A and of (6), it is sufficient to prove that $\rho[2\lambda\Omega_w g] \rightarrow 0$ as $w \rightarrow +\infty$. Since $2\lambda\Omega_w\ |g| \leq 4\lambda(AR+1)|g|$, for every $w > 0$, the assertion follows from the Lebesgue dominate convergence theorem for modular spaces (see [30], Proposition 2). □

Remark 4. By applying Theorem 3 it is possible to give also a modular convergence theorem of $T_w f$ towards f when $\eta \in L^\infty$.

Remark 5. All the previous results can be applied to every translation-invariant Köthe function spaces with absolutely continuous and absolutely finite norm. In particular we obtain corresponding results for any $L^p(G)$, $1 \leq p < +\infty$, with Haar measure.

5 Rates of Modular Convergence in Modular Lipschitz Classes

Let $\mathcal{T}$ be the class of all the functions $\tau : G \to \mathbb{R}_0^+$ with $\tau(\theta) = 0$, $\tau(t) \neq 0$ for $t \neq \theta$.

Let ρ be a measurable modular on $L^0(G)$. For a fixed $\tau \in \mathcal{T}$, we define the class (see also [10]):

$$\mathrm{lip}_\tau(\rho) = \{f \in L^\rho(G) : \exists \alpha > 0 \text{ with } \rho[\alpha|f(t+\cdot) - f(\cdot)|] = \mathcal{O}(\tau(t)),\ t \to \theta\}, \quad (8)$$

where, for any two functions $f, g \in L^0(G)$, $f(t) = \mathcal{O}(g(t))$, $t \to \theta$, means that there is a constant $B > 0$ and $U \in \mathcal{U}$ such that $|f(t)| \leq B|g(t)|$, for $t \in U$.

Here, we will assume that the involved kernels are η-homogeneous functions. The general case is easily obtained with an additional assumption (see Remark 6 below).

Let $\mathcal{X}$ be the class of all functions $\xi : \mathbb{R}_0^+ \to \mathbb{R}_0^+$ such that $\xi(0) = 0$ and $\xi(u) > 0$, $u > 0$. For a given measurable function $\eta : G \to \mathbb{R}^+$, let $\mathbb{K} = (K_w)_{w>0}$ be a family of η−homogeneous functions. Thus we have $r = R = 1$. Let $\xi \in \mathcal{X}$ be fixed. Then we will say that $\mathbb{K}$ is ξ-singular if $A_w = 1$, for every $w > 0$, and for every $U \in \mathcal{U}$ $A_w(U) = \mathcal{O}(\xi(w^{-1}))$, $w \to +\infty$.

We have the following:

Theorem 5. *Let $\xi \in \mathcal{X}$ and $\tau \in \mathcal{T}$ be fixed. Let $\eta : G \to \mathbb{R}^+$ be a measurable function and $\mathbb{K} = (K_w)_{w>0} \subset \widetilde{\mathcal{K}}_\eta$ be a ξ-singular kernel. Let ρ be a measurable, monotone, quasi-convex and strongly τ-bounded modular on $L^0(G)$. Assume that there is $U \in \mathcal{U}$ such that:*

$$\int_U K_w(\theta, z)\ (\eta(z))^{-1}\ \tau(z)\ dz = \mathcal{O}(\xi(w^{-1})),\ w \to +\infty. \quad (9)$$

If $f \in L^\rho(G)$ is such that $g = \eta f \in lip_\tau(\rho)$, then for sufficiently small $\lambda > 0$ we have:

$$\rho[\lambda(T_w f - g)] = \mathcal{O}(\xi(w^{-1})),\ w \to +\infty. \quad (10)$$

Proof. Taking into account that $A_w = 1$, for every $w > 0$, and $r = R = 1$, we have for $\lambda > 0$ and $U \in \mathcal{U}$:

$$\begin{aligned}
&\rho[\lambda(T_w f - g)] \\
&\quad \le M \int_U K_w(\theta, z)\ (\eta(z))^{-1}\ \rho[\lambda RM(g(z+\cdot) - g(\cdot)]\ d\mu(z) \\
&\qquad + M \int_{G\setminus U} K_w(\theta, z)\ (\eta(z))^{-1}\ \rho[\lambda RM(g(z+\cdot) - g(\cdot)]\ d\mu(z) \\
&\quad = I_1 + I_2.
\end{aligned}$$

Here M is the constant from quasi-convexity of ρ. Now let $V \in \mathcal{U}$ and $\alpha > 0$ such that:

$$\rho[\alpha|g(z+\cdot) - g(\cdot)|] \le B\tau(z),$$

for every $z \in V$. Let $U \in \mathcal{U}$ be such that (9) holds. We can assume that $U \subset V$. So for $\lambda RM \le \alpha$ we have:

$$I_1 \le BM \int_U K_w(\theta, z)\ (\eta(z))^{-1}\ \tau(z)\ d\mu(z) = \mathcal{O}(\xi(w^{-1})),\ \ w \to +\infty.$$

Next, we consider I_2. By strongly τ-boundedness, using the previous notations, we deduce:

$$\begin{aligned}
I_2 &\le M \int_{G\setminus U} K_w(\theta, z)\ (\eta(z))^{-1}\ \{\rho[2\lambda RM|g(z+\cdot)] + \rho[2\lambda RMg(\cdot)]\}\ d\mu(z) \\
&\le M \int_{G\setminus U} K_w(\theta, z)\ (\eta(z))^{-1}\ \{\rho[2\lambda CRMg] + h_0\} d\mu(z) \\
&\quad + MA_w(U)\ \rho[2\lambda RMg] \\
&\le M\{\rho[2\lambda CRMg] + h_0 + \rho[2\lambda RMg]\}\ A_w(U) = \mathcal{O}(\xi(w^{-1})),
\end{aligned}$$

and so the assertion follows. □

Remark 6. We analogously can state a similar result about the order of modular approximation of $T_w f$ towards $f \in \text{lip}_\tau(\rho)$ when $\eta \in L^\infty(G)$ and $\alpha(s) = 1$, for every $s \in G$. Moreover if the kernels K_w are η-subhomogeneous, we obtain a similar result by assuming $\Omega_w = \mathcal{O}(\xi(w^{-1}))$, as $w \to +\infty$, instead of $A_w = 1$.

6 Applications to Mellin-type Convolution Operators

Here we will consider the group $G = (\mathbb{R}^+, \cdot)$ provided with its Haar measure $d\mu(t) = t^{-1}\ dt$, dt being the Lebesgue measure on $\mathbb{R}^+$. Now $\mathcal{U}$ denotes the family of the neighbourhood base of 1, the neutral element. We will consider a family $(K_w)_{w>0}$ of homogeneous functions of degree 0, (this implies that $K_w : \mathbb{R}^+ \times \mathbb{R}^+ \to \mathbb{R}_0^+$ is η-homogeneous with $\eta(t) = 1$, for every $t \in \mathbb{R}^+$) of the form:

$$K_w(s,t) = H_w(st^{-1}), \quad s,t \in \mathbb{R}^+ \tag{11}$$

where, for every $w > 0$, $H_w : \mathbb{R}^+ \to \mathbb{R}_0^+$ satisfies the following conditions:

(H.1) for each $w > 0$,

$$\int_0^{+\infty} H_w(t)\, t^{-1}\, dt = 1;$$

(H.2) for any $U \in \mathcal{U}$

$$\lim_{w \to +\infty} \int_{\mathbb{R}^+ \setminus U} H_w(t)\, t^{-1}\, dt = 0.$$

We remark that from the invariance of the Haar measure, the integrals in (H.1) and (H.2) can be written with t^{-1} in place of t. For example for (H.2) putting $U =]1-\delta,\, 1+\delta[$ for $0 < \delta < 1$, we have:

$$\int_{\mathbb{R}^+ \setminus U} H_w(t^{-1})\, t^{-1}\, dt = \int_{\mathbb{R}^+ \setminus V} H_w(t)\, t^{-1}\, dt,$$

where $V =](1+\delta)^{-1}, (1-\delta)^{-1}[\in \mathcal{U}$.

If $\{H_w\}_{w>0}$ satisfies (H.1) and (H.2) we will write $H_w \in \mathcal{H}$. For any kernel $\{K_w\}_{w>0}$ of type (11) we can define the family of Mellin-type convolution operators (see [15]):

$$(T_w f)(s) = \int_0^{+\infty} K_w(s,t)\, f(t)\, t^{-1}\, dt = \int_0^{+\infty} H_w(st^{-1})\, f(t)\, t^{-1}\, dt, \tag{12}$$

for any $f \in \boldsymbol{X} = \bigcap_{w>0} \mathrm{Dom} T_w$.

As consequences of the previous general theory, for the above operators, we can state approximation and embedding results in modular spaces, in which the modular functional satisfies the assumptions used before. In particular we take an Orlicz space generated by a function $\varphi : \mathbb{R}_0^+ \to \mathbb{R}_0^+$ such that:

(φ.1) φ is continuous and nondecreasing

(φ.2) $\varphi(0) = 0$, $\varphi(u) > 0$, for $u > 0$, and $\lim_{n \to +\infty} \varphi(u) = +\infty$.

(φ.3) φ is quasi-convex (see [22], [5]), i.e. there is $M \geq 1$ such that, for any $p \in L^1_\mu(\mathbb{R}^+)$ with $\|p\|_1 = 1$, and $p \geq 0$

$$\varphi\left(\int_0^{+\infty} |f(t)|\, p(t)\, t^{-1}\, dt\right) \leq M \int_0^{+\infty} \varphi(M|f(t)|)\, p(t)\, t^{-1}\, dt, \quad (13)$$

for any $f \in L^0(\mathbb{R}^+)$.

If φ satisfies ($\varphi.i$), $i = 1, 2, 3$, we write $\varphi \in \Phi$.

We remark that the notion of quasi-convexity is much more general than classical Jensen-convexity, and (13) is equivalent to the following:

$$\varphi\left(\sum_{j=1}^{N} a_j u_j\right) \leq M \sum_{j=1}^{N} a_j \varphi(M u_j),$$

where $a_j \geq 0$, $j = 1, \ldots, N$, $\sum_{j=1}^{n} a_j = 1$, and $u_1, \ldots u_N \in \mathbb{R}^+$.

For a fixed $\varphi \in \Phi$ we define the Orlicz space generated by φ by means of:

$$L^\varphi_\mu(\mathbb{R}^+) = \{f \in L^0(\mathbb{R}^+) : I_\varphi(\lambda f) < +\infty, \text{ for some } \lambda > 0\}, \quad (14)$$

where:

$$I_\varphi(f) = \rho(f) = \int_0^{+\infty} \varphi(|f(t)|)\, t^{-1}\, dt$$

is the corresponding modular. As we remarked before, it is easy to show that I_φ is a measurable, monotone, absolutely finite, absolutely continuous, strictly τ-bounded (with $h \equiv 0$) and quasi-convex modular, with the constant $M \geq 1$, given in (φ.3). Moreover for $\{K_w\}_{w>0}$ defined by (11), we have $S_w = 0$ for every $w > 0$, where S_w is defined in Section 4. We have the following:

Proposition 2. *Let $\varphi \in \Phi$ and let $\{K_w\}_{w>0}$ be a kernel defined by (11), with $H_w \in \mathcal{H}$. Then for any $w > 0$ we have $L^\varphi_\mu(\mathbb{R}^+) \subset Dom T_w$.*

Proof. The measurability of $T_w f$ follows by standard arguments. So we will prove that for any $w > 0$, $T_w f$ is finite a.e. in $\mathbb{R}^+$. Let $f \in L^\varphi_\mu(\mathbb{R}^+)$. For a sake of simplicity we can assume that $I_\varphi(Mf) < +\infty$, where $M \geq 1$ is the constant of quasi-convexity of φ. By Fubini-Tonelli theorem, quasi-convexity of φ, and the invariance of the Haar measure μ we have:

$$\begin{aligned} M I_\varphi(Mf) &= M \int_0^{+\infty} H_w(t^{-1})\, I_\varphi(Mf)\, t^{-1}\, dt \\ &= M \int_0^{+\infty} H_w(t^{-1}) \left\{\int_0^{+\infty} \varphi(M|f(ts)|) \frac{ds}{s}\right\} \frac{dt}{t} \\ &\geq \int_0^{+\infty} \varphi(T_w|f|) \frac{ds}{s} \end{aligned}$$

Thus $\varphi((T_w|f|)(s)) < +\infty$ for a.e. $s \in I\!R^+$ and from the properties of φ this implies that $|T_w f|$ is finite for a.e. $s \in I\!R^+$. □

Corollary 1. *Let $\{K_w\}_{w>0}$ be a kernel defined by (11), with $H_w \in \mathcal{H}$. Let $\varphi \in \Phi$. Then for every $f \in L^{\varphi}_{\mu}(I\!R^+)$ there is $\lambda > 0$ such that:*

$$\lim_{w \to +\infty} I_{\varphi}(\lambda(T_w f - f)) = 0.$$

Analogously we can state also a result about the order of modular approximation for functions belonging to modular Lipschitz classes. As a consequence of Theorem 5 we have the following:

Corollary 2. *Let $\xi \in \mathcal{X}$, and $\tau \in \mathcal{T}$ be fixed. Let $\varphi \in \Phi$. Suppose that for any neighbourhood U of 1*

$$\int_{I\!R^+ \setminus U} H_w(t^{-1})\, t^{-1}\, dt = \mathcal{O}(\xi(w^{-1})), \quad w \to +\infty.$$

and there is $U \in \mathcal{U}$ such that

$$\int_U H_w(t^{-1})\, \tau(t)\, t^{-1}\, dt = \mathcal{O}(\xi(w^{-1})), \quad w \to +\infty.$$

Then if $f \in lip_{\tau}(I_{\varphi})$ we have:

$$I_{\varphi}[\lambda(T_w f - f)] = \mathcal{O}(\xi(w^{-1})), \quad w \to +\infty$$

for sufficiently small $\lambda > 0$.

Example 5. Let $\tau(t) = |1-t|^{\alpha}$, and $\xi(t) = t^{\alpha}$ with $0 < \alpha < 1$. Then

$$m_{\alpha}(H_w) = \int_0^{+\infty} H_w(t^{-1})\, |1-t|^{\alpha}\, t^{-1}\, dt$$

is the absolute α-moment of H_w for $w > 0$. Thus if $m_{\alpha}(H_w) = \mathcal{O}(w^{-\alpha})$, under the other assumptions of Corollary 2 we obtain:

$$I_{\varphi}[\lambda(T_w f - f)] = \mathcal{O}(w^{-\alpha}), \quad w \to +\infty.$$

Remark 7. The results of this section may be extended to Musielak-Orlicz spaces generated by a function $\varphi : I\!R^+ \times I\!R^+_0 \to I\!R^+_0$ depending on a parameter $t \in I\!R^+$, such that:

(i) $\varphi(\cdot, u)$ is measurable for every $u \geq 0$.

(ii) $\varphi(t, 0) = 0, \quad \varphi(t, u) > 0$, for $u > 0$.

(iii) $\varphi(t, \cdot)$ is, continuous, non decreasing and quasi-convex, for every $t \in \mathbb{R}^+$ with a fixed constant $M \geq 1$.

(iv) φ is strongly τ-bounded, i.e. there are a constant $C \geq 1$ and a measurable function $F : \mathbb{R}^+ \times \mathbb{R}^+ \to \mathbb{R}_0^+$ such that:

$$\varphi(t\tau^{-1}, u) \leq \varphi(t, Cu) + F(t, \tau),$$

for every $t, \tau \in \mathbb{R}^+$ and

$$\sup_{\tau \in \mathbb{R}^+} \int_0^{+\infty} F(t, \tau)\, t^{-1}\, dt < +\infty.$$

For references to above properties see e.g. [29], [6].

Indeed in this case the modular:

$$\widetilde{I}_\varphi(f) = \int_0^{+\infty} \varphi(t, |f(t)|)\, t^{-1} dt,$$

for $f \in L^0(\mathbb{R}^+)$, satisfies all the assumptions (a)-(f) of Section 2.

7 A Particular Case: The Moment Operator

Here we discuss a particular Mellin-type operator, which was studied by several authors in connection with its properties in Calculus of Variations (see [34], [1], [2], [8], [21]), approximation ([1], [36], [17], [2], [6]), functional inequalities ([11], [20], [26], [27]) and Fractional Calculus ([19], [24], [7], [28]). This operator, called *moment operator* or *average operator* is defined by means of the *moment kernel* $\mathbb{M} = \{M_n\}_{n \in \mathbb{N}}$,

$$M_n(s, t) = n s^{-n} t^n \chi_{]0,s[}(t), \quad n = 1, 2, \ldots$$

For each $n \in \mathbb{N}$, M_n is homogeneous of degree 0, and so we can take $\eta(t) = 1$, for every $t \in \mathbb{R}^+$. Moreover for any $n \geq 1$,

$$A_n = \int_0^{+\infty} M_n(1, t)\, t^{-1}\, dt = n \int_0^1 t^{n-1}\, dt = 1.$$

Moreover for any $\delta > 0$,

$$A_n(\delta) = \int_{\mathbb{R}^+ \setminus]1-\delta, 1+\delta[} M_n(1,t)\, t^{-1}\, dt = n \int_0^{1-\delta} t^{n-1}\, dt \to 0, \quad \text{as } n \to +\infty$$

More precisely $A_n(\delta) = (1-\delta)^n = \mathcal{O}(n^{-\alpha})$, $n \to +\infty$, for any $\alpha > 0$.

Take now $\tau(t) = |1-t|^\alpha$, $0 < \alpha < 1$. Then for any $\delta > 0$,

$$\int_{1-\delta}^{1+\delta} M_n(1,t)\, |1-t|^\alpha\, t^{-1}\, dt = n \int_{1-\delta}^{1} t^{n-1} (1-t)^\alpha\, dt,$$

for $n = 1, 2, \ldots$. If $\xi(t) = t^\alpha$, for $0 < \alpha < 1$, we have:

$$\xi(n) \int_{1-\delta}^{1+\delta} M_n(1,t)\, |1-t|^\alpha\, t^{-1}\, dt$$

$$= n^{\alpha+1} \int_{1-\delta}^{1} t^{n-1}\, (1-t)^\alpha\, dt \leq n^{\alpha+1} B(n, \alpha+1),$$

where B is the Euler Beta function. Now it is well known (see [31]) that:

$$\lim_{n \to +\infty} n^{\alpha+1} B(n, \alpha+1) = \Gamma(\alpha+1),$$

where Γ is the Euler Gamma function. Thus (9) is satisfied for any neighbourhood base of 1.

For $\mathbb{M} = (M_n)_{n \in \mathbb{N}}$, we define the moment operators:

$$(\mathcal{M}_n f)(s) = \int_0^{+\infty} M_n(s,t)\, f(t)\, t^{-1}\, dt,$$

for $f \in \boldsymbol{X}$.

Let $\varphi \in \Phi$ and let I_φ be the corresponding modular. Let $L^\varphi_\mu(\mathbb{R}^+)$ be the Orlicz space generated by I_φ. We have the following:

Corollary 3. ([6]). *Let $f \in L^\varphi_\mu(\mathbb{R}^+)$. Then*

$$I_\varphi[\lambda(\mathcal{M}_n f - f] \to 0,$$

for sufficiently small $\lambda > 0$.

Corollary 4. *Let $\tau(t) = |1-t|^\alpha$, $\xi(t) = t^\alpha$, $t \in \mathbb{R}^+$ $0 < \alpha < 1$. If $f \in lip_\tau(I_\varphi)$, then:*

$$I_\varphi[\lambda(\mathcal{M}_n f - f)] = \mathcal{O}(n^{-\alpha}), \quad n \to +\infty,$$

for sufficiently small $\lambda > 0$.

Remark 8. Corresponding results in Musielak-Orlicz spaces, or in more general modular spaces, can be obtained.

8 Applications to a Class of Fractional Integrals

Let, as before, $G = (\mathbb{R}^+, \cdot)$ provided with the Haar measure $d\mu(t) = t^{-1}\ dt$. For each $n \in \mathbb{N}$, let us consider the family of functions, homogeneous of degree 0, $\{H_n^\alpha\}_{n\in\mathbb{N}}$, defined, for $\alpha \in]0,1[$, by:

$$H_n^\alpha(s,t) = \frac{\Gamma(n+1-\alpha)}{(n-1)!\ \Gamma(1-\alpha)}\ (s-t)^{-\alpha} s^{\alpha-n}\ t^n \chi_{]0,s[}(t), \quad (s,t) \in \mathbb{R}^+ \times \mathbb{R}^+.$$

By means of the family $\{H_n^\alpha\}_{n\in\mathbb{N}}$, we define the family of Erdélyi-Kober fractional operators (see [19], [24], [32]):

$$(T_n f)(s) \equiv f_{n,1-\alpha}(s) = \int_0^{+\infty} H_n^\alpha(s,t)\ f(t)\ \frac{dt}{t},$$

for $f \in \boldsymbol{X} = \cap_n \mathrm{Dom} T_n$.

We have the following properties of the functions H_n^α.

Proposition 3. *The family $\{H_n^\alpha\}_{n\in\mathbb{N}}$, is a ξ-singular kernel with respect to $\xi(t) = t^\gamma$, for $\gamma > 0$, and (9) holds with $\tau(t) = |1-t|^\gamma$, $0 < \gamma < 1$.*

Proof. Firstly we have:

$$A_n = \int_0^{+\infty} H_n^\alpha(1,t)\ t^{-1}\ dt = \frac{\Gamma(n+1-\alpha)}{(n-1)!\ \Gamma(1-\alpha)}\ B(1-\alpha, n) = 1,$$

for $n \in \mathbb{N}$. Next, putting:

$$C_{n,\alpha} = \frac{\Gamma(n+1-\alpha)}{(n-1)!\ \Gamma(1-\alpha)},$$

we have, for $\delta \in]0,1[$,

$$\begin{aligned} A_n(\delta) &= C_{n,\alpha} \int_{\mathbb{R}^+ \setminus]1-\delta, 1+\delta[} t^n (1-t)^{-\alpha}\ \chi_{]0,1[}\ \frac{dt}{t} \\ &= C_{n,\alpha} \int_0^{1-\delta} t^{n-1}(1-t)^{-\alpha}\ dt \\ &\le \frac{C_{n,\alpha}}{\delta^\alpha} \int_0^{1-\delta} t^{n-1}\ dt = \frac{C_{n,\alpha}}{n\delta^\alpha}(1-\delta)^n. \end{aligned}$$

Now it is well known that (see [31]), that $C_{n,\alpha}$ is infinite for $n \to +\infty$, with order $1-\alpha$ with respect to n. Thus $A_n(\delta) = \mathcal{O}(n^{-\gamma})$, for any $\gamma > 0$. Finally, let $\tau(t) = |1-t|^\gamma$, $0 < \gamma < 1$. Then for $\delta \in]0,1[$, we have:

$$C_{n,\alpha} \int_{1-\delta}^1 (1-t)^{\gamma-\alpha}\ t^{n-1}\ dt \le C_{n,\alpha} B(1+\gamma-\alpha, n).$$

But using a result given in [31, pg. 11], we have:

$$\begin{aligned}
&\lim_{n\to+\infty} n^{\gamma}\, C_{n,\alpha}\, B(1+\gamma-\alpha,n)\\
&= \lim_{n\to+\infty} \frac{n^{\gamma+1-\alpha}}{\Gamma(1-\alpha)}\, \frac{\Gamma(n+1-\alpha)}{(n-1)!\, n^{1-\alpha}}\, B(1+\gamma-\alpha,n)\\
&= \frac{\Gamma(1+\gamma-\alpha)}{(\Gamma(1-\alpha))^2},
\end{aligned}$$

and so (9) holds. □

As a consequence of Proposition 3 we can obtain approximation results for the operators $f_{n,1-\alpha}$ in modular spaces. In particular, for an Orlicz space L^{φ}_{μ} generated by a function $\varphi \in \Phi$, we obtain the following corollaries:

Corollary 5. *Let* $f \in L^{\varphi}_{\mu}(\mathbb{R}^+)$. *Then:*

$$I_{\varphi}[\lambda(f_{n,1-\alpha} - f)] \to 0, \quad n \to +\infty,$$

for sufficiently small $\lambda > 0$.

Corollary 6. *Let* $\tau(t) = |1-t|^{\gamma}$, $\xi(t) = t^{\gamma}$, $t \in \mathbb{R}^+$, $0 < \gamma < 1$. *If* $f \in lip_{\tau}(I_{\varphi})$, *then:*

$$I_{\varphi}[\lambda(f_{n,1-\alpha} - f)] = \mathcal{O}(n^{-\gamma}), \quad n \to +\infty,$$

for sufficiently small $\lambda > 0$.

Remark 9. As before, corresponding results for Musielak-Orlicz spaces, or for more general modular spaces, can be obtained.

Aknowledgements. We wish to thank Professor P.L. Butzer for his encouragements and the referee for his interesting and helpful comments in the matter.

References

1. Barbieri, F., Approssimazione mediante nuclei momento, *Atti Sem. Mat. Fis. Univ. Modena*, 32, 308-328 (1983).

2. Bardaro, C., On Approximation Properties for Some Classes of Linear Operators of Convolution Type, *Atti Sem. Mat. Fis. Univ. Modena*, 33, 329-356 (1984).

3. Bardaro, C., Musielak, J., and Vinti, G., On absolute continuity of a modular connected with strong summability, *Commentationes Math*, 34, 21-33 (1994).

4. Bardaro, C., Musielak, J., and Vinti, G., On the definition and properties of a general modulus of continuity in some functional spaces, *Math. Japonica*, 43, 445-450 (1996).

5. Bardaro, C., Musielak, J., and Vinti, G., Some modular inequalities related to Fubini-Tonelli theorem, *Proc. A.Radmadze Math. Inst.*, 118, 3-19 (1998).

6. Bardaro, C. and Vinti, G., Modular convergence in generalized Orlicz spaces for moment type operators, *Applicable Analysis*, 32, 265-276 (1989).

7. Bardaro, C. and Vinti, G., Some estimates of certain integral operators in generalized fractional Orlicz classes, *Numer.Funct.Anal.Optimiz.*, 12, 443-453 (1991).

8. Bardaro, C. and Vinti, G., A general convergence theorem with respect to Cesari variation and applications, *J. Nonlinear Analysis, Theory and Appl.*, 22, 505-518 (1994).

9. Bardaro, C. and Vinti, G., Modular estimates and modular convergence for linear integral operators, in *Mathematical Analysis, Wavelets, and Signal Processing* International Conference in Honor of Professor P.L. Butzer, Cairo, January 3 - 9, 1994, *Contemporary Math.*, 190, 95-105 (1995).

10. Bardaro, C. and Vinti, G., On the order of modular approximation for nets of integral operators in modular Lipschitz classes, special volume dedicated to Prof. J. Musielak, *Functiones & Approximatio*, 26, 135-151 (1998).

11. Butzer, P.L. and Fehér, F., Generalized Hardy and Hardy-Littlewood inequalities in rearrangiament invariant spaces, *Commentationes Math.*, Tomus specialis in Honor of L. Orlicz I, 41-64 (1978).

12. Butzer, P.L. and Jansche, S., A direct approach to the Mellin transform, *J. Fourier Anal. Appl.*, 3, 325-376 (1997).

13. Butzer, P.L. and Jansche, S., Mellin Tranforms theory and the role of its differential and integral operators, in *Proceedings of the Workshop on Transform Methods and Special Function*, Varna, 1996 (Eds. P.Rusev, I. Dimovski and V. Kiryakova), Science, Culture, Technology, Singapore (1997), in press.

14. Butzer, P.L. and Jansche, S., The finite Mellin transform, Mellin-Fourier series and the Mellin-Poisson summation formula, in *Proceedings of 3rd. Int. Conference on Functional Analysis and Approximation Theory*, Maratea, 1996, (Ed. F. Altomare), *Rend. Circ. Mat. Palermo*, in press.

15. Butzer, P.L. and Jansche, S., Mellin Tranforms, the Mellin-Poisson Summation Formula and the Exponential Sampling Theorem, special issue dedicated to Prof. C. Vinti, *Atti sem. Mat. Fis. Univ. Modena*, suppl. Vol. 46, 99-122 (1998).

16. Butzer, P.L. and Nessel, R.J., *Fourier Analysis and Approximation* Academic Press, New York, 1971.

17. Cattelani, F. Degani, Nuclei di tipo distanza che attutiscono i salti in una o piú variabili, *Atti Sem. Mat. Fis. Univ. Modena*, 30, 299-321 (1981).

18. Cattelani, F. Degani Approssimazione del perimetro di una funzione mediante nuclei di tipo momento, *Atti Sem. Mat. Fis. Univ. Modena*, 34, 145-168 (1985-86).

19. Erdelyi, A., On fractional integration and its application to the theory of Hankel transforms, *Quart. J. Math.*, 11, 293 (1940).

20. Fehér, F., A generalized Schur - Hardy inequality on normed Köthe spaces, in *General Inequalities II, Prooceedings of second Int. Conference on General Inequalities*, Oberwolfach, 1978, Birkhauser, Basel, 1980, 277-285.

21. Fiocchi, C., Variazione di ordine α e dimensione di Hausdorff degli insiemi di Cantor, *Atti Sem. Mat. Fis. Univ. Modena*, 34, 649-667 (1991).

22. Gogatishvili, A. and Kokilashvili, V., Criteria of weighted inequalities in Orlicz classes for maximal functions defined on homogeneous type spaces, *Proc. Georgian Acad. Sci. Math.*, 6, 617-645 (1993).

23. Hewitt, E. and Ross, K.A., *Abstract Harmonic Analysis*, Springer-Verlag, Berlin-Göttingen 1963.

24. Kober, H., On fractional integrals and derivatives, *Quart. J. Math*, 11, 193, (1940).

25. Kozlowski, W.M., *Modular Function Spaces* (Pure Appl. Math.) Marcel Dekker, New York and Basel, 1988.

26. Love, E.R., Some inequalities for fractional integrals, in *Linear Spaces and Approximation, Proceedings Conf. Math. Research Institute*, Oberwolfach, 1977, 177-184, International Series of Numerical Mathematics, **40**, Birkhauser Verlag, Basel, Stuttgart, 1978.

27. Love, E.R., Links between some generalizations of Hardy's integral inequality, *General Inequalities IV, Proceedings 4th Int. Conf. on General Inequalities*, Oberwolfach, 47-57 (1984).

28. Mantellini, I. and Vinti, G., Modular estimates for nonlinear integral operators and applications in fractional calculus, *Numer. Funct. Anal. Optimiz.*, 17, 143-165 (1996).

29. Musielak, J., *Orlicz Spaces and Modular Spaces* Lecture Notes in Math., 1034, Springer-Verlag, (1983).

30. Musielak, J., Nonlinear approximation in some modular function spaces I, *Math. Japonica*, 38, 83-90 (1993).

31. Rainville, E.D., *Special functions* McMillan Co., New York, 1960.

32. Sneddon, I.N., The use in mathematical physisc of Erdélyi-Kober operators and of their generalizations, *Proc. Int. Conf. Univ. New Haven, 1974*, Lecture Notes in Math., 457, 37-79 (1975).

33. Vinti, C., Perimetro - Variazione, *Annali Scuola Norm. Sup. Pisa*, serie III, 43, 201-231 (1964).

34. Vinti, C., Sull'approssimazione in perimetro e in area, *Atti Sem. Mat. Fis. Univ. Modena*, 13, 187-197 (1964).

35. Vinti, C., A Survey on Recent Results of the Mathematical Seminar in Perugia, inspired by the Work of Professor P.L. Butzer, *Result.Math.*, 34, 32-55 (1998).

36. Zanelli, V., Funzioni momento convergenti dal basso in variazione di ordine non intero, *Atti Sem. Mat. Fis. Univ. Modena*, 30, 355-369 (1981).

ON THE SIMULTANEOUS APPROXIMATION OF FUNCTIONS AND THEIR DERIVATIVES

Theodore Kilgore
Department of Mathematics, Auburn University
Auburn University, AL 36849-5307
E-mail: Kilgota@banach.math.auburn.edu

The topic of this article is the simultaneous approximation of a function f and its derivatives by an algebraic polynomial and its derivatives. The article is a survey of the topic, its theoretical significance, and some practical aspects. This article will also touch on related areas of approximation theory, such as polynomial inequalities for derivatives and the simultaneous approximation properties of trigonometric functions, which are closely related to the simultaneous approximation properties of algebraic polynomials. This article also contains a brief introduction to the basic questions and the basic results of approximation theory. Inasmuch as the methods for simultaneous approximation of derivatives presented here are extensions and refinements of many of these basic results, the survey is intended to make the article as self-contained as possible, in order for it to be readily accessible to non-specialists in approximation theory. For these non-specialists, it is also hoped that a basic survey of approximation theory will be of general interest and help.

1 Introduction

The simultaneous approximation of derivatives by an approximant and its respective derivatives has both theoretical and practical aspects. The theory is a refinement and completion of results basic in approximation theory, and the practice deals with approximation methods and algorithms accompanied by error estimates. This article will discuss results in and related to both of these areas. Although this is not a treatise on applied mathematics as such, the treatment is oriented toward applications. Also, some of the fundamentals of approximation theory are surveyed and summarized in order to provide background. The treatment will be self-contained as far as it seems possible in a brief survey.

The approximation methods presented here will be based upon traditional methods of approximation, using algebraic polynomials or trigonometric polynomials as approximants. Linear approximation methods and techniques will be given, insofar as possible accompanied by error estimates or at least by some indication of how the error estimates may be obtained. Some of the methods described here have been combined most efficaciously with rapid computational methods and employed to develop actual computer implementations. Also computer comparisons on various test functions have been done using these methods, along with testing of other methods, such as spline approximation. These results are interesting in themselves, and they will be discussed further, under the topic of "Effectiveness and Efficiency." However, problems of computer programming, roundoff or truncation error, overflow error, or numerical stability in general are not the topic of this article.

2 Background

Approximation of both numbers and functions is a basic necessity in mathematical applications. In a typical approximation problem, we know a thing in principle or by description, but we do not have a way to give it explicitly. From this arises the basic question of approximation.

1. We have a problem to solve, which is either inconvenient or impossible to solve in closed form. Thus we desire an answer which can be given to arbitrary accuracy.

We try to develop a method to attack the problem. Assuming that we have been successful in constructing a method, we have further questions to answer.

2. Will our method work, in the sense of converging to a correct answer? This is inescapably a theoretical question, which in its very nature cannot be answered by numerical experimentation.

3. Is there a method to estimate error? If not, we do not know how far to carry out our procedure, or how many steps, in order to be assured of some prescribed degree of accuracy.

4. If there is more than one method available for which the previous questions are answered, can we compare the methods for efficiency, in order to make an intelligent choice between them?

The mathematical subject called approximation theory deals more properly with the approximation of functions. For such an endeavor, an obvious place to start is with the approximation of functions defined on some interval. Now still another question is appropriate, in addition to those already listed:

5. Out of many ways to measure or estimate error when one proposes to approximate one function with another, which way makes sense in the context of the given problem?

If for example we want to approximate well the function value(s) at some more or less random point(s), then we want to measure error in the uniform norm. If on the other hand our original intent is to approximate the integral of the function, then perhaps an integral norm is more relevant. There are a myriad of possibilities. We introduce some of the possibilities in the next section. Some others will be seen later on in this article.

3 Some Basic Approximation Theory

3.1 Linear spaces and norms

For reasons which should be fairly obvious, it is more convenient to pose, consider, and if possible to solve problems of approximation in a linear space (vector space) than to work in some strange environment where the ordinary laws of algebra do not work. Most approximation is done in linear spaces of functions, and that kind of approximation is discussed here.

A norm in a linear function space gives a way of describing distance which is compatible with the algebraic structure. For notation, let f be a function defined on an interval. Then the norm of f can be written as $\|f\|$ (the reader is hereby warned that we will later on use this notation to denote a particular norm, so this usage is temporary). Let us assume that we have wished to define $\|f\|$ inside of some large linear function space, such as the set of all functions defined on some given interval. If the norm which we propose obeys the defining properties of a norm but is not defined for all functions in the large space, we can confine ourselves to that subset X of our very large linear space, on which our norm everywhere makes sense. It will be seen from the defining properties of a norm (see the remark after the definition) that our subset X is in fact a itself a linear space, a subspace of the original space.

The defining properties of a norm upon X are

(a) $\|f\| \geq 0$ for all $f \in X$, with $\|f\| = 0$ if and only if f is the zero function.

(b) $\|\alpha f\| = |\alpha| \cdot \|f\|$ for all $f \in X$ and for all scalars α.

(c) $\|f + g\| \leq \|f\| + \|g\|$ for all f and g in X.

Remark. Note that if X is defined to be the set on which a proposed norm is defined, inside of some larger linear space, then X is closed under addition by (c) and is closed under scalar multiplication by (b) and is thus itself a linear space. This justifies the standard practice of defining a norm and then speaking of the space of functions for which the norm makes sense.

Now, if X is a linear space with a norm on it (a *normed linear space*), then the distance between two functions f and g in X is very naturally defined as $\|f - g\|$.

Some examples of normed spaces and norms:

- The space of continuous functions defined upon a closed and bounded interval is denoted $C[a, b]$. It is clearly possible for this purpose to define a standard closed and bounded interval. Our choice is $[-1, 1]$. The usual norm in $C[-1, 1]$ is defined by

$$\|f\| = \sup_{-1 \leq x \leq 1} |f(x)|,$$

  which is defined for every continuous function. This norm is called by the various names "supremum norm," "uniform norm," "maximum norm." Note that the notation $\|f\|$ has from henceforward a particular meaning.

- The space of all functions f defined on $[-1, 1]$ for which

$$\|f\|_p = \left(\int_{-1}^{1} |f(x)|^p \, dx \right)^{\frac{1}{p}}$$

  is defined and finite is called $L^p[-1, 1]$. The permissible values of p are $1 \leq p < \infty$. Here, the integral is defined in the sense of Lebesgue. A first requirement for this is, of course, that the function is measurable in the sense of Lebesgue. The reader who is unfamiliar with these concepts may be assured that the Lebesgue integral extends the Riemann integral of calculus and in no way overturns the methods of evaluating integrals which were presented there. To deal with the technicalities of the Lebesgue integral and the Lebesgue measure is not needed here and is beyond the scope of this article.

- If the function f is measurable on $[-1, 1]$ and if there is M such that $|f(x)| \leq M$ except on a set of measure zero, then we say that f is essentially bounded and is in the space $L^\infty[-1, 1]$. The norm of f is then defined as the infimum (greatest lower bound) of the set of all such aforementioned M. In symbols:

$$\|f\|_\infty = \text{ess} \sup_{-1 \leq x \leq 1} |f(x)|.$$

  Clearly, every function in $C[-1, 1]$ is also in $L^\infty[-1, 1]$, and

$$\|f\| = \|f\|_\infty$$

for every continuous function f. However, a bounded function can be far from continuous.

We can take spaces from the set of periodic functions with some fixed period as well as from the functions defined upon a closed and bounded interval. As we have taken the standard interval to be $[-1, 1]$, we take the standard period to be 2π. As we have defined certain normed linear function spaces on $[-1, 1]$, we can analogously define the normed linear spaces of 2π-periodic functions $C(2\pi)$, $L^p(2\pi)$ for $1 \leq p < \infty$, and $L^\infty(2\pi)$.

3.2 Approximation – basic terminology and some basic results

The notation $E_n(f)$ denotes the least possible error incurred when $f \in C[-1, 1]$ is approximated by an algebraic polynomial of degree at most n. Formally, with P_n signifying an arbitrary algebraic polynomial of degree at most n,

$$E_n(f) = \inf_{P_n} \|f - P_n\|. \tag{1}$$

In the context of $C(2\pi)$, the notation $E_n^*(\phi)$ denotes the error incurred in approximating the 2π-periodic function ϕ by a trigonometric polynomial of degree at most n. The formal definition is quite similar to that in (1).

Later, we will introduce some weighted norms, but right now we want to mention some of the basic results in the theory of approximation.

First among these is the theorem of Weierstrass, which says that, given any function $f \in C[-1, 1]$ and given an prescribed error tolerance $\epsilon > 0$, there is an algebraic polynomial P_n (of some degree n) for which the error $\|f - P_n\| < \epsilon$. That is, the theorem of Weierstrass says, for every $f \in C[-1, 1]$,

$$\lim_{n \to \infty} E_n(f) = 0.$$

In case that we want to discuss $C(2\pi)$ and approximation by trigonometric polynomials, the analogous statement of convergence also holds. In other words, the Weierstrass Theorem for $C(2\pi)$ says that for every $f \in C(2\pi)$,

$$\lim_{n \to \infty} E_n^*(f) = 0.$$

The Weierstrass theorem does not in itself answer some of our basic questions about approximation. Typically, a proof of the theorem proceeds by methods which are less useful in applications than in theory (here seems to be one of the tradeoffs of approximation: methods which always work usually are extremely inefficient).

Before continuing with others of the basic results on approximation, we need to introduce the modulus of continuity of a function f (periodic or on a closed interval):

$$\omega(f,t) = \sup_{|x-y|\le t} |f(x) - f(y)|. \tag{2}$$

It should be clear that $\omega(f,t) \to 0$ as $t \to 0$ if and only if f is a continuous function. Other properties of a modulus of continuity function are that it is non-negative and increasing and satisfies the algebraic property $\omega(f,t_1+t_2) \le \omega(f,t_1) + \omega(f,t_2)$.

The theorems of D. Jackson [29] give the feasible rates of convergence to a given continuous function. As many of the basic results in approximation theory, these results are stated first, in their simplest and strongest form, for trigonometric polynomial approximation of 2π-periodic functions. The results of Jackson, in the versions later developed by Favard [25], Achieser and Krein [1], and Korniecuk [42], say that for any 2π-periodic function f:

I. $E_n^*(f) \le a_k(n+1)^{-k}\|f^k\|$ for all $f \in C^k(2\pi)$, the space of all k times continuously differentiable 2π-periodic functions. The numbers a_k are precisely known, and all a_k satisfy $a_k \le a_1 = \frac{\pi}{2}$. Also note that this statement clearly implies $E_n^*(f) \le a_k(n+1)^{-k}E_n^*(f^k)$.

II. $E_n^*(f) \le \omega\left(\dfrac{\pi}{n+1}\right)$ for all $f \in C(2\pi)$.

There are versions of these results for algebraic polynomial approximation. The algebraic polynomial results are derived from the versions here, via $x \leftrightarrow \cos\theta$. Note that Jackson I is thereby considerably weakened. To see this, consider a function $f \in C[-1,1]$ and let $P_n(x)$ be its best approximant. Via $x \leftrightarrow \theta$, we note that both $f(\cos\theta)$ and $P_n(\cos\theta)$ are 2π-periodic and even, and clearly $P_n(\cos\theta)$ is the best approximant of $f(\cos\theta)$. Hence, by Jackson I we have

$$E_n(f(x)) = E_n^*(f(\cos\theta)) \le \frac{\pi}{2(n+1)}\|f'(\cos\theta)\sin\theta\| \le \frac{\pi}{2(n+1)}\|f'(x)\|.$$

If one insists, as might seem natural, upon estimating $E_n(f)$ with $f'(x)$, then the precision of Jackson I is significantly degraded. Though less obvious, similar problems also happen with Jackson II.

One method for dealing with the problems outlined in the previous paragraph is a variable or weighted modulus of continuity. A. F. Timan [56], was the first to introduce a variable modulus, and in 1951 he proved that for any function $f \in C^q[-1,1]$ $(q \geq 0)$ there exists a polynomial P_n of degree at most n, such that

$$|f(x) - P_n(x)| \leq C \left(\frac{\sqrt{1-x^2}}{n} + \frac{1}{n^2} \right)^q \omega \left(f^{(q)}, \frac{\sqrt{1-x^2}}{n} + \frac{1}{n^2} \right). \quad (3)$$

His proof did not make it easy actually to estimate the constant C. Note, however, that for the case $q = 0$ in particular, this result does overcome the degradation of the error estimate obtained in Jackson I.

To close our brief survey of some of the basics of approximation theory, we reiterate that the development of the theory of approximation on a finite closed interval, which can obviously be standardized as the interval $[-1,1]$, relies heavily upon related results which hold true for simultaneous approximation by the "nicer" trigonometric polynomials in the "nicer" space of 2π-periodic functions. The theorems of Jackson are a good example of this phenomenon at work. The correspondence $x \leftrightarrow \cos\theta$ then gives a natural mapping $f(x) \leftrightarrow f(\cos\theta)$ between functions on $[-1,1]$ and 2π-periodic even functions. By way of this correspondence, many crucial aspects of both the theory and the practice of algebraic polynomial approximation have trigonometric antecedents. Similar observations pertain even more strongly to the topic of simultaneous approximation. We have seen already that differentiation does not behave in exactly the same way in the two function spaces, even though functions on $[-1,1]$ can be naturally made to represent 2π-periodic even functions and vice versa. These facts cause special complications both in theory and in practice for simultaneous approximation of derivatives. If viewed correctly, they also provide opportunities for better understanding.

3.3 Linear methods for approximation – bounded linear operators and linear projections

As linear spaces are more pleasant environments in which to do approximation, so linear methods are more easily applied. Let L be an operator (a function which "operates" upon functions) between a linear space of functions X and a linear space Y. Then L is itself called linear if $L(\alpha f) + L(\beta g) = \alpha L(f) + \beta L(g)$ for arbitrary scalars α and β and arbitrary functions f and g from X. Linear operators used for approximation should preferably be continuous, that is, bounded. This is even more strongly true because many of the phenomena we wish to study are described by operators which are already not continuous. For example, differentiation can be validly viewed as a linear operator, which is highly discontinuous. The operator L is bounded if its norm is finite, and its norm is given by

$$\|L\| = \sup_{\|f\|_X \leq 1} \|Lf\|_Y. \tag{4}$$

Unfortunately, the operation of mapping a function to its best approximant is not linear.

Yet another property for an approximation operator is desirable. Continuing the discussion of the operator L, let us suppose the special case that Y, the range, is a subspace of the domain, X. It is then esthetically satisfying and computationally convenient if $Lf = f$ for every f in Y. If so, then L is a linear projection. Relative to the theoretically feasible least possible error $E_n(f)$ (or $E_n^*(f)$), approximation by a linear projection L_n into the algebraic (or respectively trigonometric) polynomials of degree at most n may be closely estimated by an inequality of Lebesgue. If we let I denote the identity operator on the domain space, then, noting that $p = L_n p$ for any polynomial p in the range, we have for any such p

$$\|f - L_n f\| = \|(I - L_n)(f - p)\| \leq \|I - L_n\| \cdot \|f - p\|. \tag{5}$$

Hence,

$$\|f - L_n f\| \leq \|I - L_n\| E_n(f) \leq (1 + \|L_n\|) E_n(f)$$

Thus, the norm of L_n is related intimately to the quality of approximations arising from it. Unfortunately, it is also true that, now matter how a sequence of linear projection operators L_n mapping into the set of polynomials of degree at most n is constructed, one has $\|L_n\| \to \infty$ as $n \to \infty$. The same is true for a sequence of projection operators mapping the continuous 2π-periodic functions into the space of trigonometric polynomials of degree at most n. In this case, the projection of least norm for each n is actually known to be the Fourier expansion truncated at n (cf. Section 3.5). The proof of this is due to Marcinkiewicz and can be found for example in the book of Cheney [19]. It is quite elegant.

3.4 Interpolation

An often-used method for the approximation of a function $f \in C[-1, 1]$ is Lagrange interpolation (it will be clear from its construction that Lagrange interpolation cannot be used if continuity is absent). The result is to approximate f by a function of degree at most n. To approximate by this procedure, one chooses or is given a set of points $x_0, \ldots, x_n$ strictly ordered consecutively in the interval $[-1, 1]$ (usually from left to right, sometimes from right to left). These points are called the nodes of interpolation. The reader should also be aware that there is not complete unanimity among those who deal with Lagrange interpolation concerning the numbering of the nodes. The problem, vexing in a minor way, arises because the set of polynomials of degree at most n is, most inconveniently, an $n+1$-dimensional space. Thus on occasion the nodes are numbered from 1 to n instead of from 0 to n, and then the resulting polynomials are of degree at most $n-1$ instead of n. For each of these numbering schemes there are situations in which it seems to be the most natural; thus this author is not even unanimous with himself on the issue. Right now we will assume that there are $n+1$ nodes, and the approximating polynomials are of degree at most n.

Assuming agreement for now about the numbering of the nodes, we construct the Lagrange interpolation operator. For $i \in \{0, \ldots, n\}$ polynomials $\ell_i(x)$ are easily constructed which satisfy $\ell_i(x_i) = 1$ and for all $j \in \{0, \ldots, n\}$ satisfy $\ell_i(x_j) = 0$ whenever $j \neq i$. One such construction is

$$\ell_i(t) = \prod_{j=0,\ j\neq i}^{n} \frac{t - t_j}{t_i - t_j}. \tag{6}$$

There are other representations, which must perforce give the same polynomial ℓ_i when applied.

The Lagrange interpolant of the given function f is now a polynomial $L_n f$ of degree at most n, defined by

$$L_n f(x) = \sum_{i=0}^{n} f(x_i)\ell_i(x). \tag{7}$$

By construction, $L_n f(x_i) = f(x_i)$ and, if f is itself a polynomial of degree at most n, then $L_n f = f$. That is, L_n is a projection operator.

A similar construction is possible for a function $f \in C(2\pi)$, which then gives a trigonometric polynomial approximant of degree at most n. The natural basis here consists of
$1, \cos\theta, \ldots, \cos n\theta, \sin\theta, \ldots, \sin n\theta$, a set of $2n+1$ functions, for which $2n+1$ distinct nodes $\theta_0, \ldots \theta_{2n}$ are needed in the interval $[0, 2\pi)$ or any other convenient interval of the same length. The interval is half open deliberately. To choose a node at 0 and another at 2π will not work because of the periodicity. The basis functions $\ell_0, \ldots, \ell_{2n}$ are given by

$$\ell_i(\theta) = \prod_{j=0;\ j\neq i}^{2n} \frac{\sin\frac{\theta-\theta_j}{2}}{\sin\frac{\theta_i-\theta_j}{2}}. \tag{8}$$

The trigonometric interpolation operator is now defined for an arbitrary continuous 2π-periodic function f, with the fundamental trigonometric polynomials (8) as

$$L_n f(\theta) = \sum_{i=0}^{2n} f(\theta_i)\ell_i(x). \tag{9}$$

For the interpolation with trigonometric polynomials, two important special cases occur. If the function which we are interpolating is 2π-periodic and even, then it is only necessary to take $n+1$ nodes $\theta_0, \ldots, \theta_n$ in the interval $[0, \pi]$, and we can construct appropriately a cosine polynomial of degree at most n to interpolate f, using

$$\ell_i(\theta) = \prod_{j=0;\ j\neq i}^{n} \frac{\cos\theta - \cos\theta_j}{\cos\theta_i - \theta_j} \tag{10}$$

and obtaining

$$L_n f(\theta) = \sum_{i=0}^{n} f(\theta_i)\ell_i(x). \tag{11}$$

Also if f is an odd 2π-periodic function we can construct an interpolant of degree not exceeding n, this time using n nodes of interpolation $\theta_1, \ldots, \theta_n$ in the interval $(0, \pi)$, with fundamental polynomials given by

$$\ell_i(\theta) = \frac{\sin\theta}{\sin\theta_i} \prod_{j=1}^{n} \frac{\cos\theta - \cos\theta_j}{\cos\theta_i - \theta_j}. \tag{12}$$

The interpolation operator is then

$$L_n f(\theta) = \sum_{i=1}^{n} f(\theta_i)\ell_i(x). \tag{13}$$

By (5) the approximation properties of L_n depend upon its norm. It is easily seen for approximation by algebraic polynomials that

$$\|L_n\| = \|\sum_{i=1}^{n} |\ell_i(x)|\|, \tag{14}$$

where the norm on the right is the usual uniform norm in $C[-1, 1]$. For trigonometric polynomial interpolation of 2π-periodic functions and for the interpolation of even functions or respectively of odd functions, the norms are obtained in like manner; only the indices must be adjusted as appropriate. In all cases, the function which is normed in the right member of the equation (14) is called the *Lebesgue function* of L_n.

It is also clear that the magnitude of $\|L_n\|$ depends only upon the placement of the nodes, for the very polynomials ℓ_n themselves (whether algebraic or trigonometric) depend solely upon the location of the nodes. It is also known that, no matter what scheme is adopted for the placement of nodes for successive values of n, the norm $\|L_n\|$ will increase without bound as $n \to \infty$. This is a particular case of an already-mentioned general phenomenon, but, as a historical footnote, these undesirable properties of Lagrange interpolation were discovered first, by Faber [24]. At any rate, Lagrange interpolation as a consequence does not converge uniformly for all continuous functions. Nevertheless, if the nodes are well chosen, it does converge well for differentiable functions.

In view of the fact that the norm of L_n must increase without bound no matter what nodes are chosen for each n, it becomes interesting to know what the least bad rate of increase of the norm is, as well as to know, if possible, which are the best sets of nodes to use. As has been indicated, the norm depends upon the nodes and is the supremum norm of the Lebesgue function, whose value at each node is 1 and which is a piecewise polynomial (algebraic or respectively trigonometric) on any interval between two nodes.

If L_n is the algebraic polynomial operator, then its Lebesgue function has n local maxima between the nodes, plus possibly two endpoint maxima, if $-1 < t_0$ and/or $t_n < 1$. If L_n is the general trigonometric polynomial operator, then there are $2n+1$ local maxima between nodes, taking periodicity into account. The case that L_n is the cosine polynomial operator on $[0, \pi]$ is equivalent to the algebraic polynomial case on $[-1, 1]$. The sine polynomial case is slightly more complicated. There may be $n-1$ or n or $n+1$ local maxima, none of them occurring at endpoints, depending upon the placement of the nodes.

An old conjecture of Bernstein [15] regarding algebraic polynomial interpolation was that the norm is minimized when the values at all local maxima of the Lebesgue function are equal. The way he stated this, it included the two endpoint maxima. His conjecture has been shown factual, but it was proven by standardizing the problem, putting the two extreme nodes at the endpoints ± 1 of the interval of interpolation and dealing with only the "interior" nodes. The proof was due to Kilgore [31] and [32]. Kilgore's work in [31] was used by de Boor and Pinkus [16] to give another proof and also to prove the analogous conjecture for the best nodes of trigonometric polynomial interpolation of the 2π-periodic functions. Again based upon the techniques provided by Kilgore

[31], de Boor and Pinkus additionally proved in their article an extension of Bernstein's conjecture by Erdös [21]. The conjecture of Erdös was that the value of the norm of the best possible interpolation must always lie between the least and the greatest of the local maximum values of the Lebesgue function, no matter what nodes are used. The space of sine polynomials of degree at most n behaves differently because all functions in it are zero at all multiples of π. However, via the transformation $x \leftrightarrow \cos\theta$, it can be viewed as a weighted algebraic polynomial space. Using this fact, in Kilgore [36] the conjectures of Bernstein and Erdös have been shown to characterize the best nodes of interpolation into the space of sine polynomials. The proof also implies that there must in fact be $n+1$ (not fewer) local maxima for the Lebesgue function if its norm is minimal.

Favorable resolution of the Bernstein and Erdös conjectures was of more than theoretical interest. First, this work showed definitely that any set of equally spaced nodes (rotation should not matter and does not) minimizes the norm of the interpolation (9) of periodic functions. The norm of the trigonometric interpolation on equally spaced nodes is approximately $\frac{2}{\pi}\log n$.

For algebraic polynomial interpolation the way to construct the nodes which give interpolation of minimal norm does not seem easy. Nevertheless, the affirmative resolution of the conjectures of Bernstein and Erdös gives a way to measure the quality of interpolation on any given set of nodes: one can compare the value of the least and greatest local maxima of the Lebesgue function, and if this difference is small, then the nodes may be judged good. By this measure, interpolation on the $n+1$ roots of the Chebyshev polynomial $\cos n \arccos\theta$ is seen to be quite good. The values at the local extrema increase as one moves from the center of the interval $[-1, 1]$ to the endpoints, and the difference between the least and greatest of the local extrema is uniformly bounded for all n by $\frac{1}{2}$. Also, the Lebesgue function on the Chebyshev nodes is bounded above by $1 + \frac{2}{\pi}\log(n+1)$ (cf. Brutman [18] for details).

Before leaving the topic of polynomial Lagrange interpolation, we mention that the formula (7) is in fact but one of many representations for it. The same polynomials ℓ_j which are used in constructing this representation of L_n can be constructed in many other ways; for example they can be represented as quotients of Vandermonde determinants. Also, and perhaps even more interesting, a change in the basis of the space of polynomials (or respectively trigonometric functions) of degree at most n leads to a corresponding change in the representation of L_n, without in any way changing its essence. For example, trigonometric polynomial interpolation can be expressed in terms of the natural basis $1, \cos\theta, \ldots, \cos n\theta, \sin\theta, \ldots, \sin n\theta$, and similarly the algebraic polynomial interpolation can be expressed as an appropriate linear combination of the

Chebyshev polynomials $T_k(x)$ for $k = 0, \ldots, n$, where $T_k(x) = \cos n \arccos x$. These expressions of Lagrange interpolation are often called the fast Fourier transform, the discrete Fourier transform, the discrete cosine transform in the case of an even expansion, or the discrete sine transform in the case of an odd expansion.

Hermite interpolation denotes the construction of a polynomial approximant by interpolating the function and some of its derivatives at one or more points. The Lagrange interpolation of a function (interpolating the function only and no derivatives) and the truncated Taylor expansion of a function are the two extreme examples.

Neither pure Lagrange interpolation nor Hermite interpolation have any intrinsic meaning in the L^p spaces, since in any L^p space, $1 \leq p \leq \infty$, two functions f and g are considered equal if they differ only on a set of measure zero. And any countable, hence any finite, set (for example, the set of nodes of interpolation) is of measure zero. In spite of these facts, we shall see that one of the ways to enhance any bounded linear projection and to turn it into a good tool for simultaneous approximation, including Lagrange interpolation, is to add some strategically placed nodes for Hermite interpolation. In so doing, one must be careful of course only to interpolate continuous functions. However, any function which can be written as the definite integral of its derivative is continuous.

3.5 Truncated Fourier series and truncated Chebyshev expansion

We have already mentioned the Fourier series, which is defined for all integrable functions. Given such a function f, its Fourier series is

$$f(\theta) \sim \frac{a_0}{2} + \sum_{k=0}^{\infty} (a_k \cos k\theta + b_k \sin k\theta) \tag{15}$$

in which

$$a_k = \frac{1}{\pi} \int_{-\pi}^{\pi} f(\theta) \cos k\theta \, d\theta \qquad \text{and} \qquad b_k = \frac{1}{\pi} \int_{-\pi}^{\pi} f(\theta) \sin k\theta \, d\theta.$$

The nth partial sum of the Fourier series is denoted S_n. A most concise way to describe $S_n f$ is in terms of an integral kernel:

$$S_n f(\theta) = \frac{1}{\pi} \int_{-\pi}^{\pi} f(t) \frac{\sin(2n+1)\frac{t-\theta}{2}}{2 \sin \frac{t-\theta}{2}} \, dt. \tag{16}$$

By construction, S_n is a linear operator. And since the Fourier series of any trigonometric polynomial is obviously the same trigonometric polynomial, S_n is a linear projection.

The norm of S_n, when used as a mapping from $C(2\pi)$ with uniform norm is given by

$$\|S_n\| = \left\| \frac{1}{\pi} \int_{-\pi}^{\pi} \left| \frac{\sin(2n+1)\frac{t-\theta}{2}}{2\sin\frac{t-\theta}{2}} \right| dt \right\|. \tag{17}$$

This quantity is approximately equal to $\frac{4}{\pi^2}\log n$ and is bounded for example by $\frac{4}{\pi^2}(2+\log n)$. As previously stated, the operator S_n is the linear projection of least possible norm from $C(2\pi)$ into the trigonometric polynomials of degree at most n. If the domain space consists of $L^p(2\pi)$, however, for $1 < p < \infty$, then the norm of S_n may be similarly estimated and turns out to be uniformly bounded for all n. And if $p = 2$, then its norm is exactly one, and in the L^2 norm $S_n f$ is in fact the best approximant for f from among the trigonometric polynomials of degree at most n.

Under certain circumstances, the Chebyshev expansion gives an algebraic polynomial approximation for a function defined on the interval $[-1, 1]$. It is defined using the correspondence $x \leftrightarrow \cos\theta$, to map the function $f(x)$ to the 2π-periodic function $f(\cos\theta)$. Then, $S_n f$ is defined if $f(\cos\theta)$ is integrable, which happens if and only if the original function was integrable with weight $(1-x^2)^{-\frac{1}{2}}$. As $S_n f$ is a polynomial involving only cosines, it can be expressed as an algebraic polynomial, written in terms of the polynomials $T_k(x) = \cos k \arccos x$, the Chebyshev polynomials.

3.6 Some monotone and "near-monotone" operators

We have noted that the norm of S_n is not uniformly bounded as $n \to \infty$, and hence $S_n f$ will not converge for every continuous function f. However, the operators

$$\sigma_n = \frac{1}{n} \sum_{k=0}^{n-1} S_k,$$

known as the Fejér means, do converge. The reason for this is that σ_n may be represented as

$$\sigma_n f(\theta) = \frac{1}{2n\pi} \int_{-\pi}^{\pi} f(t) \left(\frac{\sin\frac{n(t-\theta)}{2}}{2\sin\frac{t-\theta}{2}} \right)^2 dt,$$

which has a positive kernel. Consequently, the operator σ_n is monotone, meaning that, if f and g are continuous functions with $f \le g$, then $\sigma_n f \le \sigma_n g$. Thus in particular if $\|f\| \leq 1$ we have $\sigma_n f \leq \sigma_n 1 = 1$, and so $\|\sigma_n\| = 1$. One of the standard proofs of the theorem of Weierstrass uses the monotonicity of σ_n, along with a general result of Korovkin. Korovkin's theorem says that, for any sequence of monotone linear operators L_n into the trigonometric polynomials of degree at most n, the three statements $L_n 1 \to 1$, $L_n \cos\theta \to \cos\theta$, and $L_n \sin\theta \to \sin\theta$ taken together imply $L_n f \to f$ for every continuous and 2π-periodic f. These three statements are manifestly true (with some minor adjustment on the index n) for the sequence σ_n. However, one notes that $\sigma_n \cos\theta = \frac{n-1}{n}\cos\theta$, so that convergence is excruciatingly slow, and similar is true for $\sin\theta$. This is the kind of thing which was meant earlier, when it was stated that most of the methods actually used to prove the theorem of Weierstrass are less useful for applications than in theory.

The de la Vallée-Poussin means of the truncated Fourier series are defined as

$$V_n = \frac{1}{n}\sum_{k=n}^{2n-1} S_k = 2\sigma_{2n} - \sigma_n.$$

It is clear that this guarantees $\|V_n\| \leq 3$ for all n. Even more interestingly, $V_n t_n = t_n$ for any trigonometric polynomial t_n of degree at most n. Thus V_n is a "near-projection," sharing to some extent the nice properties of projections without being itself a projection. One of these properties is that the theorem of Lebesgue (5) can be made to work, and one has for every $f \in C(2\pi)$

$$\|f - V_n f\| \leq 4E_n^*(f).$$

Another collection of monotone operators is the sequence of Bernstein polynomials, $B_n f$ defined (this time for $f \in C[0,1]$, not our usual standard interval) by

$$B_n f(x) = \sum_{k=0}^{n} \binom{n}{k} f(\frac{k}{n}) x^k (x+1)^{n-k} \tag{18}$$

The rate at which $B_n f$ converges to f is related to the behavior of the function. It is known that

$$\|f - B_n f\| \leq \frac{3}{2}\omega(f, \frac{1}{\sqrt{n}}),$$

and this rate of convergence cannot in general be improved. We will see that this is indeed a slow rate of convergence, compared to other methods. Thus, although the Bernstein polynomial does always converge and although it, as well as the operators σ_n, does have some properties of approximating derivatives, we will not deal further with it.

3.7.1. Simultaneous approximation of derivatives on a finite interval

Leaving aside the slowly-converging monotone operators of the sort used to prove the Weierstrass theorem, one of the simplest and most obvious examples of an approximation procedure which provides the simultaneous approximation of derivatives is the truncated Fourier series operator S_n in $L^2(2\pi)$, the space of square-integrable 2π-periodic functions. Given a periodic function f, not only does $S_n f$ give the best approximation to f among trigonometric polynomials of given degree, but also the derivative of the truncated Fourier series for f is the similarly truncated Fourier series for f'. Thus in $L^2(2\pi)$ we have both $\|f - S_n f\|_2 = E_{n,2}f$ and $\|f' - S_n' f\|_2 = E_{n,2}f'$. That Fourier expansion commutes with differentiation clearly holds true no matter what the norm is. Also, this property obviously inherited by the Fejér means σ_n and the de la Valleé-Poussin means V_n. Consequently, in many normed spaces of 2π-periodic functions the simultaneous approximation of derivatives is almost as automatic as in L^2. This phenomenon will be discussed in more detail. As to algebraic polynomial approximation on an interval, we have already seen that S_n becomes the truncated Chebyshev polynomial expansion. It will also be seen that S_n and related tools can be adapted and translated to the context of algebraic polynomial approximation on $[-1, 1]$ in such a manner as to give good results for simultaneous approximation of derivatives as well.

The theoretical treatment of simultaneous approximation by trigonometric polynomials also involves the application of the Bernstein inequality (see (27) and the general discussion in Section 7) for the derivative of a trigonometric polynomial. Here, too, there are analogous tools for algebraic polynomial approximation, the inequalities of Brudnyi and Dzyadyk. We will survey some new techniques for proving such inequalities, using some new variations of the identity of M. Riesz [47].

We will present some history, some recent theoretical developments, and finally some approximation methods, along with error estimates simple enough to express numerically. The methods described will be simple enough to implement using some of the standard methods, such as the discrete cosine transform.

3.7.2. Simultaneous approximation of derivatives on infinite intervals

On infinite intervals, functions in general might not be bounded, and polynomials in particular are certainly not. A weighted norm, usually an exponential weight, can be employed to handle this problem. The use of weighted norms can lead to a theory which is quite analogous to that for the finite interval, though the existing results here are not nearly so detailed as those for finite interval approximation. However, for the simultaneous approximation problem in weighted norms on unbounded intervals there are many parallels and analogies with the simultaneous approximation problem on finite intervals. The exponentially weighted spaces of functions defined on $(-\infty, \infty)$ and $[0, \infty)$ are related in a way which is similar to the relation between the 2π-periodic functions and the functions on $[-1, 1]$: a natural transformation exists between the space of even functions defined on $(-\infty, \infty)$ and the space of all functions defined on $[0, \infty)$ generated by the $t \leftrightarrow x^2$. Similar to what happens in the finite interval case with the mapping $x \leftrightarrow \cos\theta$, the mapping $t \leftrightarrow x^2$ makes it possible to study many aspects of approximation on the half-line by transforming results valid for the approximation of functions on the whole line. Again, the fact that differentiation is not preserved under the transformation leads to problems which have to be overcome.

We will give some of the basics of approximation theory in weighted spaces on the line, along with some approximation methods for derivatives.

4 A Brief History of Simultaneous Approximation by Algebraic Polynomials

4.1 The early results on the existence of polynomials of simultaneous approximation which can converge at an optimal rate

Trigub [57] is one of the first to consider the simultaneous approximation of algebraic polynomials:

Theorem 1. (see Trigub [57]) *Let $f \in C^q[-1, 1]$. Then for each $n > 2q$, there exists a polynomial P_n of degree at most n such that for $k = 0, \ldots, q$ and for $-1 \le x \le 1$*

$$|f^{(k)}(x) - P_n^{(k)}(x)| \le M \left(\frac{\sqrt{1-x^2}}{n} + \frac{1}{n^2} \right)^{q-k} \omega \left(f^{(q)}; \frac{\sqrt{1-x^2}}{n} + \frac{1}{n^2} \right) \tag{19}$$

with M independent of n and f.

Here, the size of the step length inside the modulus of continuity depends upon the location of x within the interval $[-1, 1]$, so that the modulus itself gives a pointwise estimate.

Gopengauz [27] showed that

Theorem 2. *Let $f \in C^q[-1,1]$. Then for each $n \geq 4q+5$, there exists a polynomial P_n of degree at most n such that for $k = 0, \ldots, q$ and for $-1 \leq x \leq 1$*

$$|f^{(k)}(x) - P_n^{(k)}(x)| \leq K \left(\frac{\sqrt{1-x^2}}{n} \right)^{q-k} \omega \left(f^{(q)}; \frac{\sqrt{1-x^2}}{n} \right) \tag{20}$$

with K independent of n and f.

More recently, some variations or refinements of these results have been given, which will be mentioned along with their applications.

Though impressive, have two major deficiencies from the point of view of applications. They state only the existence of certain polynomials. Actually, the approximating polynomials are shown to exist by outlining a construction for them, but the methods of construction seem ill-adapted to applications. And the constructions are convoluted to the point that it is quite impossible to give values to the M in Trigub's theorem or to the K in the theorem of Gopengauz, even for small values of q. As $n \to \infty$, therefore, we only can know that the error is decreasing at the rate of something tangible, times a constant which is known only to exist. Thus, the error estimates are useless if we want to ask such an eminently reasonable question as "To how high a degree n do we need to go when approximating this particular f, if we want five decimal places of accuracy?" Most of the rest of this article describes the still only partially successful results of efforts to get to the bottom of such problems, in order both to complete the theory and to make theory more applicable to practice. Most ideally, we would like methods which are simple to set up, combined with error estimates which can be written out explicitly as numbers.

4.2 Early applications

As the results of Trigub and Gopengauz clearly depict, the behavior of a polynomial approximant to a function is relatively different for points close to the ends of the interval, where things seem to get out of control. And the behavior for derivatives gets relatively worse there, too. Not surprisingly, for Lagrange interpolation on a completely arbitrary set of nodes the simultaneous convergence of derivatives is quite bad. K. Balázs [4] gave estimates for the error incurred in this procedure, stated in terms of the norm of the interpolation operator. These estimates are in themselves sharp. However, the theorems of Trigub and Gopengauz suggested ways in which interpolation could be modified and adapted in order to support simultaneous approximation.

N. S. Baiguzov [2] immediately applied the result of Gopengauz, in 1969. Using interpolation on the Chebyshev nodes with additional interpolation at ± 1, he obtained better convergence for the first derivative of the interpolated function.

Rather remarkably, Baiguzov did not try anything for higher derivatives. This was done by Y. Muneer [46], in 1987. He again used interpolation on the Chebyshev nodes, augmented by Hermite interpolation of order $\left[\frac{q+1}{2}\right]$ at the points ± 1 in order to approximate simultaneously any function and q of its derivatives.. Also in 1987, K. Balázs [3] used the theorem of Gopengauz in order to study the simultaneous convergence properties on the right half-line of interpolation on the Laguerre nodes, augmented by interpolation of derivatives at zero. It was these two publications which generated my own interest in the simultaneous approximation problem. The results of [3] are generalized later on in Balázs and Kilgore [11].

J. Szabados [50], constructed a pure Lagrange interpolation operator, using as a basis the Chebyshev nodes, the endpoints ± 1, and some carefully introduced additional nodes, the number of which depended upon the number of derivatives to be approximated. The details of this along with several other results on simultaneous approximation may be found in Chapter 8 of the book of Szabados and Vértesi [51].

P. Runck and P. Vértesi [48] showed how good simultaneous approximation with Lagrange interpolation operators could be obtained, based upon interpolation on nodes at the zeroes of Jacobi polynomials, augmented by additional nodes near the ends of the interval $[-1, 1]$. The precise number of added nodes depended upon the values of the two parameters α and β used in the definition of the Jacobi polynomial.

K. Balázs and T. Kilgore [5] generalized the results of Muneer to arbitrary nodes, augmented by interpolation of derivatives at ± 1 in order to give good convergence for the derivatives. And in [6] they succeeded in generalizing the results of Muneer and of Szabados and of Runck and Vértesi. Seeking the good simultaneous approximation of a function $f \in C^q[-1, 1]$, they showed that interpolation on nearly arbitrary nodes could be augmented by extra interpolations near or at the endpoints ± 1 to achieve the desired result. Specifically, let $X_n = \{x_{1,n}, \ldots, x_{n,n}\}$ be a system of nodes in natural order in $(-1, 1)$, and let L_n be the Lagrange interpolation operator constructed upon X_n. To construct the augmented set of nodes, let $r = \left[\frac{q+1}{2}\right]$. Then a permissible set of additional nodes would consist of a set of (not necessarily all distinct) points $T_n = \{t_{0,n}, \ldots, t_{r-1,n}\} \cup \{s0, n, \ldots, s_{r-1,n}\}$ satisfying for some $C > 0$ and for some integer $N \geq \sqrt{C}$ and for all $k = 0, \ldots, r$,

$$-1 \leq t_{k,n} \leq -1 + \frac{C}{(n+N)^2} < 1 - \frac{C}{(n+N)^2} \leq s_{k,n} \leq 1$$

Out of the augmented set of nodes consisting of these T_n along with the original nodes X_n, any nodes which lie upon the same point require Hermite interpolation of the requisite multiplicity at that point. Then we have:

Theorem 3. *Let q be a positive integer, and let $r = \left[\frac{q+1}{2}\right]$. Let P_n be the interpolation operator upon the node-set $X_n \cup T_n$. Then for $f \in C^q[-1,1]$ we have with $|x| \leq 1$*

(a) For q even and $i = 0, \ldots, q$

$$|f^{(i)}(x) - P_n^(i)(x)| = O(n^{i-q})E_{n-1}(f^(q))\|L_n\|$$

(b) For q odd and $i = 0, \ldots, q$

$$|f^{(i)}(x) - P_n^(i)(x)| = O(n^{i-q})E_{n-1}(f^(q))\|L_n^*\|,$$

where $L_n^* f(x) = (1-x^2)^{\frac{1}{2}} L_n\left((1-t^2)^{\frac{1}{2}} f(t)\right)(x)$.

The proof of this result followed from an adaptation of the theorems of Trigub and Gopengauz, in Balázs, Kilgore, and Vértesi [12].

Kilgore and Prestin in two papers [37] and [39] gave closer attention to the pointwise estimates possible for simultaneous approximation by interpolation and were able to get results which retained the pointwise modulus seen in the Trigub and Gopengauz theorem in their final estimates.

5 Notation and Conventions for the Study of Simultaneous Approximation on Finite Intervals

We have already seen that basic results for approximation often come in dual forms, one pertinent to periodic functions, where the approximants are trigonometric polynomials, and one pertinent to approximation on a bounded closed interval, where the approximants are algebraic polynomials. Furthermore, there is an interplay between these two situations, with the results for periodic functions often taking a primary role. Because of this, it is sometimes inconvenient even to write the most general results for simultaneous approximation with algebraic polynomials in a purely algebraic polynomial form, and several of the following results almost necessarily involve the language of periodic functions even in their statements. We should therefore adopt from now some notational conventions about such matters, in order to keep matters as straightforward as possible.

When discussing functions on finite intervals, we will consider functions of x (understood to be defined on $[-1,1]$) and functions of θ (understood to be 2π-periodic). The symbols f and F will denote functions of x. The symbols g and G will denote functions of x which obey certain stated conditions (i. e. are zero with a certain multiplicity at ± 1). Algebraic polynomials will be written as P or p; if the degree needs to be made clear, then P_n or as p_n will be used to signify polynomials of degree related to n (either at most n, or at most $2n$, at most $n+q$, where q is some fixed integer, always to be explained in context).

The symbols h and H will always denote functions of θ and will always be either even or odd. It should be totally clear in context whether such a function is even or odd.

A 2π-periodic function which is not necessarily even or odd (unless otherwise restricted) will be denoted by ϕ or Φ. A trigonometric polynomial will be denoted by T or T_n, when the degree must explicitly be related to n.

5.1 *Derivatives*

Ordinary notation for derivatives will always denote differentiation with respect to the argument of the function. Thus H' without further ado indicates a periodic function (of θ, say) which is being differentiated with respect to θ. Also, $G'(x)$ represents a function defined upon $[-1,1]$ which is differentiated by x.

However, note that if for example H is an even 2π-periodic function it is very much consistent to write

$$D_x H(\theta) := \frac{H'(\theta)}{-\sin\theta} \tag{21}$$

to indicate the differentiation of H by x.

5.2 *Weight functions and weighted L^p spaces*

We will consider some results for weighted L^p spaces $(1 \leq p \leq \infty)$ as well as for continuous functions. Specifically, let w be a weight function (i.e. a measurable function which is positive almost everywhere) on $[-1,1]$, then for $1 \leq p < \infty$ we may define the space $L^p_w[-1,1]$ to be the set of all functions f for which

$$\|f\|_{p,w} := \left(\int_{-1}^{1} |f(x)|^p w(x)\, dx\right)^{\frac{1}{p}} < \infty. \tag{22}$$

If $p = \infty$ we define $f \in L_w^\infty[-1,1]$ if its weighted norm

$$\|f\|_{\infty,w} := \text{ess sup}_{-1\le x\le 1}|f(x)|w(x) \tag{23}$$

is finite. Clearly, if the constant function $f(x) = 1$ is in the space, then w itself must be bounded.

Similarly, if W is a 2π-periodic weight function, then the space $L_W^p(2\pi)$ consists of those 2π-periodic functions ϕ for which

$$\|\phi\|_{p,W} := \left(\int_0^{2\pi} |\phi(\theta)|^p W(\theta)\, d\theta\right)^{\frac{1}{p}}, \tag{24}$$

and $L_W^\infty(2\pi)$ is also defined in a manner analogous to the definition in (23).

Given a 2π-periodic weight function W, there is a naturally related weight function on $[-1,1]$. Restricting one's attention to 2π-periodic functions ϕ which are either even or odd, then the norm-defining integral in (24) can be taken on the half-interval $[0,\pi]$ and doubled. Now, for $x \in [-1,1]$, let $x \leftrightarrow \cos\theta$ for appropriate $\theta \in [0,\pi]$. If $U(\cos\theta) = W(\theta)$, and if $w(x) = \frac{U(x)}{\sqrt{1-x^2}}$, then we have for any function f defined on $[-1,1]$

$$\int_{-1}^{1} f(x)w(x)\,dx = \int_0^\pi f(\cos\theta)w(\cos\theta)\sin\theta\,d\theta = \int_0^\pi f(\cos\theta)W(\theta)\,d\theta, \tag{25}$$

with similar relationships on appropriate subintervals of $[-1,1]$ and $[0,\pi]$ respectively. Thus, the spaces $L_w^p[-1,1]$ and the space of even functions in $L_W^p(2\pi)$ are isometrically isomorphic.

For a function $f \in L_w^p[-1,1]$ we write

$$E_n(f)_{p,w} := \inf_{\text{degree } p_n \le n} \|f - p_n\|_{p,w}.$$

If on the other hand ϕ is a continuous 2π-periodic function and T_n denotes a trigonometric polynomial of degree at most n we write

$$E_n^*(\phi)_{p,W} := \inf_{\text{degree } T_n \le n} \|\phi - T_n\|_{p,W}.$$

An important class of weight functions here will be Muckenhoupt's class of A_p weights. That the weight function W (whether even or not) is in the class A_p means that if I is any interval of length at most 2π, then for the appropriate value of p, $1 < p < \infty$, we have

$$\left(\int_I W(\theta)\,d\theta\right)\left(\int_I [W(\theta)]^{\frac{-1}{p-1}}\,d\theta\right)^{p-1} \le K_W|I|^p. \tag{26}$$

In this definition, $|I|$ signifies the length of I, and K_W denotes a constant depending only upon W. As a consequence of the definition, both W and $[W(\theta)]^{\frac{-1}{p-1}}$ are necessarily integrable on any interval of length 2π. Also, all continuous 2π-periodic functions and hence all trigonometric polynomials are in $L^p_W(2\pi)$. Moreover, if $\phi \in L^p_W(2\pi)$, then $\phi \in L^1(2\pi)$.

6 Simultaneous Approximation by Trigonometric Polynomials

Here, we describe the basic properties of trigonometric polynomials vis a vis simultaneous approximation of derivatives. It is the case (with minor restrictions on the type of norm which is imposed) that, if any trigonometric polynomial is used to approximate a given function (whether well or badly is not the point here), then the derivative of the trigonometric polynomial approximates the derivative of the given function about equally well (or badly, as the case may be).

This rather sweeping result was proven in Czipszer and Freud [20], for the weight function $W \equiv 1$ and for $1 \leq p \leq \infty$. The theorem here includes their result and generalizes it to any space of 2π-periodic functions with a weight from the Muckenhoupt class A_p.

The proof of the result, both in Czipszer and Freud and in the more general version quoted here, is based upon three tools: the uniform boundedness of some linear projection or near-projection operator (S_n if $p < \infty$, or V_n if $p = \infty$), the Bernstein inequality

$$\|T'_n\|_{p,W} \leq c(p, W)\, n\|T_n\|_{p,W}, \tag{27}$$

and the Jackson-Favard inequality (so-called because it is an extension of the results of Jackson [29], as refined later on by Favard [25])

$$E^*_n(\phi)_{p,W} \leq \frac{c(k,p,W)}{n^k}\|\phi^{(k)}\|_{p,W}. \tag{28}$$

The constant in (27) depends only upon p and W; in case that W is identically 1, the constant itself is 1. The constant in (28) depends upon k, p, and W. If W is equal to 1 then the constant depends upon k, but the values are uniformly bounded by $\frac{\pi}{2}$ (see the original statement of the Jackson theorems).

Spaces with A_p weights are introduced and discussed in Muckenhoupt [45] and in Hunt, Muckenhoupt, and Wheeden [28], where the following fundamental results are shown:

I. The weight function W is in the class A_p if and only if for every 2π-periodic function ϕ, the Hardy maximal inequality

$$\|\phi^*(\theta)\|_{p,W} \leq C_W\|\phi(\theta)\|_{p,W} \tag{29}$$

holds with a constant C_W dependent only upon W. Here,

$$\phi^*(\theta) := \sup_{y\neq\theta;\ |y-\theta|\leq 2\pi} \frac{1}{y-\theta}\int_\theta^y |\phi(\theta)|\, d\theta.$$

II. The A_p-weighted norm of the truncated Fourier expansion S_n is uniformly bounded in n.

Based on these two observations, N. X. Ky [43] has established that the Jackson-Favard (28) inequality and the Bernstein inequality (27) hold as well for the spaces of A_p-weighted 2π-periodic functions.

With these introductory remarks, it is possible to state and prove the following theorem, valid for an A_p weight W:

Theorem 4. (Kilgore [34]) *Let ϕ be a 2π-periodic function which is $q-1$ times continuously differentiable and $\phi^{(q-1)}$ is absolutely continuous. Then*

(a) Let $T_n(\theta)$ be a trigonometric polynomial of degree at most n satisfying for some constant C independent of n the inequality

$$\|\phi(\theta) - T_n(\theta)\|_{p,W} \leq C E_n^*(\phi)_{p,W}.$$

Then for $k = 1, \ldots, q$ there exist constants $\epsilon_{k,q}$ dependent only upon k and q such that

$$\|\phi^{(k)}(\theta) - T_n^{(k)}(\theta)\|_{p,W} \leq C\frac{\epsilon_{k,q}}{n^{q-k}} E_n^*(\phi^{(q)})_{p,W}.$$

(b) Let $T_n(\theta)$ be a trigonometric polynomial of degree at most n satisfying for some ϵ the inequality

$$\|\phi(\theta) - T_n(\theta)\|_{p,W} \leq \frac{\epsilon}{n^q}.$$

Then for $k = 1, \ldots, q$ there exist constants $\alpha_{k,q}$ and $\beta_{k,q}$ dependent only upon k and q such that

$$\|\phi^{(k)}(\theta) - T_n^{(k)}(\theta)\|_{p,W} \leq \frac{\alpha_{k,q}}{n^{q-k}}\epsilon + \frac{\beta_{k,q}}{n^{q-k}} E_n^*(\phi^{(q)})_{p,W}.$$

Remark 1. In Theorem 4 part (b), the quantity ϵ does not depend upon n. However, as an obvious corollary it can be made to, which can be useful as well.

Remark 2. The proof of Czipszer and Freud used the operator V_n, whose norm is uniformly bounded with respect to n in $C(2\pi)$ and in $L^p(2\pi)$ for all p, $1 \leq p < \infty$, to give a similar proof valid for those spaces.

Proof. First, we prove part (a). Using II, we note that there is some M such that $\|S_n\|_{p,W} \leq M$ for all n. Therefore, adding and subtracting $S_n\phi$, and using the fact that always $S_n\phi^{(k)} = S_n^{(k)}\phi$, and the fact that $T_n = S_nT_n$, we obtain

$$\|\phi^{(k)}(\theta) - T_n^{(k)}(\theta)\|_{p,W} \leq \|\phi^{(k)}(\theta) - S_n\phi^{(k)}(\theta)\|_{p,W} + \|S_n^{(k)}(\phi(\theta) - T_n(\theta))\|_{p,W}$$

Now, $\|S_n^{(k)}(\phi(\theta) - T_n(\theta))\|_{p,W}$ is a trigonometric polynomial of degree at most n. Therefore, by the Bernstein inequality (27),

$$\begin{aligned}\|S_n^{(k)}(\phi(\theta) - T_n(\theta))\|_{p,W} &\leq (c(p,W))^k n^k \|S_n(\phi(\theta) - T_n(\theta))\|_{p,W} \\ &\leq (c(p,W))^k n^k \|S_n\|_{p,W}\|\phi - T_n\|_{p,W}.\end{aligned}$$

Finally, using the Jackson-Favard inequality 28,

$$\begin{aligned}\|S_n\|_{p,W}\|\phi - T_n\|_{p,W} &\leq C\|S_n\|_{p,W}E_n^*(\phi)_{p,W} \\ &\leq CMc(k,p,W)n^{-k}E_n^*(\phi^{(k)})_{p,W},\end{aligned}$$

and, using the bounded linear projection properties of S_n,

$$\|\phi^{(k)}(\theta) - S_n\phi^{(k)}(\theta)\|_{p,W} \leq (1+M)E_n^*(\phi^{(k)})_{p,W}.$$

From these considerations,

$$\|\phi^{(k)}(\theta) - T_n^{(k)}(\theta)\|_{p,W} \leq (1 + M + c(p,W)CMc(k,p,W))\, E_n^*(\phi^{(k)})_{p,W},$$

and (a) follows. The proof of (b) is similar. □

To summarize: simultaneous approximation of derivatives of 2π-periodic functions by trigonometric polynomials is "automatic." If a function is approximated by a trigonometric polynomial, then the derivative of the function is approximated about equally well by the derivative of the trigonometric polynomial.

7 Trigonometric and Algebraic Polynomial Inequalities of Bernstein Type

The Bernstein inequality (27) is one of the major tools used to study trigonometric polynomial approximation of periodic functions, as well as in the study of simultaneous approximation of derivatives by trigonometric polynomials. It has of course an analogue or more properly an interpretation which holds if the polynomials are algebraic, defined on the standard closed interval $[-1,1]$.

The correspondence $x \leftrightarrow \cos\theta$ for $0 \leq \theta \leq \pi$ transforms a given algebraic polynomial $P_n(x)$ of degree at most n into the even trigonometric polynomial $P_n(\cos\theta)$. Therefore

$$|P_n'(x)| = |P_n'(\cos\theta)\sin\theta| \leq n\|P_n\|_\infty.$$

The Markov inequality

$$|P_n'(x)| \leq n^2\|P_n\|_\infty.$$

gives more meaningful results near the endpoints but performs poorly for x in the interior of $[-1, 1]$.

Many proofs of Bernstein's inequality have been given. We give two proofs here which are useful in different situations, because of future applications and because of some interesting details. The first of the two proofs is similar to that in Ky [43], and the second is the proof of M. Riesz, which we will later modify for other purposes.

7.1 A universally valid "almost" Bernstein's inequality, from which Bernstein's inequality follows in spaces with A_p weights

Given any Lebesgue measurable 2π-periodic function h, with Fourier expansion given by

$$h(\theta) \sim a_0 + \sum_{k=1}^{\infty} a_k \cos kx + b_k \sin kx.$$

Then the *conjugate function* of h is the function $\tilde{h}$ which satisfies

$$\tilde{h}(\theta) \sim \sum_{k=1}^{\infty} b_k \cos kx - a_k \sin kx.$$

It is easily seen that, if $h = T_n$, a trigonometric polynomial of order exactly n, then

$$\frac{1}{n}T_n'(\theta) + \sigma_n\tilde{T}_n(\theta) = \tilde{T}_n(\theta).$$

Let W be any integrable weight function which is integrable if $1 \leq p < \infty$ or is bounded if $p = \infty$. Then σ_n is a bounded operator (in fact, of norm 1), since it is an operator generated by a positive kernel which takes the constant function $h(x) = 1$ to itself. This observation is completely independent of the weight function. Thus

$$\frac{1}{n}\|T_n'(\theta)\|_{p,W} \leq \|1 - \sigma_n\|\|\tilde{T}_n(\theta)\|_{p,W} \leq 2\|\tilde{T}_n(\theta)\|_{p,W}. \tag{30}$$

Restricting now to only such a W which is an A_p weight, with $1 < p < \infty$, then (cf. Hunt, Muckenhoupt, and Wheeden [28]) there is a constant C depending only upon p and W, such that for every h in $L^p_W(2\pi)$

$$\|\tilde{h}\|_{p,W} \leq C\|h\|_{p,W}. \tag{31}$$

The Bernstein inequality (27) follows immediately in $L^p_W(2\pi)$ for W an A_p weight, with $1 < p < \infty$.

7.2 The identity of M. Riesz and the ensuing proof of Bernstein's inequality

Given an arbitrary trigonometric polynomial Φ_m of degree at most m, M. Riesz [47] showed that

$$\Phi'_m(z) = \frac{1}{4m}\sum_{k=1}^{2m} \Phi_m(z+t_k)\frac{(-1)^{k+1}}{(\sin\frac{1}{2}t_k)^2}, \tag{32}$$

where $t_k := \dfrac{(2k-1)\pi}{2m}$. Now, $\Phi_m(z) = \frac{1}{m}\sin mz$ and $z = 0$ yields

$$1 = \frac{1}{4m^2}\sum_{k=1}^{2m} \frac{1}{(\sin\frac{1}{2}t_k)^2}. \tag{33}$$

If now $W = 1$ we have for $1 \leq p \leq \infty$ that $\|T_n(\theta)\| = \|T_n(\theta+\phi)\|$ for arbitrary ϕ. Bernstein's inequality follows immediately, in the precise form

$$\|T'_n\|_p \leq n\|T_n\|_p,$$

with the correct constant of 1.

More generally, if W is a weight function which satisfies

$$0 < c_1 \inf_{\phi} \|h(\theta+\phi)\|_{p,W} \leq \|h(\theta+\phi)\|_{p,W} \leq c_1 \sup_{\phi} \|h(\theta+\phi)\|_{p,W},$$

then Bernstein's inequality follows with a constant of $\frac{c_2}{c_1}$. This may happen if W is for example bounded above and below by positive constants.

7.3 Further remarks on the identity of M. Riesz

Recently, it has been seen that a host of new identities can be obtained from (32) (cf. Balázs and Kilgore [10], [11], Kilgore [35], and Felten and Kilgore [26]) by substituting specific choices of m and Φ_m. For example, we may set

$m = (r+2)\, n$ for any fixed but arbitrary $r = 0, \ldots$. In order to compute $T_n'(\theta)$, we choose

$$\Phi_m(z) = T_n(\theta + z) \left(\frac{\sin \frac{nz}{2}}{n \sin \frac{z}{2}} \right)^{2r+2},$$

noting that $\Phi_m'(0) = T_n'(\theta)$. Thus

$$T_n'(\theta) = \frac{1}{4m} \sum_{k=1}^{2m} T_n(\theta + t_k) \left(\frac{\sin \frac{nt_k}{2}}{n \sin \frac{t_k}{2}} \right)^{2r+2} \frac{(-1)^{k+1}}{(\sin \frac{1}{2} t_k)^2}. \tag{34}$$

The identities (34) can be used to give rather simple proofs of several other polynomial inequalities involving derivatives, both for trigonometric polynomials and for algebraic polynomials, via the transformation $x \leftrightarrow \cos\theta$. For example, the proof of the following result from Balázs and Kilgore [11] is fairly straightforward, giving another "almost" Bernstein inequality which holds in a very wide context:

Theorem 5. *Let T_n denote an arbitrary trigonometric polynomial of degree at most n, and let μ be an arbitrary Borel measure. Then for $1 \leq p < \infty$ we have*

$$||T_n'||_{p,\mu} \leq 6n\, \omega_{p,\mu}(T_n; \frac{\pi}{4n}). \tag{35}$$

Here, we have defined

$$\|f\|_{p,\mu} = \left(\int_0^{2\pi} |f(\theta)|^p \; d\mu \right)^{\frac{1}{p}}$$

and

$$\omega_{p,\mu}(f; h) := \sup_{|s-t| \leq h} \|f(\theta + s) - f(\theta + t)\|_{p,\mu},$$

Stechkin [49] has obtained in the unweighted uniform norm the similar inequality

$$||T_n'||_\infty \leq \left(\frac{n}{2 \sin \frac{nh}{2}} \right) ||\Delta_h(T_n)||_\infty,$$

in which $\Delta_h(f, t) = f(t + h) - f(t)$, with only a slightly better constant.

Brudnyi's inequality (cf. [17]) gives a "local" estimate for the magnitude of derivatives of an algebraic polynomial $P_n(x)$ and is used for example in the proofs of results such as those of Trigub and Gopengauz. The proof of this inequality using (34) is much simpler than the original proofs. In fact, it is (so far as the author is aware) the first proof which is simple enough to yield actual numerical estimates for the constants C_q.

7.4 Markov-Bernstein inequalities in weighted spaces

The problems with the Bernstein inequality for algebraic polynomials on $[-1,1]$ have already been mentioned, and also the partial solution which can be found in the Markov inequality. An inequality of the sort

$$|P_n'(x)| \leq Cn \frac{\|P_n\|}{\frac{1}{n} + \sqrt{1-x^2}},$$

where C is some appropriate positive constant, is valid on the entire domain $[-1,1]$ and combines in an effectual manner the information in the Bernstein and Markov inequalities. Such inequalities are commonly called inequalities of Markov-Bernstein type.

For the spaces of 2π-periodic functions with A_p weight W, there are related spaces with a related weight w, via $x \leftrightarrow \cos\theta$, as indicated in subsection 5.2. For the spaces $L_w^p[-1,1]$ we have from Kilgore [34] the following, in which the constants depend upon the weight w:

Theorem 6. *Let w be a weight on $[-1,1]$ such that the weight W associated to it by (25) is in the class A_p. Then the Markov inequality*

$$\|P_n'(x)\|_{p,w} \leq C_1\, n^2 \|P_n\|_{p,w} \tag{36}$$

and the Markov-Bernstein inequalities

$$\|(\frac{1}{n} + \sqrt{1-x^2})P_n'(x)\|_{p,w} \leq C_2\, \|P_n\|_{p,w} \tag{37}$$

and

$$\|\frac{P_n'(x)}{(\frac{1}{n} + \sqrt{1-x^2})^{k-1}}\|_{p,w} \leq C_3\, n\|\frac{P_n(x)}{(\frac{1}{n} + \sqrt{1-x^2})^k}\|_{p,w} \tag{38}$$

for $k = 1, 2, \ldots$ hold in $L_w^p[-1,1]$.

Proof. The proof is based on the Bernstein inequality (27) for trigonometric polynomials and Hardy's inequality. We omit the details here. □

8 Theory of Simultaneous Approximation of Functions on a Finite Interval by Algebraic Polynomials

8.1 Existence theorems

In Balázs and Kilgore [10], it is shown that any polynomial P_n satisfying (19) with any constant M and also satisfying

$$(f^{(k)} - P_n^{(k)})(\pm 1) = 0 \quad \text{for} \quad k = 0, \ldots, q$$

must also satisfy (20), and furthermore that the constant K in (20) depends only upon M and satisfies

$$K \leq \max\{4e^e M, 7M + 7\}$$

Kilgore and Prestin [38] have shown that a polynomial P_n exists which satisfies (19) and also interpolates on (not necessarily distinct) points $s_{0,n}, \dots, s_{q,n}, t_{0,n}, \dots, t_{q,n}$, where the points satisfy for each $n \geq 2q+1$ and each $j = 0, \dots, q$

$$-1 \leq s_{j,n} \leq -1 + \frac{1}{n^2} \text{ and} 1 - \frac{1}{n^2} \leq t_{j,n} \leq 1$$

Theorem 7. *Let $f \in C^q[-1,1]$, with $q \geq 0$. Then for each $n \geq 2q+1$, there exists a polynomial P_n of degree at most n such that for $k = 0, \dots, q$ and for $-1 \leq x \leq 1$*

$$|f^{(k)}(x) - P_n^{(k)}(x)| \leq K \left(\frac{\sqrt{1-x^2}}{n} + \frac{1}{n^2} \right)^{q-k} \omega\left(f^{(q)}; \frac{\sqrt{1-x^2}}{n} + \frac{1}{n^2} \right) \quad (39)$$

with K independent of n and f and also satisfying

$$|f(x) - P_n(x)| \leq$$

$$\frac{K}{n^q} \min_{i,j} \left(\sqrt{\frac{|Q_n(x)|}{|(x-t_i)(x-s_j)|}}\, \omega\left(f^{(q)}; \min_{\ell \neq i,\, m \neq j} \frac{\sqrt{|(x-t_\ell)(x-s_m)|}}{n} \right) \right). \quad (40)$$

Theorem 8. *Let $f \in C^q[-1,1]$ and let P_n satisfy (19) for some constant K. If P_n also interpolates f on S_n, then P_n also satisfies (40) with a constant CK, in which C is an absolute constant.*

A recent discovery was how much easier it is to prove the existence of a polynomial which simultaneously approximates a function, if the modulus in such results as the theorem of Gopengauz is replaced by $E_n(f)$. As noted by Leviatan [44], the following result can be derived from the theorem of Gopengauz. Kilgore [33] gave a new and elementary proof which depends directly upon the theory of simultaneous approximation for trigonometric polynomials and upon a very careful and economical exploitation of the transformation $x \leftrightarrow \cos\theta$. Later, it became clear that such techniques can be used to translate the theory for trigonometric polynomial simultaneous approximation systematically over to the algebraic polynomial situation:

Theorem 9. (Leviatan [44] and Kilgore [33]) *Let $f \in C^q[-1,1]$. Then there exists a sequence of polynomials P_n of degree at most n ($n > 2q$) such that for $k = 0, \ldots, q$*

$$|f^{(k)}(x) - P_n^{(k)}(x)| \leq M_{q,k} \left(\frac{\sqrt{1-x^2}}{n} \right)^{q-k} E_{n-q}(f^{(q)}),$$

with the constants $M_{q,k}$ depending only upon q and k. Moreover $f^{(q)}(\pm 1) = P_n^{(q)}(\pm 1)$.

Another very important result was given as a lemma in Kilgore [33]. We will need it later on. The lemma is couched in terms of 2π-periodic even functions, in view of the fact that under $x \to \cos\theta$ any function of x defined on $[-1,1]$ can be regarded as a 2π-periodic even function, and vice versa. The fact that the integer q in the lemma can be odd is the prime reason why the lemma has to be stated in the language of periodic functions and cannot even be stated meaningfully in terms of functions of x:

Lemma 1. (Kilgore [33], Lemma 5) *Let $G \in C^q[-1,1]$ be such that $G^{(k)}(\pm 1) = 0$ for $k = 0, \ldots, q$, and let $H(\theta) := \frac{G(\cos\theta)}{\sin^q(\theta)}$. Then for K_q depending only upon q*

$$E^*_{n-q}(D^q_\theta H) \leq K_q E_{n-q}(D^q_x G).$$

8.2 Algebraic polynomial results of universal character

Given a 2π-periodic function h and a trigonometric polynomial T_n approximating h, the relative quality of approximation can be measured in terms of a multiple of $E^*_n(h)$ (abusing the notation slightly by not going into specifics about the norm). The results of Czipszer and Freud [20] showed that the relative quality or efficiency of approximation of h' by T'_n is always about the same (up to a further constant multiple) as that for h by T_n. These results for trigonometric functions go far beyond existence proofs and are universal in character. As seen earlier in this article, it has been possible to generalize the Czipszer and Freud's results to the wide class of A_p weights. The correctly formulated algebraic polynomial analogues of the Czipszer-Freud results give great flexibility and generality in dealing with simultaneous approximation by algebraic polynomials. Here, we give the analogues concerning algebraic polynomial approximation on $[-1,1]$. We handle first the case of continuous functions with the uniform norm, and then we will deal with the analogues which arise from the weighted spaces with A_p weights. We will also present some approximation methods based upon our generalities and give error estimates.

8.2.1. Simultaneous approximation in the uniform norm on $[-1,1]$

In Kilgore and Szabados [40], we proved the following:

Theorem 10. *Let $g \in C^q[-1,1]$ be such that $g^{(k)}(\pm 1) = 0$ for $k = 0, \ldots, q-1$, and let p_n be an algebraic polynomial of degree at most $n+q$ satisfying for some ϵ independent of $x \in [-1,1]$ the inequality*

$$\left| \frac{g(x) - p_n(x)}{(\sqrt{1-x^2})^q} \right| \leq \frac{\epsilon}{n^q}.$$

Then for $|x| \leq 1$ and for $k = 1, \ldots, q$ we have

$$|g^{(k)}(x) - p_n^{(k)}(x)| \leq \left(\frac{\sqrt{1-x^2}}{n} + \frac{1}{n^2} \right)^{q-k} (\delta_{k,q} E_n(g^{(q)}) + \gamma_{k,q}\epsilon),$$

where the constants $\delta_{k,q}$ and $\gamma_{k,q}$ depend only upon k and q.

and

Corollary 1. *Let $f \in C^q[-1,1]$ and let P_n be a polynomial of degree $n + q \geq 2q - 1$ such that for some constant C and for all $x \in [-1,1]$*

$$|f(x) - P_n(x)| \leq C \left(\frac{\sqrt{1-x^2}}{n} + \frac{1}{n^2} \right)^q E_n(f^{(q)}).$$

Then there are constants $\eta_{1,q}, \ldots, \eta_{q,q}$ such that for $k = 1, \ldots, q$ and for $|x| \leq 1$

$$|f^{(k)}(x) - P_n^{(k)}(x)| \leq C\eta_{k,q} \left(\frac{\sqrt{1-x^2}}{n} + \frac{1}{n^2} \right)^{q-k} E_n(f^{(q)}).$$

The proof of these results follows the same lines as that of the theorem of Kilgore [33], with the addition of a crucial lemma which permits one to relax the more stringent assumptions in Kilgore [33] concerning the number of of the derivatives k for which $f^{k)}(\pm 1) - P_n^{(k)}(\pm 1) = 0$. The statement of this lemma was due to Kilgore, and the proof was due to Szabados. The Lemma describes some fundamentals of the relation between algebraic and trigonometric derivatives:

Lemma 2. *Let $H(\theta) \in C^{2r}(2\pi)$ be an even function. Let $D_x H(\theta)$ represent its derivative with respect to $x = \cos\theta$. Then for $r \geq 1$ the derivative $D_x^r H(\theta)$*

exists. It is in $C_{2\pi}$, and

$$\|D_x^r H(\theta)\| \leq \sum_{j=1}^{2r} A_{j,r} \|H^{(j)}\|,$$

in which $H^{(j)}$ signifies the jth derivative of H by θ and in which $A_{j,r}$ are nonnegative constants independent of H.

Remark. We note that the hypotheses of all of these results can be relaxed slightly. As was noted later in Kilgore [34], in fact only the continuity of $F^{(q-1)}$ is needed, and the results can be meaningfully stated in terms of $E_n(f^{(q)})_\infty$. It is equally true, however, that $E_n(f^{(q)})_\infty$ does not approach zero as $n \to \infty$ if $f^{(q)}$ is a discontinuous function.

Theorem 10 gives a direct statement about the quality of approximation of derivatives when any linear projection operator is suitably modified. The only restrictions are that the operator must be defined on all of $C(2\pi)$ and must preserve evenness and oddness. Via $x \leftrightarrow \cos\theta$, then, the operator becomes an algebraic polynomial operator. Examples of such operators are orthogonal expansions in terms of any of the standard orthogonal polynomials, such as Chebyshev polynomials of the first and second kind or Legendre polynomials. Let us start with such a linear projection L_n which maps $C[-1,1]$ into the algebraic polynomials of degree at most n. To construct an operator which will approximate a continuous functions and their derivatives up to the qth derivative approximately as well as L_n itself approximates continuous functions, the following steps will suffice:

1. Any function which is not zero at the ends of the interval $[-1,1]$ with multiplicity at least $[\frac{q}{2}]$ should be preconditioned, by subtracting from it a polynomial of minimal degree which interpolates f and its derivatives up to $f^{(q-1)}$ at the points ± 1. The result of this is a function g which is zero at ± 1 with multiplicity q. The polynomial subtracted is of degree at most $2q-1$, and we will call it e_{2q-1}. In fact, the construction of this polynomial is itself a linear procedure, so that we could call it $e_{2q-1}(f)$.

2. We note that the function $h(\theta) = (\sin\theta)^{-q} g(\cos\theta)$ is everywhere continuous (with removable singularities). Then the function g should be approximated by the operator $p_n(x) = \sin^q\theta L_n h(\theta)$. This operator will give simultaneous approximation of the derivatives of g by polynomials in x. We have
$$\frac{g(x) - p_n(x)}{(\sqrt{1-x^2})^q} = h(\theta) - L_n h(\theta).$$

Seeing this, from the theorem of Lebesgue we get

$$\left\| \frac{g(x) - p_n(x)}{(\sqrt{1-x^2})^q} \right\| \leq (1 + \|L_n\|)E_n^* h.$$

Now, using $x \leftrightarrow \cos\theta$ and Jackson I we get

$$E_n h \leq \frac{\pi}{2n^q} E_n^* h^{(q)},$$

and using Lemma 1, which says in the present connection that

$$E_n^*(h^{(q)}(\theta)) \leq K_q E_n(g^{(q)}(x)),$$

we get

$$E_n((\sqrt{1-x^2})^{-q} g(x)) \leq \frac{\pi}{2n^q} K_q E_n(g^{(q)}(x)).$$

Therefore, we may use the Theorem in this section with

$$\epsilon = \frac{\pi}{2} K_q (1 + \|L_n\|) E_n^*(g^{(q)})$$

to obtain for $|x| \leq 1$ and for $k = 1, \ldots, q$ we have

$$|g^{(k)}(x) - p_n^{(k)}(x)| \leq \left(\frac{\sqrt{1-x^2}}{n} + \frac{1}{n^2} \right)^{q-k}$$
$$(\delta_{k,q} E_n(g^{(q)}) + \gamma_{k,q} \frac{\pi}{2} K_q (1 + \|L_n\|) E_n(g^{(q)}),$$

where the constants $\delta_{k,q}$ and $\gamma_{k,q}$ and K_q depend only upon k and q. The term $1 + \|L_n\|$ is of course due to (5), and the norm signifies the norm of L_n as an operator defined on all of $C(2\pi)$.

3. Now we must add back to the polynomial approximant of g the polynomial $e_{2q-1}f$ of degree at most $2q-1$ which subtracted from f in order to precondition it. We notice that for $n \geq 2q-1$ we have $E_n(f) = E_n(g)$, enabling us to write

$$|f^{(k)}(x) - e_{2q-1}^{(q)}(x) - p_n^{(k)}(x)|$$
$$\leq \left(\frac{\sqrt{1-x^2}}{n} + \frac{1}{n^2} \right)^{q-k} (\delta_{k,q} E_n(f^{(q)}) + \gamma_{k,q} \frac{\pi}{2} \eta_q (1 + \|L_n\|) E_n(f^{(q)})).$$

The preceding remarks are, as stated, valid for bounded linear projections which themselves are mappings via $x \leftrightarrow \cos\theta$ of bounded linear projections defined on all of $C(2\pi)$. The described methods therefore apply without question to such methods of approximation as the truncated Chebyshev expansion, as that method is in fact no more nor no less than the application of the truncated Fourier series to an even function, using the transformation $x \leftrightarrow \cos\theta$. The method also applies to the case that L_n is an interpolation operator with nodes on the interior of the interval $[-1, 1]$ (it is usually convenient here to number the nodes from 1 to n, giving a polynomial approximant of degree at most $n-1$). Here, however, the cases that the integer q is even or respectively odd cause a branching in the method of estimating the norm. If q is even, then the estimate comes out in terms of $\|L_n\|$ itself, as we are viewing L_n as an operator on the even trigonometric functions, via $x \leftrightarrow \cos\theta$. On the other hand, if q is odd we must express the norm in terms of the odd expansion (13). As a result of this, the steps outlined above are still valid, but the expression $\|L_n\|$ in these steps gets replaced by $\|L_n^*\|$, where L_n^* is the weighted or conjugated operator defined for an odd function $h(\theta)$ by

$$L_n^* h(\theta) = \sin\theta L_n\left(\frac{h(t)}{\sin t}\right)(\theta).$$

It will be helpful to the reader in understanding how these procedures work, as well as helpful in showing how the constants which come out in the error estimates can be evaluated, if we go through the work of approximating a continuous function f and its first derivative, in the uniform norm. We will use an interpolation operator L_n defined upon nodes $-1 < x_1 < \ldots < x_n < 1$. The first step is to precondition f by subtracting $e_1(f)$ defined by

$$e_1(f)(x) = f(-1)\left(\frac{1-x}{2}\right) + f(1)\left(\frac{x+1}{2}\right),$$

obtaining g which is zero at ± 1 defined by

$$g(x) = f(x) - e_1(x).$$

Now, we let $h(\theta) = f(\cos\theta)(\sin\theta)^{-1}$, with removable singularities at all multiples of π, and we note that h thus defined is a continuously differentiable function of θ. Also, in fact $L_n^* h(\theta)$ is the sine interpolant of h given by (13). Therefore by the theorem of Lebesgue (5) we have

$$\|h(\theta) - L_n^* h(\theta)\| \leq (1 + \|L_n^*\|)E_n^* h.$$

To approximate g we use $P_n g(x) = P_n g(\cos\theta) = \sin\theta L_n^* h(\theta)$.

Using $T_n(\theta) = L_n^* h(\theta)$ for the sake of notational convenience, we have

$$g(x) - P_n g(x) = \sin\theta(h(\theta) - T_n(\theta)),$$

and therefore

$$|g(x) - P_n g(x)| = \sin\theta|h(\theta) - T_n(\theta)| \leq \sqrt{1-x^2}(1 + \|L_n^*\|)E_n^* h,$$

and by Jackson I we have

$$|g(x) - P_n g(x)| \leq \sqrt{1-x^2}(1 + \|L_n^*\|)E_n^* h \leq \sqrt{1-x^2}\frac{\pi}{2n}(1 + \|L_n^*\|)E_n^* h'$$

Now,

$$g'(x) - P_n' g(x) = -(h'(\theta) - T_n'(\theta)) - \frac{\cos\theta}{\sin\theta}(h(\theta) - T_n(\theta)),$$

whence

$$|g'(x) - P_n' g(x)| \leq \|h'(\theta) - T_n'(\theta)\| + \|\frac{\cos\theta}{\sin\theta}(h(\theta) - T_n(\theta)\|.$$

Now, as $\theta < \tan\theta$ for $0 < \theta < \frac{\pi}{2}$, with a similar estimate if $\frac{\pi}{2} < \theta < \pi$, we get

$$\|\frac{\cos\theta}{\sin\theta}(h(\theta) - T_n(\theta))\| \leq \|h'(\theta) - T_n'(\theta)\|$$

so that

$$|g'(x) - P_n' g(x)| \leq 2\|h'(\theta) - T_n'(\theta)\|.$$

Now, we get an estimate for $\|h'(\theta) - T_n'(\theta)\|$, which proceeds along the lines of the original argument of Czipszer and Freud for the supremum norm case. We have to use the "near projection" V_n (there are better choices, but to use one of them would require a deep involvement in technicalities), which has norm not more than 3, leaves all polynomials of degree at most n unmolested, and commutes with differentiation as does its parent, the Fourier expansion. We get

$$\|h'(\theta) - T_n'(\theta)\| = \|h'(\theta) - V_n h'(\theta) + V_n\left(h' - T_n'\right)(\theta)\|.$$

Now, using the fact that

$$V_n\left(h' - T_n'\right)(\theta) = V_n'\left(h - T_n\right)(\theta)$$

we get

$$\|h'(\theta) - T_n'(\theta)\| \leq \|h'(\theta) - V_n h'(\theta)\| + \|V_n'\left(h - T_n\right)(\theta)\|$$

Hence, using Bernstein's inequality to estimate the second term on the right, which is the derivative of a polynomial of degree at most $2n-1$,

$$\begin{aligned}\|h'(\theta) - T_n'(\theta)\| &\le 3E_n^* h' + (2n-1)\|V_n\,(h - T_n)\,(\theta)\| \\ &\le 3E_n^* h' + (2n-1)\|V_n\| \cdot \|h(\theta) - T_n(\theta)\|,\end{aligned}$$

so that, using the estimate which we already got for $\|h(\theta) - T_n(\theta)\|$,

$$\|h'(\theta) - T_n'(\theta)\| \le 3E_n^* h' + 3\pi(1 + \|L_n^*\|)E_n^* h'$$

whence

$$|g'(x) - P_n' g(x)| \le 6E_n^* h' + 6\pi(1 + \|L_n^*\|)E_n^* h'$$

Finally, note that

$$\frac{\cos\theta}{\sin\theta}\,(h(\theta)) = \frac{g(x)}{1-x^2}$$

whence we can get

$$E_n^* h' \le E_n g' + \frac{1}{2}E_n\left(\frac{g(x)}{1+x}\right) + \frac{1}{2}E_n\left(\frac{g(x)}{1-x}\right) \le 2E_n g'$$

and further that $E_n g' = E_n f'$, and recalling the definition of g we obtain

$$|f'(x) - e_1' f(x) - P_n' g(x)| \le 12E_n^* f' + 12\pi(1 + \|L_n^*\|)E_n^* f'$$

and

$$|f(x) - e_1 f(x) - P_n g(x)| \le \sqrt{1-x^2}(2\pi n)(1 + \|L_n^*\|)E_n^* f'$$

The reason why we have gone through this proof is in order to provide error estimates with actual numbers in them instead of some vague generalities. The argument works for any interpolation operator whose nodes are in the interior of the interval $[-1,1]$. A good choice of nodes is the Chebyshev nodes, where the norm $\|L_n^*\|$ is then of magnitude approximately $\frac{2}{\pi}\log n$.

It will be left to the reader, if it is wished for example to perform the simultaneous approximation of a polynomial of degree two, to work out the corresponding steps in order to obtain the constants. The techniques used in the proof are analogous to those laid out here. Of course, if q is even there is no need for L_n^*.

8.2.2. Analogues of the Czipszer-Freud results on $[-1,1]$, in spaces with weights which map to A_p weights

It is an example of the more fundamental nature of the results on trigonometric polynomial approximation that the weights w in this section are not themselves

A_p weights. Rather, any such w is an image under $x \leftrightarrow \cos\theta$ of an even 2π-periodic A_p weight W. The corresponding condition for w is somewhat inelegant if directly stated; but a translation of (26) using (25) would give

$$\left(\int_a^b w(x)\,dx\right)\left(\int_a^b [w(x)(\sqrt{1-x^2})^p]^{\frac{-1}{p-1}}\,dx\right)^{p-1} \le K_W(b-a)^p \qquad (41)$$

for $-1 \le a < b \le 1$. The following results appeared in Kilgore [34]:

Theorem 11. *Let $1 < p \le \infty$ and w related as in (25) to a weight W in the class A_p. Let $f \in C^{q-1}[-1,1]$ with $f^{(q-1)}$ absolutely continuous on $[-1,1]$ and with $f^{(q)} \in L^p_w[-1,1]$. Let e_{2q-1} be the algebraic polynomial of degree at most $2q-1$ which interpolates $f(\pm 1), \ldots, f^{(q-1)}(\pm 1)$. Let $B_n(\theta)$ be any trigonometric polynomial of degree at most n, even if q is even, odd if q is odd, such that for some constant C*

$$\left\| \frac{(f-e_{2q-1})(\cos\theta)}{\sin^q\theta} - B_n(\theta) \right\|_{p,W} \le C E_n^*\left(\frac{(f-e_{2q-1})(\cos\theta)}{\sin^q\theta}\right)_{p,W},$$

and let Q_n be the (algebraic) polynomial $Q_n(\cos\theta) := -e_{2q-1}(\cos\theta) + \sin^q\theta B_n(\theta)$. Then

$$\left\| \frac{f(x)-Q_n(x)}{(\sqrt{1-x^2})^q} \right\|_{p,w} \le \frac{C\,K(q,p,W)}{n^q} K_q E_n(f^{(q)})_{p,w},$$

in which the constant $K(q,p,W)$ depends only upon q, p, W, and there are constants $\eta_{1,q}, \ldots, \eta_{q,q}$ such that for $k = 1, \ldots, q$

$$\left\| \frac{f^{(k)}(x) - Q_n^{(k)}(x)}{\left(\sqrt{1-x^2}+\frac{1}{n}\right)^{q-k}} \right\|_{p,w} \le \frac{C}{n^{q-k}} \eta_{k,q} E_n(f^{(q)})_{p,w}.$$

We also showed that the estimates of Theorem 11 essentially continue to hold without the stringent requirement of exact interpolation at ± 1. It was remarkable enough in the context of L^p spaces that a good result could be obtained by requiring interpolation. In contrast, the next result requires practically nothing at all about the actual values of $f(x) - P_n(x)$ at points near ± 1, in view of the L^p norm.

Theorem 12. *Let $1 < p \le \infty$. Let $f \in C^{q-1}[-1,1]$ with $f^{(q-1)}$ absolutely continuous on $[-1,1]$ and with $f^{(q)} \in L^p_w[-1,1]$.with w as in (25). Let P_n be a polynomial of degree $n+q \ge 2q-1$ such that for some constant C*

$$\left\| \frac{f(x)-P_n(x)}{\left(\sqrt{1-x^2}+\frac{1}{n}\right)^q} \right\|_{p,w} \le \frac{C}{n^q} E_n(f^{(q)})_{p,w}.$$

Then there are constants $\eta_{1,q}, \ldots, \eta_{q,q}$ *such that for* $k = 1, \ldots, q$

$$\left\| \frac{f^{(k)}(x) - P_n^{(k)}(x)}{\left(\sqrt{1-x^2} + \frac{1}{n}\right)^{q-k}} \right\|_{p,w} \leq \frac{C}{n^{q-k}} \eta_{k,q} E_n(f^{(q)})_{p,w}.$$

These results also hold when $p = \infty$, *if in that case* w *is identically* 1.

Moving from general to particular, we mention the standard case that $p = 2$ and the weight on $[-1, 1]$ is $w(x) = (1 - x^2)^{-\frac{1}{2}}$. In this case in particular, estimates for simultaneous approximation by the appropriately modified truncated Chebyshev expansion of the function have been worked out in Kilgore and Tasche [41], where we reached the following results:

For the case $q = 1$ and $w(x) = (1 - x^2)^{-\frac{1}{2}}$ we got

Theorem 13. *Let* $g \in \mathrm{AC}(I)$ *with* $g' \in L_w^2(I)$ *and* $g(\pm 1) = 0$ *be given. Further, let*

$$h(\theta) := \begin{cases} \dfrac{g(\cos\theta)}{\sin\theta} & \text{if } \theta \in (0, \pi), \\ 0 & \text{if } \theta = 0, \pi. \end{cases}$$

The odd, 2π*-periodic extension of* h *is also denoted by* h.
Then $p \in \Pi_{n+1}$ *defined by*

$$p(x) := \sin\theta \, (S_n h)(\theta) \quad (x := \cos\theta)$$

fulfills the following estimates of simultaneous approximation

$$\|w\,(g - p)\|_w \leq \frac{4}{n+1} E_n(g')\,,$$
$$\|g' - p'\|_w \leq 3\, E_n(g')\,.$$

And for the case $q = 2$ with $w(x) = (1 - x^2)^{-\frac{1}{2}}$

Theorem 14. *Let* $g \in \mathrm{C}^1(I)$ *with* $g' \in \mathrm{AC}(I)$, $g'' \in L_w^2(I)$ *and* $g(\pm 1) = g'(\pm 1) = 0$ *be given. Further, let*

$$h(\theta) := \begin{cases} \dfrac{g(\cos\theta)}{(\sin\theta)^2} & \text{if } \theta \in (0, \pi), \\ 0 & \text{if } \theta = 0, \pi. \end{cases}$$

The even, 2π*-periodic extension of* h *is also denoted by* h.
Then $p \in \Pi_{n+2}$ *defined by*

$$p(x) := (\sin\theta)^2 \, (S_n h)(\theta) \quad (x := \cos\theta)$$

fulfills the following estimates of simultaneous approximation

$$\|w^2 (g - p)\|_w \le \frac{13}{(n+1)^2} E_n(g''),$$

$$\|(\frac{1}{w} + \frac{1}{n+1})^{-1} (g' - p')\|_w \le \frac{41}{n+1} E_n(g''),$$

$$\|g'' - p''\|_w \le 37.5\, E_n(g'').$$

The proof of these results is quite similar to that given for the approximation of a continuously differentiable function in the previous section. The differences in the constants arise for several reasons. First, as we are in L^2, the truncated Fourier series is the itself the best approximation, so that we do not need to resort to V_n. And also the operator which we are investigating is S_n itself, not some other. Also in the case of continuous functions we estimated the norms of such quantities as

$$\frac{\cos\theta}{\sin\theta}(h(\theta) - T_n(\theta)) \quad \text{and} \quad E_n\left(\frac{g(x)}{1 \pm x}\right)$$

using Rolle's theorem. Here we use the appropriate Hardy inequality, and the constants are not the same. The reader is recommended to look into the article Kilgore and Tasche [41], where the details are worked out.

8.3 Simultaneous approximation with linear projections

As far as bounded linear projections are concerned, the following result sums up what has been done in the previous sections. in which results of increasing generality were given. In this theorem, it is assumed that we have a projection operator which "lives" on the space of 2π-periodic functions , such as the Fourier expansion itself, and can be used to provide, via a restriction to the even periodic functions followed by replacing $\cos\theta$ with x. For example, the Chebyshev expansion of a function of x is of this type. The theorem then describes a method by which the operator can be modified in order to give simultaneous approximation of derivatives, with results relatively as good as the original operator could give. One starts with an operator L_n. To construct an operator P_n which does simultaneous approximation, one first preconditions the function to be approximated by subtracting e_{2q-1}, which gives a new function (call it g), which is zero with multiplicity $q-1$ at each of ± 1. To approximate the derivatives of g is then to approximate the derivatives of $f - e_{2q-1}$, and then the derivatives of f itself can be approximated in turn.

Theorem 15. (Kilgore [34]) *Let $1 < p \leq \infty$ and the weights w and W as in Theorem 11, or let $p = \infty$ with $w = 1$. Let L_n be a bounded linear projection from the space $L^p_W(2\pi)$ into the space of trigonometric polynomials of degree at most n which preserves evenness and oddness. For $f \in C[-1,1]$ such that $f^{(q-1)}$ is absolutely continuous let*

$$P_n f(x) := e_{2q-1}(\cos\theta) + \sin^q\theta L_n\left(\frac{(f - e_{2q-1})(\cos t)}{\sin^q t}\right)(\theta),$$

in which $e_{2q-1}(x)$ is the polynomial of degree at most $2q-1$ which interpolates $f = f^{(0)}, \ldots, f^{(q-1)}$ on the points ± 1. Then for $k = 0, \ldots, q$

$$\left\| \frac{f^{(k)}(x) - P_n^{(k)}(x)}{\left(\sqrt{1-x^2} + \frac{1}{n}\right)^{q-k}} \right\|_{p,w} \leq \frac{1}{n^{q-k}} E_n(f^{(q)})_{p,w}$$

$$(\delta_{k,q} + c(q,p,W) K_q \gamma_{k,q}(1 + \|L_n\|_{p,w})).$$

As a reminder, Theorem 15 does have one limitation. It is confined to operators which may be naturally extended to periodic functions, such as expansion in Chebyshev polynomials. Some operators such as S_n are already defined there, and for them the extension is obvious. Other operators, such as Lagrange interpolation, are extensible, but the consequences as to the norm must be heeded.

Finally, the constants in Theorem 15 depend on the weight function, upon the value of p, and upon the values of q and k.

8.4 Interpolation and mixed norms

It often occurs in applications that one needs to measure error of approximation in some L^p or weighted L^p norm, as in Section 8.2. However, in the "real world" we cannot in fact apply such an operator as S_n, which is based upon integration. After all, "real world" means computers, and computers are digital. We usually resort to such methods as the fast Fourier transform, which, as mentioned, is in fact nothing but interpolation under a different name. And we have also learned that interpolation does not mix well with integral norms. It is possible, however, to approximate a continuous function by interpolation and to measure the error in L^p. And often interpolation works very well when we do this.

For example, in a weighted space $L^2_w[-1,1]$ we have the result of Erdös and Turán [23]. This theorem assumes that we have obtained a sequence of polynomials $Q_0, Q_1, \ldots$ orthogonal to the weight function w, and we construct a sequence of interpolation operators $L_0, L_1, \ldots$, each operator L_n interpolating upon the roots of the polynomial Q_{n+1}. Then we have

Theorem 16. (Erdös and Turán) *For each $f \in C[-1,1]$, we have*

$$\|f - L_n f\|_{p,w} \leq \|w\|_1 E_n f,$$

where $E_n f$ denotes the error incurred in the best approximation of f in the uniform norm.

Results which are more general in another direction are available if the weight w is the Chebyshev weight $(1-x^2)^{\frac{-1}{2}}$. The theorem of Erdös and Feldheim [22] states that interpolation on the roots of the $n+1$st degree Chebyshev polynomial converges in the $L^{p,w}$ norm for $1 \leq p < \infty$. Specifically one has

$$\|f - L_n f\|_p \leq C\, E_n f.$$

Both the theorem of Erdös-Turán and the theorem of Erdös-Feldheim have extensions to simultaneous approximation of derivatives. Some details are in Balázs and Kilgore [8], but the constants which occur in the error estimates are not estimated there. However, the reader who needs to have error estimates for small values of q can find them using the methods laid out in Section 8.2 and in Section 8.2.

8.5 *Comparative effectiveness and efficiency of polynomial methods*

To compare the efficiency of two different kinds of approximation techniques is sometimes easy but is not always easy. If, for example, Method A lends itself to rapid and efficient computation methods whereas Method B does not, then it would be unwise and impractical to use method B over method A. However, the issues are not always so clear-cut.

The main choice in approximation comes down to the question of what sort of functions to use for approximation. The usual choice is one between polynomials and spline functions. Spline function approximation has a literature of its own, which it is beyond the scope of this article to review. However, it is known concerning spline approximation that it converges to every continuous function when reasonably applied, for example by interpolation on equally-spaced nodes, with the splice points of the splines at the nodes of interpolation. This is in contrast to the behavior of ordinary polynomials, for which no sequence whatsoever of interpolations on a progressively finer mesh of nodes will converge to all continuous functions. Also, it is known that $C2$ cubic splines (piecewise cubic polynomials with first and second derivatives continuous at splice points) have simultaneous approximation properties similar to those of polynomials.

In view of these facts, it would seem that spline functions have an inherent superiority over polynomial or trigonometric approximation. Nevertheless, numerical experimentation on test functions rather surprisingly indicates that this might not be true, at least when differentiable functions are approximated. G. Baszenski has written a test program which compares the rate of approximation and simultaneous approximation of a function, its first derivative, and its second derivative by polynomials of degree 2^n, for n up to 12, with the results for spline approximation with splines on a comparable number of nodes. The methods used for polynomial approximation are those of this article, combined with fast algorithms for computation. The methods of spline approximation are standard spline methods, combined with available fast algorithms as a computational technique.

In spite of conventional wisdom, the results on test functions are not very surprising, so long as the function is analytic. It would seem intuitively obvious that spline approximation would perform relatively poorly for most analytic functions. However, one example which is really surprising is the function $f(x) = (x_+)^3$, itself one of the simplest cubic splines. If this function is approximated on $[-1, 1]$ with polynomials, the results are very good, both for the function and for its first and second derivative. If it is approximated with splines, then the results are of course perfect, provided that 0 is prescribed as a node. However if 0 is *not* used as a node but always lies halfway between two adjacent nodes, then the results are very poor indeed.

The reader is referred to Tasche [52] and [53] and to Baszenski and Tasche [13] and [14] for details.

9 Simultaneous Approximation in Exponentially Weighted Spaces on $(-\infty, \infty)$ and on $[0, \infty]$

The theory of approximation in weighted spaces on the line and on the half-line is not so well worked out as is the theory of approximation on finite bounded intervals, and as a result the theory of simultaneous approximation is also not so well worked out. The broad outlines are there, but the details are not yet filled in. Nevertheless, even that which has been done may be of help or interest to someone who works with applications. We start by discussing the properties of some of the best-known and most commonly used weight functions on the line.

In a long series of papers, G. Freud worked out the theory of orthogonal polynomial, orthogonal polynomial expansions of functions, the Bernstein inequality, and the other bases of approximation theory for spaces of weighted functions on the line $(-\infty, \infty)$. His objective was to develop a unified theory of approximation which would apply both on the infinite interval and on the finite interval as well as for periodic functions. For an overview, one might consult Volume 50 (1987) of the *Journal of Approximation Theory*, called the Freud Memorial Volume, which contains a collection of articles related to his work.

Freud considered the problems of approximation on $(-\infty, \infty)$ in the presence of exponentially decaying weight functions, of the form $e^{-Q(x)}$, with certain restrictions on the function Q. His definition for the weights, called by others Freud weights, evolved over time and continues to evolve now. Rather than go into details about the definition, we recommend to the reader to look at the mentioned journal volume. The simplest examples of Freud weights are the weights $e^{-|x|^\alpha}$, where α is any number strictly greater than one. The classical example is of course the Hermite weight e^{x^2}, and related to the Hermite weight is the Laguerre weight e^t used on $[0, \infty)$.

We will see that simultaneous approximation of derivatives by polynomial approximants is automatic in the function spaces with Freud weights on the line, as it is in the spaces of 2π-periodic functions with trigonometric polynomial approximants. We will show that these results carry over to simultaneous approximation results on the half-line with corresponding weight functions. Via the transformation $t \leftrightarrow x^2$, the weight W maps to a weight w defined on the half-line, defined by $w(t) = W(x^2)$.

The main tools of approximation theory are worked out for spaces of functions on $(-\infty, \infty)$ with Freud weight W.

- The Weierstrass theorem is valid in the space $C_{0,W}(-\infty, \infty)$, the space of continuous functions with weighted limit zero at $\pm\infty$. It is also valid in all of the spaces $L^p_W(-\infty, \infty)$, for $1 \le p < \infty$. As one of the signs of the difference between the weighted infinite case and the finite case, one should note that the proof of this is a bit more work than in the finite interval case. A continuous function defined on a finite closed interval is always integrable. Here that need not be true.

- The de la Valleé Poussin means of the "Fourier series" are convergent in $C_{0,W}(-\infty, \infty)$ and in $L^p_W(-\infty, \infty)$, for $1 \le p < \infty$. We put the words Fourier series in quotes because what is meant is a generalization of the Fourier series. Namely, given the weight function there is a sequence of orthogonal polynomials associated with it. And here "Fourier series" signifies the orthogonal expansion of a function in terms of the sequence of orthogonal polynomials generated by the weight function.

- An inequality of Bernstein type is valid for polynomials in $C_{0,W}(-\infty,\infty)$ and in $L^p_W(-\infty,\infty)$, for $1 \le p < \infty$. However, the power of n in the inequality is typically different from 1. The precise value of the exponent depends upon W. Specifically for a polynomial P_n of degree at most n we have

$$\|e^{-Q(x)}P_n^{(k)}(x)\| \le A_k(\frac{n}{q_n})^k\|e^{-Q(x)}P_n(x)\|.$$

  The number q_n is defined by $q_nQ'(q_n) = n$, for n sufficiently large.

- Results of Jackson-Favard (28) type are also available in $C_{0,W}(-\infty,\infty)$ and in $L^p_W(-\infty,\infty)$, for $1 \le p < \infty$. Again the power of n is different. Using the notation

$$E_n^*(f,w) = \inf_{\text{degree}P_n \le n} \|w(x)(f(x) - P_n(x))\|_{(-\infty,\infty)},$$

  we have

$$E_n^*(f, e^{-Q(x)}) \le \alpha_k(q_n/n)^k E_{n-k}^*(f^{(k)}, e^{-Q(x)}).$$

- The results of Czipszer-Freud for trigonometric polynomial approximation hold analogously true for algebraic polynomial approximation in $C_{0,W}(-\infty,\infty)$ and in $L^p_W(-\infty,\infty)$, for $1 \le p < \infty$ with a Freud weight W. Again, the exponent on n will be different from 1 and will depend upon W. Just as in the case of a trigonometric polynomial approximating a 2π-periodic function, simultaneous approximation of derivatives is automatic.

We may suppose now that we have a weight $e^{R(t)}$ defined on $[0,\infty)$, where $R(t) = Q(x)$ with $t \leftrightarrow x^2$, and $e^{Q(x)}$ is a Freud weight. Based upon the itemized results of Freud, Balázs and Kilgore have shown

Theorem 17. (Balázs and Kilgore [7]) *Let $f \in C^q[0,\infty)$. Then for each $n > n_1$, where n_1 depends upon R and q there exists a polynomial P_n of degree at most n such that for $k = 0,\dots,q$ and for $0 \le t < \infty$*

$$e^{-R(t)}t^{\frac{k-q}{2}}|f^{(k)}(t) - P_n^{(k)}(t)| \le M_{q,k}\left(\frac{q_n}{n}\right)^{q-k} E_{n-q}(f^{(q)}; e^{-R(t)}),$$

with $M_{q,k}$ independent of n and f. Moreover $f^{(q)}(0) = P_n^{(q)}(0)$.

References

1. Achieser, N. and Krein, M., On the best approximation of periodic functions, *Dokl. Akad. Nauk SSSR* **15** (1937), 107–112.

2. Baiguzov, N., Some estimates for the derivatives of algebraic polynomials and an application to numerical differentiation, *Mat. Zametki* **5** (1969), 183–194.

3. Balázs, K., Lagrange and Hermite interpolation processes on the positive real line, *J. Approx. Theory* **50** (1987), 18–24.

4. Balázs, K., On the convergence of the derivatives of the Lagrange interpolation polynomials, *Acta Math.Hungar.* **55** (1990), 323–325.

5. Balázs, K. and Kilgore, T., On the simultaneous approximation of derivatives by Lagrange and Hermite interpolation, *J. of Approx. Theory*, 60 (1990), 231–244.

6. Balázs, K. and Kilgore, T., A discussion of simultaneous approximation of derivatives by Lagrange interpolation, *Numer. Funct. Anal. and Optim.* 11 (1990), 225–237.

7. Balázs, K. and Kilgore, T., Pointwise exponentially weighted estimates for approximation on the half-line, *Numer. Funct. Anal. and Optim.* 13 (1992), 223–232.

8. Balázs, K. and Kilgore, T., Interpolation and simultaneous mean convergence of derivatives, *Acta Math. Hungar.* **61** (1993), 63–72.

9. Balázs, K. and Kilgore, T., Approximation of functions on $[-1, 1]$ and their derivatives by polynomial projection operators, *Math. Nachr.*, 162 (1993), 39–44.

10. Balázs, K. and Kilgore, T., On some constants in simultaneous approximation, *International Journal of Mathematics and Mathematical Sciences*, 18 (1995), 279–286.

11. Balázs, K. and Kilgore, T., Some identities and inequalities for derivatives, *J. of Approx. Theory* 82 (1995), 274–286.

12. Balázs, K., Kilgore, T., and Vértesi, P., An interpolatory version of Timan's theorem on simultaneous approximation, *Acta. Math. Hungar.*, 57 (1991), 285–290.

13. Baszenski, G. and Tasche, M., Fast algorithms for simultaneous polynomial approximation, in: *Multivariate Approximation: From* CAGD *to Wavelets* (K. Jetter and F. I. Utreras (eds.)), World Scientific, Singapore, 1993, 1–16.

14. Baszenski, G. and Tasche, M., Fast polynomial multiplication and convolutions related to the discrete cosine transform, *Linear Algebra Appl.*,**252** (1997), 1–25.

15. Bernstein, S., Sur la limitation des valeurs d'un polynôme $P(x)$ de degré n sur tout un segment par ses valeurs en $(n+1)$ points du segment, *Isv. Akad. Nauk SSSR*, 7 (1931), 1025–1050.

16. de Boor, C. and Pinkus, A., Proof of the conjectures of Bernstein and Erdös concerning the optimal nodes for polynomial interpolation, *J. of Approx. Theory* 24 (1978), 289–303.

17. Brudnyi, Y., Approximation by entire functions on the exterior of an interval and of the semi-axis, *Dokl. Akad. Nauk SSSR*, **124** (1959), 739–742.

18. Brutman, L., On the Lebesgue function for polynomial interpolation, *SIAM J. Numer. Anal.*, **15** (1978), 694–704.

19. Cheney, E., "Introduction to Approximation Theory," McGraw-Hill, 1966.

20. Czipszer, J. and Freud, G., Sur l'approximation d'une fonction périodique et de ses dérivées successives, *Acta Math.* **99** (1958), 33–51.

21. Erdös, P., Some remarks on polynomials, *Bull. Amer. Math. Soc.* (1947), 1169–1176.

22. Erdös, P. and Feldheim, E., Sur le mode de convergence pour l'interpolation de Lagrange, *C. R. Acad. Sci. Paris Ser. A-B*, **203** (1936), 913–915.

23. Erdös, P. and Turán, P., On interpolation I, *Annals of Math.*, **38** (1937), 142–155.

24. Faber, G., Über die trigonometrische Darstellung stetiger Funktionen, *Jahresber. Deutsch. Math. Verein* **23** (1914), 192–210.

25. Favard, J., Sur les meilleures procédés d'approximation de certaines classes des fonctions par des polynômes trigonométriques, *Bull. Sci. Math.* **61** (1937), 209–244, 243–256.

26. Felten, M. and Kilgore, T., Some inequalities for derivatives of trigonometric and algebraic polynomials, *Results Math.* **30** (1996), 79–92.

27. Gopengauz, I., A theorem of A. F. Timan on the approximation of functions by polynomials on a finite segment, *Mat. Zametki* **1** (1967), 163–172.

28. Hunt, R., Muckenhoupt, B., and Wheeden, R., Weighted norm inequalities for the conjugate function and Hilbert transform, *Trans. Amer. Math. Soc.* **176** (1973). 227–251.

29. Jackson, D., "Über die Genauigkeit der Annäherung stetiger Funktionen durch ganz rationale Funktionen gegebenen Grades und trigonometrische Summen gegebener Ordnung," dissertation, Göttingen, 1911.

30. Kilgore, T., Optimization of the norm of the Lagrange interpolation operator, *Bull. of the Amer. Math. Soc.*, **83** (1977), 1069–1071.

31. Kilgore, T., A characterization of the Lagrange Interpolating projection with minimal Tchebycheff norm, *J. of Approx. Theory*, **24** (1978), 273–288.

32. Kilgore, T., A lower bound on the norm of interpolation with an extended Tchebycheff system, *J. of Approx. Theory* **47** (1986), 240–245.

33. Kilgore, T., An elementary simultaneous approximation theorem, Proc. Amer. Math. Soc. **118** (1993), 529–536.

34. Kilgore, T., On weighted L^p simultaneous approximation, Acta Math. Hungar. **73** (1996), 41–64.

35. Kilgore, T., Polynomial and function inequalities for derivatives, in "Approximation Theory VIII," the proceedings of the Eighth International Conference on Approximation Theory, Charles K. Chui and Larry L. Schumaker, editors, World Scientific, 1995, pp. 273–280.

36. Kilgore, T., Some remarks on weighted interpolation, in "Approximation Theory: In Memory of A. K. Varma," editors N. Govil, R. Mohapatra, Z. Nashed, A. Sharma, and J. Szabados, pp. 343–351.

37. Kilgore, T. and Prestin, J., On the order of convergence of simultaneous approximation by Lagrange-Hermite interpolation, *Math. Nachr.*, **150** (1991) 143–150.

38. Kilgore, T. and Prestin, J., A theorem of Gopengauz type with added interpolatory conditions, *Numer. Funct. Anal. Optim.*, **15** (1994), 859–868.

39. Kilgore, T. and Prestin, J., Pointwise Gopengauz estimates for interpolation, *Annales Univ. Sci. Budapestinensis, Sectio Computatorica* 16(1996), 253–261.

40. Kilgore, T. and Szabados, J., On approximation of a function and its derivatives by a polynomial and its derivatives, *Approx. Theory Appl.* **10** (1994), 93–103 .

41. Kilgore, T. and Tasche, M., *Journal of Computational Analysis and Applications*, **1** (1999), 239–261.

42. Korniecuk, N., The exact constant in D. Jackson's theorem on best uniform approximation of continuous periodic functions, *Dokl. Akad. Nauk SSSR* **145** (1962), 514–515.

43. Ky, N., On approximation by trigonometric polynomials in L^p_u-spaces, *Studia Sci. Math. Hungar.* **28** (1993), 183–188.

44. Leviatan, D., The behavior of the derivatives of the algebraic polynomials of best approximation, *J. Approx. Theory* **35** (1982), 169–176.

45. Muckenhoupt, B., Weighted norm inequalities for the Hardy maximal function, *Trans. Amer. Math. Soc.* **165** (1972), 207–226.

46. Muneer, Y., On Lagrange and Hermite interpolation. I, *Acta Math Hungar.* **49**, (1987), 293–305.

47. Riesz, M., Eine trigonometrische Interpolationsformel und einige Ungleichungen für Polynome, *Jahresber. Deutsch. Math. Verein* **23** (1914), 354–368.

48. Runck, P. and Vértesi, P., Some good point systems for derivatives of Lagrange interpolatory operators, *Acta Math. Hungar.* **56** (1990), 337–342.

49. Stechkin, S., Generalization of some inequalities of S. N. Bernstein, *Dokl. Akad. Nauk SSSR* **60** (1948), 9, 1511–1514.

50. Szabados, J., On the convergence of the derivatives of projection operators, *Analysis* **7** (1987), 349–357.

51. Szabados, J. and Vértesi, P., "Interpolation of Functions," World Scientific, 1990.

52. Tasche, M., Fast algorithms for discrete Chebyshev–Vandermonde transforms and applications, *Numer. Algorithms* **5** (1993), 453–464.

53. Tasche, M., Fast algorithms for discrete Chebyshev-Vandermonde transforms and applications, in "Algorithms for Approximation III," the proceedings of the NATO Advanced Workshop on Algorithms for Approximations, M. G. Cox and J. C. Mason, eds., Oxford, 1992, 453–464.

54. Telyakovskii, S., Two theorems on the approximation of functions by algebraic polynomials, *Mat. Sb.* **70** (112) (1966), 252–265.

55. Timan, A., "Theory of approximation of functions of a real variable," Dover, 1994 (reprint of Macmillan, 1963).

56. Timan, A., Strengthening of Jackson's theorem on the best approximation of continuous functions on a finite segment of the real axis, *Dokl. Akad. Nauk SSSR* **78** (1951), 17–20.

57. Trigub, R., Approximation of functions by polynomials with integral coefficients (in Russian), *Isv. Akad. Nauk. SSSR*, Ser. Matem. **26** (1962), 261–280.

ON THE COMPUTATION OF MINIMAL PROJECTIONS: MILLENNIUM REPORT

B. L. Chalmers, F. T. Metcalf and B. Shekhtman
Department of Mathematics, University of California, Riverside CA 92521, E-mail: blc@ucrmath.ucr.edu
Department of Mathematics, University of California, Riverside, CA 92521
Department of Mathematics, University of South Florida, Tampa, FL 33620, E-mail: boris@tarski.math.usf.edu
Contact author: B. Chalmers

Let $\tilde{P} = \sum_{i=1}^{n} \tilde{u}_i \otimes v_i : V \to V = [v_1, ..., v_n] \subset X$, where $\tilde{u}_i \in V^*$ and X is a Banach space. Let $P = \sum_{i=1}^{n} u_i \otimes v_i : X \to V$ be an extension of $\tilde{P}$ to all of X (i.e., $u_i \in X^*$) such that P has minimal (operator) norm. (E.g., if $\tilde{P} = I$, P is a minimal projection from X onto V.) In this chapter we discuss progress which has been made toward the determination of $u := u_1, ..., u_n$ and we point out several applications.

1 Introduction

1.1 *The Form of the Minimal Projection*

The problem of determining minimal projections came to our attention over twenty years ago, mostly through the work of Cheney and his colleagues (cf. [36], [44], [33]). Consequently we spent those decades of the last (twentieth) century in an effort to resolve this problem. This chapter is an attempt to share our experience of this fascinating journey in search of the form of minimal projections. The result is a mixed bag of theorems, heuristics, computational techniques and intelligent guess-work that we found instrumental in solving the problem.

Let X be a Banach space and V be an n-dimensional subspace of X. A projection P from X onto V is a linear bounded operator on X such that

$$\operatorname{Im} P = V \text{ and } P^2 = P. \tag{1}$$

Equivalently, a projection is a mapping from X onto V such that

$$Pv = v, \ \forall v \in V.$$

If V is a non-trivial subspace of X then there are infinitely many projections from X onto V. Different projections may have different norms. It is therefore customary to introduce the notion of the relative projection constant

$$\lambda(V, X) = \inf\{\|P\| : P \text{ is a projection from } X \text{ onto } V\}.$$

A projection is called minimal if

$$\lambda(V, X) = \|P\|.$$

Therefore the general problem addressed in this chapter is

Problem 1. *Given X and V, construct a minimal projection P from X onto V.*

Given an arbitrary basis $v_1, v_2, ..., v_n$ for V and an arbitrary projection P from x onto V, P can be written as

$$P = \sum_{j=1}^{n} u_j \otimes v_j$$

or, equivalently,

$$Px = \sum_{j=1}^{n} u_j(x) v_j,$$

where $u_j \in X^*$ are functionals on X such that

$$u_j(v_k) = \delta_{jk}.$$

Therefore the problem can be restated as

Problem 2. *Given a basis $v_1, v_2, ..., v_n$ for V, find the functionals $u_1, u_2, ..., u_n$ in X^* such that*

$$P = \sum_{j=1}^{n} u_j \otimes v_j$$

is a minimal projection onto V.

Of course the projection P does not depend on the choice of basis in V; it is, however, often the most difficult part of the problem to find the "most convenient" basis $v_1, ..., v_n$ for the minimal projection.

Finally, let us observe that the minimal projection is completely determined by

$$\operatorname{span}[u_j]_{j=1}^{n} = \operatorname{Im} P^*.$$

In short one can view the minimal projection problem as an optimization problem

$$\|\sum_{j=1}^{n} u_j \otimes v_j\| \to \min$$

with constraints

$$u_j \in X^*;\ u_j(v_k) = \delta_{jk}.$$

1.2 Minimal Projections and the Geometry of Banach Spaces

In the last subsection we introduced the notion of a relative projection constant $\lambda(V, X)$. It turns out that in many situations (if $X = L^1$, $X = C(K)$ or $X = \ell_\infty$) this quantity is an isometric invariant. That is, if two spaces V and V_1 are isometric, then $\lambda(V, X) = \lambda(V_1, X)$. This remarkable fact implies that the norm of a minimal projection depends only on the shape of the unit ball of the space and not on the intrinsic properties of the elements of the space. Although there are examples where the form of the minimal projection is not an isometric invariant, in many cases it is. Hence the determination of a minimal projection is often based on the geometric shape of V, i.e.

$$\{(a_j) \in \mathbf{R}^n : \|\sum a_j v_j\| \le 1\} = B(V) \subset \mathbf{R}^n.$$

The research done in the theory of Banach spaces has been very beneficial in guiding our research.

In later sections of this chapter will be presented an algorith for finding minimal projections onto some subspaces of L_1. This algorithm depends entirely on the shape of the unit ball of the space V.

1.3 First Examples

We end this introduction with a warning about the few "natural" examples of minimal projections. While there are a lot of neat ideas buried in the examples, the same examples can be misleading. In the spirit of this "new century" introduction, let us elaborate on this point with a futuristic analogy. Suppose we decide to explore Mars. We send a spaceship that finds an appropriate landing spot on the planet. We build a space-port and soon a town springs up around it. As years go by, more and more tourists visit the town, convinced that they saw Mars. Not only did they not see Mars but they saw the one spot which is precisely unlike Mars, since this is how the landing place was chosen to begin with. Undoubtedly the quest for the form of minimal projections was motivated by the simplicity of the few known examples. We found that those examples are the precise exceptions. They are unlike what is to be expected from the general form of minimal projections. Hence, upon reading this chapter, the reader can not expect to be able to write down the generic form of minimal projections onto, say, the space of polynomials of a generic degree "n". Instead, the techniques developed in this chapter will hopefully allow the reader to do so for one specific "n", provided it is not too large and enough time is available for fiddling with the project.

Finally we point out that Sections 4–6 is mostly a self-contained segment of this chapter and can be read independently.

2 Chebyshev Characterization of Polynomials of Best Approximation

This section contains a proof of the well-known theorem of Chebyshev. The theorem per se has nothing to do with minimal projections. The line of reasoning used in this section, however, is very similar to that used to find minimal projections. We hope that the readers' familiarity with the theorem will help to guide them through the rest of the chapter.

Let f be a continuous function on the interval $[-1, 1]$ and let E be the space of polynomials of degree at most n. Let $p \in E$ be the polynomial of best approximation to f. Let $X = [f] \oplus E$. Then X is an $(n+2)$-dimensional space and we can

a) Define a functional ϕ on x by

$$\phi(f - p) = \|f - p\|; \ \phi(x) = 0 \ \forall x \in E.$$

It is easy to see that $\|\phi\| = 1$.

b) By the Hahn-Banach Theorem we extend ϕ to a functional on $C([-1, 1])$ and

c) By the Riesz Representation Theorem we can indentify this extension with a regular signed-measure μ on $[-1, 1]$ such that

$$\int (f - p)\, d\mu = \|f - p\| \|\mu\| = \|f - p\| \tag{2}$$

$$\int g\, d\mu = 0 \ \forall g \in E. \tag{3}$$

d) We now use these properties to derive conclusions about the measure μ and the function $(f - p)$. First of all, observe that (2) implies that the function

$$\frac{f - p}{\|f - p\|}$$

and the measure μ are the extremals of each other. Hence

$$|f - p|(t) = \|f - p\|$$

for all $t \in \operatorname{supp} \mu$.

Secondly, for μ to annihilate E, the measure μ much change sign at least $n+2$ times. Combining these two facts we obtain

Theorem 1. *If f and p are as above, then there exist $n+2$ points $\xi_1, ..., \xi_{n+2} \in [-1,1]$ and $\lambda = \pm 1$ such that*

$$f(\xi_j) - p(\xi_j) = \lambda(-1)^j \|f - p\|. \tag{4}$$

The points $\{\xi_j\}$ form the support of μ.

How does this theorem help in finding the polynomial p of best approximation?

First of all, if one can somehow guess the "p" then the condition (4) will help to verify that the guessed "p" is in fact the polynomial of best approximation.

Secondly, it is often easier to find a polynomial that satisfies (4) than a rather "ambiguous" best approximation polynomial.

Third, if one can obtain some additional information about the measure μ (such as its support, i.e. $\{\xi_j\}$) then the problem reduces to solving a linear system

$$f(\xi_j) - \sum a_k (\xi_j)^k = \lambda(-1)^j \|f(x) - \sum a_k x^k\|, \ j = 1, ..., n+2, \tag{5}$$

for the coefficients a_k, $k = 0, ..., n+1$ and the $(n+2)$-nd unknown being $\lambda\|f - p\|$.

Finally, if the location of the points ξ_j are not known, then (5) turns into a non-linear system of $n+2$ equations with $2(n+2)$ unknowns (ξ_j are among them) and a clever choice of (ξ_j) allows one to start an interative procedure (the Remez algorithm) by means of which one can find the polynomial p.

2.1 Trace-Duality; Characterization of Minimal Projections

Let X be a Banach space and let V be a finite-dimensional subspace of X. Let P be a minimal projection from X onto V. We are seeking to characterize the minimal projection a la the Chebyshev Characterization in the last subsection. To that end we introduce a set $E \subset L(X)$ of linear bounded operators on X

$$E = \{L \in L(X) : \ker L \supset V \supset \operatorname{Im} L\}. \tag{6}$$

Observe that E is a linear subspace of $L(X)$ and for every $A \in E$ the operator $P - A$ is again a projection onto V. As in the previous subsection

a) Let ϕ be a functional defined on $[P] + E$ by

$$\phi(P) = \|P\|;\ \phi(A) = 0\ \forall A \in E. \tag{7}$$

Since 0 is the best approximation from E to P it follows that $\|\phi\| = 1$.

b) By the Hahn-Banach Theorem we extend ϕ to be a functional on $L(X)$.

c) At this juncture we are supposed to use the "Riesz Representation" to identify the functional on $L(X)$ with some well-known object. Thus the detour into "trace-duality" is in order.

Every functional ϕ on $L(X)$ is uniquely identified with a "nuclear" operator $A = A_\phi \in L(X)$ so that

$$\phi(B) = \operatorname{tr}(AB)$$

and

$$\|\phi\| = \nu(A_\phi) - \text{nuclear norm of } A_\phi.$$

In general one has to take care of the definition of the trace of an operator, but in our case the functional ϕ is only applied to the finite-dimensional operators and hence we can ignore the general case. Interested readers can find more about trace-duality, e.g. in [39], [42], and [49].

d) We now have to interpret the conditions (7) in terms of an operator A_ϕ rather than the functional ϕ. For every $v \in V$ and every $f \in V^\perp$ the operator L is given by

$$Lx = f(x)v\ \in E.$$

Hence

$$\operatorname{tr}(A_\phi L) = f(A_\phi v) = 0$$

for all $f \in V^\perp$ and thus $A_\phi v \in V$ for every $v \in V$.

To interpret the extremality condition for A_ϕ it will be convenient to represent A_ϕ as

$$A_\phi = \int_K (x^* \otimes x^{**})\, d\mu \tag{8}$$

where $K = B(X^*) \times B(X^{**})$ is a compact set in the weak-*-topology and μ is a regular measure on K. The formalism (8) simply means that

$$f(A_\phi x) = \int_K x^{**}(f) x^*(x)\, d\mu$$

for every $x \in X$ and $f \in X^*$. The fact that

$$\operatorname{tr}(A_\phi P) = \|P\|\nu(A_\phi)$$

simply means that the measure μ is positive and supported on those pairs (x^*, x^{**}) for which

$$x^*(Px^{**}) = \|P\|.$$

We now summarize all of the above into one theorem whose proof we will reproduce in the next section.

Theorem 2. *Let P be a projection from X onto V. Let $\mathcal{E}(P)$ denote the set of extremal pairs defined as*

$$\mathcal{E}(P) = \{(x^*, x^{**}) \in S(X^*) \times S(X^{**}) : \; x^{**}(P^*x^*) = \|P\|\}.$$

Then P is minimal if and only if there exists a positive measure μ on $\mathcal{E}(P)$ such that the operator

$$E_P := \int_{\mathcal{E}(P)} (x^* \otimes x^{**}) \, d\mu$$

maps V into V.

How does this theorem help in the computation of minimal projections?

Once again, if one can "guess" a minimal projection P, then one can use the theorem to verify that it is indeed minimal. In the following sections we will reproduce a few examples of such verification in order to demonstrate the power of the theorem. In particular we will illustrate the results of [46] and [16].

The other reason the theorem of this subsection has proved to be useful is as follows: In particular cases of spaces X and V one can gain some additional information about the operator E_P. As we will see in the rest of this section such information can be very instrumental in constructing minimal projections.

2.2 Minimal Projections in L^1 and in L^p

In the following sections we will also emphasize the use of M and the fact that in the case of symmetric spaces $M = I$. In addition we will show why the L^1 theory is relatively simple and more generally demonstrate the so-called "Holder equality condition" (the "($\star$-equation") to obtain minimal projections in L^p, $1 \leq p \leq \infty$.

3 General Theory and Examples

In this paper we develop first a theory providing a characterization theorem (Theorem 3) for finite-rank minimal projections. In order to demonstrate the usefulness of this characterization we provide several examples. More generally, the theory characterizes operators of minimal norm which extend a fixed linear action on a given finite-dimensional subspace. Secondly, a characterization theorem (Theorem 3) is given for minimal linear operators in a general setting. This setting includes, as examples, minimal and co-minimal projections, optimal recovery and linear estimation, linear n-widths, and best linear approximation to continuous proximity maps. These characterization theorems lead to concrete equations from which the minimal operators can be obtained and, in some important cases, described geometrically.

Let $\mathcal{B} = \mathcal{B}(X, V)$ be the space of all bounded linear operators from a real or complex normed space X into a finite-dimensional subspace V and let $\mathcal{P}$ be the family of all operators in $\mathcal{B}$ with a given fixed action on V (e.g., the identity action corresponds to the family of bounded projections onto V).

Definition 1. $(x, y) \in S(X^{**}) \times S(X^*)$ will be called an extremal pair for $Q \in \mathcal{B}$ if $\langle Q^{**}x, y\rangle = \|Q\|$, where $Q^{**} : X^{**} \to V$ is the second adjoint extension of Q to X^{**} (S denotes unit sphere).

Notation 1. Let $\mathcal{E}(Q)$ be the set of all extremal pairs for Q. To each $(x, y) \in \mathcal{E}(Q)$ associate the rank one operator $y \otimes x$ from X to X^{**} given by $(y \otimes x)(z) = \langle z, y\rangle x$ for $z \in X$.

Theorem 3. (Characterization) *P has minimal norm in $\mathcal{P}$ if and only if the closed convex hull of $\{y \otimes x\}_{(x,y)\in\mathcal{E}(P)}$ contains an operator for which V is an invariant subspace.*

Proof. The problem is equivalent to best approximating, in the operator norm, a fixed operator $P_0 \in \mathcal{P}$ from the space of operators $\mathcal{D} = \{\Delta \in \mathcal{B} : \Delta = 0 \text{ on } V\} = \text{sp}\{\delta \otimes v : \delta \in V^\perp, v \in V\}$. Let $K = B(X^{**}) \times B(X^*)$ endowed with the product topology, where $B(\cdot^*)$ denotes the unit ball with its weak* topology. Associate with any operator $Q \in \mathcal{B}$ the bilinear form $\widehat{Q} \in C(K)$ via $\widehat{Q}(x, y) = \langle Q^{**}x, y\rangle$, and let $\widehat{\mathcal{D}} = \{\widehat{\Delta} : \Delta \in \mathcal{D}\}$. Then, making use of standard duality theory for $C(K)$, K compact (see e.g., [48], Theorem 1.1 (p. 18) and Theorem 1.3 (p. 29)]), we have that $\widehat{P} = \widehat{P}_0 - \widehat{\Delta}_0$ is of minimal norm if and finite, non-zero (total mass one) signed measure $\hat{\mu}$ supported on the critical set

$$\mathcal{C}(\widehat{P}) = \{(x, y) \in S(X^{**}) \times S(X^*) : |\widehat{P}(x, y)| = \|\widehat{P}\|_\infty\}$$

such that $\text{sgn}\, \hat{\mu}\{(x, y)\} = \text{sgn}\, \widehat{P}(x, y)$ and $\hat{\mu} \in \widehat{\mathcal{D}}^\perp$, i.e.,

$$0 = \int_{\mathcal{C}(\widehat{P})} \widehat{\Delta}\, d\hat{\mu} \quad \text{for all } \widehat{\Delta} \in \widehat{\mathcal{D}}.$$

But now, since any $\widehat{Q} \in \{\widehat{P}\} \cup \widehat{\mathcal{D}}$ is a bilinear function, we can replace the signed measure $\hat{\mu}$, supported in $\mathcal{C}(\widehat{P})$, by a positive measure μ supported on $\mathcal{E}(P) \subset \mathcal{C}(\widehat{P})$ by noting that

$$\mathcal{C}(\widehat{P}) = \{(x, e^{i\theta}y) : (x,y) \in \mathcal{E}(P) \text{ and } \theta \in T\},$$

where $T = [0, 2\pi)$ in the complex case and $T = \{0, \pi\}$ in the real case, and setting

$$\mu\{(x,y)\} = |\hat{\mu}|\{(x, e^{i\theta}y) : \theta \in T\}.$$

For then $\operatorname{sgn}\mu\{(x,y)\} = \operatorname{sgn}\widehat{P}(x,y) = 1$, for $(x,y) \in \mathcal{E}(P)$ and

$$0 = \int_{\mathcal{E}(P)} \widehat{\Delta}\, d\mu \text{ for all } \Delta \in \mathcal{D},$$

since

$$\int_{\mathcal{C}(\widehat{P})} \widehat{\Delta}\, d\hat{\mu} = \int_{(x,y)\in\mathcal{E}(P),\, \theta\in T} \widehat{\Delta}(x, e^{i\theta}y)\, d\hat{\mu}(x, e^{i\theta}y)$$

$$= \int_{(x,y)\in\mathcal{E}(P),\, \theta\in T} e^{-i\theta}\, \widehat{\Delta}(x,y)\, e^{i\theta}\, d|\hat{\mu}|(x, e^{i\theta}y) = \int_{\mathcal{E}(P)} \widehat{\Delta}\, d\mu.$$

Hence,

$$0 = \int_{\mathcal{E}(P)} \widehat{\Delta}\, d\mu = \int_{\mathcal{E}(P)} \langle \Delta^{**}x, y\rangle\, d\mu(x,y) = \int_{\mathcal{E}(P)} \langle x, \delta\rangle \langle v, y\rangle\, d\mu(x,y)$$

$$= \left\langle \int_{\mathcal{E}(P)} \langle v, y\rangle x\, d\mu(x,y),\ \delta \right\rangle$$

for all $\Delta = \delta \otimes v$ ($\delta \in V^{\perp}$, $v \in V$), where, for $z \in X$, $\int_{\mathcal{E}(P)} \langle z, y\rangle x\, d\mu(x,y)$ is the element $w \in X^{**}$ defined by $\langle x^*, w\rangle = \int_{\mathcal{E}(P)} \langle z, y\rangle \langle x^*, x\rangle\, d\mu(x,y)$ for all $x^* \in X^*$. P is minimal, therefore, if and only if $\int_{\mathcal{E}(P)} \langle v, y\rangle x\, d\mu(x,y) \in (V^{\perp})^{\perp} = V$, i.e., if and only if there exists an operator (from X into X^{**})

$$E_P = \int_{\mathcal{E}(P)} y \otimes x\, d\mu(x,y) : \quad V \to V. \tag{9}$$

□

Remark 1. The identification of a minimal norm operator as the error of a best approximation problem in $C(K)$, in the proof of Theorem 3, is useful for examining various other aspects of minimal projections and extensions. For example, if in the proof of Theorem 3, we apply the Kolmogorov criterion for best approximation (see, e.g., [48], Theorem 1.16 (p. 69)]), we also get the following characterization.

Theorem 4. *P has minimal norm in $\mathcal{P}$ if and only if there does not exist $\Delta \in \mathcal{D} = \{\Delta \in \mathcal{B} : \Delta = 0$ on $V\}$ such that*

$$\sup_{(x,y)\in\mathcal{E}(P)} \operatorname{Re}\langle P^{**}x, y\rangle\overline{\langle \Delta^{**}x, y\rangle} < 0.$$

Theorem 4 was proved in [36] in the case $X = L^1$ and $\mathcal{P}$ consists of projections, where it was then used successfully in the landmark determination of a minimal projection from $L^1[-1,1]$ onto the lines (see Example 5 below). See also [45] for examples where Theorem 4 is proved and used in other settings.

Note 1. The existence of a minimal operator characterized in Theorem 3 follows from the fact that V is a finite-dimensional subspace of X and [44], where the existence of a minimal projection is shown in the more general case of V being a dual space, and by noting that the argument of [44] applies just as well to the case of minimizing over the more general class of operators in Theorem 3.

As a first example of the use of Theorem 3, we have the following sufficient condition for the adjoint of a minimal projection to be itself minimal. E_P will refer to the (not necessarily unique) operator in (9). For further discussion of the nature of E_P see Note 4 below.

Corollary 1. *Let P be a minimal projection. Then P^* is a minimal projection from X^* (onto $(\ker P)^\perp$) if $P^{**} \circ E_P = E_P \circ P$.*

Proof. (Sketch) $X = V \oplus U^\perp$, where $U^\perp = \ker P$ ($U = \operatorname{range} P^*$). Then E_P (as well as P) takes V into V, and it follows that $P^{**} \circ E_P = E_P \circ P$ if and only if E_P takes $U^\perp$ into $U^\perp$. But the latter occurs if and only if $(E_P)^*$ takes U into U. Finally, (x, y) is an extremal pair for P implies that (y, x) is an extremal pair for P^*, and hence $P^{**} \circ E_P = E_P \circ P$ implies that $(E_P)^* = E_{P^*}$. □

In the examples and discussion below it is helpful to introduce a fixed basis $v = (v_1, \ldots, v_n)$ for V; we will write $V = [v] = [v_1, \ldots, v_n]$. Then the necessary and sufficient condition (9) can be rewritten as a system of n equations

$$\int_{\mathcal{E}(P)} \langle v, y\rangle x \, d\mu(x, y) = Mv \quad \text{for some matrix } M. \tag{10}$$

Let $\sum_{i=1}^{n} u_i \otimes v_i \in X^* \check{\otimes} V$ (the injective tensor product of X^* and V) represent Q (and Q^{**}, where we set $\langle z, u_i \rangle = \langle u_i, z \rangle$ for $z \in X^{**}$), i.e., $Qx = \sum_{i=1}^{n} \langle x, u_i \rangle v_i$. Then $\mathcal{E}(Q) = \{(x,y) \in S(X^{**}) \times S(X^*) : \sum_{i=1}^{n} \langle x, u_i \rangle \langle v_i, y \rangle = \|Q\|\}$ and we can write

$$\mathcal{P} = \left\{ \sum_{i=1}^{n} u_i \otimes v_i : \ \langle v_i, u_j \rangle = A_{ij} \quad \text{for } A \text{ a given fixed } n \times n \text{ matrix} \right\}.$$

Note 2. M in (10) may be regarded as a function of A above. Hence, (10) may be regarded as determining a minimal P up to the n^2 entries of M, which are in turn determined by the n^2 entries of A.

Notation 2. If $z \in Z$ and $z^* \in Z^*$ are such that $\langle z, z^* \rangle = \|z\| \, \|z^*\| \neq 0$, then z^* is an extremal for z and we write $z^* = \operatorname{ext} z$. (Then also $z = \operatorname{ext} z^*$.) Note that $\operatorname{ext} z$ is determined only up to a non-zero scalar factor.

For purposes of illustration, observe the following simple examples (Examples 1 and 2) of Theorem 3, where minimal P has an extremal pair (x, y) with $x \in V$, and therefore we can take $E_P = y \otimes x$.

Example 1. ($\dim V = 1 = \operatorname{rank} P$) Let $P = u_1 \otimes v_1$. Then $(x, y) \in \mathcal{E}(P)$ if and only if $\langle x, u_1 \rangle \langle v_1, y \rangle = \|P\|$, i.e., $(x, y) = (\operatorname{ext} u_1, \operatorname{ext} v_1)$. Then $E_P = y \otimes x : V \to V$ if and only if $\operatorname{ext} u_1 = v_1$, i.e., $u_1 = \operatorname{ext} v_1$ (the solution given by the Hahn-Banach extension theorem). This example extends to $\dim V > 1 = \operatorname{rank} P$ by applying this $n = 1$ example to $X/V \cap \ker P$.

Example 2. (Hilbert space) Let X be a Hilbert space, let the basis v for V be orthonormal, let $\mathcal{P}$ have a fixed "diagonal" action, i.e., $A_{ij} = d_i \delta_{ij}$, where $d = (d_1, \ldots, d_n)$ is a fixed n-tuple of scalars, and let $J = \{j : |d_j| = \max |d_i|\}$. Then $P = \sum \bar{d}_i v_i \otimes v_i$ is minimal, where $E_P = y \otimes x$ for any choice of $(x, y) = (z, z)$ with z an arbitrary norm 1 element of the eigenspace corresponding to a maximum eigenvalue d_j, $j \in J$.

The minimality of the Fourier projection in the context of compact abelian groups is a simple consequence of Theorem 3 as demonstrated by the following example.

Example 3. ([16]) Let T (with "+") be a compact abelian group with Haar measure ν, $\widehat{T}$ its dual, $\{v_\gamma\}_{\gamma \in \widehat{T}}$ the set of all characters, N a finite part of $\widehat{T}$, V the linear span of the characters v_τ, $\tau \in N$, and let $X = L^p(T)$, $1 \leq p < \infty$ or $X = C(T)$, $p = \infty$. Then the Fourier projection $F = \sum_{\tau \in N} v_\tau \otimes v_\tau$ is minimal among all projections from X onto V.

Proof. (Sketch) Let (x, y) be any extremal pair for F. Then $(x_t, y_t) = (x(\cdot + t), y(\cdot + t))$ is an extremal pair for each $t \in T$. Thus, $E_F = \int_T y_t \otimes x_t \, d\nu(t) : V \to V$, since $\langle E_F v_\tau, v_\gamma \rangle \overset{\smile}{=} \langle v_\tau, y\rangle\langle x, v_\gamma\rangle\langle v_\gamma, v_\tau\rangle$ and $\langle v_\gamma, v_\tau\rangle = \delta_{\gamma\tau} = 0$ if $\tau \in N$ and $\gamma \in \widehat{T} \sim N$. $\square$

Notation 3. Write $\langle v, y\rangle \cdot u$ for $\sum_{i=1}^n \langle v_i, y\rangle u_i$ in the following.

Note 3. For $Q = \sum_{i=1}^n u_i \otimes v_i$, $(x, y) \in \mathcal{E}(Q)$ implies $x = \text{ext}\,(\langle v, y\rangle \cdot u)$ and $y = \text{ext}\,(\langle x, u\rangle \cdot v)$, which shows that, to find extremal pairs in general, we must solve the (non-linear) equation

$$d = \langle v, y\rangle = \langle v, \text{ext}\,(\langle \text{ext}\,(d \cdot u), u\rangle \cdot v)\rangle \tag{11}$$

for n-tuples of scalars $d = (d_1, \ldots, d_n) = (\langle v_1, y\rangle, \ldots, \langle v_n, y\rangle)$. For $X = C(T)$ and $L^1(T)$, respectively, however, the extremal pairs of Q have, on the support of u and the support of v, respectively, the simple forms $(x_t, y_t) = (\text{sgn}\,(v(t) \cdot u), \delta_t)$ and $(\delta_t, \text{sgn}\,(u(t) \cdot v))$, respectively, since $\|Q\| = \sup_{t \in T} L(t)$, where $L(t) = \langle x_t, v(t) \cdot u\rangle$, and $\langle u(t) \cdot v, y_t\rangle$, respectively, is the so-called "Lebesgue function" of Q. This fact makes these important cases relatively easy to consider.

That P^* in Corollary 1 is not always minimal (and that therefore E_P does not always commute with P) is shown by the following example, which also serves as an example of Theorem 3.

Example 4. ([46]) Consider $X = \ell_6^\infty$ and $V = [v_1, v_2]$, where $v_1 = (1 0 \alpha \beta \alpha \beta)$, $v_2 = (0 1 \beta \alpha \,{}^-\beta \,{}^-\alpha)$, with $\alpha = (2 + \sqrt{2})/4$, $\beta = \sqrt{2}/4$. Then the "interpolating projection" $P = \sum_{i=1}^2 u_i \otimes v_i$, where $u_i = \epsilon_i$ (the ith standard basis element in R^6), $i = 1, 2$, is minimal. This is seen by first checking that $\{(x_i, y_i)\}_{i=1}^4 \subset S(\ell_6^\infty) \times S(\ell_6^1)$ are extremal pairs, where $x_1 = (1 1 1 1 1 0)$, $x_2 = (1 1 1 1 0 \,{}^-1)$, $x_3 = (1 \,{}^-1 1 0 1 1)$, $x_4 = (1 \,{}^-1 0 \,{}^-1 1 1)$, and $y_i = \epsilon_{i+2}$, $i = 1, 2, 3, 4$. Indeed, $L(t) = \sum_{s=1}^6 |u(s) \cdot v(t)| = \langle x_{t-2}, u\rangle \cdot \langle v, y_{t-2}\rangle = \alpha + \beta$ for $t = 3, 4, 5, 6$, while $L(1) = L(2) = 1$. Secondly, check that (10) holds:

$$\sum_{i=1}^4 \langle v, y_i\rangle x_i/4 = Mv, \quad \text{where } M = \frac{\alpha + \beta}{2}\begin{pmatrix} 1 & 0 \\ 0 & 1 \end{pmatrix}.$$

Thus P is minimal, with norm $\alpha + \beta > 1$, but the projection $\sum_{i=1}^2 \epsilon_i \otimes u_i$ from ℓ_6^1 onto $[u_1, u_2]$ has norm 1, and thus $P^* : \ell_6^1 \to [u_1, u_2]$ is not minimal $(\|P^*\| = \|P\|)$.

Note 4. (See, e.g., [47] or [49] for definitions and notation.) The operator E_P of (9) can be viewed as a norm one integral operator in $(X^* \check{\otimes} V)^*$ separating

P from $\mathcal{D} = \{\Delta \in \mathcal{B} : \Delta = 0 \text{ on } V\}$, i.e.,

$$\langle P, E_P \rangle = \operatorname{tr}\left(\sum_1^n u_i \otimes E_P\, v_i\right) = \sum_1^n \langle E_P\, v_i, u_i \rangle = \sum_1^n \int \langle v_i, y \rangle \langle x, u_i \rangle \, d\mu$$

$$= \int \langle P^{**}x, y \rangle \, d\mu = \|P\|,$$

$\langle \mathcal{D}, E_P \rangle \equiv 0$ as in the proof of Theorem 3, and $\nu(E_P) \le \int_{\mathcal{E}(P)} \|y\| \|x\| \, d\mu(x,y) = 1$, where ν denotes the norm of E_P in the space of integral operators $I_1(X, X^{**})$. Note also from the above that $\|P\| = \operatorname{tr}(MA)$, for which upper bounds are determined in [23], extending those in [42] for projections ($A = I$).

By use of Note 4 and the observation that $\langle P, E_P \rangle = \operatorname{tr}(E_P|_V \circ P)$ we have the following known corollary, used for example in [42] and [39], and also used in [46] for obtaining the projection of Example 4.

Corollary 2. *The relative projection constant (of V relative to X) $\lambda(V, X) = \inf_{P \in \mathcal{P}} \|P\|$, where $\mathcal{P}$ is the family of projections, is given by*

$$\lambda(V, X) = \sup\left\{ tr(Q|_V) : \; Q \in I_1(X, X^{**}), \; Q : V \to V, \; \nu(Q) = 1 \right\}.$$

Note that if X is finite-dimensional then $I_1(X, X^{**}) = X^* \hat{\otimes} X^{**}$ (the projective tensor product of X^* and X^{**}) and ν in Corollary 2 is the nuclear norm.

Corollary 3. *Let P have minimal norm in $\mathcal{P}$. If $\mathcal{E}_1(P) = \{x : (x,y) \in \mathcal{E}(P) \cap$ supp $\mu\}$ is an independent set, then, for each $x \in \mathcal{E}_1(P)$, let $x^0 \in (\operatorname{span} \mathcal{E}_1(P))^*$ such that $\langle x, x^0 \rangle = 1$ and $\langle z, x^0 \rangle = 0$ for all $z \ne x$ in $\mathcal{E}_1(P)$, and act on (10) with x^0 to get*

$$\langle v, y \rangle \mu\{(x,y)\} = M \langle v, x^0 \rangle. \tag{12}$$

In Examples 5–10 below, we restrict ourselves to minimal projections (i.e., $A = I$).

Example 5. $X = L^1[-1,1] \supset V = [v]$, $v = (1,t)$. Use Note 3 ($(x_t, y_t) = (\delta_t, \operatorname{sgn}(u(t) \cdot v))$), Corollary 3, the clear fact that $x_t^0\big|_V = \chi_{\{t\}}$, and symmetry considerations (whence $M = \kappa \operatorname{diag}(1, m)$ for some scalar κ) to write (12) as follows (cancelling the (infinitesimal) scalar multipliers ($\mu\{(x_t, y_t)\}$ and dt)):

$$\begin{pmatrix} \langle v_1, y_t \rangle \\ \langle v_2, y_t \rangle \end{pmatrix} = \kappa \begin{pmatrix} v_1(t) \\ m\, v_2(t) \end{pmatrix} = \kappa \begin{pmatrix} 1 \\ m\, t \end{pmatrix},$$

i.e.,

$$\frac{1}{\langle v_1, y_t \rangle} = \frac{mt}{\langle v_2, y_t \rangle}, \tag{13}$$

where m is a scalar and

$$\langle v, y_t \rangle = \int_{-1}^{1} v(s)\,\mathrm{sgn}\,(u(t)\cdot v(s))ds = \epsilon(t)\left[\int_{-1}^{r(t)} - \int_{r(t)}^{1} v(s)ds\right]$$

$$= \epsilon(t)(2r(t), r^2(t) - 1),$$

where $r(t)$ is defined (uniquely since V is a Chebyshev system) by $u(t)\cdot v(r(t)) = 0$ and $\epsilon(t) = \pm 1$. Equation (13) is then easily solved for $r(t)$, yielding the admissible solution (in $[-1,1]$) $r(t) = mt - \mathrm{sgn}\,(mt)\sqrt{m^2t^2+1}$, and then u is obtained from the linear relations $u(t)\cdot v(r(t)) = 0$ and $L(t) = u(t)\cdot\langle v, y_t\rangle \equiv \lambda$, $t \in [-1,1]$. I.e., $P = \sum_{i=1}^{2} u_i \otimes v_i$ is minimal where

$$\begin{pmatrix} u_1(t) \\ u_2(t) \end{pmatrix} = \begin{pmatrix} 2r(t) & r^2(t) - 1 \\ 1 & r(t) \end{pmatrix}^{-1} \begin{pmatrix} \lambda\sigma(t) \\ 0 \end{pmatrix},$$

$\sigma(t) = \mathrm{sgn}\,((r^2(t)-1)/mt)$ and m and λ are determined to meet the remaining normality conditions, esp., $\lambda = \|P\| = 1.22040\ldots$.

This L^1-example and Example 6 which follows demonstrate how Theorem 3 provides a direct formula for P via Corollary 3. (In the L^1-cases, $\{x_t = \delta_t : t \in T\}$ is an independent set (see also [18])).

Example 6. ([18]) $X = L^1[-1,1] \supset V = [v]$, $v = (1, t, t^2)$. Analogously as in Example 5, (12) becomes

$$\frac{m_{11} + m_{13}t^2}{2(r_1(t) - r_2(t) + 1)} = \frac{t}{r_1^2(t) - r_2^2(t)} = \frac{m_{31} + m_{33}t^2}{\frac{2}{3}(r_1^3(t) - r_2^3(t) + 1)}, \tag{14}$$

where the $r_i(t)$ are defined to be the roots of the quadratic equation $u(t)\cdot v(r) = 0$, i.e., $u(t)\cdot v(r_i(t)) = 0$, $i = 1,2$. Next solve equation (14) for $r_1(t)$ and $r_2(t)$ and then u is obtained from the linear relations $u(t)\cdot v(r_i(t)) = 0$, $i = 1,2$, and $L(t) = u(t)\cdot\langle v, y_t\rangle \equiv \lambda$, $t \in [-1,1]$. I.e., $P = \sum_{i=1}^{3} u_i \otimes v_i$ is minimal where

$$\begin{pmatrix} u_1(t) \\ u_2(t) \\ u_3(t) \end{pmatrix} =$$

$$\begin{pmatrix} 2(r_1(t) - r_2(t) + 1) & r_1^2(t) - r_2^2(t) & \frac{2}{3}(r_1^3(t) - r_2^3(t) + 1) \\ 1 & r_1(t) & r_1^2(t) \\ 1 & r_2(t) & r_2^2(t) \end{pmatrix}^{-1} \begin{pmatrix} \lambda\sigma(t) \\ 0 \\ 0 \end{pmatrix},$$

$\sigma(t) = \operatorname{sgn}\left(\frac{2}{3}(r_1^3(t) - r_2^3(t) + 1)/(m_{31} + m_{33}t^2)\right)$, and $m_{11}, m_{13}, m_{31}, m_{33}$ and λ are determined to meet the remaining normality conditions, esp., $\lambda = \|P\| = 1.35948\ldots$.

In the case $X = L^1$, there is a remarkably simple geometric solution (Corollary 4 below) to the problem of minimal projections and extensions, which accounts for the relative simplicity of Examples 5 and 6 above.

Notation 4. Denote the underlying real or complex field by $\mathbf{F}$ and introduce the norm on $\mathbf{F}^n$ given by $\|a\| = \|a \cdot v\|_X$. The following proposition demonstrates a very useful geometric connection in $\mathbf{F}^n$ between the two components of an extremal pair for any $P \in \mathcal{P}$.

Proposition 1. *For any extremal pair* (x, y) *of* $P = \sum_{i=1}^n u_i \otimes v_i$,

$$\langle x, u \rangle = \|P\| \frac{\alpha\overline{\langle v, y \rangle} - c^*}{\|\alpha\overline{\langle v, y \rangle} - c^*\|}, \tag{15}$$

where α *is any positive scalar and* $c^* \in \mathbf{F}^n$ *yields* $\min\|\alpha\overline{\langle v, y \rangle} - c\|$ *subject to* $c \cdot \langle v, y \rangle = 0$.

Proof. Fix $x \in \mathcal{E}(P)$ and let $C = \{c \in \mathbf{F}^n : c \cdot \langle v, y \rangle = 0\}$. Then there exists a positive α $(= |\langle v, y \rangle|^2/\|P\|)$ and $c^* \in C$ such that $\langle x, u \rangle = \alpha\overline{\langle v, y \rangle} - c^*$. Hence, $0 = c \cdot \langle v, y \rangle = \langle c \cdot v, [\operatorname{ext}(\langle x, u \rangle \cdot v)] \rangle = \langle c \cdot v, \operatorname{ext}(\alpha\overline{\langle v, y \rangle} \cdot v - c^* \cdot v) \rangle$, for all $c \in C$, which implies that $c^* \cdot v$ is a best approximation to $\alpha\overline{\langle v, y \rangle} \cdot v$ from $\{c \cdot v : c \in C\}$ with respect to the norm of X. Hence c^* yields the minimum of $\|\alpha\overline{\langle v, y \rangle} - c\|$. Further, $\|P\| = \langle x, u \rangle \cdot \langle v, y \rangle = \langle (\alpha\overline{\langle v, y \rangle} - c^*) \cdot v, y \rangle = \|(\alpha\overline{\langle v, y \rangle} - c^*) \cdot v\|_X = \|\alpha\overline{\langle v, y \rangle} - c^*\|$, since $y = \operatorname{ext}(\langle x, u \rangle \cdot v)$. Finally, note that α can be replaced by any other positive quanitity by scaling simultaneously the numerator and denominator in (15). $\square$

Note 5. Geometrically, (15) says that $\langle x, u \rangle / \|P\|$ is a point of intersection of the unit $\|\cdot\|$-sphere in $\mathbf{F}^n$ and its tangent plane perpendicular (in the ordinary Euclidean sense) to the direction of $\langle v, y \rangle$.

For a pictorial representation of this note, see Figure A in the next section.

Theorem 5. *Under the hypotheses of Corollary 3,* $P = \sum_{i=1}^n u_i \otimes v_i$ *is minimal, where*

$$\langle x, u \rangle = \|P\| z(x),$$

with $z(x)$ *being a point of intersection of the unit* $\|\,\|$*-sphere in* $\mathbf{F}^n$ $(\|a\| = \|a \cdot v\|_X)$ *and its tangent plane perpendicular to* $M\langle v, x^0 \rangle$.

Proof. Apply (12) to (15) and use Note 5. $\square$

Corollary 4. *Let $X = L^1(T) \supset V = [v]$. Then $P = \sum_{i=1}^n u_i \otimes v_i$ is minimal where*

$$u(t) = \|P\|\, z(t), \tag{16}$$

with $z(t)$ being a point of intersection of the unit $\|\,\|$-sphere in $\mathbf{F}^n$ ($\|a\| = \|a \cdot v\|_{L^1}$) and its tangent plane perpendicular to $Mv(t)$ for some M.

Proof. Take $x_t = \delta_t$ and $x_t^0|_V = \chi_{\{t\}}$ in Theorem 5. □

Remark 2. Corollary 4 has been especially useful in [15].

By use of (10) and (3), we can rewrite Theorem 3 in a form which is constructive in terms of creating a minimal P from $\mathcal{P}$ by use only of the Banach space geometry of V as a subset of X.

Theorem 6. (Equations) $P = \sum_{i=1}^n u_i \otimes v_i$ *has minimal norm in $\mathcal{P}$ if and only if for some matrix M,*

$$\int_{\mathcal{E}} \langle v, y\rangle\, x\, d\mu(x,y) = Mv,$$

where $\mathcal{E} = ES(U^*) \times ES(V^*)$ (ES denotes the extreme points of the unit sphere S), with $U = [u] \subset X^*$, and u, x and y satisfying $y = \mathrm{ext}(\langle x, u\rangle \cdot v)$, $x = \mathrm{ext}(\langle v, y\rangle \cdot u)$, and $\langle v_i, u_j\rangle = A_{ij}$.

Corollary 5. *In Theorem 6, suppose (as in the cases $X = L^1$ or $X = C$) that the $\langle v, y\rangle = d$ are known (up to a scalar multiple). Then u (subject to the conditions $\langle v_i, u_j\rangle = A_{ij}$) is determined from the single equation $\mathrm{ext}\langle d \cdot u, u\rangle = e$, where $e = \langle x, u\rangle$ is determined from Proposition 1.*

As a first example of the use of Theorem 6, we can construct the projection of Example 4 as follows.

Example 7. Consider $X = \ell_7^\infty$ and $V = [v_1, v_2]$, where $v_1 = (0\,\beta\,\alpha\,1\,\alpha\,\beta\,0)$, $v_2 = (1\,\alpha\,\beta\,0\,{}^-\beta\,{}^-\alpha\,{}^-1)$ with $\alpha = (2+\sqrt{2})/4$ and $\beta = \sqrt{2}/4$. Write each v_i as a (convex) combination of the (signed) independent extreme points $x_1 = (1\,1\,1\,1\,1\,1\,{}^-1)$, $x_2 = (1\,1\,1\,1\,1\,{}^-1\,{}^-1)$, ..., $x_6 = (1\,{}^-1\,{}^-1\,{}^-1\,{}^-1\,{}^-1\,{}^-1)$ in $ES(\ell_7^\infty)$:

$$\frac{1}{4(\alpha+\beta)}\left[c_1x_1 + 2c_2x_2 + c_3x_3 + c_4x_4 + 2c_5x_5 + c_6x_6\right] = v$$

where $c_1 = (\alpha, \beta)$, $c_2 = (\alpha+\beta, \alpha+\beta)/2$, $c_3 = (\beta, \alpha)$, $c_4 = (-\beta, \alpha)$, $c_5 = (-\alpha-\beta, \alpha+\beta)/2$, $c_6 = (-\alpha, \beta)$. Set $\langle v, y_i\rangle = c_i$, $i = 1, \ldots, 6$, where $y_1 = \epsilon_3$, $y_2 = (\epsilon_2 + \epsilon_3)/2$, $y_3 = \epsilon_2$, $y_4 = -\epsilon_6$, $y_5 = -(\epsilon_5 + \epsilon_6)/2$, $y_6 = -\epsilon_5$. Next, construct the symmetric 12-sided ball $\|a\| = \|a \cdot v\|_{\ell_7^\infty} = 1$ and use Note 5 to conclude that $\langle x_i, u\rangle = (1,1)$, $i = 1,2,3$, and $\langle x_i, u\rangle = (-1,1)$, $i = 4,5,6$. I.e., $u_1 = \epsilon_4$ and $u_2 = \epsilon_1$.

Note 6. If $X = L^p(T)$, $1 \le p < \infty$ or $X = C(T)$, $p = \infty$, and V is piecewise continuously differentiable, then for $P = \sum_{i=1}^n u_i \otimes v_i$ minimal, we have the following necessary *linear* equation for u (obtained by the first author using different methods in the cases $p = 1, \infty$ in [3] and as a corollary of Theorem 3 of the present paper in [5])):

$$\frac{1}{p} u' \cdot Mv = \frac{1}{q} u \cdot Mv' \quad \text{on } T, \tag{$\star$}$$

where M is the matrix in (10) and $1/q + 1/p = 1$. (One may view ($\star$) as an n-dimensional version of the Hölder equality condition.) If $p = 1$, ($\star$) is an easy consequence of (10). If $p = \infty$, ($\star$) is very useful and when used in conjunction with (10), as in Example 5c below, yields a second *linear* equation for u.

Example 8. $X = C[-1,1] \supset V = [v\,]$, $v = (1,t)$. Then $P = \sum_{i=1}^2 u_i \otimes v_i$ is minimal with norm 1, where

$$\begin{pmatrix} u_1 \\ u_2 \end{pmatrix} = \begin{pmatrix} A & A \\ -C & C \end{pmatrix} \begin{pmatrix} \delta_{-1} \\ \delta_1 \end{pmatrix},$$

with $A = C = \frac{1}{2}$.

Proof. (Sketch) Check that the orthonormality conditions hold and that (10) holds with $(x_t, y_t) = (\operatorname{sgn}(\cdot - t), (\delta_1 - \delta_{-1})/2)$ and $d\mu(t) = dt/2$, $-1 \le t \le 1$, $M = \operatorname{diag}(0,1)$. ($E_P x = \frac{1}{2}[x(1) - x(-1)]v_2$ for all $x \in X$.) □

In the case of the following example, in [4] the form of P was guessed by use of the above ($\star$)-equation and local constancy of the Lebesgue function, and the minimality of P was proved by showing that its norm was the same as the norm of a known ([36]) minimal projection in an isometric L^1-setting. In the present paper we check that P is minimal by direct use of Theorem 1. Recall (Note 6 above) that the equation ($\star$) can be derived as a consequence of Theorem 3 ([5]). Also local constancy of the Lebesgue function is a consequence of using knowledge of the form of the extremal pairs in Theorem 3. In other words, Theorem 3 is being used both to obtain the formula for P and to establish its minimality.

Example 9. ([4]) $X = C[-1,1] \supset V = [v\,]$, $v = (1 - t^2, t)$. Then $P = \sum_{i=1}^2 u_i \otimes v_i$ is minimal with norm $1.220404917116354\ldots$ (same as $\|P\|$ in

Example 4a), where

$$\begin{pmatrix} u_1 \\ u_2 \end{pmatrix} = \begin{pmatrix} 0 & B & 0 \\ -C & 0 & C \end{pmatrix} \begin{pmatrix} \delta_{-1} \\ \delta_0 \\ \delta_1 \end{pmatrix} + \begin{pmatrix} b \\ cs \end{pmatrix} \frac{1}{[1+(\sigma s)^2]^{\frac{3}{2}}}, \quad -1 \le s \le 1,$$

with $(B, C, b, c) = (\lambda/2)(1, 1/\sqrt{1+\sigma^2}, 2\sigma, \sigma^2)$, $\lambda = \|P\| = -2\sigma/\log t_0$, $\sigma = (1-t_0^2)/2t_0 = (t_0^2 - t_0 - 1)\log t_0$.

Proof. (Sketch) Use $(\star)$ $(0 = u \cdot Mv')$ to determine the continuous part of $u(s)$ to be $w(s)(b, cs)$ for scalars b and c and then determine the scalar function $w(s)$ to be $[1+(\sigma s)^2]^{-3/2}$ by forcing local constancy of the Lebesgue function. Choose all the remaining parameters so that the the (ortho)normality conditions hold and that (10) holds (with $(x_t, y_t) = (\text{sgn}\,(t)\text{sgn}\,(\cdot - r(t)), \delta_t)$, $d\mu(t) = (1+t^2)dt/|t|^3$ on $t_0 \le |t| \le 1$ (μ unnormalized), where $r(t) = (t^2-1)/2\sigma t$, and $M = (4\sigma^2)\text{diag}(1, 1/\sigma))$. □

As in the previous example, the form of P in the following example is obtained with the help of the equation $(\star)$, which can be derived as a consequence of Theorem 3 ([5]) as noted in Note 6, and by use of local constancy of the Lebesgue function, again a consequence of Theorem 3. In [20] it was checked that P is minimal by direct use of Theorem 3. I.e., Theorem 3 is again being used both to obtain the formula for P and to verify its minimality.

Example 10. ([20]) $X = C[-1,1] \supset V = [v\,]$, $v = [1, t, t^2]$. This projection was first found by the authors in 1978, and may now be verified as minimal by use of Theorem 3. $P = \sum_{i=1}^{3} u_i \otimes v_i$ is minimal with norm 1.220173064217988..., where

$$\begin{pmatrix} u_1 \\ u_2 \\ u_3 \end{pmatrix} = \begin{pmatrix} A & B & A \\ -C & 0 & C \\ D & -B & D \end{pmatrix} \begin{pmatrix} \delta_{-1} \\ \delta_0 \\ \delta_1 \end{pmatrix} + \sum_{k=1}^{2} \begin{pmatrix} b_k + a_k|s| \\ c_k s \\ -b_k + d_k|s| \end{pmatrix} \frac{\chi_{S_k}(|s|)}{(1+w_k|s|)^3},$$

with $-1 \le s \le 1$ and $S_k = [s_{k1}, s_{k2}]$ $(k = 1, 2)$. For the equations (and values) for all the parameters see [20].

Proof. (Sketch) Assume extremal pairs (x_t, δ_t) and $(x_t^{(1)}, \delta_{\pm 1})$, for t in the region of constancy of $L(t)$, where $x_t = \text{sgn}\,(t)\text{sgn}\,(\cdot - r(t))$, $x_t^{(1)}(s) = \pm x_t(s)$ for $|s| \in S_1 \cup S_2$, and the sign $\pm$ is constant for $|s| \in S_k$, $k = 1, 2$. Then apply x_t^0 (x_t^0 is differentiation at $r(t)$) to (10) to obtain

$$w_1(t)v(t) + w_2(t)v(\pm 1) = Mv'(r(t)), \tag{17}$$

where w_1 and w_2 are positive scalar-valued functions. Next, "dot" both sides of (17) with $u(r)$ to get

$$w_1(t)u(r) \cdot v(t) + w_2(t)u(r) \cdot v(\pm 1) = u(r) \cdot Mv'(r).$$

But $u(r(t)) \cdot v(t) = 0$ and apply ($\star$) to obtain the linear equation $v(\pm 1) \cdot u = 0$ for u. Next use this equation, and ($\star$), to determine the continuous part of $u(s)$ to be $w(s)(b_k + a_k|s|,\, c_k s,\, -b_k + d_k|s|)$, where $c_k = (-1)^k(d_k + a_k)$, and then determine the scalar function $w(s)$ to be $1/(1 + w_k|s|)^3$, $|s| \in S_k$, $k = 1, 2$, $w_1 = (a_1 - d_1)/2b_1$, $w_2 = -w_1$, by forcing local constancy of the Lebesgue function. Finally, choose all the remaining parameters so that the orthonormality conditions hold and so that (10) holds (see [20]). □

For some further examples which could be obtained by using Theorem 3, see, e.g., [26] and [37]. Theorem 3 is a special case of Theorem 7 which follows. The proof of Theorem 7 is an easy extension of the proof of Theorem 3.

Definition 2. If X and Y are any two Banach spaces, a continuous homogeneous (not necessarily linear) operator Q from X into Y will be said to be jointly weak* continuous if Q has a continuous homogeneous extension $Q^{**} : X^{**} \to Y$ such that $\widehat{Q}(x, y) = \langle Q^{**}x, y\rangle$ is continuous on the compact set $K = B(X^{**}) \times B(Y^*)$. (E.g., $Q \in X^* \check{\otimes} Y$ or Q any compact linear operator.)

Theorem 7. *Let P_0 be a continuous homogeneous operator from a Banach space X into a Banach space Y. Let W be a subspace of X^*, V be a subspace of Y, and $\mathcal{D} = \overline{W \otimes V}$, where the closure is taken in $\mathcal{B}(X, Y)$. Then $P = P_0 - \Delta_0$ has minimal norm in $\mathcal{P} = P_0 + \mathcal{D}$ if the closed convex hull of $\{y \otimes x\}_{(x,y) \in \mathcal{E}(P)}$ contains an operator taking V into $W^\perp$, i.e., for some (total mass one) measure μ supported in $\mathcal{E}(P)$ ($\subset S(X^{**}) \times S(Y^*)$)*

$$E_P = \int_{\mathcal{E}(P)} y \otimes x \, d\mu(x, y) : V \to W^\perp, \tag{18}$$

or equivalently,

$$E_P^* = \int_{\mathcal{E}(P)} x \otimes y \, d\mu(x, y) : W \to V^\perp.$$

If P_0 is jointly weak continuous, then (18) is also necessary for P to be minimal.*

Example A. (Co-minimal (or minimal) projections (in the case V finite dimensional)) Let $Y = X$, take $P_0 = I - P_1$ (or $P = P_1$) where P_1 is a projection with (fixed) range V, and $W = V^\perp$.

$$E_P : V \to V. \tag{A}$$

As an example where P_0 is not necessarily linear, replace I above by a continuous proximity map onto V.

Example B. (Minimal extension operators (in the case V finite dimensional)) Let $V \subset S \subset X$, let P_0 be a fixed operator from S to V, and let $P : X \to V$ be a minimal-norm extension of P_0. In Theorem 7, let $Y = X$ and $W = S^\perp$.

$$E_P : V \to S. \tag{B}$$

Example C. (Optimal recovery (in the case U finite dimensional)) In Theorem 7, let $Y = X$, take $P_0 = I - P_1$ (or $P_0 = P_1$) where P_1 is a projection with (fixed) kernel $U^\perp$, $W = U$, $V = U^\perp$.

$$E_P^* : U \to U. \tag{C}$$

Corollary 6. (Geometric interpretation for optimal recovery in C) *Let $X^* = C(T)^* \supset U = [u]$. Then $P = \sum_{i=1}^n u_i \times v_i$ is minimal where*

$$v(t) = \|P\| z(t),$$

with $z(t)$ being a point of intersection of the unit $\|\cdot\|$-sphere in $\mathbf{F}^n$ ($\|a\| = \|a \cdot u\|_{C(T)^*}$) and its tangent plane perpendicular to $Mu(t)$ for some M.

Example D. (Linear optimal estimation (in the case U finite dimensional)) In Theorem 7, let P_0 be linear with (fixed) kernel $U^\perp$, $W = U$, $V = X \subset Y$.

$$E_P^* : U \to 0. \tag{D}$$

Example E. (Linear n-widths) Let P_0 be the injection of X into $Y \supset X$, let n be fixed (and finite) and let $\widetilde{\mathcal{D}} = \{\sum_{i=1}^n \delta_i \otimes \epsilon_i,\ \delta_i \otimes \epsilon_i \in X^* \otimes Y\}$. Now $\widetilde{\mathcal{D}}$ is not a subspace, but apply Theorem 7 to subspaces $\mathcal{D}_1 = X_n^* \otimes Y$ and $\mathcal{D}_2 = X^* \otimes Y_n$, where X_n^* and Y_n are n-dimensional subspaces of X^* and Y, respectively, and determine $P = P_1 = P_2$ from

$$E_{P_1}^* : X_n^* \to 0 \quad \text{and} \quad E_{P_2} : Y_n \to 0. \tag{E}$$

For some specific instances of Examples C, D, and E, and further references see [35], [40], and [41].

4 The Geometry and More Examples

For much of the following see [2]. For the sake of completeness of this section and the next we repeat some of the preliminaries mentioned in previous sections.

Let Z denote a Banach space and $\mathbf{F}$ denote the underlying real or complex field. If $z \in Z$ and $z^* \in Z^*$ are such that $\langle z, z^*\rangle = \|z\|\,\|z^*\| \neq 0$, then z^* is an *extremal* for z and we write $z^* = \operatorname{ext} z$ or $z^* = \operatorname{ext}(z)$. (Then also $z = \operatorname{ext} z^*$.) Note that any $\operatorname{ext} z$ is determined only up to a positive scalar factor. In the following we will adjust this factor for our convenience. For example if $Z = L^p$, $1 < p < \infty$, then $\operatorname{ext} z$ will denote $(\operatorname{sgn} z)|z|^{\frac{p}{q}}$ multiplied by any convenient positive scalar. ($\frac{1}{q} + \frac{1}{p} = 1$.)

Definition 3. If $u := u_1, \dots, u_n$ is an n-tuple of elements of Z then the notion of an extremal of u is defined as follows:

$$\operatorname{ext} u := \int \alpha \operatorname{ext}(\alpha \cdot u)\, d\mu(\alpha),$$

where μ is any nonnegative non-zero finite measure on $\mathbf{F}^n$. Futhermore, if $v := v_1, \dots, v_n$ is an n-tuple of elements of Z^*, the notion of a v-extremal of u, denoted

$$\operatorname{ext}_v(u),$$

will be any extremal of u such that the support of μ is contained in $\{\alpha \in S(V^*) : \|\alpha \cdot u\| = \text{maximum}\}$, where $V := [v]$. Here S denotes unit sphere and $S(V) := \{\beta \in \mathbf{F}^n : \|\beta \cdot v\| = 1\}$.

Note 7. If $w := w_1, \dots, w_n$ is an n-tuple of elements from Z^*, then, defining

$$\langle u, w\rangle := \sum_{i=1}^{n} \langle u_i, w_i\rangle,$$

we have, taking without loss $\|\operatorname{ext}(\alpha \cdot u)\| = 1$,

$$\langle u, \operatorname{ext} u\rangle = \int \langle \alpha \cdot u, \operatorname{ext}(\alpha \cdot u)\rangle\, d\mu(\alpha)$$

$$= \int \|\alpha \cdot u\|\, d\mu(\alpha).$$

Further, normalizing μ so that $\mu(S(V^*)) = 1$, we have

$$\langle u, \operatorname{ext}_v(u)\rangle = \max_{\alpha \in S(V^*)} \|\alpha \cdot u\|.$$

Thus it makes sense to define $\|u\|_v := \max_{\alpha \in S(V^*)} \|\alpha \cdot u\|$. (In the context of the operator $T = \sum_{i=1}^n u_i \otimes v_i$ defined below, $\|u\|_v = \max L(\alpha)$, where $L(\alpha) := \|\alpha \cdot u\|$ is the analogue of the classical Lebesgue function of T and thus $\|u\|_v = \|T\|$.)

Note 8. In the case $n = 1$, $\text{ext}_v(u) = \text{ext}\, u$, for all v, and the following theorem is therefore equivalent to the classical Hahn-Banach Extension Theorem. Indeed then (19) below becomes $\text{ext}\, u = v$ or, equivalently, $u = \text{ext}\, v$. For example in the case $X = L^p$, $1 < p < \infty$, then $\text{ext}\, u = (\text{sgn}\, u)|u|^{\frac{q}{p}} = v$ is equivalent to $u = (\text{sgn}\, v)|v|^{\frac{p}{q}} = \text{ext}\, v$.

The above discussion is reflected in Theorem 8 below. See also Note 9 below for further comment on the statement of Theorem 8.

Theorem 8. *Let $\tilde{P} = \sum_{i=1}^n \tilde{u}_i \otimes v_i : V \to V = [v_1, ..., v_n] \subset X$, where $\tilde{u}_i \in V^*$ and X is a Banach space. Let $P = \sum_{i=1}^n u_i \otimes v_i : X \to V$ be an extension of $\tilde{P}$ to all of X (i.e., $u_i \in X^*$) such that P has minimal (operator) norm. (E.g., if $\tilde{P} = I$, P is a minimal projection from X onto V.) Then it is necessary and sufficient that $u := u_1, ..., u_n$ is given by the formula ($v := v_1, ..., v_n$)*

$$\text{ext}_v(u) = Mv \tag{19}$$

for some $n \times n$ matrix M, where the notion of a v-extremal ("ext_v") of u is defined as follows:

$$\text{ext}_v(u) := \int \alpha\, \text{ext}\, (\alpha \cdot u)\, d\mu,$$

for μ some nonnegative non-zero finite measure on $\{\alpha \in S(V^*) : \|\alpha \cdot u\| = \|P\|\}$. ($L(\alpha) := \|\alpha \cdot u\|$ ($\leq \|P\|$) is the analogue of the classical Lebesgue function of P.)

Proof. The proof is a consequence of combining results from [19] as follows. P is a minimal-norm extension of $\tilde{P}$ means

$$\|P\| = \inf\{\|Q\| : Q : X \to V : Q|_V = \tilde{P}\}.$$

The extremal set $\mathcal{E}(P)$ of $P : X \to V$ is defined as follows:

$$\mathcal{E}(P) = \{(x, y) \in S(X^{**}) \times S(X^*) : y(P^{**}x) = \|P\|\},$$

where P^{**} denotes the second adjoint extension of P ($P^{**}x = \sum_{i=1}^n \langle u_i, x \rangle v_i$ and $\|P^{**}\| = \|P\|$). Then from [19] we have that P is a minimal-norm extension of $\tilde{P}$ if and only if there exists a probability measure μ on $\mathcal{E}(P)$ such that the operator

$$E_P = \int_{\mathcal{E}(P)} y \otimes x\, d\mu$$

maps V into V, where $y \otimes x : X \to X^{**}$ denotes the usual *dyad* operator given by $y \otimes x(z) = \langle z, y\rangle x$. I.e.,

$$\int_{\mathcal{E}(P)} \langle v, y\rangle\, x\, d\mu(x, y) = Mv,$$

for some $n \times n$ matrix M.

Furthermore, from [19], for any *extremal pair* $(x, y) \in \mathcal{E}(P)$, $\langle x, u\rangle / \|P\|$ is a point of intersection of the unit sphere $S(V)$ and its tangent plane perpendicular (in the ordinary Euclidean sense) to the direction of $\langle v, y\rangle$. (See the figure below.)

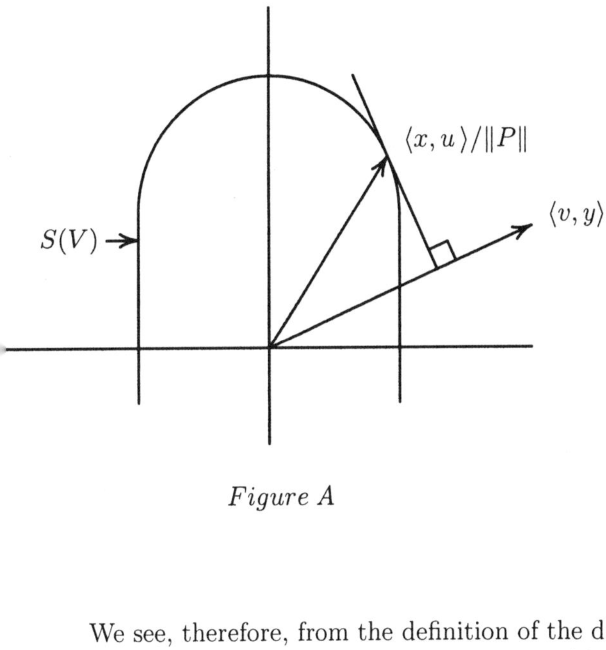

Figure A

We see, therefore, from the definition of the dual sphere $S(V^*)$, that $\alpha := \langle v, y\rangle \in S(V^*)$ is a *norming point* or *extremal* for $\langle x, u\rangle$. Thus the theorem follows by noting additionally that $\langle x, u\rangle \cdot \langle v, y\rangle = \|P\|$ implies that $x = \operatorname{ext}(\alpha \cdot u)$. $\square$

Note 9. Letting $U := [u]$, we see from Theorem 8 above that $\|P\| S(U) \cap S(V^*) = \{\alpha \in S(V^*) : \|\alpha \cdot u\| = \|P\|\}$ must be large enough so that formula (1) is possible. (See also [22].) I.e., we can interpret Theorem 8 to imply in particular that in order for P to be minimal, u must be such that a large enough part of $\|P\| S(U)$ coincides with $S(V^*)$.

(Finally, note that the Lebesgue function $L(\alpha)$ defined on $S(V^*)$ and mentioned parenthetically in Theorem 8 agrees with the classical Lebesgue function not only in the case $X = L^\infty$ ($\alpha = \langle v, y\rangle$, where $y = \delta_t$ (point-evaluation at t)) but also in the case $X = L^1$ ($\alpha = \langle v, y\rangle$, where $y = \operatorname{sgn}(u(t)\cdot v)$.)

The formula (19) above leads in many important cases to a simple geometric interpretation of minimal projections. Furthermore, by applying this formula to the case $X = L^p$, we obtain a *linear* n-dimensional analogue of the Hölder equality condition:

Corollary 7. *Let $X = L^p(T)$, $1 \le p < \infty$ or $X = C(T)$, $p = \infty$. P is a minimal-norm extension of $\tilde{P}$ implies the following* **linear** *equation holds for u:*

$$\frac{1}{p}\, u' \cdot Mv = \frac{1}{q}\, u \cdot Mv', \tag{$\star$}$$

wherever v is differentiable. (M is the matrix in (1), "$\,'\,$" denotes differentiation along an arbitrary vector field in T and $1/q + 1/p = 1$).

Proof. We will prove Corollary 7 in the case that u and v are simultaneously differentiable. The more general case is proved in [5]. From Theorem 8, P is minimal implies $\operatorname{ext} u = Mv$ for some $\operatorname{ext} u$, i.e.,

$$\int_{\mathcal{E}(P)} \alpha \operatorname{ext}(\alpha\cdot u)\, d\mu(\alpha) = Mv.$$

For ease of notation let $f(\alpha\cdot u) := \operatorname{ext}\, \alpha\cdot u$. Then we have

$$\int \alpha\, f(\alpha\cdot u(t))\, d\mu = Mv(t). \tag{20}$$

Next, "dot" both sides of the above equation with $u(t)$ to obtain

$$\int \alpha\cdot u(t)\, f(\alpha\cdot u(t))\, d\mu = u(t)\cdot Mv(t). \tag{21}$$

Assuming $g(r) := rf(r)$ is differentiable with respect to r and differentiating both sides of the equation (21) with respect to t, we have (by the chain rule and the easily verified fact that the differentiation in the left-hand side of the above equation can be taken inside the integral)

$$\int g'(\alpha\cdot u(t))\, \alpha\cdot u'(t)\, d\mu = u(t)\cdot Mv'(t) + u'(t)\cdot Mv(t). \tag{22}$$

But now, let $X = L^p$, for $1 < p < \infty$; then

$$f(r) = (\operatorname{sgn} r)|r|^{\frac{q}{p}} \text{ and } g'(r) = (rf(r))' = (\frac{q}{p}+1)(\operatorname{sgn} r)|r|^{\frac{q}{p}} = (\frac{q}{p}+1)f(r).$$

Thus equation (22) becomes

$$(\frac{q}{p}+1)\int f(\alpha\cdot u(t))\,\alpha\cdot u'(t)\,d\mu = u(t)\cdot Mv'(t)+u'(t)\cdot Mv(t). \tag{23}$$

Next, "factor out" $u'(t)$ from the left-hand side of (23) (and shift left the associated α in the integrand):

$$(\frac{q}{p}+1)u'(t)\cdot\int \alpha\, f(\alpha\cdot u(t))\,d\mu = u(t)\cdot Mv'(t)+u'(t)\cdot Mv(t).$$

Thus, we have by use of (20) that

$$(\frac{q}{p}+1)u'\cdot Mv = u\cdot Mv'+u'\cdot Mv,$$

i.e.,

$$\frac{1}{p}u'\cdot Mv = \frac{1}{q}u\cdot Mv',$$

for all $t\in T$ where u and v are simultaneously differentiable.

We obtain the result for $p=1,\infty$ either by a limiting process or by referring to [3] and [18]. □

As a prime example of the use of ($\star$) we cite the following important lemma from [7].

Notation 5. In the following we will write $(w)^r$ for $(\operatorname{sgn} w)|w|^r$.

Lemma 1. (Use of ($\star$)-Equation.) *For $1\le p<\infty$, let $P=\sum_{i=1}^{2}u_i\otimes v_i$ be the minimal projection from $L^p[-1,1]$ onto $V=[v_1,v_2]$, where $(v_1,v_2)=(1,t)$. Then*

$$u_2(t)=\kappa(t)^{p-1}\Big[-\frac{1}{m}\int^t\frac{u_1'(s)}{|s|^p}\,ds\;+\;b\Big], \tag{24}$$

where κ, m and b are constants.

Proof. Formula (24) follows by solving the following linear first-order differential equation ($\star$) for u_2 in terms of u_1 and the given $\vec{v}=(1,t)$:

$$\frac{1}{p}\vec{u}'\cdot M\vec{v}=\frac{1}{q}\vec{u}\cdot M\vec{v}'\quad\text{on } T, \tag{$\star$}$$

where M is the matrix in (10) and $1/q+1/p=1$. The fact that $M=\operatorname{diag}(1,m)$ follows from the symmetry. □

5 Minimal L^1 Projections onto π_n

The following theory and examples are taken from [17]. In [18] are derived sufficient and necessary (assuming the subspace is "smooth") equations for finite rank L^1 projections to be minimal (Theorem 9 below). (As mentioned previously, the existence of a minimal projection in this setting is proved in [44].) As an application of these conditions in [18] is obtained a sufficient condition, labeled "Prescription," for determining a minimal projection from $L^1[-1,1]$ onto a finite-dimensional subspace V. We show below that the Prescription is in fact also necessary whenever V is a Haar space, the projection (identity) action on V is generalized to an arbitrary non-singular action on V, and $L^1[-1,1]$ is generalized to $L^1([-1,1],\nu)$, where ν is an arbitrary finite nonatomic Borel measure on $[-1,1]$. (Recall that V is an n-dimensional Haar space on $[a,b]$ means that the elements of V are continuous functions on $[a,b]$ and any nonzero $v \in V$ has no more than $n-1$ zeros in $[a,b]$, or, equivalently, that any n distinct point-evaluation functionals (supported in $[a,b]$) are independent on V.) The proof is based on an application of the classical Hobby-Rice theorem, a theorem of fundamental importance in the the theory of best approximation in the L^1-norm (cf. [32]). As applications, we will use the Prescription to find numerical solutions for the minimal projections from $L^1[-1,1]$ onto $V = \pi_{n-1}$, the space of $(n-1)$-degree algebraic polynomials, $n-1 \leq 5$.

More generally, for (T,Σ,ν) a complete measure space, let $P = \Sigma u_i \otimes v_i$ be a linear operator from $L^1(T,\nu))$ onto V with $u_i \in L^\infty$ and $v_i \in V$, satisfying

$$\int_T v_i(t)u_j(t)\,d\nu(t) = a_{ij} \qquad (i,j = 1,\dots,n),$$

with the matrix $A = (a_{ij})$ fixed, but non-zero.

Additional notation which will be used in the following is

$$y = (y_1,\dots,y_n),\ y \cdot z = y_1 z_1 + \dots + y_n z_n,\ |y\,| = (y \cdot y)^{1/2},\ u \odot v = \sum u_i \otimes v_i,$$

and

$$V_i(t) = \int_T v_i(s)\operatorname{sgn} k(t,s) d\nu(s) \quad \text{where} \quad k(t,s) = \sum_{j=1}^n u_j(t)v_j(s).$$

Also, the Lebesgue function $L(t)$ for P is given by

$$L(t) = \int_T |k(t,s)|\,d\nu(s) = u\,(t) \cdot V(t), \quad tinT;$$

note that $\|P\| = \text{ess sup } L(t)$; see [36], Lemma 2. Throughout this section the notation $t \in T' \subseteq T$ will mean for almost all t out of T' relative to the measure ν.

In the following all the statements and results will refer to the operator P (and the associated "action" matrix A). Note that if P is a projection then A is the identity matrix.

The following theorem proved in [18] provides necessary and sufficient (equality) conditions for the operator P to be minimal.

Theorem 9. *Let (T, Σ, ν) be a complete measure space for which ν is strictly localizable. Let V be a finite-dimensional subspace of L^1, and let $P = u \odot v = \sum_{i=1}^{n} u_i \otimes v_i$ be an operator mapping L^1 into V with $u_i \in L^\infty$ $(i = 1, \ldots, n)$, $v = (v_1, \ldots, v_n)$ a fixed basis for V, and the matrix $A = \int_T v_i(t) u_j(t)\, d\nu(t) = a_{ij}$ $(i, j = 1, \ldots, n)$ fixed. In order that P be minimal, the following equality conditions are necessary and sufficient:*
There exists a non-zero $n \times n$ matrix M such that (a) the Lebesgue function $L(t) = \|P\|$ on $T' = \text{supp}\,(Mv)$, and

(b) there exists a positive function ϕ such that

$$\phi(t)\, V(t) = M\, v\,(t), \qquad t \in T'. \tag{25}$$

(In fact, $\phi = u \cdot Mv\, / \|P\|$.)

Notation 6. We will denote by $P_{\min}$ a minimal operator given by Theorem 9.

Recall that, in an L^1 space, the subspace V is said to be *smooth* if and only if each member of $V \backslash 0$ is almost everywhere different from 0.

Theorem 10. ([18]) *If V is smooth, then the Lebesgue function for $P_{\min}$ in Theorem A is constant on T.*

Theorem 11. ([18]) *$P_{\min}$ is unique if V is smooth and $u\,(t) \cdot v$ is determined up to a scalar factor by its roots.*

Lemma 2. (Hobby-Rice Theorem [38]) *Let V be an n-dimensional subspace of $L^1([-1, 1],\ \nu)$, where ν is a finite nonatomic Borel measure. Then there exist points $-1 = t_0 \leq t_1 \leq \cdots \leq t_{n+1} = 1$ such that*

$$\sum_{i=0}^{n} (-1)^i \int_{t_i}^{t_{i+1}} v(t) d\nu(t) = 0, \quad \forall\, v \in V. \tag{26}$$

Theorem 12. (Prescription) *Let V be an n-dimensional Haar subspace of $L^1([-1, 1], \nu)$, where ν is a finite nonatomic Borel measure. Then $P_{\min} =$*

$u \odot v = \sum_{i=1}^{n} u_i \otimes v_i$ from $L^1([-1,1],\nu)$ into V with respect to the fixed action A is given uniquely by the following prescription:

$$\begin{pmatrix} u_1(t) \\ u_2(t) \\ \cdot \\ \cdot \\ \cdot \\ u_n(t) \end{pmatrix} = \begin{pmatrix} \hat{V}_1(x(t)) & \hat{V}_2(x(t)) & \cdot\ \cdot\ \cdot & \hat{V}_n(x(t)) \\ v_1(x_1(t)) & v_2(x_1(t)) & \cdot\ \cdot\ \cdot & v_n(x_1(t)) \\ \cdot & \cdot & \cdot\ \cdot\ \cdot & \cdot \\ \cdot & \cdot & \cdot\ \cdot\ \cdot & \cdot \\ \cdot & \cdot & \cdot\ \cdot\ \cdot & \cdot \\ v_1(x_{n-1}(t)) & v_2(x_{n-1}(t)) & \cdot\ \cdot\ \cdot & v_n(x_{n-1}(t)) \end{pmatrix}^{-1} \begin{pmatrix} \lambda\sigma(t) \\ 0 \\ \cdot \\ \cdot \\ \cdot \\ 0 \end{pmatrix}, \tag{27}$$

where

$$\hat{V}_i(x) := \left(\int_{-1}^{x_1} - \int_{x_1}^{x_2} + \cdots + (-1)^{n-1} \int_{x_{n-1}}^{1} \right) v_i(s)\, d\nu(s), \quad i = 1, \ldots, n, \tag{28}$$

and $x(t)$ are solutions to

$$m_{i+1}(t)\hat{V}_i(x(t)) - m_i(t)\hat{V}_{i+1}(x(t)) = 0, \quad i = 1, \ldots, n-1, \tag{29}$$

where $m_i(t) := m_i \cdot v(t), \quad i = 1, \ldots, n$, for some $n \times n$ matrix

$$M := \begin{pmatrix} m_1 \\ \cdot \\ \cdot \\ \cdot \\ m_n \end{pmatrix}, \tag{30}$$

with

$$\sigma(t) := \operatorname{sgn}\left[\frac{\widehat{V}_n(x(t))}{m_n \cdot v(t)} \right] \tag{31}$$

and

$$\lambda := \|P\|. \tag{32}$$

Note: The n^2 parameters m_{ij} (normalized so that one of $m_{ij} = 1$) and the norm parameter λ are determined from the n^2 orthonormalization conditions

$$\int_{-1}^{+1} v_i(t) u_j(t)\, d\nu(t) = a_{ij} \qquad (i,j = 1, \ldots, n). \tag{33}$$

Proof. Keeping in mind equation (b) of Theorem 9, for a given matrix M with non-zero rows m_i $(i = 1, \ldots, n)$, we would like to solve

$$\frac{\widehat{V}_i(x(t))}{m_i \cdot v(t)} = \frac{\widehat{V}_n(x(t))}{m_n \cdot v(t)}, \quad i = 1, \ldots, n-1, \tag{34}$$

for $x(t) = (x_1(t), \ldots, x_{n-1}(t))$, with $-1 < x_1(t) < \cdots < x_{n-1}(t) < 1$ for each $t \in (-1, 1)$.

Note that (34) can be rewritten as

$$m_{i+1} \cdot v(t)\widehat{V}_i(x(t)) - m_i \cdot v(t)\widehat{V}_{i+1}(x(t)) =$$

$$\left(\int_{-1}^{x_1} - \int_{x_1}^{x_2} + \cdots + (-1)^{n-1}\int_{x_{n-1}}^{1}\right)\left[m_{i+1}(t)v_i(s) - m_i(t)v_{i+1}(s)\right] d\nu(s),$$

$$i = 1, \ldots, n-1,$$

where $m_i^t := m_i \cdot v(t)$, $i = 1, \ldots, n$. The existence of $x_i = x_i(t)$, $i = 1, \ldots, n-1$, now follows directly from Lemma [38] (the Hobby-Rice Theorem).

Next note that the invertibility of the $n \times n$ matrix in (3) is equivalent to the $n-1$ point-evaluations δ_{x_i} $(i = 1, ..., n-1)$ and the functional

$$\beta(\cdot) := \left(\int_{-1}^{x_1} - \int_{x_1}^{x_2} + \cdots + (-1)^{n-1}\int_{x_{n-1}}^{1}\right)(\cdot)(s)\, d\nu(s)$$

being independent on V. But, indeed if $0 \neq v \in V$ such that $v(x_i) = 0$, $i = 1, ..., n-1$, then, by the Haar assumption, for $x_0 := -1$ and $x_n := 1$, we have $\operatorname{sgn}(v) = \epsilon\,(-1)^i$ on (x_{i-1}, x_i), $i = 1, ..., n$, where $\epsilon = -\operatorname{sgn}(v(-1))$, and thus $|\beta(v)| > 0$, establishing the independence.

Note that $\sigma(t)\hat{V}_i(x(t))$ will be $V_i(t)$ of Theorem 9, $i = 1, \ldots, n$. The homogeneity of the system (10) allows one of the m_{ij} to be normalized to be 1, leaving exactly n^2 equations in n^2 unknowns. Finally, by the Haar condition, for each $t \in (-1, +1)$, $k(t, s) = u(t) \cdot v(s)$, as constructed according to the abov e prescription, changes sign only at $s = x_i(t)$ $(i = 1, \ldots, n-1)$ and thus $\lambda = u(t) \cdot V(x(t)) = \int_T |k(t, s)|\, d\nu(s) = L(t) > 0$, where $L(t)$ is the Lebesgue function for P. Thus Theorem 10 guarantees that we have $P_{\min}$.

Finally, by Theorem 11, $P_{\min}$ is unique. □

Remark 3. System (27) is equivalent to

$$\begin{pmatrix} v_1(x_1) & \cdots & v_{n-1}(x_1) \\ \vdots & & \vdots \\ v_1(x_{n-1}) & \cdots & v_{n-1}(x_{n-1}) \end{pmatrix} \begin{pmatrix} \psi_1 \\ \vdots \\ \psi_{n-1} \end{pmatrix} = \begin{pmatrix} -v_n(x_1) \\ \vdots \\ -v_n(x_{n-1}) \end{pmatrix}, \tag{35}$$

where $\psi_i = u_i/u_n$ ($i = 1, \ldots, n-1$), and

$$u_n = \frac{\lambda\sigma}{\widehat{V}_1(x)\psi_1 + \cdots + \widehat{V}_{n-1}(x)\psi_{n-1} + \widehat{V}_n(x)} . \tag{36}$$

Thus, if $V_{n-1} := [v_1, ..., v_{n-1}]$ is also Haar, then the $(n-1)\times(n-1)$ matrix in (35) above is invertible. In particular, in the algebraic case, $V_{n-1} := [1, t, \ldots, t^{n-2}]$ and the $(n-1) \times (n-1)$ matrix above is in fact the classical (invertible) Vandermonde matrix.

3. Applications. All the applications in this section are directed towards the determination of the minimal proj ection from $L^1[-1,1]$ onto π_{n-1} (i.e., the action $A = I$, the measure ν is standard Lebesgue measure, and $V = [1, t, \ldots, t^{n-1}]$).

The first two applications are repeated from [18] for the sake of completeness and as an aid to the reader to identify the various parts of the Prescription.

In Application 1 (n=2), for each M, the single function $x(t)$, being the root of a quadr atic, is determined explicitly, and then M :=diag$(1, m)$ and λ are determined to meet the (two) remaining (after symmetry) orthonormality conditions (via a numerical method).

In Application 2 (n=3), for each M, the two functions $x_i(t)$, $i = 1, 2$, are also determ ined explicitly (in terms of the solution of a quartic), and then M and λ are determined t o meet the (five) remaining (after symmetry) orthonormality conditions (via a numerical method).

In Applications n-1 ($4 \leq n \leq 6$), for each M, determining the $n-1$ functions $x_i(t)$, $i = 1, ..., n-1$, involves solving (for each t) an $(n-1) \times (n-1)$ system of polynomial equations of degree n and thus must be determined entirely numerically (e.g. by Newton's method). M and λ are then determined to meet the remaining (after symmetry) orthonormality conditions (via a numerical method).

In all cases we give the M matrix (up to 4 decimal places) and the projection norm λ (up to 5 decimal places).

Application 1. $V = [1, t]_{[-1,1]}$ (Franchetti-Cheney [36])

Consider the process described by the above prescription. In this case $n = 2$, $\widehat{V}_1(x) = 2x$ and $\widehat{V}_2(x) = x^2 - 1$. Using symmetry considerations, equation (5) becomes

$$\frac{2x(t)}{1} = \frac{x^2(t) - 1}{mt}, \tag{37}$$

from which the admissible solution is $x(t) = mt - \text{sgn}(mt)\sqrt{m^2t^2 + 1}$. Equations (35), (36) and (31) are then

$$\psi(t) = -x(t) \qquad\qquad (= u_1(t)/u_2(t)\,) \tag{38}$$

$$u_2(t) = \frac{\lambda\sigma(t)}{2x(t)\psi(t) + x^2(t) - 1} \tag{39}$$

$$\sigma(t) = \operatorname{sgn} x(t), \tag{40}$$

which give

$$u_1(t) = \frac{\lambda|x(t)|}{1 + x^2(t)}$$

$$u_2(t) = -\frac{\lambda \operatorname{sgn} x(t)}{1 + x^2(t)}.$$

Using the symmetry of u_1 and u_2, equations (33) become

$$\int_0^1 u_1(t)\,dt = \int_0^1 tu_2(t)\,dt = \frac{1}{2}, \tag{41}$$

which are identical to equations (32) and (34) of [36], and result in the equation

$$2\psi(1)[1 - \psi^2(1) + \psi(1)] \log|\psi(1)| + 1 - \psi^2(1) = 0 \tag{42}$$

for $\psi(1)$, and hence m. It then follows that $\lambda > 0$ and Theorem 12 guarantees we have $P_{\min}$. In this case

$$M = \begin{pmatrix} 1 & 0 \\ 0 & 1.3605 \end{pmatrix}$$

and

$$P_{\min} = \lambda = 1.22040\ldots\,.$$

Application 2. $V = [1, t, t^2]_{[-1,1]}$ (Chalmers-Metcalf [18])

In this case $n = 3$ and

$$\widehat{V}_1(x_1, x_2) = 2(x_1 - x_2 + 1), \quad \widehat{V}_2(x_1, x_2) = x_1^2 - x_2^2,$$
$$\widehat{V}_3(x_1, x_2) = \frac{2}{3}(1 + x_1^3 - x_2^3).$$

Using symmetry considerations, equations (29) become

$$\frac{2[x_1(t) - x_2(t) + 1]}{m_{11} + m_{13}t^2} = \frac{x_1^2(t) - x_2^2(t)}{t} = \frac{\frac{2}{3}[x_1^3(t) - x_2^3(t) + 1]}{m_{31} + m_{33}t^2}. \tag{43}$$

Letting

$$\gamma_1 = \frac{m_{11} + m_{13}t^2}{2t} \qquad \text{and} \qquad \gamma_3 = \frac{m_{31} + m_{33}t^2}{\frac{2}{3}t},$$

equations (43) may be rewritten

$$x_1 - x_2 + 1 = \gamma_1(x_1^2 - x_2^2), \qquad x_1^3 - x_2^3 + 1 = \gamma_3(x_1^2 - x_2^2). \tag{44}$$

Introducing the variables $y_1 = x_1 - x_2$ and $y_2 = x_1 + x_2$ leads to

$$y_1 + 1 = \gamma_1 y_1 y_2 \qquad \text{and} \qquad y_1 \frac{3y_2^2 + y_1^2}{4} + 1 = \gamma_3 y_1 y_2,$$

which reduce to the single quartic equation for y_2:

$$(\gamma_1 y_2 - 1)^2[3y_2^2 + 4(\gamma_1 - \gamma_3)y_2 - 4] + 1 = 0.$$

This equation is then solved yielding admissible x_1 and x_2 $(-1 \le x_1(t) < x_2(t) \le 1)$. The function $\sigma(t)$ (in (31)) is -1 for $|t| < t_0$ and $+1$ for $t_0 < |t|$, where $\pm t_0$ are points where the admissible solutions of the quartic equation switch from one pair of roots to another. The values of λ, m_{11}, m_{13}, m_{31}, and m_{33} are determined from the five non-trivial orthonormality conditions

$$1 = \int_{-1}^{1} u_1(t)\,dt = \int_{-1}^{1} t^2 u_3(t)\,dt = \int_{-1}^{1} t u_2(t)\,dt$$

$$0 = \int_{-1}^{1} t^2 u_1(t)\,dt = \int_{-1}^{1} u_3(t)\,dt. \tag{45}$$

The solution of these equations (for example, by the iteration method of §3 in [18] yields $t_0 = 0.45710\ldots$ and the values of λ and M given below. Hence, the $x_i(t)$ are specified, and $P_{\min} = u \odot v$, where $u\,(t)$ is given by

$$\begin{pmatrix} \hat{V}_1(x(t)) & \hat{V}_2(x(t)) & \hat{V}_3(x(t)) \\ 1 & x_1(t) & x_1^2(t) \\ 1 & x_2(t) & x_2^2(t) \end{pmatrix} \begin{pmatrix} u_1(t) \\ u_2(t) \\ u_3(t) \end{pmatrix} = \begin{pmatrix} \lambda\sigma(t) \\ 0 \\ 0 \end{pmatrix},$$

or

$$\begin{pmatrix} u_1(t) \\ u_2(t) \\ u_3(t) \end{pmatrix} = \frac{3\lambda \operatorname{sgn}(t y_2)}{y_1^3 - 2 - 3(y_1^2 - y_2^2)/2} \begin{pmatrix} (y_2^2 - y_1^2)/4 \\ -y_2 \\ 1 \end{pmatrix},$$

and $\|P_{\min}\| = \lambda$.

$$M = \begin{pmatrix} 1 & 0 & -.6675 \\ 0 & .9336 & 0 \\ -.1552 & 0 & 1.2711 \end{pmatrix}$$

and

$$P_{\min} = \lambda = 1.359484\dots .$$

Application 3. $V = [1, t, t^2, t^3]_{[-1,1]}$

$$M = \begin{pmatrix} 1 & 0 & -.4726 & 0 \\ 0 & 1.4760 & 0 & -1.1095 \\ -.0797 & 0 & 1.0520 & 0 \\ 0 & -.2648 & 0 & 1.3017 \end{pmatrix}$$

and

$$P_{\min} = \lambda = 1.46184\dots .$$

Application 4. $V = [1, t, t^2, t^3, t^4]_{[-1,1]}$

$$M = \begin{pmatrix} 1 & 0 & .2767 & 0 & -.9250 \\ 0 & 1.2925 & 0 & -.7046 & 0 \\ -.1590 & 0 & 2.3605 & 0 & -1.8019 \\ 0 & -.1523 & 0 & 1.1257 & 0 \\ -0.0126 & 0 & -.1178 & 0 & 1.0642 \end{pmatrix}$$

and

$$P_{\min} = \lambda = 1.54874\dots .$$

Application 5. $V = [1, t, t^2, t^3, t^4, t^5]_{[-1,1]}$

$$M = \begin{pmatrix} 1 & 0 & -.0466 & 0 & -.4955 & 0 \\ 0 & .8893 & 0 & .9039 & 0 & -1.4609 \\ -.0781 & 0 & 1.6151 & 0 & -.9900 & 0 \\ 0 & -.5593 & 0 & 3.5525 & 0 & -2.6180 \\ -.0743 & 0 & -.0675 & 0 & .9124 & 0 \\ 0 & -.1769 & 0 & .6439 & 0 & .2685 \end{pmatrix}$$

and

$$P_{\min} = \lambda = 1.61031\dots .$$

6 Further Applications

Formula (19) and the attendant Figure A and formula ($\star$) lead in many important cases to a simple geometric interpretation of minimal extension operators (e.g., minimal projections). In particular (see [18] for precise definitions), if $X = L^1(T)$ and V is *smooth*, it follows (since the extreme points of $S((L^1)^{**})$

are *essentially* independent) that $\langle x,\, u\rangle$ and $\langle v,\, y\rangle$ can be replaced by $u(t)$ and $Mv(t)$, respectively. (Note that then ($\star$) follows immediately since Figure A implies $u'(t) \perp Mv(t)$). See [14], [18] and [17] for applications. Furthermore, since any 2-dimensional real space is isometric to an L^1-space (see e.g., [43] and [50]) which is also a *maximal overspace* in this context ([43]; see also [50]) and since any $L^p - space, 1 < p \leq 2$, is isometric to an L^1-space (see [34]), the L^1-theory mentioned above plays an important role in a number of results in [8], [29], [30], [31], [9], [10], [12], [11], and [6]. (In retrospect, because of the essential independence of the extreme points of $S((L^1)^{**})$, it is no accident that the first *structurally non-trivial* example of a minimal projection (from $L^1[-1,1]$ onto the lines) was determined in the seminal paper [36].) More generally, in addition to the L^1-applications mentioned above, we determine minimal projections from $L^p[-1,1]$ onto π_{n-1} (the $(n-1)$-st degree algebraic polynomials), $1 \leq p \leq \infty$ (see e.g., [7], [20] and [21]).

As a final example we indicate in the following how the above theory leads to an explicit formula (see [6]) for the projection constant of ℓ_n^p. Indeed, letting $\omega = (\omega_1, ..., \omega_n)$ denote the coordinate functions of S_{n-1}, the $(n-1)$-dimensional euclidean sphere, we see that ℓ_n^p is isometric to $[\omega^{\frac{2}{q}}]$, where $\omega^{\frac{2}{q}} := ((\omega_i)^{\frac{2}{q}})_{i=1}^n$, the space of coordinate functions of $S(\ell_n^q)$ regarded as a subspace of $L^\infty(S_{n-1})$. (Here we use the notation $(a)^{\frac{2}{q}} := (\operatorname{sgn} a)|a|^{\frac{2}{q}}$.) Next we apply the $(p = \infty)$-version of formula ($\star$), i.e.

$$u \cdot Mv' = 0, \tag{46}$$

where $v = \omega^{\frac{2}{q}}$ to conclude that

$$u(s) = r(s)\omega(s)^{\frac{2}{p}}. \tag{47}$$

Indeed, $\omega \cdot \omega = 1$ implies $\omega \cdot \omega' = 0$, whence $v' = \frac{2}{q}((\omega_i)^{\frac{2}{q}-1}\omega_i')_{i=1}^n$ implies that $u(s) = r(s)\omega(s)^{\frac{2}{p}}$, since $u \cdot v' = \frac{2}{q} r\, \omega \cdot \omega' = 0$, satisfying (46).

Thus we see that (47) determines the n-tuple $u(s)$ up to a scalar multiple $r(s)$ for each $s \in S_{n-1}$. Finally, by further use of (1), we determine in [6] that

$$1/r(s) = \kappa_p \Big[\Big(\prod_{i=1}^n |\omega_i(s)|\Big)^{2-q} \sum_{i=1}^n |\omega_i(s)|^{2q-2}\Big]^{1+\frac{1}{p}},$$

for some constant κ_p.

References

1. Bacopoulos, A. and Chalmers, B. L., Vectorially minimal projections, in *Approximation Theory VIII*, C. K. Chui and L. Schumaker eds., Academic Press, New York, 1995, 15–22.

2. Chalmers, B. L., An n-dimensional Hahn-Banach extension theorem and mimimal projections, submitted.

3. Chalmers, B. L., The ($\star$)-equation and the form of the minimal projection operator, in *Approximation Theory IV*, C. K. Chui, et al, eds., Academic Press, New York, 1983, 393–399.

4. Chalmers, B. L., Absolute projection constant of the linear functions in a Lebesgue space, Constr. Approx., **4**(1988), 107–110.

5. Chalmers, B. L., The n-dimensional Hölder inequality, submitted.

6. Chalmers, B. L., The absolute projection constant of ℓ_n^p, in preparation.

7. Chalmers, B. L. and Franchetti, C., The determination of a minimal L^p projection onto the lines, in preparation.

8. Chalmers, B., Franchetti, C., and Giaquinta, M., On the self-length of two-dimensional Banach spaces, Bull. Austr. Mat. Soc., **53**(1996), 101–107.

9. Chalmers, B. L. and Lewicki, G., Minimal projections onto some subspaces of $\ell_1^{(n)}$, Functiones ed Approximatio, **26**(1998), 85–92.

10. Chalmers, B. L. and Lewicki, G., Two-dimensional real symmetric spaces with maximal projection constants, submitted.

11. Chalmers, B. L. and Lewicki, G., Symmetric spaces with maximal projection constants, in preparation.

12. Chalmers, B. L. and Lewicki, G., Minimal projections onto symmetric spaces with large projection constants, Studia Math., **134** **(2)**(1999), 119–133.

13. Chalmers, B. L., Leviatan, D., and Prophet, M. P., The Bernstein Operator is the Closest Shape-Preserving Operator to a Projection, in *Approximation Theory IX (Volume 1)*, C. K. Chui and L. Schumaker eds., Academic Press, New York, 1998, 75–82.

14. Chalmers, B. L., Leviatan, D., and Prophet, M. P., Optimal interpolating spaces preserving shape, J. Approx. Thy., **98**(1999), 354–373.

15. Chalmers, B. L. and Metcalf, F. T., A simple formula showing L^1 is a maximal overspace for two-dimensional real spaces, Ann. Polonici Math. **56**(1992), 303–309.

16. Chalmers, B. L. and Metcalf, F. T., Minimal projections and extensions for compact abelian groups, in *Approximation Theory VI*, C. K. Chui, et al, eds., Academic Press, New York, 1989, 129–132.

17. Chalmers, B. L. and Metcalf, F. T., The minimal projection from L^1 onto π_n, in *"Stochastic Processes and Functional Analysis*, Goldstein, Gretsky, Uhl eds., Marcel Dekker, Inc., New York, 1996, 61–69.

18. Chalmers, B. L. and Metcalf, F. T., The determination of minimal projections and extensions in L^1, Trans. Amer. Math. Soc., **329**(1992), 289–305.

19. Chalmers, B. L. and Metcalf, F. T., A characterization and equations for minimal projections and extensions, J. Operator Theory, **32**(1994), 31–46.

20. Chalmers, B. L. and Metcalf, F. T., Determination of a minimal projection from $C[-1,1]$ onto the quadratics, Num. Functional Anal. and Optimization, **11**(1990), 1–10.

21. Chalmers, B. L. and Metcalf, F. T., Determination of a minimal projection from $C[-1,1]$ onto π_n, in preparation.

22. Chalmers, B. L. and Metcalf, F. T., Construction of minimal projections, in *Approximation Theory VIII*, C. K. Chui and L. Schumaker eds., Academic Press, New York, 1995, 119–127.

23. Chalmers, B. L. and Pan, K. C., Finite dimensional action constants, in *Approximation Theory IX*, C. K. Chui and L. Schumaker eds., Academic Press, New York, 1998, 83–88.

24. Chalmers, B. L., Pan, K. C., and Shekhtman, B., When is the adjoint of a minimal projection also minimal, Proc. of Memphis Conf., Lect. Notes in Pure and Applied Math, **138**(1991), 217–226.

25. Chalmers, B. L. and Prophet, M. P., Minimal shape-preserving projections onto Π_n, Num. Funct. Anal. and Optimization, **18**(1997), 507–519.

26. Chalmers, B. L. and Shekhtman, B., Minimal projections and absolute projection constants for regular polyhedral spaces, Proc. Am. Math. Soc., **95**(1985), 449–452.

27. Chalmers, B. L. and Shekhtman, B., Some estimates of action constants and related parameters, Computers and Mathematics with Applications, 1997.

28. Chalmers, B. L. and Shekhtman, B., Actions that characterize $\ell_\infty^{(n)}$, Lin. Alg. and Appl., **270**(1998), 155–169.

29. Chalmers, B. L. and Shekhtman, B., A two-dimensional Hahn-Banach theorem, submitted.

30. Chalmers, B. and Shekhtman, B., Extension constants of unconditional two-dimensional operators, Lin. Alg. and Appl., **240**(1996), 173–182.

31. Chalmers, B. L. and Shekhtman, B., Actions that characterize $\ell_\infty^{(n)}$, Lin. Alg. and Appl., **270**(1998), 155–169.

32. Cheney, E. W., Applications of fixed-point theorems to approximation theory, in *Theory of Approx. with Appl.*, Academic Press Inc., New York, (1976), 1–8.

33. Minimal projections, in *Approximation Theory*, A. Talbot, ed., Academic Press, London, 1970, 261-289.

34. L. E. Dor, L. E., Potentials and isometric embeddings in L_1, Israel J. Math. **24**(1976), 260–268.

35. Fisher, S. D., *Function Theory on Planar Domains*, John Wiley & Sons, New York, 1983.

36. Franchetti, C. and Cheney, E. W., Minimal projections in L^1-space, Duke Math. J. **43**(1976), 501–510.

37. Franchetti, C. and Votruba, G., Perimeter, Macphail number and projection constant in Minkowski planes, Boll. Un. Mat. Ital. B(6), **13**(1976), 560–573.

38. Hobby, C. R. and Rice, J. R., A moment problem in L^1 approximation, Proc. Amer. Math. Soc. **16**(1965), 665–670.

39. König, K. and Tomczak-Jaegermann, N., Norms of minimal projections, J. Funct. Anal. **119**(1994), 253–280.

40. Micchelli, C. A. and Rivlin, T. J., eds. *Optimal Estimation in Approximation Theory*, Plenum, New York, 1977.

41. Micchelli, C. A. and Rivlin, T. J., Lectures on optimal recovery, in *Numerical Analysis, Lancaster 1984*, P. R. Turner, ed., Springer-Verlag, New York, 1985, 21–93.

42. König, H., Lewis, D. R., and Lin, P.-K., Finite dimensional projection constants, Studia Mathematica, **75**(1983), 341–358.

43. Lindenstrauss, J., On the extension of operators with a finite-dimensional range, Illinois J. Math. **8**(1964), 488–499.

44. Morris, P. D. and Cheney, E. W., On the existence and characterization of minimal projections, J. Reine Angew. Math., **270**(1974), 61–76.

45. Odyniec, W. and Lewicki, G., *Minimal Projections in Banach Spaces*, Springer-Verlag, Berlin, 1990.

46. Pan, K. C. and Shekhtman, B., On minimal interpolating projections and trace duality, J. of Approx. Theory, **65**(1991), 216–230.

47. Pisier, G., *Factorization of Linear Operators and Geometry of Banach Spaces*, C.B.M.S. Regional Conf. Series, **60**(1986), Am. Math. Soc.

48. Singer, I., *Best Approximation in Normed Linear Spaces by Elements of Linear Subspaces*, Springer-Verlag, Berlin, 1970.

49. Tomczak-Jaegermann, N., *Banach-Mazur Distances and Finite-dimensional Operator Ideals*, John Wiley & Sons, New York, 1989.

50. Yost, D., L_1 contains every two-dimensional normed space, Ann. Polonici Math. **49**(1988), 17–19.

COPOSITIVE POLYNOMIAL APPROXIMATION REVISITED

Yingkang Hu and Xiang Ming Yu
Department of Mathematics & Computer Science, Georgia Southern University, Statesboro, GA 30460-8093, E-mail: yhu@gasou.edu
Department of Mathematics, Southwest Missouri State University, Springfield, MO 65804, E-mail: xmy944f@mail.smsu.edu
Contact author: X. M. Yu

It is a survey of recent results in copositive polynomial approximation and related areas. Several new theorems and lemmas are added.

1 Introduction

We are interested in constrained polynomial approximation that preserves certain geometric properties of the function to be approximated. We want to know not only how to realize such an approximation but also how well it performs, comparing with its nonconstrained counterpart.

There are various types of constrained approximation, see [5, Chapter 2] for examples. Among them, monotone and convex approximation has been investigated extensively in recent years.

Let $\mathbf{\Pi}_n[a, b]$ be the set of all algebraic polynomials on $[a, b]$ of degree $\le n$, $\mathbf{C}[a, b]$ the set of all continuous functions on $[a, b]$, and $\mathbf{C}^k[a, b]$ that of all k times continuously differentiable functions, $k = 1, 2, \dots$ For $0 < p < \infty$, let $\mathbf{L}_p[a, b]$ be the set of all measurable functions on $[a, b]$ such that $\|f\|_{\mathbf{L}_p[a,b]} < \infty$, where

$$\|f\|_{\mathbf{L}_p[a,b]} := \left(\int_a^b |f(x)|^p\,dx\right)^{1/p}.$$

Throughout this paper, $\mathbf{L}_\infty[a, b]$ is understood as $\mathbf{C}[a, b]$. We denote by $\mathbf{W}_p^k[a, b]$, $p \ge 1$, the Sobolev space, the set of all functions f on $[a, b]$ such that $f^{(k-1)}$ is absolutely continuous and $f^{(k)} \in \mathbf{L}_p[a, b]$. We shall omit the interval in the notation of a function space if the interval is $[-1, 1]$, for example, $\mathbf{L}_p$ means $\mathbf{L}_p[-1, 1]$. We shall also use the short notation $\|f\|_p$ for $\|f\|_{\mathbf{L}_p[-1,1]} = \|f\|_{\mathbf{L}_p}$. Whenever p is omitted in this paper, it is understood that $p = \infty$, for instance, $\|f\| := \|f\|_\infty$.

For $f \in \mathbf{L}_p$, $0 < p \le \infty$, we define

$$E_n(f)_p := \inf_{P_n \in \mathbf{\Pi}_n} \|f - P_n\|_p.$$

For $f \in \mathbf{C}^k$ or $\mathbf{W}_p^k$, and $f^{(k)} \geq 0$, we define

$$E_n^{(k)}(f)_p := \inf_{P_n \in \mathbf{\Pi}_n, P_n^{(k)} \geq 0} \|f - P_n\|_p \tag{1}$$

for $k \geq 0$, with $f^{(0)}$ understood as f itself. It is well-known that in regard to approximation rate, nonconstrained approximation is better in general than monotone and convex approximation. Indeed, Lorentz and Zeller [25] proved the following in 1969.

Theorem A. *For each* $k = 1, 2, \ldots$, *there exists a function* $f \in \mathbf{C}^k$ *with* $f^{(k)} \geq 0$ *such that*

$$\limsup_{n \to \infty} \frac{E_n^{(k)}(f)}{E_n(f)} = \infty.$$

On the other hand, if we allow slow convergence, we can always use for $f \in \mathbf{C}[0, 1]$ the Bernstein polynomials $B_n(f, x)$ defined by

$$B_n(f, x) := \sum_{i=0}^{n} f(i/n) \binom{n}{i} x^i (1 - x)^{n-i},$$

because for $k = 1, 2, \ldots$, $B_n^{(k)}(f, x) \geq 0$ if $f^{(k)}(x) \geq 0$ on $[0, 1]$, see [5, Thm. 10.3.3].

There are many results on estimates of $E_n^{(k)}(f)$, especially for $k = 1$ and 2. We shall mention several beautiful recent results in the last section. In this paper, we focus on a different type of constrained approximation: copositive approximation. Although estimating the approximation rate of a nonnegative function $f \in \mathbf{C}$ by nonnegative polynomials is trivial, it is not so in the $\mathbf{L}_p$ space, neither is estimating the rate of *copositive approximation*, in which one approximates a function having a finite number of sign changes by polynomials that have the same sign everywhere on the interval, (see the next section for a more accurate definition). In §3 we shall present latest results on copositive polynomial approximation in $\mathbf{C}$, as well as a discussion on the construction methods used in these results. A summary of both positive and copositive approximation in the $\mathbf{L}_p$ and Sobolev spaces will be given in §4. In these two sections we shall also show several new results: Lemmas 2 and 20, Theorems 9, 11, 13 and 14. Results on other types of constrained approximation, such as intertwining approximation, will be discussed as well in this paper.

2 Notations and Definitions

Let

$$Y_s := \{y_1, \dots, y_s \mid y_0 := -1 < y_1 < y_2 < \dots < y_s < 1 =: y_{s+1}\},$$

$0 \le s < \infty$, be a point set, and

$$\delta := \min_{0 \le j \le s} (y_{j+1} - y_j).$$

We denote by $\Delta^0(Y_s)$ the set of all functions f such that $(-1)^{s-j} f(x) \ge 0$ for $x \in [y_j, y_{j+1}]$, $j = 0, \dots, s$, i.e., every $f \in \Delta^0(Y_s)$ has s sign changes at the points in Y_s and is nonnegative near 1. In particular, if $s = 0$, then $\Delta^0 := \Delta^0(Y_0)$ denotes the set of all nonnegative functions on $[-1, 1]$. A function g is said to be copositive with f if f and g belong to the same class $\Delta^0(Y_s)$. For $f \in \mathbf{L}_p \cap \Delta^0(Y_s)$, let

$$E_n^{(0)}(f, Y_s)_p := \inf_{P_n \in \mathbf{\Pi}_n \cap \Delta^0(Y_s)} \|f - P_n\|_p$$

be the degree of copositive polynomial approximation of f. In particular, if $s = 0$,

$$E_n^{(0)}(f)_p := E_n^{(0)}(f, Y_0)_p := \inf_{P_n \in \mathbf{\Pi}_n \cap \Delta^0} \|f - P_n\|_p$$

is the degree of positive approximation. Note that this notation is consistent with (1).

To estimate $E_n^{(0)}(f, Y_s)_p$ and $E_n^{(0)}(f)_p$, we shall use both the mth usual and Ditzian-Totik moduli of smoothness. The mth (usual) modulus of smoothness of $f \in \mathbf{L}_p[a, b]$ is defined by

$$\omega_m(f, t, [a, b])_p := \sup_{0 < h \le t} \|\Delta_h^m(f, \cdot, [a, b])\|_{\mathbf{L}_p[a, b]},$$

where $\Delta_h^m(f, x, [a, b]) :=$

$$\begin{cases} \sum_{i=0}^m \binom{m}{i} (-1)^{m-i} f(x - \frac{mh}{2} + ih), & \text{if } x \pm \frac{mh}{2} \in [a, b], \\ 0, & \text{otherwise.} \end{cases}$$

We write $\omega_m(f, t)_p := \omega_m(f, t, [-1, 1])_p$. The mth Ditzian-Totik modulus of smoothness is defined by

$$\omega_m^\varphi(f, t)_p := \sup_{0 < h \le t} \|\Delta_{h\varphi(\cdot)}^m(f, \cdot, [-1, 1])\|_p$$

where $\varphi(x) := \sqrt{1-x^2}$. For $f \in \mathbf{C}$, the term of $\omega_m(f, \Delta_n(x)) := \omega_m(f, \Delta_n(x))_\infty$ is also used, where $\Delta_n(x) := \frac{\sqrt{1-x^2}}{n} + \frac{1}{n^2}$.

Closely related to copositive approximation is the so-called intertwining approximation, which is sometimes considered as well. The degree of intertwining polynomial approximation of a function $f \in \mathbf{L}_p$ is defined by

$$\widetilde{E}_n(f, Y_s)_p := \inf\{\|P_n - Q_n\|_p \mid P_n, Q_n \in \mathbf{\Pi}_n; P_n - f, f - Q_n \in \Delta^0(Y_s)\}.$$

In the case $s = 0$, we use the notation $\widetilde{E}_n(f)_p := \widetilde{E}_n(f, Y_0)_p$, which is the degree of onesided approximation. If $f \in \Delta^0(Y_s)$, an intertwining polynomial P_n is also a copositive approximation to f, thus

$$E_n^{(0)}(f, Y_s)_p \le \widetilde{E}_n(f, Y_s)_p.$$

It turns out that very often a routine procedure can be applied to a locally constrained polynomial approximation to obtain a globally constrained one, that is, the usual constrained approximation. Denote by $\Delta_n^0(Y_s)$ the set of all functions f such that $(-1)^{s-j} f(x) \ge 0$ for $x \in [y_j, y_j + \frac{1}{2n}) \cup (y_{j+1} - \frac{1}{2n}, y_{j+1}]$, $j = 0, 1, \ldots, s$. The degree of local copositive polynomial approximation of a function $f \in \mathbf{C} \cap \Delta^0(Y_s)$ is defined as

$$\operatorname{loc}E_n^{(0)}(f, Y_s) := \inf_{P_n \in \mathbf{\Pi}_n \cap \Delta_n^0(Y_s)} \|f - P_n\|.$$

The degree of local intertwining polynomial approximation of a function $f \in \mathbf{C}$ is defined as

$$\operatorname{loc}\widetilde{E}_n(f, Y_s) := \inf\{\|P_n - f\| + \|f - Q_n\| \mid P_n, Q_n \in \mathbf{\Pi}_n; P_n - f, f - Q_n \in \Delta_n^0(Y_s)\}.$$

3 Copositive Approximation in C

Let us first explore ideas and methods of constructing copositive polynomials for functions $f \in \mathbf{C} \cap \Delta^0(Y_s)$. Here are some successfully used ones.

1) It is noticed that if the uniform norm is used to measure copositive approximation error, then one merely needs to construct polynomials copositive with f only in a neighborhood of diameter $1/n$ of each sign change y_j, $j = 1, \ldots, s$. This can simplify the procedure of construction. In fact, we have

Lemma 1. [15] *Let $f \in \mathbf{C} \cap \Delta^0(Y_s)$, $1 \le s < \infty$. If for each $n > 4/\delta$ there is a polynomial $p_n \in \mathbf{\Pi}_n$ which is copositive with f on $\cup_{j=1}^s (y_j - \frac{1}{2n}, y_j + \frac{1}{2n})$,* *then there exists a polynomial $P_n \in \mathbf{\Pi}_{Cn}$ which is copositive with f on $[-1, 1]$ and satisfies*

$$\|f - P_n\| \le C\|f - p_n\|, \tag{2}$$

where the constant C depends only on δ and s.

To get such a P_n, one only needs to modify p_n by adding a polynomial $\bar{p}_n$ that is copositive with f on $[-1, 1]$ and satisfies

$$\|\bar{p}_n\| \le C\|f - p_n\|,$$

$$|\bar{p}_n(x)| > \|f - p_n\|, \qquad x \in [-1, 1] \setminus \bigcup_{j=1}^{s} \left(y_j - \frac{1}{2n}, y_j + \frac{1}{2n}\right).$$

Such a $\bar{p}_n$ can be found by using a special copositive polynomial approximation to $\operatorname{sgn}(x)$. An analogue of this for intertwining approximation is given below.

Lemma 2. *Let $f \in \mathbf{C}$ and Y_s, $1 \le s < \infty$, be given. If for every $n > 4/\delta$ there is a pair of polynomials p_n and $q_n \in \mathbf{\Pi}_n$ with both $p_n - f$ and $f - q_n$ in $\Delta_n^0(Y_s)$, then there exists a pair of polynomials P_n and $Q_n \in \mathbf{\Pi}_{Cn}$ such that both $P_n - f$ and $f - Q_n$ belong to $\Delta^0(Y_s)$ and*

$$\|P_n - f\| + \|f - Q_n\| \le C(\|p_n - f\| + \|f - q_n\|), \tag{3}$$

where C depends only on s.

Proof. Let $R_n := \|p_n - f\| + \|q_n - f\|$ and $n > 4/\delta$. Note that the intervals $(y_j - \frac{1}{2n}, y_j + \frac{1}{2n})$, $j = 1, \ldots, s$, are disjoint. By DeVore [3], there is an odd polynomial $\bar{S}_N \in \mathbf{\Pi}_N$ with $\bar{S}'_N(x) \ge 0$ on $[-1, 1]$ and $\bar{S}_N(1) = 1$ such that

$$|\operatorname{sgn}(x) - \bar{S}_N(x)| < K(1 + |Nx|)^{-2}, \tag{4}$$

where K is an absolute constant. Take $N := 2n([\sqrt{2K}] + 1)$. From (4) we have for $|x| \ge \frac{1}{2n}$

$$|\bar{S}_N(x)| \ge 1 - K(1 + |Nx|)^{-2} \ge 1 - K(Nx)^{-2} \ge \frac{1}{2}.$$

The polynomial

$$S_N(x) := \prod_{j=1}^{s} \bar{S}_N(x - y_j),$$

satisfies

$$|S_N(x)| \ge 2^{-s} \qquad \text{for } x \notin \textstyle\bigcup_{j=1}^{s} (y_j - \frac{1}{2n}, y_j + \frac{1}{2n}), \tag{5}$$

$S_N \in \Delta^0(Y_s)$, and $\|S_N\| \le 1$. We now define P_n and Q_n by

$$P_n(x) := p_n(x) + 2^s R_n S_N(x)$$

and

$$Q_n(x) := q_n(x) - 2^s R_n S_N(x).$$

Since $p_n - f$ and $f - q_n \in \Delta_n^0(Y_s)$ and $S_N(x) \in \Delta^0(Y_s)$, we know $P_n - f$ and $f - Q_n$ both belong to $\Delta_n^0(Y_s)$. For $x \notin \cup_{j=1}^s (y_j - \frac{1}{2n}, y_j + \frac{1}{2n})$, we have from (5)

$$\operatorname{sgn}(P_n(x) - f(x)) = \operatorname{sgn}(p_n(x) - f(x) + 2^s R_n S_N(x)) = \operatorname{sgn}(S_N(x)),$$

therefore $P_n - f \in \Delta^0(Y_s)$. Similarly, $f - Q_n \in \Delta^0(Y_s)$. Finally, (3) follows from the definitions of P_n and Q_n:

$$\begin{aligned} \|P_n - f\| + \|f - Q_n\| &\le \|p_n - f\| + \|f - q_n\| + 2^{s+1} R_n \|S_N\| \\ &\le (2^{s+1} + 1) R_n \end{aligned}$$

□

2) Certain other types of constrained approximation will help with constructing copositive approximation. These include interpolation, simultaneous approximation, and one-sided approximation. If a polynomial P_n is copositive with a continuous function $f \in \Delta^0(Y_s)$, then $P_n(y_j) = 0 = f(y_j)$, $j = 1, \ldots, s$. This is interpolation. If f is smooth, say $f \in \mathbf{C}^2$, then naturally one may modify a simultaneous approximation P_n ($P_n \to f$, $P_n' \to f'$, and $P_n'' \to f''$) to get a copositive approximation. For a copositive approximating polynomial of lower degree on a small subinterval, Whitney's Theorem for onesided polynomial approximation is often used. Some authors have also constructed comonotone or coconvex simultaneous polynomial approximation to f with f' or $f'' \in \mathbf{C} \cap \Delta^0(Y_s)$, so the derivatives of these polynomials are indeed good copositive approximation to f' or f''.

3) For $f \in \mathbf{C} \cap \Delta^0(Y_s)$, sometimes it is an effective way to break down the whole process into two steps: constructing a smooth copositive approximation $\tilde{f}$ to f, and then approximating $\tilde{f}$ by a copositive polynomial. For example [15], a smooth function $f_n \in \mathbf{C}^2$ is constructed for each n so that it is copositive with f on $\cup_{j=1}^s (y_j - \frac{1}{2n}, y_j + \frac{1}{2n})$ and satisfies

$$\|f - f_n\| \le C\omega_2(f, 1/n),$$

$$\|f_n''\| \le Cn^2\omega_2(f, 1/n),$$

and

$$|f_n'(x)| \geq n\omega_2(f, 1/n), \qquad x \in \bigcup_{j=1}^{s} \Big(y_j - \frac{1}{2n},\, y_j + \frac{1}{2n}\Big).$$

The last inequality above guarantees a simultaneous polynomial approximation interpolating f_n at the points in Y_s to be copositive with f_n in $\cup_{j=1}^{s}\big(y_j - \frac{1}{2n},\, y_j + \frac{1}{2n}\big)$. Combining this with Lemma 1, one obtains a polynomial copositive with f on $[-1,\, 1]$ having an approximation rate of $\omega_2(f, 1/n)$.

More frequently used smooth functions in this manner are splines. In [12], [16], and [18], for examples, a copositive spline approximation was first constructed and then was converted to a copositive polynomial approximation.

4) For constructing copositive splines, another powerful tool, in addition to the constrained Whitney's Theorem, is Beatson's Blending Lemma [1]. See [12] for an example. While a polynomial approximation of $\text{sgn}(x)$ is widely used in converting a copositive spline approximation to a polynomial one, a more important and delicate method in intertwining approximation uses a polynomial approximation of the truncated $(x-a)_+^{2k}$, $-1 < a < 1$. We refer the interested reader to [12].

Let us now review some existing Jackson type estimates for copositive approximation of $f \in \mathbf{C}$. In 1983, D. Leviatan [22] first proved that if $f \in \mathbf{C} \cap \Delta^0(Y_s)$, $s \geq 1$, then

$$E_n^{(0)}(f, Y_s) \leq C\omega(f, 1/n),$$

where C is an absolute constant. S. P. Zhou showed in 1992 that it is impossible to approximate $f \in \mathbf{C} \cap \Delta^0(Y_s)$ by copositive polynomials at a rate of ω_4:

Theorem 1. [28] *There is a function $f \in \mathbf{C}^1$ with one sign change in* $(-1,\, 1)$ *such that*

$$\limsup_{n\to\infty} \frac{E_n^{(0)}(f, Y_1)}{\omega_4(f, 1/n)} = \infty. \tag{6}$$

This means that the best possible approximation rate for $E_n^{(0)}(f, Y_s)$ is ω_3. In 1993, Hu, Leviatan and Yu [15] proved that for $f \in \mathbf{C} \cap \Delta^0(Y_s)$, $s \geq 1$,

$$E_n^{(0)}(f, Y_s) \leq C\omega_2(f, 1/n),$$

where C depends on s and δ. They then made an improvement in 1995 [16] by showing that for $f \in \mathbf{C}^1 \cap \Delta^0(Y_s)$ and any $m \in \mathbb{N}$,

$$E_n^{(0)}(f, Y_s) \leq Cn^{-1}\omega_m(f', 1/n), \tag{7}$$

where C depends on s, δ, and m. A year later, Hu and Yu [18] filled the gap by estimating $E_n^{(0)}(f, Y_s)$ in terms of $\omega_3(f, 1/n)$ for $f \in \mathbf{C} \cap \Delta^0(Y_s)$. However, Kopotun obtained a better estimate in terms of $\omega_3^{\varphi}(f, 1/n)$:

Theorem 2. [21] *For $f \in \mathbf{C} \cap \Delta^0(Y_s)$, $s \geq 1$, we have*

$$E_n^{(0)}(f, Y_s) \leq C\omega_3^{\varphi}(f, 1/n), \qquad n \geq 2, \tag{8}$$

where C depends only on Y_s.

Theorems 1 and 2 together complete a satisfactory investigation on copositive approximation for $f \in \mathbf{C} \cap \Delta^0(Y_s)$. Hu, Kopotun and Yu also improved (7):

Theorem 3. [12] *For $f \in \mathbf{C}^1 \cap \Delta^0(Y_s)$, $s \geq 1$, and any $m \in \mathbb{N}$, we have*

$$E_n^{(0)}(f, Y_s) \leq Cn^{-1}\omega_m^{\varphi}(f', 1/n), \qquad n \geq C, \tag{9}$$

where C depends only on Y_s and m.

Polynomial approximation rate in $\mathbf{C}$ is often measured pointwise. The following are two pointwise estimates of $E_n^{(0)}(f, Y_s)$

Theorem 4. [12] *For $f \in \mathbf{C} \cap \Delta^0(Y_s)$, $s \geq 1$, there is a polynomial $P_n \in \mathbf{\Pi}_n$ copositive with f such that*

$$|f(x) - P_n(x)| \leq C\omega_3(f, \Delta_n(x)), \qquad n \geq C, \tag{10}$$

where C depends only on Y_s.

Theorem 5. [7, 12] *For $f \in \mathbf{C}^1 \cap \Delta^0(Y_s)$, $s \geq 1$, and any $m \in \mathbb{N}$, there is a polynomial $P_n \in \mathbf{\Pi}_n$ copositive with f such that*

$$|f(x) - P_n(x)| \leq C\Delta_n(x)\omega_m(f', \Delta_n(x)), \qquad n \geq C, \tag{11}$$

where C depends only on Y_s and m.

Positive approximation is trivial if the uniform norm is used, since one can simply add $\|f - P_n\|$ to an unconstrained approximation P_n for a positive one. It no longer works when pointwise estimate is desired. We do have the same rate as the nonconstrained case, though.

Theorem 6. [7, 12] *For a nonnegative function $f \in \mathbf{C}$, and any $m \in \mathbb{N}$, there is a nonnegative polynomial $P_n \in \mathbf{\Pi}_n$ such that*

$$|f(x) - P_n(x)| \leq C\omega_m(f, \Delta_n(x)), \qquad n \geq m - 1, \tag{12}$$

where C depends only on m.

Lemma 1 and Theorem 1 imply that even if we only consider a local copositive approximation, which seems less constrained than copositive approximation at first glance, a rate better than ω_3 is still impossible.

Theorem 7. *There is a function $f \in \mathbf{C}^1$ with one sign change in $(-1, 1)$ such that*

$$\limsup_{n\to\infty} \frac{\mathrm{loc}E_n^{(0)}(f, Y_1)}{\omega_4(f, 1/n)} = \infty. \tag{13}$$

In another word, a local constraint tends to be equally strict as a global one. A significant improvement in approximation rate can be achieved only by relaxing the constraint in a neighborhood of each sign change y_j. This leads to the consideration of various types of weak copositive approximation. In our opinion, the so-called almost copositive approximation defined in [13] is the most sensible way to achieve the same rate of $\omega_m^{\varphi}(f, 1/n)$ or $\omega_m(f, \Delta_n(x))$ for any $m \in \mathbb{N}$ as that for nonconstrained approximation. The interested reader can see [13] for details.

For the error of intertwining approximation $\widetilde{E}_n(f, Y_s)$, we have the following.

Theorem 8. [12, Thm. 5] *Let $f \in \mathbf{C}^1$, $m \in \mathbb{N}$, and Y_s be given. Then*

$$\widetilde{E}_n(f, Y_s) \le C(m, s) n^{-1} \omega_m^{\varphi}(f', 1/n), \quad n \ge C(Y_s). \tag{14}$$

Moreover, there exists an intertwining pair of polynomials $\{P_n, Q_n\} \subset \mathbf{\Pi}_n$ such that

$$|P_n(x) - Q_n(x)| \le C(m, s)\Delta_n(x)\omega_m(f', \Delta_n(x)), \quad n \ge C(Y_s). \tag{15}$$

It is shown in the same paper that for any $A > 0$, there exists an $f \in \mathbf{C}$ for which $\widetilde{E}_n(f, Y_s) > A\|f\|$. By Lemma 2, a similar result holds for local intertwining polynomial approximation.

Theorem 9. *For any $A > 0$ and $n \in \mathbb{N}$, there exists an $f \in \mathbf{C}$ such that*

$$\mathrm{loc}\widetilde{E}_n(f, Y_s) > A\|f\|. \tag{16}$$

In [13], weak versions of intertwining approximation are considered. By relaxing the constraint in a neighborhood of each y_j, one looks at *almost intertwining approximation*, whose error $\widetilde{E}_n(f, \mathrm{alm}Y_s)$ has the same order as the nonconstrained case.

Theorem 10. [13, Thm. 13] *Let $f \in \mathbf{C}$, $m \in \mathbb{N}$ and Y_s be given. Then for sufficiently large n*

$$\widetilde{E}_n(f, \text{alm}Y_s) \le C\omega_m^{\varphi}(f, 1/n). \tag{17}$$

Moreover, there exists an almost intertwining pair of polynomials P_n, $Q_n \in \mathbf{\Pi}_n$ such that

$$|P_n(x) - f(x)| + |f(x) - Q_n(x)| \le C\omega_m(f, \Delta_n(x)). \tag{18}$$

From Theorems 8 and 9, we can see that one way to improve approximation order is to relax the constraint in neighborhoods of y_js. Another way is to require more on the smoothness of f in the neighborhoods.

Theorem 11. *Let $f \in \mathbf{C}$, $m \in \mathbb{N}$ and Y_s be given. For sufficiently large n, if f' is continuous on $\cup_{j=1}^{s}[y_j - \frac{1}{2n}, y_j + \frac{1}{2n}]$, then*

$$\widetilde{E}_n(f, Y_s) \le C\big(\omega_{m+1}^{\varphi}(f, 1/n) + \frac{1}{n}\sum_{j=1}^{s}\omega_m(f', 1/n, [y_j - \frac{1}{2n}, y_j + \frac{1}{2n}])\big). \tag{19}$$

In addition, if $f \in \Delta^0(Y_s)$, then

$$E_n^{(0)}(f, Y_s) \le C\big(\omega_{m+1}^{\varphi}(f, 1/n) + \frac{1}{n}\sum_{j=1}^{s}\omega_m(f', 1/n, [y_j - \frac{1}{2n}, y_j + \frac{1}{2n}])\big). \tag{20}$$

For the sake of simplicity, we now prove the following result that is slightly weaker than Theorem 11, but a similar proof works for Theorem 11. Denote $I_j := [y_j - \frac{1}{2n}, y_j + \frac{1}{2n}]$ and $\tilde{I}_j := [y_j - \tilde{m}/n, y_j + \tilde{m}/n]$, $j = 1, \ldots, s$, where $\tilde{m} := 2m^2$.

Theorem 12. *Let $f \in \mathbf{C}$, $m \ge 2$ and Y_s be given. For sufficiently large n, if f' is continuous on $\cup_{j=1}^{s}\tilde{I}_j$, then*

$$\widetilde{E}_n(f, Y_s) \le C\big(\omega_{m+1}(f, 1/n) + \frac{1}{n}\sum_{j=1}^{s}\omega_m(f', 1/n, \tilde{I}_j)\big). \tag{21}$$

In addition, if $f \in \Delta^0(Y_s)$, then

$$E_n^{(0)}(f, Y_s) \le C\big(\omega_{m+1}(f, 1/n) + \frac{1}{n}\sum_{j=1}^{s}\omega_m(f', 1/n, \tilde{I}_j)\big). \tag{22}$$

Proof. Let f' be continuous on $\cup_{j=1}^{s}\tilde{I}_j$. By Whitney's Theorem for onesided polynomial approximation, there exist p_j and $q_j \in \mathbf{\Pi}_{m-1}$ such that

$$p_j(x) \geq f'(x) \geq q_j(x), \qquad x \in \tilde{I}_j$$

and

$$\|p_j - q_j\|_{\mathbf{C}(\tilde{I}_j)} \leq C\omega_m(f', |\tilde{I}_j|, \tilde{I}_j).$$

Define $\tilde{p}_j := p_j$ and $\tilde{q}_j := q_j$ if $(-1)^{s-j} > 0$; and $\tilde{p}_j := q_j$ and $\tilde{q}_j := p_j$ if $(-1)^{s-j} < 0$;

$$\tilde{P}_j := \int_{y_j}^{x} \tilde{p}_j(t)\, dt + f(y_j)$$

and

$$\tilde{Q}_j := \int_{y_j}^{x} \tilde{q}_j(t)\, dt + f(y_j).$$

We have for $x \in \tilde{I}_j$

$$(-1)^{s-j}(\tilde{P}_j(x) - f(x))\mathrm{sgn}(x - y_j) \geq 0,$$

$$(-1)^{s-j}(\tilde{Q}_j(x) - f(x))\mathrm{sgn}(x - y_j) \leq 0,$$

and

$$\|\tilde{P}_j - \tilde{Q}_j\|_{\mathbf{C}(\tilde{I}_j)} \leq \left\| \int_{y_j}^{x} (\tilde{p}_j(t) - \tilde{q}_j(t))\, dt \right\|_{\mathbf{C}(\tilde{I}_j)} \leq C|\tilde{I}_j|\omega_m(f', |\tilde{I}_j|, \tilde{I}_j).$$

For each of $\bar{I}_j := [y_j + \frac{1}{2n}, /, y_{j+1} - \frac{1}{2n}]$, $j = 1, \ldots, s-1$, $\bar{I}_0 := [-1,\, y_1 - \frac{1}{2n}]$ and $\bar{I}_s := [y_s + \frac{1}{2n},\, 1]$, it is well-known that there exists a spline $\tilde{S}_j \in \mathbf{C}^{m-1}(\bar{I}_j)$ of order $m+1$ such that

$$\|f - \tilde{S}_j\|_{\mathbf{C}(\bar{I}_j)} \leq C\omega_{m+1}(f, 1/n, \bar{I}_j).$$

By Beatson's Blending Lemma [1], there exists a pair of splines S_{n1} and $S_{n2} \in \mathbf{C}^{m-1}$ of order $m+1$ such that

$$S_{n1}(x) = \tilde{P}_j(x), \quad S_{n2}(x) = \tilde{Q}_j(x), \qquad x \in I_j;$$

$$S_{n1}(x) = S_{n2}(x) = \tilde{S}_j(x), \qquad x \in [y_j + \tilde{m}/n,\, y_{j+1} - \tilde{m})/n], \quad 1 \leq j < s,$$

$$\|S_{ni} - f\| \leq C(\omega_{m+1}(f, 1/n) + \frac{1}{n}\sum_{j=1}^{s} \omega_m(f', 1/n, \tilde{I}_j)), \qquad i = 1, 2.$$

Note that the first of these properties implies $S_{n1} - f$ and $f - S_{n2} \in \Delta_n^0(Y_s)$. By [17] we also have

$$\omega_m(S'_{ni}, 1/n) \le C(n\omega_{m+1}(f, 1/n) + \sum_{j=1}^{s} \omega_m(f', 1/n, \tilde{I}_j)), \qquad i = 1, 2.$$

Since $S_{ni} \in \mathbf{C}^{m-1} \subseteq \mathbf{C}^1$, from Theorem 8 there exist P_{ni} and $Q_{ni} \in \mathbf{\Pi}_n$ such that $P_{ni} - S_{ni}$, $S_{ni} - Q_{ni} \in \Delta^0(Y_s)$ and

$$\begin{aligned} \|P_{ni} - f\|,\ \|f - Q_{ni}\| &\le C\frac{1}{n}\omega_m(S'_{ni}, 1/n) \\ &\le C(\omega_{m+1}(f, 1/n) + \frac{1}{n}\sum_{j=1}^{s} \omega_m(f', 1/n, \tilde{I}_j)). \end{aligned}$$

Here both $P_{n1} - f$ and $f - Q_{n2}$ belong to $\Delta_n^0(Y_s)$. Therefore

$$\mathrm{loc}\widetilde{E}_n(f, Y_s) \le C(\omega_{m+1}(f, 1/n) + \frac{1}{n}\sum_{j=1}^{s} \omega_m(f', 1/n, \tilde{I}_j)).$$

Now (21) follows from this and Lemma 2. □

4 Positive and Copositive Approximation in $\mathbf{L}_p$

The investigation on positive and copositive approximation in $\mathbf{L}_p$, $0 < p < \infty$, was started later than that in $\mathbf{C}$. It turned out that things become much more complicated in $\mathbf{L}_p$ and, as a consequence, the positive and copositive approximation rates are significantly lower than those in $\mathbf{C}$. Indeed, it is impossible to obtain the rate in terms of ω_2 even for positive polynomial approximation in $\mathbf{L}_p$.

Theorem 13. [11, Thm. 1] *For each $n \in \mathbb{N}$, $0 < p < \infty$, $0 < \varepsilon \le 2$, and $A > 0$, there exists a nonnegative function $f \in \mathbf{C}^\infty$ such that for every polynomial $P_n \in \mathbf{\Pi}_n$ that is nonnegative at $x = 1$, the following inequality holds:*

$$\|f - P_n\|_{\mathbf{L}_p[1-\varepsilon,\,1]} > A\omega_2(f, 1)_p. \tag{23}$$

The positive results are given below.

Theorem 14. [11, Thm. 2], [19] *If $f \in \mathbf{L}_p$, $0 < p < \infty$, and $f \ge 0$, then for $\forall n \in \mathbb{N}$,*

$$E_n^{(0)}(f)_p \le C\omega^\varphi(f, 1/(n+1))_p, \tag{24}$$

where C is an absolute constant if $1 \le p < \infty$, and $C = C(p)$ if $0 < p < 1$.

Theorem 15. [11, Thm. 3] *Let Y_s be given. If $f \in \mathbf{L}_p \cap \Delta^0(Y_s)$, $0 < p < \infty$, then for $\forall n \in \mathbb{N}$,*

$$E_n^{(0)}(f, Y_s)_p \le C\omega^\varphi(f, 1/(n+1))_p, \tag{25}$$

where C depends on s and δ, and also on p if $0 < p < 1$.

Hu, Kopotun and Yu also studied copositive approximation in Sobolev spaces $\mathbf{W}_p^k[-1, 1]$, $1 \le p < \infty$. For $k = 1$ we have

Theorem 16. [12, Thm. 10] *If $f \in \mathbf{W}_p^1 \cap \Delta^0(Y_s)$, $1 \le p < \infty$, then for sufficiently large n,*

$$E_n^{(0)}(f, Y_s)_p \le Cn^{-1}\omega_2^\varphi(f', 1/(n+1))_p, \tag{26}$$

where C depends on Y_s.

For $\mathbf{W}_p^1$, the result in Theorem 16 is the best in the sense that the term $n^{-1}\omega_2^\varphi(f', 1/n)_p$ can not replaced by $\omega_3(f', 1)_p$.

For $k = 2$ we have

Theorem 17. [12, Corollary 9] *If $f \in \mathbf{W}_p^2 \cap \Delta^0(Y_s)$, $1 \le p < \infty$, then for $\forall m \in \mathbb{N}$ and sufficiently large n,*

$$E_n^{(0)}(f, Y_s)_p \le Cn^{-2}\omega_m^\varphi(f'', 1/(n+1))_p, \tag{27}$$

where C depends on Y_s and m.

The above results were proved by showing their analogues for positive and copositive spline approximation and then applying some technical lemmas to convert the results to the polynomial cases. We should emphasize that the modification is much more difficult in $\mathbf{L}_p$ than in $\mathbf{C}$, for the amount of modification is determined by the uniform norm while the approximation error is measured by the $\mathbf{L}_p$ norm. However, even for special functions such as splines their L_p norm and L_q norm ($p \ne q$) are not switchable. Let $h := 1/n$ and s be a spline of order r on $[0, 1]$ with knot sequence $T := \{x_i = ih\}$. We can see this matter from the following lemma, which is a generalization of DeVore's result [5, Theorem 5.1.2]:

$$\|s\|_p \le Ch^{1/p-1/q}\|s\|_q, \qquad 1 \le q \le p.$$

Various weak versions of this lemma have been used in construction of constrained approximation. See [17] and [10] for similar properties of moduli of smoothness of splines.

Lemma 3. *Let $1 \le r < n$, $0 \le m \le r$ and $1 \le q \le p$. Then, we have*

$$\omega_m(s,h)_p \le Ch^{1/p-1/q}\omega_m(s,h)_q. \tag{28}$$

Proof. Let $\{\mathbf{c}\}$ be the B-spline coefficient sequence of s and Δ^m the mth difference operator on sequences. By [17], we have

$$\omega_m(s,h)_p \sim h^{1/p}\|\Delta^m \mathbf{c}\|_p$$

and

$$\omega_m(s,h)_q \sim h^{1/q}\|\Delta^m \mathbf{c}\|_q.$$

Note for $1 \le q \le p$, $\|\mathbf{a}\|_p \le C\|\mathbf{a}\|_q$ holds for any sequence $\{\mathbf{a}\}$ in l_q. So the above equivalences imply the desired inequality. □

For $\mathbf{W}_p^1$, the best estimate for positive approximation belongs to Stozavova.

Theorem 18. [26] *If $f \in \mathbf{W}_p^1$, $1 < p < \infty$, and $f \ge 0$, then we have*

$$E_n^{(0)}(f)_p \le Cn^{-1}E_{n-1}(f')_p. \tag{29}$$

There have been discussions on onesided approximation, intertwining approximation, or their weaker versions in $\mathbf{L}_p$. There are estimates using the so-called τ-modulus $\tau_m(f,t)_p$, which seems more proper than the usual modulus of smoothness for these approximations in L_p. We omit details here.

5 Other Types of Constrained Approximation

Since this paper aims mainly at copositive approximation, we are not going to get deeply into other types of constrained approximation. Here, we only mention a few of recent interesting results on monotone and convex approximation. Recall from (1) that $E_n^{(1)}(f)_p$ and $E_n^{(2)}(f)_p$ denote the monotone and convex approximation errors of $f \in \mathbf{L}_p$, respectively.

In his 1981 paper, Švedov [27] proved that for any given monotone (or convex) function $f \in \mathbf{L}_p$, $0 < p \le \infty$, and $n \ge 1$, there exists a monotone (or convex) polynomial $P_n \in \mathbf{\Pi}_n$ such that

$$\|f - P_n\|_p \le C\omega_2(f,1/n)_p, \tag{30}$$

where C depends only on p. He also proved that one cannot replace $\omega_2(f,1/n)_p$ even by $\omega_3(f,1)_p$ for monotone approximation or by $\omega_4(f,1)_p$ for convex approximation.

Since then, many improvements have been made to (30), such as using in (30) a pointwise estimate or Ditzian-Totik modulus of smoothness, filling the gap for convex approximation by ω_3 or ω_3^φ, and obtaining better estimates for differentiable functions. We list some of them below.

1) For monotone approximation, DeVore and Yu [6] proved for any monotone function in $\mathbf{C}$ the estimate

$$|f(x) - P_n(x)| \le C\omega_2(f, \sqrt{1-x^2}/n).$$

2) Leviatan [23] proved for any monotone function $f \in \mathbf{C}$ the estimate

$$E_n^{(1)}(f) \le C\omega_2^\varphi(f, 1/n).$$

3) For any monotone function $f \in \mathbf{C}^k$, $k \ge 0$, the following holds:

$$E_n^{(1)}(f) \le Cn^{-k}\omega(f^{(k)}, 1/n).$$

Lorentz and Zeller [25] proved the case $k = 0$, Lorentz [24] proved the case $k = 1$, and DeVore [2], [3] obtained the rest, that is, for any $k \ge 2$.

4) In 1998, G. A. Dzyubekno, J. Gilewicz and I. A. Shevchuk [8] investigated piecewise monotone approximation for a piecewise monotone function $f \in \mathbf{C}^k$ and obtained the pointwise estimate as follows. Let $\Pi(x) := \prod_{1\le j\le s}(x - y_j)$, and let $f \in \mathbf{C}^k$ be such that $f'(x)\Pi(x) \ge 0$, $\forall x \in [-1, 1]$. Let m be any positive integer if $k > 1$, and $m \le 3$ if $k = 1$. Then for $n \ge m + k - 1$, there exists a polynomial $P_n \in \mathbf{\Pi}_n$ such that $P_n'(x)\Pi(x) \ge 0$ and

$$|f(x) - P_n(x)| \le C\left(\frac{1}{n^2} + \frac{1}{n}\sqrt{1-x^2}\right)^k \omega_m\left(f^{(k)}, \frac{1}{n^2} + \frac{1}{n}\sqrt{1-x^2}\right)$$

for $\forall x \in [-1, 1]$, where C depends only on k, m, and Y_s.

5) In 1994, Hu, Leviatan and Yu [14] obtained an estimate for convex approximation of convex functions $f \in \mathbf{C}$:

$$E_n^{(2)}(f) \le C\omega_3(f, 1/n).$$

Slight later and independently, Kopotun [20] proved a better estimate for convex functions $f \in \mathbf{C}$:

$$E_n^{(2)}(f) \le C\omega_3^\varphi(f, 1/n).$$

These closed the gap left by Švedov.

6) In 1996, DeVore, Hu and Leviatan [4] generalized the above results to $\mathbf{L}_p$. $0 < p < \infty$:

$$E_n^{(2)}(f)_p \le C\omega_3^\varphi(f, 1/n)_p,$$

where C is a constant depending at most on p as $p \to 0$.

References

1. Beatson, R. K., *Restricted range approximation by splines and variational inequalities*, SIAM J. Num. Anal., **19**, 372–380 (1982).

2. DeVore, R. A., *Degree of approximation*, Approximation Theory II, New York : Academic Press, 117–162 (1976).

3. DeVore, R. A., *Monotone approximation by polynomials*, SIAM J. Math. Anal., **8**, 891–905 (1977).

4. DeVore, R. A., Hu, Y. K., and Leviatan, D., *Convex polynomial and spline approximation in* $\mathbf{L}_p$, $0 < p < \infty$, Constr. Approx., **12**, 409–422 (1996).

5. DeVore, R. A. and Lorentz, G. G., Constructive Approximation, Springer-Verlag, Berlin, 1993.

6. DeVore, R. A. and Yu, X. M., *Pointwise estimates for monotone polynomial approximation*, Constr. Approx., **1**, 323–331 (1985).

7. Dzyubenko, G. A., *Copositive and positive pointwise approximation*, preprint.

8. Dzyubekno, G. A., Gilewicz, J., and Shevchuk, I. A., *Piecewise monotone pointwise approximation*, Constr. Approx., **14**, 311–348 (1998).

9. Hu, Y. K., *Positive and copositive spline approximation in* $\mathbf{L}_p[0, 1]$, Computers and Math. with Appl., **30**, 137-146 (1995).

10. Hu, Y. K., *On equivalence of moduli of smoothness*, J. Approx. Theory, **97**, 282–293 (1999).

11. Hu, Y. K., Kopotun, K., and Yu, X. M., *On positive and copositive polynomial and spline approximation in* $\mathbf{L}_p[-1, 1]$, $0 < p < \infty$, J. Approx. Theory, **86**, 320–334 (1996).

12. Hu, Y. K., Kopotun, K., and Yu, X. M., *Constrained approximation in Sobolev spaces*, Canad. Math. J., **49**, 74–99 (1997).

13. Hu, Y. K., Kopotun, K., and Yu, X. M., *Weak copositive and intertwining approximation*, J. Approx. Theory, **96**, 213–236 (1999).

14. Yingkang Hu, Yingkang, Leviatan, D., and Yu, Xiang Ming, *Convex polynomial and spline approximation in* $C[-1, 1]$, Constr. Approx., **10**, 31–64 (1994).

15. Hu, Y. K., Leviatan, D., and Yu, X. M., *Copositive polynomial approximation in* **C**[0, 1], J. of Analysis, **1**, 85–90 (1993).

16. Hu, Y. K., Leviatan, D., and Yu, X. M., *Copositive polynomial and spline approximation*, J. Approx. Theory, **80**, 204–218 (1995).

17. Hu, Y. K. and Yu, X. M., *Discrete modulus of smoothness of splines with equally spaced knots*, SIAM J. Num. Anal., **32**, 1428–1435 (1995).

18. Hu, Y. K. and Yu, X. M., *The degree of copositive approximation and a computer algorithm*, SIAM J. Num. Anal., **33**, 388–398 (1996).

19. Ivanov, K. G., *On a new characteristic of functions. II. Direct and converse theorems for the best algebraic approximation in* $C[-1,1]$ *and* $L_p[-1,1]$, PLISKA Stud. Math. Bulgar. , **5**, 151–163 (1983).

20. Kopotun, K. A., *Pointwise and uniform estimates for convex approximation of functions by algebraic polynomials*, Constr. Approx., **10**, 153–178 (1994).

21. Kopotun, K. A., *On copositive approximation by algebraic polynomials*, Analysis Math., **21**, 269–283 (1995).

22. Leviatan, D., *The degree of copositive approximation by polynomials*, Proc. of AMS, **88**, 101–105 (1983).

23. Leviatan, D., *Monotone and comonotone polynomial approximation revisited*, J. Approx. Theory, **53**, 1–16 (1988).

24. Lorentz, G. G., *Monotone approximation*, Inequalities III. New York : Academic Press, 201–215 (1972).

25. Lorentz, G. G. and Zeller, K., *Degree of approximation by monotone polynomials I*, J. Approx. Theory, **1**, 501–504 (1968).

26. Stojanova, M., *The best onesided algebraic approximation in* $L_p[-1,1]$, $(1 \leq p \leq \infty)$, Math. Balkanica, **2**, 101–113 (1988).

27. Švedov, A. S., *Orders of coapproximation of functions by algebraic polynomials*, Mat. Zametki, **29**, 117–130 (1981); English tranl. in Math. Notes, **29**, 63–70 (1981).

28. Zhou, S. P., *A counterexample in copositive approximation*, Israel J. Math., **78**, 75–83 (1992).

29. Zhou, S. P., *On copositive approximation*, Approx. Theory Appl. **9**, 104–110 (1993).

NEURO-FUZZY ADAPTIVE CONTROL: STRUCTURE, ALGORITHMS, AND PERFORMANCE

Augustine O. Esogbue
Intelligent Systems and Controls Lab, School of Industrial and Systems Engineering, Georgia Institute of Technology
E-mail: augustine.esogbue@isye.gatech.edu

1 Introduction

Solution to control problems in which there is considerable uncertainty or lack of knowledge about the process is the focal concern of our research mission. The class of problems used as the leitmotif of our work may be subsumed under the rubric of what is generally known as the general set-point regulator problem. This group encompasses the usual dynamical system plants as well as discrete state space problems which arise in the control of operations and manufacturing. Our work has focused on the development and demonstration of the potential of the proposed Statistical Fuzzy Associative Learning Controller in simulated applications to three representatives of this class of problems, namely, process control of particular second-order dynamical systems, message routing in communication networks and the power system stabilization and control problem. In each, there are complexities and uncertainties which call for intelligence, but intelligence which can be automated for speed and repeatability.

Classical analytical and optimization approaches generally require some type of model of the plant and specific knowledge of what constitutes desirable plant behavior. For cases where the lack of knowledge is of a high order, a variety of intelligent methods have been suggested to provide more acceptable performance than can be achieved with the classical methods. In our review of existing approaches (Esogbue and Murrell 1994), it was shown that many of them consist of off-line empirical modeling using a training set of data. There are many instances in which these approaches work well for control of uncertain systems. However, a training set generally requires some knowledge of what constitutes good control, and in uncertain systems, this is often lacking. Additionally, many systems are based on an explicit gradient-type of approach, such as the well-known backpropagation neural networks with their attendant drawbacks. It is also noteworthy that some of these problems are framed in terms of discrete spaces, do not have objective functions which are analytically differentiable, or any known objective function formula at all. Hence, gradient methods which proliferate in the literature may not be applicable.

An alternative approach to control which has had notable success is that of fuzzy controllers. These methods mathematically mimic the imprecise reasoning processes which are an important part of what makes humans so successful at solving problems with large uncertainties. However, the recurrent problem had hitherto been to find effective ways to capture or acquire the knowledge or skill required for a particular problem. Recent research has been devoted to ways of integrating learning algorithms with fuzzy control to automate the knowledge acquisition. Most of the attempts to combine the power of fuzzy reasoning with learning capability have taken either of two approaches. One approach has been to implement fuzzy reasoning with AI-type rule-based methods. The resulting controller models can become quite complicated and cumbersome as they are scaled up to larger systems. The second approach is to use gradient-type neural networks combined with fuzzy logic. Although these have been used successfully in some applications, they share the limitations of gradient methods.

It is now believed that learning systems (in the sense of mathematical learning theory) may be the most appropriate for the problems with a high order of uncertainty. A reinforcement learning trial system can learn on-line from its own experience with the process it is designed to control. Although such systems have been suggested since the 1960's, their applicability had been significantly inhibited by the almost exponential growth in complexity as it is scaled up to larger systems due to the necessity of discretizing the spaces. Combining these methods with fuzzy discretization and fuzzy logic can make learning trial methods a viable, practical approach. However, until the research first reported in Esogbue and Murrell (1993), and then furthered by Murrell (1994) and Esogbue, Hearnes and Song (1995), very little development of this idea had been seen in the literature.

These problems are second-order linear and nonlinear dynamical systems of the type which often occur in industrial process control. In these problems, uncertainty about the true plant model is often introduced by the environment, or by frequent changes in configuration necessitated for example by flexible manufacturing practices. Additionally, unknown, time-varying and nonlinear dynamics complicate the use of the plant model required by classical control theory methods. These concerns, necessitate the pursuit of novel approaches of the sort dissussed in the sequel.

2 The Self-Learning Fuzzy-Neuro Controller

In response to the quest for an intelligent controller that can learn online, does not require an exact model of the plant, operates without the benefit of what is considered optimal, does not need a set of predefined control rules, and uses feedback in an instructive way without the attendant expensive computational costs, Esogbue and Murrell [10] reported the development of a self-learning fuzzy-neuro controller with unique features and capabilities. This controller consists of five components: (1) the Statistical Fuzzy Discretization Network (SFDN) which employs a variation of the Kohonen self-organizing map (SOM) to fuzzify the state space of the plant; (2) the Fuzzy Correlation Network (FCN) which implements the learned fuzzy control rules as fuzzy relation; (3) the Stochastic Learning Correlation Network (SLCN) which maps a particular fuzzy state to a set of fuzzy control actions through an adaptive stochastic algorithm; (4) the Control Activation Network (CAN) which defuzzifies the fuzzy control to a crisp control signal; and (5) the Performance Evaluation System (PES) which provides feedback reinforcement signals to the learning algorithm based on the effectiveness of the control action. A block diagram of the controller is shown in Figure 1.

2.1 Statistical Fuzzy Discretization Network (SFDN)

This subsystem consists of a network of automata nodes ("neurons") arranged in a grid. Each node receives as its input the current process state vector. Every time a vector is input, each node computes an output that represents the degree of membership in the fuzzy subset of the input space that corresponds to that node. This output, called the node activation, is a measure of the degree of similarity of the input state vector to the ideal or prototype member of that fuzzy set. It is computed as some combination of the state vector and a location parameter vector associated with that node which represents the prototype process state for the corresponding fuzzy set. Also associated with each node is a parameter that encodes the degree of dispersion or spread of its fuzzy set membership function used to calculate the node activation. For a particular membership function form, the location parameters and spread parameter together define a fuzzy set. The SFDN thus performs a fuzzy discretization, inducing a fuzzy partition of the state space X into reference fuzzy subsets $X_1, X_2, \ldots, X_r$, each represented by a node in the grid.

The network described here is an extension of Kohonen's self-organizing feature map to fuzzy characterization of dynamic plant states. The output of the ith node in the map is

$$a_i = \exp(-\| x - m_i \| / s_i) \tag{1}$$

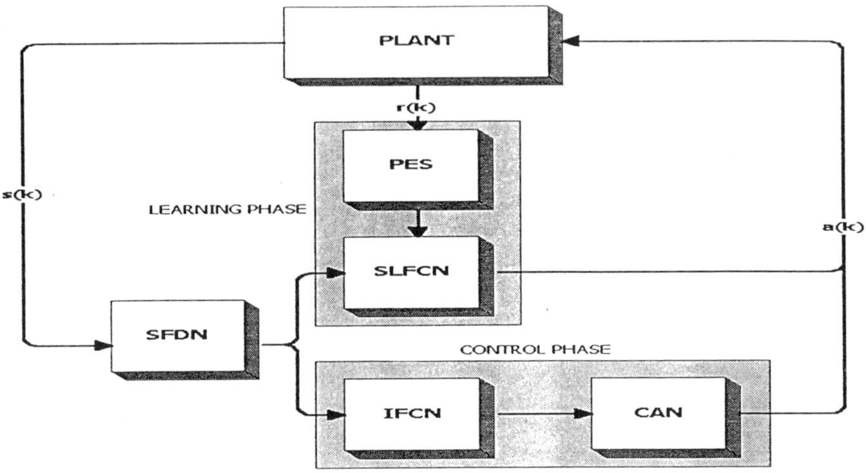

Figure 1: Block Diagram of Fuzzy-Neuro Controller

where x is the state vector input, m_i is the vector of location parameters and s_i the spread parameter for the ith node. and it is assumed that the choice of similarity measure is the Euclidean metric and the functional form of the membership functions is a gaussian function. Thus, the vector a of node activations is the fuzzy hyperstate due to input state vector x:

$$a = (\pi_{X_1}(x), \ldots, \pi_{X_r}(x))^T \tag{2}$$

where $\pi_{X_i}(\bullet)$ is the membership function for fuzzy set X_i given by the foregoing equation.

A sequence of state vectors is input to the network over time, and an adjustment algorithm adapts the location parameters to reflect the actual clustering of the state vectors by which the aggregation into fuzzy subsets is determined. A simplified version of the update rule of the jth component of m_i for this example is

$$m_{ij,k+1} = \begin{cases} m_{ij,k} + \alpha_k(x_{j,k} - m_{ij,k}) & \text{for } i \in T_{c_k} \\ m_{ij,k} & \text{otherwise} \end{cases} \tag{3}$$

where k indexes the time step of the algorithm, T_{c_k} is a small neighborhood of nodes in the grid within a radius c_k around the node most activated by x_k, and α_k and c_k are decreasing functions of k. The basic concept for updating the spread parameter is given by

$$s_{i,k+1} = \begin{cases} s_{i,k} + \eta_k(|x_{j,k} - m_{ij,k}|^2 - s_{i,k}) & \text{for } i \in T_{c_k} \\ s_{i,k} & \text{otherwise} \end{cases} \tag{4}$$

where η_k is also a decreasing function of k.

The SFDN provides a means of aggregating similar plant states, thus permitting implementation of the control as a discrete relation. The adaptation update equations as initially configured are of the simple delta-rule type, which is more easily implemented in real time than clustering algorithms and permits the parallel distributed computation of neural networks.

2.2 Fuzzy Correlation Network (FCN)

The Fuzzy Correlation Network implements the fuzzy control rules as a fuzzy relation G (learned by the SLCN) which associates the collection of fuzzy sets $X_1, \ldots, X_r$ for input vectors $x \in X$ with the fuzzy sets $U_1, \ldots, U_s$ for the controls $u \in U$. This is accomplished with a fuzzy associative memory (FAM) or correlation network. The ith node on one side represents the degree to which X_i has been selected, given by the SFDN node output a_i for state x. Each of these is linked to every node on the other side. The output b_j of the jth node on the other side indicates the degree to which fuzzy output set U_j is the correct choice given the activations of the X_i's, or the "firing strength" of each rule for which that fuzzy control is the consequent. The connection weight parameter g_{ij} indicates the degree to which X_i relates to U_j. The control rule "If x is X_i then u is U_j" is represented by a strong link g_{ij} between node X_i and node U_j. The network connection weight matrix $G = \{g_{ij}\}$ specfies a fuzzy relation on $\{X_1, \ldots, X_r\} \times \{U_1, \ldots, U_s\}$. Thus, given input activation a, the fuzzy hyperstate, the fuzzy control vector is given by

$$b^T = \sigma(a^T G) \tag{5}$$

where σ is a vector-valued function whose components are $b_j = \sigma_j(a^T G\bullet, j)$, $G\bullet, j$ is the jth column of G, and each σ_j is some type of limiting function, (i.e., σ_i satisfies $\sigma_i(\alpha) \to 0$ as $\alpha \to -\infty$ and $\sigma_i(\alpha) \to 1$ as $\alpha \to \infty$), such as a sigmoid function. Thus, this implementation uses the product and limited-sum logic operators which is more easily implemented with a neural network associative memory than the common min-max logic.

2.3 Stochastic Learning Correlation Network (SLCN)

The purpose of this subsystem is to test and learn the effectiveness of pairing a particular control vector fuzzy set with each given state vector fuzzy set, using the performance evaluation provided by the Performance Evaluation System (PES), then use this knowledge base to generate the fuzzy control relation used by the FCN. Both the SLCN and the FCN receive as input the fuzzy hyperstate a output by the state map grid of the SFDN. The first phase of operation of the controller is a learning phase in which the fuzzy control vector b is generated by the SLCN. In the second phase of operation, the fuzzy relation learned by the SLCN is used by the FCN to generate b.

Initially, nothing is known about what control vector gives the best response when the process is in a given state. So, a fuzzy output is just picked and the performance measure indicates how good the selection was. If it was not that good, the controller will be less inclined to pick that control again the next time the system enters that fuzzy state. If the selection was good, that control action is reinforced, so that it is more likely to be selected next time. The network that implements this is akin to Narendra's stochastic learning automata (SLA).

The SLCN consists of a matrix of nodes where each row corresponds to a particular fuzzy input state and each column to a particular fuzzy control action. The degree of activation of a node indicates the fuzzy degree to which it selects the control fuzzy subset to which it is assigned. Each node has a spread parameter h_{ij} for the ith fuzzy state and the jth fuzzy control. The location parameters λ_{ij} (scalar) are adjusted so that they always fall at the center of the spread function. In this case, a box-shaped function defined on a bounded interval is used. Thus, the output of the node for the ith fuzzy state and the jth fuzzy control is given by

$$c_{ij,t} = \begin{cases} 1 & \text{if } \lambda_{ij,t} - h_{ij,t} < \zeta_t < \lambda_{ij,t} + h_{ij,t} \\ 0 & \text{otherwise} \end{cases} \tag{6}$$

$$\lambda_{ij,t} = \sum_{k=1}^{j-1} h_{ij,t} + \frac{1}{2}\, h_{ij,t} \tag{7}$$

where $h_{ij,t}$ and $\lambda_{ij,t}$ are location and spread parameters, respectively, at time t for node (i,j), and $\zeta_t \in [0,1]$ is generated by a chaotic or pseudo-random process and serves as the input to the node. The node with the largest spread parameter is then the one that will have the maximum activation most often.

The fuzzy control vector **b** is given by

$$b_{j,t} = \begin{cases} c_{ij,t} & \text{for } i = \text{ argmax } (a_i) \\ 0 & \text{otherwise} \end{cases} \tag{8}$$

When i is the index of the most activated input fuzzy set, then the most activated node in the ith row of the SLCN node matrix selects the control fuzzy set. The update algorithm for h_{ij} is given as

$$h_{ij,t+1} = (h_{ij,t} + r_t a_{i,t} b_{j,t} \gamma_{ij,t})/(1 + \sum_{j^r t} a_{i,t} b_{j,t} \gamma_{ij,t}) \tag{9}$$

$$\gamma_{ij,t} = \begin{cases} h_{ij,t} & \text{if } r_t < 0 \\ 1 - h_{ij,t} & \text{if } r_t \geq 0 \end{cases} \tag{10}$$

where r_t is the reinforcement which is a function of the performance measure p_t, and $a_{i,t}$ and $b_{i,t}$ are the input and output activations, respectively, for the ith fuzzy state and j^{th} fuzzy control at time t. The product $a_{i,t}\ b_{j,t}$ is the association or correlation between the state and control fuzzy sets. The quantities p_t and r_t are computed by the PES subsystem.

2.4 Control Activation Network (CAN)

The input to the Control Activation Network is the fuzzy control vector b. This fuzzy control is defuzzified to produce a crisp output quantity u, which is a vector for multivariable control processes. Each CAN node has its location parameter vector set to the desired control vector prototype levels. Its input is the vector b, and its output is a crisp control vector u. The nodes can be set up as a map network to adapt the fuzzy sets according to the control vectors that are actually being output from the controller.

Using the max criterion defuzzification method, the B_i node with the largest activation (degree of truth) triggers the activation of the CAN node whose output is the prototype value $\bar{u}_j$ corresponding to the fuzzy set B_j. Alternatively, in the center of area method, u is calculated as the normalized weighted sum over the fuzzy sets B_i in which the weights are the activations b_j, given as

$$u = \left(\sum_{k=1}^{|U|} b_k s_k \bar{u}_k \right) / \sum_{k=1}^{|U|} b_k \tag{11}$$

where s_k are the spread parameters for the output fuzzy set membership functions. If the membership function form is symmetrical, then the effect of the spread is trivial.

2.5 Performance Evaluation System (PES)

The particular nature of the plant or process to be controlled and the characterization of the desired performance dictate the details of how the performance evaluation network is configured. When a performance measure is available which is an analytical function of the plant states or output, then the reinforcement signal r_t is simply a normalization of the performance measure p_t to lie in the interval [-1,1] or [0,1]. It is often the case, however, that complex processes which have no known well-defined plant model also do not have a well-defined formula for computing performance. Rather, there is a certain qualitative goal or objective to be reached, but it is not known what the values of the plant variables should be when that goal is reached. Even when a formula in terms of the variables is known, there is often an unknown delay between the control action taken and the effect on the plant variables, so that the result of the current control is not known until some future time. In such cases, various methods of estimating the performance evaluation function must be used. Our investigation into performance function estimation methods are reported elsewhere. In particular, we note the use of various dynamic programming-like algorithms such as the temporal difference (TD) algorithm by Murrell (1994) and both TD and Q-Learning by Esogbue, et al. (1995).

The most straight-forward approach to the situation in which a qualitative determination of reaching the goal is given only after a period of many time intervals is described here. Reaching the goal is indicated by $p_t = 1$ (success) and reaching forbidden states (such as plant shutdown due to variables out-of-bounds) is indicated by -1 (failure). For each state entered at each time, a control action is taken. The average performance over time of this state-control pair is computed and updated whenever there is a success or failure. The reinforcement signal r_t can then simply be the current value of this average for the current state-control pair that just occurred. This method was used with success in the earliest version of this controller.

3 Complexity Analysis of Controller Algorithms

To determine the size of the problem, one must consider the number of state space dimensions p and the number of control space dimensions q. Recall that the size of the spaces must be used in deciding the number of fuzzy subsets utilized in the fuzzy discretization of these spaces. As a consequence, the size of the problem in terms of the complexity of the controller algorithms, must also be expressed in terms of the number s of state space terms (i.e., the number of state space nodes N_i) and the number r of control terms.

The state space node map approach permits increasing the dimension of the input state vectors without an exponential increase in controller complexity. We note that other fuzzy discretization schemes generate a complete set of fuzzy subset terms for each dimension of the state, thereby greatly increasing the number of fuzzy rules as the dimension increases. In general, the usual number of rules is $l^P \bullet m^q$ with l and l the number of state terms per dimension and the number of control terms per dimension respectively. In the proposed method, the number of state terms s is the number of map nodes. This is, in general, selected according to the size of the state space p so that every dimension of the state space can be adequately covered. However, s need not be set in such a way that it grows exponentially with p, since we can elect to cover the state space more sparsely than with a full factorial lattice. Since fuzzification is of the state space vectors rather than merely within each state dimension separately, the generalizing and interpolating characteristics of the fuzzy inference can exploit any smoothness or redundancy in information in all the dimensions simultaneously. Also, the adaptation of the location vectors and spreads tends to move the coverage of the membership function to regions where it is most needed in the state space. The number of rules is determined by the product sr, which can be chosen to grow with p and q more slowly than exponentially. Whether the number of nodes needs to be increased as the dimension increases is a matter of discretion involving a trade-off between the precision of the fuzzification and an increase in the number of nodes. In discussing the size of a problem , we must also consider the number of input data points. For on-line control, there is no fixed number of input data points, rather there is a potentially infinite sequence of data. However, there are some types of algorithms, such as recursive least squares or some types of adaptive clustering algorithms which would require either the complete past history or the history for some period into the past to be stored and used in the computations performed at each time step as a new data point becomes available. The usual clustering algorithms, for example, would require a pass through a set of past data for each additional point, greatly adding to the complexity of the computations as time advanced. Incremental parametric algorithms, including many neural network algorithms and the algorithms proposed here, do not have this drawback. Others such as those recently developedby Esogbue and Liu have attractive features that can be explored in future versions of the controller. Let us briefly address the complexity issues associated with the computation for one time step of the SFAL controller.

We begin with the SFDN algorithms. Computation of a_i requires s sets of calculations, one for each node; each of these includes operations that must be performed on each state dimension, i.e., p times. It has been determined that for one step of the SFDN algorithms, the number of computations is bounded above by $Ms + Nsp + L$ (for some integer l, M, and N), and thus the order of complexity is $O(sp)$.

In the SLFCN algorithm, while in principle, the quantities γ_{ij} require, sr computations, the simplified algorithm utilized by Murrell(1994) with γ_{ij} as an indicator, needs only r computations. On the average, even this can be made smaller. Obtaining all the b_j similarly requires sr, or simply r computations, depending on whether or not γ_{ij} indicates more than one node N_i. The c_{ij} require sr, or actually $s'(k) \bullet r$ computations, where again $s'(k)$ decreases with time down from s. The SLFCN therefore requires $Mr + Nsr + L$ operations (L, M, N some integers, different from above),and as such has complexity $O(sr)$.

The IFCN requires only a few arithmetic operations performed sr times to obtain the g_{ij} from the c_{ij}, and sr computations to obtain $=^T$ leading to a complexity of $O(sr)$. The CAN similarly needs only r sets of simple arithmetic operations, while the PES needs only 1 small set of arithmetic operations per time step.

Thus, one iteration of the entire controller algorithm has a complexity of $O(sp + sr)$. Let us next address the problem of the number of iterations that may be required.

To learn the control law as a fuzzy relation matrix, requires that s pieces of information be learned or estimated. This is accomplished by a stochastic search of r^s combinations which could be of formidable complexity. However, the algorithm does not try each combination one at a time. Each SFDN node is a learning unit which explores only r possibilities, in parallel sequences of learning trials with all the other nodes. If the reinforcement signals were based on performance scores that were absolutely accurate and certain, then in principle, each control choice for each state need be visited exactly once, hence requiring only sr iterations of the controller algorithm. However, since the information is uncertain in a *Statistical* as well as a fuzzy sense, some multiple of sr is required to obtain an adequate sample of information. A conservative analysis shows that , with a probability greater than 0.98, the controller can generate a complete set of performance predictions in approximately $4r$ plant runs.

4 Implementation Issues for the Application of the Controller

In order to utilize this controller in its current configuration, the following conditions must be met:

1. Ability to characterize the process at any given time by its process state.

2. Ability of the control inputs to affect the sequence of input states.

3. Availability of a good performance measure which is related to the system goal

4. There must be some topological nearness measure (e.g., a metric) for both the state space $\mathbf{X}$ and control space $\mathbf{U}$ which additionally satisfies the smoothness assumptions listed below:

   (a) For every $\mathbf{x} \in \mathbf{X}$ there exists a control $\mathbf{u}^\star$ such that $y(\mathbf{x}, \mathbf{u}^\star) = \sup_{\mathbf{u} \in \mathbf{U}}\{y(\mathbf{x}, \mathbf{u}\}$.

   (b) If the optimal control for $\mathbf{x}_i$ is $\mathbf{u}^\star$ then the optimal control for every $\mathbf{x}$ in a neighborhood of $\mathbf{x}_i$ is near $\mathbf{u}^\star$.

5. There must be a direct relationship between the degree to which a control action contributes to the final computed control and the probability of that action being successful when applied purely.

6. The process is either intrinsically recurrent or can be repeatedly restarted at random initial states.

We note that the nearness structure of the spaces stipulated in the foregoing conditions 4 can be imposed or arrived at via some transformation of the original spaces. In particular, it is neither necessary for the smoothness to be analytical differentiability nor the nearness to be in terms of a metric computed by a formula.

Let us now outline the steps to be followed when trying to apply the controller to a system or process which has met the conditions stipulated earlier. These are succintly summarized in the flowchart for the algorithm 2.

As usual, we begin with the set-up and initialization of the controller for the particular process as outlined below:

1. Set the state space and control space boundaries.

2. Set the state map size and initial location and spread parameters.

3. Set the control terms (control space map).

4. Set initial correlation parameters.

5. Determine the sampling interval for the process.

6. Determine process events to be used to index algorithm steps for each algorithm:

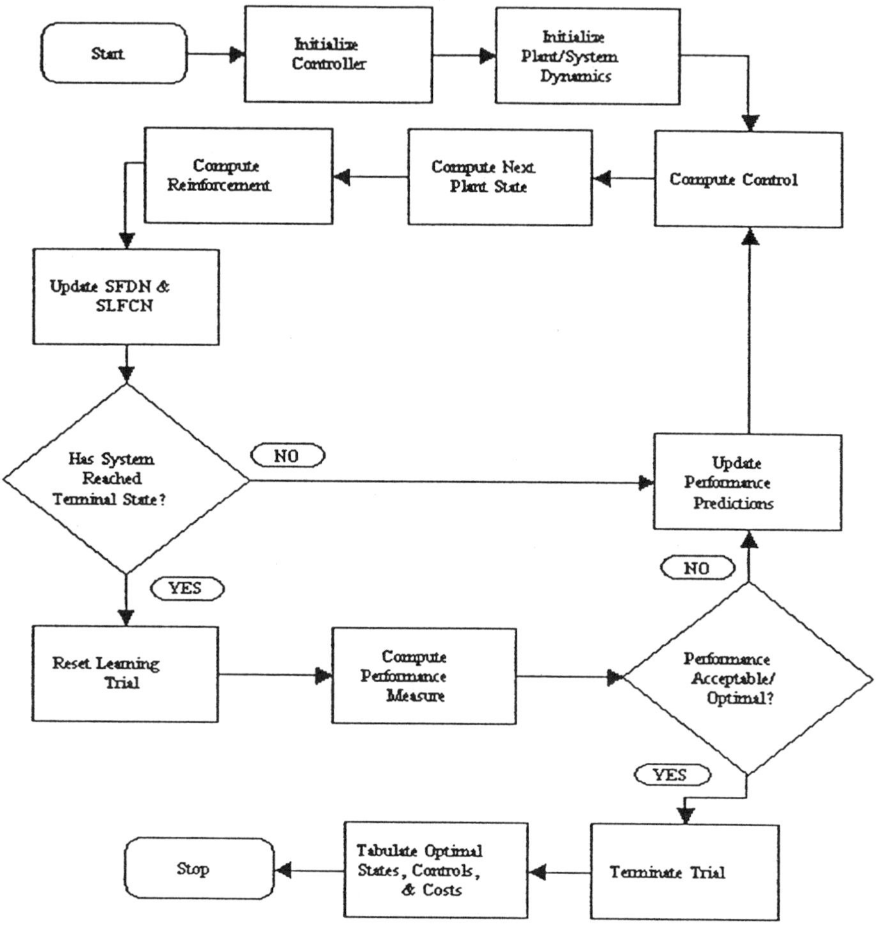

Figure 2: Flowchart for Implementation of Fuzzy-Neuro Controller

(a) Map node location update.

(b) Map node location neighborhood decay.

(c) Map node location update rate decay.

(d) Spreads update.

(e) Correlation parameter update.

(f) Correlation neighborhood decay.

(g) Performance evaluation update.

7. Set controller learning parameters.

8. Determine a normalized performance measure to be computed at terminal states.

As in most algorithms, it is beneficial to have some basic knowledge or information about the process such as its state and control variables as well their ranges. Using this knowledge, the state and control space limits used by the controller are determined. Determining the number of nodes, the size of the crisp neighborhoods, and the basic level of fuzziness or dispersion for the membership functions requires some reasoning about what level of precision is appropriate for the process. The size of the state space region to be covered and what level of resolution is required to resolve the control switching surfaces must be considered. Ability to partition both the state and control spaces so that they can be fuzzy discretized is essential. This relates to the issues of continuity and nearness discussed above. According to condition 4b the control behavior for a locality in the state space must be generalizable throughout the neighborhood.

When no prior information is available, the map node locations are set at the points of a uniformly spaced grid. If it is impractical to fill all the points of the grid, then design of experiments methods can be used to determine a good subset of the full factorial coverage. The nodes can be placed with greater density in regions where more coverage is needed if there is any prior information. Similarly, if there is no prior information about the control relation, then one can set the correlation parameter vectors to be uniform probability distributions. Any prior information can be encoded by biasing the distribution vectors towards favored control actions.

The operation then consists of repeated cycles of computing controls to input to the plant, plant state transitions, computing reinforcements, and performing learning updates. Whenever a terminal state is reached, such as a goal state or set-point of the problem, then a performance measure is computed and provided to the controller Performance Evaluation System in order to update its performance prediction algorithm. The system is reset—i.e., a new initial state is determined and any algorithm counters, etc., are reset. The operation terminates if it is determined that the attained performance is acceptable or near optimal.

In general, the most ideal problem situations for applying the controller are those in which there exist a high tolerance for mistakes to be made for a brief period early in the learning phase or which allow for initial learning with a simulated on-line system.

5 Controller Application and Performance on Sample Problems

5.1 Application to Inverted Pendulum Problem

A classic testbed problem for nonlinear controllers is the inverted pendulum problem. One of the simplest inherently unstable systems known, yet it has a broad base for comparison throughout the literature. The system is shown in the following Figure 3.

Model of Process

Mathematically, the system is described as follows:

Inputs: State vector—$\mathbf{x} = [\theta, \Delta\theta]$.

Outputs: Force applied (in Newtons)—$\mathbf{u} = [F]$ where $F > 0$ denotes a force applied in the positive x direction.

Equations of Motion: These equations, it must be wmphasized, serve only to simulate the system and are not used in the derivation of the control law:

$$\ddot{\theta} = \frac{g \sin\theta + \cos\theta \left(\frac{-F - m_p l \dot{\theta}^2 \sin\theta + \mu_p \mathrm{sgn}(\dot{x})}{m_p + m_c} \right) - \frac{\mu_p \dot{\theta}}{m_p l}}{l \left(\frac{4}{3} - \frac{m_p \cos^2\theta}{m_p + m_c} \right)} \tag{12}$$

$$\ddot{x} = \frac{F + m_p l \left(\dot{\theta}^2 \sin\theta - \ddot{\theta}\cos\theta \right) - \mu_c \mathrm{sgn}\left(\dot{x}\right)}{m_p + m_c} \tag{13}$$

where the variables are defined as g: acceleration due to gravity, m_c: mass of the cart (kg), m_p: mass of the pole (kg), l: half-length of pole (m), μ_c: coefficient of friction for cart (N), and μ_p is the coefficient of friction for pole (N) with values shown in Table 1.

$g = 9.8\text{m/s}^2$	$m_c = 1.0\text{kg}$
$m_p = 0.1\text{kg}$	$l = 0.5\text{m}$
$\mu_c = 0.0\text{N}$	$\mu_p = 0.0\text{N}$

Table 1: Parameters of Inverted Pendulum System Simulation.

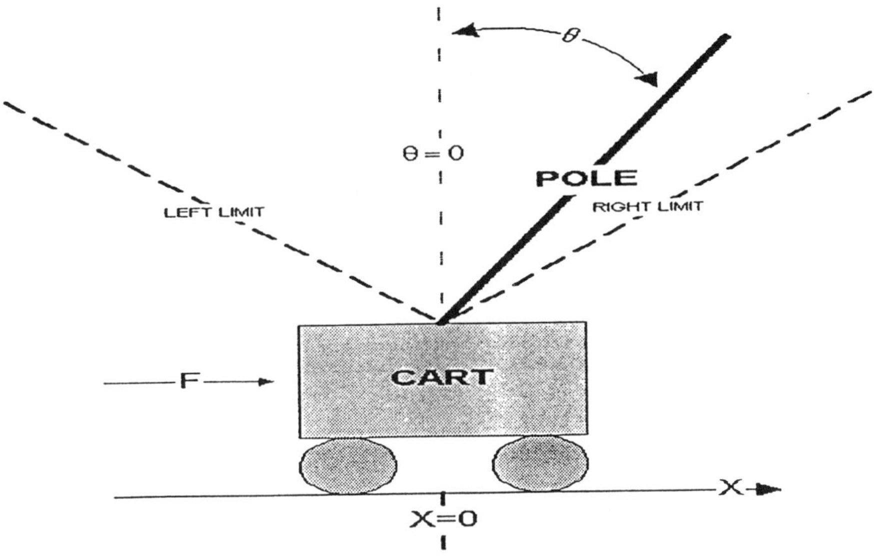

Figure 3: The Pole Cart System

Success and Failure: The setpoint for the inverted pendulum problem is $[0, 0]$. Failure occurs at either of the following conditions:

$$| \theta | > 12^\circ$$
$$| \Delta\theta | > 25^\circ/\mathrm{s}$$

In a dynamical system problem such as the inverted pendulum under investigation, the system is always reset upon failure (exceeding state space boundaries), and may be reset upon reaching the goal state; in all the experiments reported in this study for this type of system, reset is not performed upon reaching a goal state. Also, new initial states of the system are selected randomly. The results of a representative run of the simulations are depicted in the following figures of the trajectory 4 and the control surface 5.

5.2 Application to Communication Networks

The controller's performance was further tested on the communications network problem of great interest in the literature. An instructive example involving learning is the problem discussed in [27]. Initially, the validation of the learning capabilities of our controller was effected via a classic three node problem and then a larger ten node problem. The configuration for the later is shown in figure 6. This problem basically involves the routing of arriving messages from several origins to their destinations. Network characteristics and performance index are as in the classic literature. Here, reset is upon arrival of the current message to its destination, and determining a new initial state consists of observing where the next message arrives in the network. No failure state is defined for the routing problem since every message eventually reaches its destination (network protocols prevent endless cycling). The networks were successfully simulated and optimal routing controls learned by the controller.

Comparison with Other Network Methods

The performance of the controller in the sample networks was compared to several other message routing approaches known in the literature. A rigorous statistical analyis of the results indicate acceptable performance which is equal to and superior in instances to the approaches in question. These results are reported in detail in Murrell [29]. As an example of the comparative performance of our controller on these sample problems, we present a graphical representation of these results where the 89 percent and 95 percent confidence intervals show that both the controller and the dynamic shortest path algorithms have statistically equal performance which is better than that attained by both the SLA and random controllers. This is depicted in figure 7.

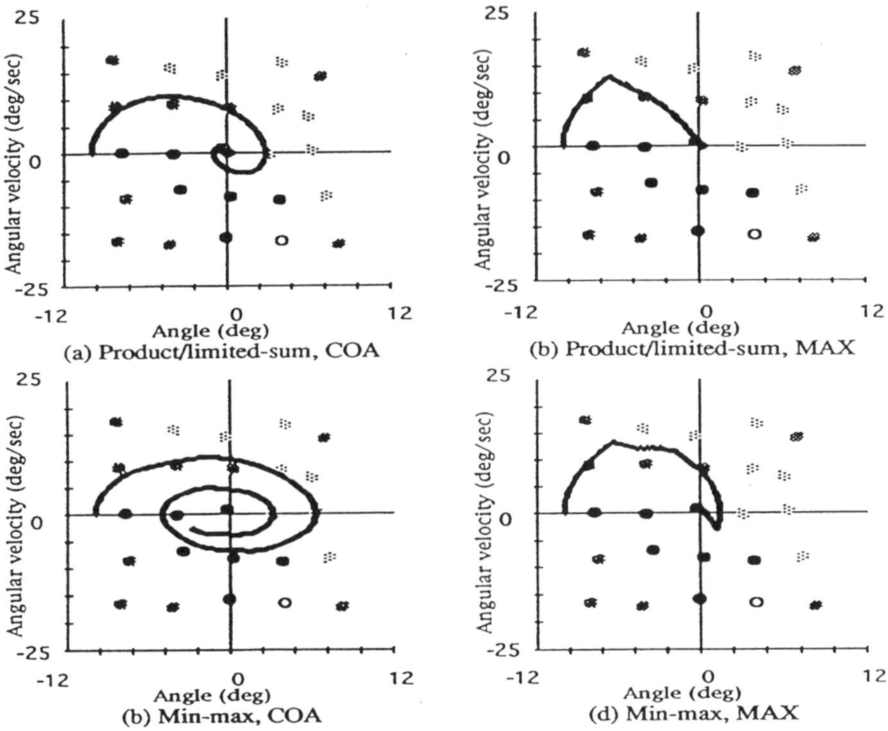

Figure 4: Control Trajectory for the Pole Cart System

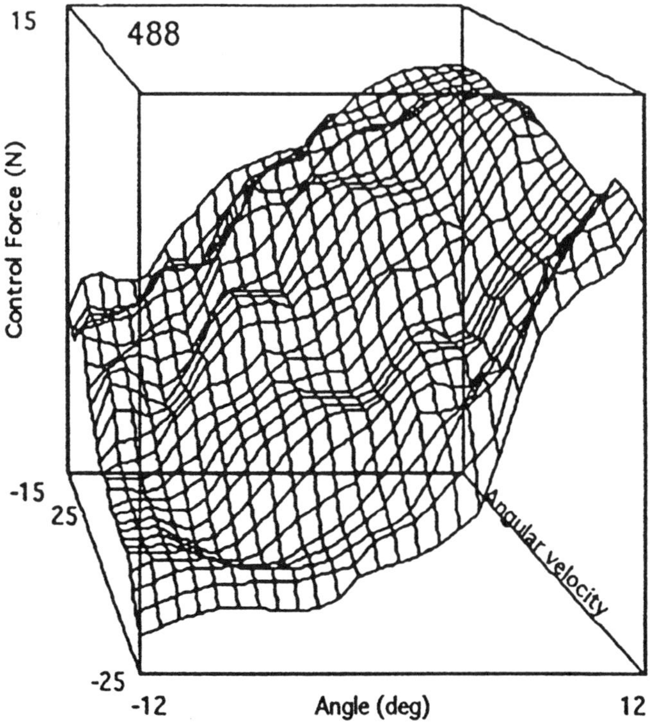

Figure 5: The Control Surface for the Pole Cart System

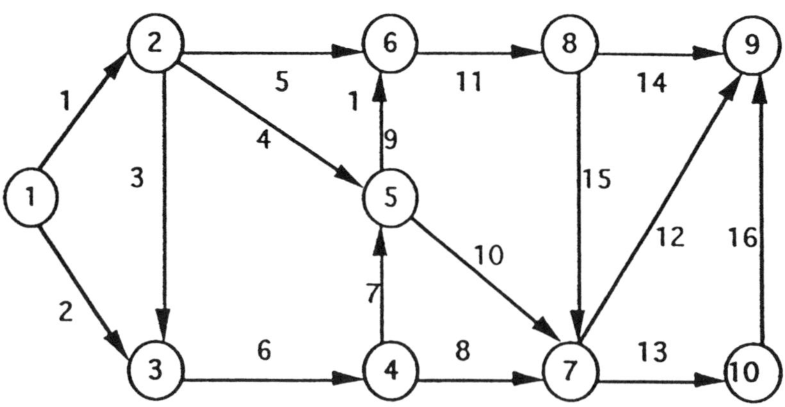

Figure 6: A Ten Node Communication Network Problem

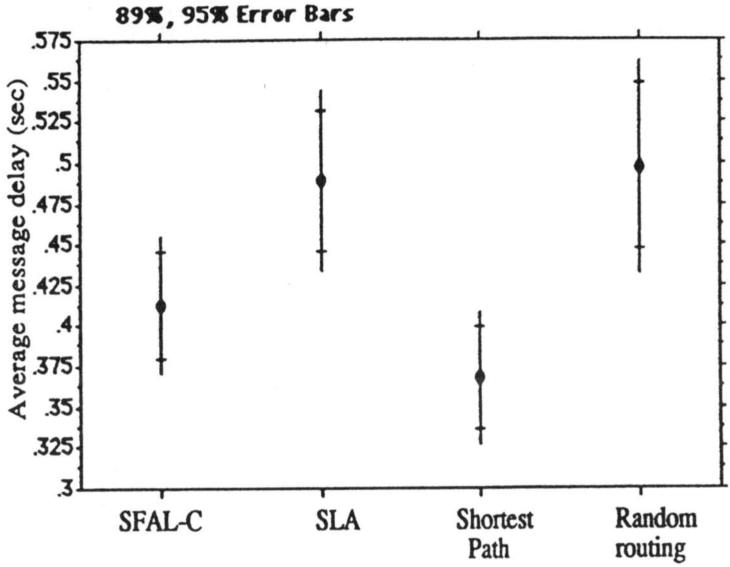

Figure 7: Average Delay Statistics of 4 Controllers for the Network Problem

5.3 Application to Power System Stabilization Problems

One of the most intriguing and frequently investigated problem areas for deploying potent and novel tools of control engineering is the power system stabilization problem. Many different control strategies as well as controllers have been tested on this problem. Part of this interest is engendered by both the challenge and intractability of power systems which are characterized by the existence of power inherently complex, nonlinear, time-varying and indeterminable elements, simple controllers which work well in one situation may not perform equally well in another. As part of the experimental ivestigations with the SFAL controller, we explored its ability to learn a robust control law to stabilize the power system under various operating conditions.

The earliest stabilizers consisted of a lead-lag analog circuit with the speed as the input. Such a simple controller cannot satisfy the high standards of the power system. PID controllers [18] perform better than the lead-lag circuit. Yet, unless their parameters are tuned automatically as operating conditions change, PID controllers in general do not work satisfactorily within a wide range of conditions. The self-tuning controller [6. 14, 24, 23] and the adaptive controller [4, 5, 20] are designed for this purpose. By continuously identifying the model of the plant, the self-tuning controller adjusts its parameters to achieve optimal performance, while the adaptive controller adjusts its parameters based on the knowledge of the plant model. These two controller paradigms are time-consuming in design but can perform well under different operating conditions. Most recently, fuzzy logic controllers [16, 17, 19, 23, 34] have been successfully applied to stabilize power systems. It has been found that the fuzzy logic controller performs as well as the self-tuning controller in power system stabilization [23] and shows great potential for application in power systems. When the system is of large scale and of high complexity, however, it is not easy to extract the control rules from human expert(s) [21] and, even if this can be done, the expert's experience is still limited. Therefore, it is necessary to design a controller that can "learn" the control law via its own experience.

6 Mathematical Models of the Power System

The system considered here is composed of a synchronous machine with an exciter and a stabilizer connected to an infinite bus. The dynamics of the synchronous machine can be expressed as follows using the linearized incremental model [18]. To reiterate, these equations serve only to simulate the system and are not used in the derivation of the control law.

$$\begin{aligned}
\Delta\dot{\omega} &= \frac{1}{M}(\Delta T_m - \Delta T_e - \\
&\quad \Delta T_L - D\Delta\omega) && (14)\\
\Delta\dot{\delta} &= 377\Delta\omega && (15)\\
\Delta T_e &= K_e\Delta\delta + K_2\Delta e_q && (16)\\
\Delta\dot{e}_q &= \frac{1}{K_3 T_{de}}(K_3\Delta e_{fd} - \\
&\quad K_3K_4\Delta\delta - \Delta e_q) && (17)\\
\Delta V_t &= K_5\Delta\delta + K_6\Delta e_q && (18)\\
\Delta\dot{V}_f &= \frac{1}{T_F}(K_f\Delta\dot{e}_{fd} - \Delta V_F) && (19)\\
\Delta\dot{e}_{fd} &= \frac{1}{T_E}(\Delta V_A - K_E\Delta e_{fd}) && (20)\\
\Delta\dot{V}_A &= \frac{1}{T_A}(K_A\Delta V_{ref} - K_A\Delta V_F + K_A u - \\
&\quad K_AK_6\Delta e_q - K_AK_5\Delta\delta - \Delta V_A) && (21)
\end{aligned}$$

$$| u | \leq u_{\max} \tag{22}$$

where

V_{ref}	constant reference input voltage
ΔV_t	terminal voltage change,
ΔV_o	infinite bus voltage change
Δe_{fd}	equivalent excitation voltage change
Δe_q	q-axis component voltage behind transient reactance change
ΔV_F	stabilizing transformer voltage change
u	stabilizer output
ΔT_m	mechanical input change
ΔT_e	energy conversion torque change
ΔT_L	load demand change
$\Delta \delta$	torque angle deviation,
$\Delta \omega$	angular velocity deviation
K_A, K_E	voltage regulator gains
T_A, T_E	voltage regulator time constants
K_F	stabilizing transformer gain
T_F	stabilizing transformer time constant
$K_1, \ldots, K_6$	constants of the linearized model of synchronous machine
T_{do}	d-axis transient open circuit time constant
M	inertia coefficient
D	damping coefficient
T_s	sampling period

The objective of the controller is to drive the state of the system $\vec{x}$ to $[0, 0]$ via the stabilizer output u. The values for the above parameters are given in Table 2 below.

7 Simulation Results

The experiments run on the power system stabilization problem consisted of multiple replications of the learning phase of the controller on a simulated power system written in C. The inputs to the controller are $\Delta\omega$ and $\Delta\dot{\omega}$. Thus, the controller in effect mimics a PD-like controller with unknown structure. The state space is defined as $\Delta\omega \in [-0.012, 0.012]$ and $\Delta\dot{\omega} \in [-0.025, 0.025]$. The number of nodes for the SFDN is set at $N = 25$ and there are 5 refer-

$K_1 = 1.4479$	$K_2 = 1.3174$	$K_3 = 0.3072$
$K_4 = 1.8050$	$K_5 = 0.0294$	$K_6 = 0.5257$
$K_A = 400$	$T_F = 1.0$	$T_A = 0.05$
$D = 0$	$T_{do} = 5.9$	$K_E = -0.17$
$M = 4.74$	$T_E = 0.95$	$K_F = 0.025$
$\Delta T_m = 0$	$\Delta V_{ref} = 0$	$T_s = 0.01$

Table 2: Parameters of Simulation.

ence control fuzzy sets defined for $u \in [-0.12, 0.12]$. Once the controller has completed the learning phase, it is used as a stabilizer in the system.

Several experiments were run and an example of the resulting controller is shown in the figures below. Figure 8 shows the transient process of $\Delta\omega$ when the load increases 0.05 pu and 0.3 pu, respectively. It takes about 2 seconds for the speed deviation $\Delta\omega$ to vanish for the 0.05 pu load change and about 3 seconds for the 0.3 pu load change. Figure 9 shows the learned control surface using product–limited sum inference and center-of-area defuzzification.

7.1 Comparison to Existing Controllers

The simulation results clearly showed that the controller can learn an effective control law to stabilize the system under varying load conditions. However, the results, are not optimal with regard to settling time. The settling time obtained with our controller was slightly longer than the results obtained using an existing PID controller [18] and a fuzzy controller [19], but comparable to or shorter than the settling time for other fuzzy controllers [16, 17, 34] reported in the literature. The optimality issue (see Section 7.2) is currently under investigation, but the relative ease of developing an "efficient" controller via the self-learning controller with respect to the existing methods illustrates the potential of this approach.

Despite the foregoing, the advantages of our controller over other controllers that have been applied to the power systems stabilization problem are very significant and should be noted:

1. The controller successfully learned the control law via its own experience. It did not require the analytic solution of a dynamical model, the tuning of parameters as in PID control, and it did not rely on existing expert knowledge about the control of the process.

2. The learning phase of the controller took less than 5 minutes to complete.

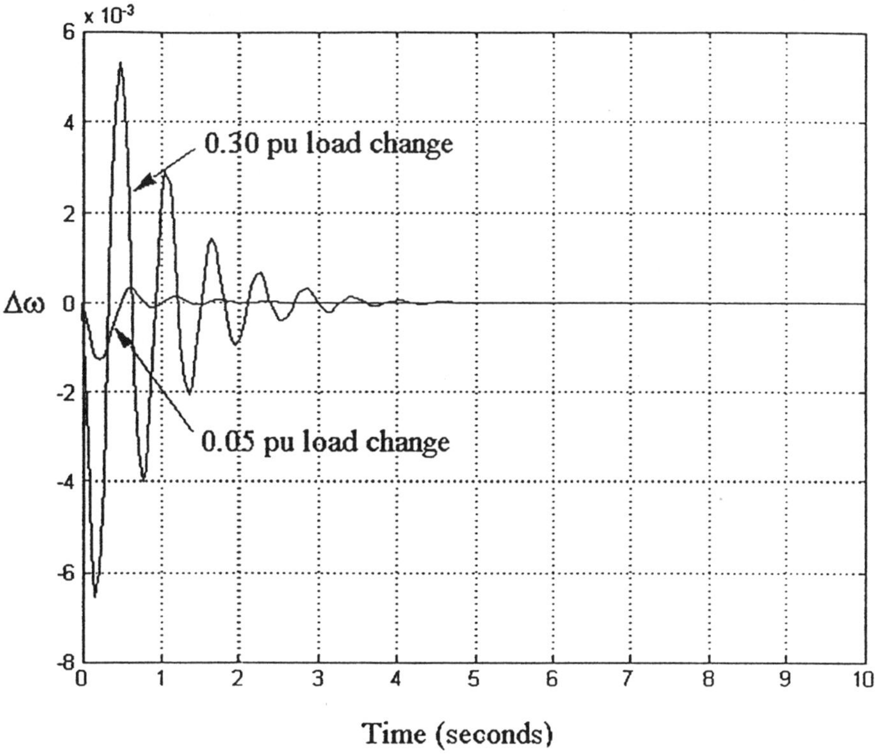

Figure 8: Transient Process for Selected Load Changes.

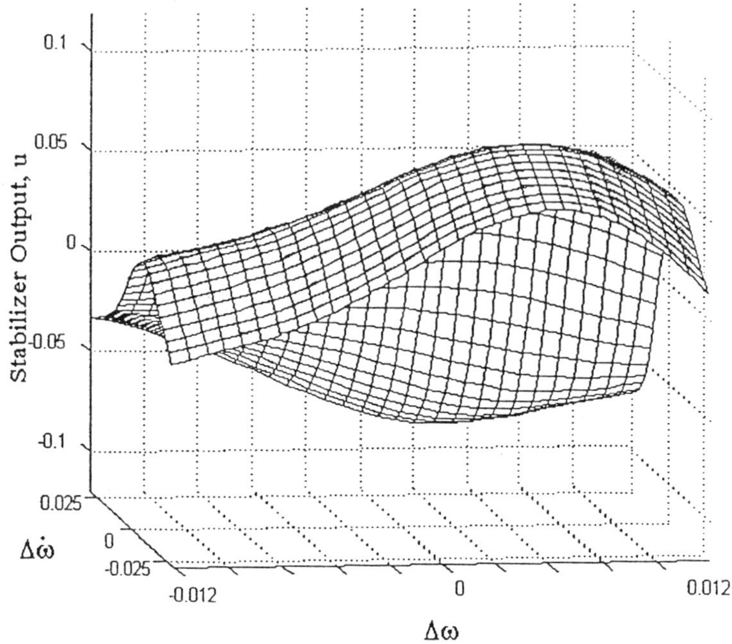

Figure 9: Learned Control Surface for Stabilization.

Thus, there is a huge time savings in development over the existing controllers.

3. The internal controller parameters (that control the learning and other attributes) are very robust—the parameters used for the inverted pendulum problem were used for the power system stabilization problem. No tuning of these parameters was performed, although doing so may improve the resulting control.

4. The resulting control is robust—the controller can handle a more extreme range of load changes than the PID controller [18].

7.2 Reinforcement Learning and Optimal Control

The lack of optimality with respect to settling time is not unexpected since the controller, which learns via reinforcements. is only given one goal: *Drive the power system to its set point* $\vec{x} = [0, 0]$. The only reinforcements that are given are upon *success* or *failure* of the plant and a desired trajectory through the state space is not defined. These external reinforcements are used to update an internal prediction function for each maximally activated node n_t at time t via $P_{t+1}(n_t) = P_t(n_t) + \alpha(P_t(n_{t+1}) - P_t(n_t))$, similar to Sutton's method of temporal differences. Thus, only by trial and error does the controller learn to drive the system to the set point, and it does so without requiring (and therefore without necessarily satisfying) a performance objective function.

The two controllers that performed better [19, 18] both utilized external information about the plant or the behavior of the controller. This additional information about the plant dynamics and desired control behavior could possibly be used by our controller to develop near-optimal control automatically. For example, knowing the desired trajectory of the power system through the state space allows the controller to give itself *additional* external reinforcement about its behavior, thereby changing $P_t(n_t)$ and the learned control law.

8 Unique Features

This controller has several unique features mentioned earlier which we summarize here. It is adaptive and well suited to the control of complex processes. In particular, it is capable of learning effective control using process data and improving its control through on-line adaption. The controller performs a fuzzy discretization of the state and control spaces and learns the fuzzy relations for these fuzzy subsets using a variation of the TD method with its dynamic programming inspirations. While it adapts both the membership functions and the control rule state-control association, the controller primarily learns the control rule associations, unlike many other methods which fix the rules and adjust the membership functions. Most important, it does so for the entire state space. Additionally, no training data sets nor any error signal derived from knowledge of the desired plant trajectory are needed. This self-learning controller has been successfully applied to the inverted pendulum problem, the DC servomotor position control problem, the switching problem in a distributed communications network, and the power system stabilization problem.

9 Conclusions

We have reported the development an intelligent controller with many unique features and successfully applied it, with various modifications, to an array of problems. Of note are: Esogbue and Murrell, 1993a; Esogbue and Murrell, 1993b; Murrell, 1993; Esogbue and Murrell, 1994; Esogbue and Hearnes,

1995; Esogbue, Hearnes and Song, 1995). To repeat the summarization presented in Murrell (1994), one of the principal developments of this research includes a feasible and practical tool for the application of fuzzy generalization to the discrete rules or associations of action -response utilized in reinforcement learning methods. Further investigations and extensions are underway. These include the use of more intelligent reinforcement strategies especially those that are DP-based algorithms useful in learning real-time control strategies. Of interest are Watkin's Q-learning algorithm and its variations as well as the hybridization of the controller involving dynamic switching between the reinforcement controller and a stabilizing controller.

10 Acknowledgements

The contributions of several graduate and undergraduate students including James Murrell, Warren Hearnes, and Qiang Song who worked on various aspects of this project that are reported here, under funding provided principally by the National science Foundation under Grant No. ECS 91216004 and the Electic Power Research Institute under Grant No RP 8030-16 are gratefully acknowledged.

References

1. Barto, A.G., Bradtke, S.J., and Singh, S.P. Learning to act using real-time dynamic programming. *Artificial Intelligence*, **72:1-2**, 1995, 81–138.

2. Berenji, H.R. Fuzzy Q-learning: a new approach for fuzzy dynamic programming. *Proceedings of the Third IEEE Conference on Fuzzy Systems*, Orlando, FL, June 26-29, 1994, 486–491.

3. Berenji, H.R. and Ralescu, A.L. Fuzzy reinforcement learning and dynamic programming. *Proceedings of the Second Workshop on Fuzzy Logic in Artificial Intelligence*, Chamberry, France, August 28, 1993, 1–9.

4. Cheng, C.-H. and Hsu, Y.-H. Damping of generator oscillations using an adaptive static VAR compensator. *IEEE Transactions on Power Systems*, **7:2**, 1992, 718–725.

5. Cheng, S.-J., Chow, Y.S., Malik, O.P and Hope, G.S. An adaptive synchronous machine stabilizer. *IEEE Transactions on Power Systems*, **PWRS-1:3**, 1986, 101–107.

6. Cheng, S.-J., Malik, O.P and Hope, G.S. Self-tuning stabilizer for a multimachine power system. *IEE Proceedings*, **133-C:4**, 1986, 176–185.

7. Esogbue, A.O. Optimal clustering of fuzzy data via fuzzy dynamic programming. *Fuzzy Sets and Systems*, **18**, 1986, 283–298.

8. Esogbue, A.O. and Hearnes, W.E. Constructive experiments with a new fuzzy adaptive controller. *NAFIPS/IFIS/NASA '94. Proceedings of the First International Joint Conference of the North American Fuzzy Information Processing Society Biannual Conference. The Industrial Fuzzy Control and Intelligent Systems Conference, and the NASA Joint Technology Workshop on Neural Networks and Fuzzy Logic*, San Antonio, TX, December 18-21, 1994, 377–380.

9. Esogbue, A.O., Hearnes, W.E.and Song, Q. A reinforcement learning fuzzy controller for set-point regulator problems. *Proceedings of the Fifth IEEE Conference on Fuzzy Systems*, New Orleans, LA, September 8-11, 1996, pp. 2136–2142.

10. Esogbue, A.O. and Murrell, J.A. A fuzzy adaptive controller using reinforcement learning neural networks. *Proceedings of Second IEEE International Conference on Fuzzy Systems*, March 28–April 1, 1993, 178–183.

11. Esogbue, A.O. and Murrell, J.A. Advances in fuzzy adaptive control. *Computers & Mathematics with Applications*, **27:9-10**, 1994, 29–35.

12. Esogbue, A.O. and Song, Q. Optimal defuzzification and applications. *Proceedings of the International Joint Conference on Information Science, Fourth Annual Conference on Fuzzy Theory & Technology*, Wrightsville Beach, NC, September 28–October 1, 1995.

13. Esogbue, A.O., Song, Q. and Hearnes, W. E. Application of a self-learning controller to the power system stabilization problem. *Proceedings of the 1995 World Conference on Neural Networks* Washington, D.C. **II** 699–700.

14. Ghandakly, A.A. and Farhoud, A.M. A parametrically optimized self-tuning regulator for power system stabilizers. *IEEE Transactions on Power Systems*, **7:3**, 1992, 1245–1250.

15. Glorennec, P.Y. Fuzzy Q-learning and dynamical fuzzy Q-learning. *Proceedings of the Third IEEE Conference on Fuzzy Systems*, Orlando, FL, June 26-29, 1994, 474–479.

16. Hassan, M.A.M., Malik, O.P and Hope, G.S. A fuzzy logic based stabilizer for a synchronous machine. *IEEE Transactions on Energy Conversion*, **6:3**, 1991, 407–413.

17. Hiyama, T. and Sameshima, T. Fuzzy logic control scheme for on-line stabilization of multi-machine power system. *Fuzzy Sets and Systems*, **39**, 1991, 181–194.

18. Hsu, Y.-Y. and Hsu, C.-Y. Design of a proportional-integral power system stabilizer. *IEEE Transactions on Power Systems*, **PWRS-1:2**, 1986, 46–53.

19. Hsu, Y.-Y. and Cheng, C.-H. A fuzzy controller for generator excitation control. *IEEE Transactions on Systems, Man and Cybernetics*, **23:2**, 1993, 532–539.

20. Irving, E., Barret, J.P., Charcossey, C. and Monville, J.P. Improving power network stability and unit stress with adaptive generator control. *Automatica*, **15**, 1979, 31–46.

21. Jang, J.-S.R. ANFIS: Adaptive-network-Based fuzzy inference system. *IEEE Transactions on Systems, Man and Cybernetics*, **23:3**, 1993.

22. Kaymak, U. and Babuska, R. Compatible cluster merging for fuzzy modelling. *Proceedings of 1995 IEEE International Conference on Fuzzy Systems. The International Joint Conference of the Fourth IEEE International Conference on Fuzzy Systems and The Second International Fuzzy Engineering Symposium*, Yokohama, Japan, March 20-24, 1995, 897–904.

23. Lim, C.M. and Hiyama, T. Comparison study between a fuzzy logic stabilizer and a self-tuning stabilizer. *Computers in Industry*, **21**, 1993, 199–215.

24. Lim, C.M. and Hiyama, T. Self-tuning control scheme for stability enhancement of multimachine power systems. *IEE Proceedings*, **137-C:4**, 1990, 269–275.

25. Lin, C.-T. and Lee, C.S.G. Reinforcement structure/parameter learning for neural-network-based fuzzy logic control systems. *IEEE Transactions on Fuzzy Systems*, **2:1**, 1994, 46–63.

26. Liu, B. and Esogbue, A.O. Optimal fuzzy criterion clustering based on fuzzy prototypes. *Proc. 1995 IEEE International Conference on Systems, Man, and Cybernetics*, Vancouver, Canada, 1995, 4702–4705.

27. Mars, P. and Narendra, K.S. Routing,flow control and learning algorithms. *First IEEE National Conference on UK Telecommunication Networks - Present and Future*, London, England, June 2-3, 1987, 78–83.

28. Maeda, M., Sato, T., and Murakami, S. Design of the self-tuning fuzzy controller. *Proceedings of the International Conference on Fuzzy Logic and Neural Networks*, Iizuka, Japan, 1990, 393–396.

29. Murrell, J. A. *A statistical fuzzy associative learning approach to intelligent control.* Ph.D. Thesis, Georgia Institute of Technology, Atlanta, Georgia, December 1993.

30. Nakanishi, S., Takagi, T., Unehara, K., and Gotoh, Y. Self-organizing fuzzy controllers by neural networks. *Proceedings of the International Conference on Fuzzy Logic and Neural Networks*, Iizuka, Japan, 1990, 187–191.

31. Nedzelnitsky, O. V., and Narenda, K. S. Nonstationary models of learning automata routing in data communication networks. em IEEE Trans. Syst. Man Cybern. **17:6**, 1987, 1004–1015.

32. Patrikar, A. and Provence, J. A self-organizing controller for dynamic processes using neural networks. *Proceedings of the International Joint Conference on Neural Networks*, **3**, 1990, 359–364.

33. Procyk, T.J. and Mamdani, E.H. A linguistic self-organizing process controller. *Automatica*, **15**, 1979, 15–30.

34. Shi, J., Herron, L.H. and Kalam, A. A fuzzy logic controller applied to power system stabilizer for a synchronous machine power system. *Proceedings of IEEE Region 10 Conference, Tencon 92*, Melbourne, Australia, November 11–13, 1992.

35. Sutton, R.S. Learning to predict by the method of temporal differences. *Machine Learning*, **3**, 1988, 9–44.

36. Watkins, C.J.C.H. *Learning from delayed rewards.* Ph.D. Thesis, Cambridge University, Cambridge, England, 1989.

37. Watkins, C.J.C.H. and Dayan, P. Q-Learning. *Machine Learning*, **8**, 1992, 279–292.

38. Werbos, P.J. An overview of neural networks for control. *IEEE Control Systems Magazine*, **11:1**, 1991, 40–41.

39. Whitehead, S.D. and Lin, L.-J. Reinforcement learning of non-Markov decision processes. *Artificial Intelligence*, **73:1-2**, 1995, 271–306.

40. Yamaoka, M. and Mukaidono, M. A learning method of fuzzy inference rules with a neural network. *Proceedings of the International Fuzzy Systems Association, Fourth World Congress*, July, 1991.

41. Zadeh, L.A. Outline of a new approach to the analysis of complex systems and decision processes. *IEEE Transactions on Systems, Man, and Cybernetics*, **SMC-1**, 1973, 28–44.

42. Zeng, S. and He, Y. Learning and tuning fuzzy logic controllers through genetic algorithm. *Proceedings of the 1994 IEEE International Conference on Neural Networks*, Orlando, FL, June 27-July 2, 1994, 1632–1637.

43. Zomaya, A.Y. Reinforcement learning for the adaptive control of nonlinear systems. *IEEE Transactions on Systems, Man and Cybernetics*, **24:2**, 1994, 357–363.

SAMPLING THEORY AND FUNCTION SPACES

Hans–Jürgen Schmeisser and Winfried Sickel
Mathematisches Institut, F.–Schiller–Universität Jena,
D–07743 Jena, Germany
E-mail: mhj@minet.uni-jena.de, sickel@minet.uni-jena.de
Contact author: W. Sickel

We investigate the convergence and the rate of convergence in $\| \cdot |L_p\|$, $0 < p \leq \infty$ of the classical Shannon series and its generalizations. This will be done within the framework of function spaces of Nikol'skij type using Fourier analytical characterizations as well as characterizations by different average moduli of continuity. Various examples of generalized kernels are treated.

1 Introduction

The aim of this survey article is to bring together sampling theory and function spaces of Besov-Nikol'skij type. We consider sampling from the point of view of approximation processes. The Kotel'nikov-Shannon theorem says that each function bandlimited in $[-\pi W, \pi W]$ and belonging to L_p, $p < \infty$, can be represented for all t as

$$S_W f(t) = \sum_{k=-\infty}^{\infty} f(\frac{k}{W}) \operatorname{sinc}(Wt - k) . \tag{1}$$

If f is not bandlimited, in particular (as in real life) if f is timelimited, one is forced to consider the approximation process

$$S_W f \longrightarrow f \qquad (W \to \infty) \tag{2}$$

for certain classes of functions. More general, if replacing the sinc-kernel in (1) by an appropriate function Ψ, which has a faster decay at infinity, one achieves at

$$S_W^{\Psi} f(t) = \sum_{k=-\infty}^{\infty} f(\frac{k}{W}) \, \Psi(Wt - k) \tag{3}$$

and the approximation process

$$S_W^{\Psi} f \longrightarrow f \qquad (W \to \infty) . \tag{4}$$

Hundreds of papers have been published dealing with this topic and related questions during the last 25 years, stimulated by practical (signal analysis) as well as by mathematical reasons. An excellent survey on the research carried out by the Butzer School at Aachen has been given in [21]. A comprehensive treatment of different aspects has been given by Higgins [39]. We refer also to Marks [48, 49] and Jerri [44]. With respect to approximation most of the papers deal with uniform or pointwise convergence in (2) and (4) for spaces of Hölder-continuous and differentiable functions and with construction of optimal kernels. Much less seems to be known for L_p–convergence, $0 < p < \infty$, see Rahman and Vertesi [57] and Fang Gensun [33]. In some sense we try to fill this gap or to complete the picture looking for Fourier analytical methods which can be used to investigate the rate of convergence with respect to L_p–(quasi)–norms in (2) (for $1 < p < \infty$) and in (4) (even for $0 < p < \infty$). The main ideas come from the theory of function spaces of Sobolev-Nikol'skij-Besov type as developed in the books Nikol'skij [51], Peetre [54], Triebel [79, 80, 81] or Schmeisser and Triebel [66] and its natural interplay with approximation theory. For the latter aspect we refer also to Tikhomirov [76], Löfström [47] and de Vore and Lorentz [27].

The paper is organized as follows. Section 2 deals with properties of bandlimited functions, i.e. entire analytic functions of exponential type, in the context of Bernstein spaces. This is of interest of its own but here mainly for later use. To make the article more readable we shall give full proofs of inequalities of Nikol'skij, Bernstein and Plancherel-Polya type and of representations of Shannon type. In Section 3 we recall the function spaces we shall work with. Our approach will be based on the Fourier analytical definition of these classes. But we shall collect properties such as characterizations via derivatives and differences (moduli of continuity) and embeddings. Connections to approximation theory are discussed in great detail. The last section is devoted to the rate of L_p–convergence in (2) and (4) for non-bandlimited signals. Here we use the function spaces $B^s_{p,\infty}$ of Nikol'skij type (extended to parameters p, $0 < p \le \infty$; Zygmund type if $p = \infty$) to describe the smoothness and the approximation properties of such signals. We subdivide into three parts:

(i) approximation by the Shannon series $S_W f$;

(ii) approximation by generalized sampling series $S^\Psi_W f$ with bandlimited kernel Ψ;

(iii) approximation by generalized sampling series $S^\Psi_W f$ with timelimited kernel Ψ.

Let us emphasize that we concentrate on the approximation (aliasing) aspect. Moreover, we restrict ourselves to uniform sampling. We do not touch such topics as truncation and jitter errors, sampling of derivatives, prediction of signals, multivariate sampling, irregular sampling and Kramer's sampling theorem. Here we refer to the papers and books Butzer, Splettstößer and Stens [21], Higgins [39], Marks [48, 49], Feichtinger and Gröchenig [35, 36] and Zayed [84].

The paper is essentially self-contained except basic theory of Fourier transform and a few facts about real interpolation (which will be used very rarely). Comments to the literature will be given at the end of each section.

In some sense the survey is an extended version of parts of lectures on "Fourier Analysis" held by the first named author at the University of Jena in the academic year 1997/98.

This survey is part of the project "Abtastreihen". Both authors take the opportuinity to thank the DFG for supporting this project.

Notation

As usual, ℓ_p denotes the sequence spaces equipped with the quasi-norm

$$\| \{a_j\}_{j\in J} \,|\ell_p\| = \| \, a_j \,|\ell_p\| = \left(\sum_{k\in J} |a_j|^p \right)^{1/p} ,$$

and

$$\| \{a_j\}_{j\in J} \,|\ell_\infty\| = \| \, a_j \,|\ell_\infty\| = \sup_{j\in J} |a_j| \, ,$$

(here J denotes a countable index set, which one will be clear from the context).

Throughout the paper we restrict ourselves to the one-dimensional case. $\mathcal{S}$ and $\mathcal{S}'$ denote the Schwartz space of infinitely differentiable and rapidly decreasing functions and its dual the space of tempered distributions, respectively. If $0 < p \leq \infty$ we put $L_p = L_p(\mathbb{R})$ equipped with the quasi-norms

$$\| \, f \,|L_p\| = \left(\int_{-\infty}^{\infty} |f(t)|^p dt \right)^{1/p} \qquad \text{and} \qquad \| \, f \,|L_\infty\| = \operatorname*{ess\,sup}_{t\in\mathbb{R}} |f(t)| \, .$$

Moreover, $C^r = C^r(\mathbb{R})$, $r = 0, 1, \ldots$, stands for the space of all functions f on $\mathbb{R}$ which are r–times differentiable such that $f^{(j)}$ is uniformly continuous and bounded for all $j = 0, 1, \ldots, r$, normed by

$$\| \, f \,|C^r\| = \sum_{j=0}^{r} \| \, f^{(j)} \,|L_\infty\| \, .$$

The Fourier transform and its inverse on $\mathcal{S}'$ are denoted by $\mathcal{F}$ and $\mathcal{F}^{-1}$, respectively. If $f,\, g \in L_1$ we have

$$\mathcal{F}f(\omega) = \frac{1}{2\pi} \int_{-\infty}^{\infty} e^{-i\omega t}\, f(t)\, dt\,,$$

and

$$\mathcal{F}^{-1}g(\omega) = \int_{-\infty}^{\infty} e^{i\omega t}\, g(t)\, dt\,.$$

We put $f * g$ for the convolution of $f \in \mathcal{S}'$ and $g \in \mathcal{S}'$ (whenever it makes sense). Under certain assumptions we have

$$(f*g)(t) = \int_{-\infty}^{\infty} f(\tau)\, g(t-\tau)\, d\tau\,, \qquad a.e..$$

In particular, this holds if $g \in L_1$ and $f \in L_p$, $1 \le p \le \infty$. The following properties of $\mathcal{F}$ and $\mathcal{F}^{-1}$ will be frequently used:

$$\begin{aligned}
\mathcal{F}^{-1}f(\omega) &= 2\pi\, \mathcal{F}f(-\omega), & \mathcal{F}[f(t-t_0)](\omega) &= e^{-it_0\omega}\, \mathcal{F}f(\omega),\\
\mathcal{F}[f(dt)](\omega) &= (1/d)\, \mathcal{F}f(\omega/d), & \mathcal{F}[f^{(j)}](\omega) &= (i\omega)^j\, \mathcal{F}f(\omega),\\
\mathcal{F}(f*g)(\omega) &= 2\pi\, \mathcal{F}f(\omega)\, \mathcal{F}g(\omega), & \mathcal{F}[f\, g](\omega) &= (\mathcal{F}f * \mathcal{F}g)(\omega)\,.
\end{aligned}$$

A major role will be played by the linear combinations of the translates and dilates of the sinc-function given by

$$\operatorname{sinc} t = \begin{cases} \frac{\sin \pi t}{\pi t} & \text{if} \quad t \neq 0\,, \\ 1 & \text{if} \quad t = 0 \end{cases}$$

Its Fourier transform is given by

$$\mathcal{F}\mathrm{sinc}\,(\omega) = \begin{cases} \frac{1}{2\pi} & \text{if} \quad |\omega| < \pi\,, \\ \frac{1}{4\pi} & \text{if} \quad |\omega| = \pi\,, \\ 0 & \text{otherwise}\,. \end{cases}$$

Let us recall the famous Riesz theorem. Let $\chi_{(a,b)}$, $-\infty < a < b < \infty$ be the characteristic function of the interval (a,b). If $1 < p < \infty$, then it holds

$$\| \mathcal{F}^{-1}[\chi_{(a,b)}(\omega)\, \mathcal{F}f(\omega)](t)\, |L_p\| \le c_p\, \|\, f\, |L_p\| \tag{5}$$

for all $f \in \mathcal{S}$. The constant c_p does not depend on a, b and f. For the basics in Fourier analysis we refer to Hörmander [40], Vladimirov [82], Stein and Weiss [72], Strichartz [75], or Butzer and Nessel [20]. We finish this collection of notations by the following agreement. At various places we shall deal with pointwise properties of functions $f \in L_p$. Of course, in such a situation we suppose that f is pointwise given and will be identified with its equivalence class $[f]$ which belongs to L_p.

2 Properties of Bandlimited Functions

Section 2 and 3 are of preparatory character. To keep the paper mainly self-contained we shall give full proves.

2.1 Bernstein spaces

One of the most important concepts in the field of bandlimited functions is that of the Bernstein space.

Definition 1. Let $0 < p \leq \infty$ and let $\Omega > 0$ be a real number. We put

$$B_\Omega^p = \left\{ f \in \mathcal{S}' \cap L_p \ : \mathrm{supp}\, \mathcal{F}f \subset [-\Omega, \Omega] \right\}. \tag{6}$$

The class B_Ω^p is called Bernstein space of Ω-bandlimited functions.

Remark 1. By the Paley–Wiener–Theorem (cf. [40, p. 181]) the functions in B_Ω^p are infinitely differentiable (even more, they can be extended to a function holomorphic in the complex plane) and of at most polynomial growth: that means, there exist positive constants c_α and m_α such that

$$|D^\alpha f(t)| \leq c_\alpha (1 + |t|^{m_\alpha}) \tag{7}$$

for all $\alpha = 0, 1, \ldots$ and all $t \in \mathbb{R}$.

A useful tool in investigating Bernstein spaces will be the following explicit approximation procedure.

Theorem 1. *Let* $\varphi \in \mathcal{S}$ *such that* $\mathcal{F}\varphi(\omega) \geq 0$, $\mathrm{supp}\, \mathcal{F}\varphi \subset [-1, 1]$ *and* $\varphi(0) = 1$.
(i) *If* $f \in B_\Omega^p$, $0 < p < \infty$, *and* $\varepsilon > 0$, *then* $\varphi(\varepsilon t)\, f(t) \in B_{\Omega+\varepsilon}^p \cap \mathcal{S}$ *and* $[\varphi(\varepsilon \cdot)\, f] \underset{\varepsilon \to 0}{\longrightarrow} f$ *in* L_p.
(ii) *If* $f \in B_\Omega^\infty$ *and* $\varepsilon > 0$, *then* $\varphi(\varepsilon t)\, f(t) \in B_{\Omega+\varepsilon}^\infty \cap \mathcal{S}$ *and* $\|\, \varphi(\varepsilon t)\, f(t)\, |L_\infty\| \to \|\, f\, |L_\infty\|$.

Proof. Part (i) follows easily from Lebesgue's Dominated Convergence Theorem and the representation

$$\mathcal{F}[\varphi(\varepsilon \cdot)\, f] = [\frac{1}{\varepsilon}\, \mathcal{F}\varphi(\frac{\cdot}{\varepsilon})] * \mathcal{F}f$$

which makes sense for all $f \in \mathcal{S}'$. To prove (ii) we choose t such that $\|\, f\, |L_\infty\| - |f(t)| < \frac{\delta}{2}$ and then ε_0 such that $|\, \varphi(\varepsilon t)\, f(t)\, | - |\, f(t)\, | \leq \frac{\delta}{2}$ for all $\varepsilon \leq \varepsilon_0$. Then we have

$$
\begin{aligned}
|\varphi(\varepsilon t)\, f(t)\,| &\leq \|\, f\,|L_\infty\| \left| \int_{-\infty}^{\infty} \mathcal{F}\varphi(\omega)\, e^{i\varepsilon\omega t}\, d\omega \right| \\
&\leq \|\, f\,|L_\infty\| \int_{-1}^{1} \mathcal{F}\varphi(\omega)\, d\omega \;=\; \|\, f\,|L_\infty\|
\end{aligned}
$$

and therefore

$$
0 \leq \|\, f\,|L_\infty\| - \|\varphi(\varepsilon\cdot)\, f\,|L_\infty\| \leq \|\, f\,|L_\infty\| - \|\,\varphi(\varepsilon t)\, f(t)\,\| \leq \delta\,.
$$

This proves the theorem. □

Further properties of Bernstein spaces will be derived during the course of this section, in particular in Subsections 3 and 4.

2.2 Inequalities I

Because of the importance and convenience for the reader we give proofs of inequalities of Nikol'skij, Bernstein and Plancherel–Polya.

Theorem 2. (Nikol'skij's inequality). *Let $0 < p \leq q \leq \infty$. Then there exists a constant c (independent of f and Ω) such that*

$$
\|\, f\,|L_q\| \leq c\,\Omega^{\frac{1}{p}-\frac{1}{q}}\, \|\, f\,|L_p\| \tag{8}
$$

holds for all f in B^p_Ω.

Proof. By homogeneity arguments and the transform $f \mapsto f(\frac{\cdot}{\Omega})$ we can restrict ourselves to the case $\Omega = 1$. Let $q = \infty$ and let $f \in B^p_1$. We chose $\varphi \in C_0^\infty$ such that $\varphi(\omega) = 1$ on $[-1, 1]$. Then we have (with $\psi = \mathcal{F}^{-1}\varphi$)

$$
f(t) = \frac{1}{2\pi} \int_{-\infty}^{\infty} f(\tau)\, \psi(t-\tau)\, d\tau \tag{9}
$$

for all $t \in \mathbb{R}$. If $p \geq 1$ we obtain from (9)

$$
|f(t)| \leq \frac{1}{2\pi}\, \|\,\psi\,|L_{p'}\|\, \|\, f\,|L_p\|
$$

by Hölder's inequality. If $0 < p < 1$ we have

$$
\begin{aligned}
|f(t)| &\leq \frac{1}{2\pi} \int_{-\infty}^{\infty} |f(\tau)|^p |f(\tau)|^{1-p}\, |\psi(t-\tau)|\, d\tau \\
&\leq \frac{\|\,\psi\,|L_\infty\|}{2\pi}\, \|\, f\,|L_\infty\|^{1-p}\, \|\, f\,|L_p\|^p\,.
\end{aligned}
$$

Taking $\sup_{t\in\mathbb{R}}$ on both sides and dividing by $\| f \,| L_\infty\|^{1-p}$ gives (8) with $q = \infty$ and $0 < p < 1$. The general case $0 < p < q < \infty$ follows by

$$\| f \,| L_q\|^q = \int_{-\infty}^{\infty} |f(t)|^{q-p} |f(t)|^p \, dt \leq \| f \,| L_\infty\|^{q-p} \, \| f \,| L_p\|^p$$

and the (just proved) estimate (8) for $q = \infty$. □

Remark 2. As a corollary of Theorem 2 we get the continuous imbeddings

$$B_\Omega^p \hookrightarrow B_\Omega^q \hookrightarrow B_\Omega^\infty \tag{10}$$

if $0 < p < q < \infty$. In particular, any function belonging to a Bernstein space is bounded.

Lemma 1. *Let $f \in B_\pi^\infty \cap \mathcal{S}$. Then it holds*

$$f'(t) = \frac{1}{\pi} \sum_{k=-\infty}^{\infty} \frac{(-1)^{k+1}}{(k-1/2)^2} \, f(t - \frac{1}{2} + k)\,, \tag{11}$$

where the convergence is in the L_p-sense for all p, $1 \leq p \leq \infty$.

Proof. It is clear that the series on the right-hand side converges in L_p, $(1 \leq p \leq \infty)$. We have to show equality in $\mathcal{S}'$. One has

$$f'(t + \frac{1}{2}) = \mathcal{F}^{-1}[e^{i\omega/2} \, i\omega \, \mathcal{F}f](t)\,.$$

Let us denote the 2π-periodic extension of the function $g(\omega) = e^{i\omega/2} \, i\omega$ (restricted to $[-\pi, \pi]$) by $g_*(\omega)$. Its Fourier series converges uniformly and its Fourier coefficients can be easily calculated as

$$\hat{g}_*(k) = \frac{(-1)^{k+1}}{\pi} \frac{1}{(k-1/2)^2}\,, \qquad k \in \mathbb{Z}\,. \tag{12}$$

Because of $\mathcal{F}f \in \mathcal{S}$ and $\operatorname{supp} \mathcal{F}f \subset [-\pi, \, \pi]$ we obtain

$$\begin{aligned} f'(t + \frac{1}{2}) &= \mathcal{F}^{-1}\Bigg[\sum_{k=-\infty}^{\infty} \frac{(-1)^{k+1}}{\pi \, (k-1/2)^2} \, e^{ik\omega} \mathcal{F}f \Bigg](t) \\ &= \frac{1}{\pi} \sum_{k=-\infty}^{\infty} \frac{(-1)^{k+1}}{(k-1/2)^2} \, f(t-k) \end{aligned}$$

in $\mathcal{S}'$. This proves the lemma. □

Remark 3. We have

$$e^{i\omega/2}\, i\omega = \frac{1}{\pi} \sum_{k=-\infty}^{\infty} \frac{(-1)^{k+1}}{(k-1/2)^2}\, e^{ik\omega}\,, \qquad |\omega| \le \pi$$

by formula (12). Putting $\omega = \pi$ this leads to the equality

$$\pi^2 = \sum_{k=-\infty}^{\infty} \frac{1}{(k-1/2)^2}\,. \tag{13}$$

Theorem 3. (Bernstein's inequality). *Let* $1 \le p \le \infty$ *and let* $\Omega > 0$. *Then the inequality*

$$\|\, f' \,|L_p\| \le \Omega \,\|\, f \,|L_p\| \tag{14}$$

holds for all $f \in B^p_\Omega$.

Proof. Again by homogeneity it is sufficient to consider a special case. Let $\Omega = \pi$ and suppose $f \in B^p_\pi \cap \mathcal{S}$. As a consequence of Lemma 1 and formula (13) we obtain

$$\|\, f' \,|L_p\| \le \frac{1}{\pi} \sum_{k=-\infty}^{\infty} \frac{1}{(k-1/2)^2} \,\|\, f \,|L_p\| = \pi \,\|\, f \,|L_p\|$$

and hence (14) for all $f \in B^p_\pi \cap \mathcal{S}$ and all $\Omega > 0$. If f is an arbitrary function belonging to B^p_Ω we use the approximation procedure of Theorem 1 and apply (14) to the functions $\varphi(\varepsilon t)\, f(t) \in B^p_{\Omega+\varepsilon} \cap \mathcal{S}$. The limit process $\varepsilon \to 0$ finishes the proof of the theorem. □

Remark 4. The simple example $f(t) = e^{i\Omega t}$ shows the sharpness of (14) in the case $p = \infty$. Furthermore, by iteration of (14) it follows

$$\|\, D^\alpha f \,|L_p\| \le \Omega^\alpha \,\|\, f \,|L_p\|$$

for all $\alpha = 1, 2, \dots$ and $f \in B^p_\Omega$.

Next we deal with the equivalence of continuous and discrete L_p-norms for $f \in B^p_\Omega$.

Lemma 2. *Let* $1 \le p < \infty$, $h > 0$, *and let* $t_k \in [hk,\, h(k+1)]$ *for all* $k \in \mathbb{Z}$. *Then the inequality*

$$\left(h \sum_{k=-\infty}^{\infty} |f(t_k)|^p \right)^{1/p} \le (1 + \Omega h) \,\|\, f \,|L_p\| \tag{15}$$

holds for all $f \in B^p_\Omega$.

Proof. There exist natural numbers $\tau_k \in [hk,\, h(k+1)]$, $k \in \mathbb{Z}$, such that

$$\int_{-\infty}^{\infty} |f(t)|^p\, dt = h \sum_{k=-\infty}^{\infty} |f(\tau_k)|^p .$$

Now, let $t_k \in [hk,\, h(k+1)]$, $k \in \mathbb{Z}$, be arbitrary real numbers. With the help of the above identity and the triangle inequality for ℓ_p-norms we find

$$\begin{aligned}\left(h \sum_{k=-\infty}^{\infty} |f(t_k)|^p\right)^{1/p} &\le h^{1/p}\, \|\, f(\tau_k)\, |\ell_p\| + h^{1/p}\, \|\, f(\tau_k) - f(t_k)\, |\ell_p\| \\ &= \|\, f\, |L_p\| + h^{1/p}\, \|\, f(\tau_k) - f(t_k)\, |\ell_p\| . \end{aligned} \tag{16}$$

Obviously, it holds

$$|f(\tau_k) - f(t_k)| \le \int_{hk}^{h(k+1)} |f'(t)|\, dt \le h^{1/p'} \left(\int_{hk}^{h(k+1)} |f'(t)|^p\, dt\right)^{1/p}$$

by Hölder's inequality. Consequently, the second summand on the right-hand side of (16) can be estimated by

$$h^{1/p}\, \|\, f(\tau_k) - f(t_k)\, |\ell_p\| \le h\, \|\, f'\, |L_p\| \tag{17}$$

and (15) follows from (16) and Bernstein's inequality (Theorem 3). □

Remark 5. Observe, that the left-hand side of (15) can be replaced by

$$\sup_{u\in\mathbb{R}} \left(h \sum_{k=-\infty}^{\infty} |f(hk-u)|^p\right)^{1/p} .$$

This represents the usual formulation of this inequality, sometimes also called Nikol'skij's inequality. Furthermore, notice that the case $p = \infty$ is trivial.

Lemma 3. *Let $1 \le p < \infty$. Let $f \in B^p_\Omega$ for some $\Omega > 0$. Let $h \le h_0 < 1/\Omega$ and let $hk \le t_k \le h(k+1)$, $k \in \mathbb{Z}$. Then it holds*

$$\|\, f\, |L_p\| \le \frac{1}{1 - h_0\Omega} \left(h \sum_{k=-\infty}^{\infty} |f(t_k)|^p\right)^{1/p} . \tag{18}$$

Proof. Let τ_k, $hk \le \tau_k \le h(k+1)$, $k \in \mathbb{Z}$, have the same meaning as in proof of Lemma 2. Then it follows by triangle inequality, formula (17), and Bernstein's inequality (14) that

$$\| f \,| L_p \| \le h^{1/p} \, \| f(\tau_k) - f(t_k) \,| \ell_p \| + \left(h \sum_{k=-\infty}^{\infty} |f(t_k)|^p \right)^{1/p}$$

$$\le h \,\Omega \, \| f \,| L_p \| + \left(h \sum_{k=-\infty}^{\infty} |f(t_k)|^p \right)^{1/p} .$$

This leads to (18) because we assumed $h\Omega \le h_0 \Omega < 1$. □

Theorem 4. *Let $\Omega > 0$ and $0 < \varepsilon < 1$.*
(i) *Suppose $1 \le p < \infty$. If $0 < h \le (1-\varepsilon)/\Omega$ and if t_k, $hk \le t_k \le h(k+1)$, $k \in \mathbb{Z}$, are arbitrary real numbers then it holds*

$$\frac{1}{2-\varepsilon} \left(h \sum_{k=-\infty}^{\infty} |f(t_k)|^p \right)^{1/p} \le \| f \,| L_p \| \le \frac{1}{\varepsilon} \left(h \sum_{k=-\infty}^{\infty} |f(t_k)|^p \right)^{1/p} \tag{19}$$

for all $f \in B_\Omega^p$.
(ii) *Let $0 < h < (1-\varepsilon)/\Omega$ and let t_k, $hk \le t_k \le h(k+1)$, $k \in \mathbb{Z}$ be arbitrary real numbers. Then it holds*

$$\sup_{k \in \mathbb{Z}} |f(t_k)| \le \| f \,| L_\infty \| \le \frac{1}{\varepsilon} \sup_{k \in \mathbb{Z}} |f(t_k)| \tag{20}$$

for all $f \in B_\Omega^\infty$.

Proof. Part (i) follows from Lemma 2 and Lemma 3 (with $h_0 = (1-\varepsilon)/\Omega$). To prove (ii) we argue as follows. Let $\delta > 0$ and let $f \in B_\Omega^\infty$. We find a real number τ such that

$$\| f \,| L_\infty \| \le (1+\delta) \, |f(\tau)| \, .$$

For some $k \in \mathbb{Z}$ we obtain by Bernstein's inequality (14) for $p = \infty$

$$\| f \,| L_\infty \| \le (1+\delta) \, |f(\tau) - f(t_k)| + (1+\delta) \, |f(t_k)|$$

$$\le (1+\delta) \, h \, \| f' \,| L_\infty \| + (1+\delta) \sup_{k \in \mathbb{Z}} |f(t_k)|$$

$$\le (1+\delta) \, h \, \Omega \, \| f \,| L_\infty \| + (1+\delta) \sup_{k \in \mathbb{Z}} |f(t_k)| \, .$$

Letting $\delta \to 0$ it follows that

$$\| f \,| L_\infty \| \le h \, \Omega \, \| f \,| L_\infty \| + \sup_{k \in \mathbb{Z}} |f(t_k)| \, .$$

This yields (20) because of $h\Omega \le 1 - \varepsilon$. □

Lemma 4. *Let $g \in B^1_{2\pi}$. Then it holds*

$$\int_{-\infty}^{\infty} g(t)\,dt = \sum_{k=-\infty}^{\infty} g(k) \tag{21}$$

where the sum is absolutely convergent.

Proof. The absolute convergence of the series $\sum_{k=-\infty}^{\infty} g(t+k)$ follows immediately from Lemma 2. Its sum $g_*(t) = \sum_{k=-\infty}^{\infty} g(t+k)$ is 1-periodic and belongs to $L_1([0,1])$. The Fourier coefficients can be calculated as

$$\begin{aligned}\hat{g}_*(\ell) &= \int_0^1 \sum_{k=-\infty}^{\infty} g(t+k)\,e^{-i2\pi\ell t}\,dt = \sum_{k=-\infty}^{\infty} \int_k^{k+1} g(\tau)\,e^{-i2\pi\ell\tau}\,d\tau \\ &= \int_{-\infty}^{\infty} g(\tau)\,e^{-i2\pi\ell\tau}\,d\tau = 2\pi\,\mathcal{F}g(2\pi\ell)\,.\end{aligned}$$

The function $\mathcal{F}g$ is continuous and has support in $[-2\pi,\,2\pi]$, which means that all except one of these coefficients are vanishing. Moreover, we have

$$|g_*(t) - g_*(t_0)| \le \sum_{k=-\infty}^{\infty} |g(t+k) - g(t_0+k)| \le \sum_{k=-\infty}^{\infty} |g'(t_k)|\,|t-t_0|$$

where $|t_k - k| \le |t - t_0|$. Hence g_* is Lipschitz continuous by Bernstein's inequality and Lemma 2. Consequently,

$$g_*(t) = 2\pi \sum_{\ell=-\infty}^{\infty} \mathcal{F}g(2\pi\ell)\,e^{i2\pi\ell t} = 2\pi\,\mathcal{F}g(0)$$

holds for all t. This implies the desired result. □

Theorem 5. *Let $f \in B^p_\pi$, $g \in B^{p'}_\pi$, where $1 \le p \le \infty$ and $1/p + 1/p' = 1$. Then it holds*

$$(f * g)(t) = \sum_{k=-\infty}^{\infty} f(k)\,g(t-k) = \sum_{k=-\infty}^{\infty} g(k)\,f(t-k)\,, \tag{22}$$

where the series converge absolutely for all t and uniformly on $\mathbb{R}$.

Proof. The function $h(\tau) = f(\tau)\,g(t-\tau)$ belongs to the Bernstein space $B^1_{2\pi}$. This can be easily verified. Lemma 4 implies

$$\int_{-\infty}^{\infty} f(\tau)\,g(t-\tau)\,d\tau = \sum_{k=-\infty}^{\infty} f(k)\,g(t-k)$$

(absolute convergence). Changing the roles of f and g and taking into account the commutativity of convolution it follows (22). It remains to show uniform convergence with respect to t. We can restrict ourselves to $1 \le p < \infty$ (if $p = \infty$ we change the roles of f and g). Let $\varepsilon > 0$. Because of Lemma 2 we can find $N_0 \in \mathbb{N}$ such that

$$\left(\sum_{|k| \ge N} |f(k)|^p \right)^{1/p} \le \varepsilon$$

for all $N \ge N_0$. Using Hölder's inequality and Lemma 2 again we obtain

$$\left| \sum_{|k| \ge N} f(k)\, g(t-k) \right| \le \left(\sum_{|k| \ge N} |f(k)|^p \right)^{1/p} \left(\sum_{k=-\infty}^{\infty} |g(t-k)|^{p'} \right)^{1/p'}$$
$$\le \varepsilon\,(1+\pi)\, \| g(t-\cdot)\,|L_{p'}\| = \varepsilon\,(1+\pi)\, \| g\,|L_{p'}\|$$

for all $t \in \mathbb{R}$. This implies uniform convergence. □

Now we are in a position to formulate and prove the main assertion of this subsection.

Theorem 6. *Let* $1 < p < \infty$ *and let* $\Omega > 0$.
(i) *(Plancherel-Polya inequalities). There exist positive constants* A_p *and* B_p *(independent of* Ω *and* f*) such that*

$$A_p \,\| f\,|L_p\| \le \left(\frac{\pi}{\Omega} \sum_{k=-\infty}^{\infty} |f(\frac{\pi}{\Omega} k)|^p \right)^{1/p} \le B_p \,\| f\,|L_p\| \tag{23}$$

holds for all $f \in B^p_\Omega$.
(ii) *(Shannon Sampling Theorem). Any function* $f \in B^p_\Omega$ *can be represented as*

$$f(t) = \sum_{k=-\infty}^{\infty} f(\frac{\pi}{\Omega} k)\, \mathrm{sinc}\,(\frac{\Omega}{\pi} t - k)\,, \tag{24}$$

where the convergence is absolute for all t *and in the sense of* L_p.

Proof. Step 1. The identity (24) is a consequence of Theorem 5 if we put $g(t) = \mathrm{sinc}\,(t)$, which belongs to $B^{p'}_\pi$ for all p, $1 < p < \infty$. Indeed, (22) gives

$$\sum_{k=-\infty}^{\infty} f(k)\, \mathrm{sinc}\,(t-k) = \sum_{k=-\infty}^{\infty} f(t-k)\, \mathrm{sinc}\,(k) = f(t)\,,$$

where the convergence is absolute. The general case $\Omega \neq \pi$ follows by homogeneity. The L_p-convergence will be shown in Step 3.

Step 2. We prove (23). Again we may restrict ourselves to $\Omega = \pi$. If $f \in B_\pi^p$, then it follows from Lemma 2 (with $h = 1$, $\Omega = \pi$, $t_k = k$) the estimate

$$\left(\sum_{k=-\infty}^{\infty} |f(k)|^p \right)^{1/p} \leq (1+\pi) \, \| f \, | L_p \| \,. \tag{25}$$

To prove the estimate from below we shall use duality arguments. Let $f \in B_\pi^p$. By the well-known duality $(L_{p'})' = L_p$ we have

$$\| f \, | L_p \| = \sup_{g \in L_{p'}} \frac{| \int_{-\infty}^{\infty} f(t) \, g(t) \, dt \, |}{\| g \, | L_{p'} \|} \,. \tag{26}$$

We put

$$f_N(t) = \sum_{|\ell| \leq N} f(\ell) \, \mathrm{sinc}\,(t - \ell) \in B_\pi^p \,.$$

The previous theorem yields $\lim_{N \to \infty} f_N(t) = f(t)$ uniformly in t and

$$f_N(k) = \begin{cases} f(k) & \text{if } \ |k| \leq N \,, \\ 0 & \text{if } \ |k| > N \,. \end{cases} \tag{27}$$

Let $g \in \mathcal{S}$ and let χ be the characteristic function of $[-\pi, \pi]$. By Plancherel's Theorem we get

$$\begin{aligned} \left| \int_{-\infty}^{\infty} f_N(t) \, g(t) \, dt \right| &= \frac{1}{2\pi} \left| \int_{-\infty}^{\infty} (\mathcal{F} f_N)(\omega) \, \overline{\mathcal{F} \overline{g}(\omega)} \, d\omega \right| \\ &= \frac{1}{2\pi} \left| \int_{-\infty}^{\infty} (\mathcal{F} f_N)(\omega) \, \overline{\chi(\omega) \, \mathcal{F} \overline{g}(\omega)} \, d\omega \right| \\ &= \left| \int_{-\infty}^{\infty} f_N(t) \, \overline{\mathcal{F}^{-1} [\chi(\omega) \, \mathcal{F} \overline{g}(\omega)](t)} \, dt \right| . \end{aligned}$$

By Riesz's Theorem χ is a Fourier multiplier in $L_{p'}$, $1 < p' < \infty$. Hence, $h(t) = \overline{\mathcal{F}^{-1} [\chi(\omega) \, \mathcal{F} \overline{g}(\omega)](t)} \in B_\pi^{p'}$ and we obtain

$$\begin{aligned} \left| \int_{-\infty}^{\infty} f_N(t) \, g(t) \, dt \right| &= \left| \int_{-\infty}^{\infty} f_N(t) \, h(t) \, dt \right| \\ &= \left| \sum_{k=-\infty}^{\infty} f_N(k) \, h(-k) \right| \end{aligned}$$

$$\leq \left(\sum_{|k| \leq N} |f_N(k)|^p \right)^{1/p} \left(\sum_{k=-\infty}^{\infty} |h(k)|^{p'} \right)^{1/p'}$$

$$\leq \left(\sum_{|k| \leq N} |f(k)|^p \right)^{1/p} (1+\pi) \, \| \, h \, | L_{p'} \|$$

using Theorem 5, Hölder's inequality, (27) and (25) with p' in place of p. Denoting by C the norm of the operator $g \mapsto \mathcal{F}^{-1}[\chi(\omega) \, \mathcal{F} g(\omega)]$ in $\mathcal{L}(L_{p'})$ we obtain

$$\left| \int_{-\infty}^{\infty} f_N(t) \, g(t) \, dt \right| \leq C \, (1+\pi) \left(\sum_{k=-\infty}^{\infty} |f(k)|^p \right)^{1/p} \| \, g \, | L_{p'} \|$$

for all $g \in L_{p'}$. If N tends to infinity this implies that

$$\left| \int_{-\infty}^{\infty} f(t) \, g(t) \, dt \right| \leq C \, (1+\pi) \left(\sum_{k=-\infty}^{\infty} |f(k)|^p \right)^{1/p} \| \, g \, | L_{p'} \| \, .$$

Together with (26) and (25) this completes the proof of (23).

Step 3. To show the L_p-convergence of (24) for $f \in B_\pi^p$ let us consider the functions

$$f_{M,N}(t) = \sum_{M \leq |\ell| \leq N} f(\ell) \, \mathrm{sinc} \, (t - \ell) \in B_\pi^p \, .$$

It holds

$$f_{M,N}(k) = \begin{cases} f(k) & \text{if} \quad M \leq |k| \leq N \, , \\ 0 & \text{otherwise} \, . \end{cases}$$

Together with (23) this leads to

$$\| \, f_{M,N} \, | L_p \| \leq c \left(\sum_{M \leq |k| \leq N} |f(k)|^p \right)^{1/p}$$

and to L_p-convergence. □

Thanks to the identities

$$\int_{-\infty}^{\infty} \mathrm{sinc} \, (\frac{\Omega}{\pi} t - k) \, \mathrm{sinc} \, (\frac{\Omega}{\pi} t - m) \, dt = \frac{\pi}{\Omega} \frac{1}{2\pi} \int_{-\pi}^{\pi} e^{-i(k-m)\omega} \, d\omega = \frac{\pi}{\Omega} \, \delta_{k,m}$$

($\delta_{k,m}$ denotes the Kronecker symbol) it is clear that the set $\{\sqrt{\pi/\Omega}\,\mathrm{sinc}\,(\Omega/\pi t - k)\}_{k=-\infty}^{\infty}$ is an orthonormal basis for B^2_Ω. By means of Theorem 6 this generalizes as follows.

Corollary 1. *Let $1 < p < \infty$ and let $\Omega > 0$. Then the system of functions $\{\mathrm{sinc}\,\frac{\Omega}{\pi}t - k\}_{k\in\mathbb{Z}}$ is an unconditional Schauder basis for B^p_Ω.*

Remark 6. For the notion of an unconditional Schauder basis we refer to Wojtaszczyk [83].

Remark 7. Further immediate consequences are (under the same conditions on p and Ω):

(i) the Bernstein space B^p_Ω is isomorphic to ℓ_p;

(ii) Bernstein spaces are Banach spaces.

The second assertion extends to $p = 1$ and $p = \infty$. Moreover, B^p_Ω is a quasi-Banach space for $0 < p < 1$.

Corollary 2. *If $1 < p < \infty$, $1/p + 1/p' = 1$, then we have $(B^p_\Omega)' = B^{p'}_\Omega$ for the dual spaces. The norm of $f \in B^{p'}_\Omega$ as a functional is equivalent to $\| f \,|L_{p'}\|$.*

Proof. Of course the duality has to be understood via the duality pairing

$$f(g) = \int_{-\infty}^{\infty} f(x)\, g(x)\, dx\,.$$

It will be enough to consider $\Omega = \pi$. If $f \in B^p_\pi$ and $g \in B^{p'}_\pi$, then $f\,g \in B^1_{2\pi}$ and

$$\left| \int_{-\infty}^{\infty} f(x)\, g(x)\, dx \right| = \left| \sum_{k=-\infty}^{\infty} f(k)\, g(-k) \right| \le B_p\, B_{p'}\, \| f \,|L_p\|\, \| g \,|L_{p'}\|\,.$$

Vice versa, if $T \in (B^p_\pi)'$, then $T \circ I \in (\ell_p)'$ with

$$I: \quad \{\alpha_k\}_{k=-\infty}^{\infty} \mapsto \sum_{k=-\infty}^{\infty} \alpha_k\, \mathrm{sinc}\,(t-k)\,.$$

Hence, there exists a sequence $\{\beta_k\}_{k=-\infty}^{\infty} \in \ell_{p'}$ such that

$$(T \circ I)(\alpha) = \sum_{k=-\infty}^{\infty} \alpha_k\, \beta_k$$

for all $\alpha = \{\alpha_k\}_{k=-\infty}^{\infty} \in \ell_p$. By Theorem 6

$$g(t) := \sum_{k=-\infty}^{\infty} \beta_k \,\mathrm{sinc}\,(t-k) \in B_\pi^{p'}$$

and from Theorem 5 we conclude

$$\int_{-\infty}^{\infty} f(x)\, g(x)\, dx = \sum_{k=-\infty}^{\infty} f(k)\, g(-k) = (T \circ I)(\{f(k)\}_k) = T(f)$$

for all $f \in B_\pi^p$. From this identity, Theorem 6, and $(\ell_p)' = \ell_{p'}$ we derive

$$\begin{aligned} \| T \| &= \sup_{f \in B_\pi^p,\, \| f \,|L_p\| \le 1} \left| \sum_{k=-\infty}^{\infty} f(k)\, g(-k) \right| \\ &= \sup_{\| \{\alpha_k\}_k \,|\ell_p\| \le B_p} \left| \sum_{k=-\infty}^{\infty} \alpha_k\, g(-k) \right| \\ &= B_p \left(\sum_{k=-\infty}^{\infty} |g(k)|^{p'} \right)^{1/p'} \\ &\ge B_p\, A_{p'} \, \| g \,|L_{p'}\| \,. \end{aligned}$$

This proves the claim. □

2.4 Fourier multipliers for Bernstein spaces

Theorem 7. (Convolution algebras). *Let $0 < p \le 1$ and let $\Omega > 0$. If $f, g \in B_\Omega^p$, then $g * f \in B_\Omega^p$ and there exists a constant independent of f, g, and Ω such that*

$$\| g * f \,|L_p\| \le c\, \Omega^{\frac{1}{p}-1} \, \| g \,|L_p\| \, \| f \,|L_p\| \,. \tag{28}$$

Proof. By Nikol'skij's inequality (Theorem 2) B_Ω^p is imbedded into B_Ω^1 and the convolution integral makes sense. Moreover, the function $h_t(\tau) := f(\tau)\, g(t-\tau)$ is in $B_{2\Omega}^p$ for all $t \in \mathbb{R}$. Hence, by Nikol'skij's inequality again,

$$|\,(g * f)(t)| \le c\, \Omega^{\frac{1}{p}-1} \left(\int_{-\infty}^{\infty} |f(\tau)|^p \, |g(t-\tau)|^p \, d\tau \right)^{1/p} .$$

Integrating with respect to t and Fubini's theorem imply (28). □

Theorem 8. (Fourier multipliers). *Let $0 < p \leq \infty$, $\Omega > 0$, and $\tilde{p} = \min(1,p)$. Let $m(\omega) \in \mathcal{S}'$ such that $\mathcal{F}^{-1}m \in B_\Omega^{\tilde{p}}$. Then it holds*

$$\| \mathcal{F}^{-1}[m\, \mathcal{F}f] \,|L_p\| \leq c\, \Omega^{\frac{1}{\tilde{p}}-1} \, \| \mathcal{F}^{-1}m \,|L_{\tilde{p}}\| \; \| f \,|L_p\| \tag{29}$$

for all $f \in B_\Omega^p$. The constant c does not depend on Ω and m.

Proof. Let $0 < p \leq 1$. By Nikol'skij's inequality both f and $\mathcal{F}^{-1}m$ belong to L_1. Hence $\mathcal{F}f$ and m are continuous functions with support contained in $[-\Omega,\, \Omega]$. Moreover, $\mathcal{F}^{-1}[m(\omega)\, \mathcal{F}f(\omega)]$ makes sense and represents a bounded function in B_Ω^∞. It holds

$$\mathcal{F}^{-1}[m\, \mathcal{F}f](t) = \frac{1}{2\pi} \int_{-\infty}^{\infty} (\mathcal{F}^{-1}m)(t-\tau)\, f(\tau)\, d\tau = \frac{1}{2\pi}\, (\mathcal{F}^{-1}m) * f(t)\,.$$

Now, (29) follows from Theorem 7. If $1 < p < \infty$ (29) reads as

$$\| \mathcal{F}^{-1}[m\mathcal{F}f] \,|L_p\| \leq c \, \| \mathcal{F}^{-1}m \,|L_1\| \; \| f \,|L_p\|$$

and this is a well-known convolution inequality. □

Proposition 1. *Let $0 < p < \infty$ and let $\Omega > 0$. The set*

$$S_\Omega = \Big\{ f \in \mathcal{S} : \quad \operatorname{supp} \mathcal{F}f \subset (-\Omega,\, \Omega) \Big\}$$

is dense in B_Ω^p.

Proof. We choose two functions ψ_1 and $\psi_2 \in \mathcal{S}$ such that

$$\operatorname{supp} \psi_1 \subset (-2\Omega,\, \frac{\Omega}{2})\,, \qquad \operatorname{supp} \psi_2 \subset (-\frac{\Omega}{2},\, 2\Omega)$$

and $\psi_1(t) + \psi_2(t) = 1$ on $[-\Omega, \Omega]$. It holds

$$f = \mathcal{F}^{-1}[\psi_1\, \mathcal{F}f] + \mathcal{F}^{-1}[\psi_2\, \mathcal{F}f] = f_1 + f_2$$

for $f \in B_\Omega^p$. By Theorem 8 we see that $f_i \in B_\Omega^p$, $i = 1,2$. Moreover,

$$\operatorname{supp} \mathcal{F}f_1 \subset [-\Omega,\, \frac{\Omega}{2}) \qquad \text{and} \qquad \operatorname{supp} \mathcal{F}f_2 \subset (-\frac{\Omega}{2},\, \Omega)\,.$$

Let us consider f_1 (the function f_2 may be treated analogously). Put $f_{1,\nu}(t) = f_1(t)\, e^{i\nu\Omega t}$. Because of the uniform convergence of $e^{i\nu\Omega t} - 1$ we have $\lim_{\nu\to 0} f_{1,\nu} = f_1$ in the L_p-sense. If $\nu \leq 1/3$, then

$$\operatorname{supp} \mathcal{F}f_{1,\nu} = \operatorname{supp} (\mathcal{F}f_1)(\omega - \nu\Omega) \subset (-\Omega,\, \frac{5}{6}\,\Omega)\,.$$

Now, it suffices to approximate $f_{1,\nu}$ by means of the approximation procedure of Theorem 1. The functions $\varphi(\varepsilon t)\, f_{1,\nu}(t)$ belong to S_Ω if $\varepsilon \leq \varepsilon(\nu)$ and approximate $f_{1,\nu}(t)$ in L_p. This completes the proof. □

2.5 *Inequalities II: The case* $0 < p < 1$

We prove Bernstein and Plancherel-Polya inequalities in case $0 < p < 1$.

Theorem 9. (Bernstein's inequality). *Let* $0 < p \leq 1$ *and let* $\Omega > 0$. *Then there exists a constant* c *independent of* Ω *and* f *such that*

$$\| f' \,|L_p\| \leq c\,\Omega\, \| f \,|L_p\| \tag{30}$$

holds for all $f \in B^p_\Omega$.

Proof. We can choose $\Omega = 1$ by homogeneity. Let $g \in \mathcal{S}$ with $\operatorname{supp} \mathcal{F}g \subset [-2,\, 2]$ and $(\mathcal{F}g)(\omega) = 1/(2\pi)$ if $\omega \in [-1,\, 1]$. Then $f * g = f$ and $f'(t) = (f * g')(t)$. Theorem 7 implies

$$\| f' \,|L_p\| \leq c_1 \| g' \,|L_p\| \, \| f \,|L_p\| = c_2 \| f \,|L_p\| \,.$$

This poves (30). □

To prove the analogue of Theorem 6 we need the following preparation. Let $f \in B^p_2$, $0 < p < 1$. We choose $g \in \mathcal{S}$ such that $\mathcal{F}g(\omega) = 1/(2\pi)$ if $|\omega| \leq 2$ and $\operatorname{supp} \mathcal{F}g \subset [-\pi,\, \pi]$. If $0 < h \leq 1$, then

$$\operatorname{supp} \mathcal{F}[f(h\cdot)] \subset [-2,\, 2] \qquad \text{and} \qquad f(ht) = (f(h\cdot) * g)(t)\,.$$

By Theorem 5 it follows

$$f(ht) = \sum_{k=-\infty}^{\infty} f(hk)\, g(t-k)\,, \qquad 0 < h \leq 1\,. \tag{31}$$

Lemma 5. *Let* $0 < p < 1$ *and let* $t_k \in \mathbb{R}$ *such that* $k \leq t_k \leq k+1$, $k \in \mathbb{Z}$. *Then there exists a constant* c *such that*

$$\left(\sum_{k=-\infty}^{\infty} |f(t_k)|^p \right)^{1/p} \leq c \,\| f \,|L_p\| \tag{32}$$

for all $f \in B^p_2$.

Proof. Let $t \in [k, k+1]$ be an arbitrary number and let g have the meaning of (31). It follows (with $h = 1$) from (31) that

$$f(t_k) = (f * g)(t_k) = (f(\cdot + t) * g)(t_k - t) = \sum_{\ell=-\infty}^{\infty} f(\ell + t)\, g(t_k - t - \ell)\,.$$

Because of $0 < p < 1$ and $g \in \mathcal{S}$

$$|f(t_k)|^p = \sum_{\ell=-\infty}^{\infty} |f(\ell+t)|^p \sup_{|\tau|\le 1} |g(\tau-\ell)|^p$$

$$\le c_M \sum_{\ell=-\infty}^{\infty} |f(\ell+t)|^p \sup_{|\tau|\le 1} \frac{1}{(1+|\tau-\ell|)^{Mp}}$$

$$\le 2^{Mp}\, c_M \sum_{\ell=-\infty}^{\infty} |f(\ell+t)|^p (1+|\ell|)^{-Mp} \,,$$

where $M > 0$ is at our disposal. We choose $M = 2/p$. Integration with respect to $t \in [k, k+1]$ and summation over $k \in \mathbb{Z}$ leads to

$$\sum_{k=-\infty}^{\infty} |f(t_k)|^p \le c_1 \sum_{\ell=-\infty}^{\infty} \sum_{k=-\infty}^{\infty} \int_k^{k+1} |f(\ell+t)|^p \, dt \, (1+|\ell|)^{-2} \le c_2 \, \| f \,|L_p\|^p \,.$$

This proves (32). □

Lemma 6. *Let $0 < p < 1$ and let $f \in B_2^p$. Then there exist positive real numbers $h_0 < 1$ and $c > 0$ such that for all $h < h_0$ and all $t_k \in [hk,\, h(k+1)]$, $k \in \mathbb{Z}$*

$$\| f \,|L_p\| \le c \left(h \sum_{k=-\infty}^{\infty} |f(t_k)|^p \right)^{1/p} \tag{33}$$

holds. Here the numbers h_0 and c may be chosen independent of f.

Proof. Using (31) (with $h = 1$) and the mean-value theorem one obtains for arbitrary $t \in \mathbb{R}$

$$|f(ht)|^p \le \sum_{k=-\infty}^{\infty} |f(hk)|^p \, |g(t-k)|^p$$

$$\le \sum_{k=-\infty}^{\infty} \left[|f(t_k)|^p + |f'(\tau_k)|^p |t_k - hk|^p \right] |g(t-k)|^p$$

$$\le \sum_{k=-\infty}^{\infty} |f(t_k)|^p |g(t-k)|^p + h^p \sum_{k=-\infty}^{\infty} |f'(\tau_k)|^p |g(t-k)| \,,$$

where $\tau_k \in [hk,\, h(k+1)]$. Integration with respect to t implies

$$\| f \,|L_p\|^p \le h \sum_{k=-\infty}^{\infty} |f(t_k)|^p \| \, g \,|L_p\|^p + h^{p+1} \sum_{k=-\infty}^{\infty} |f'(\tau_k)|^p \| \, g \,|L_p\|^p$$

$$\leq c_1 \left(h \sum_{k=-\infty}^{\infty} |f(t_k)|^p + h^{p+1} \sum_{k=-\infty}^{\infty} |f'(\tau_k)|^p \right) . \tag{34}$$

We use Lemma 5 with $f'(h\,\cdot)$ in place of f and Theorem 9 to see that

$$\sum_{k=-\infty}^{\infty} |f'(\tau_k)|^p \leq c_2 \, \| \, f'(h\,\cdot) \, |L_p\|^p = c_2 \, h^{-1} \, \| \, f' \, |L_p\|^p$$
$$\leq c_3 \, h^{-1} \, \| \, f \, |L_p\|^p \, .$$

Substituting this estimate into (34) leads to

$$\| \, f \, |L_p\|^p \leq c_1 \, h \sum_{k=-\infty}^{\infty} |f(t_k)|^p + c_4 \, h^p \, \| \, f \, |L_p\|^p \, .$$

It follows (33) if we choose $c_4 \, h_0^p < 1$. □

Theorem 10. *Let* $0 < p < 1$, $\Omega > 0$, *and let* $f \in B_\Omega^p$. *Then there exist constants* h_0, c_1, c_2 *such that for all* $h \leq h_0/\Omega$ *and all* t_k, $hk \leq t_k \leq h(k+1)$, $k \in \mathbb{Z}$

$$c_1 \, \| \, f \, |L_p\| \leq \left(h \sum_{k=-\infty}^{\infty} |f(t_k)|^p \right)^{1/p} \leq c_2 \, \| \, f \, |L_p\| \, . \tag{35}$$

The constants h_0, c_1, *and* c_2 *are independent of* $f \in B_\Omega^p$.

Proof. Combining Lemmas 5 and 6 proves the case $\Omega = 2$. The general case follows by the transform $f \mapsto f(\frac{2t}{\Omega})$. □

Comments

As mentioned at the beginning the results of this section are well-known. There exist different proofs. We followed mainly [51], [78, 80] and [66].

As far as the inequalities are concerned we refer also to [50] (Nikol'skij inequality for $0 < p \leq q \leq \infty$). The proof of Theorem 3 (Bernstein's inequality for $1 \leq p \leq \infty$) is adopted from [51] but first proofs have been given by Bernstein [3]. Inequalities of Plancherel-Polya type (Theorems 4 and 10) go back to [56], see also [6]. In the case $p \geq 1$ we followed again [51]. The case $p < 1$ can be found in [78, 80]. Our proof uses ideas from [34], see also [60] for the periodic case. Lemma 4, Theorem 5 and the Shannon formula (24) are well-known in sampling theory. For proofs and historical remarks we refer to [21, 22], [38, 39], [29], and [33]. Properties of Bernstein spaces (Theorems 1, 6, 7, and 8, Corollary 1 and 2 as well as Proposition 1) have been proved in a more general form (multivariate case, weighted L_p–quasi-norms) in [78, 80, 66]. However, also in Achieser [1], Timan [77], Nikol'skij [51], Peetre [54], and Higgins [39] these topics are treated.

3 Function Spaces

This section deals with function spaces of several types, but mainly with Sobolev spaces as well as Nikol'skij-Besov type on the real line. The only exception will be Subsection 5. There spaces defined by means of the τ–modulus as well as spaces of functions of bounded p–variation will be investigated.

The theory of Nikol'skij-Besov spaces has been systematically developed in several books. We refer to [51], [54], [4], and [80, 81]. We are interested in its close and natural connection with problems arising in approximation theory, in particular sampling theory of non-bandlimited signals. The interplay of approximation and smoothness is well-known, cf. e.g., [51], [77], [4], [27] and [76]. The aim of this section is twofold. On the one hand we recall some well-known facts (characterizations, imbeddings) without proofs. On the other hand we give a more detailed description of approximation procedures in function spaces. Our approach is strictly Fourier analytical and relies on the so-called decomposition method, cf. [54] and [80].

3.1 Sobolev spaces

All the derivatives $D^\alpha f = f^{(\alpha)} \frac{d^\alpha f}{dt^\alpha}$, $\alpha = 1, 2, \ldots$ ($D^0 f := f$) must be understood in the distributional sense.

Definition 2. (i) Let $1 \leq p < \infty$ and $r = 1, 2, \ldots$. Then we define the Sobolev space

$$W_p^r = \left\{ f \in L_p : \quad D^\alpha f \in L_p \right\}$$

and the Sobolev norm

$$\| f \,| W_p^r \| = \sum_{\alpha=0}^{r} \| D^\alpha f \,| L_p \| \,.$$

(ii) Let $1 < p < \infty$ and $-\infty < s < \infty$. Then we define the fractional Sobolev (or Bessel potential) space

$$H_p^s = \left\{ f \in \mathcal{S}' : \quad \mathcal{F}^{-1}[(1+\omega^2)^{s/2} \, \mathcal{F} f(\omega)] \in L_p \right\}$$

and the corresponding norm

$$\| f \,|H_p^s\| = \| \mathcal{F}^{-1}[(1+\omega^2)^{s/2}\, \mathcal{F} f] \,|L_p\| \,.$$

Remark 8. The Sobolev spaces W_p^r could be alternatively defined as follows: $f \in W_p^r$ if and only if $f \in L_p$ and there exists a function $\tilde{f}$ satisfying $f = \tilde{f}$ a.e., $\tilde{f}$ has derivatives (in the ordinary sense) of order less than or equal to $r-1$, $\tilde{f}^{(r-1)}$ is absolutely continuous and its derivative $\tilde{f}^{(r)}$ (which exists almost everywhere) belongs to L_p, cf. e.g. [46, 5.6]. Both W_p^r and H_p^s are Banach spaces and

$$W_p^r = H_p^r \,, \qquad \text{if} \quad 1 < p < \infty \,, \tag{36}$$

in the sense of equivalent norms, cf. e.g., [54] or [79].

3.2 Nikol'skij–Besov spaces

Let h and t be real numbers. Then we put

$$\begin{aligned} \Delta_h f(t) &= \Delta_h^1 f(t) = f(t+h) - f(t) \\ \Delta_h^m f(t) &= \Delta_h(\Delta^{m-1} f)(t) \,, \qquad m = 2, 3, \ldots \,. \end{aligned}$$

If $0 < p \leq \infty$, $m = 1, 2, \ldots$ the m–th modulus of continuity is denoted by

$$\omega_p^m(f, \tau) = \sup_{|h| \leq \tau} \| \Delta_h^m f(t) \,|L_p\| \,.$$

Based on these characteristics of a function f one may introduce the following smoothness classes (in the spirit of generalized Lipschitz spaces).

Definition 3. Let $0 < p < \infty$, $0 < q < \infty$ and $s > 0$. Let m be a natural number with $m > s$. Then we put

$$\Lambda_{p,q}^s = \left\{ f \in L_p : \quad [\tau^{-s} \omega_p^m(f,\tau)]^q / \tau \in L_1([0,1]) \right\}$$

equipped with the quasi-norm

$$\| f \,|\Lambda_{p,q}^s\| = \| f \,|L_p\| + \left(\int_0^1 [\tau^{-s}\, \omega_p^m(f,\tau)]^q \frac{d\tau}{\tau} \right)^{1/q}$$

and

$$\Lambda_{p,\infty}^s = \left\{ f \in L_p : \quad \sup_{0<\tau<1} [\tau^{-s} \omega_p^m(f,\tau)] < \infty \right\}$$

equipped with the quasi-norm

$$\| f \,|\Lambda^s_{p,\infty}\| = \| f \,|L_p\| + \sup_{0<\tau<1} \left[\tau^{-s}\, \omega^m_p(f,\tau)\right].$$

Remark 9. Taking $p = q = \infty$ then it becomes clear that $\Lambda^s_{\infty,\infty}$ is a Hölder-Zygmund space. In particular, we have for $0 < s < 1$

$$\Lambda^s_{\infty,\infty} = \Lambda^s = \left\{ f \in C : \quad \sup_{u \neq v} \frac{|f(u) - f(v)|}{|u - v|^s} < \infty \right\}$$

In case $s = 1$ the modification reads as follows

$$\Lambda^1_{\infty,\infty} = \Lambda^1 = \left\{ f \in C : \quad \sup_{h \neq 0} \frac{|f(t+2h) - 2\, f(t+h) + f(t)|}{|h|} < \infty \right\},$$

which is a Zygmund-type space.

At the first glance (e.g. the Jackson and Bernstein theorems) this definition looks most natural (in comparison with what will follow) and simple. However, for $p < 1$ we are confronted with certain difficulties. Consider

$$f_{\alpha,\delta}(t) = \psi(t)\, |t|^{-\alpha}\, |\log|t|\,|^{-\delta}, \qquad \alpha > 0, \quad \delta \geq 0,$$

where ψ is a smooth cut-off function supported around the origin. Then elementary calculations show $f_{\alpha,\delta} \in L_1$ if and only if either $\alpha < 1$ or $\alpha = 1$ and $\delta > 1$. Further, if $\delta > 0$, then $f_{\alpha,\delta} \in \Lambda^s_{p,q}$ if and only if either $0 < s < 1/p - \alpha$ or $s = 1/p - \alpha$ and $q\delta > 1$. In case $\delta = 0$ this reads as $f_{\alpha,0} \in \Lambda^s_{p,q}$ if and only if either $0 < s < 1/p - \alpha$ or $s = 1/p - \alpha$ and $q = \infty$. All together this means that $\Lambda^s_{p,q}$ contains functions which are not distributions as long as either $s < 1/p - 1$ or $s = 1/p - 1$ and $q > 1$. Since one of our main tools will be the Fourier transform this case would be not appropriate. For that reason we introduce a second scale based on Fourier analytical methods. To prepare this we start with a dyadic decomposition of unity. Let

$$Q_j = [-2^j,\, 2^j], \qquad j = 0, 1, \ldots \tag{37}$$

and put

$$I_0 = Q_0, \qquad \text{and} \qquad I_j = Q_j \setminus Q_{j-1}, \quad j = 1, 2, \ldots. \tag{38}$$

Now, let $\varphi_0 \in \mathcal{S}$ such that $\varphi_0(\omega) = 1$ on Q_0 and $\operatorname{supp} \varphi_0 \subset Q_1$. We put

$$\varphi(\omega) := \varphi_0(\omega) - \varphi_0(2\omega), \tag{39}$$

$$\varphi_j(\omega) := \varphi(2^{-j}\omega) = \varphi_0(2^{-j}\omega) - \varphi_0(2^{-j+1}\omega), \quad j = 1, 2, \ldots. \tag{40}$$

Clearly, it follows for $\ell = 0, 1, 2, \ldots$

$$\sum_{j=0}^{\ell} \varphi_j(\omega) = \varphi_0(\omega) + (\varphi_0(2^{-1}\omega) - \varphi_0(\omega)) + \ldots + (\varphi_0(2^{-\ell}\omega) - \varphi_0(2^{-\ell+1}\omega))$$
$$= \varphi_0(2^{-\ell}\omega)$$

and hence

$$\sum_{j=0}^{\infty} \varphi_j(\omega) = 1\,, \qquad \operatorname{supp} \varphi_j \subset I_j \cup I_{j+1} = \{\omega : \quad 2^{j-1} < |\omega| \leq 2^{j+1}\}\,,$$

$j = 1, 2, \ldots$. Observe, if $f \in \mathcal{S}'$, then multiplication of $\mathcal{F}f$ with φ_j makes sense in $\mathcal{S}'$ and by means of the Paley-Wiener theorem $\mathcal{F}^{-1}[\varphi_j(\omega)\,\mathcal{F}f(\omega)](t)$ is a smooth function on $\mathbb{R}$ which will be abbreviated by $f_j(t)$.

Definition 4. Let $0 < p \leq \infty$, $0 < q < \infty$ and $-\infty < s < \infty$.
(i) We define the Nikol'skij spaces

$$B^s_{p,\infty} = \left\{ f \in \mathcal{S}' : \quad \sup_{j=0,1,\ldots} 2^{js}\, \| \mathcal{F}^{-1}[\varphi_j\, \mathcal{F}f]\, |L_p\| < \infty \right\}$$

and the Nikol'skij quasi-norm

$$\| f\, |B^s_{p,\infty}\| = \sup_{j=0,1,\ldots} 2^{js}\, \| \mathcal{F}^{-1}[\varphi_j\, \mathcal{F}f]\, |L_p\|\,.$$

(ii) We define the Besov spaces

$$B^s_{p,q} = \left\{ f \in \mathcal{S}' : \quad \sum_{j=0}^{\infty} 2^{jsq}\, \| \mathcal{F}^{-1}[\varphi_j\, \mathcal{F}f]\, |L_p\|^q < \infty \right\}$$

and the Besov quasi-norm

$$\| f\, |B^s_{p,q}\| = \left(\sum_{j=0}^{\infty} 2^{jsq}\, \| \mathcal{F}^{-1}[\varphi_j\, \mathcal{F}f]\, |L_p\|^q \right)^{1/q}\,.$$

We are going to compare these two scales $\Lambda^s_{p,q}$ and $B^s_{p,q}$. By means of the properties of the family $f_{\alpha,\delta}$ and the definition of $B^s_{p,q}$ (in fact $B^s_{p,q} \subset \mathcal{S}'$ is required) it follows that $B^s_{p,q} \neq \Lambda^s_{p,q}$ if either $s < 1/p - 1$ or $s = 1/p - 1$ and $q > 1$. It is a non-trivial fact that these are almost all cases where the spaces do not coincide.

Theorem 11. *Let $0 < p \le \infty$, $0 < q \le \infty$, and $s > 0$.*
(i) *The spaces $B^s_{p,q}$ are quasi-Banach spaces continuously imbedded in $\mathcal{S}'$.*
(ii) *Suppose $s > \max(1/p - 1, 0)$. Then*

$$B^s_{p,q} = \Lambda^s_{p,q} \qquad \textit{(in the sense of equivalent quasi-norms)}. \tag{41}$$

Not only for completeness but also for later use we state some further equivalent characterizations of $B^s_{p,q}$ or $\Lambda^s_{p,q}$, respectively. Let $m \in \mathbb{N}$ and $0 < u \le \infty$. Then we introduce the following means of m–th order differences

$$d^m_{\tau,u} f(t) := \left(\frac{1}{\tau} \int_{-\tau}^{\tau} |\Delta^m_h f(t)|^u \, dh \right)^{1/u} \tag{42}$$

$$d^m_{\tau,\infty} f(t) := \sup_{|h| \le \tau} |\Delta^m_h f(t)| \,. \tag{43}$$

Theorem 12. *Let $0 < p \le \infty$ and $0 < q \le \infty$,*
(i) *Suppose $\max(1/p - 1, 0) < s < m$. Then $f \in B^s_{p,q}$ if and only if $f \in L_p$ and*

$$\| f \,|\Lambda^s_{p,q}\|^{\triangle}_m := \| f \,|L_p\| + \left(\int_{-\infty}^{\infty} |h|^{-sq} \, \| \Delta^m_h f(t) \,|L_p\|^q \frac{dh}{|h|} \right)^{1/q} < \infty \tag{44}$$

(in case $0 < q < \infty$) and

$$\| f \,|\Lambda^s_{p,\infty}\|^{\triangle}_m := \| f \,|L_p\| + \sup_{h \in \mathbb{R}} |h|^{-s} \, \| \Delta^m_h f(t) \,|L_p\| < \infty \tag{45}$$

(in case $q = \infty$). Moreover, all quasi-norms $\| f \,|\Lambda^s_{p,q}\|^{\triangle}_m$ are equivalent to $\| f \,|B^s_{p,q} \|$.
(ii) *Suppose $\max(1/p - 1, 0) < s - k < m$ for some $k \in \mathbb{N}$. Then $f \in B^s_{p,q}$ if and only if $D^\alpha f \in L_p$, $0 \le \alpha \le k$, and*

$$\| f \,|\Lambda^s_{p,q}\|^{\clubsuit}_{m,k} :=$$
$$\sum_{\alpha=0}^{k} \| D^\alpha f \,|L_p\| + \left(\int_{-\infty}^{\infty} |h|^{-(s-k)q} \, \| \Delta^m_h D^k f(t) \,|L_p\|^q \frac{dh}{|h|} \right)^{1/q} < \infty$$

(modification if $q = \infty$). Moreover, all quasi-norms $\| f \,|\Lambda^s_{p,q}\|^{\clubsuit}_{m,k}$ are equivalent to $\| f \,|B^s_{p,q} \|$.

(iii) *Suppose* $1 \le r \le \infty$, $0 < u \le r$, *and* $\max(1/p - 1/r, 0) < s < m$. *Then* $f \in B^s_{p,q}$ *if and only if* $f \in L_{\max(p,r)}$ *and*

$$\| f \,|\Lambda^s_{p,q}\|^{\spadesuit}_{m,u} := \| f \,|L_{\max_{p,r}}\| + \left(\int_0^1 \tau^{-sq} \, \| d^m_{\tau,u} f(t) \,|L_p\|^q \frac{d\tau}{\tau} \right)^{1/q} < \infty .$$

Moreover, all quasi-norms $\| f \,|\Lambda^s_{p,q}\|^{\spadesuit}_{m,u}$ *are equivalent to* $\| f \,|B^s_{p,q}\|$.

Remark 10. Again the case $p = q = \infty$ is that one which is most transparent. Let $0 < s - k < 1$. Then

$$\| f \,|\Lambda^s_{\infty,\infty}\|^{\clubsuit}_{m,k} = \sum_{\alpha=0}^{k} \sup_{t \in \mathbb{R}} |\, D^\alpha f(t) \,| + \sup_{u \neq v} \frac{|D^k f(u) - D^k f(v)|}{|u - v|^{s-k}}$$

and hence $\Lambda^s_{\infty,\infty}$ coincides with C^s if s is not a natural number.

Remark 11. As a consequence of the above theorems we see that for $1 \le p < \infty$, $r < s < r + 1$ the function f belongs to $B^s_{p,\infty}$ if and only if $f \in W^r_p$ and $\omega^1_p(\tau, f) = O(\tau^{s-r})$.

3.3 Imbedding theorems for Nikol'skij-Besov spaces

We concentrate on $B^s_{p,q}$. From the monotonicity of the ℓ_q-quasi-norms and the summability of the sequence $2^{-j\varepsilon q}$, $j = 0, 1, \ldots$ for any $q > 0$ and any $\varepsilon > 0$ one derives immediately the following.

Lemma 7. (Elementary imbeddings). *If* $s_0 > s_1 \ge s_2$, $q_1 \le q_2$ *and without further restrictions on* p *and* q_0, *then it holds* $B^{s_0}_{p,q_0} \hookrightarrow B^{s_1}_{p,q_1} \hookrightarrow B^{s_2}_{p,q_2}$.

Employing Littlewood-Paley arguments one can compare Besov spaces and fractional order Sobolev spaces.

Lemma 8. *Let* $-\infty < s < \infty$ *and* $0 < q_1, q_2 \le \infty$.
(i) *Let* $1 < p < \infty$. *Then*

$$B^s_{p,1} \hookrightarrow B^s_{p,\min(p,2)} \hookrightarrow H^s_p \hookrightarrow B^s_{p,\max(p,2)} \hookrightarrow B^s_{p,\infty} .$$

(ii) *(Diversity). Let* $1 < p < \infty$. *It holds* $H^{s_1}_{p_1} = B^{s_2}_{p_2,q_2}$ *if and only if* $s_1 = s_2$ *and* $p_1 = p_2 = q_2 = 2$.
(iii) *(Diversity). Let* $0 < p \le \infty$. *It holds* $B^{s_1}_{p_1,q_1} = B^{s_2}_{p_2,q_2}$ *if and only if* $s_1 = s_2$, $p_1 = p_2$, *and* $q_1 = q_2$.
(iv) *(Lift property). Let* $0 < p \le \infty$ *and let* $m \in \mathbb{N}$. *It holds* $f \in B^s_{p,q}$ *if and only if* $f \in B^{s-m}_{p,q}$ *and* $D^\alpha f \in B^{s-m}_{p,q}$, $0 \le \alpha \le m$.

The main part of this subsection is taken by imbedding theorems for different metrics, also called Sobolev imbeddings. The first assertion is nothing than a consequence of Nikol'skij's inequality.

Theorem 13. *Let $s_0, s_1 \in \mathbb{R}$, $0 < q_0, q_1 \le \infty$, and $0 < p_0 < p_1 \le \infty$. Then it holds*

$$B^{s_0}_{p_0,q_0} \hookrightarrow B^{s_1}_{p_1,q_1} \quad \textit{if and only if} \quad \textit{either} \quad s_0 > s_1 + \left(\frac{1}{p_0} - \frac{1}{p_1}\right)$$

$$\textit{or} \quad s_0 = s_1 + \left(\frac{1}{p_0} - \frac{1}{p_1}\right) \quad \textit{and} \quad q_0 \le q_1 \,. \tag{46}$$

Next we want to compare the Besov spaces with the Lebesgue spaces. Because of $H^0_p = L_p$ some information is contained in Lemma 8. The counterpart of Theorem 13 reads as follows.

Theorem 14. *Let $s \in \mathbb{R}$, $0 < p_0 < p_1 < \infty$, $1 \le p_1$ and $0 < q \le \infty$. Then the following assertions are equivalent:*

(a) $\quad B^s_{p_0,q} \hookrightarrow L_{p_1}$;

(b) $\quad$ *either* $s > 1/p_0 - 1/p_1$ *or* $s = 1/p_0 - 1/p_1$ *and* $0 < q \le p_1$.

Remark 12. Let $0 < p < 1$. Let $f \in B^{\frac{1}{p}-1}_{p,1}$. Then it follows from Nikol'skij's inequality (Theorem 1 with $\Omega = 2^{j+1}$) that

$$\| \sum_{j=M}^{N} \mathcal{F}^{-1}[\varphi_j \, \mathcal{F}f] \, |L_1\| \le c \sum_{j=M}^{N} 2^{j(\frac{1}{p}-1)} \| \, \mathcal{F}^{-1}[\varphi_j \, \mathcal{F}f] \, |L_1\| \,.$$

Hence, by completeness the series $\sum_{j=0}^{\infty} \mathcal{F}^{-1}[\varphi_j \, \mathcal{F}f] = f$ converges in L_1. On the other hand one has

$$\| \sum_{j=M}^{N} \mathcal{F}^{-1}[\varphi_j \, \mathcal{F}f] \, |L_p\|^p \le \sum_{j=M}^{N} \| \, \mathcal{F}^{-1}[\varphi_j \, \mathcal{F}f] \, |L_p\|^p$$

and therefore, by completeness of L_p the series $\sum_{j=0}^{\infty} \mathcal{F}^{-1}[\varphi_j \, \mathcal{F}f] = g$ converges in L_p. But this implies $f = g$ a.e., and thus $f \in B^{\frac{1}{p}-1}_{p,1}$ can be interpreted as a function in $L_p \cap L_1$ if $0 < p < 1$. $B^{\frac{1}{p}-1}_{p,1}$ represents the limit case for such an assertion. Indeed, if either $s < 1/p - 1$ or $s = 1/p - 1$ and $q > 1$, then

$B^s_{p,q}$ contains singular distributions. In particular, the δ–distribution belongs to $B^{\frac{1}{p}-1}_{p,\infty}$ for all p, [71].

Of peculiar interest with respect to sampling theory are imbeddings into spaces of continuous functions.

Theorem 15. *Let $s \in \mathbb{R}$, $0 < p \leq \infty$, and $0 < q \leq \infty$. Then the following assertions are equivalent:*

(a) $B^s_{p,q} \hookrightarrow L_\infty$;

(b) $B^s_{p,q} \hookrightarrow C$;

(c) either $s > 1/p$ or $s = 1/p$ and $0 < q \leq 1$.

Remark 13. The implication $(c) \to (b)$ follow by means of the elementary embeddings (cf. Lemma 7) and

$$B^{1/p}_{p,1} \hookrightarrow B^0_{\infty,1}\,. \tag{47}$$

The imbedding (47) is an immediate consequence of Definition 4 and Nikol'skij's inequality.

Theorems 14 and 15 can be illustrated by Figure 1 below.

If $f \in B^s_{p,q}$ is a continuous function one can ask for the behaviour of discrete values $\{f(t_k)\}_k$. We prove the following proposition.

Proposition 2. *Let $0 < p < \infty$ and let $\tilde{p} = \min(1, p)$. Then there exists a constant c such that*

$$\left(\sum_{k=-\infty}^{\infty} |f(k)|^p \right)^{1/p} \leq c \, \| \, f \, | B^{1/p}_{p,\tilde{p}} \| \tag{48}$$

holds for all $f \in B^{1/p}_{p,\tilde{p}}$.

Proof. If $f \in B^{1/p}_{p,\tilde{p}}$, then the series

$$f(t) = \sum_{j=0}^{\infty} \mathcal{F}^{-1}[\varphi_j(\omega)\, \mathcal{F}f(\omega)](t)$$

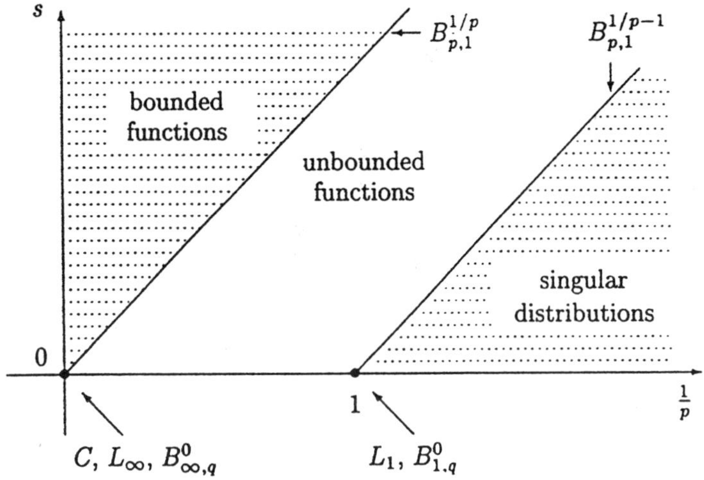

Figure 1:

converges absolutely and uniformly on $\mathbb{R}$. This is an immediate consequence of the embedding $B^{1/p}_{p,\bar{p}} \hookrightarrow B^0_{\infty,1}$ (see (47)). Let $1 \le p \le \infty$. Then

$$
\begin{aligned}
\left(\sum_{k=-\infty}^{\infty} |f(k)|^p \right)^{1/p} &= \| \sum_{j=0}^{\infty} \mathcal{F}^{-1}[\varphi_j\, \mathcal{F}f](k)\, |\ell_p\| \\
&\le \sum_{j=0}^{\infty} \left(\sum_{k=-\infty}^{\infty} |\mathcal{F}^{-1}[\varphi_j\, \mathcal{F}f](k)|^p \right)^{1/p} \\
&\le \sum_{j=0}^{\infty} \left(\sum_{k=-\infty}^{\infty} |\mathcal{F}^{-1}[\varphi_j\, \mathcal{F}f](2^{-j}k)|^p \right)^{1/p} \\
&\le c \sum_{j=0}^{\infty} 2^{j/p} \| \mathcal{F}^{-1}[\varphi_j\, \mathcal{F}f]\, |L_p\| = c\, \|\, f\, |B^{1/p}_{p,1}\|
\end{aligned}
$$

where the last estimate follows from Lemma 2 with $\Omega = 2^{j+1}$ and $h = 2^{-j}$. Now, let $0 < p < 1$. We have

$$
\sum_{k=-\infty}^{\infty} |f(k)|^p \le \sum_{j=0}^{\infty} \sum_{k=-\infty}^{\infty} |\mathcal{F}^{-1}[\varphi_j\, \mathcal{F}f](k)|^p
$$

$$\leq c \sum_{j=0}^{\infty} 2^j \| \mathcal{F}^{-1}[\varphi_j \, \mathcal{F} f] \, | L_p \|^p = c \, \| \, f \, | B^{1/p}_{p,p} \|^p$$

where the last estimate follows from Lemma 5 with $f(2^{-j}t)$ in place of f and $t_k = k$. Thus, Proposition 2 is proved. □

Corollary 3. *Let $0 < p < \infty$, $\tilde{p} = \min(1,p)$ and let $\lambda > 0$. Suppose $\lambda k \leq \tau_k \leq \lambda(k+1)$, $k \in \mathbb{Z}$. Then there exists a constant c (independent of λ and the sequence $\{\tau_k\}_k$) such that*

$$\left(\sum_{k=-\infty}^{\infty} |f(\tau_k)|^p \right)^{1/p} \leq c \, (1 + \lambda^{-1/p}) \, \| \, f \, | B^{1/p}_{p,\tilde{p}} \| \tag{49}$$

holds for all $f \in B^{1/p}_{p,\tilde{p}}$.

Proof. A small modification of the proof of Proposition 2 yields (49) with $\lambda = 1$. The general case can derived now from the inequality

$$\| \, f(\lambda \, \cdot \,) \, | B^{1/p}_{p,\tilde{p}} \| \leq c \, (1 + \lambda^{-1/p}) \, \| \, f \, | B^{1/p}_{p,\tilde{p}} \|$$

which is a simple consequence of (41) and the definition of $\Lambda^{1/p}_{p,\tilde{p}}$. □

3.4 Approximation by bandlimited functions

In this subsection we concentrate on the characterization of the Nikol'skij spaces $B^s_{p,\infty}$ by means of approximation processes. Although well-known (at least if $p \geq 1$) we give complete proofs for the convenience of the reader and for use in the next section.

Definition 5. (i) Let $f \in \mathcal{S}$ and $W > 0$. We put

$$M_W f(t) := \mathcal{F}^{-1}[\chi_{[-\pi W, \, \pi W]}(\omega) \, \mathcal{F} f(\omega)](t) \, . \tag{50}$$

(ii) Let $f \in \mathcal{S}$, $W > 0$ and let ψ be a continuous function on $\mathbb{R}$ with compact support and $\psi(0) = 1$. We define

$$M^{\psi}_W f(t) := \mathcal{F}^{-1}[\psi(\omega/W) \, \mathcal{F} f(\omega)](t) \, . \tag{51}$$

(iii) Let $0 < p \leq \infty$ and let $W > 0$. Let $f \in \mathcal{S} \cap L_p$. We denote by

$$E_p(W, f) := \inf \left\{ \| \, f - g \, | L_p \| \, : \quad g \in B^p_W \right\} \tag{52}$$

the best approximation of f in L_p by bandlimited functions from the Bernstein space B^p_W.

Remark 14. It follows easily that

$$M_W f \to f \qquad \text{and} \qquad M^\psi_W f \to f \qquad \text{in } \mathcal{S}' \quad \text{as} \quad W \to \infty .$$

By means of density arguments these approximation processes can be extended to quasi-Banach spaces contained in $\mathcal{S}'$. In signal analysis these procedures correspond to low pass filters (ideal low pass in the case of $M_W f(t)$). We have

$$M_W f(t) = W \int_{-\infty}^{\infty} \operatorname{sinc}\,(W(t-\tau))\, f(\tau)\, d\tau \tag{53}$$

and

$$M^\psi_W f(t) = \frac{W}{2\pi} \int_{-\infty}^{\infty} (\mathcal{F}^{-1}\psi)(W(t-\tau))\, f(\tau)\, d\tau . \tag{54}$$

Hence, (50) makes sense for all $f \in L_p$, $1 < p < \infty$, and (51) can be extended to all $f \in L_p$, $1 \le p \le \infty$, if $\mathcal{F}^{-1}\psi \in L_1$.

Theorem 16. *Let* $0 < p \le \infty$ *and let* $\max(1/p - 1, 0) < s < \infty$. *Then it holds* $f \in B^s_{p,\infty}$ *if and only if* $f \in L_p$ *and* $E_p(W, f) = O(W^{-s})$. *Furthermore*

$$\| f \,|N^s_p\| := \| f \,|L_p\| + \sup_{W \ge 2} W^s\, E_p(W, f) \tag{55}$$

is an equivalent quasi-norm in $B^s_{p,\infty}$.

Proof. By monotonicity properties of $E_p(W, f)$ it is sufficient to consider the dyadic subsequence $W = 2^\ell$, $\ell = 1, 2, \ldots$. Let us put

$$f^\ell(t) := \sum_{j=0}^{\ell} \mathcal{F}^{-1}[\varphi_j\, \mathcal{F} f](t) = \mathcal{F}^{-1}[\varphi_0(2^{-\ell}\omega)\mathcal{F} f(\omega)](t)$$

where φ_0 and φ_j have the meaning from (39). If $f \in B^s_{p,\infty}$, then $f^\ell \in B^p_{2^{\ell+1}}$ and

$$\begin{aligned}
\| f - f^\ell \,|L_p\|^{\tilde p} &= \| \sum_{j=\ell+1}^{\infty} \mathcal{F}^{-1}[\varphi_j\, \mathcal{F} f] \,|L_p\|^{\tilde p} \\
&\le \sum_{j=\ell+1}^{\infty} \| \mathcal{F}^{-1}[\varphi_j\, \mathcal{F} f] \,|L_p\|^{\tilde p} \\
&\le \sum_{j=\ell+1}^{\infty} 2^{-js\tilde p}\, \| \mathcal{F}^{-1}[\varphi_j\, \mathcal{F} f] \,|L_p\|^{\tilde p} \\
&\le c\, 2^{-\ell s \tilde p}\, \| f \,|B^s_{p,\infty} \|^{\tilde p} .
\end{aligned}$$

Hence $E_p(2^\ell, f) = O(2^{-\ell s})$. Remark 12 and the elementary imbeddings (Lemma 7) show that $f \in L_p$ ($L_p \cap L_1$ if $0 < p < 1$) if $f \in B^s_{p,\infty}$, $s > \max(1/p - 1, 0)$. Moreover,

$$\| f \,|L_p\|^{\tilde{p}} \le \sum_{j=0}^{\infty} \| \mathcal{F}^{-1}[\varphi_j \, \mathcal{F} f] \,|L_p\|^{\tilde{p}} \le \| f \,|B^0_{p,\tilde{p}}\|^{\tilde{p}} \le c \, \| f \,|B^s_{p,\infty} \|^{\tilde{p}} \,.$$

To prove the converse direction we choose functions $g_{\ell-1} \in B^p_{2^\ell}$ such that

$$\| f - g_{\ell-1} \,|L_p\| \le 2 \, E_p(2^\ell, f) \,, \qquad \ell = 1, 2, \ldots \,.$$

Then, by $s > 0$

$$f = \lim_{\ell \to \infty} g_\ell = g_0 + \sum_{j=1}^{\infty} (g_{j+1} - g_j) = \sum_{j=0}^{\infty} a_j$$

in L_p. It is easily checked that $a_j \in B^p_{2^{j+1}}$ and

$$\begin{aligned}
\sup_{j=0,1,\ldots} \; 2^{sj} \, \| a_j \,|L_p\| &\le c_1 \left(\| f \,|L_p\| + \sup_{\ell=1,2,\ldots} \; 2^{\ell s} \, \| f - g_{\ell-1} \,|L_p\| \right) \\
&\le c_2 \left(\| f \,|L_p\| + \sup_{\ell=1,2,\ldots} \; 2^{\ell s} \, E_p(2^\ell, f) \right) .
\end{aligned}$$

It is clear that $f \in \mathcal{S}'$ if $1 \le p \le \infty$. If $0 < p < 1$ and $s > \max(1/p - 1, 0)$, then by Nikol'skij's inequality

$$\sum_{j=0}^{\infty} \| a_j \,|L_1\| \le c \sum_{j=0}^{\infty} 2^{j(\frac{1}{p}-1)} \, \| a_j \,|L_p\| \le C \sum_{j=0}^{\infty} 2^{j(\frac{1}{p}-1-s)} < \infty \,.$$

Hence, the series $\sum_{j=0}^{\infty} a_j$ converges in L_1 and $f \in L_1 \hookrightarrow \mathcal{S}'$. As a consequence,

$$\mathcal{F}^{-1}[\varphi_\ell \, \mathcal{F} f] = \sum_{j=\ell-1}^{\infty} \mathcal{F}^{-1}[\varphi_\ell \, \mathcal{F} a_j] \,, \qquad \ell = 1, 2, \ldots$$

in $\mathcal{S}'$. Let $\sigma_p = \max(0, 1/p - 1)$. The Fourier multiplier criterion from Theorem 8 (with $\Omega = 2^{j+1}$) yields

$$\begin{aligned}
\| \mathcal{F}^{-1}[\varphi_\ell \, \mathcal{F} a_j] \,|L_p\| &\le c \, 2^{j \, \sigma_p} \| \mathcal{F}^{-1} \varphi_\ell \,|L_{\tilde{p}}\| \; \| a_j \,|L_p\| \\
&= c \, 2^{(j-\ell) \, \sigma_p} \| \mathcal{F}^{-1} \varphi \,|L_{\tilde{p}}\| \; \| a_j \,|L_p\| \,.
\end{aligned}$$

Hence, for $\ell = 1, 2, \ldots$

$$2^{\ell s\tilde{p}} \, \| \, \mathcal{F}^{-1}[\varphi_\ell \, \mathcal{F} f] \, | L_p \|^{\tilde{p}} \leq c \sum_{j=\ell-1}^{\infty} 2^{(\ell-j)\,(s-\sigma_p)\tilde{p}} \, 2^{js\tilde{p}} \, \| \, a_j \, | L_p \|^{\tilde{p}}$$

$$\leq c \, (\sup_{j=0,1,\ldots} 2^{js} \, \| \, a_j \, | L_p \|)^{\tilde{p}} \, .$$

On the same way we get

$$\| \, \mathcal{F}^{-1}[\varphi_0 \, \mathcal{F} f] \, | L_p \|^{\tilde{p}} \leq c \sum_{j=0}^{\infty} 2^{-j(s-\sigma_p)\tilde{p}} \, 2^{js\tilde{p}} \, \| \, a_j \, | L_p \|^{\tilde{p}}$$

$$\leq c \, (\sup_{j=0,1,\ldots} 2^{js} \, \| \, a_j \, | L_p \|)^{\tilde{p}} \, .$$

This completes the proof of the theorem. □

In the first part of the proof we showed that the means $M_{2^\ell}^{\varphi_0} f$ approximate the elements of $B^s_{p,\infty}$ in the correct order. This can be improved for particular ψ. By $L_1 + L_\infty$ we mean the set of functions $f = f_0 + f_1$ with $f_0 \in L_1$ and $f_1 \in L_\infty$.

Proposition 3. *Let $\psi_0 \in \mathcal{S}$ be a function such that $\psi(0) = 1$. Let $m \in \mathbb{N}$. Define*

$$\psi(\omega) = (-1)^{m+1} \sum_{j=0}^{m-1} \binom{m}{j} (-1)^j \, \psi_0((m-j)\omega) \, . \tag{56}$$

Then the means $M_W^\psi f$ satisfy the following inequalities.
(i) *Let $1 \leq p \leq \infty$. Then there exists a constant c such that*

$$\| \, f - M_W^\psi f \, | L_p \| \leq c \, \omega_p^m (1/W, f) \tag{57}$$

holds for all $f \in L_p$ and all $W \geq 1$.
(ii) *Let $0 < p < 1$. Suppose $f \in L_1 + L_\infty$ and satisfies*

$$\| \, d^m_{\tau,1} f \, | L_p \| = O(\tau^s) \qquad (\tau \to 0) \, ,$$

for some $s > 0$. Then

$$\| \, f - M_W^\psi f \, | L_p \| = O(W^{-s}) \qquad (W \to \infty) \, . \tag{58}$$

Proof. We use a tricky argument of Nikol'skij. Based on $\psi_0(0) = 1$ elementary calculations yield the following identity

$$f(t) - M_W^\psi f(t) = \frac{1}{2\pi} (-1)^{m+1} \int_{-\infty}^{\infty} \Delta^m_{W^{-1}v} f(t) \, (\mathcal{F}^{-1}\psi_0)(v) \, dv$$

This identity holds as long as f belongs to L_p with $p \geq 1$ or more general, if $f \in L_1 + L_\infty$. Let $I_0 = [-1,1]$ and $I_n = [-2^{n+1}, -2^n] \cup [2^n, 2^{n+1}]$, $n = 1, 2, \ldots$. We assume $2^\ell \leq W < 2^{\ell+1}$. The generalized Minkowski inequality together with $|\mathcal{F}^{-1}\psi_0(v)| \leq c_M \, (1 + |v|)^{-M}$ (here M is at our disposal) show that

$$\begin{aligned}
\| \, f - M_W^\psi f \, | L_p \| &\leq c_M \int_{-\infty}^{\infty} \| \, \Delta^m_{W^{-1}v} f(t) \, | L_p \| \, (1 + |v|)^{-M} \, dv \\
&\leq c \sum_{n=0}^{\infty} 2^{-n(M-1)} \sup_{2^n \leq |v| \leq 2^{n+1}} \| \, \Delta^m_{W^{-1}v} f(t) \, | L_p \| \\
&\leq c \sum_{n=0}^{\infty} 2^{-n(M-1)} \omega_p^m(f, 2^{n-\ell}) \, .
\end{aligned}$$

The inequality (57) follows now from

$$\omega_p^m(f, \, \lambda\tau) \leq (\lambda + 1)^m \, \omega_p^m(f, \, \tau) \, ,$$

cf. e.g. [27, 2.7/(8)], and the fact that we can choose $M > m + 1$.
The case $0 < p < 1$ is more complicated. For brevity we put $f^\ell = M^\psi_{2^\ell} f$. Similar as above we find

$$\begin{aligned}
\| \, f^{\ell+1} - f^\ell \;\; | L_p \|^p &\leq 2 \, \max_{j=0,1} \, \| \int_{-\infty}^{\infty} \Delta^m_{2^{-\ell-j}v} f(t) \, \mathcal{F}^{-1}\psi_0(v) \, dv \, | L_p \|^p \\
&\leq c_M \, \max_{j=0,1} \sum_{n=0}^{\infty} 2^{-n(M-1)p} \, \| \, 2^{-n} \int_{I_n} |\Delta^m_{2^{-\ell-j}v} f(t)| \, dv \, | L_p \|^p \\
&\leq c \sum_{n=0}^{\infty} 2^{-n(M-1)p} \, \| \, d^m_{2^{n-\ell},1} f \, | L_p \|^p \, .
\end{aligned}$$

Next we use

$$\| \, d^m_{2\tau,1} f \, | L_p \|^p \leq 2^m \, \| \, d^m_{\tau,1} f \, | L_p \|^p$$

which can be derived from the identity

$$\Delta^m_{2h} f(t) = \sum_{i_1=0}^{1} \cdots \sum_{i_m=0}^{1} \Delta^m_h f(t + i_1 h + i_2 h + \ldots + i_m h) \, ,$$

cf. [67, 1.1/(5)]. Hence

$$\| f^{\ell+1} - f^{\ell} \,|L_p\|^p \le c \,\| d^m_{2^{-\ell},1} f \,|L_p\|^p \,,$$

for $M > m+1$. But this implies

$$\begin{aligned} \| f^{\ell+n} - f^{\ell} \,|L_p\|^p &\le c \sum_{j=0}^{n-1} \| d^m_{2^{-\ell-j},1} f \,|L_p\|^p \\ &\le c\, 2^{-\ell s p} \,, \end{aligned}$$

with c independent of ℓ. By our assumption the sequence f^{ℓ}, $\ell = 0, 1, \ldots$ converges in L_p and, of course, in S' (here one uses $f \in L_1 + L_\infty$). The arguments used in Remark 12 apply also in this situation here. Consequently, these limits coincide almost everywhere and for $n \to \infty$ the claim follows. □

Remark 15. Let $m \in \mathbb{N}$. Let ψ_0 and ψ be as in Proposition 3. If we define a class of functions by

$$f \in L_1 \,, \qquad \sup_{0<\tau<1} \tau^{-s} \| d^m_{\tau,1} \,|L_p\| < \infty \,,$$

$s > 0$ and $0 < p < 1$, then the same proof as above yields

$$\| f - M^{\psi}_W f \,|L_p\| \le c\, W^{-s} \sup_{0<\tau<1} \tau^{-s} \| d^m_{\tau,1} \,|L_p\|$$

(c is independent of W and f). By Theorem 12 these spaces coincide with $B^s_{p,\infty}$ if $1/p - 1 < s < m$. Observe, if s is to large in comparison to m, the space becomes trivial. With more effort one can obtain the estimate

$$E_p(W, f) \le c\, \omega^m_p(W^{-1}, f) \,, \qquad W \ge 1 \,,$$

also for $p < 1$, cf. E.A. Storozenko and P. Oswald [74] for the periodic situation. As a consequence we get $f \in \Lambda^s_{p,\infty}$ implies $E_p(W, f) = O(W^{-s})$, $(W \to \infty)$ for all $s > 0$ also if $p < 1$. This again turns into a characterization of $\Lambda^s_{p,\infty}$. This follows from the close relations between approximation scales and interpolation scales, cf. [27, 26], and the characterization of the real interpolation spaces between L_p and $\Lambda^s_{p,\infty}$, namely $(L_p, \Lambda^s_{p,\infty})_{\Theta,\infty} = \Lambda^{s\Theta}_{p,\infty}$, cf. [28]. Since we tried to avoid interpolation theory this extension of Theorem 16 lies a bit outside of our scope.

Theorem 16 shows that the Nikol'skij spaces $B^s_{p,\infty}$ are the appropriate spaces to describe the asymptotic behaviour of the approximation error with respect to (50) and (51), at least if $s > \max(0, 1/p - 1)$.

Theorem 17. *Let $1 < p < \infty$, $s > 0$ and $m \in \mathbb{N}$.*
(i) *There exists a constant c such that*

$$\| f - M_W f \,|L_p\| \le c\,\omega_p^m(1/W, f) \tag{59}$$

holds for all $f \in L_p$ and all $W \ge 1$.
(ii) *We have $f \in B^s_{p,\infty}$ if and only if $f \in L_p$ and $\| f - M_W f \,|L_p\| = O(W^{-s})$. Moreover,*

$$\| f \,|N^s_p\|^\chi := \| f \,|L_p\| + \sup_{W \ge 2} W^s \,\| f - M_W f \,|L_p\| \tag{60}$$

is an equivalent norm in $B^s_{p,\infty}$.

Proof. Let $\psi_0 \in \mathcal{S}$ be a function such that $\operatorname{supp}\psi_0 \subset \{\omega : \ |\omega| \le 3/2\}$ and $\psi_0(\omega) = 1$ if $|\omega| \le 1$. Let ψ be the function defined in (56). Then $\psi(\omega) = 1$ if $|\omega| \le 1/m$ and $\operatorname{supp}\psi \subset \{\omega : \ |\omega| \le 3/2\}$. As above we put $f^\ell = M^\psi_{2^\ell} f$. Let $2^\ell \le W < 2^{\ell+1}$. Now, using the projection property of M_W we see that

$$M_W(M^\psi_{2^\ell} f) = M^\psi_{2^\ell} f\,, \qquad \ell = 0, 1, \ldots\,.$$

Using this identity, Riesz theorem and the previous proposition we derive

$$\begin{aligned} \| f - M_W f \,|L_p\| &\le \| f - M^\psi_{2^\ell} f \,|L_p\| + \| M_W(f - M^\psi_{2^\ell}) \| \\ &\le c\,\omega_p^m(f, 2^{-\ell})\,. \end{aligned}$$

Hence (59) follows.
To prove part (ii) we observe that Theorem 16 yields

$$\| f \,|B^s_{p,\infty} \| \le c \,\| f \,|N^s_p\|^\chi\,.$$

The converse estimate is implied by (59) and the coincidence of $B^s_{p,\infty}$ and $\Lambda^s_{p,\infty}$. □

Remark 16. The theorem does not extend to $p = \infty$ and $p \le 1$. That may be seen immediately by testing M_W on functions $f(\omega) = \mathcal{F}^{-1}\varphi_0(\lambda\omega) \in \mathcal{S}$. For appropriate λ we have $M_W f(t) = \operatorname{sinc} Wt$. Hence, $p \le 1$ is excluded. If $p = \infty$, we may concentrate on $W = 1$. Then we consider the family of piecewise linear (with respect to the intervals $[k, k+1]$, $k \in \mathbb{Z}$) and continuous functions f_N defined by

$$f_N(t) = \begin{cases} 0 & t \le 0\,, \\ 1 & t = 2k-1\,,\ k = 1, 2, \ldots, N\,, \\ -1 & t = 2k\,,\ k = 1, 2, \ldots, N\,, \\ 0 & t \ge 2N+1\,. \end{cases}$$

Then f_N is uniformly bounded in C but

$$\sum_k f_N(k)\text{sinc}\,(1/2-k) = \frac{1}{\pi}\sum_{k=1}^{2N}\frac{1}{k-1/2} \geq \log(2N)\,.$$

Next we investigate the approximation by M_W^ψ for more general functions ψ. We need some preparations.

Proposition 4. (Szasz theorem). *Let $0 < p \leq 2$ and let $\tilde{p} = \min(1,p)$. Then there exists a constant c such that*

$$\|\,\mathcal{F}m\,|L_p\| \leq c\,\|\,m\,|B_{2,\tilde{p}}^{\frac{1}{p}-\frac{1}{2}}\| \tag{61}$$

Proof. Let $m \in B_{2,\tilde{p}}^{\frac{1}{p}-\frac{1}{2}}$. Because of $0 < p \leq 2$ we have the imbedding $B_{2,\tilde{p}}^{\frac{1}{p}-\frac{1}{2}} \hookrightarrow L_2$ and hence, $\mathcal{F}m$ is a regular distribution. If $0 < p \leq 1$ we claim (using Hölder's inequality)

$$\begin{aligned}\int_{-\infty}^{\infty}|\mathcal{F}m(\omega)|^p d\omega &\leq \sum_{j=0}^{\infty}\int_{\text{supp}\,\varphi_j}|\,\varphi_j(\omega)\,\mathcal{F}m(\omega)\,|^p d\omega\\ &\leq c\sum_{j=0}^{\infty}2^{j(1-p/2)}\,\|\,\varphi_j\,\mathcal{F}m\,|L_2\|^p = c\,\|\,m\,|B_{2,p}^{\frac{1}{p}-\frac{1}{2}}\|^p\,.\end{aligned}$$

If $1 < p \leq 2$, then

$$\begin{aligned}\|\mathcal{F}m|L_p\| &\leq \sum_{j=0}^{\infty}\|\,\varphi_j\,\mathcal{F}m\,|L_p\|\\ &= \sum_{j=0}^{\infty}\left(\int_{\text{supp}\,\varphi_j}|\,\varphi_j(\omega)\,\mathcal{F}m(\omega)|^p d\omega\right)^{1/p}\\ &\leq c\sum_{j=0}^{\infty}2^{j(1-p/2)\,\frac{1}{p}}\,\|\,\varphi_j\,\mathcal{F}m\,|L_2\| = c\,\|\,m\,|B_{2,1}^{\frac{1}{p}-\frac{1}{2}}\|\,.\end{aligned}$$

This proves the proposition. □

Proposition 5. *Let $0 < q < \infty$ and let σ be a real number with*

$$\sigma + \frac{1}{q} > \max(\frac{1}{q}-1,0)\,.$$

Let $\varphi_0 \in C_0^\infty$ be a function equal to 1 in a neighbourhood of the origin. Then

$$\varphi_0(x)\,|x|^\sigma \in B^{\sigma+\frac{1}{q}}_{q,\infty}\,. \tag{62}$$

Proof. Clearly, $\varphi_0(x)\,|x|^\sigma \in L_q$ because of $\sigma + 1/q > 0$. Hence, by Theorem 12, formula (45) it is sufficient to show that

$$\|\,\Delta_h^m \varphi_0(x)\,|x|^\sigma\,|L_q\|^q \le c\,|h|^{\sigma q+1} \tag{63}$$

for large $m > \sigma + 1/q$ and small $|h| < \varepsilon$, where m and ε are at our disposal. Let $m > \sigma + 1/q$ be fixed. There exists a $\delta > 0$ such that

$$\varphi_0(x+jh) = 1 \qquad \text{for all} \quad |x| < \delta \quad \text{and all} \quad j = 0, 1, \ldots, m\,.$$

We choose $\varepsilon > 0$ such that $2m|h| < \delta$ if $|h| \le \varepsilon$. Taking into account

$$\Delta_h^m f(x) = \sum_{j=0}^m \binom{m}{j} (-1)^{m-j}\, f(x+jh)$$

we obtain

$$\int_{|x|\le 2m|h|} |\,\Delta_h^m \varphi_0(x)\,|x|^\sigma\,|^q\,dx \le c \int_{|y|\le 3m|h|} |y|^{\sigma q}\,dy \le C\,|h|^{\sigma q+1}\,. \tag{64}$$

We split

$$\int_{|x|\ge 2m|h|} |\,\Delta_h^m \varphi_0(x)\,|x|^\sigma\,|^q\,dx = \sum_{j=1}^J \int_{I_j} |\,\Delta_h^m \varphi_0(x)\,|x|^\sigma\,|^q\,dx\,,$$

where J is a finite number depending on m, h, and $\operatorname{supp}\varphi_0$ and

$$I_j = \{x : \quad 2mj|h| \le |x| \le 2m(j+1)|h|\,\}\,.$$

Let $x \in I_j$. Then we use

$$|\,\Delta_h^m \varphi_0(x)\,|x|^\sigma\,| \le c\,|h|^m \sup_{|x-y|\le m|h|} \left| \frac{d^m}{dy^m}\left(\varphi_0(y)\,|y|^\sigma\right)\right|.$$

If $x \in I_j$ and $|x-y| \le m|h|$, then $(2j-1)m|h| \le |y| \le (2j+3)m|h|$ and hence

$$\left| \frac{d^m}{dy^m}\left(\varphi_0(y)\,|y|^\sigma\right)\right| \le c_1\,|y|^{\sigma-m} \le c_2\,|jh|^{\sigma-m}\,.$$

As a consequence we get

$$\begin{aligned}\int_{|x|\geq 2m|h|} |\,\Delta_h^m \varphi_0(x)\,|x|^\sigma\,|^q\,dx &\leq c \sum_{j=1}^{J} \int_{I_j} (|h|^m\,|jh|^{\sigma-m})^q\,dx \\ &\leq c \sum_{j=1}^{J} j^{(\sigma-m)q}\,|h|^{\sigma q+1} \\ &\leq c\,|h|^{\sigma q+1}\,, \qquad (65)\end{aligned}$$

because of $(\sigma - m)q < -1$. Finally, (64) and (65) prove (63). □

Remark 17. Proposition 4 and 5 show that

$$\mathcal{F}[\varphi_0(x)\,|x|^\sigma](\omega) \in L_p\,, \qquad 0 < p \leq 1\,, \tag{66}$$

if $\sigma > 1/p - 1$. Indeed, we find that

$$\|\,\mathcal{F}[\varphi_0(x)\,|x|^\sigma](\omega)|L_p\| \leq c\,\|\,\varphi_0(x)\,|x|^\sigma\,|B_{2,p}^{\frac{1}{p}-\frac{1}{2}}\,\| \leq c\,\|\,\varphi_0(x)\,|x|^\sigma\,|B_{2,\infty}^{\sigma+\frac{1}{2}}\,\|\,.$$

Furthermore, let us recall the imbedding $B_{1,p}^{1/p} \hookrightarrow B_{2,p}^{\frac{1}{p}-\frac{1}{2}}$, see Theorem 13. It follows that a function $m(\omega)$ with compact support is a Fourier multiplier in the Bernstein space B_Ω^p in the sense of Theorem 8 if it belongs to the space $B_{1,\tilde{p}}^{1/\tilde{p}}$, $\tilde{p} = \min(1,p)$.

Theorem 18. *Let $0 < p \leq \infty$, $\sigma > 0$, and let ψ be a continuous function with compact support. We put $\tilde{p} = \min(1,p)$ and suppose that*

$$\mathcal{F}^{-1}\psi \in L_{\tilde{p}} \tag{67}$$

and

$$\|\,\mathcal{F}^{-1}\big[|\omega|^{-\sigma}\,(1-\psi(\omega))\,\eta(\omega)\big]\,|L_{\tilde{p}}\| < \infty \tag{68}$$

for some function $\eta \in C_0^\infty$ equal to 1 in a neighbourhood of the origin. If $\max(1/p-1,0) < s < \sigma$, then it holds $f \in B_{p,\infty}^s$ if and only if $f \in L_p$ and

$$\|\,f - M_W^\psi f\,|L_p\| = O(W^{-s}) \qquad (W \to \infty). \tag{69}$$

Moreover,

$$\|\,f\,|N_p^s\|^\psi := \|\,f\,|L_p\| + \sup_{W\geq 2} W^s\,\|\,f - M_W^\psi f\,|L_p\| \tag{70}$$

is an equivalent quasi-norm in $B_{p,\infty}^s$.

Proof. Let $\operatorname{supp}\psi \subset [-R, R]$ for some $R > 0$. Then the definition of $M_W^\psi f$ shows that

$$E_p(RW, f) \le \| f - M_W^\psi f \,|L_p\| \,.$$

Assume that $f \in L_p$ and that (69) holds. Then Theorem 16 yields $f \in B^s_{p,\infty}$ and $\| f \,|B^s_{p,\infty} \| \le c \| f \,|N^s_p\| \le c \| f \,|N^s_p\|^\psi$. In order to prove the converse direction it is sufficient to show that

$$\sup_W W^s \| f - M_W^\psi f \,|L_p\| \le c \| f \,|B^s_{p,\infty} \| \,, \tag{71}$$

where the supremum can be taken with respect to $W \ge W_0$, $W_0 > 0$. Let $W = 2^j\tau$, where $j = 0, 1, \ldots,$ $1 \le \tau \le 2$. We put $\psi_\tau(\omega) = \psi(\omega/\tau)$. Then $M_W^\psi f(t) = M_{2^j}^{\psi_\tau} f(t) = \mathcal{F}^{-1}[\psi_\tau(2^{-j}\omega)\,\mathcal{F}f(\omega)](t)$. Further, there exists a natural number L such that $\operatorname{supp}\psi_\tau \subset [-L, L]$, $1 \le \tau \le 2$. Obviously, the following estimates are true

$$\begin{aligned} W^{s\tilde{p}} \| f - M_W^\psi f|L_p\|^{\tilde{p}} &\le c\, 2^{js\tilde{p}} \sum_{\ell=L+j+1} \| \mathcal{F}^{-1}[\varphi_\ell\, \mathcal{F}f] \,|L_p\|^{\tilde{p}} \\ &+ c\, 2^{js\tilde{p}} \sum_{\ell=0}^{L+j} \| \mathcal{F}^{-1}[(1 - \psi_\tau(2^{-j}\omega)\, \varphi_\ell(\omega)\, \mathcal{F}f(\omega)] \,|L_p\|^{\tilde{p}} \,. \end{aligned}$$

Because of $s > 0$ the first term on the right-hand side can be estimated by $c \| f \,|B^s_{p,\infty} \|^{\tilde{p}}$. Hence, it remains to prove that

$$\sup_{j=0,1,\ldots} 2^{js\tilde{p}} \sum_{\ell=0}^{L+j} \|\mathcal{F}^{-1}[(1-\psi_\tau(2^{-j}\omega)\varphi_\ell(\omega)\mathcal{F}f(\omega)] \,|L_p\|^{\tilde{p}} \le c \| f \,|B^s_{p,\infty} \|^{\tilde{p}} \,, \tag{72}$$

where c is independent of τ, $1 \le \tau \le 2$. It holds the identity

$$\begin{aligned} &\| \mathcal{F}^{-1}[(1 - \psi_\tau(2^{-j}\omega)\varphi_\ell(\omega)\, \mathcal{F}f(\omega)] \,|L_p\| \\ &= 2^{(\ell-j)\sigma} \| \mathcal{F}^{-1}\left[|2^{-j}\omega|^{-\sigma}\, (1 - \psi_\tau(2^{-j}\omega)\, \mathcal{F}[\mathcal{F}^{-1}|2^{-\ell}\xi|^\sigma\, \varphi_\ell(\xi)\, \mathcal{F}f(\xi)]\right] \,|L_p\| \\ &= 2^{(\ell-j)\sigma} \| \mathcal{F}^{-1}\left[\varrho_\tau(2^{-j}\omega)\, \varphi_0(2^{-j-L-1}\omega)\, \mathcal{F}\tilde{f}_\ell(\omega)\right] \,|L_p\| \end{aligned} \tag{73}$$

where

$$\varrho_\tau(\omega) = |\omega|^{-\sigma}\, (1 - \psi(\frac{\omega}{\tau}))$$

and

$$\tilde{f}_\ell(t) = \mathcal{F}^{-1}[|2^{-\ell}\omega|^\sigma\, \varphi_\ell\,(\omega)\mathcal{F}f(\omega)](t) \,,$$

(φ_0 has the meaning of Subsection 2.1). Using the Fourier multiplier criterion stated in Theorem 8 we see that for $\ell = 0, 1, \ldots, j + L$

$$\begin{aligned} &\| \mathcal{F}^{-1}[\varrho_\tau(2^{-j}\omega)\ \varphi_0(2^{-j-L-1}\omega)\, \mathcal{F}\tilde{f}_\ell]\, |L_p\| \\ &\qquad \le c_L \, \| \mathcal{F}^{-1}[\varrho_\tau(\omega)\, \varphi_0(2^{-L-1}\omega)]\, |L_{\tilde{p}}\| \ \| \tilde{f}_\ell \, |L_p\| \, . \end{aligned} \tag{74}$$

We have $\varphi_0(\omega) = \varphi_0(\omega/(2\tau))\, \varphi_0(\omega)$ for $1 \le \tau \le 2$. Applying Theorem 8 again we get

$$\begin{aligned} \| \mathcal{F}^{-1}[\varrho_\tau(\omega)\, \varphi_0(2^{-L-1}\omega)]\, |L_{\tilde{p}}\| &\le c \, \| \mathcal{F}^{-1}[\varrho_\tau(\omega)\, \varphi_0(2^{-L-2}\omega/\tau)]\, |L_{\tilde{p}}\| \\ &\le c \, \| \mathcal{F}^{-1}[\varrho_1(\omega)\, \varphi_0(2^{-L-2}\omega)]\, |L_{\tilde{p}}\| \, . \end{aligned}$$

Without loss of generality we may assume that the function η in (68) has support in $[-1, 1]$. Therefore, it follows from (67), (68), and Theorem 8 that

$$\begin{aligned} &\|\mathcal{F}^{-1}[\varrho_1(\omega)\, \varphi_0(2^{-L-2}\omega)]\, |L_{\tilde{p}}\| \\ &\quad \le c\, (\| \mathcal{F}^{-1}[\varrho_1\, \eta]\, |L_{\tilde{p}}\| + \|\mathcal{F}^{-1}[\varrho_1(\omega)(1 - \eta(\omega))\, \varphi_0(2^{-L-2}\omega)]\, |L_{\tilde{p}}\|) \\ &\quad \le c\, (\| \mathcal{F}^{-1}[\varrho_1\, \eta]\, |L_{\tilde{p}}\| + \|\mathcal{F}^{-1}[|\omega|^{-\sigma}\, (1 - \eta(\omega))\, \varphi_0(2^{-L-2}\omega)]\, |L_{\tilde{p}}\| \\ &\qquad\qquad + \| \mathcal{F}^{-1}\psi\, |L_{\tilde{p}}\|) < \infty \, . \end{aligned}$$

Substituting into (74) and (73) we find

$$\| \mathcal{F}^{-1}[(1 - \psi_\tau(2^{-j}\omega)\, \varphi_\ell(\omega)\, \mathcal{F}f(\omega)]\, |L_p\| \le c\, 2^{(\ell-j)\sigma} \, \| \tilde{f}_\ell \, |L_p\| \, ,$$

where c is independent of τ, j, and ℓ. The left-hand side of (72) can be estimated from above by

$$\begin{aligned} &c \sup_{j=0,1,\ldots} 2^{js\tilde{p}} \sum_{\ell=0}^{L+j} 2^{(\ell-j)\sigma\tilde{p}} \, \| \tilde{f}_\ell \, |L_p\|^{\tilde{p}} \\ &\qquad \le c \sum_{\ell=0}^{L+j} 2^{(\ell-j)(\sigma-s)\tilde{p}} \, (\sup_{\ell=0,1,\ldots} 2^{\ell s} \| \tilde{f}_\ell \, |L_p\|)^{\tilde{p}} \\ &\qquad \le c_L \sup_{\ell=0,1,\ldots} 2^{\ell s} \, \| \mathcal{F}^{-1}[|2^{-\ell}\omega|^\sigma \, \varphi_\ell(\omega)\, \mathcal{F}f(\omega)]\, |L_p\|^{\tilde{p}} \, . \end{aligned} \tag{75}$$

In the last inequality we used the assumption $s < \sigma$. By means of Remark 17 we conclude from Theorem 8

$$\| \mathcal{F}^{-1}[|\omega|^\sigma \varphi_0(\omega)\, \mathcal{F}f(\omega)]\, |L_p\| \le c\, (\| \mathcal{F}^{-1}[\varphi_0 \, \mathcal{F}f]\, |L_p\| + \| \mathcal{F}^{-1}[\varphi_1 \, \mathcal{F}f]\, |L_p\|) \, .$$

Moreover, for $\ell = 1, 2, \ldots$,

$$\| \mathcal{F}^{-1}[|2^{-\ell}\omega|^{\sigma} \varphi_{\ell}(\omega) \, \mathcal{F}f(\omega)] \, |L_p\| \le c \sum_{\nu=-1}^{1} \| \mathcal{F}^{-1}[\varphi_{\nu+\ell} \, \mathcal{F}f] \, |L_p\|$$

also follows from Theorem 8. Together with (75) and (72) this completes the proof. □

Remark 18. The method of proof in Theorem 17 which is based on a projection property and Proposition 3 applies to M_W^{ψ} if ψ is equal to 1 in a neigbourhood of 0, at least if $1 \le p \le \infty$ and one gets

$$\| f - M_W^{\psi} f \, |L_p\| \le c \, \omega_p^m(f, 1/W) \, .$$

3.5 The τ–modulus and related function spaces

As we shall see later on the classes of functions introduced above are not sufficient for our purposes. The approximation properties of sampling sums in L_p–norms are connected in natural way with functions of bounded p–variation. For low rates of approximation by sampling sums it seems that the role played by the modulus of continuity ω_p^m in the previous subsection is taken over by an averaged version of it, the so-called τ–modulus.

Here in this subsection we shall give definitions and collect some properties of interest for us.

Functions with bounded p–variation

Definition 6. Let $1 \le p < \infty$. A function f is said to be of bounded p–variation if

$$\| f \, |V_p\| = \sup_{\mathcal{Z}} \left(\sum_k |f(t_{k+1}) - f(t_k)|^p \right)^{1/p}$$

where $\mathcal{Z}$ denotes a finite collection of points $\{t_k\}_k$ such that $t_k < t_{k+1}$ for all k, is finite. The class of all functions having bounded p–variation will be denoted by V_p.

Sometimes it will be convenient for us to deal with $V_p \cap L_p$. Then we shall equip this space with the norm

$$\| f \, |V_p \cap L_p\| = \| f \, |L_p\| + \| f \, |V_p\| \, .$$

Whereas V_p becomes a Banach space modulo constants only $V_p \cap L_p$ itself is a Banach space.

Lemma 9. *Let* $1 \le p,\, p_0,\, p_1 < \infty$.
(i) *For* $p_0 < p_1$ *it follows* $V_{p_0} \cap L_{p_0} \hookrightarrow V_{p_1} \cap L_{p_1}$.
(ii) *We have the continuous imbeddings*

$$B^{1/p}_{p,1} \hookrightarrow V_p \cap L_p \hookrightarrow B^{1/p}_{p,\infty}\,. \tag{76}$$

Proof. We select a point y such that

$$|f(y)|^p \le \int_{-\infty}^{\infty} |f(t)|^p\, dt\,.$$

Then

$$|f(t)| \le |f(t) - f(y)| + |f(y)| \le \|\, f\,|V_p\| + \|\, f\,|L_p\|\,.$$

Hence, $V_p \cap L_p$ consists of bounded functions only. Hölder's inequality yields

$$\|\, f\,|V_{p_1}\| \le \|\, f\,|V_{p_0}\|^{p_0/p_1}\, \big(2 \sup_{t\in\mathbb{R}} |\, f(t)\,|\big)^{1-p_0/p_1}$$

and

$$\|\, f\,|L_{p_1}\| \le \|\, f\,|L_{p_0}\|^{p_0/p_1} \sup_{t\in\mathbb{R}} |\, f(t)\,|^{1-p_0/p_1}\,.$$

This proves part (i). Finally, (ii) is proved in [54, Theorem 5.7]. □

The τ–modulus

Let $f : \mathbb{R} \to \mathbb{R}$ be pointwise given. Let $m \in \mathbb{N}$. The local modulus of smoothness is defined to be

$$\omega^m(f,t,\delta) = \sup\{\, |\Delta_h^m f(y)| \,:\, t - \frac{m\delta}{2} \le y,\, y + mh \le t + \frac{m\delta}{2}\}\,, \qquad \delta > 0\,.$$

Based on this notion we can introduce the τ–modulus of order m:

$$\tau_p^m(f,\delta) := \|\, \omega^m(f,t,\delta)\,|L_p\|\,.$$

As usual, if $m = 1$, then we drop the exponent 1 and write $\tau_p(f,\delta)$ instead of $\tau_p^1(f,\delta)$. This τ-modulus and ω_p^m can be compared through an inequality due to Ivanov [41]: for $1 \le p < \infty$

$$\tau_p^m(f,\delta) \le c\,\delta^{1/p} \int_0^{\delta} \omega^m(f,t)_p\, t^{-1/p}\, \frac{dt}{t}\,, \tag{77}$$

valid for continuous f and with c independent of f and δ, and the elementary inequality

$$\omega_p^m(f,\delta) \le \tau_p^m(f,\delta)\,.$$

Corresponding to the τ–modulus one can introduce classes of functions in the same way as done for ω_p^m.

Definition 7. Let $0 < p \le \infty$ and $s > 0$. Let $m \in \mathbb{N}$ and $m > s$. Then we put

$$A^s_{p,\infty} = \left\{ f \in L_p : \quad \| f \,| A^s_{p,\infty} \| = \| f \,| L_p \| + \sup_{0<\delta<\infty} \delta^{-s}\, \tau_p^m(f,\delta) < \infty \right\}.$$

These classes are known to be important in several fields of approximation theory, e.g. in connection with one-sided approximation or with the approximation order of quadrature rules, cf. e.g., [67]. The notation indicates that there is a full scale $A^s_{p,q}$ as in case of the Besov spaces. But for us only $q = \infty$ turns out to be of interest. We collect a few properties of these classes and compare them with the Besov spaces and the classes $V_p \cap L_p$.

Lemma 10. *Let $0 < p \le \infty$ and $s > 0$.*
(i) *It holds*

$$\sup_{t\in\mathbb{R}} |f(t)| \le c\,(\tau_p(f,1) + \| f \,| L_p \|)$$

with c independent of f.
(ii) *Always we have $A^s_{p,\infty} \hookrightarrow \Lambda^s_{p,\infty}$.*
(iii) *We have $A^s_{p,\infty} = \Lambda^s_{p,\infty}$ if and only if $s > 1/p$.*
(iv) *For $1 \le p < \infty$ it holds $\Lambda^{1/p}_{p,q} \hookrightarrow A^{1/p}_{p,\infty}$ if and only if $q \le 1$.*
(v) *Suppose $1 \le p < \infty$. The space $V_p \cap L_p$ is continuously imbedded into $A^{1/p}_{p,\infty}$.*

Proof. Assertions (i) and (ii) are elementary. The third one and partly the fourth one (sufficiency) are consequences of the inequality (77). Part (iii) for $p < 1$ has been proved in [25]. Necessity of $q \le 1$ follows from the fact that $\Lambda^{1/p}_{p,q}$ contains unbounded functions if $q > 1$, cf. e.g. [61, 2.3.1] for some explicit examples. Finally, (v) becomes a consequence of

$$\begin{aligned}\int_{-\infty}^{\infty} |\omega(f,t,\delta)|^p\, dt &\le 2\,\delta \sum_{k=-\infty}^{\infty} |f(t_{W,k}) - f(y_{W,k})|^p \\ &\le 4\,\delta \sum_{k=-\infty}^{\infty} |f(z_{W,k+1}) - f(z_{W,k})|^p\end{aligned}$$

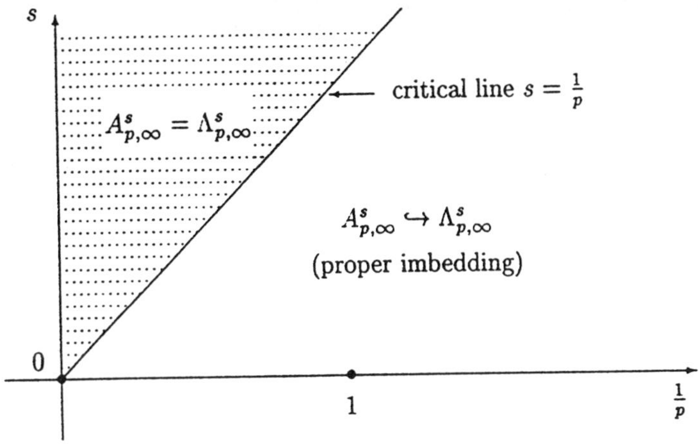

Figure 2:

where we have chosen $t_{W,k}$, $y_{W,k}$ such that

$$(k-1)\delta \leq t_{W,k},\, y_{W,k} \leq (k+2)\delta\,,$$

$$2\,|f(t_{W,k}) - f(y_{W,k})| \geq \sup_{(k-1)\delta \leq x \leq (k+2)\delta} f(x) - \inf_{(k-1)\delta \leq x \leq (k+2)\delta} f(x)\,,$$

and $z_{W,k}$ belongs to a common refinement of $\{t_{W,k}\}_k$ and $\{y_{W,k}\}_k$. □

Remark 19. Of particular interest for us are the spaces $A^s_{p,\infty}$, $s \leq 1/p$. These spaces contain bounded functions but they can be discontinuous. The most simple examples are characteristic functions of intervals.

The relations between $A^s_{p,\infty}$ and $\Lambda^s_{p,\infty}$ are described in Figure 2.

Comments

Theorem 11 and Theorem 12 have been proved in [43] and [80, 81]. Partial results may be found also in [51], [54], and [4]. Proofs of the imbedding and comparison theorems 13-15 are given, for example, in [42] and [80]. For the sharpness see [42] and [71]. Proposition 4 is due to Peetre [54]. The presentation of the approximation results of Theorems 16-18 follows the periodic case as treated in [66], [64, 65], and [62]. However, it should be mentioned that a lot of papers and books have dealt with these problems. We refer to [4], [20], [27], [47], [51], [52], [54], [55], [68], and [80, 81]. The connection between spaces defined by best or even "near best" approximation and scales of interpolation spaces, mentioned in Remark 15, have been highlighted in [4], [55], [27] and [23].

The τ-modulus has been made popular by the activities of the Bulgarian school around B. Sendov and V. Popov. It can be used to characterize best one-side approximation. In introduction is given in [67], but we refer also to [25] and [41].

4 Sampling of Non-bandlimited Functions

4.1 Uniform convergence problems

Let f be a function on $\mathbb{R}$, pointwise given, and let $W > 0$ be a real number. We put formally

$$S_W f(t) := \sum_{k=-\infty}^{\infty} f(\frac{k}{W}) \operatorname{sinc}(Wt - k)\,. \tag{78}$$

It follows from Theorem 6 the projection property

$$S_W f = f \qquad \text{(convergence in } L_p \quad \text{and uniformly)} \tag{79}$$

if $f \in B^p_{\pi W}$, $p < \infty$. Moreover, we have the interpolation property

$$S_W f(\frac{k}{W}) = f(\frac{k}{W})\,. \tag{80}$$

Let either $f \in B^\infty_{\pi W}$ or let f be non-bandlimited (which means $\operatorname{supp} \mathcal{F} f$ is not compact). We shall discuss mainly two problems:

(1) the uniform convergence of the series

$$\sum_{k=-M}^{N} f(\frac{k}{W}) \operatorname{sinc}(Wt - k) \;=:\; S_{W,M,N} f(t)\,, \tag{81}$$

(2) the convergence/divergence of $\sup_{t\in\mathbb{R}} |\, f(t) - S_W f(t)\,|$ if $W \to \infty$, provided the right-hand side of (78) makes sense.

Let us denote by

$$\mathcal{A} = \left\{ f \in \mathcal{S}' : \quad \mathcal{F} f \in L_1 \right\} \tag{82}$$

the Wiener algebra. The following theorem is classical and well-known.

Theorem 19. (Brown [16], Higgins [39, 11.7]).
(i) *If $f \in \mathcal{A}$, then the symmetric partial sums*

$$S_{W,N}f(t) := \sum_{k=-N}^{N} f(\frac{k}{W}) \operatorname{sinc}(Wt-k) \qquad (W>0 \quad \textit{fixed}) \tag{83}$$

are uniformly convergent on each compact subset of the real line. In particular, (78) makes sense for all $t \in \mathbb{R}$.
(ii) *If $f \in \mathcal{A}$, then we have the estimate*

$$\sup_{t\in\mathbb{R}} |\, f(t) - S_W f(t)\,| \le 2 \int_{|\omega|\ge \pi W} |\mathcal{F}f(\omega)|\, d\omega\,. \tag{84}$$

In particular, $S_W f$ tends to f uniformly on $\mathbb{R}$ if $W \to \infty$.

Next we quote some results by Rahman, Vertesi, and H. Boche which show that the problem of uniform convergence is rather sophisticated.

Theorem 20. (i) (Boche [11]). *Let $W > 0$ be fixed. There exists a function $f \in \mathcal{A} \cap B^\infty_{\pi W}$ such that*

$$\overline{\lim_{N\to\infty}} \sup_{t\in\mathbb{R}} |S_{W,N}f(t)| = \infty\,.$$

(ii) (Boche [7, 8], Boche and Schreiber [13]). *Let $W > 0$ be fixed. There exists a function $f \in \mathcal{A} \cap B^\infty_{\pi W}$ such that*

$$\overline{\lim_{N,\,M\to\infty}} \sup_{0<t<1/W} |S_{W,N}f(t)| = \infty\,.$$

(iii) (Boche [9], Rahman and Vertesi [57]). *Let $0 < s < 1/2$. There exists a continuous function $f \in H^s_2$ with compact support such that*

$$\overline{\lim_{W\to\infty}} \sup_{t\in\mathbb{R}} |S_W f(t)| = \infty\,.$$

Remark 20. Part (i) shows that one cannot expect uniform convergence on the whole axis for symmetric finite sampling series. For nonsymmetric finite sampling series we do not have uniform convergence on compact subsets.. Both problems can be removed by oversampling, see e.g. Higgins [39, 6.2.1] and Boche [11]. On the other hand, even for time-limited continuous signals in general there is no uniform convergence of $S_W f$ if $W \to \infty$.

Remark 21. Comparing Part (i) of Theorem 20 with Part (ii) of Theorem 19 we see that uniform convergence of $S_W f$ against f $(W \to \infty)$ has nothing to do with uniform convergence of $S_{W,N} f$ against $S_W f$ $(N \to \infty)$.

In view of the preceding remark it is desirable to have classes of functions as large as possible such that both happens that means uniform convergence of the finite sampling sums $S_{W,M,N}f$ and uniform convergence of the the sampling sums $S_W f$. If $f \in B^{1/2}_{2,1}$ (or $f \in H^s_2$, $s > 1/2$), then f belongs to the Wiener algebra $\mathcal{A}$ (see Proposition 4) and (84) applies to f. According to Theorems 13 and 15 it is natural to ask about the behaviour of $S_W f$ for $f \in B^{1/p}_{p,1}$, especially if p is large.

Theorem 21. *Let* $0 < p < \infty$ *and let* $f \in B^{1/p}_{p,1}$.
(i) *The non-symmetric finite sampling sums* $S_{W,M,N}f$ *are uniformly convergent on* $\mathbb{R}$ *for each fixed* $W > 0$.
(ii) *It holds*

$$\lim_{W\to\infty} \sup_{t\in\mathbb{R}} |\, f(t) - S_W f(t)| = 0\,. \tag{85}$$

Proof. If $p_0 < p_1$, then $B^{1/p_0}_{p_0,1} \hookrightarrow B^{1/p_1}_{p_1,1}$ (see Theorem 13). Hence it will be sufficient to deal with p large. We assume $p > 1$. In view of Corollary 3 we know

$$\sum_{k=-\infty}^{\infty} |f(\frac{k}{W})|^p < \infty\,. \tag{86}$$

Moreover, it holds $\operatorname{sinc}(Wt - \cdot) \in B^{p'}_{\pi}$, $1/p + 1/p' = 1$, and

$$\sum_{k=-\infty}^{\infty} |\operatorname{sinc}(Wt-k)|^{p'} \le c\|\operatorname{sinc}(Wt-\cdot)\,|L_{p'}\|^{p'} = c\,\|\operatorname{sinc} y\,|L_{p'}\|^{p'} < \infty \tag{87}$$

by Lemma 2 (with $h = 1$, $\Omega = \pi$, and $t_k = k$). As a consequence of (86), (87), and Hölder's inequality the right-hand side of (78) is absolutely convergent. We obtain that ($W > 0$ fixed)

$$\begin{aligned}
&|\, S_W f(t) - S_{W,M,N}f(t)\,| \\
&\le |\sum_{k=-\infty}^{-M-1} f(\frac{k}{W}) \operatorname{sinc}(Wt-k)\,| + |\sum_{k=N+1}^{\infty} f(\frac{k}{W}) \operatorname{sinc}(Wt-k)\,| \\
&\le c\left(\sum_{k=-\infty}^{-M-1} |f(\frac{k}{W})|^p\right)^{1/p} + c\left(\sum_{k=N+1}^{\infty} |f(\frac{k}{W})|^p\right)^{1/p}.
\end{aligned}$$

This implies part (i) of the theorem according to (86). To prove part (ii) let $2^\ell \le W \le 2^{\ell+1}$, $\ell \in \mathbb{N}$. We put

$$f^\ell(t) := \sum_{j=0}^{\ell} \mathcal{F}^{-1}[\varphi_j(\omega)\,\mathcal{F}f(\omega)](t) = \mathcal{F}^{-1}[\varphi_0(2^{-\ell}\omega)\,\mathcal{F}f(\omega)](t)\,,$$

cf. Subsection 3.2. From $f \in B^{1/p}_{p,1}$ we derive $f^\ell \in B^p_{2^{\ell+1}} \hookrightarrow B^p_{\pi W}$, and

$$S_W f^\ell = f^\ell\,,$$

cf. the projection property (79). It follows

$$\begin{aligned} |\,f(t) - S_W f(t)\,| &\le |\,f(t) - f^\ell(t)\,| + |\,S_W(f - f^\ell)(t)\,| \\ &\le \sum_{j=\ell+1}^{\infty} |f_j(t)| + \sum_{j=\ell+1}^{\infty} |S_W(f_j)(t)|\,, \end{aligned} \tag{88}$$

where $f_j(t) = \mathcal{F}^{-1}[\varphi_j(\omega)\,\mathcal{F}f(\omega)](t) \in B^{1/p}_{p,1}$. Clearly, $S_W(f_j) \in B^\infty_{\pi W}$ and we get

$$\begin{aligned} |(S_W f_j)(t)| &\le c_1\, W^{1/p}\,\|\,S_W f_j\,|L_p\| \\ &\le c_2\,\|\,S_W f_j(\frac{k}{W})\,|\ell_p\| \\ &= c_2\,\|\,f_j(\frac{k}{W})\,|\ell_p\| \\ &\le c_3\, 2^{j/p}\,\|\,f_j\,|L_p\|\,, \end{aligned}$$

using successively Nikol'skij's inequality (Theorem 2), Plancherel-Polya inequality (Theorem 6), the interpolation property (80), and Lemma 2 (with $\Omega = 2^{j+1}$ and $h = 2^{-j}$). Substitution into (88) and Nikol'skij's inequality again lead to the estimate

$$\sup_{t\in\mathbb{R}} |\,f(t) - S_W f(t)\,| \le c \sum_{j=\ell+1}^{\infty} 2^{j/p}\,\|\,f_j\,|L_p\|\,. \tag{89}$$

This implies part (ii). □

Corollary 4. *Let $0 < p < \infty$ and let $s > 1/p$. Then there exists a constant c such that*

$$\sup_{t\in\mathbb{R}} |\,f(t) - S_W f(t)\,| \le c\,W^{-(s-1/p)}\,\|\,f\,|B^s_{p,\infty}\| \tag{90}$$

for all $f \in B^s_{p,\infty}$.

Proof. (90) is an immediate consequence of (89) and the definition of $B^s_{p,\infty}$ if $1 < p < \infty$. For $0 < p_0 < 1$ we apply (89) for some p, $1 < p < \infty$ and then we continue with Nikol'skij's inequality to switch from the L_p–norms of f_j to the L_{p_0}–quasi–norms of f_j. This proves (89) also for $p \leq 1$. □

4.2 Shannon sampling: L_p–convergence

We shall give necessary and sufficient conditions for the L_p–convergence of the sampling sums under a mild additional condition on f.

Proposition 6. *Let* $1 < p < \infty$ *and let* $|f|$ *be locally integrable in the Riemannian sense.*
(i) *The non-symmetric partial sums* $S_{W,M,N} f$ *are convergent in* L_p *for fixed* $W > 0$ *if and only if*

$$\sum_{k=-\infty}^{\infty} |f(\frac{k}{W})|^p < \infty . \tag{91}$$

(ii) *Let* $f \in L_p$. *It holds that* $\lim_{W\to\infty} \| f - S_W f \,|L_p\| = 0$ *if and only if (91) is satisfied for all* $W > 0$.

Proof. Step 1. Obviously, we have $S_{W,M,N} f \in B^p_{\pi W}$ for all M, N. Using Theorem 6(i) and the interpolation properties of $S_{W,M,N} f$ we obtain

$$A_p \, \| S_{W,M,N} f \,|L_p\| \leq \left(\frac{1}{W} \sum_{k=-M}^{N} |f(\frac{k}{W})|^p \right)^{1/p} \leq B_p \, \| S_{W,M,N} f \,|L_p\| \tag{92}$$

with A_p, B_p independent of f, W, M and N. This proves part (i).
Step 2. Let $f \in L_p$. Assume (91) for all $W > 0$. Let $\varepsilon > 0$. We choose $\Omega > 0$ and $g \in B^p_\Omega$ such that $\| f - g \,|L_p\| < \varepsilon$. Because of Lemma 2 and our assumption on f the function $(f - g)$ satisfies (91), too and $|f - g|^p$ is locally integrable in the Riemannian sense. Therefore

$$\| f - g \,|L_p\| = \lim_{W\to\infty} \left(\frac{1}{W} \sum_{k=-\infty}^{\infty} |(f-g)(\frac{k}{W})|^p \right)^{1/p} . \tag{93}$$

This may be seen as follows. By definition of the Lebesgue and Riemann integral we have

$$\int_{-\infty}^{\infty} |f(t)|^p \, dt = \lim_{N\to\infty} \lim_{W\to\infty} \frac{1}{W} \sum_{|k|\leq NW} |f(\frac{k}{W})|^p$$

$$\leq \lim_{W\to\infty} \frac{1}{W} \sup_{N=1,2,\dots} \sum_{|k|\leq NW} |f(\frac{k}{W})|^p .$$

On the other hand

$$\int_{-\infty}^{\infty} |f(t)|^p \, dt \geq \lim_{W\to\infty} \frac{1}{W} \sum_{|k|\leq NW} |f(\frac{k}{W})|^p$$

for all N. Hence (93).

If W is sufficiently large, then we have on the one hand $S_W g = g$ (projection property) and on the other hand

$$\left(\frac{1}{W} \sum_{k=-\infty}^{\infty} |(f-g)(\frac{k}{W})|^p \right)^{1/p} \leq 2 \| f - g \,|L_p\| \leq 2\varepsilon .$$

Together with Theorem 6(i) and the interpolation property of $S_W f$ we obtain

$$\begin{aligned} \| f - S_W f \,|L_p\| &\leq \| f - g \,|L_p\| + \| S_W(f-g) \,|L_p\| \\ &\leq \varepsilon + A_p^{-1} \left(\frac{1}{W} \sum_{k=-\infty}^{\infty} |(f-g)(\frac{k}{W})| \right)^{1/p} \\ &\leq (1 + 2A_p^{-1})\, \varepsilon \end{aligned}$$

for $W \geq W_0(\varepsilon)$. This proves part (ii) of the theorem. □

In view of this proposition we introduce the following classes of functions.

Definition 8. Let $1 \leq p < \infty$. Then we put

$$\mathcal{R}_p = \Big\{ f \in L_p : \ |f| \text{ is Riemann integrable on each finite interval in } \mathbb{R},$$
$$\| f \,|\mathcal{R}_p\| = \sup_{W\geq 1} \left(\frac{1}{W} \sum_{k=-\infty}^{\infty} |f(\frac{k}{W})|^p \right)^{1/p} < \infty \Big\} .$$

Of course, for $f \in \mathcal{R}_p$ we have convergence of $S_W f$ in the L_p–sense.

Lemma 11. *Let* $1 \leq p, p_0, p_1 < \infty$.
(i) *The functions in* $\mathcal{R}_p$ *are bounded and for* $p_0 < p_1$ *it follows* $\mathcal{R}_{p_0} \hookrightarrow \mathcal{R}_{p_1}$.
(ii) *The space* $V_p \cap L_p$ *is continuously imbedded into* $\mathcal{R}_p$.

Proof. Let $k \le t \le k+1$ with $k \in \mathbb{Z}$. Then there exist a number W, $1 \le W < 2$ such that either $Wt = k+1$ (if $k \ge 0$) or $Wt = k$ (if $k \le 0$). Hence

$$|f(t)| \le W^{1/p} \, \| f \,|\mathcal{R}_p\| \le 2^{1/p} \, \| f \,|\mathcal{R}_p\| \, .$$

The monotonicity of $\mathcal{R}_p$ with respect to p follows from this and Hölder's inequality. Indeed, let $(1/p_0) = (1-\Theta)/p_1$. Then

$$\begin{aligned} W^{-1/p_1} \, \| \{f(k/W)\}_k \ \ |\ell_{p_1}\| & \\ & \le W^{-\frac{1-\Theta}{p_0}} \, \| \{f(k/W)\}_k \,|\ell_{p_0}\|^{1-\Theta} \| \{f(k/W)\}_k \,|\ell_\infty\|^{\Theta} \\ & \le 2^{\Theta/p} \, \| f \,|\mathcal{R}_p\| \, . \end{aligned}$$

Concerning part (ii) we select points $y_{W,k}$ such that $k \le W\, y_{W,k} \le (k+1)$ and

$$|f(y_{W,k})| \le \left(W \int_{k/W}^{(k+1)/W} |f(t)|^p \, dt \right)^{1/p} .$$

Then

$$\left(\frac{1}{W} \sum_{k=-\infty}^{\infty} |f(\frac{k}{W})|^p \right)^{1/p} \le W^{-1/p} \, \| f \,|V_p\| + \| f \,|L_p\| \, .$$

This proves (ii). □

As an immediate consequence of Lemma 9(i), Lemma 11(ii) and Proposition 6 we obtain the following corollary.

Corollary 5. *For $f \in V_1 \cap L_1$ it holds $\lim_{W \to \infty} \| f - S_W f \,|L_p\| = 0$ for all p, $1 < p < \infty$.*

Approximation of order $s \ge 1/p$

In view of Corollary 3 condition (91) is satisfied for all $W > 0$ if f belongs to $B^{1/p}_{p,1}$, $1 < p < \infty$, in particular if $f \in B^s_{p,\infty}$ where $s > 1/p$. In this case f is also bounded and uniformly continuous (cf. Theorem 15), hence Proposition 6 can be applied. The following theorem deals with the rate of convergence.

Theorem 22. *Let $1 < p < \infty$.*
(i) *If $f \in B^{1/p}_{p,1}$, then the non-symmetric partial sums $S_{W,M,N} f$ converge in L_p for any fixed $W > 0$. Moreover, it holds*

$$\| f - S_W f \,|L_p\| = o(W^{-1/p}) \qquad (W \to \infty) \, . \tag{94}$$

(ii) *Let $s > 1/p$. The function f belongs to $B^s_{p,\infty}$ if and only if f is continuous function belonging to L_p and satisfying*

$$\| f - S_W f \,|L_p\| = O(W^{-s}) \qquad (W \to \infty)\,. \tag{95}$$

Proof. Let $f \in B^{1/p}_{p,1}$ and let $2^\ell \le W < 2^{\ell+1}$. By the same arguments as in the proof of Theorem 21 (and using the same abbreviations) we get

$$\begin{aligned} \| f - S_W f \,|L_p\| &\le \| f - f^\ell \,|L_p\| + \| S_W(f - f^\ell) \,|L_p\| \\ &\le \sum_{j=\ell+1}^{\infty} \| f_j \,|L_p\| + \sum_{j=\ell+1}^{\infty} \| S_W f_j \,|L_p\| \\ &\le (1 + A_p^{-1} B_p) \sum_{j=\ell+1}^{\infty} 2^{(j-\ell)/p} \| f_j \,|L_p\|\,. \end{aligned} \tag{96}$$

This implies (94). Further, we conclude from (96)

$$\begin{aligned} W^s \| f - S_W f \,|L_p\| &\le c_1\, 2^{\ell s} \sum_{j=\ell+1}^{\infty} 2^{(j-\ell)/p} \| f_j \,|L_p\| \\ &\le c_1 \| f \,|B^s_{p,\infty}\| \sum_{j=\ell+1}^{\infty} 2^{(j-\ell)(\frac{1}{p}-s)} \\ &\le c_2 \| f \,|B^s_{p,\infty}\| \end{aligned}$$

for some constant c_2 independent of f and W. Hence (95). The converse direction of part (ii) is a consequence of Theorem 16, and the fact that $S_W f \in B^p_{\pi W}$. □

Approximation of order $s \le 1/p$

For order of L_p–approximation $0 < s \le 1/p$ we shall use the τ-modulus.

Theorem 23. *Let $1 < p < \infty$ and suppose $0 < s \le 1/p$.*
(i) *There exists a constant c such that*

$$\| f - S_W f \,|L_p\| \le c\, \tau_p(f, 1/W) \tag{97}$$

holds for all $f \in A^s_{p,\infty}$ and all $W \ge 1$.
(ii) We have

$$\| f - S_W f \,|L_p\| = O(W^{-s}) \tag{98}$$

for all $f \in A^s_{p,\infty}$.

Proof. Step 1. To prove (97) we proceed as in proof of Theorem 22. Also we use the same abbreviations as there. Let $2^\ell \le W < 2^{\ell+1}$. Then $S_W f^\ell = f^\ell$ and Proposition 3 together with Theorem 6 yield

$$\begin{aligned} \| f - S_W f \mid L_p\| &\le \| f - f^\ell \,|L_p\| + \| S_W(f - f^\ell) \,|L_p\| \\ &\le \| f - f^\ell \,|L_p\| + c \left(\frac{1}{W} \sum_{k=-\infty}^{\infty} |(f - f^\ell)(\frac{k}{W})|^p \right)^{1/p} . \end{aligned} \tag{99}$$

In the next step we use

$$|g(\frac{k}{W})| \le |g(\frac{k}{W}) - g(y_k)| + \left(W \int_{k/W}^{(k+1)/W} |g(t)|^p \, dt \right)^{1/p}$$

for y_k, $k \le W\, y_k \le (k+1)$, chosen appropriate. With $g = f - f^\ell$ this leads to

$$\begin{aligned} &\frac{1}{W} \sum_{k=-\infty}^{\infty} |(f - f^\ell)(\frac{k}{W})|^p \\ &\qquad \le c \left(\frac{1}{W} \sum_{k=-\infty}^{\infty} |(f - f^\ell)(\frac{k}{W}) - (f - f^\ell)(y_k)|^p + \| f - f^\ell \,|L_p\|^p \right) \\ &\qquad \le c \left(\tau_p(f - f^\ell, 2/W)^p + \| f - f^\ell \,|L_p\|^p \right) . \end{aligned}$$

Inserting this into (99), using $A^s_{p,\infty} \hookrightarrow B^s_{p,\infty}$, cf. Lemma 10(ii), and Proposition 3 we arrive at

$$\| f - S_W f \,|L_p\| \le c \left(\omega_p(f, 2^{-\ell}) + \tau_p(f - f^\ell, 2^{-\ell}) \right) .$$

Using the convolution structure of f^ℓ and $p \ge 1$ then the generalized Minkowski inequality yields

$$\tau_p(f^\ell, \delta) \le c\, \tau_p(f, \delta)$$

for some constant c independent of f, ℓ, and δ. This proves (i). Part (ii) is an immediate consequence of (i). □

In contrast to the situation $s > 1/p$ we do not know inverse estimates for $s \le 1/p$. But a partial answer will be given in the next lemma. To compare this assertion with our results derived before, recall $V_1 \cap L_1 \hookrightarrow V_p \cap L_p \hookrightarrow A^{1/p}_{p,\infty}$.

Lemma 12. *Let* $1 < p < \infty$. *The characteristic function of an finite interval* χ_I *belongs to* $V_1 \cap L_1$ *and satisfies*

$$\| \chi_I - S_W \chi_I \,|L_p\| \ge c\, W^{-1/p} \tag{100}$$

for some positive c independent of $W > 0$.

Proof. Elementary calculations yield $\chi_I \in V_1$ which is enough for our purpose. To derive the estimate (100) we restrict us to the case $I = [0,1]$ and drop the subscript I. Furthermore, we take $W \in \mathbb{N}$ (otherwise the same arguments apply with the integer part of W instead of W itself). We obtain

$$\begin{aligned}
\| \chi_I - S_W \chi_I \,|L_p\|^p &\geq \int_{-\infty}^{0} |\sum_{k=0}^{W} \mathrm{sinc}\,(Wt-k)|^p \, dt \\
&\geq W^{-1} \int_{1/4}^{1/2} |\sum_{k=0}^{W} \mathrm{sinc}\,(y+k)|^p \, dy \\
&\geq W^{-1} \Big(\frac{1}{\pi\sqrt{2}}\Big)^p \int_{1/4}^{1/2} |\sum_{k=0}^{W} (-1)^k \frac{1}{y+k} |^p \, dy \\
&\geq W^{-1} \Big(\frac{1}{\pi\sqrt{2}}\Big)^p \int_{1/4}^{1/2} |\frac{1}{y^2+y} - \frac{1}{y^2+(2W+1)y+W^2+W} |^p \, dy \\
&\geq W^{-1} \Big(\frac{1}{\pi\sqrt{2}}\Big)^p \int_{1/4}^{1/2} |\frac{2y+2}{(y^2+y)(y^2+3y+2)} |^p \, dy \\
&\geq c\, W^{-1}
\end{aligned}$$

which proves our claim. □

Remark 22. The Lemma shows that the estimate (97) cannot be improved for functions of bounded variation. In particular, it makes clear that the capital "O" in Theorem 23(i) can not be replaced by a small "o" for $s = 1/p$. This should be compared with Theorem 22(i).

4.3 Approximation by generalized sampling series: the case of bandlimited kernels

In the following we consider generalized sampling series defined (formally) by

$$S_W^{\Psi}(t) := \sum_{k=-\infty}^{\infty} f(\frac{k}{W})\, \Psi(Wt-k)\,, \tag{101}$$

where Ψ belongs (at least) to L_1 in contrast to $\Psi(t) = \mathrm{sinc}\,(t)$. Moreover, we assume that

$$\sup_{t\in\mathbb{R}} \sum_{k=-\infty}^{\infty} |\Psi(t-k)| < \infty\,. \tag{102}$$

Then (101) makes sense for all bounded functions f defined on the whole real axis. We shall study the rate of convergence of $\| f - S_W^\Psi f \,|L_p\|$ for $W \to \infty$. Here we distinguish two cases:

(i) Ψ is bandlimited ($\Psi \in B^1_{2\pi}$);

(ii) Ψ is time-limited (supp Ψ is compact) and continuous.

The first assertion is probably more restrictive as it looks like, e.g. it cancels the Jackson kernels. However, we shall use it here to keep the conditions on the kernels more transparent (think on the classical Strang-Fix conditions for the approximation order of principal shift-invariant spaces, cf. e.g. [59], [24], or [45]).

The following assertion is well-known.

Proposition 7. *If Ψ is continuous and satisfies (102), then the following statements are equivalent:*

$$\text{(i)} \quad \| f - S_W^\Psi f \,|L_p\| \to 0 \qquad (W \to \infty) \quad \textit{for all } f \in C\,; \tag{103}$$

$$\text{(ii)} \quad \sum_{k=-\infty}^{\infty} \Psi(t-k) = 1 \qquad \textit{for all} \quad t \in \mathbb{R}\,; \tag{104}$$

$$\text{(iii)} \quad \mathcal{F}^{-1}\Psi(2\pi k) = \begin{cases} 1 & k = 0\,; \\ 0 & k \neq 0\,. \end{cases} \tag{105}$$

In this subsection we assume that $\Psi \in B^1_{2\pi}$ and $\mathcal{F}^{-1}\Psi(0) = 1$. Then Ψ satisfies (102) by Lemma 2 and, obviously (using the continuity of $\mathcal{F}^{-1}\Psi$) we have (105).

Proposition 8. *Let $0 < a < 1$. Suppose $g \in B^\infty_{(1-a)\pi}$ and $\Psi \in B^1_{(1+a)\pi}$. Then it holds*

$$\sum_{k=-\infty}^{\infty} g(t-k)\,\Psi(k) = \sum_{k=-\infty}^{\infty} g(k)\,\Psi(t-k) = \int_{-\infty}^{\infty} g(\tau)\,\Psi(t-\tau)\,, \quad t \in \mathbb{R}\,, \tag{106}$$

where the series are absolutely convergent.

Remark 23. Formula (106) can be proved as Lemma 4, cf. also Theorem 5. One of the nice consequences is as follows: If we assume $\Psi \in B^1_{(1+a)\pi}$ and the interpolation property

$$\Psi(\ell) = \begin{cases} 0 & \ell \neq 0\,, \\ 1 & \ell = 0\,, \end{cases}$$

then $S_1^\Psi g = g$ if $g \in B^\infty_{(1-a)\pi}$ (projection property).

Define $\psi = \mathcal{F}^{-1}\Psi$ (hence also $\mathcal{F}\psi = \Psi$). Recall the definition of the means $M_W^\psi f$ (cf. (51)). A consequence of (106) is

$$S_1^\Psi g(t) = M_1^\psi g(t)\,, \qquad g \in B^\infty_{(1-a)\pi}\,.$$

By homogeneity arguments and coordinate transform we achieve

$$S_W^\Psi f(t) = M_W^\psi f(t)\,, \qquad f \in B^\infty_{(1-a)\pi W}\,. \tag{107}$$

If additionally $\psi(\omega) = 1$ on $[-(1-a)\pi,\,(1-a)\pi]$, then the projection property of M_W^ψ yields

$$S_W^\Psi f(t) = f(t)\,, \qquad f \in B^\infty_{(1-a)\pi W}\,. \tag{108}$$

Lemma 13. *Let* $0 < p \le \infty$ *and let* $\Psi \in B^{\tilde p}_{2\pi}$, *where* $\tilde p = \min(1,p)$. *Let* W, $c_0 > 0$.
(i) *There exists a constant* c *such that*

$$\|\, M_W^\psi f \,|L_p\| \le c \left(\frac{\Omega}{W}\right)^{\max(0,\frac{1}{p}-1)} \|\, f \,|L_p\| \tag{109}$$

holds for all $f \in B^p_\Omega$ *and all* Ω *with* $\Omega \ge W\, c_0$.
(ii) *There exists a constant* c *such that*

$$\|\, S_W^\Psi f \,|L_p\| \le c \left(\frac{\Omega}{W}\right)^{\frac{1}{p}} \|\, f \,|L_p\| \tag{110}$$

holds for all $f \in B^p_\Omega$ *and all* Ω *with* $\Omega \ge W\, c_0$.

Proof. Step 1. We prove (109). We have $(M_W^\psi f)(t) = [M_1^\psi f(\frac{\cdot}{W})](Wt)$. Because of $f(\frac{\cdot}{W}) \in B^p_{\Omega/W}$ and homogeneity arguments it is sufficient to consider the case $W = 1$. Furthermore we may assume $\Omega \ge 2\pi$. Now, (109) is a consequence of Theorem 8.

Step 2. To show (110) it is also sufficient to consider the case $W = 1$. This follows from $(S_W^\Psi f)(t) = [S_1^\Psi f(\frac{\cdot}{W})](Wt)$ as above. By means of Theorem 4 (if $1 \le p \le \infty$) and Theorem 10 (if $0 < p < 1$) we see that

$$\|\, S_1^\Psi f \,|L_p\| \le c \left(\sum_{m=-\infty}^{\infty} |(S_1^\Psi f)(h_0 m)|^p\right)^{1/p} \tag{111}$$

if $h_0 = h_0(p)$ is sufficiently small (modification if $p = \infty$).
Let $1 \le p \le \infty$. By the definition of $S_1^\Psi f$ and by Hölder's inequality we obtain from (111) that

$$\| S_1^\Psi f \ |L_p\|^p \le c \sum_{m=-\infty}^{\infty} \left[\sum_{k=-\infty}^{\infty} |f(k)| \, |\Psi(h_0 m - k)| \right]^p$$

$$\le c \sum_{m=-\infty}^{\infty} \left[\sum_{k=-\infty}^{\infty} |f(k)|^p \, |\Psi(h_0 m - k)| \right] \left[\sum_{k=-\infty}^{\infty} |\Psi(h_0 m - k)| \right]^{p/p'} .$$

Using Lemma 2 and $\Psi \in B^1_{2\pi}$ we see that

$$\sum_{k=-\infty}^{\infty} |\Psi(h_0 m - k)| \le c \, \| \Psi(h_0 m - \cdot\,) \, |L_1\| = c \, \| \Psi \, |L_1\|$$

and

$$\sum_{m=-\infty}^{\infty} |\Psi(h_0 m - k)| \le c \, \| \Psi(h_0 \, \cdot \, -k) \, |L_1\| = c \, h_0^{-1} \, \| \Psi \, |L_1\| \, .$$

It follows

$$\| S_1^\Psi f \, |L_p\|^p \le c \sum_{k=-\infty}^{\infty} |f(k)|^p \, \| \Psi \, |L_1\|^{1+p/p'} \, . \tag{112}$$

Next we choose $j \in \mathbb{N}$ such that $2^j \le \Omega < 2^{j+1}$. Using Lemma 2 one obtains from (112)

$$\| S_1^\Psi f \, |L_p\| \le c_1 \left(\sum_{k=-\infty}^{\infty} |f(2^{-j} k)|^p \right)^{1/p}$$

$$\le c_2 \, 2^{j/p} \, \| f \, |L_p\| \le c_2 \, \Omega^{1/p} \, \| f \, |L_p\|$$

for $1 \le p < \infty$. The case $p = \infty$ is obvious. If $0 < p < 1$, then (111) and Lemma 5 lead to

$$\| S_1^\Psi f \, |L_p\|^p \le c_1 \sum_{k=-\infty}^{\infty} |f(k)|^p \sum_{m=-\infty}^{\infty} |\Psi(h_0 m - k)|^p$$

$$\le c_2 \, \| \Psi \, |L_p\|^p \sum_{k=-\infty}^{\infty} |f(k)|^p$$

$$\le c_3 \sum_{k=-\infty}^{\infty} |f(2^{-j} k)|^p$$

$$\le c_4 \, 2^j \, \| f \, |L_p\|^p \le c_4 \, \Omega \, \| f \, |L_p\|^p \, .$$

This completes the proof of (110). $\square$

Approximation order $s \geq 1/p$

As in case of the classical sampling we have to split our considerations into two parts (high and low approximation rates). Also as before our results have some final character for $s > 1/p$.

Theorem 24. *Let $0 < p \leq \infty$, $\sigma > 0$, and $0 < a < 1$. We put $\tilde{p} = \min(1,p)$ and suppose that $\Psi \in B^{\tilde{p}}_{(1+a)\pi}$ and*

$$\| \mathcal{F}^{-1}[|\omega|^{-\sigma}(1-\psi(\omega))\,\eta(\omega)]\,|L_{\tilde{p}}\| < \infty$$

for some function $\eta \in C_0^\infty$ equal to 1 in a neighbourhood of the origin. If $1/p < s < \sigma$, then it holds $f \in B^s_{p,\infty}$ if and only if f is continuous, belongs to L_p and satisfies

$$\| f - S^\Psi_W f\,|L_p\| = 0(W^{-s})\,. \tag{113}$$

Moreover,

$$\| f\,|L_p\| + \sup_{W \geq 1} \| f - S^\Psi_W f\,|L_p\| \tag{114}$$

is an equivalent quasi-norm on $B^s_{p,\infty}$.

Proof. Step 1. Preliminaries. By assumption

$$\sum_{k=-\infty}^{\infty} |\Psi(t-k)|^{\tilde{p}} \leq c\,\| \Psi\,|L_{\tilde{p}}\|^{\tilde{p}}\,,$$

cf. Lemmata 2 and 5. In particular, (102) holds. Furthermore, $\Psi \in L_1$ and consequently $\mathcal{F}^{-1}\Psi = \psi$ is a continuous function with compact support. From the definition of the Fourier transform and Nikol'skij's inequality we derive

$$\begin{aligned}\sup_{\omega\in\mathbb{R}} \Big| |\omega|^{-\sigma}(1-\psi(\omega))\,\eta(\omega)\Big| &\leq \| \mathcal{F}^{-1}[|\omega|^{-\sigma}(1-\psi(\omega))\,\eta(\omega)]\,|L_1\| \\ &\leq c\,\| \mathcal{F}^{-1}[|\omega|^{-\sigma}(1-\psi(\omega))\,\eta(\omega)]\,|L_{\tilde{p}}\|\end{aligned}$$

which means $\psi(0) = 1$, hence (105).

Let $f \in B^s_{p,\infty}$, $s > 1/p$. Then $f \in C$ by Theorem 15. Now Proposition 6 shows that $S^\Psi_W f$ itself is an uniformly convergent series in C, hence a continuous function. Moreover, we also have convergence in L_p because of

$$\| \sum_{k=-\infty}^{\infty} f(\frac{k}{W})\Psi(Wt-k)\,|L_p\| \leq c\left(\frac{1}{W}\sum_{\ell=-\infty}^{\infty}\Big|\sum_{k=-\infty}^{\infty} f(\frac{k}{W})\Psi(\ell-k)\Big|^p\right)^{1/p}$$

$$\leq c\, W^{-1/p} \, \| \, \{\Psi(k)\}_k \, |\ell_1\| \, \| \, \{f(\frac{k}{W})\}_k \, |\ell_p\|$$

$$\leq C_W \, \| \, f \, | B^{1/p}_{p,1} \| \tag{115}$$

for $1 < p < \infty$ (using Theorem 6(i), Corollary 3, and a classical convolution inequality in ℓ_p) and

$$\int_{-\infty}^{\infty} \Big| \sum_{k=-\infty}^{\infty} f(\frac{k}{W}) \, \Psi(Wt-k) \Big|^p \, dt \leq \sum_{k=-\infty}^{\infty} |f(\frac{k}{W})|^p \, \| \, \Psi(Wt-k) \, |L_p\|^p$$

$$\leq C_W \, \| \, f \, | B^{1/p}_{p,p} \|^p \tag{116}$$

for $0 < p \leq 1$ (using Corollary 3).

Step 2. We shall prove that (113) holds if f belongs to $B^s_{p,\infty}$. Let W be large enough ($W \geq 2/((1-a)\pi)$). We choose $\ell \in \mathbb{N}$ such that $2^{\ell+1} \leq (1-a)\pi W < 2^{\ell+2}$ and put

$$f^\ell = \sum_{j=0}^{\ell} \mathcal{F}^{-1}[\varphi_j \, \mathcal{F} f] \qquad \text{and} \quad f_j = \mathcal{F}^{-1}[\varphi_j \, \mathcal{F} f] \, .$$

Then it holds

$$S^\Psi_W f^\ell = M^\psi_W f^\ell$$

by (107) and we get

$$\begin{aligned} \| \, f - S^\Psi_W f \, |L_p\| &\leq c_p \Big(\| \, f - M^\psi_W f \, |L_p\| + \| \, M^\psi_W (f^\ell - f) \, |L_p\| \\ &\qquad + \| \, S^\Psi_W (f^\ell - f) \, |L_p\| \Big) \\ &\leq c_p \Big(\| \, f - M^\psi_W f \, |L_p\| + \Big(\sum_{j=\ell+1} \| \, M^\psi_W f_j \, |L_p\|^{\tilde p} \Big)^{1/\tilde p} \\ &\qquad + \Big(\sum_{j=\ell+1}^{\infty} \| S^\Psi_W f_j | L_p\|^{1/\tilde p} \Big)^{\tilde p} \Big) . \end{aligned} \tag{117}$$

The first summand on the right-hand side can be estimated by Theorem 18. According to Lemma 13(i) we have

$$\| \, M^\psi_W f_j \, |L_p\| \leq c_1 \, (2^j W^{-1})^{\frac{1}{\tilde p}-1} \, \| \, f_j \, |L_p\| \leq c_2 \, 2^{(j-\ell)/p} \, \| \, f_j \, |L_p\| \tag{118}$$

and

$$\| S_W^\Psi f_j \,|L_p\| \le c_1 \,(2^j W^{-1})^{1/p} \,\| f_j \,|L_p\| \le c_2 \,2^{(j-\ell)/p} \,\| f_j \,|L_p\| \,. \tag{119}$$

Substituting this into (117) yields

$$\begin{aligned}\| f - S_W^\Psi f \,|L_p\| &\le c_1 \left(W^{-s} \,\| f \,|B^s_{p,\infty} \| + \left(\sum_{j=\ell+1}^{\infty} 2^{(j-\ell)\tilde{p}/p} \,\| f_j \,|L_p\|^{\tilde{p}} \right)^{1/\tilde{p}} \right) \\ &\le c_1 \left(W^{-s} \,\| f \,|B^s_{p,\infty} \| + 2^{-\ell s} \left(\sum_{j=\ell+1}^{\infty} 2^{(j-\ell)(\frac{1}{p}-s)\tilde{p}} \right)^{1/\tilde{p}} \| f \,|B^s_{p,\infty} \| \right) \\ &\le c_2 \, W^{-s} \,\| f \,|B^s_{p,\infty} \| \end{aligned} \tag{120}$$

because of $s > 1/p$. If W is small, then we use the estimates

$$\begin{aligned}\| f - S_W^\Psi f \,|L_p\| &\le c \left(\| f \,|L_p\| + \| S_W^\Psi f \,|L_p\| \right) \\ &\le c \left(\| f \,|L_p\| + \left(\frac{1}{W} \sum_{k=-\infty}^{\infty} |f(\frac{k}{W})|^p \right)^{1/p} \right) \\ &\le c \,\| f \,|B^{1/p}_{p,\tilde{p}} \| \,, \end{aligned}$$

(cf. Theorem 16 and Corollary 3).

Step 3. The converse direction follows immediately from Theorem 16 and $\| f - S_W^\Psi f \,|L_p\| \le E_p((1+a)\pi W, f)$. □

Remark 24. The proof of (120) and a modification of the proof of Theorem 18 show that

$$\| f - S_W^\Psi f \,|L_p\| = o(W^{-1/p}) \,, \qquad f \in B^{1/p}_{p,\tilde{p}} \,. \tag{121}$$

Remark 25. The Theorem can be simplified if S_W^Ψ has the projection property (108). That is the case, e.g. if we assume the interpolation property for Ψ (cf. Remark 23) or if we assume that ψ is 1 on $[-(1-a)\pi, \,(1-a)\pi]$. Then (117) reduces to

$$\| f - S_W^\Psi f \,|L_p\| \le c_p \left(\| f - f^\ell \,|L_p\| + \| S_W^\Psi (f - f^\ell) \,|L_p\| \right) .$$

Corollary 6. *Let* $0 < p \le \infty$. *Suppose that*

$$\Psi \in B^{\tilde{p}}_{(1+a)\pi} \,, \qquad 0 < a < 1 \,, \quad \tilde{p} = \min(1, p)$$

and the projection property (108). Then the statements (113) and (114) of Theorem 24 remain true for all $1/p < s < \infty$.

Approximation order $s \leq 1/p$

We can proceed similar as in proof of Theorem 23.

Theorem 25. *Let $0 < p < \infty$, $\sigma > 0$, and $0 < a < 1$. Let $\tilde{p} = \min(1,p)$. Let $\Psi \in B^{\tilde{p}}_{(1+a)\pi}$.*
(i) *Suppose*

$$\| \mathcal{F}^{-1}[|\omega|^{-\sigma}(1-\psi(\omega))\,\eta(\omega)]\,|L_{\tilde{p}}\| < \infty$$

for some function $\eta \in C_0^\infty$ equal to 1 in a neighbourhood of the origin. If $\max(0, 1/p - 1) < s < \min(1, 1/p, \sigma)$, then

$$\| f - S_W^\Psi f\,|L_p\| = 0(W^{-s}), \qquad (W \to \infty), \tag{122}$$

holds for all $f \in A^s_{p,\infty}$.
(ii) *If in addition S_W^Ψ has the projection property (108) (cf. also Remark 23) and if $0 < s < \min(1, 1/p)$, then*

$$\| f - S_W^\Psi f\,|L_p\| \leq c\,\tau_p(f, 1/W), \tag{123}$$

holds for all $f \in A^s_{p,\infty}$ and all $W \geq 1$.

Proof. Step 1. We shall give a proof of part (i). Recall Lemma 10(i). So, $f \in A^s_{p,\infty}$ is bounded and belongs to L_p. Hence, by arguing as in Step 1 of the proof of Theorem 24 (cf. (115) and (116)) we find

$$\begin{aligned}\| S_W^\Psi f\,|L_p\| &\leq c\left(\frac{1}{W}\sum_{k=-\infty}^{\infty}|f(\frac{k}{W})|^p\right)^{1/p}\\ &\leq c\,(\tau(f, 1/W)_p + \| f\,|L_p\|)\,.\end{aligned}$$

Next we employ the same splitting as in (117). For $0 < s < \sigma$ we may apply Theorem 18 and Proposition 3 to estimate the term $\| f - M_W^\psi f\,|L_p\|$. Furthermore, we use (118) and obtain similar to (120)

$$\| M_W^\psi (f - f^\ell)\,|L_p\| \leq c\,W^{-s}\,\| f\,|B^s_{p,\infty}\,\|\,,$$

which is, in view of Lemma 10(ii), sufficient for us. Finally, we have to deal with $\| S_W^\Psi (f - f^\ell)\,|L_p\|$. By the same arguments as in proof of Theorem 23 we arrive at

$$\| S_W^\Psi (f - f^\ell)\,|L_p\| \leq c\,\tau_p(f - f^\ell, 1/W)\,, \qquad 2^\ell \leq W < 2^{\ell+1}\,,$$

also if $p \leq 1$. It remains to estimate the τ–modulus on the right-hand side. We concentrate on $p < 1$ because for $p \geq 1$ it follows from the generalized

Minkowski inequality. Let $I_0 = [-2^{-\ell}, 2^{-\ell}]$ and $I_m = [-2^{m-\ell}, -2^{m-1-\ell}] \cup [2^{m-1-\ell}, 2^{m-\ell}]$, $m = 1, 2, \ldots$. We obtain

$$\begin{aligned}
\tau_p(f^\ell, 1/W)^p &\le \int_{-\infty}^{\infty} \left| \int_{-\infty}^{\infty} 2^\ell \, |(\mathcal{F}^{-1}\varphi_0)(2^\ell y)| \, \omega(f, t-y, 1/W) \, dy \right|^p dt \\
&\le \sum_{m=0}^{\infty} 2^{\ell p} \, 2^{-mMp} \int_{-\infty}^{\infty} \left| \int_{I_m} \omega(f, t-y, 2^{-\ell}) \, dy \right|^p dt \\
&\le \sum_{m=0}^{\infty} 2^{\ell p} \, 2^{-mMp} \, |I_m|^p \int_{-\infty}^{\infty} \omega(f, t, 2^{-\ell+m})^p \, dt \\
&\le c \, \tau_p(f, 2^{-\ell}) \,,
\end{aligned}$$

using

$$\tau_p(f, 2\delta) \le 2^{1/p} \, \tau_p(f, \delta) \,,$$

see [67]. This proves the claim also for $p < 1$. □

Examples

Example 1. (Fejér kernel). Let

$$\Psi(t) = \left[\frac{\sin(t/2)}{(t/2)} \right]^2 .$$

Then

$$\psi(\omega) = \mathcal{F}^{-1}\Psi(\omega) = (1 - |\omega|)_+ = \begin{cases} 1 - |\omega| & |\omega| \le 1 \,, \\ 0 & \text{otherwise} \,. \end{cases}$$

It follows easily that $\Psi \in B_1^p$ for all $p > 1/2$ and that (68) is satisfied with $\sigma = 1$. It holds (113) for all $f \in B_{p,\infty}^s$ if $1 < p \le \infty$, $1/p < s < 1$, and (121) for all $f \in B_{p,1}^{1/p}$ if $1 < p \le \infty$.

Example 2. (de la Vallée Poussin kernel). Let

$$\Psi(t) = \frac{2 \, \sin(t/2) \, \sin(3t/2)}{\pi \, t^2)} .$$

Then

$$\psi(\omega) = \mathcal{F}^{-1}\Psi(\omega) = \begin{cases} 1 & |\omega| \le 1 \,, \\ 2 - |\omega| & 1 < |\omega| \le 2 \,, \\ 0 & \text{otherwise} \,. \end{cases}$$

In this case $\Psi \in B_2^p$ for all $p > 1/2$ and (68) is satisfied for all $\sigma > 0$. Consequently (113) holds for all $f \in B_{p,\infty}^s$ if $1/2 < p \leq \infty$ and $s > 1/p$. We have (121) if $f \in B_{p,\tilde{p}}^{1/p}$ where $1/2 < p \leq \infty$.

Example 3. As before χ_I denotes the characteristic function of the set I. Let $0 < a < 1$ and put

$$\Psi_{a,m}(t) = \left[\operatorname{sinc}\left(\frac{at}{m}\right)\right]^m \operatorname{sinc} t\,, \qquad m \in \mathbb{N}\,.$$

Then

$$\psi_{a,m}(\omega) = \left(\frac{m}{2\pi a}\right)^m \Big[\underbrace{\chi_{[-(a\pi)/m,(a\pi)/m]} * \cdots * \chi_{[-(a\pi)/m,(a\pi)/m]}}_{m\,-\,\text{times}}\Big] * \chi_{[-\pi,\pi]}$$

In particular, we have $\psi_{a,m} = 1$ if $|\omega| \leq (1-a)\pi$ and $\operatorname{supp}\psi_{a,m} \subset [-(1+a)\pi, (1+a)\pi]$. Thus $\Psi \in B_{(1+a)\pi}^p$ for all $p > 1/(m+1)$. It follows (113) for all $f \in B_{p,\infty}^s$ if $1/(m+1) < p \leq \infty$ and $1/p < s < \infty$ by Corollary 6.

Example 4. Let

$$\Psi(t) = \frac{2}{\pi}\,\frac{\cos(\pi t)}{1-4t^2}\,.$$

Then

$$\psi(\omega) = \mathcal{F}^{-1}\Psi(\omega) = (\cos\frac{\omega}{2})_+ = \begin{cases} \cos(\omega/2) & |\omega| \leq \pi\,, \\ 0 & \text{otherwise}\,. \end{cases}$$

is the cosine-kernel. Obviously, $\Psi \in B_\pi^p$ for all $p > 1/2$ and (68) is satisfied with $\sigma = 2$. Theorem 24 applies to $1/2 < p \leq \infty$ and $1/p < s < 2$.

Example 5. Let

$$\Psi(t) = \frac{1}{2\pi}\,\frac{\sin(\pi t)}{t(1-t^2)} = \operatorname{sinc} t\,\frac{1}{2\,(1-t^2)}\,.$$

Then

$$\psi(\omega) = \mathcal{F}^{-1}\Psi(\omega) = \frac{1}{2}\,(1+\cos\omega)_+ = \frac{1}{2}\begin{cases} 1+\cos\omega & |\omega| \leq \pi\,, \\ 0 & \text{otherwise}\,. \end{cases}$$

We have $\Psi \in B_\pi^p$ for all $p > 1/3$ and (68) is satisfied with $\sigma = 2$. Theorem 24 applies to $1/3 < p \leq \infty$ and $1/p < s < 2$.

Example 6. (Boche [10], Boche and Schreiber [13], Boche and Fischer [12]). (Cosine-Roll-Off kernels). Let $0 < a < 1$ and put

$$\Psi_a(t) = \operatorname{sinc} t \, \frac{\cos(\pi a t)}{1 - 4(at)^2}\,.$$

Then

$$\begin{aligned}
\psi_a(\omega) &= \frac{1}{2\pi}\,(\mathcal{F}^{-1}\operatorname{sinc}) * \left(\mathcal{F}^{-1}\frac{\cos(\pi a t)}{1-4(at)^2}\right) \\
&= \chi_{[-\pi,\,\pi]} * \left(\frac{1}{4a}\cos(\omega/(2a))\right)_+ \\
&= \begin{cases} 1 & |\omega| \le (1-a)\pi\,, \\ \cos^2\left(\frac{1}{4a}(\omega - \pi) + \frac{\pi}{4}\right) & (1-a)\pi < |\omega| \le (1+a)\pi\,, \\ 0 & \text{otherwise}\,. \end{cases}
\end{aligned}$$

If $a = 0$ we have $\Psi_0(t) = \operatorname{sinc} t$ and if $a = 1$ we easily see that

$$\psi_1(\omega) = \begin{cases} \cos^2(\omega/4) & |\omega| \le 2\pi\,, \\ 0 & \text{otherwise}\,, \end{cases} = \frac{1}{2}\begin{cases} 1 + \cos(\omega/2) & |\omega| \le 2\pi\,, \\ 0 & \text{otherwise}\,, \end{cases}$$

is related to Example 5. In the case $0 < a < 1$ the assumptions of Theorem 24 are satisfied for $p > 1/2$ and $1/p < s < \infty$ by Corollary 6.

Example 7. We have seen above that kernels are easily constructed multiplying the sinc-kernel with appropriate bandlimited functions. If we put

$$\Psi(t) = \vartheta(t)\operatorname{sinc} t$$

with $\vartheta \in \mathcal{S}$, $\vartheta(0) = 1$, $\operatorname{supp}\mathcal{F}\vartheta \subset (-a\pi,\, \pi a)$ for some $0 < a < 1$, then our theorem applies to all p, $0 < p \le \infty$ and all s, $1/p < s < \infty$. Moreover, $S_W^\Psi f$ has the interpolation property.

Example 8. (Riesz kernels). Let

$$\psi_{\alpha,\beta}(\omega) = (1 - |\omega|^\alpha)_+^\beta = \begin{cases} (1 - |\omega|^\alpha)^\beta & |\omega| \le 1\,, \\ 0 & \text{otherwise}\,, \end{cases}$$

where $0 < \alpha$, $\beta < \infty$ are real numbers. We have $\mathcal{F}\psi_{\alpha,\beta} \in B_1^p$, $p > 1/(\beta+1)$ and (68) for $\sigma = \alpha$, cf. [72, Theorem 4.4.15]SW.

4.4 Approximation by generalized sampling series: the case of timelimited kernels

In this subsection we consider sampling series $S_W^\Psi f$, $W > 0$, where Ψ is a continuous function with compact support. One of the advantages of these kernels consists in the fact that existence of the sampling sums is always satisfied. One has only to deal with the approximation problem itself.

Let us assume that

$$\operatorname{supp}\Psi \subset [-a, a]\,, \tag{124}$$

$$\sum_{k=-\infty}^{\infty} \Psi(t-k) = 1\,, \qquad t \in \mathbb{R}\,,$$

$$\sum_{k=-\infty}^{\infty} (t-k)^j\,\Psi(t-k) = 0\,, \quad j = 1, 2, \ldots, \sigma - 1\,,\ t \in \mathbb{R}, \tag{125}$$

where $\sigma \in \mathbb{N}$ is a given number. Because of (124) we have

$$c_j(\Psi) := \sup_{t \in \mathbb{R}} \sum_{k=-\infty}^{\infty} |t-k|^j\,|\Psi(t-k)| \leq \sup_{0 \leq t \leq 1} \sum_{|k| < a+1} |t-k|^j\,|\Psi(t-k)| < \infty\,.$$

In analogy to (105) one can easily show (computing Fourier coefficients) that the discrete moment condition (125) is equivalent to

$$(\mathcal{F}^{-1}\Psi)^{(j)}(2k\pi) = 0 \qquad \text{for all} \quad k \in \mathbb{Z}\,. \tag{126}$$

This gives the possibility to construct kernels Ψ as linear combinations of translates of B–splines, cf. [21], [2], [32], [39], and [59]. The most simple case is $\Psi(t) = (1 - |t|)_+$ which satisfies (125) for $j = 1$.

Let us consider the sampling sums $S_W^\Psi f(t)$. Obviously, it is well-defined for each $t \in \mathbb{R}$ and it is continuous on $\mathbb{R}$ if $\{f(k/W)\}_{k=-\infty}^{\infty}$ makes sense. It is bounded and it belongs to L_p, $0 < p < \infty$, if (91) holds. Because of the identity $S_W^\Psi f(t) = (S_1^\Psi f(\frac{\cdot}{W}))(Wt)$ we can restrict ourselves to the case $W = 1$. Recall, (115) and (116). Whereas the latter one applies without any change we have to modify the estimate for $p \geq 1$. However, Hölder's inequality yields

$$\int_{-\infty}^{\infty} \left| \sum_{k=-\infty}^{\infty} f(k)\,\Psi(t-k) \right|^p dt \tag{127}$$

$$\leq \int_{-\infty}^{\infty} \sum_{k=-\infty}^{\infty} |f(k)|^p\,|\Psi(t-k)| \left(\sum_{k=-\infty}^{\infty} |\Psi(t-k)| \right)^{p/p'} dt$$

$$\leq \sup_{t\in\mathbb{R}} \left(\sum_{k=-\infty}^{\infty} |\Psi(t-k)| \right)^{p/p'} \| \Psi \,|L_1\| \sum_{k=-\infty}^{\infty} |f(k)|^p .$$

Now, assuming (91) and $f \in L_p$, it makes sense to ask for the rate of convergence of $\| f - S_W f \,|L_p\|$ $(W \to 0)$.

Recall the definition of the means d,

$$d^1_{\varrho,1} f(t) = \left(\frac{1}{\varrho} \int_{-\varrho}^{\varrho} |\Delta_h f(t)| \, dh \right) , \qquad \text{(see (42))},$$

$$d^1_{\varrho,\infty} f(t) = \sup_{|h|\leq\varrho} |\Delta_h f(t)| , \qquad \text{(see (43))}.$$

Proposition 9. (Pointwise estimates).
(i) Suppose Ψ satisfies (124) and (104). Then it holds

$$| f(t) - S^{\Psi}_W f(t) | \leq c_0(\Psi) \, d^1_{aW^{-1},\infty} f(t) \tag{128}$$

provided that the right-hand side and f are defined.
(ii) Suppose that Ψ satisfies (124), (104), and (125) for $j = 1, 2, \ldots, r$, $(r \in \mathbb{N})$. Let $f^{(j)} \in C$, $j = 0, 1, \ldots, r-1$, and let $f^{(r)}$ be locally integrable ($f^{(r)}$ has to be taken in the distributional sense here). Then it holds almost everywhere

$$| f - S^{\Psi}_W f(t) | \leq c_0(\Psi) \, \frac{(aW^{-1})^r}{(r-1)!} \, d^1_{aW^{-1},1} \, f^{(r)}(t) . \tag{129}$$

Proof. Step 1. Let t be fixed. We have

$$| f(t) - S^{\Psi}_W f(t) | \leq \sum_{|Wt-k|<a} |f(t) - f(\frac{k}{W})| \, |\Psi(Wt-k)|$$

$$\leq \sup_{|h|<aW^{-1}} |\Delta_h f(t)| \sum_{k=-\infty}^{\infty} |\Psi(Wt-k)| .$$

This leads to (128).

Step 2. Under the assumptions of part (ii) it holds Taylor's formula

$$f(x) = \sum_{j=0}^{r-1} \frac{f^{(j)}(x_0)}{j!} (x - x_0)^j + \int_{x_0}^{x} f^{(r)}(u) \, \frac{(x-u)^{r-1}}{(r-1)!} \, du . \tag{130}$$

Let t, W be fixed. We use (104) and (130) with $k/W = x$, $x_0 = t$ and see that

$$f - S_W^\Psi f(t) = \sum_{k=-\infty}^{\infty} \left[f(t) - f(\frac{k}{W})\right] \Psi(Wt-k) \tag{131}$$

$$= \sum_{j=1}^{r-1} \frac{f^{(j)}(t)}{j!} \sum_{k=-\infty}^{\infty} (\frac{k}{W} - t)^j \Psi(Wt-k) \tag{132}$$

$$+ \sum_{k=-\infty}^{\infty} \int_t^{k/W} f^{(r)}(u) \frac{(\frac{k}{W} - u)^{r-1}}{(r-1)!} du \, \Psi(Wt-k),$$

where, indeed, the sum extends over all k such that $|t - k/W| \le aW^{-1}$. It holds almost everywhere

$$f^{(r)}(t) \frac{(\frac{k}{W} - t)^r}{r!} = \int_t^{k/W} f^{(r)}(t) \frac{(\frac{k}{W} - u)^{r-1}}{(r-1)!} du.$$

Substituting into (131) and using the discrete moment conditions (125) gives

$$\begin{aligned} & f(t) - S_W^\Psi f(t) \\ &= \sum_{k=-\infty}^{\infty} \left[\int_t^{k/W} (f^{(r)}(u) - f^{(r)}(t)) \frac{(\frac{k}{W} - u)^{r-1}}{(r-1)!} du\right] \Psi(Wt-k) \\ &= \sum_{k=-\infty}^{\infty} \left[\int_0^{k/W-t} (\Delta_h f^{(r)}(t)) \frac{(\frac{k}{W} - t - h)^{r-1}}{(r-1)!} dh\right] \Psi(Wt-k). \end{aligned}$$

This yields the estimate

$$|\, f(t) - S_W^\Psi f(t)\,| \le \frac{(aW^{-1})^{r-1}}{(r-1)!} \int_{|h| \le aW^{-1}} |\Delta_h f^{(r)}(t))|\, dh \sum_{k=-\infty}^{\infty} |\Psi(Wt-k)|$$

and the desired inequality (129) follows for almost all $t \in \mathbb{R}$. □

Theorem 26. *Let $1/2 < p \le \infty$ and suppose that Ψ satisfies (124), (104), and (125) for $j = 1, 2, \dots, \sigma - 1$, where $\sigma \ge 1$ is a natural number. If $1/p < s < \sigma$, then it holds*

$$\|\, f - S_W^\Psi f \,|L_p\| = 0(W^{-s}) \tag{133}$$

for all $f \in B_{p,\infty}^s$.

Proof. Step 1. Let $1/p < s < 1$ (and $1 < p \le \infty$). It follows from part (i) of Proposition 9 that

$$\|\, f - S_W^\Psi f(t) \,|L_p\| \le c_0(\Psi)\, W^{-s}\, a^s \left[(aW^{-1})^{-s} \,\|\, d^1_{aW^{-1},\infty} f \,|L_p\|\right]. \tag{134}$$

This implies (133) applying Theorem 12(iii) with $r = u = \infty$, $m = 1$, and $q = \infty$ for $\sigma = 1$.

Step 2. Let $1/2 < p \le 1$ and let $1/p < s < 2$. If $f \in B^s_{p,\infty}$, then $f \in B^{s-1/p}_{\infty,\infty} = \Lambda^{s-1/p}_{\infty,\infty}$ and $f' \in B^{s-1/p}_{1,\infty} \hookrightarrow L_1$, see Theorems 11 and 13 and using Bernstein's inequality. Hence, Proposition 9(ii) can be applied with $r = 1$. As a consequence we get the estimate

$$\begin{aligned} \| f - S^{\Psi}_W f \,| L_p \| &\le c_0(\Psi)\, a\, W^{-1}\, \| d^1_{aW^{-1},1} f' \,| L_p \| \\ &\le c_0(\Psi)\, a^s\, W^{-s} \left[(aW^{-1})^{-s+1} \| d^1_{aW^{-1},1} f' \,| L_p \| \right]. \end{aligned} \tag{135}$$

By means of Theorem 12(iii) with $r = u = 1$, $m = 1$ and $q = \infty$ we obtain

$$\| f - S^{\Psi}_W f \,| L_p \| \le c_1\, W^{-s}\, \| f' \,| B^{s-1}_{p,\infty} \| \le c_2\, W^{-s}\, \| f \,| B^s_{p,\infty} \| \tag{136}$$

for $W \ge W_0$, where the last estimate is a consequence of Bernstein's inequality.

Step 3. Let $1 < p \le \infty$ and let $\sigma - 1 < s < \sigma$. If $f \in B^s_{p,\infty}$, then $f^{(j)} \in C$, $j = 0, 1, \ldots, \sigma - 2$, and $f^{(\sigma-1)} \in L_p$ by imbedding arguments as in Step 2. As a consequence of Proposition 9(ii) with $r = \sigma - 1$ and Theorem 12 we get analogously to Step 2 the estimate

$$\| f - S^{\Psi}_W f \,| L_p \| \le c_1\, W^{-s}\, \| f^{(\sigma-1)} \,| B^{s-(\sigma-1)}_{p,\infty} \| \le c_2\, W^{-s}\, \| f \,| B^s_{p,\infty} \| . \tag{137}$$

Step 4. Let $1/2 < p \le 1$ and let $\sigma - 2 + 1/p < s < \sigma$. If $f \in B^s_{p,\infty}$, then as in Step 3 $f^{(j)} \in C$, $j = 0, 1, \ldots, \sigma - 2$, and $f^{(\sigma-1)} \in L_1$ as in Step 2. Analogously to Step 3 we obtain (137).

Step 5. It remains to show (136) respectively (133) for $s = r = 1, \ldots, \sigma - 1$ if $1 < p \le \infty$ and for s, $2 \le s \le \sigma - 2 + 1/p$ if $1/2 < p \le 1$. In both cases we use an interpolation technique. If p is fixed and if s is given as above, then we find numbers s_0 and s_1 with $s_0 < s < s_1$, $s_0 < 1$, $s_1 > \sigma - 1$ if $1 < p < \infty$ and with $1/p < s_0 < 2$, and $\sigma - 2 + 1/p < s_1 < \sigma$ if $1/2 < p \le 1$, see Figure 3. For fixed W we consider the operator $T_W : f \mapsto f - S^{\Psi}_W f$ which maps $B^{s_i}_{p,\infty}$ into L_p according to the estimate (136) where the norms are less than $c\, W^{-s_i}$, $i = 0, 1$. Now, by real interpolation, cf. e.g. [80], we have $(B^{s_0}_{p,\infty}, B^{s_1}_{p,\infty})_{\Theta,\infty} = B^s_{p,\infty}$ for appropriate Θ and T_W maps $B^s_{p,\infty}$ into L_p with operator norm less than $c\, W^{-s}$. This finishes the proof. □

Remark 26. A slight modification of the proof shows that the approximation order σ can be realized in $\| \cdot \,| L_\infty \|$ if $f \in C^\sigma$. In other words, it holds

$$\sup_{t \in \mathbb{R}} | f(t) - S^{\Psi}_W f(t) | = 0(W^{-\sigma}), \qquad f \in C^\sigma ,$$

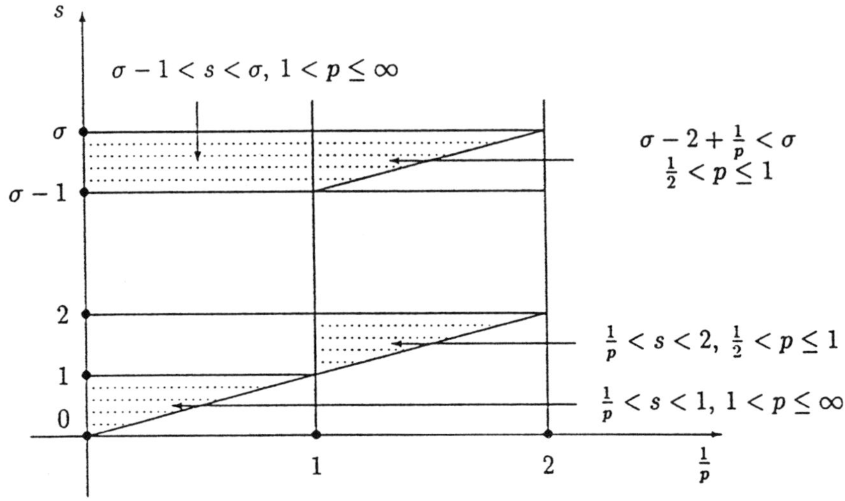

Figure 3:

if Ψ satisfies (124), (104), and (125).

Example 9. As mentioned at the beginning of this subsection kernels Ψ can be constructed by means of spline functions. Let us quote two variants. The (central) B–splines $M_n(t)$, $n = 2, 3, \ldots$ can be defined by Fourier transform

$$M_n(t) = \frac{1}{\pi} \int_0^\infty (\operatorname{sinc} \omega/2)^n \cos(\omega t)\, d\omega \tag{138}$$

or, equivalently, by

$$(\mathcal{F}^{-1} M_n)(\omega) = (\operatorname{sinc} \omega/2)^n \,.$$

In [32], see also [21, 4.3], kernels Ψ of the form

$$\Psi(t) = \sum_{j=1}^{m} \gamma_j\, M_{n_j}(t)$$

are constructed with $\sigma = \min_j n_j$. The most simple example is $\Psi(t) = M_2(t)$, obviously with $\sigma = 2$. As further examples may serve

$$\begin{aligned} \Psi(t) &= 4\, M_3(t) - 3\, M_4(t) \qquad \text{(quadratic and cubic splines)}\,, \\ \Psi(t) &= 21\, M_5(t) - 35\, M_6(t) + 15\, M_7(t)\,, \end{aligned}$$

which satisfy (125) with $\sigma = 3$ and $\sigma = 5$, respectively.
A second set of admissible kernels is obtained by considering

$$\Psi(t) = \alpha_0 \, M_n(t) + \sum_{j=1}^{[n-1/2]} \alpha_j \left(M_n(t+j) + M_n(t-j) \right).$$

The coefficients $\alpha_0, \alpha_1, \ldots$ can be chosen in such a way that the assumptions of the theorem are satisfied with $\sigma = n$. We refer to [21] and [2]. Simple examples are

$$\Psi(t) = -\frac{1}{8} M_3(t+1) + \frac{5}{4} M_3(t) - \frac{1}{8} M_3(t-1) \qquad \text{(quadratic splines)},$$
$$\Psi(t) = -\frac{1}{6} M_4(t+1) + \frac{4}{3} M_4(t) - \frac{1}{6} M_4(t-1) \qquad \text{(cubic splines)},$$

which satisfy (125) with $\sigma = 3$ and $\sigma = 4$, respectively.

A modification of this setting leads to prediction of non-bandlimited signals, see [21, 5.5] and literature quoted there. For example, the kernels

$$\Psi(t) = 2\, M_2(t-1) - M_(t-2) \qquad \text{(piecewise linear)},$$
$$\Psi(t) = \frac{17}{4} M_3(t - \frac{3}{2}) - 5\, M_3(t - \frac{5}{2}) + \frac{7}{4} M_3(t - \frac{7}{2}) \qquad \text{(quadratic splines)},$$
$$\Psi(t) = \frac{28}{3} M_4(t-2) - \frac{109}{6} M_4(t-3) + \frac{40}{3} M_4(t-4) - \frac{7}{2} M_4(t-5) \qquad \text{(cubic splines)},$$

lead to prediction and satisfy our assumptions with $\sigma = 2$, $\sigma = 3$, $\sigma = 4$, respectively.

Comments

Estimates of the aliasing error for (generalized) sampling series of non-bandlimited functions have attracted much attention in the literature. Most of the results deal with pointwise estimates or estimates in the sup–norm. Appropriate function spaces are the spaces Λ^s of Hölder-Zygmund type. We refer to [21, 17, 18, 19, 22], [39], [73] and the papers quoted there. Convergence results or error estimates in the L_p–norm seem to be less known, cf. [15, 69, 70]. R.L. Stens presented results about L_p–convergence in a talk at the SAMPTA'99 held in Loen (Norway). In particular, we learned from him about the papers [57] and [33] and about the use of the τ–modulus with respect to these problems. Theorem 21 and 22 go back to [70], see also [15, 33]. Concerning Proposition 6 we refer in addition to [57]. Theorems 23 and 25 seem to be new. Dryanov [30, 31] has characterized $A^s_{p,\infty}$ by approximation properties of S_W, but using either a different norm (instead of $\| \cdot |L_p\|$) or dealing with one-side approximation. Also he investigated generalized Jackson kernels and studied corresponding saturation problems. Theorems 24 and 26 are well-known in the case $p = \infty$, see the references quoted above. The extension to parameters p, $0 < p < \infty$, has been announced in [63], see also [5]. There can be found quite a lot of examples of kernels for generalized sampling series. Our choice is not complete and not representative. Let us refer again to [21], [2], [37], [58], [10, 14] as well as to [53].

5 Sampling of Unbounded Signals

Our aim is to indicate in a vary rough way how one could weaken the conditions on f near infinity. All the sufficient conditions for convergence of the sampling series have involved global boundedness of the signal. Now we remove this restriction and allow polynomial growth near infinity.

5.1 Bandlimited functions

Suppose f is a tempered distribution satisfying $\operatorname{supp} \mathcal{F} f \subset [-\pi, \pi]$. Hence, by the Paley-Wiener theorem f is a smooth function such that

$$|f(t)| \le c\,(1 + |t|^2)^{\alpha/2}\,, \qquad t \in \mathbb{R}\,, \tag{139}$$

for some $\alpha \ge 0$ and $c > 0$. Assume (139). Choosing

$$g(t) = \left(\operatorname{sinc} \frac{at}{m}\right)^m, \qquad 0 < a < 1\,, \quad t \in \mathbb{R}\,,$$

then

$$\operatorname{supp} \mathcal{F}(f\,g) \subset [-(1+a)\pi,\, (1+a)\pi]$$

and

$$f(t)\,g(t) \in L_p\,, \qquad \text{if} \quad m > \frac{1}{p} + \alpha - 1\,.$$

Consequently, we have

$$f(t)\,g(t) = \sum_{k=-\infty}^{\infty} f(\frac{k}{1+a}) \operatorname{sinc}((1+a)t - k)$$

or, simply rewriting this expression,

$$f(t) = \sum_{k=-\infty}^{\infty} f(\frac{k}{1+a}) \frac{\text{sinc}\,((1+a)t-k)}{(\text{sinc}\,(at/m))^m}$$

if $t \neq (\ell m)/a$ by Theorem 6.

5.2 Non-bandlimited functions

Assume that $f(t)\,(1+|t|^2)^{-\alpha/2} \in B^s_{p,\infty}$, $s > 1/p$. Let Ψ be an appropriate kernel, cf. Theorems 22 (Theorem 24 and Theorem 26). Then

$$\|\, f(t)\,(1+|t|^2)^{-\alpha/2} - S^{\Psi}_W(f(\cdot)\,(1+|\cdot|^2)^{-\alpha/2})(t)\,|L_p\| = 0(W^{-s})$$

holds ($s < \sigma$ in Theorems 24 and 26). But

$$S^{\Psi}_W(f(\cdot)\,(1+|\cdot|^2)^{-\alpha/2})(t) \;=\; \sum_{k=-\infty}^{\infty} \frac{f(\frac{k}{W})}{(1+|k/W|^2)^{\alpha/2}}\,\Psi(Wt-k)\,.$$

Now we get a reformulation in terms of weighted function spaces. Let $w(t) > 0$ be a measurable function. Then we put

$$\|\, f\,|L_{p,w}\| = \left(\int_{-\infty}^{\infty} |f(t)\,w(t)|^p\,dt\right)^{1/p}.$$

We have

$$\|f(t) - (1+|t|^2)^{\alpha/2} \sum_{k=-\infty}^{\infty} \frac{f(k/W)}{(1+|k/W|^2)^{\alpha/2}}\Psi(Wt-k)\,|\;\; L_{p,(1+|t|^2)^{-\alpha/2}}\|$$
$$= 0(W^{-s})$$

if and only if $f \in B^s_{p,\infty}((1+|t|^2)^{-\alpha/2})$ normed by

$$\|\, f(t)\,|L_{p,(1+|t|^2)^{-\alpha/2}}\| + \sup_{|h|\leq 1}\; |h|^{-s}\|\,\Delta^M_h f(t)\,|L_{p,(1+|t|^2)^{-\alpha/2}}\|$$

where $M > s > 1/p$. For the latter reformulation of $f(t)\,(1+|t|^2)^{-\alpha/2} \in B^s_{p,\infty}$, see [66, Theorem 5.1.4(iii)].

References

1. N. L. Achieser, *Vorlesungen über Approximationstheorie*, Akademie-Verlag, Berlin, 1967.

2. H. Babovsky, T. Beth, H. Neunzert, and M. Schulz-Reese, *Mathematische Methoden in der Systemtheorie: Fourier Analysis*, Teubner, Stuttgart, 1987.

3. S. N. Bernstein, *On the absolute convergence of trigonometric series*, Coll. papers, Vol. I, 217–223.

4. J. Bergh and J. Löfström, *Interpolation spaces. An introduction*, Springer, Berlin 1976.

5. L. Bittermann, *Irreguläres Sampling: die Transformationsmethode*, Diploma-Thesis, FSU Jena, 1992, 67 pages.

6. R. P. Boas, *Entire functions*, Academic Press, New York, 1954.

7. H. Boche, *Neuere Beiträge zur Theorie der Funktionaltransformationen und ihre Anwendungen*, VDI-Forschungsberichte, Reihe 21, Bd. **158**, 126 pages, VDI-Verlag Düsseldorf, 1994.

8. H. Boche, *Bemerkungen zum Rekonstruktionsverhalten endlicher Shannonscher Abtastreihen*, Frequenz **51** (1997), 11–12, 289–291.

9. H. Boche, *Verhalten des Aliasing-Fehlers bei der Abtastung nicht bandbegrenzter Signale*, Frequenz **52** (1998), 3–4, 71–75.

10. H. Boche, *Neuere Untersuchungen zur Approximierbarkeit und Rekonstruierbarkeit bandbegrenzter Signale*, Forsch. Ingenieurwes. **63** (1998), 321–337.

11. H. Boche, *Konvergenzverhalten der Shannonschen Abtastreihe für Funktionen aus dem Wiener-Raum und Einfluß der Überabtastung.* Preprint, 12 pages, Berlin, 1997

12. H. Boche and J. Fischer, *Analyse von Abtastreihen zur Signalrekonstruktion*, Zeitschrift Ingenieur in der Kommunikationstechnik, Verlag Technik GmbH, Berlin, **46** (1996), 26–29.

13. H. Boche and H. Schreiber, *The behaviour of finite Shannon sampling series.* Proc. SAMPTA '97, Univ. de Aveiro, 1997, 419–425.

14. H. Boche, and H. Schreiber, *Rekonstruktionsverhalten von Abtastreihen mit Kosinus-roll-off-Kernen*, Kleinheubacher Berichte, Deutsche Telekom AG, **40** (1997), 665–675.

15. J. H. Bramble and S. R. Hilbert, *Estimation of linear functionals on Sobolev spaces with application to Fourier transforms and spline interpolation*, SIAM J. Numer. Anal. **7** (1970), 112–124.

16. J. L. Brown, *On the error in reconstructing a non-bandlimited function by means of the bandpass sampling theorem*, J. Math. Anal. Appl. **18** (1967); Erratum, same journal, **21** (1968), p. 699.

17. P. L. Butzer, W. Engels, S. Ries and R. L. Stens, *The Shannon sampling series and reconstruction of signals in terms of linear, quadratic and cubic splines*, SIAM J. Appl. Math. **46** (1986), 299–323.

18. P. L. Butzer, A. Fischer and R. L. Stens, *Generalized sampling approximation of multivariate signals: general theory*, Proc. 4 th Meeting on Real Anal. and Measure Th., Capri, 1990.

19. P. L. Butzer, A. Fischer and R. L. Stens, *Generalized sampling approximation of multivariate signals: theory and some applications*, Note di Matematica (1990), 173–191.

20. P. L. Butzer and R. Nessel, *Fourier analysis and approximation*, Birkhäuser, Basel, 1971.

21. P. L. Butzer, W. Splettstößer and R. L. Stens, *The sampling theorem and linear prediction*, Jahresberichte Dt. Math.-Verein. **90** (1988), 1–70.

22. P. L. Butzer, R. Stens, *Sampling theory for non-bandlimited functions: a historical overview*, SIAM Reviews **34** (1992), 40–53.

23. A. Cohen, R. de Vore, and R. Hochmuth *Restricted approximation*, Research Report IMI **6** (1997), Univ. of South Carolina, 27 pages.

24. C. de Boor, R. de Vore and A. Ron, *Approximation from shift-invariant subspaces of $L_2(\mathbb{R}^d)$*, Trans. Amer. Math. Soc. **341** (1994), 787–806.

25. L.T. Dechevski, *τ-moduli and interpolation.* In: Function spaces and applications, Lect. Notes in Math. **1302**, 177-190, Springer, Berlin 1988.

26. R.A. de Vore, *Nonlinear approximation.* Acta Numerica **7** (1998), 51–150.

27. R.A. de Vore and G.G. Lorentz, *Constructive Approximation*, Springer, Berlin, 1993.

28. R.A. de Vore and R.C. Sharpley, *Besov spaces on R^d*, Trans. Amer. Math. Soc. **335** (1993), 843–864.

29. Dinh-Dung, *The sampling theorem, L_q^T-approximation and ε–dimension*, J. Approx. Theory **70** (1992), 1–15.

30. D. P. Dryanov, *On the convergence and saturation problem of a class of discrete linear operators of entire exponential type in $L_p(-\infty, \infty)$ spaces*, In: Proc. Conf.:Constructive theory of functions '84, 312–318, Publ. House of the Bulg. Acad. of Science, Sofia, 1984.

31. D. P. Dryanov, *Equiconvergence and equiapproximation for entire functions*, In: Proc. Conf.:Constructive theory of functions, Varna '91, 123-136, Publ. House of the Bulg. Acad. of Science, Sofia, 1992.

32. W. Engels, E. L. Stark and L. Vogt, *Optimal kernels for a generalized sampling theorem*, J. Approx. Theory **50** (1987), 69–83.

33. Fang Gensun, *Whittaker-Kotel'nikov-Shannon sampling theorem and aliasing error*, J. Approx. Theory **85** (1996), 115–131.

34. M .Frazier and B. Jawerth, *Decomposition of Besov spaces*, Indiana Univ. Math. J. **34** (1985), 777–799.

35. H.-G. Feichtinger and K. Gröchenig, *Irregular sampling theorems and series expansions of band-limited functions*, J. Math. Anal. Appl. **167** (1992), 530–556.

36. H.-G. Feichtinger and K. Gröchenig, *Error analysis in regular and irregular sampling theory*, Applicable Analysis **50** (1993), 167–169.

37. R. Gervais, Q. I. Rahman and G. Schmeisser, *A bandlimited function simulating a duration limited one*, Approx. Theory and Funct. Anal., Anniversary Vol., Proc. Conf. Oberwolfach 1983, ISNM **65** (1984), 355–362.

38. J. R. Higgins, *Five short stories about cardinal series*, Bull. Amer. Math. Soc. **12** (1985), 45–89.

39. J.R. Higgins, *Sampling theory in Fourier and signal analysis. Foundations*, Clarendon Press, Oxford, 1996.

40. L. Hörmander, *The analysis of linear partial differential operators. I.*, Springer, Berlin, 1990.

41. K.G. Ivanov, *On the behaviour of two moduli of smoothness.* Compt. Rend. Acad. Bulg. Sci. **38** (1985), 539–542.

42. B. Jawerth, *Some observations on Besov and Triebel-Lizorkin spaces*, Math. Scand. **40** (1977), 94–104.

43. B. Jawerth, *The trace of Sobolev and Besov spaces if* $0 < p < 1$, Studia Math. **62** (1978), 65–71.

44. A. Jerri, *The Shannon sampling theorem – its various extensions and applications; a tutorial review*, Proc. IEEE **65** (1977), 1565–1596.

45. R.-Q. Jia and J. Lei, *Approximation by multiinteger translates of functions having global support*, J. Approx. Theory **72** (1993), 2–23.

46. A. Kufner, O. John, and S. Fucik, *Function spaces*, Academia, Prague, 1977.

47. J. Löfström, *Besov spaces in the theory of approximation*, Ann. Mat. Pura Appl. **85** (1970), 93–184.

48. R. J. Marks, *Introduction to Shannon sampling and interpolation theory*, Springer, New York, 1991.

49. R. J. Marks (ed.), *Advanced topics in Shannon sampling and interpolation theory*, Springer, New York, 1993.

50. R. Nessel and G. Wilmes, *Nikol'skij type inequalities for trigonometric polynomials and entire functions of exponential type*, J. Austral. Math. Soc. **25** (1978), 7–18.

51. S.M. Nikol'skij, *Approximation of functions of several variables and imbedding theorems*, Springer, Berlin, 1975.

52. P. Oswald, *Multilevel finite element approximation; theory and applications*, Teubner, Stuttgart 1995.

53. A. Papoulis, *Signal analysis*, McGraw-Hill, New York, 1977.

54. J. Peetre, *New thoughts on Besov spaces*, Duke Univ. Press, Durham, 1976.

55. A. Pietsch, *Approximation spaces*, J. Approx. Theory **32** (1981), 115–134.

56. M. Plancherel and G. Polya, *Fonctions entieres et integrales de Fourier multiples*, Comment. Math. Helv. **9** (1937), 224–248; ebenda **10** (1938), 110–163.

57. Q. I. Rahman and P. Vertesi, *On the L^p convergence of Lagrange interpolating entire functions of exponential type*, J. Approx. Theory **69** (1992), 302–317.

58. S. Ries, *Approximation stetiger und unstetiger Funktionen durch verallgemeinerte Abtastreihen*, Doctoral Dissertation, 96 pages, RWTH Aachen, 1984.

59. S. Ries and R.L. Stens, *Approximation by generalized sampling series*, In: Constr. Theory of Functions '84, 746-756, Publ. House of the Bulg. Acad. of Science, Sofia, 1984.

60. K. Runovski and H.-J. Schmeisser, *On Marcinkiewicz-Zygmund type inequalities for irregular knots in L_p–spaces,* $0 < p \leq \infty$, Math. Nachr. **189** (1998), 209–220.

61. T. Runst and W. Sickel, *Nemytskij operators, Sobolev spaces of fractional order and nonlinear partial differential equations.* de Gruyter, Berlin, 1996.

62. H.-J. Schmeisser, *Characterizations of periodic function spaces of Besov-Hardy-Sobolev type via approximation processes and relations to the strong summability of Fourier series*, Banach Center Publ. **22**, 341–361, PWN Polish Scientific Publishers, Warsaw, 1989.

63. H.-J. Schmeisser, *Abtasttheoreme aus der Sicht der Theorie der Funktionenräume*, In: Wavelet-Approximation und Anwendungen, 48–52, Lübeck, 1995.

64. H.-J. Schmeisser and W. Sickel, *On the strong summability of multiple Fourier series and approximation of periodic functions*, Math. Nachr. **133** (1987), 211–236.

65. H.-J. Schmeisser and W. Sickel, *Characterization of periodic function spaces via means of Abel-Poisson and Bessel-Potential type*, J. Approx. Theory **61** (1990), 239–262.

66. H.-J.Schmeisser and H. Triebel, *Topics in Fourier analysis and function spaces*, Wiley, Chichester, 1987.

67. B. Sendov and V.A. Popov, *The averaged moduli of smoothness*, Wiley, Chichester, 1988.

68. H.S. Shapiro, *Topics in approximation theory*, Lect. Notes in Math. **187**, Springer, Berlin, 1971.

69. W. Sickel, *Some remarks on trigonometric interpolation on the n-torus*, ZAA **10** (1991), 551–562.

70. W. Sickel, *Characterizations of Besov-Triebel-Lizorkin spaces via approximation by Whittaker's cardinal series and related unconditional Schauder bases*, Constr. Approx. **8** (1992), 257–274.

71. W. Sickel and H. Triebel, *Hölder inequalities and sharp embeddings in function spaces of $B^s_{p,q}$ and $F^s_{p,q}$ type*, ZAA **14** (1995), 105–140.

72. E. M. Stein and G. Weiss, *Introduction to Fourier analysis on Euclidean spaces*, Princeton Univ. Press, Princeton, 1971.

73. R. L. Stens, *Error estimates for sampling sums based on convolution integrals*, Inform. and Control **45** (1980), 37–47.

74. E.A. Storozenko and P. Oswald, *Jackson's theorem in the spaces $L_p(R^k)$*, $0 < p < 1$. Sibirsk. Math. J. **19** (1978), 888–901.

75. R. Strichartz, *A guide to distribution theory and Fourier transforms*, CRC Press, Boca Raton, 1994.

76. V.M. Tikhomirov, *Approximation theory*, in Encyclopedia of Math. Science: Analysis II (R.V. Gamkrelidze, ed.), Vol. **14**, Springer, Berlin, 1990, 93–244.

77. A.F. Timan, *Theory of approximation of functions of a real variable*, MacMillan, New York, 1963.

78. H. Triebel, *Fourier analysis and function spaces*, Teubner-Texte Math. **7**, Teubner, Leipzig 1977.

79. H. Triebel, *Interpolation theory, function spaces, differential operators*, Deutscher Verlag der Wissenschaften, Berlin, 1978.

80. H. Triebel, *Theory of Function Spaces*, Birkhäuser, Basel, 1983.

81. H. Triebel, *Theory of Function Spaces. II*, Birkhäuser, Basel, 1992.

82. V. S. Vladimirov, *Gleichungen der mathematischen Physik*, VEB Deutscher Verlag der Wissenschaften, Berlin, 1971.

83. P. Wojtasczczyk, *A mathematical introduction to wavelets.* Cambridge Univ. Press, Cambridge, 1997.

84. A. Zayed, *Advances in Shannon sampling theory*, CRC Press, Boca Raton, 1993.

COMPUTATIONAL ISSUES IN STABLE FINANCIAL MODELING

Carlo Marinelli and Svetlozar T. Rachev
Department of Economics, University of California at Santa Barbara, Santa Barbara, CA 93106. E-mail: carlo@pstat.ucsb.edu
Institute of Statistics and Mathematical Economics, School of Economics, University of Karlsruhe, Kollegium am Schloss, Bau II, 20.12, R210, Postfach 6980, D-76128, Karlsruhe, Germany.
E-mail: rachev@lsoe.uni-karlsruhe.de
Contact author: C. Marinelli

We overview the definition and basic properties of stable laws and processes, and discuss how to estimate simple univariate stable models. In particular, we show how to perform maximum likelihood estimation of stable parameters and how to extend this procedure to more sophisficated conditional models. Furthermore, we present two applications: regression with stable disturbances, and stable option pricing.

1 Introduction

Many theories in Finance are based upon distributional assumptions, among which the most widely used is the normal. A strong and plausible argument in favor of this assumption comes from the Central Limit Theorem, which states that (suitably normalized) sums of i.i.d. random variables with finite variance converge in distribution to a gaussian random variable. As a consequence, a variable that results as the sum of many random components can assumed to be approximately Gaussian. However, since the path-breaking studies of Mandelbrot and Fama in the 60's, it is now clear that the empirical distribution of many observed economic and financial data deviates from the ideal gaussian law, since they often exhibit skewness (asymmetry) and fat tails. The two authors proposed the α-stable (Paretian) assumption, on the basis of its attractive modeling properties: stable laws not only provide a better empirical fit, but they also possess heavy tails and result as limit distribution of sums of i.i.d. random variables under very general conditions. Moreover, they have domains of attraction, which make them robust models: in fact, since the distributional model is only an approximation of the exact distribution underlying the observed data, it is reasonable to assume that a slight modification in the data will give a distribution that still lies in the same domain of the attraction. Another attractive property of stable laws is the so-called stability property: the sum of two stable variables with the same index α is again stable with stability index α.

In the following section we briefly recall the definition and main properties of stable laws, with emphasis on the univariate case. Section 3 deals with parameters estimation of stable laws and with numerical methods for computer simulation of stable random variates.

In Section 4 we overview the basic properties of stable processes, in particular of Lévy and fractional stable motions.

Section 5 discusses the application of the stable assumption in two classical problems from statistics and mathematical finance: regression with stable errors, and option pricing with stable distributed returns.

2 The Stable Distribution

The theory of univariate stable distributions was mainly developed in the 1920's and 1930's by Paul Lévy and Aleksander Kinchine. Since then it has been used with very interesting results in many fields, ranging from geophysics to economics and engineering (especially in communications and for modeling $1/f$ noise — see Nikias and Shao (1996)). Stable distributions have also been applied to financial modeling, following the path outlined by Mandelbrot and Taylor (1967) in their seminal paper.

For accounts of the classical theory of stable laws, two excellent references are Gnedenko and Kolmogorov (1954) and Feller (1971). A very comprehensive treatment can be found in Zolotarev (1986). Nikias and Shao (1996) is an introduction to the applications of symmetric stable distributions to digital signal processing and other fields of electrical engineering. Rachev and Mittnik (2000) is a survey of the state of the art in stable financial modeling.

As we will see in more detail later on, stable random variables are very well suited to model the contribution of many small effects, as are Gaussian variables, with the significative difference that stable laws are more general, since we may say, loosely speaking, that these small effects can have infinite variance. Therefore, the stable distribution permits more variations than the normal, and so it is a better option to handle with extreme variations, since its density admits heavy tails.

2.1 Definitions

The following theorem fully characterizes a random variable with stable distribution.

Theorem 1. *Let X be a random variable. The following conditions are equivalent:*

(i) Let $a, b \in \mathbb{R}^+$ and X_1, X_2 be independent copies of X. There exist $c \in \mathbb{R}^+$ and $d \in \mathbb{R}$ such that

$$aX_1 + bX_2 \stackrel{d}{=} cX + d \tag{1}$$

where $\stackrel{d}{=}$ denotes equality in distribution.

(ii) Let n be a positive integer, $n \geq 2$, and $X_1, X_2, \ldots, X_n$ be independent copies of X. There exist $c_n \in \mathbb{R}^+$ and $d_n \in \mathbb{R}$ such that

$$X_1 + X_2 + \cdots + X_n \stackrel{d}{=} c_n X + d_n \tag{2}$$

(iii) X has a domain of attraction, i.e. there are a sequence of i.i.d. random variables $\{Y_i\}_{i\in\mathbb{N}}$, a real positive sequence $\{a_i\}_{i\in\mathbb{N}}$ and a real sequence $\{b_i\}_{i\in\mathbb{N}}$ such that

$$\frac{1}{a_n}\sum_{i=1}^{n} Y_i - b_n \stackrel{d}{\longrightarrow} X$$

where "$\stackrel{d}{\longrightarrow}$" stands for convergence in distribution.

(iv) The characteristic function of X admits the following form:

$$Ee^{iXt} = \begin{cases} \exp\left(-\sigma^\alpha |t|^\alpha \left(1 - i\beta(\operatorname{sgn} t)\tan\frac{\pi\alpha}{2}\right) + i\mu t\right) & \text{if } \alpha \neq 1 \\ \exp\left(-\sigma |t| \left(1 + i\beta\frac{2}{\pi}(\operatorname{sgn} t)\log|t|\right) + i\mu t\right) & \text{if } \alpha = 1. \end{cases} \tag{3}$$

where the parameters satisfy the following contraints: $\alpha \in]0,2]$, $\sigma \in \mathbb{R}_0^+$, $\beta \in [-1,1]$, $\mu \in \mathbb{R}$.

Proof. It is straightforward to show, by induction, that (i) implies (ii). A proof of the inverse implication can be found in Feller (1971). Taking Y_i's as independet copies of X, one easily proves that (ii) implies (iii). The converse is proved, for example, in Gnedenko and Kolmogorov (1954). Here we only show that (iv) implies (ii): for $\alpha \neq 1$ we can write

$$Ee^{it(X_1+X_2+\cdots+X_n)} = \exp\left(-n\sigma^\alpha |t|^\alpha \left(1 - i\beta(\operatorname{sgn} t)\tan\frac{\pi\alpha}{2}\right) + in\mu t\right).$$

and also

$$\begin{aligned} Ee^{it(c_n X + d_n)} &= e^{itd_n} Ee^{i(tc_n)X} \\ &= e^{itd_n}\exp\left(-\sigma^\alpha |c_n t|^\alpha \left(1 - i\beta(\operatorname{sgn} c_n t)\tan\frac{\pi\alpha}{2}\right) + i\mu c_n t\right). \end{aligned}$$

Choosing $c_n = n^{1/\alpha}$ and $d_n = \mu(n - n^{1/\alpha})$, it follows

$$Ee^{it(c_nX+d_n)} = Ee^{it(X_1+X_2+\cdots+X_n)}.$$

Therefore, by the uniqueness of the characteristic function, we have

$$X_1 + X_2 + \cdots + X_n \stackrel{d}{=} c_nX + d_n$$

where, in particular, the constant c_n and d_n have been explicitly calculated. The proof for $\alpha = 1$ is very similar. We give next a sketch of the proof of the converse: one can show that the characteristic function of a stable variable admit the so-called *Lévy-Khinchine representation*:

$$Ee^{iXt} = \begin{cases} \exp\left(iMt + P\int_0^\infty \psi(t,x)\frac{dx}{x^{1+\alpha}} + Q\int_{-\infty}^0 \psi(t,x)\frac{dx}{|x|^{1+\alpha}}\right) & \text{if } \alpha < 2 \\ \exp\left(iMt - \sigma^2t^2\right) & \text{if } \alpha = 2 \end{cases} \tag{4}$$

where $M \in \mathbb{R}$, σ, P, $Q \in \mathbb{R}_0^+$,

$$\psi(t,x) = e^{itx} - 1 - \frac{itx}{1+x^2},$$

and the measure

$$L(dx) = \frac{P}{x^{1+\alpha}}H(x)\,dx + \frac{Q}{|x|^{1+\alpha}}H(-x)\,dx$$

in (4) is called *Lévy measure* ($H(x)$ is the Heavyside function, i.e. $H(x) = 1$ $\forall x \in \mathbb{R}^+$ and $H(x) = 0$ elsewhere).

Then, in the case $\alpha < 2$, that is when X is non-degenerate ($P + Q > 0$), one sets

$$\beta = \frac{P-Q}{P+Q}$$

and evaluates the integral in (4), eventually arriving at (3). For the detail of the proof, which is rather involved, we refer the interested reader to Gnedenko and Kolmogorov (1954), Section 34. □

Definition 1. A random variable X is said to be stable if it satisfies (one of) the four equivalent conditions expressed by Theorem 1.

A stable random variable X is called *strictly* stable if $d = 0$ in (1), and *symmetric* if X and $-X$ have the same distribution.

The parameter α is called the *index of stability* or *characteristic exponent* of the random variable X, which in turn is called α-stable. Moreover, the following proposition holds, which contributes to the characterization of stable random variables.

Proposition 1. *Let X be a stable random variable. Then there exist α', $\alpha'' \in (0,2]$ such that*

(i) the constants appearing in (1) are related by

$$c^{\alpha'} = a^{\alpha'} + b^{\alpha'};$$

(ii) in (2) it holds

$$c_n = n^{1/\alpha''};$$

(iii) the constants α' and α'' are the index of stability, i.e.

$$\alpha' = \alpha'' = \alpha.$$

Proof. See Feller (1971), Section VI.1. □

We denote by $S_\alpha(\sigma, \beta, \mu)$ the stable distribution with parameters α, β, σ and μ, and with the notation

$$X \sim S_\alpha(\sigma, \beta, \mu)$$

we mean that the distribution function of the random variable X is $S_\alpha(\sigma, \beta, \mu)$.

The first two equivalent conditions of Theorem 1 express the stability property: the family of independent stable distributions is closed under convolution.

The third conditions relates stable distributions with a generalized form of the Central Limit Theorem (CLT). In particular, it states that stable distributions are the only distributions that can be obtained as limits of normalized sums of i.i.d. random variables. Note that if the Y_i's are i.i.d. with finite variance, taking $a_n = n^{1/2}$, we obtain a statement of the ordinary central limit theorem. We may say that relaxing the assumption that the Y_i's have finite variance in the ordinary CLT, we "discover" stable distributions as "replacement" of the Gaussian limit.

2.2 Stable Central Limit Theorem and Domain of Attraction of Stable Laws

We state here an alternative formulation of the third condition in Theorem 1 in terms of the domain of attraction (DA) of a random variable, and report a result which characterizes the stable DA.

Let $\{Y_i\}_{i\in\mathbb{N}}$ be i.i.d. random variables with common distribution function $F(x)$. If there exist constants a_n and b_n such that the normalized sums

$$S_n = \frac{1}{a_n}\sum_{i=1}^{n} Y_i - b_n$$

converge in distribution to a random variable X with characteristic function $G(x)$, then we say that $F(x)$ is *attracted* to $G(x)$. Moreover, the totality of distribution functions attracted to $G(x)$ is called the *domain of attraction* of the law $G(x)$.

A concise formulation of the Stable Central Limit Theorem can now be stated.

Theorem 2. *The domain of attraction of a distribution function $G(x)$ is nonempty if and only if $G(x)$ is a stable law.*

For a proof, see Gnedenko and Kolmogorov (1954), Section 33.

There exist in the literature many results about the characterization of distribution functions attracted to stable laws. For a thorough treatment of this topic we refer, among others, to Gnedenko and Kolmogorov (1954), and for recent developments to Aaronson and Denker (1998) and references therein.

Here we only give a theorem which states some properties of the distribution function of a sequence of i.i.d. random variables attracted to a stable law.

Theorem 3. *Let $(Y_i)_{i\in\mathbb{N}}$ be i.i.d. random variables with distribution function $F(x)$. Necessary and sufficient condition for the existence of sequences $(a_n)_{n\in\mathbb{N}} \in \mathbb{R}^+$ and $(b_n)_{n\in\mathbb{N}} \in \mathbb{R}$ such that*

$$\frac{1}{a_n}\sum_{i=1}^{n} Y_i - b_n \xrightarrow{d} X \sim S_\alpha(1,\beta,0)$$

is that

$$L(x) = x^\alpha(1 - F(x) + F(-x))$$

is slowly varying at infinity and that it holds

$$\lim_{x\to\infty} \frac{F(-x)}{1 - F(x) + F(-x)} = \frac{1-\beta}{2}.$$

A proof of this theorem can be found in Mijnheer (1975), where it is also shown that the sequence a_n must satisfy

$$\lim_{n\to\infty}\frac{nL(a_n)}{a_n^\alpha}=\begin{cases}\Gamma(1-\alpha)\cos\dfrac{\pi\alpha}{2} & \text{if } \alpha\in(0,1)\\ \frac{2}{\pi} & \text{if } \alpha=1\\ \dfrac{\Gamma(2-\alpha)}{\alpha-1|\cos\frac{\pi\alpha}{2}|} & \text{if } \alpha\in(1,2),\end{cases}$$

and the sequence b_n can be chosen as

$$b_n=\begin{cases}0 & \text{if } \alpha\in(0,1)\\ na_n\displaystyle\int_{\mathbb{R}}\sin\frac{x}{a_n}\,dF(x) & \text{if } \alpha=1\\ n\displaystyle\int_{\mathbb{R}}x\,dF(x) & \text{if } \alpha\in(1,2).\end{cases}$$

Moreover, Mijnheer (1975) shows that $a_n=n^{1/\alpha}L_0(n)$ with L_0 slowly varying at infinity.
We recall that a function $L:\mathbb{R}\to\mathbb{R}_0^+$ is called slowly varying at infinity if it satisfies

$$\lim_{t\to\infty}\frac{L(xt)}{L(x)}=1\quad\forall t\in\mathbb{R}^+.$$

2.3 Parametrizations

While we shall almost always use the parametrization of the one-dimensional characteristic function given in (3), several others have been proposed (see Zolotarev (1986) and Nolan (1998)), some of which are well suited for computational purposes.

As a matter of fact, a major drawback of the S parametrization of the characteristic function of a stable random variable X, that we report here for convenience:

$$Ee^{iXt}=\begin{cases}\exp\Big(-\sigma^\alpha|t|^\alpha\Big(1-i\beta(\operatorname{sgn}t)\tan\frac{\pi\alpha}{2}\Big)+i\mu t\Big) & \text{if } \alpha\neq1\\ \exp\Big(-\sigma|t|\Big(1+i\beta\frac{2}{\pi}(\operatorname{sgn}t)\log|t|\Big)+i\mu t\Big) & \text{if } \alpha=1,\end{cases}$$

is that is it not continous in the four parameters α, β, σ and μ.

The following parametrization, proposed by Zolotarev (1986), and called the (M) parametrization,

$$Ee^{iXt}=\exp\Big(\sigma^\alpha\Big(-|t|^\alpha+it\beta(|t|^{\alpha-1}-1)\tan\frac{\pi\alpha}{2}\Big)+i\mu_1t\Big)\qquad(5)$$

has the distinctive advantage of being continuous with respect to the parameters α, β, σ and μ_1, as a consequence of the relation

$$\lim_{\alpha\to 1} \tan\frac{\pi\alpha}{2}(|t|^{1-\alpha}-1) = \frac{2}{\pi}\log|t|.$$

This makes (5) better numerically conditioned, although μ_1 looses the meaning of location parameter. In fact, it holds

$$\mu_1 = \begin{cases} \mu + \beta\sigma\tan\frac{\pi\alpha}{2} & \text{if } \alpha \neq 1 \\ \mu & \text{if } \alpha = 1. \end{cases}$$

A similar parametrization, proposed by Nolan (1998), defines

$$X^0 = \sigma Z + \mu^0,$$

where Z is a standard stable random variable in Zolotarev's (M) parametrization. The resulting so-called S^0 parametrization is such that α, β and σ have the same meaning as in the S parametrization, while

$$\mu = \begin{cases} \mu^0 - \beta\sigma\tan\frac{\pi\alpha}{2} & \text{if } \alpha \neq 1 \\ \mu^0 - \beta\sigma\frac{2}{\pi}\log\sigma & \text{if } \alpha = 1. \end{cases}$$

In the S^0 parametrization, σ and μ^0 are true scale and location parameters, meaning that if $X \sim S_\alpha(\sigma,\beta,\mu^0)$, then

$$aX + b \sim S_\alpha\Big(|a|\sigma, \frac{|a|}{a}\beta, a\mu^0 + b\Big)$$

for any real constants $a \neq 0$ and b. Furthermore, the characteristic function in the S^0 parametrization (and hence the density and the distribution functions) are jointly continuous in all four parameters $\alpha, \beta, \sigma, \mu^0$.

In all the algorithms that we have implemented, for simulation of stable random variates, calculation of the stable density function and stable parameter estimation, we relied either on Zolotarev's (M) or on Nolan's S^0 parametrizations. For example, interpolation on a lattice $(x_i, \alpha_j, \beta_k)_{ijk}$ of precomputed values of the stable density, for example, makes sense only if the density is continuous, and similarly, the continuity in the parameter α of the density function in a neighborhood of $\alpha = 1$ is essential for correct parameter estimation.

Note that, when $\alpha = 2$, the characteristic function (3) becomes

$$Ee^{itX} = e^{-\sigma^2 t^2 + i\mu t},$$

that is the characteristic function of a Gaussian random variable with mean μ and variance $2\sigma^2$, i.e. σ is not equal to the standard deviation. Note also that although the value of β is not specified, it is customary to take $\beta = 0$.

2.4 Properties

In this section we will prove some basic properties of stable random variables, which will also give an interpretation to the parameters α, β, σ and μ.

Stable random variables have continuous probability density fuctions (even better, it can be proved that they are smooth functions, i.e. belonging to $C^\infty(\mathbb{R})$), but unfortunately they do not admit a closed form representation (see Zolotarev (1986)), except for a few exceptions:

1. the Gaussian distribution $S_2(\sigma, 0, \mu) = N(\mu, 2\sigma^2)$, whose density function is
$$f_2(x;\sigma,0,\mu) = \frac{1}{2\sigma\sqrt{\pi}} e^{-\frac{(x-\mu)^2}{4\sigma^2}}.$$
2. The Cauchy distribution $S_1(\sigma, 0, \mu)$, whose density function is
$$f_1(x;\sigma,0,\mu) = \frac{\sigma}{\mu} \frac{1}{(x-\mu)^2+\sigma^2}.$$
3. The Lévy distribution $S_{1/2}(\sigma, 1, \mu)$, whose density function is
$$f_{1/2}(x;\sigma,1,\mu) = \left(\frac{\sigma}{2\pi}\right)^{1/2} \frac{1}{(x-\mu)^{3/2}} e^{-\frac{\sigma}{2(x-\mu)}}$$
and is defined on (μ, ∞).

Proposition 2. *Let X_1, X_2 be independent α-stable random variables, with $X_i \sim S_\alpha(\sigma_i, \beta_i, \mu_i)$, $i = 1, 2$. Then $X = X_1 + X_2$ is a stable random variable with distribution $S_\alpha(\sigma, \beta, \mu)$, where*
$$\sigma = (\sigma_1^\alpha + \sigma_2^\alpha)^{1/\alpha}$$
$$\beta = \frac{\beta_1\sigma_1^\alpha + \beta_2\sigma_2^\alpha}{\sigma_1^\alpha + \sigma_2^\alpha}$$
$$\mu = \mu_1 + \mu_2 .$$

Proof. As we did before, we prove this statement only for $\alpha \neq 1$, leaving the case $\alpha = 1$ to the reader. Since X_1 and X_2 are independent we have
$$\begin{aligned}
&\log Ee^{it(X_1+X_2)} \\
&= \log Ee^{itX_1}Ee^{itX_2} = \log Ee^{itX_1} + \log Ee^{itX_2} \\
&= -(\sigma_1^\alpha + \sigma_2^\alpha)|t|^\alpha + i|t|^\alpha(\operatorname{sgn} t)\tan\frac{\pi\alpha}{2}(\beta_1\sigma_1^\alpha + \beta_2\sigma_2^\alpha) + it(\mu_1+\mu_2) \\
&= -(\sigma_1^\alpha + \sigma_2^\alpha)|t|^\alpha \left(1 - i\frac{\beta_1\sigma_1^\alpha + \beta_2\sigma_2^\alpha}{\sigma_1^\alpha + \sigma_2^\alpha}(\operatorname{sgn} t)\tan\frac{\pi\alpha}{2}\right) + it(\mu_1+\mu_2).
\end{aligned}$$

□

The following proposition gives an interpretation of the parameter μ, which is often called *shift parameter.*

Proposition 3. *Let* $X \sim S_\alpha(\sigma, \beta, \mu)$ *and* $a \in \mathbb{R}$. *Then* $X+a \sim S_\alpha(\sigma, \beta, \mu+a)$.

Proof. It is an easy consequence of the form of the characteristic function of X.
□

Proposition 4. *Let* $X \sim S_\alpha(\sigma, \beta, \mu)$ *and* $a \in \mathbb{R} \setminus \{0\}$. *Then*

$$\begin{array}{ll} aX \sim S_\alpha(|a|\sigma, \operatorname{sgn}(a)\beta, a\mu) & \text{if } \alpha \neq 1 \\ aX \sim S_1(|a|\sigma, \operatorname{sgn}(a)\beta, a\mu - \frac{2}{\pi}a(\log|a|)\sigma\beta) & \text{if } \alpha = 1. \end{array}$$

Proof. For $\alpha \neq 1$ we have

$$\begin{aligned} \log Ee^{it(aX)} &= -\sigma^\alpha |ta|^\alpha \left(1 - i\beta(\operatorname{sgn} at)\tan\frac{\pi\alpha}{2}\right) + i\mu(t\alpha) \\ &= -(\sigma|a|)^\alpha |t|^\alpha \left(1 - i\beta(\operatorname{sgn} a)(\operatorname{sgn} t)\tan\frac{\pi\alpha}{2}\right) + i(\mu a)t. \end{aligned}$$

The proof for $\alpha = 1$ is similar. □

As a consequence of this result, the parameter σ is called *scale parameter.* However, note that when $\alpha = 1$, multiplication by a constant coefficient turns out in a nonlinear transformation of the shifting parameter μ. Therefore, when $\alpha = 1$ and $\beta \neq 0$, this would be an abuse of language.

We can give now two propositions that give to the parameter β the interpretation of *skewness parameter.*

Proposition 5. *Let* $\alpha \in (0,2)$ *and* $X \sim S_\alpha(\sigma, \beta, 0)$. *Then*

$$-X \sim S_\alpha(\sigma, -\beta, 0).$$

Proof. Very easy to verify using the characteristic function of X. □

Proposition 6. *The stable variable* $X \sim S_\alpha(\sigma, \beta, \mu)$ *is symmetric if and only if* $\beta = 0$ *and* $\mu = 0$. *It is symmetric around* μ *if and only if* $\beta = 0$.

Proof. Recall that a random variable is symmetric if and only if its characteristic function is real. Now observe that (3) is real if and only if $\beta = \mu = 0$. The second result follows from Proposition 3. □

The following propositions tell us some interesting facts about strictly stable random variables.

Proposition 7. *Let* $X \sim S_\alpha(\sigma, \beta, \mu)$ *with* $\alpha \neq 1$. *Then* X *is strictly stable if and only if* $\mu = 0$.

Proof. Let X_1, X_2 be independent copies of X, a, $b \in \mathbb{R}^+$. Then it holds

$$aX_1 + bX_2 \sim S_\alpha\big(\sigma(a^\alpha + b^\alpha)^{1/\alpha}, \beta, \mu(a+b)\big).$$

Setting $c = (a^\alpha + b^\alpha)^{1/\alpha}$ we can write

$$cX + d \sim S_\alpha\big(\sigma(a^\alpha + b^\alpha)^{1/\alpha}, \beta, \mu(a^\alpha + b^\alpha)^{1/\alpha} + d\big).$$

Therefore we have

$$aX_1 + bX_2 \stackrel{d}{=} cX + d$$

with $d = 0$ if and only if $\mu = 0$. □

Corollary 1. *Let* $X \sim S_\alpha(\sigma, \beta, \mu)$ *with* $\alpha \neq 1$. *Then* $X - \mu$ *is strictly stable.*

Proof. Since $X - \mu \sim S_\alpha(\sigma, \beta, 0)$ by Proposition 3, the result follows from the previous proposition. □

Proposition 8. *Let* $X \sim S_1(\sigma, \beta, \mu)$. *Then* X *is strictly stable if and only if* $\beta = 0$.

Proof. Let X_1, X_2 be independent copies of X, a, $b \in \mathbb{R}^+$. Then it holds

$$aX_1 + bX_2 \sim S_1\Big((a+b)\sigma, \beta, (a+b)\mu - \frac{2}{\pi}\sigma\beta(a\log a + b\log b)\Big)$$

and

$$(a+b)X + d \sim S_1\Big((a+b)\sigma, \beta, (a+b)\mu - \frac{2}{\pi}\sigma\beta(a+b)\log(a+b) + d\Big).$$

Therefore $aX_1 + bX_2 \stackrel{d}{=} (a+b)X$ (i.e. $d = 0$) if and only if

$$\beta(a\log a + b\log b) = \beta(a+b)\log(a+b)$$

for any a, $b \in \mathbb{R}^+$, hence if and only if $\beta = 0$. □

Proposition 9. *Let* $X_1, X_2, \ldots, X_n$ *be i.i.d. with distribution* $S_\alpha(\sigma, \beta, \mu)$. *Then we have*

$$X_1 + X_2 + \cdots + X_n \stackrel{d}{=} n^{1/\alpha}X_1 + \mu(n - n^{1/\alpha})$$

if $\alpha \neq 1$ *and*

$$X_1 + X_2 + \cdots + X_n \stackrel{d}{=} nX_1 + \frac{2}{\pi}\sigma\beta n \log n$$

if $\alpha = 1$.

Proof. It was given in the proof of Theorem 1 for $\alpha \neq 1$, and the case $\alpha = 1$ is analogous. But it can also be proved directly, noting that $X_1 + X_2 + \cdots + X_n \sim S_\alpha(\sigma', \beta', \mu')$ with

$$\begin{aligned} \sigma' &= (\sigma^\alpha + \sigma^\alpha + \cdots + \sigma^\alpha)^{1/\alpha} \\ &= n^{1/\alpha}\sigma \\ \beta' &= \frac{\beta\sigma^\alpha + \beta\sigma^\alpha + \cdots + \beta\sigma^\alpha}{\sigma^\alpha + \sigma^\alpha + \cdots + \sigma^\alpha} \\ &= \beta \\ \mu' &= n\mu. \end{aligned}$$

Since it also holds

$$n^{1/\alpha}X_1 \sim S_\alpha(n^{1/\alpha}\sigma, \beta, n^{1/\alpha}\mu)$$

and

$$n^{1/\alpha}X_1 + \mu(n - n^{1/\alpha}) \sim S_\alpha(n^{1/\alpha}\sigma, \beta, n\mu),$$

then $n^{1/\alpha}X_1 + \mu(n - n^{1/\alpha}) \sim S_\alpha(\sigma', \beta', \mu')$, i.e. $X_1 + X_2 + \cdots + X_n \stackrel{d}{=} n^{1/\alpha}X_1 + \mu(n - n^{1/\alpha})$. □

We need to introduce here some terminology. The distribution $S_\alpha(\sigma, \beta, \mu)$ is said to be *skewed to the right* if $\beta > 0$ and *skewed to the left* if $\beta < 0$. It is said to be *totally skewed to the right* if $\beta = 1$ and *totally skewed to the left* if $\beta = -1$. Unfortunately, the terminology is confusing, since it does not refer to the support of the distribution, but to the parameters P and Q of the Lévy measure. However, in the case $\alpha < 1$ and $\beta = 1$, the support of the distribution $S_\alpha(\sigma, 1, 0)$ is actually the positive real half-line, i.e. $\mathbb{R}^+$. This result can be proved by representing the random variable $X \sim S_\alpha(\sigma, 1, 0)$ as the limit of a sequence of positive compound Poisson random variables (see Samorodnitsky and Taqqu (1994), Proposition 1.2.11).

The random variable $X \sim S_\alpha(\sigma, 1, 0)$ with $\alpha \in (0, 1)$ is called a *stable subordinator*. We will make use of subordinators in the modeling of arrival times of exchange rates in foreign exchange markets. More generally, stable subordinators are often used to model the intrinsic time in stochastically subordinated processes.

The following proposition shows how stable random variables that are totally skewed to the right can be regarded as "building blocks" of general stable variables.

Proposition 10. *Let $X \sim S_\alpha(\sigma, \beta, 0)$ with $\alpha < 2$. Then there exist two i.i.d. stable random variables Y_1, Y_2 with common distribution $S_\alpha(\sigma, 1, 0)$ such that*

$$X \stackrel{d}{=} \left(\frac{1+\beta}{2}\right)^{1/\alpha} Y_1 - \left(\frac{1-\beta}{2}\right)^{1/\alpha} Y_2$$

if $\alpha \neq 1$, and

$$X \stackrel{d}{=} \left(\frac{1+\beta}{2}\right) Y_1 - \left(\frac{1-\beta}{2}\right) Y_2 + \sigma \left(\frac{1+\beta}{\pi} \log \frac{1+\beta}{2} - \frac{1-\beta}{\pi} \log \frac{1-\beta}{2}\right)$$

if $\alpha = 1$.

Proof. Apply Propositions 6 through 8. □

As a corollary, we could show that, for $\alpha < 1$ and fixed σ, the family of distributions $S_\alpha(\sigma, \beta, 0)$ is stochastically ordered in $\beta \in [-1, 1]$, that is, if $X_\beta \in S_\alpha(\sigma, \beta, 0)$ and $\beta_1 \le \beta_2$, then $P(X_{\beta_1} \ge x) \le P(X_{\beta_2} \ge x)$. Moreover, the support of $S_\alpha(\sigma, \beta, 0)$ is $\mathbb{R}$ for $\beta \in (-1, 1)$. For a proof, see Samorodnitsky and Taqqu (1994), Proposition 1.2.14.

The following proposition gives a description of the asymptotic behavior of the tail probabilities, which is one of the key features of stable laws.

Proposition 11. *Let $X \sim S_\alpha(\sigma, \beta, \mu)$ with $\alpha \in]0, 2[$. Then we have*

$$\lim_{\lambda \to +\infty} \lambda^\alpha P(X > \lambda) = C_\alpha \frac{1+\beta}{2} \sigma^\alpha \tag{6}$$

$$\lim_{\lambda \to +\infty} \lambda^\alpha P(X < -\lambda) = C_\alpha \frac{1-\beta}{2} \sigma^\alpha \tag{7}$$

where

$$C_\alpha = \left(\int_0^\infty x^{-\alpha} \sin x \, dx\right)^{-1}.$$

Proof. We give only a sketch of the proof here. For more details, see Feller (1971), Theorem XVII.5.1.

In the case $X \sim S_\alpha(\sigma, 1, 0)$, one can show that it holds

$$Ee^{-\gamma X} = e^{-a^\alpha \gamma^\alpha}$$

with $a^\alpha = \frac{\sigma^\alpha}{\cos \frac{\pi\alpha}{2}}$. It is also possible to write

$$\int_0^\infty e^{-\gamma\lambda} P(X > \lambda) \, d\lambda = \frac{1 - Ee^{-\gamma X}}{\gamma} \approx a^\alpha \gamma^{\alpha-1}.$$

As $\gamma \to 0$, applying the Tauberian theorem cited in Feller (1971), Theorem XIII.5.4, we obtain

$$\lim_{\lambda\to\infty} P(X > \lambda) \approx \frac{a^\alpha}{\Gamma(1-\alpha)}\lambda^{-\alpha} = \sigma^\alpha C_\alpha \lambda^{-\alpha}.$$

Then we can extend this result to all the cases $\alpha < 1$, $\beta \in [-1,1]$ by means of Proposition 10. Finally, it can be shown that if (6) is valid for $\alpha < 1$, then it holds also for $\alpha \in [1,2)$: see for example Samorodnitsky and Taqqu (1994), Exercise 1.28. The derivation of (7) is very similar and follows the same reasoning. □

Since it holds $E|X|^p = \int_0^\infty P(|X|^p > \lambda)\, d\lambda$, we easily obtain the result contained in this corollary.

Corollary 2. *Let $X \sim S_\alpha(\sigma,\beta,\mu)$ with $\alpha \in (0,2)$. Then $E|X|^p$ exists for any $0 < p < \alpha$, and $E|X|^p = \infty$ for any $p \geq \alpha$.*

Since α-stable random variables with $\alpha < 2$ do not have a finite second order moment makes impossible to use the standard techniques developed for the Gaussian case $\alpha = 2$. Moreover, note that also $E|X|^\alpha$ does not exist, and if $\alpha < 1$ (as in the case of subordinators, for example), even the mean is not finite.

2.5 Multivariate Stable Laws

In analogy to the unidimensional case, an $\mathbb{R}^k$–valued random vector $\mathbf{X}$ is said to follow a multivariate stable distribution if for any positive real numbers a and b there exist a positive real number c and a vector $\mathbf{d} \in \mathbb{R}^k$ such that

$$a\mathbf{X}_1 + b\mathbf{X}_2 = c\mathbf{X} + \mathbf{d}, \tag{8}$$

where $\mathbf{X}_1$ and $\mathbf{X}_2$ are independent copies of $\mathbf{X}$.
Moreover, $\mathbf{X}$ is stable if and only if it holds

$$\mathbf{X}_1 + \mathbf{X}_2 + \cdots + \mathbf{X}_n \stackrel{d}{=} n^{1/\alpha}\mathbf{X} + \mathbf{d}_n$$

for any $n \geq 2$, with $\mathbf{X}_1, \ldots, \mathbf{X}_n$ independent copies of $\mathbf{X}$ and $\mathbf{d}_n$ vector in $\mathbb{R}^k$. A stable vector $\mathbf{X}$ is also uniquely specified by its characteristic function

$$\log Ee^{i(\mathbf{t},\mathbf{X})} = \begin{cases} -\int_{\mathbb{S}^k} |(\mathbf{t},\mathbf{s})|^\alpha \Big(1 - i\,\mathrm{sgn}((\mathbf{t},\mathbf{s}))\tan\frac{\pi\alpha}{2}\Big)\Gamma(d\mathbf{s}) + i(\mathbf{t},\boldsymbol{\mu}) & \text{if } \alpha \neq 1 \\ -\int_{\mathbb{S}^k} |(\mathbf{t},\mathbf{s})|\Big(1 + i\frac{2}{\pi}\,\mathrm{sgn}((\mathbf{t},\mathbf{s}))\log|(\mathbf{t},\mathbf{s})|\Big)\Gamma(d\mathbf{s}) + i(\mathbf{t},\boldsymbol{\mu}) & \text{if } \alpha = 1, \end{cases}$$

where $(\cdot,\cdot)$ denotes the scalar product in $\mathbb{R}^k$, Γ is a finite measure on the unit ball $\mathbb{S}^k$ of $\mathbb{R}^k$, and $\boldsymbol{\mu}$ is a shifting vector. Note that the pair $(\Gamma, \boldsymbol{\mu})$ in the expression of the multivariate stable characteristic function is unique (see Samorodnitsky and Taqqu (1994) for a proof). Moreover, the spectral measure Γ determines the dependence structures among the components of the vector $\mathbf{X}$: for example, a stable vector $\mathbf{X}$ in $\mathbb{R}^k$ has independent components if and only if its spectral measure is concentrated on the canonical base of $\mathbb{R}^k$, i.e. in $(1,0,...,0)$, $(0,1,0,\ldots,0)$, ..., $(0,\ldots,0,1)$. More generally, the spectral measure of an $\mathbb{R}^k$–valued stable vector is concentrated on a finite number of points of the unit ball $\mathbb{S}^k$ if and only if it holds

$$\mathbf{X} = A\mathbf{Y}\ ,$$

where $\mathbf{Y}$ is an $\mathbb{R}^m$–valued stable vector with independent components and A is a $k \times m$ matrix.

3 Estimation and Simulation of Stable Random Variables

There are several methods to estimate the parameters of stable distribution $S_\alpha(\sigma, \beta, \mu)$, most of which have been proposed in the past 25 years. We will not consider here estimation procedures valid only for peculiar classes of stable distributions (for example, symmetric, or totally skewed distributions), while we will discuss a quantile-based method, due to McCulloch (1986), and an approximated maximum likelihood (ML) method, originally proposed in a slightly different form by DuMouchel (1971).

Before presenting the two algorithms, we give in the following section a short account of some tail index estimation procedures that have been proposed for estimating the index of stability α.

3.1 Tail Estimators of the Stability Index α

We know from the previous chapter that if X is a stable random variable, then it holds

$$\lim_{\lambda \to \infty} \lambda^\alpha P(X > \lambda) = C_\alpha (1 + \beta)\sigma^\alpha. \tag{9}$$

This tail behavior suggests a quick estimate of α by plotting the empirical distribution function of the data on a logarithmic scale, and measuring the slope of the straight line approached by the tail. In fact, if the data are stable, then the tail should converge to a straight line with slope $-\alpha$ in logarithmic scale.

This method, although interesting for its ease, presents at least two major drawbacks: McCulloch (1997) shows that this procedure leads to overstimation of α when applied to stable data sets with $\alpha \in (1, 2)$, while Fofack and Nolan (1998) show that , for α close to 2, the tail of the distribution actually resembles the algebraic asymptotic behavior only very far out on the tail, i.e. only if the data set is very large. Moreover, they show that the point where the algebraic decay starts to appear depends heavily on the parametrization, and is a complicated function of α and β.

A better method for estimating the index of stability α is the Hill estimator, which is given by

$$\hat{\alpha} = \frac{1}{\frac{1}{k}\sum_{j=1}^{k} \log X_{n+1-j:n} - \log X_{n-k:n}} \tag{10}$$

where $X_{j:n}$ represents the j-th order statistics of the sample $X_1, X_2, \ldots, X_n$. If the right tail of the distribution shows an asymptotic algebraic behavior, as predicted by (9), then (10) gives an estimate of α, provided that we choose k appropriately. The choice of k is not very easy, and a good discussion can be found in Danielsson and de Vries (1997).

Unfortunately, also the Hill estimator tends to overstimate the index of stability α, as shown in many simulation based studies, while it performs quite well when applied to Student's t and Pareto distributions.

Several variants have been proposed, like the so called Pickands estimator (which still performs rather poorly on stable data sets), further improved by Mittnik and Rachev (1997), who proposed the so-called Modified Unconditional Pickands (MUP) tail index estimator, which is based on Bergström's asymptotic expansion of the α-stable distribution, and eliminates the dependence on k.

3.2 Quantile-Based Estimation

We describe in this section a method for estimating the parameters of a stable distribution that is based on the calculation of sample quantiles and interpolation to tabulated population quantiles. This method is essentially identical to McCulloch (1986), except for the use of a different interpolation algorithm.

Let $\{x_j\}$, $j = 1, 2, \ldots, n$ be independent samples of a stable random variable $X \sim S_\alpha(\sigma, \beta, \mu)$, with α, β, σ and μ to be estimated. Let x_p be the p-th population quantile, i.e. x_p is such that $F(x_p) = p$, where F is the distribution function of X, and let $\hat{x}_p$ be the p-th sample quantile corrected for continuity. Let us explain in details what we mean by this: if we identify the i-th order statistics with $\hat{x}_{q(i)}$, where $q(i) = \frac{2i-1}{2n}$, and an integer i^* exists such that $p = q(i^*)$, then $\hat{x}_p = \hat{x}_{q(i^*)}$. Otherwise, we can always find an $i' \geq n$ such that

$$q(i') < p \leq q(i' + 1).$$

In fact it is enough to solve for i the equation $q(i) = p$, which yields

$$i = \frac{2np+1}{2},$$

and put $i' = \lfloor i \rfloor$. We define $\hat{x}_p$ as the linearly interpolated value to $\hat{x}_{q(i')}$ and $\hat{x}_{q(i'+1)}$, that are known. Since $q(i+1) - q(i) = 1/n$ for every i, as it is easy to verify, we get

$$\hat{x}_p = n(\hat{x}_{q(i'+1)} - \hat{x}_{q(i')})(p - \hat{x}_{q(i')}).$$

Without this correction we would introduce spurious skewness in the estimated values.

It is evident that $\hat{x}_p$ is a consistent estimator of x_p.

We define now the following index:

$$\nu_\alpha = \frac{x_{0.95} - x_{0.05}}{x_{0.75} - x_{0.25}}.$$

It can be proved that ν_α depends only on α and β, while it does not depend on σ and μ. Moreover, ν_α is a strictly decreasing function of α.

McCulloch (1986) calculated ν_α as a function of α and β using DuMouchel's (1971) tables of the stable distribution.

Since $\hat{x}_p$ is a consistent estimator of x_p for every p, then the statistics

$$\hat{\nu}_\alpha = \frac{\hat{x}_{0.95} - \hat{x}_{0.05}}{\hat{x}_{0.75} - \hat{x}_{0.25}} \tag{11}$$

is a consistent estimator of the index ν_α.

Let us now define a second index

$$\nu_\beta = \frac{x_{0.95} + x_{0.05} - 2x_{0.5}}{x_{0.95} - x_{0.05}}$$

and the corresponding statistics $\hat{\nu}(\beta)$ in analogy with (11). Also ν_β does not depend on σ and μ, but it is a strictly increasing function of β.

The relationships between ν_α, ν_β and α, β can be formalized saying that there exists a one-to-one function Φ defined by

$$\begin{aligned} \Phi : \mathbb{R}^2 &\longrightarrow \mathbb{R}^2 \\ (\alpha, \beta) &\longmapsto (\nu_\alpha, \nu_\beta), \end{aligned}$$

which henceforth admits an inverse Ψ defined by

$$\begin{aligned} \Psi : \mathbb{R}^2 &\longrightarrow \mathbb{R}^2 \\ (\nu_\alpha, \nu_\beta) &\longmapsto (\alpha, \beta). \end{aligned}$$

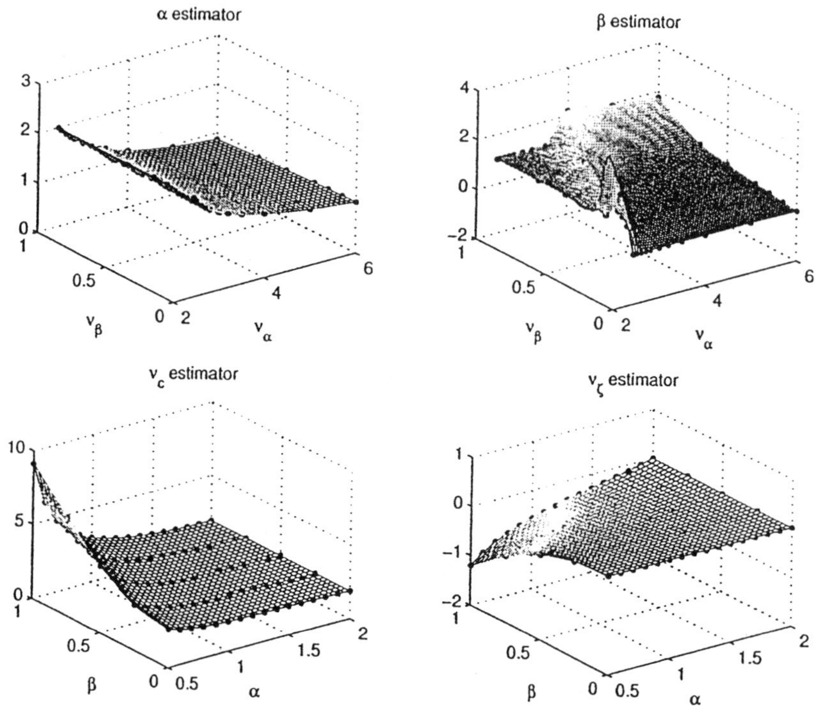

Figure 1: Quantile-based estimators

Given the tabulated values of ν_α and ν_β, we know Ψ on a discrete (finite) subset of $\mathbb{R}^2$. Written $\Psi = (\psi_1, \psi_2)$, i.e.

$$\psi_1 : \mathbb{R}^2 \longrightarrow \mathbb{R}$$
$$(\nu_\alpha, \nu_\beta) \longmapsto \alpha,$$

and analogously for ψ_2, we show a surface plot of ψ_1 and ψ_2, obtained by two-dimensional bicubic interpolation of the known tabulated values.

The estimation procedure is now clear: after calculating $\hat{\nu}_\alpha$ and $\hat{\nu}_\beta$ from the sample, we get the estimates

$$\hat{\alpha} = \hat{\psi}_1(\hat{\nu}_\alpha, \hat{\nu}_\beta)$$
$$\hat{\beta} = \hat{\psi}_2(\hat{\nu}_\alpha, \hat{\nu}_\beta)$$

where $\hat{\psi}_i$, $i = 1, 2$ are the functions obtained by interpolation of the tabulated values.

Let us define the index

$$\nu_\sigma = \frac{x_{0.75} - x_{0.25}}{\sigma}.$$

It can be shown that ν_σ does not depend on μ, and can be written as $\nu_\sigma = \phi_3(\alpha, \beta)$. Thus, since $\hat{\alpha}$, $\hat{\beta}$, $\hat{x}_{0.75}$ and $\hat{x}_{0.25}$ are all consistent estimators, we obtain the following estimator for σ:

$$\hat{\sigma} = \frac{\hat{x}_{0.75} - \hat{x}_{0.25}}{\phi_3(\hat{\alpha}, \hat{\beta})}. \tag{12}$$

Similarly as we did before, we build by two-dimensional bicubic interpolation the function $\hat{\phi}_3$, 'discretized' version of ϕ_3, and subsequently, again by the same interpolation procedure, we calculate $\hat{\phi}_3(\hat{\alpha}, \hat{\beta})$ and then plug it in (12).

In order to get a good estimate of μ, we need to introduce a new parameter, ζ, defined as

$$\zeta = \begin{cases} \mu + \beta\sigma \tan \frac{\pi\alpha}{2} & \text{for } \alpha \neq 1 \\ \mu & \text{for } \alpha = 1. \end{cases}$$

Note that the parameter ζ is μ_1 in Zolotarev's (M) parametrization (see Section 2 for more details).

Let us now define the index

$$\nu_\zeta = \frac{\zeta - x_{0.5}}{\sigma}$$

which can be shown to depend only on α and β, i.e. there exists a function ϕ_4 such that

$$\nu_\zeta = \phi_4(\alpha, \beta).$$

Following exactly the same procedure we presented for the estimation of σ, we are now able to get an estimate of ζ:

$$\hat{\zeta} = \hat{x}_{0.5} + \hat{\sigma}\phi_4(\hat{\alpha}, \hat{\beta})$$

and hence

$$\hat{\mu} = \hat{\zeta} - \hat{\beta}\hat{\sigma} \tan \frac{\pi\hat{\alpha}}{2}.$$

Asymptotic Properties of the Estimators

Since the method presented here is a modification of the original McCulloch's (1986) algorithm (we adopt two-dimensional nonlinear interpolation instead of its linear counterpart as suggested by McCulloch), it is reasonable to suppose the asymptotic behavior of the two estimators to be quite similar. We do not concentrate here on asymptotic covariances and efficiency, but instead we refer the interested reader to the original papers of Fama and Roll (1971) and McCulloch (1986). We only recall that McCulloch's estimator is consistent and asymptotically normal, with computable asymptotic standard errors.

3.3 Maximum Likelihood Estimation

Let $\{x_i\}$, $i = 1, 2, \ldots, N$ be independent samples of a stable random variable $X \sim S_\alpha(\sigma, \beta, \mu)$, with density function $f(x; \alpha, \beta, \sigma, \mu)$. It is well known that the maximum likelihood estimates of the parameters α, β, σ, μ is given by the solution to the following maximization problem:

$$\max_{\theta \in \Theta} \prod_{i=1}^{N} f(x_i; \theta)$$

where $\theta = (\alpha, \beta, \sigma, \mu)$ and $\Theta = (0, 2] \times [-1, 1] \times \mathbb{R}^+ \times \mathbb{R}$.

The main difficulty of this problem is that there is no closed form expression for the density of stable variables. Therefore it is necessary to rely on numerical approximations of the stable PDF. There are two main approaches to numerical computation of the stable density: one relies on Zolotarev's integral representation, and another one on the Fast Fourier Transform (FFT) applied to the characteristic function. From the point of view of computational complexity, it can be shown that the integration method for computing $\{f(x_i)\}_{i=1,\ldots,N}$ is of order $O(N^2)$, while its counterpart based on the FFT is of order $O(N \log N)$. While generally faster, the FFT method computes the stable PDF on a uniformly spaced grid, so it cannot calculate the value of f at a generic point $x \in \mathbb{R}$. To do so, we interpolate on the grid of values where f is known. Mittnik, Doganoglu and Chenyao (1999) have shown that the combination of FFT and linear interpolation yields virtually the same results as the integration method, therefore without any significant loss in the complexity of computing $f(x_i)$.

In the following we briefly explain how to implement the FFT method to calculate the stable PDF. Let $\psi(t)$ be the characteristic function of a standardized stable random variable X, i.e. such that $X \sim S_\alpha(1, \beta, 0)$. Then it holds

$$f(x) = \frac{1}{2\pi} \int_{\mathbb{R}} \psi(t) e^{-itx} \, dt = \frac{1}{2\pi} \hat{\psi}(x),$$

where $\hat{\psi}$ is the Fourier transform of ψ. In order to obtain an approximation of f by the FFT, we need to compute ψ on a finite grid on a finite support on $\mathbb{R}$.

Therefore, the critical choices are the length of the support (the "bandwidth"), and the number of samples, which therefore determines the sampling period. In particular, if we compute $\psi(t)$ on $(-T_p/2, T_p/2]$, then the sampling period of $f(x)$ will be $F_s = T_p^{-1}$, and if T_s is the sampling period of $\psi(t)$, then the support of $f(x)$ will have length $F_p = T_s^{-1}$. Note that if M is the number of points on the grid for ψ, i.e. $T_p = MT_s$, it will also hold $F_p = MF_s$, that is we obtain a uniformly spaced grid for $f(x)$ with M points, extending from $-F_p/2$ to $F_p/2$. The best choices for M are of the type $M = 2^Q$, as the FFT algorithms achieve their best performance when the length of the input vector is a power of two. Once f is known on a grid, $f(x)$ can be easily approximated on arbitrary values of x by linear interpolation. The ML estimation is then performed by numerical maximization of the log likelihood function

$$\log \mathcal{L}(\theta) = \sum_{i=1}^{N} \log \frac{1}{\sigma} f\left(\frac{x_i - \mu}{\sigma}\right)$$

over the parameter space $\Theta = (0,2] \times [-1,1] \times \mathbb{R}^+ \times \mathbb{R}$. As initial value one can use the quantile–based estimate discussed above. Loosely speaking, if the true parameter θ_0 is far from the boundary of Θ, an unconstrained optimization routine usually converges rapidly, while a more complex constrained optimization procedure is required when θ_0 is close to the boundary of Θ.

Another approach to stable ML estimation is represented by the STABLE program of Nolan (1999), which uses spline interpolation over a lattice of pre-computed values of the stable densitiy for different values of α, β and x. The log likelihood is then maximized by direct or gradient search.

DuMouchel proved that if the true value θ_0 belongs to the interior of the parameter space Θ, then the maximum likelihood estimator is consistent and asymptotically normal with

$$\sqrt{N}\left(\hat{\theta} - \theta_0\right) \xrightarrow{d} N(0, I(\theta_0)^{-1}),$$

where $I(\theta_0)$ is the Fisher information matrix. I can be evaluated either as

$$\hat{I}_{ij} = -N^{-1} \sum_{i=1}^{N} \frac{\partial^2 \mathcal{L}(\hat{\theta})}{\partial \theta_i \partial \theta_j}, \tag{13}$$

or by numerical approximation of

$$I_{ij} = \int_{\mathbb{R}} \frac{\partial f}{\partial \theta_i} \frac{\partial f}{\partial \theta_j} \frac{1}{f}.$$

Note that the estimate of I given in (13) can easily be obtained by the Hessian arising in the maximization of the log likelihood if we use, for example, a quasi–Newton algorithm. From the estimate of I one can derive large sample confidence intervals for each of the parameters as

$$\hat{\theta}_i \pm z_{\alpha/2} \frac{\sigma_{\hat{\theta}_i}}{\sqrt{N}},$$

where $\sigma_{\hat{\theta}_i}$ are the square roots of the diagonal entries of I^{-1}.

Unfortunately, when θ_0 is near, or even worse on the boundary of Θ, the behavior of the estimator is unknown. So far, it is only known that if θ_0 is on the boundary of Θ, then the asymptotic distribution of the estimators tends to a degenerate distribution and the maximum likelihood estimators are super–efficient, in the sense that they have a zero asymptotic standard deviation.

Maximum likelihood estimation, apart of providing asymptotic efficient estimates, can also be extended to more sophisticated stable models, where, for instance, we assume only the conditional distribution of the data to be stable. An example of such models is the stable GARCH(1,1) process specified by

$$\begin{aligned} x_t &= \mu + c_t \varepsilon_t \\ c_t^\delta &= \omega + a|\varepsilon_t|^\delta + b c_{t-1}^\delta, \end{aligned}$$

where ε_t are i.i.d. with distribution $S_\alpha(1, \beta, 0)$. Mittnik et al. showed that this model has a unique stationary solution if $\alpha \in (1, 2]$, $\delta \in (0, \alpha)$, $\omega > 0$, $a \geq 0$, $b \geq 0$, and

$$E|\varepsilon_t|^\delta a + b \leq 1,$$

where $E|\varepsilon_t|^\delta$ admits a closed–form representation $\lambda(\alpha, \beta, \delta)$.

The likelihood function in this case is given by

$$L(\theta) = \prod_{t=1}^{N} c_t^{-1} f\left(\frac{x_t - \mu}{c_t}\right),$$

where f is the density of a standardized stable variable and

$$\theta = (\mu, c_0, \omega, a, b, \alpha, \beta, \delta).$$

ML estimation of the stable GARCH(1,1) model is then obtained by numerical maximization of $\mathcal{L}(\theta) = \log L(\theta)$, subject to the constraints on θ guaranteeing the existence of a stationary solution. The problem is therefore significantly more complicated in terms of dimension and constraints, and does not have either an easy choice of initial estimates. However, Mittnik, Paolella and Rachev

(1999) have shown, through a Monte Carlo study, that whenever $\alpha > 1.4$, $|\beta| < 0.7$, $|\mu| < 0.2$, $\omega > 0$, $a > 0$, and $b > 0.2$, an unconstrained optimization routine (namely, the BFGS algorithm) is insensitive to the initial values and converges relatively fast.

3.4 Simulation of Stable Random Variates

It is necessary in empirical Monte Carlo simulations of stable models to have a fast generator of pseudorandom stable variates. Unfortunately, there are no analytic expression of the inverse of the stable distribution function, except for a few cases, and therefore the inverse transform method cannot be applied.

Chamber, Mallows and Stuck (1976) developed a method to simulate stable random variables based on an integral representation of the stable density function due to Zolotarev (1986). More recently, Weron (1996) has proved the equality in distribution of a general stable random variables with a nonlinear transformation of a uniform and an exponential random variable, independent of eachother.

We need to define another parametrization of the stable characteristic function, that was originally introduced by Zolotarev (1986) for its analytic properties. In particular, a random variable X is stable if (and only if) its characteristic function admits the following parametrization:

$$Ee^{iXt} = \begin{cases} \exp\Big(- \sigma_2^\alpha |t|^\alpha \exp\Big(- i\beta_2 \operatorname{sgn}(t) \frac{\pi}{2} K(\alpha)\Big) + i\mu t\Big) & \text{for } \alpha \neq 1 \\ \exp\Big(- \sigma_2 |t| \Big(\frac{\pi}{2} + i\beta_2 \operatorname{sgn}(t) \log|t|\Big) + i\mu t\Big) & \text{for } \alpha = 1, \end{cases} \tag{14}$$

where

$$K(\alpha) = \alpha - 1 + \operatorname{sgn}(1-\alpha) = \begin{cases} \alpha & \text{for } \alpha < 1 \\ \alpha - 2 & \text{for } \alpha > 1. \end{cases}$$

While the stability index α and the location parameter μ coincide in this and the S parametrization, σ_2 and β_2 are related to the corresponding σ and β of the S parametrization by the following relations:

$$\sigma_2 = \sigma\left(1 + \beta^2 \tan^2 \frac{\pi\alpha}{2}\right)^{1/(2\alpha)}$$

$$\tan\left(\beta_2 \frac{\pi K(\alpha)}{2}\right) = \beta \tan \frac{\pi\alpha}{2},$$

for $\alpha \neq 1$, and

$$\sigma_2 = \frac{2}{\pi}\sigma$$

$$\beta_2 = \beta,$$

for $\alpha = 1$. The fundamental result underlying the simulation method we describe here is contained in the following theorem.

Theorem 4. *Let γ be a uniformly distributed random variable on $(\pi/2, \pi/2)$, W an exponential random variable with mean 1 independent of γ, and*

$$\gamma_0 = -\frac{\pi}{2}\beta_2\frac{K(\alpha)}{\alpha}.$$

Let us define a random variables X as follows:

$$X = \begin{cases} \frac{\sin\alpha(\gamma-\gamma_0)}{(\cos\gamma)^{1/\alpha}}\left(\frac{\cos(\gamma-\alpha(\gamma-\gamma_0))}{W}\right)^{\frac{1-\alpha}{\alpha}} & \text{for } \alpha \neq 1 \\ \left(\frac{\pi}{2} + \beta_2\gamma\right)\tan\gamma - \beta_2\log\left(\frac{W\cos\gamma}{\frac{\pi}{2}+\beta_2\gamma}\right) & \text{for } \alpha = 1. \end{cases}$$

Then X is stable with parameters α, β_2, $\sigma_2 = 1$, $\mu = 0$ in the parametrization (14).

For a proof, see Weron (1996).

Relying on this result, we can derive the following algorithm for simulating stable random variables in the usual S parametrization:

1. Simulate a random variable γ uniformly distributed on $(-\pi/2, \pi/2)$, and a standard exponential random variable W, independent of γ.

2. If $\alpha \neq 1$, compute

$$X = S_{\alpha,\beta}\frac{\sin\alpha(\gamma + B_{\alpha,\beta})}{(\cos\gamma)^{1/\alpha}}\left(\frac{\cos\left(\gamma - \alpha(\gamma + B_{\alpha,\beta})\right)}{W}\right)^{\frac{1-\alpha}{\alpha}},$$

where

$$B_{\alpha,\beta} = \frac{\arctan\left(\beta\tan\frac{\pi\alpha}{2}\right)}{\alpha}$$

and

$$S_{\alpha,\beta} = \left(1 + \beta^2\tan^2\frac{\pi\alpha}{2}\right)^{\frac{1}{2\alpha}}.$$

If $\alpha = 1$, compute

$$X = \frac{2}{\pi}\left(\left(\frac{\pi}{2} + \beta\gamma\right)\tan\gamma - \beta\log\frac{\frac{\pi}{2}W\cos\gamma}{\frac{\pi}{2} + \beta\gamma}\right).$$

Note that $B_{\alpha,\beta}$ accounts for the skewness parameter change, and $S_{\alpha,\beta}$ for the scale parameter change.

The resulting X will be distributed like $S_\alpha(1, \beta, 0)$, therefore we proceed as follows.

3. For arbitrary $\sigma \in \mathbb{R}^+$ and $\mu \in \mathbb{R}$, the random variable

$$Y = \begin{cases} \sigma X + \mu & \text{for } \alpha \neq 1 \\ \sigma X + \frac{2}{\pi}\beta\sigma \log \sigma + \mu & \text{for } \alpha = 1 \end{cases}$$

will be $S_\alpha(\sigma, \beta, \mu)$-distributed.

The original Fortran routine `rstab` listed in Chambers, Mallows and Stuck (1976) and its updated version included in S-Plus generate standard α-stable pseudorandom variates in Zolotarev's (M) parametrization, not in the usual S parametrization. Consequently, they generates pseudorandom variates X distributed like

$$X \sim \begin{cases} S_\alpha(1, \beta, -\beta \tan \frac{\pi\alpha}{2}) & \text{for } \alpha \neq 1 \\ S_\alpha(1, \beta, 0) & \text{for } \alpha = 1. \end{cases}$$

A further updated version of `rstab` is listed in Samorodnitsky and Taqqu (1994), while a Matlab implementation of Weron's (1996) algorithm, simulating general $S_\alpha(\sigma, \beta, \mu)$-distributed random variables, is available from the authors.

4 Stable Processes

In this section we define stable stochastic processes and give a few simple examples, that extend to the stable setting some well–known Gaussian processes.

4.1 Definition

Definition 2. A stochastic process $\{X(t), t \in T\}$ is stable if its finite-dimensional distributions

$$\Big(X(t_1), X(t_2), \ldots, X(t_n)\Big)$$

are stable for any $n \in \mathbb{N}$ and $t_1, t_2, \ldots, t_n \in T$.

It can be proved that the finite-dimensional distributions of a stable process must have the same index of stability α. Moreover, if $\alpha \in [1, 2]$, then a necessary a sufficient condition for $X(t)$ to be α-stable is that all linear combinations of the type

$$\sum_{i=1}^{n} c_i X(t_i)$$

are α-stable for any $n \in \mathbb{N}$, $t_1, t_2, \dots, t_n \in T$, and $c_1, c_2, \dots, c_n \in \mathbb{R}$.

4.2 Stable Motions

The simplest example of a stable process is the symmetric α-stable Lévy motion, i.e. a stochastic process $X(t)_{t \in \mathbb{R}_0^+}$ that satisfies the following conditions:

1. $X(0) = 0$ almost surely;
2. $X(t)$ has independent increments;
3. $X(t) - X(s) \sim S_\alpha\Big(|t-s|^{1/\alpha}, 0, 0\Big)$ for some $\alpha \in]0, 2]$.

Note that the Lévy motion $X(t)$ is the most natural extension to the stable setting of the Brownian motion, which can be defined as a Gaussian process $B(t)$ with mean 0 and autocovariance function

$$EX(t)X(s) = \min(t, s).$$

In fact, the increments of the α-stable Lévy motion $X(t)$ are stationary, and $X(t)$ is a Brownian motion if $\alpha = 2$.

Brownian motion and α-stable Lévy motion are also the simplest examples of self-similar processes in the Gaussian and stable settings respectively. The basic idea of self-similarity for stochastic processes can be seen as the invariance in distribution under suitable scaling of time scale. Self-similar processes occur in a natural way, among other disciplines, in probability theory (in connection to limit theorems) and physics (renormalization groups of statistical mechanics).

A formal definition of self-similarity for stochastic processes reads as follows.

Definition 3. The $\mathbb{R}$-valued stochastic process $X(t)_{t \in \mathbb{R}}$ is said to be self-similar with index $H > 0$ if, for all $c > 0$, the finite-dimensional distributions of $X(ct)_{t \in \mathbb{R}}$ are identical to the finite-dimensional distributions of $c^H X(t)_{t \in \mathbb{R}}$. That is, if, for all $c > 0$ and for any $N > 1$, $t_1, t_2, \dots, t_N \in \mathbb{R}$ it holds

$$\Big(X(ct_1), X(ct_2), \dots, X(ct_N)\Big) \stackrel{d}{=} \Big(c^H X(t_1), c^H X(t_2), \dots, c^H X(t_N)\Big).$$

We will often adopt the simpler notation $X(ct)_{t\in\mathbb{R}} \stackrel{d}{=} c^H X(t)_{t\in\mathbb{R}}$.

To see that a Brownian motion $B(t)$ is a self-similar process, note that for all $c > 0$ and $t, s \in \mathbb{R}$ it holds

$$EB(ct)B(cs) = \min(ct, cs) = c\min(t,s) = E\Big(c^{1/2}B(t)\Big)\Big(c^{1/2}B(s)\Big),$$

that is $B(t)$ is self-similar with index $H = 1/2$.

Similarly, an α-stable Lévy motion $X(t)$ is self-similar with index $H = 1/\alpha$. In fact, since this process has independent increments, it is sufficient to show that $X(ct) - X(cs) \stackrel{d}{=} c^H\Big(X(t) - X(s)\Big)$:

$$X(ct) - X(cs) \sim S_\alpha\Big(c^{1/\alpha}|t-s|^{1/\alpha}, 0, 0\Big) = c^{1/\alpha}S_\alpha\Big(|t-s|^{1/\alpha}, 0, 0\Big)$$

i.e. $X(ct) - X(cs) \stackrel{d}{=} c^H\Big(X(t) - X(s)\Big)$ with $H = 1/\alpha$.

Moreover, Brownian motion and α-stable Lévy motion share the property of having stationary increments. Formally, an $\mathbb{R}$-valued stochastic process $X(t)_{t\in\mathbb{R}}$ is said to have stationary increments if

$$\Big(X(t+h) - X(t)\Big)_{t\in\mathbb{R}} \stackrel{d}{=} \Big(X(t) - X(0)\Big)_{t\in\mathbb{R}}$$

for all $h \in \mathbb{R}$. So $B(t)$ and $X(t)$ are so-called H-s.s.s.i. processes, i.e. self-similar with index H and with stationary increments. In particular, $H = 1/2$ for Brownian motion, and $H = 1/\alpha$ for α-stable motion.

In general, it can be proved that for a non-degenerate α-stable H-s.s.s.i. process, it holds:

$$\alpha < 1 \Longrightarrow H \in]0, 1/\alpha]$$
$$\alpha \geq 1 \Longrightarrow H \in]0, 1].$$

Moreover, it is possible to show that for each permissible pair (H, α) we can find an α-stable H-s.s.s.i. process, and that for fixed α and H there are often several such processes. However, in the Gaussian case, i.e. for $\alpha = 2$, the only H-s.s.s.i. process is fractional Brownian motion (FBM).

The following proposition gives a characterization of fractional Brownian motion.

Proposition 12. *Let $H \in (0,1)$ and $\sigma^2 = EX(1)^2$. The following statements are equivalent:*

1. *$X(t)$ is Gaussian and H-s.s.s.i.*

2. $X(t)$ is a fractional Brownian motion with self-similarity index H.

3. $X(t)$ is Gaussian, has mean zero and autocovariance function

$$\Sigma(t_1, t_2) = \frac{1}{2}\sigma^2\Big(|t_1|^{2H} + |t_2|^{2H} - |t_1 - t_2|^{2H}\Big).$$

Note that FBM reduces to Brownian motion when $H = 1/2$, and that for $H = 1$ it degenerates ($X(t) = tX(1)$ a.s. $\forall t \in \mathbb{R}$). Moreover, when $H > 1/2$, the increment process of FBM (which are usually called fractional Gaussian noise) exhibits long-range dependence, that is its autocorrelation function r is such that $\sum_{k\in\mathbb{Z}} r(k) = \infty$. Long-range dependence, or long memory, is a characteristic displayed by observed data in many fields, ranging from geophysics to network traffic analysis.

In the more general setting of α-stable processes, there are several ways to generalize fractional Brownian motion. A process which admits a particularly simple representation is the so-called well-balanced linear fractional stable motion, which can be written as

$$X(t) = \int_{\mathbb{R}} \left(|t - x|^{H-1/\alpha} - |x|^{H-1/\alpha}\right) M(dx),$$

where $H \in (0, 1)$, $H \neq 1/\alpha$, and M is a symmetric α-stable measure. The well-balanced LFSM is self-similar with index H and has stationary increments. Moreover, it serves as a paradigm for long-range dependent stable processes: in fact, in analogy to the Gaussian case, we say that the increments of $X(t)$ are long-range dependent if $H > 1/\alpha$. For more details, we refer to Samorodnitsky and Taqqu (1994).

5 Two Applications

In this section we present a couple of applications of stable modeling in econometrics and mathematical finance. In particular, we study the problem of univariate regression with stable errors, and derive a formula for option pricing when the price of the underlying is described by a stable Lévy motion.

5.1 Regression with Stable Disturbances

Let the process Y_t belong to the class

$$Y_t = \beta_0 + \beta_1 X_t + U_t$$

with $(X_t)_{t=1,\dots,n}$ nonstochastic regressors and U_t i.i.d. random variables in the domain of attraction of symmetric α-stable random variables, i.e. having characteristic function $\psi(t) = e^{-|\sigma t|^{\alpha}}$ with $\alpha \in (0,2]$. The Ordinary Least Squares (OLS) estimators for β_0 and β_1 are defined by

$$\hat{\beta}_{1n} = \frac{n\sum_{t=1}^{n} X_t Y_t - \sum_{t=1}^{n} X_t \sum_{t=1}^{n} Y_t}{n\sum_{t=1}^{n} X_t^2 - (\sum_{t=1}^{n} X_t)^2}$$

and

$$\hat{\beta}_{0n} = \frac{1}{n}(\sum Y_t - \hat{\beta}_{1n}\sum X_t).$$

It can be shown that, if $\alpha > 1$, the estimators are consistent, although they are inefficient in comparison with Maximum Likelihood (ML) estimators. Moreover, the corresponding OLS residuals

$$\hat{U}_t^{(n)} = Y_t - \hat{\beta}_{0n} - \hat{\beta}_{1n} X_t$$

are consistent for all α (see McCulloch (1998)).

Since the disturbances U_t are distributed like $S_\alpha(\sigma,0,0)$, we also need an estimator of σ: let us define the quantities:

$$C(p,\alpha) = 2^{p+1}\frac{\Gamma(\frac{p+1}{2})\Gamma(\frac{-p}{\alpha})}{\alpha\sqrt{\pi}\Gamma(\frac{-p}{2})}$$

$$\hat{m}_p = \frac{1}{n}\sum_{i=1}^{n}|\hat{U}_i|^p$$

$$\hat{\sigma}_p = \left(\frac{\hat{m}_p}{C(p,\alpha)}\right)^{1/p},$$

where $p \in]0,\alpha[$, $\Gamma(x) = \int_0^\infty t^{x-1}e^{-t}\,dt$ is the Gamma function and $\hat{m}_p$ is the sample absolute p-th moment. By the Law of Large Numbers, as $n \to \infty$, it holds

$$\hat{\sigma}_p \xrightarrow{d} \sigma$$

(see Samorodnitsky and Taqqu (1994).)
Furthermore, the following result holds.

Theorem 5. *Let the regressors X_t satisfy some mild regularity conditions, the coefficients β_0, β_1 be independent of time t, $p \in]0,\alpha[$, and define the stochastic process*

$$\hat{B}_{p,n}(x) = \frac{1}{\hat{\sigma}_p n^{1/\alpha}}\sum_{t=1}^{\lfloor nx \rfloor}\hat{U}_{t,n}$$

on $x \in [0,1]$. Then, in the Skorohod space $D[0,1]$, it holds, as n tends to ∞,

$$\left(\hat{B}_{p,n}(x)\right)_{x\in[0,1]} \xrightarrow{w} \left(LB_\alpha(x)\right)_{x\in[0,1]},$$

where $LB_\alpha(x)$ is an α-stable Lévy bridge, and "$\xrightarrow{w}$" stands for weak convergence.

For a proof, see Kim, Mittnik and Rachev (1997).

An α-stable Lévy bridge is the solution of the following stochastic differential equation

$$LB_\alpha(x) = \int_0^x \frac{LB_\alpha(s)}{s-1} + \int_0^x dL_\alpha(s)$$

where $L_\alpha(s)$ is a standard α-stable Lévy motion. For an account of this stochastic differential equation, we refer to Protter (1990) and Janicki and Weron (1994).

We recall that the Skorohod space $D[0,1]$ is the set of right-continuous real function on $[0,1]$ with left-hand limits (cádlág functions), endowed with the so-called Skorohod metric. For further details, see Billingsley (1968).

An improvement of the previous result that does not depend on the stability index α, is stated in the following theorem.

Theorem 6. *Let $\xi \in [0,1]$, $n \geq 1$, and let us define the sequence of processes $X_n(\xi)$ as*

$$X_n(\xi) = \frac{\sum_{i=1}^{[n\xi]} U_i^{(n)}}{\left(\sum_{i=1}^{n} (U_i^{(n)})^2\right)^{1/2}}.$$

Let the process $X_\infty(\xi)$ be defined as

$$X_\infty(\xi) = \frac{\sum_{i=1}^{\infty} \delta_i \Gamma_i^{-1/\alpha}\left(\mathbf{1}_{\{V_i\leq\xi\}} - \xi\right)}{\left(\sum_{i=1}^{\infty} \Gamma_i^{-2/\alpha}\right)^{1/2}},$$

where the sequences V_i, δ_i and Γ_i are independent, V_i are uniformly distributed on $[0,1]$, Γ_i are the arrivals of a standard Poisson process and $\delta_i \in \{1,-1\}$ with $P(\delta_i = 1) = p$, $p = \lim_{t\to\infty} \frac{P(U_i>t)}{P(|U_i|<t)}$. If $\alpha \in (1,2)$ and $EU_1 = 0$, or if $\alpha = 1$ and $\lim_{n\to\infty} \int_{-n}^{n} x\, dF_{U_1}(x) = 0$, then, in the Skorohod space $D[0,1]$, it holds

$$X_n(\xi) \xrightarrow{w} X_\infty(\xi)$$

as $n \to \infty$, where "$\xrightarrow{w}$" stands for weak convergence.

For a proof see Mittnik, Rachev and Samorodnitsky (1998).

5.2 Stable Option Pricing

One of the most important topics in Mathematical Finance is derivative pricing, which includes as a fundamental case option pricing. Recall that an option is a contract defined as the right to buy (*call option*) or sell (*put option*) an underlying asset within a predetermined period of time, at a fixed price. The most important characteristics of an options are:

- the maturity (or expiration) time T, that is the latest admissible time to exercise the option;[a]
- the exercise (or strike) price K, i.e. the price at which the option holder can buy or sell the underlying asset;
- the price P to be paid to buy the option itself, i.e. to acquire the right to buy or sell an underlying asset.

Note that the price P is the only parameter of the option allowed to change over time.

Black–Scholes Theory

The classical Black–Scholes model is based on the assumption that the underlying stock price process S_t follows the stochastic differential equation

$$dS_t = \mu S_t\,dt + \sigma S_t\,dB_t, \tag{15}$$

where r is the risk–free interest rate, B_t is a standard Brownian motion, μ is the expected rate of return of the underlying, and σ is the volatility. As a consequence, this model assumes (log) price increments to be independent, stationary and normally distributed.

The value of a call option with maturity T and strike price K at time T is given by

$$C_T = \max(S_T - K, 0),$$

while the value of a put option with the same expiration time and strike price at time T would be

$$P_T = \max(K - S_T, 0).$$

Here we consider the problem of calculating the price of a Europan call option only, being the put option pricing completely similar, while we refer to specialized texts such as Duffie (1996) or Hull (1999) for the problem of American and path-dependent option pricing.

---

[a]Recall that so-called *European* options can be exercised only at time T, while *American* options can be exercised at any time before or at time T.

Let $V_t = V(S_t, t)$ be the price of an option on the underlying asset with price S_t. Then the Black–Scholes equation

$$\frac{\partial V}{\partial t} + \frac{1}{2}\sigma^2 S_t^2 \frac{\partial^2 V}{\partial S_t^2} + r S_t \frac{\partial V}{\partial S_t} - r V_t = 0. \tag{16}$$

admits the following closed-form solution for the price of a European call option:

$$\begin{aligned} C_t = {} & S_t \Phi\left(\frac{1}{\sigma\sqrt{T-t}} \log \frac{S_t}{Ke^{-r(T-t)}} + \frac{1}{2}\sigma\sqrt{T-t}\right) \\ & - Ke^{-r(T-t)} \Phi\left(\frac{1}{\sigma\sqrt{T-t}} \log \frac{S_t}{Ke^{-r(T-t)}} - \frac{1}{2}\sigma\sqrt{T-t}\right). \end{aligned}$$

Note that in this expression the drift μ does not appear, i.e. the only influential parameter in the diffusion model on which the Black and Scholes theory relies is the volatility σ.

In general, there are no closed-form solution to the problem of European option pricing if the underlying asset price S_t follows a different stochastic differential equation than (15).

α-Stable Option Pricing

It would be intuitively appealing to model asset prices with a diffusion-like equation, where we replace the Brownian motion integrator with an α-stable Lévy process:

$$dS_t = \mu S_t \, dt + \sigma S_t \, dL_t^{(\alpha)},$$

and use the same no-arbitrage arguments as in the original Black and Scholes paper. However, this approach is unfeasible, and does not lead to a closed-form solution of the problem.

Hurst, Platen and Rachev (1999) provide an elegant solution for the valuation of European options on an asset whose price process is driven by a stable motion. In their approach, the asset price S_t is generated by a process of the type

$$\log S(t) - \log S(t_0) = \mu(t - t_0) + \rho(T(t) - T(t_0)) + \sigma\big(Z(t) - Z(t_0)\big), \tag{17}$$

where the driving process $Z(t)$ is a subordinated process specified by

$$Z(t) = W \circ T\,(t) = W(T(t)). \tag{18}$$

In (18), $W(t)$ is a standard Brownian motion and $T(t)$ is a scaled $\alpha/2$-stable subordinator, i.e. $T(t)$ is a Lévy process with independent increments distributed like

$$T(t) - T(s) \sim S_{\alpha/2}\Big(c|t-s|^{\alpha/2}, 1, 0\Big)$$

where c is chosen so that increments are closed under convolution, i.e.

$$c = 2\cos\left(\frac{\pi\alpha}{4}\right)^{2/\alpha}.$$

Note that increments are non-negative, since the support of the stable distribution with stability index $\alpha < 1$, skewness $\beta = 1$ and $\mu = 0$ is $\mathbb{R}_0^+$.

Under these assumptions on $W(t)$ and $T(t)$, the compound process $Z(t) = W(T(t))$ turns out to be a standard symmetric α-stable Lévy process, with

$$Z(t) - Z(s) \sim S_\alpha\Big(|t-s|^{1/\alpha}, 0, 0\Big).$$

If we think of $W(t)$ as indexed by the physical time t, the process $T(t)$ can be interpreted as a stochastic deformation on the time scale indexing the process $W(t)$. The process $T(t)$ is usually called the intrinsic time process, or the market time process. Note that $Z(t)$ is indexed on the physical time and exhibits heavy-tailed increments, while its increments indexed on the intrinsic time are Gaussian.

This construction of $T(t)$ translates in mathematical terms the observation that market time (i.e. the time at which the markes operates) evolves at different rates. In fact, as Clark (1973) noted, the flow of information available to traders is nonuniform in time, and the "speed" at which the price process evolves depends on the "quantity" of new information arriving: more news usually imply more intense trading and faster price movements.

For a treatment of subordinated stable processes in a more general setting, see Samorodnitsky and Taqqu (1994), and Hurst, Platen and Rachev (1997).

In (17), μ is the drift in the physical time scale, ρ is the drift in the intrinsic time scale, and $\sigma > 0$ is the volatility. We also assume $t_0 \in [0, t]$. Note that the logarithmic price $L(t) = \log S(t)$ can be written, after the normalization $S(t_0) = 1$ (without loss of generality), as

$$L(t) = \mu t + \rho T(t) + \sigma Z(t).$$

Therefore, the log price $L(t)$ itself is an α-stable Lévy process, its increments are stationary and independent with

$$L(t) - L(s) \sim S_\alpha\Big(\sigma|t-s|^{1/\alpha}, 0, \mu|t-s|\Big).$$

Note that the increments of the log price (i.e. the returns) do not admit moments of order equal or grater than the index of stability α.

This model is known in the literature as *log-stable*. It can be shown that for $\alpha \to 2$ the intrinsic time process $T(t)$ asymptotically tends to the physical time, leading us back to the classical lognormal model.

In this setting, the value at time t of a European call option with exercise price K and time to maturity τ is

$$C_t = S_t F_-\left(\log \frac{S_t}{\overline{K}_{r,t,\tau}}\right) - \overline{K}_{r,t,\tau} F_+\left(\log \frac{S_t}{\overline{K}_{r,t,\tau}}\right),$$

where

$$\overline{K}_{r,t,\tau} = K e^{-r(\tau - t)},$$

r is the risk-free interest rate,

$$F_\pm(x) = \int_0^\infty \Phi\left(\frac{x \mp \frac{1}{2}y}{y^{1/2}}\right) dF_Y(y),$$

σ is the volatility, Φ is the standard normal distribution function, and F_Y is the distribution function of the random variable $Y = \sigma^2(T(\tau) - T(t))$. Note that it holds $Y = \lambda V$, with

$$\lambda = 2\sigma^2 \cos\left(\frac{\pi\alpha}{4}\right)^{2/\alpha} (\tau - t)^{2/\alpha}$$

and $V \sim S_{\alpha/2}(1,1,0)$. Therefore $F_\pm(x)$ admits the representation

$$F_\pm(x) = \int_0^\infty \Phi\left(\frac{x \mp (1/2)\lambda v}{\sqrt{\lambda v}}\right) f_V(v)\, dv, \tag{19}$$

where f_V is the density function of V. Note that both f_V (which can be approximated either using the FFT or the integration methods discussed above) and the normal distribution Φ are continuous functions, hence the integral in (19) can be numerically calculated rather easily.

Note that the European call option price at time t depends on the asset price S_t, the strike price K, the time to maturity τ, the risk-free interest rate r, the volatility σ and the index of stability α of the driving process $Z(t)$.

6 Conclusion

Stable laws and processes represent the most natural way to generalize Gaussian models, as suggested by the central limit theorem. Apart of accounting for heavy tails, stable models can also exhibit rich dependence structures, making them far more flexible their Gaussian counterpart. On the other hand, stable laws, and thus processes, have often been considered too difficult to deal with, as they do not admit simple analytical representations. However, the wide and cheap availability of computing power has made them more attractive, as demonstrated by the growing interest for stable models in engineering, physics, and other sciences.

We have briefly described here how to estimate simple univariate stable models, and how to solve two classical problems in statistics and mathematical finance in the stable setting. We believe that there is still a great potential to be discovered in stable models, especially in the multivariate case, and that their attractive theoretical properties, combined with the fast development of hardware and computational techniques, will favor their adoption in the applied sciences and among practitioners.

References

1. Aaronson, J. and Denker, M.. Characteristic functions of random variables attracted to 1–stable laws. *Annals of Probability*, 26, 399–415, 1998.

2. R. Adler, R. *The Geometry of Random Fields.* Wiley & Sons, New York, 1981.

3. Adler, R., Feldman, R., and Taqqu, M. S. *A Practical Guide to Heavy Tails - Statistical Techniques and Applications.* Birkäuser, Boston, 1998.

4. Akgiray, V. and Lamoureux, C. G. Estimation of stable-law parameters: A comparative study. *Journal of Business and Economic Statistics*, **7**, 85–93, 1989.

5. Bachelier, L. Théorie de la Spéculation. *Annales de l'Ecole Normale Supérieure*, **3**, 1900. English translation in Cootner (1964).

6. Baillie, B. T. Long memory processes and fractional integration in econometrics. *Journal of Econometrics*, **73**, 5–59, 1996.

7. Beirlant, J., Vynckier, P., and Teugels, J. L. Tail Index Estimation, Pareto Quantile Plots, and Regression Diagnostics. *Journal of the American Statistical Association*, **91**, 1659–1667, 1996.

8. Beran, J. *Statistics for Long–Memory Processes.* Chapman and Hall, New York, 1994.

9. Bertoin, J. *Lévy Processes.* Number 121 in Cambridge Tracts in Mathematics. Cambridge UP, 1998.

10. Billingsley, P. *Convergence of Probability Measures.* Wiley & Sons, New York, 1968.

11. Black, F. and Scholes, M. The pricing of options and corporate liabilities. *Journal of Political Economics*, **3**, 637–654, 1973.

12. Bochner, S. *Harmonic analysis and the theory of probability.* University of California Press, Berkeley, 1955.

13. Bollerslev, T. Generalized Autoregressive Conditional Heteroscedasticity. *Journal of Econometrics*, **31**, 307–327, 1986.

14. Bollerslev, T., Chou, R., and Kroner, K. ARCH Modeling in Finance: A Review of Theory and Empirical Evidence. *Journal of Econometrics*, **52**, 5–59, 1992.

15. Borodin, A. N. and Ibragimov, I. A. *Limit Theorems for Functionals of Random Walks*, volume 195 of *Proceedings of the Steklov Institute of Mathematics.* American Mathematical Society, 1995.

16. Bouleau, N. and Lépingle, D. *Numerical Methods for Stochastic Processes.* Wiley & Sons, New York, 1994.

17. Brachet, M., Taflin, E., and Tcheou, J. M. Scaling transformation and probability distributions for financial time series. Technical report, Laboratoire de Physique Statistique, CNRS URA 1306, France, 1997.

18. Breiman, L. *Probability.* Addison-Wesley, Reading, MA, 1968.

19. Brockwell, P. J. and Davis, R. A. *Time Series: Theory and Methods.* Springer, New York, 2 edition, 1991.

20. Buckle, D. J. Bayesian Inference for Stable Distributions. *Journal of the American Statistical Association*, **90**, 605–613, 1995.

21. Campbell, J. Y., Lo, A. W., and MacKinlay, A. C. *The Econometrics of Financial Markets.* Princeton UP, Princeton, NJ, 1997.

22. Chamberlain, G. A Characterization of the Distributions that Imply Mean Variance Utility Functions. *Journal of Economic Theory*, **29**, 985–988, 1983.

23. Chambers, J. M., Mallows, C. L., and Stuck, B. W. A Method for Simulating Stable Random Variables. *Journal of the American Statistical Association*, **71**, 340–344, 1976.

24. Chan, G. and Wood, A. T. A. Simulation of Multifractional Brownian Motion. Technical report, Department of Statistics, University of New South Wales, 1998.

25. Cheng, B. N. and Rachev, S. T. Multivariate Stable Future Prices. *Mathematical Finance*, **5**, 133–153, 1995.

26. Connor, G. A Unified Beta Pricing Theory. *Journal of Economic Theory*, **34**, 13–31, 1984.

27. Cutland, N. J., Kopp, P. E., and Willinger, W. Stock price returns and the Joseph effect: a fractional version of the Black–Scholes model. In E. Bolthausen, M. Dozzi, and F. Russo, editors, *Seminar on Stochastic Analysis, Random Fields, and Applications*, pages 327–351, Boston, 1995. Birkäuser.

28. Daníelsson, J. and de Vries, C. G. Tail index and quantile estimation with very high frequency data. *Journal of Empirical Finance*, **4**, 241–257, 1997.

29. Devroye, L. *Non-uniform Random Variate Generation.* Springer, New York, 1986.

30. DuMouchel, W. H. *Stable distributions in statistical inference.* PhD thesis, University of Ann Arbor, Ann Arbor, MI, 1971.

31. DuMouchel, W. H. On the Asymptotic Normality of the Maximum-Likelihood Estimate when Sampling from a Stable Distribution. *Annals of Statistics*, **1**, 948–957, 1973.

32. DuMouchel, W. .H. Stable Distributions in Statistical Inference 2: Information From Stably Distributed Samples. *Journal of the American Statistical Association*, **70**, 386–393, 1975.

33. DuMouchel, W. H. Estimating the stable index α in order to measure tail thickness: a critique. *Annals of Statistics*, **11**, 1019–1031, 1983.

34. Evertsz, C. J. G. Fractal geometry of financial time series. *Fractals*, **3**, 609–616, 1995.

35. Fama, E. The behavior of stock market prices. *Journal of Business*, **38**, 34–105, 1965.

36. Fama, E. Risk, Return and Equilibrium. *Journal of Political Economy*, **78**, 30–55, 1970.

37. Fama, E. and Roll, R. Some Properties of Symmetric Stable Distributions. *Journal of the American Statistical Association*, **63**, 817–836, 1968.

38. Fama, E. and Roll, R. Parameter Estimates for Symmetric Stable Distributions. *Journal of the American Statistical Association*, **66**, 331–339, 1971.

39. Feller, W. *An Introduction to Probability Theory and Its Applications*, volume 2nd. Wiley, New York, 2nd edition, 1971.

40. Fofack, H. and Nolan, J. P. Tail Behavior, Modes and Other Characteristics of Stable Distributions, 1998. Preprint.

41. Gamrowski, G. and Rachev, S. T. Stable Models in Testable Asset Pricing. In *Approximation, Probability and Related Fields*, pages 315–320. Plenum Press, New York, 1994.

42. Gamrowski, B. and Rachev, S. T. Financial Models Using Stable Laws. In Y. V. Prohorov, editor, *Applied and Industrial Mathematics*, pages 556–604. 1995.

43. Geman, H. and Ané, T. Stochastic Subordination. *Risk*, **9**, 146–149, 1996.

44. Gnedenko, B. V. and Kolmogorov, A. N. *Limit distributions for sum of independent random variables.* Addison–Wesley, Reading, MA, 1954.

45. Gouriéroux, C. *ARCH Models and Financial Applications.* Springer, New York, 1997.

46. Harrison, J. M. and Pliska, S. R. Martingales and Stochastic Integrals in the Theory of Continuous Trading. *Stochastic Processes and their Applications*, **11**, 215–260, 1981.

47. Hill, B. M. A Simple General Approach about the Tail of a Distribution. *Annals of Statistics*, **3**, 1163–1174, 1975.

48. Hull, J. *Option, Futures, and Other Derivatives.* Prentice Hall, 3rd edition, 1997.

49. Hull, J. and White, A. The Pricing of Options on Assets with Stochastic Volatilities. *Journal of Finance*, **2**, 281–300, 1987.

50. Hurst, H. E. Long–term storage capacity of reservoirs. *Transactions of the American Society of Civil Engineers*, **116**, 770–808, 1951.

51. Hurst, S. R., Platen, E., and Rachev, S. T. Subordinated Market Index Models: A Comparison. *Finan. Engin. Japan. Markets*, **4**, 97–124, 1997.

52. Hust, S. R., Platen, E., and Rachev, S. T. Option Pricing for a Logstable Asset Price Model. In S. Mittnik and S. T. Rachev, editors, *Stable Models in Finance.* Pergamon Press, 1999.

53. Janicki, A. and Weron, A. *Simulation and Chaotic Behavior of α-stable Stochastic Processes.* Marcel Dekker, New York, 1994.

54. Jansen, D. W. and de Vries, C. G. On the Frequency of Large Stock Returns: Putting Booms and Busts into Perspective. *Review of Economics and Statistics*, **73**, 18–24, 1983.

55. Jarrow, R. A. and Rudd, A. Approximate Option Valuation for Arbitrary Stochastic Processes. *Journal of Financial Economics*, **10**, 347–369, 1982.

56. Karandikar, R. L. and Rachev, S. T. A Generalized Binomial Model and Option Formulae for Subordinated Stock–Price Processes. *Probability and Mathematical Statistics*, **15**, 427–446, 1995.

57. Karandikar, R. L. and Rachev, S. T. A Generalized Binomial Model and Option Formulae for Subordinated Stock Price Processes. *Probability and Mathematical Statistics*, **15**, 427–446, 1997.

58. Khinchin, A. Y. *Limit Laws for Sums of Independent Random Variables.* ONTI, Moscow, 1938.

59. Kim, J. R., Mittnik, S., and Rachev, S. T. Econometric Modelling in the presence of heavy–tailed innovations. *Communications in Statistics – Stochastic Models*, **13**, 841–886, 1997.

60. Kon, S. J. Models of Stock Returns: A Comparison. *Journal of Finance*, **39**, 147–165, 1984.

61. Koutrouvelis, I. A. Regression–type Estimation of the Parameters of Stable Laws. *Journal of the American Statistical Association*, **75**, 918–928, 1980.

62. Koutrouvelis, I. A. An Iterative Procedure for the Estimation of the Parameters of Stable Laws. *Commun. Statist. - Simula.*, **10**, 17–28, 1981.

63. Kwapień, S. and Woycziński, W. A. *Random Series and Stochastic Integrals - Single and Multiple.* Springer, New York, 1992.

64. Lee, M. L. T., Rachev, S. T., and Samorodnitsky, G. Dependence of Stable Random Variables. In *Stochastic Inequalities*, IMS Lecture Notes 22, pages 219–234. 1993.

65. Lévy-Vehel, J. Fractal Approaches in Signal Processing. *Fractals*, **3**, 755–775, 1995.

66. Lo, A. W. Long-term memory in stock market prices. *Econometrica*, **59**, 1279–1313, 1991.

67. Mandelbrot, B., Fisher, A., and Calvet, L. A Multifractal Model of Asset Returns. Technical Report 1164, Cowles Foundation, 1997.

68. Mandelbrot, B. B. The Variation of Certain Speculative Prices. *Journal of Business*, **26**, 394–419, 1963.

69. Mandelbrot, B. B. and van Ness, J. W. Fractional Brownian motions, fractional noises and applications. *SIAM Review*, **10**, 422–437, 1968.

70. Mantegna, R. N. and Stanley, H. E. Scaling behavior in the dynamics of an economic index. *Nature*, **376**, 46–49, 1995.

71. McCulloch, J. H. Simple consistent estimators of stable distribution parameters. *Commun. Statist. - Simula.*, **15**, 1109–1136, 1986.

72. McCulloch, J. H. Measuring Tail Thickness in Order to Estimate the Stable Index α: A Critique. *Journal of Business and Economic Statistics*, **15**, 74–81, 1997.

73. McCulloch, J. H. Linear regression with stable disturbances. In R. Adler, R. Feldman, and M. Taqqu, editors, *A Practical Guide to Heavy Tails*, Boston, 1998. Birkäuser.

74. Merton, R. Option Pricing when Underlying Stock Returns Are Discontinuous. *Journal of Financial Economics*, **3**, 125–144, 1976.

75. Mijnheer, J. L. Sample path properties of Stable Processes, 1975.

76. Mittnik, S., Paolella, M. S., and Rachev, S. T. Modeling the Persistence of Conditional Volatilities with GARCH–stable Processes. Technical report, University of California, Santa Barbara, 1997.

77. Mittnik, S., Paolella, M. S., and Rachev, S. T. Diagnosing and Treating the Fat Tails in Financial Return Data. Technical report, Institute of Statistics and Econometrics, University of Kiel, 1999.

78. Mittnik, S. and Rachev, S. T. Alternative Multivariate Stable Distributions and Their Applications to Financial Modeling. In S. Cambanis, editor, *Stable Processes and Related Topics*, pages 107–119. Birkäuser, Boston, 1991.

79. Mittnik, S. and Rachev, S. T. Modeling Asset Returns with Alternative Stable Distributions. *Econometric Reviews*, **12**, 261–330, 1993.

80. Mittnik, S. and Rachev, S. T. Tail Estimation of the Stable Index α. *Applied Mathematics Letters*, **9**, 53–56, 1996.

81. Mittnik, S., Rachev, S. T., and Samorodnitsky, G. Testing for Structural Breaks in Time Series Regressions with Heavy–tailed Disturbances. Technical report, Department of Statistics and Mathematical Economics, University of Karlsruhe, 1998.

82. Nelson D. Stationarity and Persistence in the GARCH(1,1) model. *Econometric Theory*, **6**, 318–334, 1990.

83. Nikias, C. L. and Shao, M. *Signal Processing with Alpha-Stable Distributions.* Wiley, 1996.

84. Nolan, J. P. Numerical calculation of stable densities. *Stochastic Models*, **4**, 1997.

85. Nolan, J. P. Parametrizations and modes of stable distributions. *Statistics and Probability Letters*, **38**, 187–195, 1998.

86. Panorska, A. K., Mittnik, S., and Rachev, S. T. Stable GARCH models for Financial Time Series. *Applied Mathematics Letters*, **8**, 33–37, 1995.

87. Petrov, V. V. *Limit Theorems of Probability Theory.* Oxford UP, Oxford, 1995.

88. Pollard, D. *Convergence of Stochastic Processes.* Springer, New York, 1984.

89. Press, W. H., Teukolsky, S. A., Vetterling, W. T., and Flannery, B. P. *Numerical recipes in C.* Cambridge UP, 1992.

90. Protter, P. *Stochastic Integration and Differential Equations – A New Approach.* Springer, New York, 1990.

91. Rachev, S. T. and Mittnik, S. *Stable Paretian Models in Finance.* Wiley, New York, 2000. Forthcoming.

92. Rachev, S. T. and Samorodnitsky, G. Option Pricing Formulae for Speculative Prices Modelled by Subordinated Stochastic Processes. *Pliska,* **19**, 175–190, 1993.

93. Rachev, S. T. and Xin, H. Test on Association of Random Variables in the Domain of Attraction of Multivariate Stable Law. *Probability and Mathematical Statistics,* **14**, 125–141, 1993.

94. Rossi, P. E., editor. *Modelling Stock Market Volatility – Bridging the Gap to Continuous Time.* Academic Press, New York, 1996.

95. Samorodnitsky, G. and Taqqu, M. S. *Stable non-Gaussian Random Processes: Stochastic Models with Infinite Variance.* Chapman and Hall, New York, 1994.

96. Shao, M. and Nikias, C. L. Signal Processing with fractional lower order moments: stable processes and applications. *Proceedings of the IEEE,* **81**, 984–1010, 1993.

97. Silverman, B. W. *Density Estimation for Statistics and Data Analysis.* Chapman and Hall, New York, 1986.

98. Tapia, R. A. and Thompson, J. R. *Nonparametric Function Estimation, Modelling, and Simulation.* SIAM, Philadelphia, 1990.

99. Taqqu, M. S. and Teverovsky, V. Robustness of Whittle–type estimates for time series with long–range dependence. *Stochastic Models,* **13**, 723–757, 1997.

100. Taylor, S. J. *Modelling Financial Time Series.* Wiley & Sons, Chichester, 1986.

101. Weron, R. On the Chambers–Mallows–Stuck method for simulating skewed stable random variables. *Statistics and Probability Letters*, **28**, 165–171, 1996.

102. Willinger, W., Taqqu, M. S., and Teverovsky, V. Stock market prices and long–range dependence. *Finance and Stochastics*, **3**, 1–13, 1999.

103. Yamazato, M. Unimodality of infinitely divisible distributions of class *l*. *Annals of Probability*, **6**, 523–531, 1978.

104. Zenios, S. A. *Financial Optimization.* Cambridge University Press, 1993.

105. Ziemba, W. T. Choosing investment portfolios when the returns have stable distributions. In P. L. Hammer and G. Zoulendijl, editors, *Mathematical Programming in Theory and Practice*, pages 443–482. North-Holland, 1974.

106. Ziemba, W. T. and Vickson, R. G. *Stochastic Optimization Models in Finance.* Academic Press, New York, 1975.

107. Zolotarev, V. M. On representation of stable laws by integrals. *Selected Translations in Mathematical Statistics and Probability*, **6**, 84–88, 1966.

108. Zolotarev, V. M. *One-dimensional stable distributions.* American Mathematical Society, Providence, RI, 1986.

PERFORMANCE MEASUREMENTS: THE STABLE PARETIAN APPROACH

G. Götzenberger, S. T. Rachev, and E. Schwarz
Institute of Statistics and Mathematical Economics, School of Economics, University of Karlsruhe, Germany.
E-mail:Gero.Goetzenberger@daimlerchrysler.com
Institute of Statistics and Mathematical Economics, School of Economics University of Karlsruhe, Karlsruhe, Germany.
Anderson School of Management, University of California at Los Angeles, Los Angeles, CA.
Contact author: G. Götzenberger

In the first part of the paper we review recent results on performance measurements in finance. In the second part we introduce stable Paretian distributions to asset pricing models in view of the vast empirical evidence that stable Paretian models out perform the classical Gaussian models. Consequently, we develop performance measures based on stable non-Gaussian distribution for asset returns, and present a methodology describing the those new performance measures based on regression models with heavy tailed distributed innovations. We develop estimators that enable us to assume different indices of stability (that is different tail behavior) for the market and asset return distributions. To check our stable Paretian model, we stimate the t-statistics, quantifying our assumptions on the indices of stability for the market and the asset returns. We expect that this new methodology will provide more realistic models for performance measurement than was previously possible.

1 Introduction

Over the last thirty years, the problem of investment performance measurement has lost none of its allure or importance for the financial community. Just consider two reasons, which will appeal immediately to every investor. At a time when more than fifty percent of all financial assets in North America are controlled by pension or mutual funds, a lot of people apparently employ someone else to manage their money. Also, they are obviously willing to pay high fees or expenses for these services and are naturally very interested in how well "their" funds are performing. On the other hand, insights on performance measurement allow us to draw conclusions about the notion of efficient markets, an area that has occupied scholars and the "real-world" alike. If certain investors would realize consistently "abnormal" returns by trading on knowledge or information that is available to all other investors in the market, then the hypothesis of efficient markets could be rejected (assuming the model of the real world employed by the test is correct). That in turn would have substantial consequences on investment strategies. Recent studies have shown that the use of predetermined information variables such as lagged interest rates, dividend yields, and others are useful in predicting asset returns. These insights should certainly have an impact on performance measurement. Instead of rejecting the efficient market hypothesis, which would be the consequence of these results, it may be the performance measures that have to be adjusted. In addition, no investor should be willing to pay high fees for simple mechanical forecasts that exhibit superior performance under conventional measures when he/she could replicate the payoffs easily himself/herself.

Mean returns of securities are positively related to their risk. Therefore, performance measurement is all about relating the returns of the portfolios of investors in some way or another to the risks that those investors were willing to bear through their portfolios. In general, this is done as follows. The returns of the investigated, actively managed portfolio are compared to a passive portfolio that exhibits the same level of risk. There are two ways to accomplish this task: i) based on the observation of the returns of the investigated portfolio as well as the returns of a benchmark portfolio plus a risk-free asset and ii) based on the composition of the investigated portfolio and on the observation of the returns of that portfolio. In this paper, we analyze performance measures that benefit from both methods.

Recently, research has made advances in other fields as well. It has long been known that returns are in general not normally distributed. In fact, the variance of return distributions is most probably not even finite. Following Mandelbrot and others, researchers have made great advances in the field and extended the work of Fama (1965) and Ross (1976) to develop asset pricing models, and consequently, performance measures that allow return distributions in L_p with $1 < p < 2$ ($p = 2$ implies a Gaussian distribution). This paper presents performance measures and empirical evidence based on these works. We extend the conventional measures based on normal distributions to a general paretian stable case.

In Section 2, we review the theory of asset pricing and performance measurement. Section 3 gives an overview of relevant empirical studies in the field. We concentrate on studies based on the capital asset pricing model (CAPM) and the asset pricing theory (APT) as well as event-study measures. Given the vast number of material on the subject, it is obvious that we cannot consider all of the existing papers. Section 4 introduces and develops performance measures based on pareto-stable distributions. We present the regression methodology for the stable paretian approach in Section 5.

The summary in Section 6 six concludes the paper.

2 Review of the Theory on Performance Measurement Studies

Fama (1972) identified two possibilities for investors to realize "superior" returns. Microforecasting, or selectivity, is the strategy with which the investor identifies mispriced securities. Hence, he will be able to make an excess profit if the market recognizes the true value at a later time. Alternatively macroforecasting, or market timing, refers to forecasts of price movements of the market as a whole. If the investor perceives a hausse in the next period, he will increase the risk exposure of his portfolio and vice versa realizing abnormal returns along the way. We consider performance measures based on both notions of "superior" returns.

2.1 Performance Measurement Based on the Capital Asset Pricing Model

2.1.1. The Capital Asset Pricing Model (CAPM)

The widely used capital asset pricing model was developed by Sharpe (1964) and Lintner (1965) based on the works of Markowitz (1959). It states that there is a linear relationship between risk and return. Assets are subject to two kinds of risk, diversifiable and non-diversifiable risk. Both types together represent the total risk of the asset, which can be measured as the variance of the asset's return. On the other hand, it may be argued that the relevant type of risk is not total risk but non-diversifiable or market risk. By diversifying his portfolio, an investor can reduce his risk-exposure while still earning the same return. Assuming no transaction costs it becomes clear that rational investors diversify their portfolios. Risk can then be measured in the portfolio's beta:

$$\beta_p = \frac{\text{cov}(r_p, r_m)}{\text{var}(r_m)},$$

where r_p is the return on the portfolio, and r_m is the return on the market. Therefore, there are two ways to measure investment performance. First, the security market line (SML) which plots expected returns of assets or portfolios against β, assuming that non-diversifiable risk is relevant:

$$E(r_i) = r_f + (E(r_m) - r_f)\beta_i,$$

where $E(r_i)$ is the expected return on the asset or the portfolio, r_f is the average risk-free rate, $E(r_m)$ is the expected return on the market, and β_i is the beta of the asset or the portfolio. Second, the capital market line (CML) which plots expected returns against the standard deviation, assuming that total risk is relevant:

$$E(r_p) = r_f + (E(r_m) - r_f)\sigma_p,$$

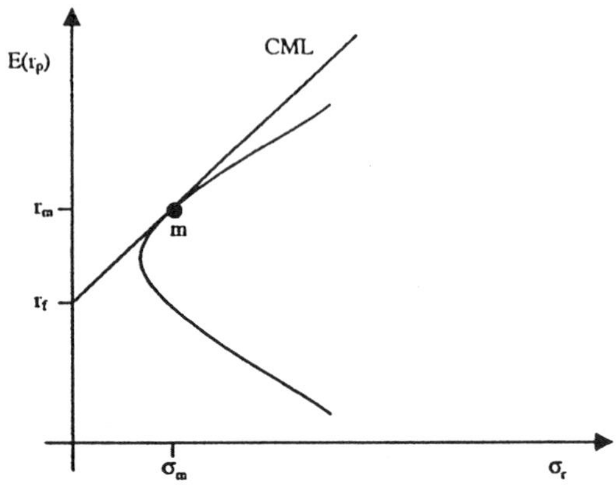

Figure 1: Portfolio frontier and the capital market line.

where σ_p is the standard deviation of the returns on the portfolio.

2.1.2. Performance Measures Based on the CAPM

The Treynor ratio Treynor (1965) developed the Treynor ratio, also known as the reward-to-volatility ratio. The average excess return of the portfolio is divided by its risk expressed through its β_p. Here the non-diversifiable risk is assumed to be relevant, therefore, the SML is the benchmark. The Treynor ratio is defined as

$$\mathrm{RVOL}_p = \frac{\bar{r}_p - \bar{r}_f}{\beta_p}.$$

Treynor assumed the CAPM to be true and used the RVOL_m of the mean/variance efficient market portfolio ($\beta = 1$) as a benchmark. Portfolios that have greater ratios than $\bar{r}_m - \bar{r}_f$ lie consequently above the ex post SML, indicating that they have outperformed the market and vice versa. Though not used very often today, it was one of the first attempts to obtain some measure that would link the return and the risk exposure of portfolios.

The Sharpe ratio Sharpe (1966) introduced the reward-to-variability ratio as a measure of performance. It is constructed similarly to the Treynor ratio. However, instead of beta or non-diversifiable risk it utilizes the variance or total risk of the portfolio, and as a result the CML is considered. This might be more appropriate if the investor is not very diversified and therefore worries about the total risk of his assets. The Sharpe ratio is defined as

$$\mathrm{RVAR}_p = \frac{\bar{r}_p - \bar{r}_f}{\sigma_p}.$$

The Jensen measure Introduced in Jensen (1968), this measure has become the most widely accepted measure of investment performance. Assuming the CAPM is valid, Jensen extended this one period model into a multiperiod model simply by inferring that it is valid at every point in time. Consider the regression

$$r_{pt} = a_p + \beta_p r_{mt} + u_{pt}$$

where r_{pt} is the excess return on the portfolio over the prevailing risk-free rate, r_{mt} is the excess return on the market over the risk-free rate, u_{pt} is the error term, and β_p is the measure for the portfolio's systematic risk. If certain investors choose portfolios that perform consistently better than the market, the a_ps in the above regression will be significantly greater than zero for these portfolios. Jensen and Treynor based their benchmarks on the SML, assuming that it represents the world correctly, a_ps different from zero or Treynor ratios greater than $\bar{r}_m - \bar{r}_f$ would hence indicate abnormal performance. Both, Jensen and Treynor accounted for the systematic risk of portfolios. However, in contrast to the return-to-volatility ratio the a_ps are an absolute measure of performance. Since the two measures refer to a different kind of risk than the Sharpe ratio, the latter can yield inconsistent results with the classification of portfolios into "winners" and "losers".

Market timing measures of Treynor and Mazuy As described above, the notion of market timing implies that the investor enhances the risk exposure of his portfolio if he perceives an upward or booming market and vice versa. In contrast, passive strategies should produce fairly constant risk exposures of the portfolio implying that the relationship between excess market returns and excess portfolio returns is linear. Hypothesizing the extreme case of market timing in which the investor switches his wealth between assets with very high risk exposure (during a market rise) and "default free" bonds, the relationship between market excess returns and portfolio excess returns is no longer linear. The slope of the relation would be greater for positive excess market returns and lower for negative ones. The methodology of Treynor and Mazuy (1966) tries to pick up the above effect by adding an additional quadratic term into the regression specification utilized by Jensen:

$$r_{pt} = a_p + \beta_{p1} r_{mt} + \gamma_p {r_{mt}}^2 + u_{pt}$$

where γ_p describes the ability of fund managers to anticipate turns in the stock market.

Market timing measures of Henriksson and Merton Merton (1981) developed another model to test for market timing, assuming that investors do try to predict market movements, but they do not try to predict the magnitude of excess returns, leading to the extreme case of market timing described above. The benefit of this methodology is that, given the investors predictions are observable, a non-parametric test, which does not require any assumptions about the joint return distribution or the return generating process, can be employed to measure market timing ability. For the case that the forecasts are not available, Merton developed a parametric test, however he had to put restrictions on the return generating process, assuming either the CAPM or an APT model.

Henriksson and Merton (1981) presented two statistical techniques to perform these tests. In the parametric test the fund manager is assumed to have two target betas, one for a market rise and one for a market decline. Suppose he switches his portfolio's beta between the two target betas according to his forecast. The consequent regression is written as:

$$r_{pt} = a_p + \beta_{dp} r_{mt} + \gamma_p \max(0, r_{mt}) u_{pt}$$

where r_{pt} is the realized return on the portfolio, r_{mt} is the realized excess return on the market and u_{pt} is the error term. Obviously, the selection ability of the investor is measured by a_p, whereas the market timing ability is expressed by γ_p. Hence, if γ_p is significantly different from zero, the investor is capable of "timing the market".

2.1.3. Criticisms of Performance Measures Based on the CAPM

Roll's critique What are "abnormal" returns, "abnormal" relative to what? In order to measure the quality of returns, most of the testing literature has employed a benchmark with which the observable returns are compared. For all the measures based on the CAPM this would be the market portfolio. Richard Roll's (1977 and 1978) severe criticisms of the utility of the CAPM and consequently the market portfolio in performance measurement, had serious impacts on the testing literature. The power of Roll's arguments comes from the fact that they are derived from the mathematics of the CAPM and are not based on econometrics or empirical evidence. His arguments could be summarized in the phrase: "*securities plot on the security market line (SML) if and only if the market portfolio is efficient*". He pointed out that Jensen's alpha and all the other measures do not provide any evidence on whether some securities exhibit superior or inferior performance; they merely test whether or not the market portfolio is efficient. Due to the fact that the true market portfolio is not observable, the utility of all the previous literature may be doubted. Furthermore, Roll pointed out that if the market portfolio would be observable and mean-variance efficient, the securities would plot on the security market line, and consistently positive or consistently negative deviations should not occur. What would then be the point in testing?

Roll (1978) provided a hypothetical example, in which he showed that the application of three different benchmarks to calculate the betas provided completely different rankings for the performance of the assets in his universe. (Of course, the choice of the true mean-variance efficient market portfolio provided no superior or inferior performance at all, all securities plotted on the SML. Hence, no statements about performance could be inferred.) Consequently, any desired performance could be achieved by choosing the appropriate benchmark. In the next section Roll examined the CAPM in a multiperiod setting. He distinguished two cases. First, he assumed that the mean returns and the covariance matrix were stationary. If a mean-variance efficient index is chosen, no investor will be able to achieve consistently abnormal returns, since every asset is neither a consistent loser nor a consistent winner. If the number of examined periods is large enough, the performance of all investors will be the same, no matter which rules they applied to build their portfolios. Second, if the means and covariance matrix were not stationary, some investment strategies could yield more superior returns than others. However, the SML would not able to detect this superior investment performance.

The robustness of the SML Green (1986) evaluated the robustness of the Security Market Line relationship when the benchmark portfolio is not mean-variance efficient. He derived how the location of an asset in mean-variance space determines its alpha and proved that this "benchmark error" will lie within a certain interval. The size of the benchmark error for mean efficient portfolios is linearly dependent on the size of the return on the portfolio. Green ultimately presented some severe inconsistencies for the relative rankings of portfolios according to a certain proxy for the market portfolio. An arbitrarily close (in mean-variance space) alternative proxy can exactly reverse the rankings of assets and portfolios.

Further ambiguity when performance is measured by the SML Grauer (1991) also concentrated his criticism on the CAPM and the SML. For the single period he developed three ideas that are inconsistent with the application of the SML in performance measurement. First, if investors do not choose solely on the basis of mean and variance but have linear risk tolerance utility functions, the SML criterion might be completely irrelevant. Second, if mean-variance investors faces active constraints, the SML is no longer a line (see also Brennan, 1971). Therefore, the SML does not characterize a market equilibrium. Assets no longer plot on the SML, but on a hyperplane. Third, if we calculate the betas against an inefficient index portfolio, assuming this is the true market portfolio, we assign to each asset a new mean, and thereby make the inefficient index portfolio efficient. Since an infinite number of mean vectors will make our benchmark efficient, it is possible to get almost any ranking we like as we keep the benchmark portfolio's mean the same. Further ambiguities appear when time enters the analysis. Most of the testing literature calculated the betas from time series, assuming them to stay constant. However, as soon as the weights in the benchmark portfolio change, the betas of the individual assets change, as do their expected returns. Grauer provided a real-world example. Over the period from 1935 to 1979 he examined a portfolio of high-beta stocks of the NYSE. He found the portfolio had an average return on 19.6% per annum, whereas Jensen's alpha was 1.5%. However, allowing for changing weights in the benchmark portfolio, the SML consistent means ranged from 16% to 30% per annum.

Informed investors Dybvig and Ross (1985) and others were more and more concerned about the content and the quality of the information available to the investor (also see Admati, Bhattacharya, Pfleiderer and Ross, 1986). Since the claim of superior information is more or less the nature of portfolio management, funds managers will extend their efforts to obtain this information. However, according to Dybvig and Ross the performance of their portfolios should not be measured by the SML, but rather from the informed manager's perspective, since the SML more or less reflects the uninformed majority of investors. Similar arguments led to the development of conditional performance measures.

2.2 Performance Measurement Based on the Arbitrage Pricing Theory

2.2.1. The Arbitrage Pricing Theory

Partly in response with the criticisms of the CAPM, Ross (1976) developed his "Arbitrage Theory of Capital Asset Pricing". The Arbitrage Pricing Theory (APT) was born. Ross argued that in order to derive the CAPM, overly severe assumptions have to be made. For investors to choose portfolios exclusively on the base of mean and variance, either the asset returns have to be multivariate normally distributed or the investors must have quadratic utility functions — two assumptions that are most probably incorrect. Therefore, he drew on another fairly intuitive argument, the absence of arbitrage. It is a common approach to assume that ex post returns are generated by some stochastic relation, where the returns are generated by some factor, like

$$R_i = E_i + \beta_i \delta + \varepsilon_i,$$

where ε_i is a mean zero noise error term, E_i is a constant representing ex ante expected return, β_i is the ex ante sensitivity to movements of the factor (factor loading, factor beta), and δ is the risk premium for the exposure to this factor.

First, we form an arbitrage portfolio, η, of the n assets. The ith component of η represents the weight invested in asset i, $\eta \in \Re^n$. Since an arbitrage portfolio uses no wealth, we get

$$\eta^T \mathbf{e} = 0,$$

where $\mathbf{e}$ is the unit vector. Then the return on η will be

$$\begin{aligned} R_\eta &= \eta^T \mathbf{x} = \eta^T \mathbf{E} + (\eta^T \beta)\delta + \eta^T \varepsilon \\ &\approx \eta^T \mathbf{E} + (\eta^T \beta)\delta, \end{aligned}$$

when the arbitrage portfolio is sufficiently well diversified to permit the use of the law of large numbers to approximately eliminate the noise term. If the arbitrage portfolio with a risk-free position is chosen,

$$\eta^T \beta = 0.$$

The return on this special arbitrage portfolio is therefore

$$R_\eta = \eta^T \mathbf{E} + (\eta^T \beta)\delta = \eta^T \mathbf{E}.$$

The portfolio η uses no wealth and is risk-free. With the absence of arbitrage this implies

$$\eta^T \mathbf{E} = 0.$$

But this is simply the algebraic statement that all vectors η, orthogonal to $\mathbf{e}$ and β are orthogonal to $\mathbf{E}$. Hence, $\mathbf{E}$ must be a linear combination of $\mathbf{e}$ and β. Therefore, there are constants such that

$$\mathbf{E} = E_0 \mathbf{e} + a\beta.$$

When considering the market portfolio η_m, δ can be normalized so that $\eta_m \beta = 1$, then the above equation becomes

$$\mathbf{E} = E_0 \mathbf{e} + (E_m - E_0)\beta.$$

No assumption about equilibrium was necessary to derive the result, only the consideration that portfolios that cost nothing, and are not subject to any risks should not earn any return. The same argument yields that E_0 must represent the return on a zero-beta portfolio (a risk-free asset). This basic arbitrage argument easily generalizes to the k-factor case, provided that the number of common factors is significantly less than the number of assets. The generating model takes the form

$$R_i = E_i + \beta_{i1}\delta_1 + \cdots + \beta_{ik}\delta_k + \varepsilon_i,$$

where the β_{ij} are the asset's sensitivities to movements in the k factors (factor loadings or factor betas) and the δ_j represent the k factors. The basic arbitrage condition takes the form

$$E_i = E_0 + (E_m - E_0)[\gamma_1\beta_{i1} + \cdots + \gamma_k\beta_{ik}],$$

where γ_l are normalized constants so that $\gamma\mathbf{e} = 1$. Defining $\rho_l = (E_m - E_0)\gamma_l$, E_i becomes

$$E_i = E_0 + \rho_1\beta_{i1} + \cdots + \rho_k\beta_{ik},$$

where ρ_l can be interpreted as that part of the risk premium (expected return above the risk-free rate) of the market that is due to factor l. Replacing E_i in the generating model, it becomes

$$R_i = E_0 + (\delta_1 + \rho_1)\beta_{i1} + \cdots + (\delta_k + \rho_k)\beta_{ik} + \varepsilon_i.$$

Two basic theories have evolved from this work. The asymptotic APT (see Huberman (1982)), which utilizes the notion of the nth economy as one with n assets. However, the distance between the vector of real returns and the vector of explained returns is bounded. As n increases this distance, the distance between the vector of real returns and the vector of explained returns tends to zero.

Another development of the APT is the work of Connor (1984), who derived a competitive equilibrium version of the APT which seemed to avoid most of the problems associated with CAPM and common APT models. However, his results came at two costs: i) the market portfolio must be perfectly diversified and ii) the simple intuitive arbitrage argument that makes the APT so appealing is abandoned in favor of arguments based on utility maximization and the assumption of equilibrium. Of course, there were numerous other authors who developed refinements of the above findings.

2.2.2. Performance Measures

Connor and Korajczyk (1986) and others derived performance measures similar to Jensen's α_p's and Treynor ratio that are theoretically consistent with Connor's derivation and the APT itself. The most widely used performance measure based on the APT is the equivalent of the Jensen's measure. It is derived from the regressions

$$r_{pt} = a_p + \beta_{p1}(\rho_{1t} + \delta_{1t}) + \cdots + \beta_{pk}(\rho_{kt} + \delta_{kt}) + u_{pt}$$

where r_{pt} is the excess return on the portfolio over the prevailing risk-free rate, $(\rho_{jt} + \delta_{jt})$ is the excess return on the factor j over the risk-free rate, u_{pt} is the error term, and β_{pj} is the measure for the portfolio's sensitivity to movements of the factor j. If certain investors choose portfolios that perform consistently better than the market, the a_ps in the above regression will be significantly greater than zero for these portfolios. Connor and Koraczyk (1986) showed that in an equilibrium arbitrage pricing model only informed investors, who have information in addition to what is commonly known about the market, would be able to choose portfolios with positive a_ps.

2.2.3. Criticisms of Performance Measures Based on the APT

One of the most severe criticisms of the arbitrage theory came from Shanken (1982). Similar to Roll's critique of the CAPM, the paper develops thoughts about what the previous tests of the arbitrage pricing theory really meant. Empirical investigations of the APT have attempted to test the following statement: *If a set of asset returns conforms to a k-factor model, then the expected return vector is equal to a linear combination of a unit vector and the factor loading vectors.* Shanken pointed out that the rejection of the hypothesis could not be equated with the rejection of the theory, however its acceptance would be consistent with the theory and might be interpreted as evidence in favor of the APT. Shanken criticized that a theory that cannot be rejected might not necessarily be preferable to the CAPM.

Nevertheless, Shanken's most important argument concerns the uniqueness of the factors in the model. He considered two sets of securities as equivalent if the corresponding sets of obtainable portfolio returns are the same. Intuitively, the same pervasive forces in the economy should move two equivalent sets of securities. These forces should be represented by the same factors. Shanken showed with a simple example that this is not so. He repackaged a set of two securities that conformed originally to a one-factor model. After repackaging the two securities conformed to a zero-factor model, although the two sets, the original and the repackaged, were equivalent. As Ross and Roll (1980) recognized earlier, it is possible to transform the set of factors by an arbitrary invertible transformation to another basis for the same factor space. Two equivalent sets of securities may confirm to very different factor structures. This alone does not really present a problem. However, the fact that the number of factors in the same respective model need not be the same does imply a severe problem. Shanken showed that the empirical formulation of the APT is consistent if and only if all securities have the same expected return. Therefore, the empirical formulation cannot be considered adequate. It rules out the expected return differentials, which the theory tries to explain.

Next Shanken was concerned about the conclusions drawn from the various empirical studies. He claimed that whenever some of the factors in a given factor model representation are highly correlated with the return of a mean-variance efficient portfolio, it is not very surprising to find these factors priced. The economic significance of such a result is therefore questionable.

Finally, Shanken investigates the work of Connor, who would later summarize his findings (Connor (1984)). His equilibrium APT requires that all the idiosyncratic risk, defined relative to a given factor structure, is completely diversified away in the market portfolio. This would be a possibility to validate the equilibrium ATP. Without the observation of the market portfolio it is unlikely that the diversification condition can ever be verified in practice. It seems that the equilibrium APT is subject to the same difficulties as the CAPM. When reconsidering the 0-factor model, then all risk is idiosyncratic risk, which should be diversified away in the market portfolio according to Connor. Therefore, the variance of the market portfolio should be zero. Considering the substantial variation in the used market proxies, this condition can be rejected. As a result, the usual empirical formulation of the APT is very questionable.

Another argument, much in line with the thoughts above, is the fact that in general, the factors must be identified in one way or another. The most common approach is to identify portfolios that seem to mimic the effects of the factors. This is, however, very similar to the CAPM approach; there the only factor is the market factor and the difficulties of obtaining a proxy for this factor are well known. Therefore, one can argue whether or not it is easier to find a proxy for only one factor or a number of them.

Concluding, as the CAPM can be restated as "the market portfolio is mean variance efficient", the APT pricing restriction can be restated as "any linear combination of factor portfolios is mean-variance efficient".

2.3 Conditional Performance Measurement

2.3.1. Performance Measures

Recent studies (e.g. Fama and French (1992)) showed some evidence that stock returns are at least partly predictable by using predetermined variables such as the lagged dividend yield or lagged Treasury bill yields. These variables represent public information and are available to everybody. The fact that public information seems to help in predicting stock prices might be interpreted as standing in harsh contrast to the efficient market hypothesis. Instead of dismissing the hypothesis, Ferson and Schadt (1996) proposed new performance measures, which take these recent insights into account. They developed a modification of the Jensen measure of selectivity as well as modifications for the Henriksson-Merton test and the Treynor-Mazuy test for market timing.

So far all studies assumed that the betas of stock would stay constant over time. As Best and Grauer (1990) and others have pointed out, this is hardly possible. Assuming a CAPM environment, if one of the parameters changes, the rest of them including the betas would have to change as well. Ferson and Schadt introduced conditional betas defining them as linear functions of predetermined (lagged) instruments, which are proven to be useful for predicting stock prices. These instruments include lagged dividend yield, lagged one month Treasury bill yield, information about the term structure and the bond market as well as a dummy variable to account for the January effect. The conditional models attempt to bring the realm of performance measurement up to date with the current level of research and are able to incorporate dynamic behavior of the stock market.

Conditional models assume the betas of assets are linear functions of predetermined information variables. Ferson and Schadt (1996) employed the lagged risk-free rate and the lagged dividend yield on the value-weighted CRSP index in order to represent the conditioning information. They used a Taylor series expansion, hence the beta of the portfolio becomes:

$$\beta_p = \beta_{0p} + \beta_{1p} Tb_{t-1} + \beta_{2p}\, dy_{t-1} \quad \text{for all } t,$$

where Tb_{t-1} is the Treasury bill yield lagged one period, and dy_{t-1} is the dividend yield on the value-weighted CRSP index, also lagged one period. This allows conditional models to account for dynamic environments, and also it takes insights into account that asset returns are somewhat predictable using these predetermined information variables.

Substituting the expression for the beta into the original Jensen equation, we find that the conditional Jensen test is a straightforward modification of the original version:

$$r_{pt} = \alpha_p + \beta_{0p} r_{mt} + \beta_{1p}[Tb_{t-1} r_{mt}] + \beta_{2p}[dy_{t-1} r_{mt}] + u_{pt}.$$

The excess return on the portfolio is regressed against the excess return on the market and on the product of the excess return on the market and the information variables. α_p is still the measure of selectivity.

Ferson and Schadt also developed a conditional market timing measure. The idea is to distinguish between "mechanical market timing" based on public information from market timing using information that is superior to the predetermined information variables. Replacing the beta in the original Treynor-Mazuy test with the linear function, the regression specification for the conditional Treynor-Mazuy test becomes:

$$r_{pt} = \alpha_p + \beta_{0p} r_{mt} + \beta_{1p}[Tb_{t-1} r_{mt}] + \beta_{2p}[dy_{t-1} r_{mt}] + \gamma_p {r_{mt}}^2 + u_{pt}$$

where all returns are excess returns over the risk-free rate, β_{1p}, β_{2p} capture the effect of the information variables on the portfolio's beta, and γ_p is the measure of market timing based on information superior to the information variables.

The conditional Henriksson-Merton modification assumes that the investor uses the predetermined variables to obtain the conditional mean of the return on the market. The forecast of the conditional mean can obviously be conducted in many different ways. Ferson and Schadt obtain it by a simple regression of the return on the market against the lagged Treasury bill rate and the lagged dividend yield. Assuming the investor attempts to forecast whether or not the actual excess return on the market will be greater or less than the excess conditional mean return, he will shift the β_{0p} of his portfolio accordingly, increasing the beta if the forecast is negative and vice versa. Hence, the following modification for the conditional Henriksson-Merton regression specification is derived:

$$\begin{aligned} r_{pt} = {} & \alpha_p + \beta_{0p} r_{mt} + \beta_{1p}[Tb_{t-1} r_{mt}] + \beta_{2p}[dy_{t-1} r_{mt}] + \gamma_p {r^*}_{mt} \\ & + {\beta^*}_{1p}[Tb_{t-1} {r^*}_{mt}] + {\beta^*}_{2p}[dy_{t-1} {r^*}_{mt}] + u_{pt} \end{aligned}$$

where all returns are excess returns, and ${r^*}_{mt}$ is the product of the index excess return and an indicator dummy for positive values of the difference between the index excess return and the conditional mean of the excess return, where the conditional mean is estimated by a linear regression on Tb_{t-1} and dy_{t-1}.

A somewhat different approach can be examined in Jagannathan and Wang (1996). They also favor a conditional approach; however their framework is slightly different. The CAPM was developed as a one period model. When researchers extended the model, they had to make assumptions, one of which was the assumption that betas remain constant over time. In contrast, recent studies (e.g. Ferson and Harvey (1991, 1993)) have shown that neither the betas nor the risk premium stay constant over time. Hence, betas and expected returns do most likely depend on the amount and content of the information available at any given point in time. Jagannathan and Wang assumed that the conditional CAPM holds. For each asset i and in each period

$$E[R_{it} \mid I_{t-1}] = \gamma_{0t-1} + \gamma_{1t-1}\beta_{it-1},$$

where R_{it} is the return on security i in t, I_{t-1} is the available information at $t-1$, γ_{0t-1} is the conditional expected return on a zero-beta portfolio, γ_{1t-1} is the conditional market risk premium and β_{it-1} is the conditional beta of the security i defined as

$$\beta_{it-1} = \text{cov}(R_{it}, R_{mt} \mid I_{t-1})/\text{var}(R_{mt} \mid I_{t-1}).$$

From the conditional CAPM they derived an unconditional version of the CAPM by taking the unconditional expectations,

$$E[R_{it}] = \gamma_0 + \gamma_1\overline{\beta}_1 + \text{Cov}(\gamma_{1t-1}, \beta_{it-1}),$$

where $\gamma_1 = E[\gamma_{1t-1}]$ is the expected market risk premium, $\gamma_0 = E[\gamma_{0t-1}]$, and $\overline{\beta}_1 = E[\beta_{it-1}]$ is the expected beta. The main difference from the conventional static CAPM is the factor that includes the covariance between the conditional beta and the conditional market risk premium. When that covariance is zero, then the unconditional CAPM resembles the static CAPM. As Jagannathan and Wang pointed out, that is unlikely since during bad economic times the risk premium is quite high and distressed firms are likely to have high betas as well, indicating that the conditional beta and conditional market risk premium is correlated.

Consequently, Jagannathan and Wang showed that the $\text{cov}(\gamma_{1t-1}, \beta_{it-1})$ depends only on the part of the conditional beta that is in the linear span of the market risk premium. They break down the conditional beta into three orthogonal parts. The first part is the expected beta, which is a constant. The second part is the beta-prem sensitivity, which is perfectly correlated with the market risk premium. While the third part is the residual beta, which is on average zero and uncorrelated with the market risk premium. Since only the first two parts do affect the unconditional expected return, and since they cannot be observed directly, Jagannathan and Wang defined the following two betas

1. The market beta $\beta_i := \text{Cov}(R_{it}, R_{mt})/\text{Var}(R_{mt})$,
2. The premium beta $\beta_{iprem} := \text{Cov}(\gamma_{1t-1}, r_{it})/\text{Var}(\gamma_{1t-1})$,

where β_i measures market risk and β_{iprem} measures beta-instability. Furthermore, they showed that under fairly mild assumptions the unconditional expected return is

$$E[r_{it}] = a_0 + a_1\beta_i + a_2\beta_{iprem}.$$

They also assumed the market risk premium γ_{1t-1} is a linear function of the yield spread between BAA- and AAA-rated bonds, and the return on the market r_{mt} is a linear function of the return on the value-weighted stock index and the growth rate in per capita labor income.

In addition, they followed an argument of Mayers (1972) that states that human capital represents a substantial part of the total capital in the economy. Hence, the inclusion of the growth rate in per capita income will yield a benchmark portfolio that is somewhat more appropriate. Finally, they derived the regression specification for the equivalent of the Jensen's measure as

$$r_{pt} = a_p + c_{vwt}\beta_{ivw} + c_{premt}\beta_{iprem} + c_{labort}\beta_{ilabor},$$

where r_{pt} is the excess return on the observed portfolio, a_p is the equivalent of the Jensen measure for their conditional CAPM, c_{vw} is the risk premium on the value-weighted stock index, c_{prem} is the risk premium denoted to beta-instability and c_{labor} is the risk premium on human capital.

Ferson and Harvey (1999) developed performance measures in a multi-beta environment. Similar to Ferson and Schadt (1996), they applied a number of predetermined information variables. In other words, they used the spread between Moody's Baa and Aaa corporate bond yields, the difference between the lagged returns on a three-month and a one-month Treasury bill plus the spread between a ten-year and a one-year Treasury bond yield. In addition to those, they applied the same variables as in Ferson and Schadt described above. These instruments were proposed earlier in a number of studies. Their empirical framework assumes the following return generating process,

$$r_{it+1} = E_t(r_{it+1}) + \beta_{it}'\{\mathbf{r}_{pt+1} - E_t(\mathbf{r}_{pt+1})\} + \varepsilon_{it+1},$$

where r_{it+1} is the excess return for any stock or portfolio i at time t (in excess of the return on a one-month Treasury bill), $\mathbf{r}_{pt+1}$ is the vector of excess returns on the risk factor mimicking portfolios and ε_{it+1} is the error term, where $E_t(\varepsilon_{it+1}) = E_t(\varepsilon_{it+1}\, r_{fpt+1}) = 0$ for all factors f. Furthermore, the model for the conditional expected returns and the betas is assumed as

$$E_t(r_{it+1}) = \mathbf{a}_{it} + \beta_{it}' E_t(\mathbf{r}_{pt+1}),$$

with

$$\begin{aligned}\beta_{it} &= b_{0i} + \mathbf{b}_{1i}'\mathbf{Z}_t,\\ \mathbf{a}_{it} &= a_{0i} + \mathbf{a}_{1i}'\mathbf{Z}_t,\end{aligned}$$

where $\mathbf{Z}_t$ is the vector of mean zero information variables known at time t. Modeling the betas as a linear function of the predetermined instruments, they followed Shanken (1990) and Ferson and Schadt (1996). Combining the equations the econometric model becomes

$$r_{it+1} = a_{0i} + \mathbf{a}'_{1i}\mathbf{Z}_t + (b_{0i} + \mathbf{b}'_{1i}\mathbf{Z}_t)\mathbf{r}_{pt+1} + \varepsilon_{it+1}.$$

The a_{0i} can be interpreted as a conditional multi-beta version of the Jensen measure. It can also be interpreted as the multi-beta extension of the study of Ferson and Schadt (1996). If a_{0i} is consistently greater than zero, the measured portfolio is able to earn consistently higher returns than what is predicted by the model.

2.3.2. Criticisms of Conditional Performance Measures

The latest frenzy about time-varying betas has led Ghysels (1998) to examine the usefulness of conditional betas. He suspected that the methods applied to model the betas had significant inconsistencies and were not suited to lead to more reliable results. He concluded that the misspecification in beta risk dynamics is almost always revealed by the non-constancy of the beta risk model parameters (just as the misspecification of the CAPM is revealed by the time variation in the betas). Therefore, he tested whether or not there are structural shifts in the parameters of conditional CAPM models. He found significant evidence for the existence of these breaks. Arguing that the time-varying betas do not capture the temporal dynamics very well, he concluded that they misprice risk. Ghysels suspected the reasons for this in the fact that betas change very slowly over time. However, all the parameters used to predict the betas were very volatile and hence not suited very well. In fact, the currently applied methods to model time-varying factor loadings in conditional CAPM or APT models lead to results that imply larger errors than the conventional fixed factor loading models. Therefore, more attention should be paid to a more satisfying specification.

2.4 Performance Measurement Based on Event Studies

2.4.1. Performance Measures

It seems that the shortcomings of the above models were accepted after some time. However, due to the lack of better models or testing methodologies, the problems were acknowledged and then more or less ignored. Nonetheless, there have been some attempts to overcome these difficulties and find alternative ways to measure performance. Since the above logical arguments are not easily dismissed, as early as 1979 Cornell (1979) suggested an event study methodology that is based on the knowledge of the composition of the evaluated portfolio. His rationale was that the assets in the portfolio should have higher returns when included in the portfolio than when excluded, assuming informed portfolio managers. Cornell obtained his performance measure by replacing a benchmark with the portfolio's asset mean returns in the evaluation period. He compared the portfolio's asset mean returns just prior to the test period with those achieved during this time span. This procedure is repeated, and significantly greater returns during the test period relative to those in the comparison period indicate superior performance. Copeland and Mayers (1982) proposed a similar Event Study Measure (ESM), where the average return on a security in a later period is used as the proxy for the period t expected return.

$$\text{ESM} = \sum_i \sum_t \frac{w_{jt}(r_{jt} - \overline{r_{jt+k}})}{T}$$

where w_{jt} is the portfolio's weight of security j at time t, r_{jt} is the return on security j at time t, and $\overline{r_{jt+k}}$ is the future average return.

Grinblatt and Titman (1993) introduced a new measure that uses a security's portfolio weight in an earlier period as the proxy for the expected portfolio weight of the security. That measure was called the Portfolio Change Measure:

$$\text{PCM} = \sum_i \sum_t \frac{r_{jt}(w_{jt} - w_{jt-k})}{T}$$

where r_{jt} is the return on security j at time t, w_{jt} and w_{jt-k} are the portfolio's weights of security j at time t and time $t-k$, the event period and the comparison period, respectively. Intuitively, this measure compares the returns of an actively managed portfolio with those of a passive portfolio formed in an earlier time period. When the portfolio manager decreases the weights of securities with positive returns and increases the weights of those with negative returns, the Performance Change Measure (PCM) will be negative. Therefore, a positive PCM attests returns higher than those earned by an uninformed investor and vice versa.

In addition, as mentioned above, Henriksson and Merton (1981) developed a non-parametric measure in order to detect systematic market timing. That measure requires that the forecast of the observed portfolio manager is observable. Henriksson and Merton define two conditional probabilities of a correct forecast: p_1, given that the risk-free asset is more profitable than the market, and p_2, given that the return on the market is higher than the return on the risk-free asset. They then show that p_1 and p_2 follow a binomial distribution. Merton (1981) showed that the forecast is only valuable to other investors if and only if the conditional probabilities p_1 and p_2 sum up to more than one. If the sum is less than one than the forecast has negative value. Hence, trading against the forecast would be profitable. Assuming that forecasts of rational managers will have positive value, the non-parametric test is a one-tailed test of the hypothesis

$$H_0 : p_1 + p_2 = 1$$

where p_1 is the probability of a correct forecast in a down market and p_2 the probability of a correct forecast in an up market.

2.4.2. Criticisms of Performance Measurement Based on Event Studies

First of all, while professional portfolio managers do have access to the portfolio weights of most mutual funds, they are very difficult and expensive for academics to obtain. To date there exist only a few papers concerning this manner of performance measurement. Also, the number of funds examined is quite small and the time periods are relatively short. Hence, the empirical evidence and concluding remarks on these performance measures are somewhat limited.

One of the critical assumptions of the above measures is that the proxy for the expected portfolio weights be independent of the security return. Often, this assumption is certainly not valid. As described in Section 3, a lot of portfolio managers implement momentum strategies, which imply that they persistently buy past winners. Hence, there is a positive relation between security returns and expected security weights, which will downwardly bias the PCM. Similarly, the results of Cornell (1979) would be downwardly biased by this relation, since he used past returns as a benchmark for expected returns.

It is obvious that the PCM and the ESM also rely on the stationary nature of returns for the correct determination of portfolio performance. But then, the same is true for basically all performance measures investigated so far. What will happen is that both measures will exhibit positive performance when the portfolio manager systematically chooses securities which are temporarily very risky.

Another weakness of the ESM is its exposure to survivorship bias. If an asset currently included in an investigated portfolio fails to survive in the short-term, the holdings of that asset cannot be used to assess the performance of the portfolio manager. Especially when managers tend to hold very risky stocks that are in severe distress, the survivorship bias becomes critical. On the other hand, the PCM cannot be influenced by survivorship bias by construction.

The non-parametric Henriksson-Merton test has been subjected to a number of criticisms. It assumes a "black or white" view of the world. For example, it does not take the extent of the difference between the risk-free rate and the return on the market into account, but only the sign of the difference. In addition, the forecast of a portfolio manager is not easily observed. In general, it has to be constructed from the portfolio weights of the managed portfolio. There are several ways to accomplish that each might yield different results.

2.5 Performance Measurement Based on Characteristics

2.5.1. Performance Measures Based on Characteristics

Daniel, Grinblatt, Titman and Wermers (1997) developed new performance measures that are based on characteristics rather than on the covariance structure of returns. These performance measures are motivated by the argument that mechanical investment strategies based on characteristics should not lead to superior abnormal performance. The conditional approach presented in an earlier section is supported by a similar argument. However, the construction of the performance measures presented here is somewhat different. Daniel, Grinblatt, Titman and Wermers (1997) construct benchmark portfolios for every single available security by sorting the universe of securities according to certain characteristics. Then the performance of every asset can be matched with a portfolio of other assets, which belong to the same class as the examined asset according to its characteristics. They separated the performance of funds into three different parts: i) characteristic selectivity (CS), ii) characteristic timing (CT), and iii) average style (AS). CS detects whether of not mutual fund managers are able to pick stocks that did better than those that would have been chosen on the base of the stated characteristics. The month t component of the CS measure is defined as

$$\mathrm{CS}_t = \sum_{j=1}^{N} w_{jt-1}(r_{jt} - r_{b(jt-1)t})$$

where w_{jt-1} is the portfolio weight on stock j at the end of month $t-1$, r_{jt} the month t return of stock j, and $r_{b(jt-1)t}$ is the month t return of the characteristics based passive portfolio that is matched to stock j during month $t-1$. The time series average, over all months that a fund exists, gives the CS measure for that fund. CT is positive when managers are able to pick high-book-to-market stocks when the returns on these stocks is especially high (this definition of timing ability is different from the traditional market timing presented earlier in the text). The month t CT measure is defined as

$$\text{CT}_t = \sum_{j=1}^{N} (w_{jt-1} r_{b(jt-1)t} - w_{jt-13} r_{b(jt-13)t}).$$

The AS measures the returns earned by a fund due to that fund's tendency to hold stocks with certain characteristics. The month t component of the AS measure equals

$$\text{AS}_t = \sum_{j=1}^{N} (w_{jt-13}\, r_{b(jt-13)t}).$$

The sum of the three measures equals the total return of the fund.

2.5.2. Criticisms of Performance Measures Based on Characteristics

Unfortunately, to our knowledge no papers concerning any criticisms of these measures have been published so far. That is certainly due to the fact that these measures have been developed only recently. However, it is obvious that when expected returns are indeed based on characteristics rather than on risk, the implications for performance measurement and portfolio analysis are tremendous. Portfolios could be constructed that achieve higher Sharpe ratios than efficient markets would allow for. It implies that the characteristics model is inconsistent with the Modilgliani and Miller (1958) theorem. Hence, if the characteristics based benchmarks are appropriate — which means that returns are based on characteristics rather than risk — the consequences for corporate finance would also be substantial.

3 Review of Empirical Evidence on Performance Measurement

3.1 Empirical Evidence on Performance Measures Based on the CAPM

3.1.1. Empirical Evidence on Selectivity Measures

Sharpe (1966) argued that differences in the performance of mutual funds were due to either differences in fees and expenses or superior skills of some fund managers to identify mispriced securities. He used data on 34 mutual funds in the time period from 1954 to 1963 and concluded that fees account for a substantial part of the differences in performance. Furthermore, fees were strongly negatively correlated to performance. When compared to the Dow-Jones Industrial Average, none of the mutual funds performed significantly better then the average, hence supporting the efficient market hypothesis. Consequently, investors should choose the fund with the lowest expense ratio or, today, invest directly in index funds.

In one of the most influential studies on performance measurement Jensen (1968) examined the performance of 115 open-end mutual funds over the period from 1945 to 1964. His findings show that mutual funds were not able to predict future securities prices; the $a_p s$ were on average slightly negative over the whole period. Since professional fund managers are supposed to be at least as well informed and educated as the general market, these findings also support the notion of efficient markets and are consistent with Sharpe's results.

Grinblatt and Titman (1989) employed an extension of the Jensen measure to analyze the performance of 279 mutual funds over the period from 1974 to 1984. They constructed their own benchmark of eight portfolios (P8), arguing that it generates the $a_p s$ that are closest to zero when applied to passive portfolios. The rationale underlying this sort of benchmark is that portfolios constructed on the basis of securities characteristics are correlated with their stock's factor loadings, and can therefore be used as a proxy. A benchmark constructed of all relevant factors is considered to be a better market portfolio. Assuming characteristics such as size, dividend yield, interest rate sensitivity, etc. are relevant factors, Grinblatt and Titman found that aggressive and small cap funds displayed superior performance. However, after accounting for transaction costs these funds no longer exhibited abnormal returns. In an extension of their study Grinblatt and Titman (1992) measured the persistence of mutual fund performance and found evidence for positive persistence, indicating that the past performance of funds provides useful information for the future.

Ippolito (1989) employed Jensen's aps on data of mutual funds over the period from 1965 to 1984. His findings are significantly different from the ones above. As an appropriate benchmark portfolio, the S&P 500 index was applied. Two important conclusions can be drawn from his test. First, he found positive $\alpha_p s$ even after accounting for transaction costs which led him to the conclusion that the money managers had superior selection skills, and their fees and expenses were well deserved. Second, his findings did not associate inferior performance with higher management fees or expenses. Considering these results Ippolito concluded that he had found evidence against the efficient market hypothesis.

Elton, Gruber, Das and Hlavka (1993) examined Ippolito's results more closely. In their study they investigated mutual funds for the period 1945 to 1964 (Jensen's period) and 1965 to 1984 (Ippolito's period). In their opinion the S&P 500 index is not the correct benchmark when examining the performance of mutual funds, because funds will always hold assets like smaller stocks or bonds that are not included in the S&P 500. Therefore, they accounted for the impact of non-S&P assets and found that the α_p's were no longer positive, but slightly negative and consistent with earlier findings. Their modified regression model is:

$$r_{pt} - r_{ft} = a_p + \beta_{p1}(r_{mt} - r_{ft}) + \beta_{p2}(r_{st} - r_{ft}) + \beta_{p3}(r_{bt} - r_{ft}) + u_{pt}$$

where r_{pt} is the return on portfolio p, r_{mt} the return on the S&P 500 index, r_{st} the return on a non S&P index orthogonal to the S&P index, r_{bt} the return on a bond index orthogonal to the former two and β_{pk} the sensitivity of the portfolio p to the factor k. In order to prove their point they regressed the CRSP value-weighted and equal-weighted index as well as the CRSP small stock index against the S&P index. For example, they found the a_p in the period from 1945 to 1964 was -4.04 for the small stock index. The a_p was significantly lower but still negative for the value-weighted index. For the second period the two indices showed $a_p s$ of 10.06 (significantly positive) and 0.57 (slightly positive), respectively. It is clear that any portfolio including unmanaged ones would exhibit positive $a_p s$ over the period from 1965 to 1984 as long as part of the money was invested in small stocks and non-S&P assets.

Reilly and Akhtar (1995) examined the CAPM in a global setting. Special attention was paid to the several alternative benchmarks that are possible when the setting is changed from the US to the whole world. Considering dominant stock market indices for four major countries in addition to a world equity index and a multi-asset index, they were interested in the benchmark return and risk relation as well as the correlation between the different benchmarks. Reilly and Akhtar calculated the domestic SML's for the three periods from 1983 to 1988, from 1989 to 1994, and from 1983 to 1994. They showed apparent inconsistencies of the different SML's, especially in the two subperiods. They concluded that the problem identified by Roll (1977), i.e. the unknown true market portfolio, only increases with global investing, making the use of a domestic stock index as a proxy for the true market portfolio very questionable. The analysis of the systematic risk measures indicates significant differences depending which benchmark is employed. Interestingly, the betas are significantly larger if a highly diversified benchmark is chosen compared to a domestic one. In fact, the betas are highly sensitive to whether a domestic stock index, a global stock index, or a well-diversified global stock and bond portfolio is chosen. Reilly and Akhtar argued that in a world where global investment is the norm rather than the exception, the financial community should start to think of systematic risk measures that employ global benchmarks.

3.1.2. Empirical Evidence on Market Timing

Treynor and Mazuy (1966) applied their market timing measure to the data of 56 mutual funds from the period 1953–1962. For the examined period, however, Treynor and Mazuy did not find any evidence of market timing, indicating that markets were efficient.

Henriksson (1984) carried out the parametric test, employing data on 116 mutual funds over the period from 1968 to 1980. His findings provided no evidence of market timing ability on behalf of fund managers. In fact, the results show a strong negative correlation between the estimated selectivity and the timing coefficients, which Henriksson explained as a possible misspecification.

Since other researchers obtained similar results, several explanations were forwarded. Jagannathan and Korajczyk (1986) used an approach which characterizes the stocks as options to explain these findings. Securities can be seen as a call option where the debt reflects the strike price and the companies' assets the underlying value. On the strike date the shareholders can either exercise their option (sell the assets, pay the debt and receive the difference) or refuse to exercise (the debtors are left with the assets). Obviously, the shareholders will only exercise if the value of the assets exceeds the debt. The more leveraged a company is, the more extensive is the option character of the stock. Jagannathan and Korajczyk showed that funds with a considerable amount of these stocks will show positive market timing and negative selectivity ability, whereas funds exposed to lower proportions of these stocks will show reverse results. Applying their findings on Merton's and Henriksson's work, they empirically showed that the parametric test results will exhibit market timing and reversed selectivity even when none exist. They proved the existence of artificial timing by running a simple regression of the CRSP equally weighted index against the value-weighted index and vice versa.

The evidence on market timing is not consistent. Vandell and Stevens (1989) and Lee and Rahman (1990) found positive evidence. Vandell and Stevens (1989) examined Wells-Fargo's model, which is based on the spread between bond yields, and the equity returns. They discovered that Wells-Fargo's strategy results in a significantly higher return than a buy-and-hold strategy. However, Chen, Lee, Rahmann and Chan (1992) obtained the same results as Henriksson (1984).

3.2 Empirical Evidence on Performance Measures Based on the APT

Lehmann and Modest (1987) used data on 130 mutual funds over the period from 1968 to 1982 to test whether the application of different benchmarks changes relative performance. They contributed to the discussion by applying the usual CAPM benchmarks as well as a variety of APT benchmarks to the same set of data which enabled them to compare the two measures directly. Roll (1978) and Copeland and Mayers (1982) had done similar research previously. However, their findings suggested that alternative risk-adjustment procedures should lead to few substantive differences in performance measures. On the other hand, Lehmann and Modest found mutual fund rankings to be very sensitive to the specific APT-model chosen to measure "normal" performance. Different implementations of the model would also yield substantial differences in the absolute and relative performance measures, although the number of relevant factors did not seem to be crucial. Crucial was the choice of the relevant model. CAPM and APT benchmarks yielded considerably different measures of performance.

First of all, these findings could be interpreted as a criticism of the APT itself. A model that allows the construction of different benchmarks, which yield significantly different performance measures, might be of little value. Second, these findings stress the importance of choosing the correct model and benchmark (two choices!) of normal performance. If this choice had been irrelevant, it would not have affected the relative and absolute performance significantly, which it did. This evidence stands in sharp contrast to many conventional assumptions in the literature.

Connor and Korajczyk (1991) examined 130 US mutual funds in the time period from 1968 to 1982. The data set consists of monthly returns. They apply performance measures based on the CAPM and the APT. When considering the CAPM, they observed the size effect described above. Also, APT based performance measures exhibit size effects, although the effect is smaller for the APT measures. They apply the Jensen measure for both models. Acknowledging the problems associated with APT models, they identify statistically significant factors, and rotate them with the help of economic variables, so that the factors are significant and have an economic meaning. In addition, they examined the timing measures of Henriksson and Merton (1981). They showed theoretically that when some funds inhibit an option character, the conventional timing measure gives misleading results, and strongly negatively correlated timing and selectivity coefficients. Since their empirical results seem to support their claim, they developed a new measure that combines the selectivity and the timing ability and is consistent with the fact that funds do indeed include put-option payoffs in their returns.

3.3 Recent Empirical Evidence on Multibeta Models

In the years before 1992 all the shortcomings of the models of the APT and especially the CAPM have been acknowledged, however no other or better models were in sight, and hence the criticisms were more or less ignored. Results like those of Banz (1981), who discovered the size effect, and Bhandari (1988), who found a positive relationship between leverage and average return, did not have a lasting impact on the discussion in the financial community. However, Fama and French (1992) showed results that seriously questioned the existence of the CAPM.

Black, Jensen and Scholes (1972) and Fama and MacBeth (1973) found that in the years prior to 1969 there was a positive relationship between the β and the average stock return. In the period from 1963 to 1990 Fama's and French's results no longer exhibit a relation between β and the return. In contrast, they found that the univariate relations between average return and size, leverage, earnings/price (E/P) and book-to-market equity (B/M) are strong. In multivariate tests only book-to-market equity and size keep their explanatory power; it seems that book-to-market equity absorbs the effects of leverage and E/P. It seems that β does not help to explain the cross-section of average returns. In fact, it seems that characteristics such as size and B/M do a much better job at least for the sample period of 1963 to 1990.

French and Fama (1993) extended their study a year later and included bonds data. Also, they expanded the set of variables used to explain returns. They find three common factors in the stock market: i) an overall market factor ii) firm size and iii) book-to-market equity. In addition, there are two bond market factors related to maturity and default risk. In view of these results, Fama and French concluded that the CAPM model derived by Sharpe, Lintner and Black is not at all suited to describe the last 50 years of average stock returns. However, they fail to deliver economic reasons that explain the story behind the size and the book-to-market effect.

Two years later Fama and French (1995) tried to overcome this problem. Based on their previous work they concluded that when assets are priced rationally, then systematic differences in stock returns are due to changes in risks. Since size and book-to-market equity seem to explain the cross-section of expected returns very well, there must be common risk factors associated with these two characteristics. Moreover, the patterns of size and book-to-market equity must be explained by the behavior of the earnings. Firms with a high book-to-market equity ratio tend to be distressed. Also, their earnings tend to be lower than those of firms with low book-to-market equity (growth stocks). French and Fama find that this is true for at least 11 years around the for-mation of the tested portfolios. The relation between size and profitability is largely due to the 1980s. The economic depression of 1981/1982 hit small companies harder than big firms. Also, small stocks did, on average, not participate as much in the boom of the following years. Summarizing their results, they found that size and book-to-market equity are related to profitability. In an effort to document that the common variation in stock returns is driven by the common factors in earnings, they achieve mixed results. Market and size factors in earnings help explain the market and size factors in returns. However, the book-to-market equity factor in earnings has little explanatory power for the factor that seems to drive stock returns.

Since the results of Fama and French are not easily dismissed, Kothari, Shanken and Sloan (1995) had another look on the conclusions of their study. In spite of their colleagues they used annual data to estimate the βs and somehow obtained different results. Suddenly, market risk or β gets priced significantly. Also, the relation between book-to-market equity is much lower than that in Fama and French (1992). Moreover, they conclude that the former results are biased significantly by the survivorship bias in the COMPUSTAT database affecting the high book-to-market equity stocks' performance. When using an alternative data source (S&P industry-level data from 1947 to 1987) they find that book-to-market equity is at best weakly correlated with average stock returns. On the other hand, they also find evidence of a size effect. Other researchers were also concerned about possible bias in the data of French and Fama.

Kim (1997) investigated the COMPUSTAT selection bias. In order to achieve independent results, he collected most of the missing data from Moody's Manuals. However, even with aggregated data the results with the COMPUSTAT sample do not change significantly. The selection bias is not so severe that the relationship between average returns and book-to-market equity is significantly affected.

Kim also compensated for another bias, i.e. the errors-in-variables (EIV) bias. Since true βs are unknown, estimated βs serve as a proxy for the unobservable true βs and this involves an EIV problem. Several studies showed that the EIV problem causes an underestimation of the price of β-risk and an overestimation of the other cross-sectional regressors associated with idiosyncratic variables that are observed without errors such as firm size book-to-market equity, and earnings per share price. Kim and others developed a correction method for that issue. After correcting for the EIV problem, market βs had economically and statistically significant force, regardless of the absence or presence of firm size, book-to-market equity, and earnings price ratio. Interestingly, the intercept estimate is insignificant when market β alone is used.

Size is slightly significant when monthly returns are used, while in the case of quarterly returns firm size is insignificant. On the other hand, the influence of book-to-market equity is unimpressed by the correction for the EIV problem or different return measurement intervals. Therefore, Kim concludes that this characteristic gives stronger evidence in favor of a misspecification of the CAPM than does firm size.

Carhart (1997) analyzed monthly data on diversified equity funds from January 1962 to December 1993. Due to Carhart, the data is free of survivorship bias. He applied Jensen measures of the traditional CAPM, a three factor model suggested by Fama and French (1993), and his own four factor model. The latter is identical to the three factor model except for a factor for the prior year return to capture the effects of the momentum effect of Jegadeesh and Titman (1993). The mean absolute errors from the CAPM, the three-factor, and the four-factor model are 0.35 percent, 0.31 percent, and 0.14 percent respectively. The argument that his four-factor model seems to explain the cross-section in average stock returns rather well receives further support from the fact that most of the patterns in pricing errors disappear when the model is applied.

Carhart was especially concerned about the persistence of mutual fund performance. He showed that buying last year's top-decile investment funds and selling the bottom-decile funds yields a return of 8% per year. The largest part (4.6%) is due to momentum effects. The rest is explained by differences in expense ratios and transaction costs. In fact, expense ratios, portfolio turnover, and load fees are significantly and negatively related to performance. Load funds underperform no-load funds on average by an additional 0.8%. Furthermore, Carhart found that funds with the highest Jensen measures exhibit also the highest expenses, which he interpreted as evidence in line with market efficiency. On the other hand, he admitted that buying last year's winners is a feasible strategy for capturing the one-year momentum effect. He claimed that after that strategy has been widely followed, mutual funds will charge higher transaction fees to incoming and outgoing investors in order to reestablish their performance.

French and Fama (1998) again examined stocks with high book-to-market equity ratios or earnings to price ratios, which they named value stocks. Lakonishok, Shleifer, and Vishny (1994) showed that in the US there exists a strong value premium in average returns. French and Fama interpreted the value premium to be associated with relative distress. In their latest paper the authors sought the answer on two questions: i) does the value premium exist outside the United States, and ii) if it exists does a risk factor for relative distress explain this risk premium in the same way it explains the premium in the United States. They examined annual data on market, value and growth portfolios for thirteen major markets in America, Europe, Australia and the Far East from 1974 to 1994. Indeed, they found evidence for an international value premium. The difference between the average annual return from a high and a low book-to-market equity portfolio was 7.68%, which was statistically significant. As far as their second question is concerned, they found that an international CAPM did not explain the observed value premium. However, a factor for relative distress in Merton's (1973) intertemporal CAPM or an APT version did a good job in explaining the differences in returns.

3.4 Empirical Evidence on Conditional Performance Measures

Ferson and Schadt applied their conditional performance measures to the data of 67 open mutual funds from January 1968 to December 1990. They divided the mutual funds into four categories: income, growth, growth-income and maximum gain funds. For the unconditional Jensen test, it seems that the funds have more negative than positive alpha's. When the conditional model is used, this tendency disappears. The significant alphas are split exactly in half between negative and positive ones. Analyzing their market timing results when applying a naive buy and hold strategy, Ferson and Schadt found that for both the unconditional Treynor-Mazuy and the unconditional Henriksson-Merton test the alphas are significantly positive and the market timing coefficient is significantly negative — obviously a misspecification of the models. Jagannathan and Korajczyk (1986) obtained the same result and explained the opposite signs for timing coefficient and alphas by showing that naive results can exhibit option characteristics. These results disappear as well when the conditional model is used, thereby further promoting the conditional approach. For the Treynor-Mazuy test, the unconditional model delivers mostly negative market timing coefficients, which is consistent with earlier findings. The results change completely when the conditional model is used. Suddenly, the vast majority of the significant timing coefficients is positive. The unconditional Henriksson-Merton test provides very similar results to the unconditional Treynor-Mazuy test. However, when the conditional test is performed, all significant forms of market timing vanish and the coefficients tend to be positive.

Jagannathan and Wang (1996) also developed conditional performance measures described above. However, they primarily tried to justify empirically their derivation of the unconditional implications of the conditional CAPM. In order to achieve this, they used the returns to stocks of non-financial firms listed in NYSE and AMEX from 1962 to 1990. When forming their portfolios they followed the procedure of Fama and French (1992). Next, they showed that the results of French and Fama (1992) could be consistent with their conditional CAPM. They further suggested that the results of French and Fama were due to the fact that they took a poor proxy for the true market portfolio. In order to get a better proxy for the return on aggregate wealth, they included a measure of return on human capital. When they tested their model empirically, they found that it explained more than fifty percent of the cross-sectional variation in average returns; the data failed to reject their model. Furthermore, size and book-to-market equity have little additional explanatory power.

Similar reasoning comes from Ferson and Harvey (1999). They examined factor models such as that of Fama and French (1993) or Elton, Gruber and Blake (1995), who concluded that predetermined variables explain the cross-section of expected returns. While these researchers claimed that the characteristics are proxies for common risk factors, others wrote that the market systematically misprices securities with these characteristics (Lakonishok, Shleifer and Vishny (1994), Daniel and Titman (1997)). In addition, the third party expressed the opinion that the aforementioned results were due to data snooping and various biases (Kothari, Shanken and Sloan (1995), Kim (1997)). The Fama and French model was developed to explain unconditional mean average returns. Ferson and Harvey test it on conditional returns and use a set of lagged economy-wide predictor variables similar to that of Ferson and Schadt (1996) as a model of dynamic patterns in returns. They utilized monthly data of returns on US common stock portfolios from July 1963 to December 1994. The portfolios were formed with a method similar to that of Fama and French (1993). First of all, there seems to be strong evidence for time-varying betas. In the following, the unconditional regression of the portfolio excess returns over time on the three Fama and French (1992) factors basically delivers the same results as those of Fama and French. However, when the excess returns are regressed on the three factors plus the predetermined information variables, there is strong evidence against the Fama and French model. Ferson and Harvey admitted that this test might be biased against the tested model, since the betas were fixed and not time-varying. In a third regression they allow for time-varying betas that depend on the instruments. However, even the conditional version of the Fama and French model could be rejected. The alphas were time-varying and significant. Not only the intercept, but also the loads on the lagged instruments were highly significant. They concluded that these loads reveal information that is not captured by the popular factors for the cross-section of expected returns. Therefore, they reasoned that the conditional Fama and French model and the performance measures derived therefrom are not appropriate, since even funds that implement a mechanical strategy based on the predetermined information variables can outperform these measures. When exploring the four-factor model of Elton, Gruber and Blake they basically obtained the same results. Given the fact that practitioners and researchers already utilize the Fama and French model for performance measurement, risk analysis and cost of capital calculations, the authors doubt the correctness of the results of these applications.

3.5 Empirical Evidence on Performance Measurement Based on Event Studies

Copeland and Mayers (1982) used the event study measure in their study of the Value Lines Investment Survey recommendations made between 1965 and 1978. They found that investors who followed Value Line's recommendations earned significantly abnormal positive returns. The event study measure utilized by Copeland and Mayers provides an estimate of the sum of the time-series covariances between the portfolios weights and the subsequent returns of each asset included in the evaluated portfolio. However, they admitted that their methodology might be biased, since the estimation period for the benchmark and the testing period are not the same, and therefore the results might be biased. In addition, survivorship problems can systematically bias the results quite substantially.

Grinblatt and Titman (1993) employed the portfolio change measure that avoids some of these issues and, in addition, has some computational advantages for statistical interference. They reexamined the data on 155 mutual funds in the period from 1974 to 1984. The results of the study indicate abnormal positive performance implying superior information and/or superior ability on behalf of the fund managers. Aggressive growth funds achieved the highest performance. However, on average all mutual funds exhibited abnormal positive performance. The persistence of these results is interesting. Successful funds showed consistent superior performance. However, private investors can not profit from the ability of investment managers by investing in their funds, since expenses and fees account for a large part of the abnormal returns. In fact, the net performance of mutual funds is on average zero. Which is consistent with the notion of efficient markets.

Grinblatt, Russ and Wermers (1995) suggested that the persistence of the abnormal positive performance is due to momentum strategies of investment funds. A momentum strategy states that the fund buys stocks that were successful in the past and are likely to stay successful in the short-term. Jegadeesh and Titman (1993) documented this momentum effect and suggested that it is used by most mutual funds as a stock selection criteria.

Daniel and Titman (1997) gave a new interpretation to the results of previous studies, which found that characteristics did help explain the cross-section of expected returns. They argued that it is the characteristics that explain the cross-sectional variation in stock returns rather than the covariance structure of returns. They started with a discussion of the results of Fama and French (1992, 1993, 1995, 1996)). As the economic story behind their findings, Fama and French suggested that size and book-to-market equity are proxies for the firm's loadings on priced common risk factors. Hence, they assume that the "true" model is a multifactor model, such as an APT version or Merton's (1973) Multibeta CAPM. On the contrary, as stated in the previous section, Lakonishok, Shleifer, and Vishny (1994) suggested that the rationale underlying the results is different. They thought that investors are overly optimistic about firms that have done well in the past and overly pessimistic about those which have done poorly. They did not argue about the possibility that there are common risk factors, however they claimed that the return premia associated with these factor portfolios are just too large and the covariances with macro factors are just too low. In the opinion of Daniel and Titman (1997) their evidence is compelling. Therefore, Daniel and Titman (1997) tested for two issues: i) whether there really are pervasive factors that are directly associated with size and book-to-market equity, and ii) whether there are risk premia associated with these factors. They tested whether the high returns of high book-to-market and small size stocks can be attributed to their factor loadings. The results of the study indicate that there is no separate risk factor associated with high or low book-to-market equity. Moreover, Daniel and Titman could not find a return premium associated with any of the three factors identified by Fama and French (1993). They found that high book-to-market stocks covary strongly with other high book-to-market stocks. However, the covariances do not result from there being particular risks associated with distress, but rather from the fact that those firms often have similar properties (industries, business cycles, etc.). Distressed firms' covariances were equally strong before they became distressed. It seems that there is no evidence of a separate distress factor. In addition, once controlled for the characteristics, expected returns no longer seem to be positively related to the loadings on the market or size or book-to-market equity. Hence, the factor loadings do not explain average returns beyond the extent to which they act as proxies for these characteristics. The authors concluded that when expected returns are indeed based on characteristics rather than on risk, the implications are inconsistent with the Modilgliani and Miller (1958) theorem — one of the basics of corporate finance.

Daniel, Grinblatt, Titman and Wermers (1997) argued that no abnormal performance should be associated with mechanical strategies based on stock characteristics such as size, book-to-market equity or momentum (prior year return). Picking up on the above findings, they developed new performance measures that are based on characteristics rather than on the covariance structure of returns. They did not form factor portfolios that are based on characteristics of sorted portfolios, which are then regressors in a traditional multifactor regression model — the standard testing procedure. Instead, they compared the returns of portfolios of stocks with equivalent characteristics and decomposed the performance of each fund into the three measures characteristic selectivity (CS), characteristic timing (CT) and average style (AS). Daniel, Grinblatt, Titman and Wermers (1997) applied their measures to an extensive new database of over 2500 equity funds from 1974 to 1994. They claimed that their database has been the largest and most complete sample of mutual fund holdings so far and that the database has been free from survivorship bias. One drawback is the fact that for this method the mutual fund holdings must be known. They used characteristics (size, book-to-market, prior-year-return) to form 125 (5*5*5) benchmark portfolios (after separating the funds in quintiles for each criterion). The appropriate benchmark for each component stock is chosen by directly matching the characteristics of the component stock with those of the 125 benchmark portfolios. The CS measure states positive abnormal performance of investment funds over the full period. However, that performance is relatively small and close to the funds' expenses. Aggressive-growth and growth funds exhibit the highest performance, on the other hand, they produce the highest costs. This is consistent with other findings which that suggest that informed traders are able to outperform the market just enough to earn back the costs of obtaining the information. The CT measure provides no evidence that fund managers are style timers.

Moreover, Daniel, Grinblatt, Titman and Wermers apply three additional performance measures i) the Grinblatt and Titman (1993) measure (GT), ii) the Carhart (1997) measure that is based on the Fama and French (1993) factor model, and iii) the traditional CAPM-based Jensen's measure. The results for the GT measure show that the average fund has consistently provided abnormal returns. However, they stated that momentum effects most likely bias the GT measure, since funds that apply a momentum strategy outperform funds that don't on average under the GT measure. The Carhart measure of selectivity exhibits small positive abnormal performance statistically significant at the ten percent level. However, the average benchmark-adjusted returns are about the same size as the typical management fee. According to the authors that evidence is consistent with equilibrium.

4 The Stable Paretian Approach

4.1 Empirical Evidence

It is a standard assumption in theoretical and empirical research in finance that stock returns follow multivariate normal distributions. In fact, a number of asset pricing models has its roots in the multivariate normal assumption. For example, it is well-known that if stock returns would indeed follow a multivariate normal distribution, rational investors would choose mean-variance efficient portfolios and the CAPM would hold. Chamberlain (1983) showed that the hypothesis of normality could be replaced by one that assumes finite variance. However, this is still equivalent to the proposition that stock returns are in the normal domain of attraction of a Gaussian law. By the Central Limit Theorem, assuming the variance exists,

$$\frac{\sum_{i=1}^{m}(r_j^i - E(r_j^i))}{\sqrt{m}} \xrightarrow{d} N \quad \text{for all } j = 1, 2, \ldots, n,$$

where $\mathbf{r}^i$ is the ith observation of the vector of gross returns $\mathbf{r}$, $E(\mathbf{r})$ is the vector of mean gross returns. Hence, $\mathbf{N}$ is an n-dimensional normal random vector with zero mean. So far there has been considerable focus on the question as to whether or not this assumption is correct.

The hypothesis was has not been questioned seriously before Mandelbrot (1963) began his seminal work on certain speculative prices. He analyzed cotton prices and found that the empirical distributions were leptokurtic, which implies that too many values are close to the mean and too many values are out in the tails. Earlier studies had simply excluded the outliers in order to promote normal distributions. Alternatively, in another classical procedure it was assumed that observations are generated by a mixture of two normal distributions. One is considered to be a random contaminator since it is assumed to have a large variance but a small weight. Nevertheless, Mandelbrot proposed a special class of distributions, which he has labeled as stable Paretian distributions. In an extension to his paper Mandelbrot (1968) also examined wheat prices, railroad stocks, and various interest and exchange rates and received similar results. Fama (1965) took up the work of Mandelbrot and analyzed the daily returns of the thirty stocks of the Dow Jones Industrial Average. He found evidence that the distributions of stock returns also, too, exhibited significant leptokurtosis. The departures from normality are in the direction of the Mandelbrot stable Paretian hypothesis. Furthermore, Fama estimated the α of the underlying distributions, the index of stability. For $\alpha = 2$ the distribution is normal. The more weight in the tails of the distribution, the lower the α — therefore, α is also a measure for the tail thickness. Fama concluded that changes in the log price of stocks of large mature companies follow stable Paretian distributions with index of stability close to two, but still less than two.

Generally, the hypothesis of normally distributed stock returns has been rejected. These conclusions, however, have been based on univariate tests of normality. Fama (1976) rejects the normal distribution for one half of the companies in the Dow Jones Industrial Index over the 1951–1968 sample period. However, as Fama pointed out, these results are difficult to interpret, since the research has investigated the multivariate normality of stock returns using tests based on the marginal distribution of returns. Due to the fact that returns are contemporaneously correlated, the results might not be unambiguous. Richardson and Smith (1993) developed a test for multivariate normality and applied it on the companies of the Dow Jones Industrial Average. They found highly significant evidence that stock market returns and market model residuals are non-normal. The non-normality appears in both the marginal and the joint distribution of the assets. Of course, these findings through doubt on put results into doubt that are based on the assumption of normality.

More recent evidence comes from Mittnik, Rachev and Paollela (1997), who have provided empirical evidence that return distributions belong to L_P with p equal to or greater than one and less than two. They paid close attention to three financial time series, the daily AMEX Composite index, the daily AMEX OIL index, and the daily DM/US$ exchange rate. For our purposes especially the former two are of particular interest. The Composite index is often used in performance measurement as a market portfolio, and an OIL index is often considered as a factor portfolio in APT models. Mittnik, Rachev and Paollela estimated the parameters for the time series assuming either a normal distribution or an α-stable distribution. In all cases the estimated index of stability is well below two. Furthermore, the two goodness-of-fit measures are applied. The log-likelihood values and the Kolmogorov distances clearly indicate clearly that the α-stable distribution dominates the normal distribution.

4.2 Stable Paretian Distributions

4.2.1. Basic Properties of Pareto-Stable Laws

A stable Paretian distribution is any distribution that is invariant under addition, which means that the sums of independent identically distributed (i.i.d.) stable Paretian random variables are stable Paretian themselves. In fact, it can be shown that Paretian distributions are the only possible limiting distributions for sums of i.i.d. random variables. This is to say that if for a sequence of m i.i.d observations $(r^{(i)})_{i>0}$ there exist normalizing constants $a^{(m)}$ and $b^{(m)}$, $m \geq 1$, such that

$$\frac{1}{a^{(m)}} \sum_{i=1}^{m} r^{(i)} + b^{(m)} \xrightarrow{d} X, \quad \text{as } m \to \infty,$$

then the limiting random variable X is an α-stable random variable. Substantial parts of the theory of these distributions has long been well-known, see Feller (1966), Zolotarev (1951) and Samorodnitsky and Taqqu (1994). Nevertheless, we review the most important facts on stable laws. A random variable X has a stable distribution if its characteristic function admits to the form

$$\Phi(t) = E(e^{itX}) = \begin{cases} \exp\{-\sigma^\alpha |t|^\alpha (1 - i\beta(\text{sign}(t))) \tan\left(\frac{\pi\alpha}{2}\right) + it\mu\} & \text{if } \alpha \neq 1 \\ \exp\{-\sigma|t| \left(1 - i\beta\frac{2}{\pi}(\text{sign}(t))\right) + it\mu\} & \text{if } \alpha = 1, \end{cases}$$

where sign(t) is defined as $t/|t|$, and $\alpha \in (0, 2]$ is the so-called index of stability of X. When $\alpha = 2$, X is the normally distributed. The index of stability α can be viewed as a measure of tail thickness. In fact, the asymptotic behavior of the right tail is

$$\lim_{\lambda \to \infty} \lambda^\alpha P(X > \lambda) = C_\alpha \frac{1+\beta}{2} \sigma^\alpha,$$

and that of the left tail is

$$\lim_{\lambda \to \infty} \lambda^\alpha P(X < -\lambda) = C_\alpha \frac{1-\beta}{2} \sigma^\alpha,$$

where

$$C_\alpha = \begin{cases} \frac{1-\alpha}{\Gamma(2-\alpha)\cos(\pi\alpha/2)} & \text{if } \alpha \neq 1 \\ \frac{2}{\pi} & \text{if } \alpha = 1. \end{cases}$$

In the situation observed in real asset return data; for daily and weekly stock returns, the index of stability α is around 1.6–1.8 and 1.7–1.9, respectively.

The parameter $\beta \in [-1, 1]$ is an index of skewness. Whenever $\beta > 0$ the distribution is skewed to the right and vice versa. When $\beta = 0$, the distribution is symmetric, and in this case the characteristic function has the simple form:

$$\Phi(t) = E(e^{itX}) = \exp(-\sigma^\alpha |t|^\alpha + it\mu)$$

Furthermore, when X is symmetric α-stable ($S\alpha S$), the characteristic function further reduces to

$$\Phi(t) = E(e^{itX}) = \exp(-\sigma^\alpha |t|^\alpha),$$

since the parameter μ is equal to zero.

The parameter μ is a location parameter. Whenever $\alpha \in [1,2]$, μ is the expected value or mean of the distribution.

The parameter σ is the scale parameter, the index of dispersion. If $\alpha = 2$, the square of σ equals one half-the variance.

Assume that the random variables X_i, $i = 1, \ldots, n$, are α-stable distributed with the same index of stability α ($\alpha \in [1,2]$), but possibly different location and scale parameters. Then, every linear combination of the X_i is again α-stable. Given $\alpha \in [1,2]$, this implies that the vector $\mathbf{X}$ of the X_i is a α-stable random vector (see Samorodnitsky and Taqqu (1994)). In fact, when the X_i are $S\alpha S$, then $\mathbf{X}$ is an $S\alpha S$ random vector. If $\alpha \neq 1$, the characteristic function of a α-stable random vector becomes

$$\begin{aligned}\Phi_\alpha(\mathbf{t}) &= E\left[\exp\left\{i\sum_{k=1}^{n} t_k X_k\right\}\right] \\ &= \exp\left\{-\int_{S_n} |(\mathbf{t},\mathbf{s}|^\alpha(1 - i\,\mathrm{sign}(\mathbf{t},\mathbf{s})\tan(\pi\alpha/2)\Gamma_X(ds) + i(\mathbf{t},\mu)\right\},\end{aligned}$$

where Γ_X is called the spectral measure of the vector $\mathbf{X}$, and its support is concentrated on the n-dimensional unit sphere S_n. If $\mathbf{X}$ is a $S\alpha S$ random vector, the characteristic function simplifies to

$$\Phi_\alpha(\mathbf{t}) = \exp\left\{-\int_{S_n} |(\mathbf{t},\mathbf{s})|^\alpha \Gamma_X(ds)\right\}.$$

Furthermore, the concept of the "covariation" will be needed. The "covariation" is related to the covariance in the finite variance case and is designed to replace the covariance in the case $1 < \alpha < 2$. Let X_1 and X_2 be jointly symmetric α-stable distributed, which is to say that $\mathbf{X} = (X_1, X_2)$ is a symmetric α-stable random vector, then the "covariation" of X_1 on X_2 is the real number

$$[X_1, X_2]_\alpha = \int_{S_2} s_1 s_2^{\langle \alpha-1 \rangle} \Gamma(ds),$$

where the "signed power" $a^{\langle p \rangle}$ is $|a|^p\,\mathrm{sign}(a)$. The measure Γ is the spectral measure of the random vector (X_1, X_2). The variation of X_1 is then defined as

$$V_\alpha(X_1) = [X_1, X_1]_\alpha = \int_{S_n} |s_1|^\alpha \Gamma_X(ds) = \sigma_1^\alpha,$$

where Γ_X is the spectral measure of (X_1, X_2). Note that for $\alpha = 2$ the covariation becomes one half the covariance and the variation becomes one half the variance.

Assume that all the gross returns r_i are α-stable distributed with the same index of stability α ($\alpha \in [1,2]$), but possibly with a different location and scale parameters. Then, every linear combination of the r_i is again α-stable. Hence, every portfolio's gross return $\mathbf{w}'\mathbf{r}$ has an α stable distribution $S_\alpha(\sigma_w, \beta_w, \mu_w)$, where $\mathbf{w} = (w_1, \ldots, w_n)$ are the portfolio weights of the assets j (the w_j sum up to one), and $\mathbf{r} = (r_1, \ldots, r_n)$ the vector of gross returns on the assets j, $j = 1, \ldots, n$. The scale parameter σ_w and the skewness parameter β_w are given by

$$\sigma_w = \left(\int_{S_n} |w^T s|^\alpha \Gamma(ds) \right)^{1/\alpha}, \quad \beta_w = \frac{\int_{S_n} |w^T s|^\alpha \text{sign}(w^T s) \Gamma(ds)}{\sigma_w^\alpha}.$$

If the relevant distributions are symmetric ($\beta_j = 0$), then σ_w becomes:

$$\sigma_w = \left(\sum_{j=1}^{n} (\sigma_j)^\alpha |w_j|^\alpha \right)^{1/\alpha},$$

where σ_j is the scale parameter and w_j is the portfolio weight of r_j. Also, in the symmetric case, the variation of the return on the portfolio becomes

$$[r_m, r_m]_\alpha = \int_{S_n} \left| \sum_{j=1}^{n} w_{mj} s_j \right|^\alpha \Gamma_r(d\mathbf{s}).$$

The sub-Gaussian random vectors represent an important class of stable random vectors. Their importance is two-fold. First, sub-Gaussian distributions have a finite number of parameters, and thus have desired forecasting properties. Second, the set of all weak limits of sub-Gaussian distributions coincides with the class of stable distributions. Choose

$$A \sim S_{\alpha/2}((\cos(\pi\alpha/4))^{2/\alpha}, 1, 0),$$

and let

$$\mathbf{G} = (G_1, G_2, \ldots, G_n)$$

be a zero mean Gaussian vector independent of A. Then the random vector

$$\mathbf{X} = (A^{1/2}G_1, A^{1/2}G_2, \ldots, A^{1/2}G_n)$$

is a sub-Gaussian $S\alpha S$ random vector with underlying Gaussian vector $\mathbf{G}$. The characteristic function of a sub-Gaussian $S\alpha S$ random vector becomes

$$\Phi_\alpha(\mathbf{t}) = \exp\{-2^{-\alpha/2}\sigma^\alpha|\mathbf{t}|^\alpha\} = \exp\left\{ -\left| \frac{1}{2} \sum_{j=1}^{n} \sum_{k=1}^{n} t_j t_k \operatorname{cov}(G_j, G_k) \right| \right\},$$

since the spectral measure Γ_X is uniform if and only if $\mathbf{X}$ is sub-Gaussian $S\alpha S$.

4.3 The Principle of Diversification

Another property of stable random variables is that the variance is not necessarily defined. In fact, let $X \sim S_\alpha(\sigma, \beta, \mu)$ with $0 < \alpha < 2$. Then

$$\begin{aligned} E(|X|^p) &< \infty \quad \text{for any } 0 < p < \alpha, \\ E(|X|^p) &= \infty \quad \text{for any } p \geq \alpha. \end{aligned}$$

This implies that the second moment is infinite for all stable random variables that are not Gaussian. Hence, many techniques applied above in Section 4.2 are no longer valid or are no longer meaningful for α-stable random variables with $\alpha < 2$. For example, both the CAPM model and the APT model argue that risk can be measured by the dispersion of asset returns expressed by the variance. However, not the total variance is relevant, but only a certain part of it, since investors can reduce the dispersion of the distribution of the return on a portfolio by diversification. In order to identify the relevant part, the covariance is used extensively. Although both measures, variance and covariance are no longer meaningful, the same cannot be said about the principle of diversification. The following argument is basically taken from Fama (1965b).

Assume that the returns r_j on all assets are related to a common underlying factor, otherwise, they are independent of each other. Then, the returns are generated by

$$r_j = b_j M + e_j,$$

where e_j expresses random error or noise in the relation of r_j and M or the idiosyncratic component of the return, and b_j is a measure of the relationship between r_j and the factor variable M, which is itself a random variable. Henceforth, without loss of generality, we assume that all random variables follow α-stable symmetric distributions. Let the $e_j \sim S_\alpha(\sigma_{ej}, 0, 0)$ and $M \sim S_\alpha(\sigma, 0, 0)$. Hence, r_j is an α-stable random variable, exactly $r_j \sim S_\alpha(\sigma_j, 0, 0)$, where $\sigma_j = ((\sigma_{ej})^\alpha + (b_j \sigma_m)^\alpha)^{1/\alpha}$. Furthermore, assume for the sake of simplicity that $w_j = 1/n$, which means that the same amount is invested in every available asset. The $(\sigma_w)^\alpha$ of the portfolio becomes

$$\sigma_w^\alpha = \sum_{j=1}^{n} (\sigma_{ej}^\alpha + |b_j|^\alpha \sigma_m^\alpha)(1/n)^\alpha = (1/n)^\alpha \sum_{j=1}^{n} \sigma_{ej}^\alpha + |\bar{b}_n|^\alpha \sigma_m^\alpha).$$

Therefore, when $\alpha = 1$,

$$\sigma_w = \bar{\sigma}_{ej} + |\bar{b}_n| \sigma_m.$$

However, when $\alpha > 1$,

$$\sigma_w^\alpha < \bar{\sigma}_{ej}^\alpha + |\bar{b}_n|^\alpha \sigma_m^\alpha,$$

since

$$(1/n)^\alpha \sum_{j=1}^{n} \sigma_{ej}^\alpha < \overline{\sigma}_{ej}^\alpha.$$

In fact, the limit becomes

$$\lim_{n\to\infty} \sigma_{ej}^\alpha = |\overline{b}_j|^\alpha \sigma_m^\alpha.$$

When the number of the securities in the portfolio reaches infinity, then the dispersion of the portfolio consists only of the component due to the common underlying factor. The component due to noise or idiosyncratic factors isgets diversified away completely in the case $\alpha > 1$. It is obvious that the opposite is true when $\alpha < 1$. In this case the dispersion of the portfolio growsincreases proportionally stronger and no rational investor would diversify his portfolio. In empirical financial data there have been no reported observations that are in line with an index of stability below one. Given that result, performance measures based on stable distributions should compare returns exclusively to that part of the dispersion that cannot be diversified away.

4.4 Performance Measures Based on a α-Stable Version of the CAPM

4.4.1. The CAPM in the Case of Stable Distributions

Since the CAPM was never really satisfactory when empirically tested, Fama (1971) considered the case when the asset returns follow α-stable distributions. Although there are many different alternative methods, the derivation of the risk - return relationship commenced by Fama and completed by Gamrowski and Rachev (1994) is probably the most intuitive. Fama assumed that asset returns r_j are distributed symmetrically α-stable ($S\alpha S$) distributed and generated by

$$r_j = a_j + b_j M + \varepsilon_j,$$

where a_j is some constant, b_j is a measure of the relationship to the random variable M, and ε_j represents random disturbances. M and the ε_j are supposed to be mutually independent $S\alpha S$-distributed random variables. Furthermore, Fama supposed that consumers are all risk-averse and expected utility maximizers. Consumers must allocate their resources to current consumption and a portfolio, with a whose market value in the next period which determines consumption in period two. In addition, the market is in equilibrium. First, Fama derived that the alternative investment decisions can be ranked on the basis of current consumption, expected portfolio return μ_w, and dispersion of portfolio returns σ_w alone. In a next step, he proved that the expected utility maximizing portfolio for consumers that are risk-averse must be return - dispersion efficient, given a concave utility function. Every consumer will choose an optimal portfolio on the efficient set. Suppose for a certain consumer that is point e. The efficient portfolio e is given by the solution to the following maximization problem:

$$\max \mu_w = \max \sum_{j=1}^{n} w_j \mu_j$$

$$\text{s.t. } \sigma_w = \left(\sigma_m^\alpha \left| \sum_{j=1}^{n} w_j b_j \right|^\alpha + \sum_{j=1}^{n} \sigma_{\varepsilon j}^\alpha |w_j|^\alpha \right)^{1/\alpha} = \sigma_e$$

and

$$\sum_{j=1}^{n} w_j = 1.$$

Using the Lagrangian method, it can be shown that after some elementary calculations, the solution of the programming problem also satisfies the following risk - return relation,

$$\mu_i - \mu_j = \lambda_e \left[\frac{\partial \sigma_e}{\partial w_{ej}} - \sigma_e \right],$$

where λ_e is the Lagrangian multiplier representing the slope of the efficient set at the point e, w_{ej} is the proportion invested in asset i in the efficient portfolio, and σ_e is the dispersion of the efficient portfolio. Points on *bed* represent the efficient set in Figure 2. Next, Fama introduces a risk-free asset. Portfolios consisting of the risk-free asset and any portfolio a of risky assets lie on a straight line according to

$$r_p = wr_f + (1 - w)r_a \qquad \text{(see Figure 3).}$$

Rotating the straight line from r_f upward until it can moved no further without leaving the feasible set yields the new efficient set. In equilibrium, all investors with the same ex ante beliefs must hold the same fund of risky assets. Sharpe (1964) and Lintner (1965) were the first to recognize that this implies that the efficient fund of risky assets, m, is the same as the market portfolio. Separation takes place and the optimal portfolio for every investor consists of some combinations of the risk-free asset and the market portfolio m. Therefore, the new efficient set will be the straight line according to

$$r_p = wr_f + (1 - w)r_m.$$

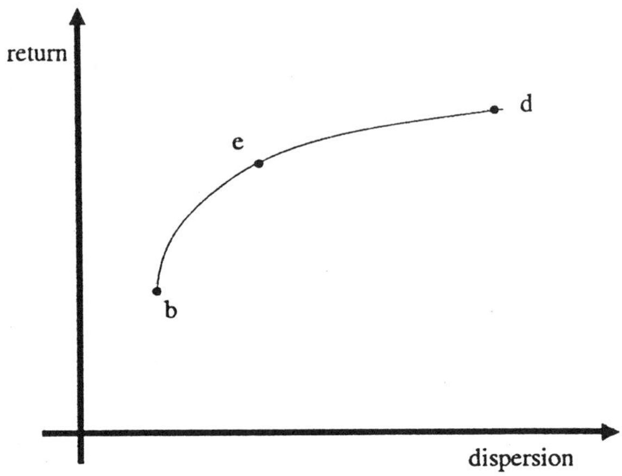

Figure 2: Return - dispersion efficient set.

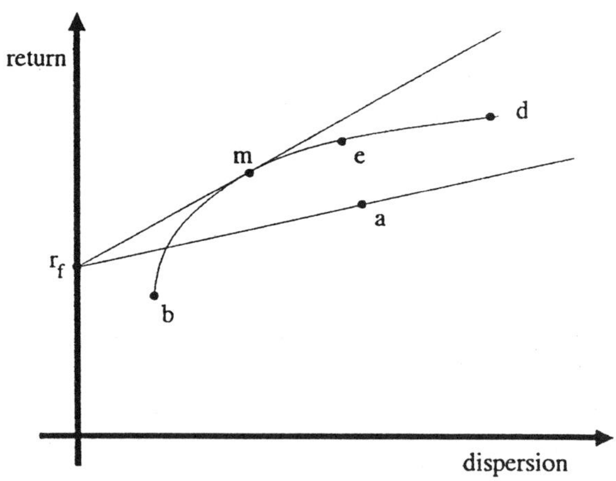

Figure 3: Risk - return relation with a risk-free asset.

The slope of that straight line (see Figure 3) is

$$\lambda_m = [\mu_m - r_f]/\sigma_m.$$

Since the market portfolio is efficient, the risk - dispersion relation from above must also hold for m. Combined with the equation for λ_m that relation becomes

$$\mu_j = r_f + \left[\frac{\mu_m - r_f}{\sigma_m}\right]\frac{\partial \sigma_m}{\partial w_{mj}},$$

where μ_j and μ_m are the expected returns on asset j and the market portfolio, respectively, r_f is the return on the risk-free asset, σ_m is the dispersion of the market portfolio, and w_{mj} is the portfolio weight in the market portfolio of asset j. Thus, the above equation must hold in equilibrium for the risk - dispersion relationship.

As stated above, $\sigma_m = V_\alpha(r_m)^{1/\alpha}$, where $V_\alpha(\,)$ denotes the variation. As a result

$$\frac{1}{\sigma_m}\frac{\partial \sigma_m}{\partial w_{mj}} = \frac{1}{V_\alpha(r_m)}\frac{\partial V_\alpha(r_m)}{\partial w_{mj}}.$$

Recalling that $V_\alpha(r_m) = \int_{S_n}\left|\sum_{j=1}^n w_{mj}s_j\right|^\alpha \Gamma_r(ds)$, where Γ_r is the spectral measure of the vector of asset returns $\mathbf{r}' = (r_1, \ldots, r_n)$. Hence, the partial derivative of $V_\alpha(r_m)$ becomes

$$\frac{\partial V_\alpha(r_m)}{\partial w_{mj}} = \alpha \int_{S_n} s_j \left(\sum_{i=1}^n w_{mi}s_i\right)^{\langle \alpha - 1\rangle} \Gamma_r(ds) = \alpha[r_j, r_m]_\alpha.$$

Therefore, we can rewrite the risk - dispersion relation derived by Fama (1971) and Gamrowski and Rachev (1994) as

$$\mu_j = r_f + (\mu_m - r_f)\frac{[r_j, r_m]_\alpha}{V_\alpha(r_m)},$$

which represents the version of the traditional CAPM assuming that asset returns are jointly α-stable distributed with the index of stability α.

There are a number of different ways, in which the above result can be derived. Indeed, the methodway used by of Fama shown here, represents a restricted case. Fama requires M and the ε_j to be mutually independent, whereas Gamrowski and Rachev (1994) derive the result with the help of Ross' (1978) mutual fund separation theorem under the less restricting hypothesis of $E(\varepsilon_j \mid M) = 0$.

It is obvious that the stable CAPM is still subject to some of the criticisms stated above. Roll's critique is not answered by the assumption of stable distributions, the issue of the efficient market portfolio is still not resolved. However, it seems that the stable case is nevertheless able to better explain the empirical data. In fact, one of the biggest drawbacks is the assumption of a constant, common index of stability for all of the return distributions. Empirical data clearly draws a different picture. For example, the index of stability is in generally larger for an index than for the individual assets included in this index. As long as there does not exist an no inner product on the set of all stable distributions exists, there is little hope ofto formulatinge a better theory. For the empirical work there seems to be a solution when concentrating on the estimator in the regression equation. We refer the reader to Section 5. In contrast to this problem, the restriction due to the assumption of symmetry seems to be of minor importance.

Performance Measures Based on α-Stable Distributions — The CAPM Case

The Treynor ratio As described in Section 2.1.2, the Treynor ratio is based on the Security Market Line. Considering the results of the previous chapter, the SML for a CAPM based on α-stable distributions resembles the traditional result. The main difference is the expression for the sensitivity β_j of an asset j towards the market risk. We will call the new measure $\beta_{\alpha p}$, which is defined as

$$\beta_{\alpha p} = [r_p, r_m]_\alpha / V_\alpha(r_m),$$

where r_p and r_m are the return on portfolio p and the market portfolio m, respectively. $[\ ,\ ]_\alpha$ denotes the covariation and $V_\alpha(\ )$ denotes the variation. Thus, the Treynor ratio becomes

$$\mathrm{RVOL}_p = \frac{\overline{r}_p - \overline{r}_f}{\beta_{\alpha p}} = \frac{\overline{r}_p - \overline{r}_f}{[r_p, r_m]_\alpha} V_\alpha(r_m).$$

The Sharpe ratio The Sharpe ratio assumes that total risk is relevant. Hence, the Capital Market Line is considered. Of course, the Capital Market Line looks somewhat different under the assumption of jointly symmetric α-stable distributed asset returns in the case $\alpha < 2$. Under that condition, the CML becomes

$$\mu_p = r_f + (\mu_m - r_f)\sigma_p = r_f + (\mu_m - r_f)(V_\alpha(r_p))^{1/\alpha},$$

where μ_p and μ_m are the expected returns on portfolio p and the market portfolio m, respectively. It should be noted, however, that in the case $\alpha = 2$, the σ_p in the above equation does not equal the standard deviation on the return on portfolio p utilized in Section 2.1.2. In fact, it differs by the factor $\sqrt{2}/2$. On the other hand, as Bawa (1975) points out, "the scale parameter σ and not the variance is the natural measure of dispersion...." Hence, the Sharpe ratio becomes

$$\mathrm{RVAR}_p = \frac{\overline{r}_p - \overline{r}_f}{\sigma_p} = \frac{\overline{r}_p - \overline{r}_f}{V_\alpha(r_p)^{1/\alpha}}\,.$$

The Jensen measure Like the Treynor ratio, the Jensen measure is based on the SML. Assuming the same preliminaries as for that ratio, the regression specification of the Jensen measure for the CAPM becomes

$$r_{pt} = a_p + \beta_{\alpha p} r_{mt} + u_{pt},$$

where a_p is the Jensen measure, r_{pt} is the excess return on the portfolio over the prevailing risk-free rate, r_{mt} is the excess return on the market over the risk-free rate, u_{pt} is the error term, an $S\alpha S$ distributed random variable, and $\beta_{\alpha p}$ is the measure for the portfolio's systematic risk. In Section 5, we will derive appropriate estimators for a_p and $\beta_{\alpha p}$.

4.5 Performance Measures Based on an α-Stable Version of the APT

4.5.1. The APT in the Case of Stable Distributions

In the following, we will present an α-stable version of the asymptotic APT following the work of Huberman (1982) and Gamrowski and Rachev (1994). The notion of the nth economy simply states that there are n assets in this economy. Suppose that the n assets in the nth economy are generated by the following k-factor model

$$r_i^n = E_i^n + \beta_{i1}^n \delta_1^n + \cdots + \beta_{ik}^n \delta_k^n + \varepsilon_i^n,$$

where the β_{ij}^n are the asset's sensitivities to movements in the k factors (factor loadings or factor betas) and the δ_j^n represent the k factors and

$$E(\delta_j^n) = 0.$$

The ε_j^ns are $S\alpha S$-distributed and assumed to be independent. Furthermore, for every ε_i^n

$$V_\alpha(\varepsilon_i^n) \le A, \quad A \in \Re^n.$$

In vector notation we get

$$\mathbf{r}^n = \mathbf{E}^n + \beta^n \delta^n + \varepsilon^n.$$

From linear algebra we know that it is possible to express the vector of expected returns $\mathbf{E}^n$ as

$$\mathbf{E}^n = \mathbf{E}_0^n + \sum_{j=1}^{k} \beta_j^n \delta^n + \mathbf{c}^n,$$

where $\mathbf{c}^n \in \Re^n$ and $\mathbf{c}^n$ is a vector orthogonal to β_j^n for all j,

$$\beta_j^n \mathbf{c}^n = 0.$$

In addition, choose the $\mathbf{c}^n$, which is also orthogonal to the vector of ones $\mathbf{e}^n$, namely

$$\mathbf{e}^n \mathbf{c}^n = 0.$$

Note that $\mathbf{c}^n$ can be interpreted as an arbitrage portfolio, since it uses no wealth. In the case of arbitrage, a subsequence n' would exist, so that,

$$\lim_{n' \to \infty} E(\mathbf{r}^n \, \mathbf{c}^n) = +\infty,$$

and

$$\lim_{n' \to \infty} V_\alpha(\mathbf{r}^n \, \mathbf{c}^n) = 0,$$

which is equivalent to an almost sure profit that increases to infinity by scaling without investing anything. In a next step, Rachev and Gamrowski (1994) prove the following theorem for the $S\alpha S$ case, following the proof in the traditional case of Huberman (1982).

Theorem. *Suppose that the returns satisfy the conditions above and that there is no arbitrage. Then for* $n = 1, 2, \ldots$ *there exists* E_0^n, $\gamma_1^n, \ldots, \gamma_k^n$, *and* A *such that*

$$\sum_{i=1}^{n} \left| E_i^n - E_0^n - \sum_{j=1}^{k} (\beta_{ij}^n \gamma_j^n) \right|^\alpha \leq A, \quad \textit{for } n = 1, 2, \ldots .$$

The interpretation of the theorem is that for most assets in a large economy, the mean return is approximately linearly correlated to the β_{ij}^n, the sensitivities of the ith asset towards economy-wide risk factors. Huberman states that the β_{ij}^n would be the covariances of the asset with these common factors. In the symmetric α-stable setting here, these would be proportional to the covariations. As the number of assets becomes larger the linear approximation approves and most of the asset returns are almost exact linear functions of the sensitivities.

4.5.2. *Performance Measures Based on α-Stable Distributions — The APT Case*

Using the results of the preceding section and the findings of Connor and Korajczyk (1986), the modification of the Jensen measure for the α-stable case is a straightforward modification. If we interpret γ_j above as

$$\gamma_j^n = E(R_j^n - r_{\text{free}}^n),$$

and δ_j^n as

$$\delta_j^n = R_j^n - E(R_j^n),$$

then the regression equation for the Jensen measure becomes

$$\begin{aligned} r_{pt} &= \alpha_p + (\gamma_{1t}^n + \delta_{1t}^n)\beta_1^n + \cdots + (\gamma_{kt}^n + \delta_{kt}^n)\beta_k^n + \varepsilon_p, \\ r_{pt} &= \alpha_p + \beta_1^n r_{1t}^n + \cdots + \beta_k^n r_{kt}^n + \varepsilon_p \end{aligned}$$

where r_{it} is the excess return over the risk-free rate r_{free} of portfolio p, α_p is the equivalent of the Jensen measure, r_{jt} is the excess return on factor j.

5 Regression Methodology in the Stable Case

This section aims to develop the necessary econometric tools in order to apply the aforementioned performance measures in the α-stable case. In Section 5.1 we will consider the regression properties when all return distributions have the same index of stability, whereas in Section 5.2 we will drop this assumption.

5.1 *Jointly SαS Distributed Asset Returns*

It must be remembered that when asset returns are jointly symmetric α-stable ($S\alpha S$) distributed, the return on each individual asset r_{pt} is $S\alpha S$ distributed. Since the excess return on the benchmark portfolio r_{mt}, for example the return on the market, is a linear combination of the excess returns of the individual assets, it is also $S\alpha S$ distributed. All of the above models assume a linear relationship between return and risk. Therefore, linear regressions are most important for the realm of performance evaluation.

5.1.1. *Simple Regressions*

A regression of a random variable X_1 on a random variable X_2 is linear if there exists a constant c exists such that

$$E(X_1 \mid X_2) = cX_2 \quad \text{a.s.}$$

It is well known that in the Gaussian case the conditional expectation is linear in X_2. Samorodnitsky and Taqqu (1994) show that when $\mathbf{X} = (X_1, X_2)$ is a $S\alpha S$ random vector and $\mathbf{X} \in \Re^2$, and $\alpha \in [1, 2]$ so that $E|X_1|$ and $E|X_2|$ exist, then

$$E(X_1 \mid X_2) = cX_2,$$

where

$$c = [X_1, X_2]_\alpha / V_\alpha(X_2) = [X_1, X_2]_\alpha / \sigma_{X_2}^\alpha.$$

Consider the non-t homogenous regression for the Jensen measure

$$r_{pt} = a + \beta r_{mt} + u_t,$$

where a is the Jensen measure. It is expected to be equal to zero for the average unmanaged asset. Assume that the excess returns on the assets are jointly $S\alpha S$ distributed, then so are r_{mt} and r_{at}. Hence, due to by the above result the β is equivalent to

$$\beta_\alpha = [r_p, r_m]_\alpha / V_\alpha(r_m).$$

In the Gaussian case, the well-known ordinary least squares (OLS) estimator is applied, which is identical to the best linear unbiased estimator (BLUE) when $\alpha = 2$:

$$\hat{\beta} = \frac{\sum_t r_{mt} r_{pt}}{\sum_t r_{mt}^2},$$
$$\hat{a} = \bar{r}_p - \bar{r}_f \hat{\beta}.$$

When $\alpha < 2$ this estimator is no longer the BLUE. Blattberg and Sargent (1971) presented a straightforward derivation of the appropriate BLU estimator conditionally based on the known sequence r_{mt},

$$\hat{\beta}_\alpha = \frac{\sum_t r_{mt}^{\langle 1/(\alpha-1) \rangle} r_{pt}}{\sum_t |r_{mt}|^{\alpha/(\alpha-1)}},$$

which coincides with $\hat{\beta}$ when $\alpha = 2$.

5.1.2. Multiple Regressions

Whereas multiple regressions are always linear in the Gaussian case, this is not so for the case $\alpha < 2$. In fact, it can be shown that multiple regression is only guaranteed to be linear in two cases. In the first case all the X_i must have to be linearly independent from each other. When considering asset returns, this assumption is quite unlikely to be correct. Therefore, we will concentrate on the second case, which assumes that the X_i belong to the sub-Gaussian family.

Let $\mathbf{X}$ be a sub-Gaussian $S\alpha S$ random vector. In that case

$$E(X_n \mid X_1, \ldots, X_{n-1}) = c_1 X_1 + \cdots + c_{n-1} X_{n-1},$$

where

$$c_i = [X_n, X_i]_\alpha / V_\alpha(X_i) = \operatorname{cov}(G_n, G_i)/\operatorname{var}(G_i),$$

where

$$G = (G_1, \ldots, G_n)$$

is the underlying mean zero Gaussian vector of $\mathbf{X}$. Therefore, it is obvious that the well-known OLS estimators can be applied. Consider, for example, the regression formula for the Jensen measure in an APT framework,

$$r_{pt} = a + \beta_1 r_{f1t} + \cdots + \beta_k r^n_{fkt} + u_t.$$

In vector notation we get

$$\mathbf{r}_p = \mathbf{r}_f \beta + u_t,$$

where $\beta^T = (a\,\beta 1\,\beta 2\,\ldots\,\beta k)$ and $\mathbf{r}_f = (\mathbf{1}\,\mathbf{r}_{f1}\,\mathbf{r}_{f2}\,\ldots\,\mathbf{r}_{fk})$ is a $t \times k+1$ matrix. The estimator for β becomes

$$\hat{\beta} = (\mathbf{r}'_f, \mathbf{r}_f)^{-}\mathbf{r}'_f \mathbf{r}_p.$$

5.2 Alternative Indices of Stability for Assets and Benchmarks

Researchers feel that the most severe restriction in performance measurement and asset pricing in the stable case is the assumption of a common index of stability for all assets — individual securities and portfolios alike. As we know that asset returns are not normally distributed, we also know that they are not distributed with the same index of stability.

Consider the stable CAPM and let r_a, the excess return on the asset a, be $S\alpha S(\sigma)$ distributed. Let r_m, the excess return on the market, be $S\gamma S(\delta)$ distributed, exactly,

$$(r_{mt}) \sim S_\gamma(\delta, 0, 0),$$

where $\gamma \in (1, 2]$. Considering the regression for the Jensen measure

$$r_{at} = a + \beta r_{mt} + u_t,$$

where r_{at} is the excess return on asset a in period t, r_{mt} is the excess return on the market in period t, and u_t is the error term. Consequently, u is $S\alpha S$ distributed, exactly

$$(u_t) \sim S_\alpha(\sigma, 0, 0),$$

where $\alpha \in (1,2]$. The sequence $(r_{mt})_{t>0}$ are i.i.d. $S\gamma S(\delta)$ random variables and the sequence $(u_t)_{t>0}$ are i.i.d. $S\alpha S(\sigma)$ random variables.

When $\alpha = \gamma = 2$ it is well-known, that β becomes

$$\beta = \mathrm{cov}(r_a, r_m)/\mathrm{var}(r_m).$$

However, when $\alpha < 2$ and $\gamma < 2$, the covariance and the variance in the above equation do not exist any longer. Then the first question is as to whether the standard estimators for β and μ in the regression model

$$y_t = \mu + \beta x_t + u_t,$$

where

$$x_t \sim S\gamma S(\delta),$$

and

$$u_t \sim S\alpha S(\sigma),$$

preserve the unbiasedness. Also, the asymptotic properties of those standard estimators are important.

We will consider the well-known OLS estimators

$$\hat{\mu}_n = \frac{\sum_{t=1}^n x_t^2 \sum_{t=1}^n y_t - \sum_{t=1}^n x_t \sum_{t=1}^n y_t x_t}{n\sum_{t=1}^n x_t^2 - (\sum_{t=1}^n x_t)^2}$$

and

$$\hat{\beta}_n = \frac{n\sum_{t=1}^n y_t x_t \sum_{t=1}^n x_t \sum_{t=1}^n y_t}{n\sum_{t=1}^n x_t^2 - (\sum_{t=1}^n x_t)^2}.$$

In addition, we will consider the corresponding t-statistics for these estimators

$$t_{\hat{\mu}} = \frac{\hat{\mu} - \mu}{s_{\hat{\mu}}} \quad \text{and} \quad t_{\hat{\beta}} = \frac{\hat{\beta} - \beta}{s_{\hat{\beta}}},$$

where

$$s_{\hat{\mu}}^2 = \frac{\hat{\sigma}_u^2 \sum_{t=1}^n x_t^2}{n\sum_{t=1}^n x_t^2 - (\sum_{t=1}^n x_t)^2}, \quad s_{\hat{\beta}}^2 = \frac{n\hat{\sigma}_u^2}{n\sum_{t=1}^n x_t^2 - (\sum_{t=1}^n x_t)^2},$$

$\hat{\sigma}_u^2 := n^{-1}\sum_{t=1}^n \hat{u}_t^2$ and $\hat{u}_t^2 := y_t - \hat{\mu}_n - \hat{\beta}_n x_t$. The t-statistics are important, since they provide us with a way to check for the heavy-tailedness and to determine the unknown indices of stability. In the model we let xt and ut be γ-stable and α-stable distributed, however, we have no guarantee that α and γ are the correct right values. The t-statistics converge to values that are dependent of α and γ. We can calculate those values and order them in a grid for certain values of α and γ. After we have run the regression we compare the t-statistics with those values that should occur when the assumed α and γ are correct. It is possible to define confidence conifdence intervals for the t-statistics. Consequently we have a test, whether or not our assumption of α and γ was reasonable.

In the following, we will consider four different cases.

$$
\begin{array}{lll}
\text{Case 1:} & \alpha = 2, \gamma = 2 & \Rightarrow u_t \sim N(0,\sigma^2) \text{ and } x_t \sim N(0,\delta^2). \\
\text{Case 2:} & 1 < \alpha < 2, \text{ and } \gamma = 2 & \Rightarrow u_t \sim S\alpha S(\sigma) \text{ and } x_t \sim N(0,\delta^2). \\
\text{Case 3:} & \alpha = 2, \text{ and } 1 < \gamma < 2 & \Rightarrow u_t \sim N(0,\sigma^2) \text{ and } x_t \sim S\gamma S(\delta). \\
\text{Case 4:} & 1 < \alpha < 2, \text{ and } 1 < \gamma < 2 & \Rightarrow u_t \sim S\alpha S(\sigma) \text{ and } x_t \sim S\gamma S(\delta).
\end{array}
$$

Case 1. This case is well-known and there exists extensive literature on the case is available. Therefore, we will only present the results.

Theorem 1. *Suppose* $\alpha = 2$, $\gamma = 2$, $u_t \sim N(0,\sigma^2)$ *and* $x_t \sim N(0,\delta^2)$. *Assume* (u_t) *is independent of* (x_t). *Then as* $n \to \infty$,

$$\left(\sqrt{n}(\hat{\mu} - \mu), \sqrt{n}(\hat{\beta} - \beta), t_{\hat{\mu}}, t_{\hat{\beta}}\right) \xrightarrow{d} \left(\sigma U, \frac{\sigma}{\delta} U, U, U\right),$$

where $U \sim N(0,1)$.

For the proof refer to Appendix A.

Case 2. In this case the single asset is assumed to follow an α-stable distribution and the benchmark or the explaining variable is normally distributed.

Theorem 2. *Suppose* $1 < \alpha < 2$, $\gamma = 2$, $u_t \sim S\alpha S(\sigma)$ *and* $x_t \sim N(0,\delta^2)$. *Assume* (u_t) *is independent of* (x_t). *Then as* $n \to \infty$,

$$
\begin{aligned}
&\left(n^{1-1/\alpha}(\hat{\mu} - \mu), n^{1-1/\alpha}(\hat{\beta} - \beta), t_{\hat{\mu}}, t_{\hat{\beta}}\right) \\
&\xrightarrow{d} \left(\sigma U, \frac{\sigma}{\delta}(E|N(0,1)|^{\alpha})^{1/\alpha} U, \frac{U}{\sqrt{[L_\alpha](1)}}, \frac{(E|N(0,1)|^{\alpha})^{1/\alpha} U}{\sqrt{[L_\alpha](1)}}\right)
\end{aligned}
$$

where $U \sim S\alpha S(1)$ *is a standard* α*-stable random variable and* $[L_\alpha](1)$ *is a "bracket process", which is defined for the strictly stable random variable* $L_\alpha \sim S\alpha S(\sigma)$ *as*

$$[L_\alpha](1) = L_\alpha^2 - \int_0^1 L_\alpha(s-)\, dL_\alpha(s).$$

For further details refer to Rachev (1999).

The sketch of the proof is presented in Appendix B. For the complete proof and further results, refer to Götzenberger, Rachev and Schwarz (1999).

Case 3. The single asset is assumed to be normally distributed and the benchmark or the explaining variable follows an γ-stable distribution.

Theorem 3. *Suppose* $2 = \alpha$, $1 < \gamma < 2$, $u_t \sim N(0, \sigma^2)$ *and* $x_t \sim S\gamma S(\delta)$. *Assume* (u_t) *is independent of* (x_t). *Then as* $n \to \infty$,

$$(\sqrt{n}(\hat{\mu} - \mu), n^{1/\gamma}(\hat{\beta} - \beta), t_{\hat{\mu}}, t_{\hat{\beta}}) \xrightarrow{d} \left(\sigma U, \frac{\sigma U}{\delta\sqrt{[L_\gamma](1)}} U, U\right)$$

where $U \sim N(0,1)$ *a standard normal random variable and* $[L_\gamma](1)$ *is a "bracket process", which is defined for the strictly stable random variable* $L_\gamma \sim S\gamma S(\delta)$ *as*

$$[L_\gamma](1) = L_\gamma^2 - \int_0^1 L_\gamma(s-)\, dL_\gamma(s).$$

For further details, refer to Rachev (1999).

The sketch of the proof is presented in Appendix C. For the complete proof and further results, refer to Götzenberger, Rachev and Schwarz (1999).

Case 4. Both the single asset and the explaining variable are assumed to follow symmetric stable distributions.

Theorem 4. *Suppose* $1 < 2 < \alpha$, $1 < \alpha < 2$, $u_t \sim S\alpha S(\sigma)$ *and* $x_t \sim S\gamma S(\delta)$. *Assume* (u_t) *is independent of* (x_t). *Then as* $n \to \infty$,

$$(n^{1-1/\alpha}(\hat{\mu} - \mu), n^{1/\gamma}(\hat{\beta} - \beta), t_{\hat{\mu}}, n^{1/2-1/\alpha} t_{\hat{\beta}})$$
$$\xrightarrow{d} \left(\sigma U, \frac{\sigma U\, V^{1/\alpha}}{\delta[L_\gamma](1)}, \frac{U}{\sqrt{[L_\gamma](1)}}, \frac{V^{1/\alpha}\, U}{\sqrt{[L_\gamma](1)}\sqrt{[L_\alpha](1)}}\right)$$

where $U \sim S\alpha S(1)$ *a standard* α*-stable random variable,* V *is a random variable in the normal domain of attraction of an* α*-stable random variable, and* $[L_\gamma](1)$ *is a "bracket process", which is defined for the strictly stable random variable* $L_\gamma \sim S\gamma S(\delta)$ *as*

$$[L_\gamma](1) = L_\gamma^2 - \int_0^1 L_\gamma(s-)\, dL_\gamma(s).$$

$[L_\alpha](1)$ *is defined in the same way.*

For further details, refer to Rachev (1999).

The sketch of the proof is presented in Appendix D. For the complete proof and further results, refer to Götzenberger, Rachev and Schwarz (1999).

6 Conclusions

In the first part of this paper we have presented an overview of the field of performance measurement. Also, we reflected the current state of research. Then, we shifted our focus to stable distributions. This area of statistics has been neglected for some thirty years, mostly due to the lack of theory and the difficulties associated with stable Paretian distributions. However, since recent research has shown the inferiority of the standard approach, we promote the application of stable theory. After presenting asset pricing models in the stable case, we develop stable performance measures. In addition, we overcome one of the biggest drawbacks of regressions in the stable case. We present estimators that allow different indices of stability for the regressor and error term. As previous studies have shown (e.g. Mittnik, Rachev and Paollela (1997), this approach should lead to performance measures that are closer to the realty of the market than the standard methodology in performance measurement. Further results in this direction are to be published in a forthcoming paper. Also, empirical results are to be expected in an additional paper.

Appendix A

Sketch of the proof It can easily be verified that

$$(\hat{\mu}_n - \mu) = \frac{\sum_{t=1}^n x_t^2 \sum_{t=1}^n u_t - \sum_{t=1}^n x_t \sum_{t=1}^n u_t x_t}{n \sum_{t=1}^n x_t^2 - (\sum_{t=1}^n x_t)^2}$$

and

$$(\hat{\beta}_n - \beta) = \frac{n \sum_{t=1}^n u_t x_t - \sum_{t=1}^n x_t \sum_{t=1}^n u_t}{n \sum_{t=1}^n x_t^2 - (\sum_{t=1}^n x_t)^2}.$$

Define

$$N_\mu := \sum_{t=1}^n x_t^2 \sum_{t=1}^n u_t - \sum_{t=1}^n x_t \sum_{t=1}^n u_t x_t,$$

$$N_\beta := n \sum_{t=1}^n u_t x_t - \sum_{t=1}^n x_t \sum_{t=1}^n u_t$$

and

$$D := n \sum_{t=1}^n x_t^2 - \left(\sum_{t=1}^n x_t \right)^2.$$

Let us first consider N_μ. We must determine the dominant term in order to see the behavior for $n \to \infty$. To this end, we condition on the known sequence

x_t,

$$\begin{aligned} N_\mu &= \sum_{t=1}^{n} x_t^2 \sum_{t=1}^{n} u_t - \sum_{t=1}^{n} x_t \sum_{t=1}^{n} u_t x_t \\ &=^d \left(\frac{n}{n}\sum_{t=1}^{n} x_t^2\right)\left(\frac{\sqrt{n}\sigma}{\sqrt{n}\sigma}\sum_{t=1}^{n} u_t\right) - \left(\frac{\sqrt{n}\sigma}{\sqrt{n}\sigma}\sum_{t=1}^{n} x_t\right)\sqrt{\sum_{t=1}^{n} x_t^2}\sigma U, \end{aligned}$$

where $U \sim N(0,1)$. Thus,

$$N_\mu =^d n^{3/2}\sigma\left(\frac{1}{n}\sum_{t=1}^{n} x_t^2\right) U - \sqrt{n}\delta\sigma X\sqrt{\frac{n}{n}\sum_{t=1}^{n} x_t^2} U,$$

where $X \sim S\gamma S(1)$. Then as $n \to \infty$,

$$N_\mu \xrightarrow{d} n^{3/2}\sigma\delta^2 U - n\delta^2\sigma X U.$$

Therefore, the dominant part is $n^{3/2}\sigma\delta^2 U$. Then, as $n \to \infty$,

$$n^{-3/2} N_\mu \overset{n\to\infty}{\longrightarrow} \sigma\delta^2 U.$$

Next, let us consider N_β.

$$\begin{aligned} N_\beta &= n\sum_{t=1}^{n} u_t x_t - \sum_{t=1}^{n} x_t \sum_{t=1}^{n} u_t \\ &=^d n\sqrt{\frac{n}{n}\sum_{t=1}^{n} x_t^2}\sigma U - \left(\frac{n^{1/2}\delta}{n^{1/2}\delta}\sum_{t=1}^{n} x_t\right)\left(\frac{\sqrt{n}\sigma}{\sqrt{n}\sigma}\sum_{t=1}^{n} u_t\right), \end{aligned}$$

where $U \sim N(0,1)$. Thus,

$$N_\beta =^d n^{3/2}\sqrt{\frac{1}{n}\sum_{t=1}^{n} x_t^2}\sigma U - n\delta\sigma X U,$$

where $X \sim S\gamma S(1)$. Then as $n \to \infty$,

$$N_\beta \xrightarrow{d} n^{3/2}\sigma\delta U - n\delta\sigma X U.$$

Therefore, the dominant term is $n^{3/2}\sigma\delta U$. Then, as $n \to \infty$,

$$n^{-3/2} N_\beta \overset{n\to\infty}{\longrightarrow} \sigma\delta U.$$

Next, let us consider D.

$$\begin{aligned} D &= n\sum_{t=1}^{n} x_t^2 - \left(\sum_{t=1}^{n} x_t\right)^2 \\ &=^d n\left(\frac{n}{n}\sum_{t=1}^{n} x_t^2\right) - \left(\frac{n^{1/2}\delta}{n^{1/2}\delta}\sum_{t=1}^{n} x_t\right)^2 \\ &\xrightarrow{d} n^2\delta^2 - n\delta^2 X^2, \end{aligned}$$

where $X \sim N(0,1)$.

Therefore, the dominant part is $n^2\delta^2$. Then, as $n \to \infty$,

$$n^{-2}D \overset{n\to\infty}{\longrightarrow} \delta^2.$$

So far, we have shown the convergence of the univariate marginal distributions of (N_μ, N_β, D). Using more advanced techniques of stochastic processes (see Paulauskas and Rachev (1998)) one can show that the joint distribution of (N_μ, N_β, D) converges as well. In summary we get

$$\frac{n^{-3/2}N_\mu}{n^{-2}D} = \sqrt{n}(\hat{\mu}_n - \mu) \overset{n\to\infty}{\longrightarrow} \frac{\sigma\delta^2}{\delta^2}U = \sigma U,$$

and

$$\frac{n^{-3/2}N_\beta}{n^{-2}D} = \sqrt{n}(\hat{\beta}_n - \beta) \overset{n\to\infty}{\longrightarrow} \frac{\sigma\delta}{\delta^2}U = \frac{\sigma}{\delta}U.$$

From now on X and U are defined as above.

For the t-statistics, we must look at

$$\begin{aligned} t_{\hat{\mu}} &= \frac{\hat{\mu} - \mu}{s_{\hat{\mu}}} \quad \text{and} \quad t_{\hat{\beta}} = \frac{\hat{\beta} - \beta}{s_{\hat{\beta}}} \\ &= \frac{N_\mu}{D s_{\hat{\mu}}} \quad \text{and} \quad\quad = \frac{N_\beta}{D s_{\hat{\beta}}}, \end{aligned}$$

where

$$\begin{aligned} s_{\hat{\mu}}^2 &= \frac{\hat{\sigma}_u^2 \sum_{t=1}^n x_t^2}{n\sum_{t=1}^n x_t^2 - (\sum_{t=1}^n x_t)^2}, & s_{\hat{\beta}}^2 &= \frac{n\hat{\sigma}_u^2}{n\sum_{t=1}^n x_t^2 - (\sum_{t=1}^n x_1)^2} \\ s_{\hat{\mu}} &= \frac{\hat{\sigma}_u\sqrt{\sum_{t=1}^n x_t^2}}{\sqrt{D}}, & s_{\hat{\beta}} &= \frac{\sqrt{n}\hat{\sigma}_u}{\sqrt{D}}. \end{aligned}$$

Therefore,

$$\begin{aligned}
t_{\hat{\mu}} &= \frac{N_\mu}{Ds_{\hat{\mu}}} & \text{and} \quad t_{\hat{\beta}} &= \frac{N_\beta}{Ds_{\hat{\beta}}} \\
&= \frac{N_\mu \sqrt{D}}{D\hat{\sigma}_u \sqrt{\sum_{t=1}^n x_t^2}} & \text{and} \quad &= \frac{N_\beta \sqrt{D}}{D\sqrt{n}\hat{\sigma}_u} \\
&= \frac{N_\mu}{\sqrt{D}\sqrt{\sum_{t=1}^n x_t^2}\hat{\sigma}_u} & \text{and} \quad &= \frac{N_\beta}{\sqrt{D}\sqrt{n}\hat{\sigma}_u},
\end{aligned}$$

where

$$\hat{\sigma}_u^2 := n^{-1}\sum_{t=1}^n \hat{u}_t^2 \quad \text{and} \quad \hat{u}_t^2 := y_t - \hat{\mu}_n - \hat{\beta}_n x_t.$$

For both $t_{\hat{\mu}}$ and $t_{\hat{\beta}}$ the term $\hat{\sigma}_u^2 := n^{-1}\sum_{t=1}^n \hat{u}_t^2$ is important.

$$\begin{aligned}
\hat{\sigma}_u^2 &= n^{-1}\sum_{t=1}^n \hat{u}_t^2 = n^{-1}\sum_{t=1}^n (y_t - \hat{\mu}_n - \hat{\beta}_n x_t)^2 \\
&= n^{-1}\sum_{t=1}^n (\mu + \beta x_t + u_t - \hat{\mu}_n - \hat{\beta}_n x_t)^2 \\
&= n^{-1}\sum_{t=1}^n (u_t - (\hat{\mu}_n - \mu) - (\hat{\beta}_n - \beta)x_t)^2 \\
&\xrightarrow{d} n^{-1}\sum_{t=1}^n \left(u_t - \left(\frac{\sigma U}{\sqrt{n}}\right) - \left(\frac{\sigma U}{\delta\sqrt{n}}\right)x_t\right)^2 \\
&=^d n^{-1}\sum_{t=1}^n \left(u_t^2 + \frac{\sigma^2 U^2}{n} + \frac{\sigma^2 U^2}{\delta^2 n}x_t^2 - 2\frac{\sigma U}{\sqrt{n}}u_t \right. \\
&\quad \left. - 2\frac{\sigma U}{\delta\sqrt{n}}u_t x_t + 2\frac{\sigma^2}{\delta n}U^2 x_t\right) \\
&=^d \frac{1}{n}\sum_{t=1}^n u_t^2 + \frac{\sigma^2 U^2}{n} + \frac{\sigma^2 U^2}{\delta^2 n}\frac{1}{n}\sum_{t=1}^n x_t^2 \\
&\quad - 2\frac{\sigma U}{n}\frac{\sigma}{\sqrt{n}\sigma}\sum_{t=1}^n - 2\frac{\sigma U}{\delta\sqrt{n}}\sqrt{\frac{n}{n}\sum_{t=1}^n x_t^2}\sigma U + 2\frac{\sigma^2 U^2}{n^{3/2}}\frac{1}{n^{1/2}}\delta\sum_{t=1}^n x_t \\
&\overset{n\to\infty}{\longrightarrow} \sigma^2 + \frac{\sigma^2 U^2}{n} + \frac{\sigma^2 U^2}{n} - 2\frac{\sigma^2 U^2}{n} - 2\frac{\sigma^2}{n}U^2 + 2\frac{\sigma^2}{n^{3/2}}U^2 X
\end{aligned}$$

Therefore σ^2 is the most dominant term. Then, as $n \to \infty$

$$\hat{\sigma}_u^2 \stackrel{n\to\infty}{\longrightarrow} \sigma^2.$$

First we will focus on $t_{\hat{\mu}}$. It must be remembered that $t_{\hat{\mu}}$ becomes

$$\begin{aligned} t_{\hat{\mu}} &= \frac{N_\mu}{\sqrt{D}\sqrt{\sum_{t=1}^n x_t^2 \hat{\sigma}_u}} \\ &=^d \frac{N_\mu}{\sqrt{D}\sqrt{\frac{n}{n}\sum_{t=1}^n x_t^2 \hat{\sigma}_u}} \\ &\xrightarrow{d} \frac{n^{3/2}\sigma\delta^2 U}{n\delta\sqrt{n}\,\delta\sigma} \\ &= U. \end{aligned}$$

Second, we must remember that $t_{\hat{\beta}}$ becomes

$$\begin{aligned} t_{\hat{\beta}} &= \frac{N_\beta}{\sqrt{D}\sqrt{n}\,\hat{\sigma}_u} \\ &= \frac{n^{3/2}\delta\sigma U}{n\delta\sqrt{n}\,\sigma} \\ t_{\hat{\beta}} &= U, \end{aligned}$$

which concludes the sketch of the proof.

Appendix B

Sketch of the proof It can easily be verified that

$$(\hat{\mu}_n - \mu) = \frac{\sum_{t=1}^n x_t^2 \sum_{t=1}^n u_t - \sum_{t=1}^n x_t \sum_{t=1}^n u_t x_t}{n\sum_{t=1}^n x_t^2 - (\sum_{t=1}^n x_t)^2}$$

and

$$(\hat{\beta}_n - \beta) = \frac{n\sum_{t=1}^n u_t x_t - \sum_{t=1}^n x_t \sum_{t=1}^n u_t}{n\sum_{t=1}^n x_t^2 - (\sum_{t=1}^n x_t)^2}.$$

Define

$$N_\mu := \sum_{t=1}^n x_t^2 \sum_{t=1}^n u_t - \sum_{t=1}^n x_t \sum_{t=1}^n u_t x_t,$$

$$N_\beta := n\sum_{t=1}^n u_t x_t - \sum_{t=1}^n x_t \sum_{t=1}^n u_t$$

and

$$D := n\sum_{t=1}^{n} x_t^2 - \left(\sum_{t=1}^{n} x_t\right)^2.$$

Let us first consider N_μ. We must determine the dominant term in order to see the behavior for $n \to \infty$. To this end, we condition on the known sequence x_t, then

$$\begin{aligned} N_\mu &= \sum_{t=1}^{n} x_t^2 \sum_{t=1}^{n} u_t - \sum_{t=1}^{n} x_t \sum_{t=1}^{n} u_t x_t \\ &=^d \left(\frac{n}{n}\sum_{t=1}^{n} x_t^2\right)\sum_{t=1}^{n}\left(\frac{n^{1/\alpha}\sigma}{n^{1/\alpha}\sigma}\sum_{t=1}^{n} u_t\right) \\ &\quad - \left(\frac{\sqrt{n}\delta}{\sqrt{n}\delta}\sum_{t=1}^{n} x_t\right)\left(\left(\sum_{t=1}^{n}|x_t|^\alpha\right)^{1/\alpha}\sigma U\right), \end{aligned}$$

where $U \sim S\alpha S(1)$. Thus,

$$N_\mu =^d n^{1+1/\alpha}\sigma\left(\frac{1}{n}\sum_{t=1}^{n} x_t^2\right)U - \sqrt{n}\delta\sigma XU\left(\left(\sum_{t=1}^{n}|x_t|^\alpha\right)^{1/\alpha}\right),$$

where $X \sim N(0,1)$, and as $n \to \infty$,

$$\begin{aligned} N_\mu &\xrightarrow{d} n^{1+1/\alpha}\sigma\delta^2 U - n^{1/2+1/\alpha}\delta\sigma XU(E|N(0,\delta^2)|^\alpha)^{1/\alpha} \\ &= n^{1+1/\alpha}\sigma\delta^2 U - n^{1/2+1/\alpha}\delta^2\sigma XU(E|N(0,1)|^\alpha)^{1/\alpha}. \end{aligned}$$

Therefore, the dominant part is $n^{1+1/\alpha}\sigma\delta^2 U$. Then, as $n \to \infty$,

$$n^{-1-1/\alpha}N_\mu \overset{n\to\infty}{\longrightarrow} \sigma\delta^2 U.$$

Next, let us consider N_β. To this end, we condition on the known sequence x_t, then

$$\begin{aligned} N_\beta &= n\sum_{t=1}^{n} u_t x_t - \sum_{t=1}^{n} x_t \sum_{t=1}^{n} u_t \\ &=^d n\left(\left(\sum_{t=1}^{n}|x_t|^\alpha\right)^{1/\alpha}\sigma U\right) - \left(\frac{\sqrt{n}\delta}{\sqrt{n}\delta}\sum_{t=1}^{n} x_t\right)\left(\frac{n^{1/\alpha}\sigma}{n^{1/\alpha}\sigma}\sum_{t=1}^{n} u_t\right), \end{aligned}$$

where $U \sim S\alpha S(1)$. Thus,

$$N_\beta =^d n \left(\frac{n}{n} \left(\sum_{t=1}^{n} |x_t|^\alpha \right)^{1/\alpha} \sigma U \right) - n^{1/2+1/\alpha} \delta \sigma X U,$$

where $X \sim N(0,1)$. Then as $n \to \infty$,

$$N_\beta \xrightarrow{d} n^{1+1/\alpha} \sigma \delta (E|N(0,1)|^\alpha)^{1/\alpha} U - n^{1/2+1/\alpha} \delta \sigma X U.$$

Therefore, the dominant term is $n^{1+1/\alpha} \sigma \delta (E|N(0,1)|^\alpha)^{1/\alpha} U$. Then, as $n \to \infty$,

$$n^{-1-1/\alpha} N_\beta \overset{n \to \infty}{\longrightarrow} \sigma \delta (E|N(0,1)|^\alpha)^{1/\alpha} U.$$

Next, let us consider D.

$$\begin{aligned} D &= n \sum_{t=1}^{n} x_t^2 - \left(\sum_{t=1}^{n} x_t \right)^2 \\ &=^d n \left(\frac{n}{n} \sum_{t=1}^{n} x_t^2 \right) - \left(\frac{\sqrt{n}\delta}{\sqrt{n}\delta} \sum_{t=1}^{n} x_t \right)^2 \\ &\xrightarrow{d} n^2 \delta^2 - n \delta^2 X^2, \end{aligned}$$

where $X \sim N(0,1)$.

Therefore, the dominant part is $n^2\delta^2$. Then, as $n \to \infty$,

$$\frac{1}{n^2} D \overset{n \to \infty}{\longrightarrow} \delta^2.$$

So far, we have shown the convergence of the univariate marginal distributions of (N_μ, N_β, D). Using more advanced techniques of stochastic processes (see Paulauskas and Rachev (1998)) one can show that the joint distribution of (N_μ, N_β, D) converges as well. In summary we get

$$\frac{n^{-1-1/\alpha} N_\mu}{n^{-2} D} = n^{1-1/\alpha} (\hat{\mu}_n - \mu) \overset{n \to \infty}{\longrightarrow} \frac{\sigma \delta^2 U}{\delta^2} = \sigma U$$

and

$$\begin{aligned} \frac{n^{-1-1/\alpha} N_\beta}{n^{-2} D} &= n^{1-1/\alpha} (\hat{\beta}_n - \beta) \overset{n \to \infty}{\longrightarrow} \frac{\sigma \delta (E|N(0,1)|^\alpha)^{1/\alpha} U}{\delta^2} \\ &= \frac{\sigma}{\delta} (E|N(0,1)|^\alpha)^{1/\alpha} U. \end{aligned}$$

For the t-statistics, we must look at

$$\begin{aligned} t_{\hat{\mu}} &= \frac{\hat{\mu}-\mu}{s_{\hat{\mu}}} \quad \text{and} \quad & t_{\hat{\beta}} &= \frac{\hat{\beta}-\beta}{s_{\hat{\beta}}} \\ &= \frac{N_\mu}{Ds_{\hat{\mu}}} \quad \text{and} \quad & &= \frac{N_\beta}{Ds_{\hat{\beta}}}, \end{aligned}$$

where

$$\begin{aligned} s_{\hat{\mu}}^2 &= \frac{\hat{\sigma}_u^2 \sum_{t=1}^n x_t^2}{n\sum_{t=1}^n x_t^2 - (\sum_{t=1}^n x_t)^2}, \quad & s_{\hat{\beta}}^2 &= \frac{n\hat{\sigma}_u^2}{n\sum_{t=1}^n x_t^2 - (\sum_{t=1}^n x_t)^2} \\ s_{\hat{\mu}} &= \frac{\hat{\sigma}_u \sqrt{\sum_{t=1}^n x_t^2}}{\sqrt{D}}, & s_{\hat{\beta}} &= \frac{\sqrt{n}\hat{\sigma}_u}{\sqrt{D}}. \end{aligned}$$

Therefore,

$$\begin{aligned} t_{\hat{\mu}} &= \frac{N_\mu}{Ds_{\hat{\mu}}} & \text{and} \quad t_{\hat{\beta}} &= \frac{N_\beta}{Ds_{\hat{\beta}}} \\ &= \frac{N_\mu\sqrt{D}}{D\hat{\sigma}_u\sqrt{\sum_{t=1}^n x_t^2}} & \text{and} \quad &= \frac{N_\beta\sqrt{D}}{D\sqrt{n}\hat{\sigma}_u} \\ &= \frac{N_\mu}{\sqrt{D}\sqrt{\sum_{t=1}^n x_t^2}\hat{\sigma}_u} & \text{and} \quad &= \frac{N_\beta}{\sqrt{D}\sqrt{n}\hat{\sigma}_u}, \end{aligned}$$

where

$$\hat{\sigma}_u^2 := n^{-1}\sum_{t=1}^n \hat{u}_t^2 \quad \text{and} \quad \hat{u}_t^2 := y_t - \hat{\mu}_n - \hat{\beta}_n x_t.$$

For both $t_{\hat{\mu}}$ and $t_{\hat{\beta}}$ the term $\hat{\sigma}_u^2 := n^{-1}\sum_{t=1}^n \hat{u}_t^2$ is important.

$$\begin{aligned} \hat{\sigma}_u^2 &= n^{-1}\sum_{t=1}^n \hat{u}_t^2 = n^{-1}\sum_{t=1}^n (y_t - \hat{\mu}_n - \hat{\beta}_n x_t)^2 \\ &= n^{-1}\sum_{t=1}^n (\mu + \beta x_t + u_t - \hat{\mu}_n - \hat{\beta}_n x_t)^2 \\ &= n^{-1}\sum_{t=1}^n (u_t - (\hat{\mu}_n - \mu) - (\hat{\beta}_n - \beta)x_t)^2 \\ &\xrightarrow{d} n^{-1}\sum_{t=1}^n \left(u_t - \left(\frac{\sigma U}{n^{1-1/\alpha}}\right) - \left(\frac{\frac{\sigma}{\delta}(E|N(0,1)|^\alpha)^{1/\alpha}U}{n^{1-1/\alpha}}\right) x_t \right)^2 \end{aligned}$$

$$
\begin{aligned}
&=^d n^{-1}\sum_{t=1}^{n}\Bigg(u_t^2+\frac{\sigma^2U^2}{n^{2-2/\alpha}}+\frac{(\frac{\sigma}{\delta}(E|N(0,1)|^\alpha)^{1/\alpha})^2U^2}{n^{2-2/\alpha}}x_t^2-2\frac{\sigma U}{n^{1-1/\alpha}}u_t\\
&\quad -2\frac{\frac{\sigma}{\delta}(E|N(0,1)|^\alpha)^{1/\alpha}U}{n^{1-1/\alpha}}u_tx_t-2\frac{\frac{\sigma^2}{\delta}(E|N(0,1)|^\alpha)^{1/\alpha}U^2}{n^{2-2/\alpha}}x_t\Bigg)\\
&=^d \frac{1}{n^{1-2/\alpha}}\frac{\sigma^2}{n^{2/\alpha}\sigma^2}\sum_{t=1}^{n}u_t^2+\frac{\sigma^2U^2}{n^{2-2/\alpha}}+\frac{(\frac{\sigma}{\delta}(E|N(0,1)|^\alpha)^{1/\alpha})^2U^2}{n^{2-2/\alpha}}\frac{1}{n}\sum_{t=1}^{n}x_t^2\\
&= -2\frac{\sigma U}{n^{2-2/\alpha}}\frac{\sigma}{n^{1/\alpha}\sigma}\sum_{t=1}^{n}u_t-2\frac{\frac{\sigma}{\delta}(E|N(0,1)|^\alpha)^{1/\alpha}U}{n^{2-1/\alpha}}\left(\sum_{t=1}^{n}|x_t|^\alpha\right)^{1/\alpha}\sigma U\\
&\quad +2\frac{\frac{\sigma^2}{\delta}(E|N(0,1)|^\alpha)^{1/\alpha}U^2}{n^{2.5-2/\alpha}}\frac{\delta}{\sqrt{n}\delta}\sum_{t=1}^{n}x_t\\
&\xrightarrow{d} \frac{\sigma^2}{n^{1-2/\alpha}}[L_\alpha](1)+\frac{\sigma^2U^2}{n^{2-2/\alpha}}+\frac{(\sigma(E|N(0,1)|^\alpha)^{1/\alpha})^2U^2}{n^{2-2/\alpha}}\\
&\quad -2\frac{\sigma^2U^2}{n^{2-2/\alpha}}-2\frac{\frac{\sigma^2}{\delta}(E|N(0,1)|^\alpha)^{1/\alpha}U^2}{n^{2-2/\alpha}}(E|N(0,1)|^\alpha)^{1/\alpha}\\
&\quad +2\frac{\sigma^2(E|N(0,1)|^\alpha)^{1/\alpha}U^2}{n^{2.5-2/\alpha}}.
\end{aligned}
$$

Therefore $\frac{\sigma^2}{n^{1-2/\alpha}}[L_\alpha](1)$ is the most dominant term. Then, as $n\to\infty$

$$\hat{\sigma}_u^2 \overset{n\to\infty}{\longrightarrow} \frac{\sigma^2}{n^{1-2/\alpha}}[L_\alpha](1).$$

First we will focus on $t_{\hat{\mu}}$. Remember that $t_{\hat{\mu}}$ becomes

$$
\begin{aligned}
t_{\hat{\mu}} &= \frac{N_\mu}{\sqrt{D}\sqrt{\sum_{t=1}^n x_t^2}\hat{\sigma}_u}\\
&=^d \frac{N_\mu}{n^{1/2}\sqrt{D}\sqrt{\frac{1}{n}\sum_{t=1}^n x_t^2}\hat{\sigma}_u}\\
&\xrightarrow{d} \frac{n^{1+1/\alpha}\sigma\delta^2U}{n\delta n^{1/2}\delta n^{-1/2+1/\alpha}\sigma\sqrt{[L_\alpha](1)}}\\
&= \frac{U}{\sqrt{[L_\alpha](1)}}.
\end{aligned}
$$

Second, remember that $t_{\hat{\beta}}$ becomes

$$t_{\hat{\beta}}=\frac{N_\beta}{\sqrt{D}\sqrt{n}\,\hat{\sigma}_u}$$

$$= \frac{n^{1+1/\alpha}\delta\sigma(E|N(0,1)|^{\alpha})^{1/\alpha}U}{n\delta\sqrt{n}n^{-1/2+1/\alpha}\sigma\sqrt{[L_\alpha](1)}} \cdot$$

$$= \frac{(E|N(0,1)|^{\alpha})^{1/\alpha}U}{\sqrt{[L_\alpha](1)}} .$$

which concludes the sketch of the proof.

Appendix C

Sketch of the proof It can easily be verified that

$$(\hat{\mu}_n - \mu) = \frac{\sum_{t=1}^n x_t^2 \sum_{t=1}^n u_t - \sum_{t=1}^n x_t \sum_{t=1}^n u_t x_t}{n\sum_{t=1}^n x_t^2 - (\sum_{t=1}^n x_t)^2}$$

and

$$(\hat{\beta}_n - \beta) = \frac{n\sum_{t=1}^n u_t x_t - \sum_{t=1}^n x_t \sum_{t=1}^n u_t}{n\sum_{t=1}^n x_t^2 - (\sum_{t=1}^n x_t)^2} .$$

Define

$$N_\mu := \sum_{t=1}^n x_t^2 \sum_{t=1}^n u_t - \sum_{t=1}^n x_t \sum_{t=1}^n u_t x_t,$$

$$N_\beta := n\sum_{t=1}^n u_t x_t - \sum_{t=1}^n x_t \sum_{t=1}^n u_t$$

and

$$D := n\sum_{t=1}^n x_t^2 - \left(\sum_{t=1}^n x_t\right)^2 .$$

Let us first consider N_μ. We must determine the dominant term in order to see the behavior for $n \to \infty$. To this end, we condition on the known sequence x_t,

$$\begin{aligned} N_\mu &= \sum_{t=1}^n x_t^2 \sum_{t=1}^n u_t - \sum_{t=1}^n x_t \sum_{t=1}^n u_t x_t \\ &=^d \left(\frac{n^{2/\gamma}\delta^2}{n^{2/\gamma}\delta^2}\sum_{t=1}^n x_t^2\right)\left(\frac{\sqrt{n}\sigma}{\sqrt{n}\sigma}\sum_{t=1}^n u_t\right) \\ &\quad - \left(\frac{n^{1/\gamma}\delta}{n^{1/\gamma}\delta}\sum_{t=1}^n x_t\right)\sqrt{\sum_{t=1}^n x_t^2}\sigma U, \end{aligned}$$

where $U \sim N(0,1)$. Thus,

$$N_\mu =^d n^{2/\gamma+1/2}\sigma\delta^2 \left(\frac{1}{n^{2/\gamma}\delta^2}\sum_{t=1}^n x_t^2\right) U - n^{1/\gamma}\delta^2\sigma X \sqrt{\frac{1}{n^{2/\gamma}\delta^2}\sum_{t=1}^n x_t^2 U},$$

where $X \sim S\gamma S(1)$. Then as $n \to \infty$,

$$N_\mu \xrightarrow{d} n^{2/\gamma+1/2}\sigma\delta^2[L_\gamma](1)U - n^{2/\gamma}\delta^2\sigma X\sqrt{[L_\gamma](1)U}.$$

Therefore, the dominant part is $n^{2/\gamma+1/2}\sigma\delta^2[L_\gamma](1)U$. Then, as $n \to \infty$,

$$n^{-2/\gamma-1/2}N_\mu \stackrel{n\to\infty}{\longrightarrow} \sigma\delta^2[L_\gamma](1)U.$$

Next, let us consider N_β.

$$\begin{aligned} N_\beta &= n\sum_{t=1}^n u_t x_t - \sum_{t=1}^n x_t \sum_{t=1}^n u_t \\ &=^d n\sqrt{\sum_{t=1}^n x_t^2}\sigma U - \left(\frac{n^{1/\gamma}\delta}{n^{1/\gamma}\delta}\sum_{t=1}^n x_t\right)\left(\frac{\sqrt{n}\sigma}{\sqrt{n}\sigma}\sum_{t=1}^n u_t\right), \end{aligned}$$

where $U \sim N(0,1)$.

$$N_\beta =^d n\sqrt{\frac{n^{2/\gamma}\delta^2}{n^{2/\gamma}\delta^2}\sum_{t=1}^n x_t^2}\sigma U - n^{1/\gamma+1/2}\delta\sigma XU,$$

where $X \sim S\gamma S(1)$. Then as $n \to \infty$,

$$N_\beta \xrightarrow{d} n^{1/\gamma+1}\sqrt{[L_\gamma](1)}\sigma\delta U - n^{1/\gamma+1/2}\delta\sigma XU.$$

Therefore, the dominant term is $n^{1/\gamma+1}\sqrt{[L_\gamma](1)}\sigma\delta U$. Then, as $n \to \infty$,

$$n^{-1/\gamma-1}N_\beta \stackrel{n\to\infty}{\longrightarrow} \sqrt{[L_\gamma](1)}\sigma\delta U.$$

Next, let us consider D.

$$\begin{aligned} D &= n\sum_{t=1}^n x_t^2 - \left(\sum_{t=1}^n x_t\right)^2 \\ &=^d n\left(\frac{n^{2/\gamma}\delta^2}{n^{2/\gamma}\delta^2}\sum_{t=1}^n x_t^2\right) - \left(\frac{n^{1/\gamma}\delta}{n^{1/\gamma}\delta}\sum_{t=1}^n x_t\right)^2 \\ &\xrightarrow{d} n^{1+2/\gamma}\delta^2[L_\gamma](1) - n^{2/\gamma}\delta^2 X^2, \end{aligned}$$

where $X \sim N(0,1)$.

Therefore, the dominant part is $n^{1+2/\gamma}\delta^2[L_\gamma](1)$. Then, as $n \to \infty$,

$$n^{-1-2/\gamma} D \overset{n\to\infty}{\longrightarrow} \delta^2[L_\gamma](1).$$

So far, we have shown the convergence of the univariate marginal distributions of (N_μ, N_β, D). Using more advanced techniques of stochastic processes (see Paulauskas and Rachev (1998)) one can show that the joint distribution of (N_μ, N_β, D) converges as well. In summary we get

$$\frac{n^{-1/2-2/\gamma}N_\mu}{n^{-2/\gamma-1}D} = \sqrt{n}(\hat{\mu}_n - \mu) \overset{n\to\infty}{\longrightarrow} \frac{\sigma\delta^2[L_\gamma](1)}{\delta^2[L_\gamma](1)}U = \sigma U,$$

and

$$\frac{n^{-1-1/\gamma}N_\beta}{n^{-2/\gamma-1}D} = n^{1/\gamma}(\hat{\beta}_n - \beta) \overset{n\to\infty}{\longrightarrow} \frac{\sigma\delta\sqrt{[L_\gamma](1)}}{\delta^2[L_\gamma](1)}U = \frac{\sigma}{\delta\sqrt{[L_\gamma](1)}}U.$$

From now on X and U are defined as above.

For the t-statistics, we must look at

$$\begin{aligned} t_{\hat{\mu}} &= \frac{\hat{\mu}-\mu}{s_{\hat{\mu}}} \quad \text{and} \quad & t_{\hat{\beta}} &= \frac{\hat{\beta}-\beta}{s_{\hat{\beta}}} \\ &= \frac{N_\mu}{Ds_{\hat{\mu}}} \quad \text{and} & &= \frac{N_\beta}{Ds_{\hat{\beta}}}, \end{aligned}$$

where

$$\begin{aligned} s_{\hat{\mu}}^2 &= \frac{\hat{\sigma}_u^2 \sum_{t=1}^n x_t^2}{n\sum_{t=1}^n x_t^2 - (\sum_{t=1}^n x_t)^2}, & s_{\hat{\beta}}^2 &= \frac{n\hat{\sigma}_u^2}{n\sum_{t=1}^n x_t^2 - (\sum_{t=1}^n x_t)^2} \\ s_{\hat{\mu}} &= \frac{\hat{\sigma}_u\sqrt{\sum_{t=1}^n x_t^2}}{\sqrt{D}}, & s_{\hat{\beta}} &= \frac{\sqrt{n}\hat{\sigma}_u}{\sqrt{D}}. \end{aligned}$$

Therefore,

$$\begin{aligned} t_{\hat{\mu}} &= \frac{N_\mu}{Ds_{\hat{\mu}}} & \text{and} \quad t_{\hat{\beta}} &= \frac{N_\beta}{Ds_{\hat{\beta}}} \\ &= \frac{N_\mu\sqrt{D}}{D\hat{\sigma}_u\sqrt{\sum_{t=1}^n x_t^2}} & \text{and} \quad &= \frac{N_\beta\sqrt{D}}{D\sqrt{n}\hat{\sigma}_u} \\ &= \frac{N_\mu}{\sqrt{D}\sqrt{\sum_{t=1}^n x_t^2}\hat{\sigma}_u} & \text{and} \quad &= \frac{N_\beta}{\sqrt{D}\sqrt{n}\hat{\sigma}_u}, \end{aligned}$$

where

$$\hat{\sigma}_u^2 := n^{-1}\sum_{t=1}^{n}\hat{u}_t^2 \quad \text{and} \quad \hat{u}_t^2 := y_t - \hat{\mu}_n - \hat{\beta}_n x_t.$$

For both $t_{\hat{\mu}}$ and $t_{\hat{\beta}}$ the term $\hat{\sigma}_u^2 := n^{-1}\sum_{t=1}^{n}\hat{u}_t^2$ is important.

$$\begin{aligned}
\hat{\sigma}_u^2 &= n^{-1}\sum_{t=1}^{n}\hat{u}_t^2 = n^{-1}\sum_{t=1}^{n}(y_t - \hat{\mu}_n - \hat{\beta}_n x_t)^2 \\
&= n^{-1}\sum_{t=1}^{n}(\mu + \beta x_t + u_t - \hat{\mu}_n - \hat{\beta}_n x_t)^2 \\
&= n^{-1}\sum_{t=1}^{n}(u_t - (\hat{\mu}_n - \mu) - (\hat{\beta}_n - \beta)x_t)^2 \\
&\xrightarrow{d} n^{-1}\sum_{t=1}^{n}\left(u_t - \left(\frac{\sigma U}{\sqrt{n}}\right) - \left(\frac{1}{n^{1/\gamma}}\frac{\sigma}{\delta\sqrt{[L_\gamma](1)}}U\right)x_t\right)^2 \\
&=^d n^{-1}\sum_{t=1}^{n}\Bigg(u_t^2 + \frac{\sigma^2 U^2}{n} + \frac{1}{n^{2/\gamma}}\left(\frac{\sigma}{\delta\sqrt{[L_\gamma](1)}}\right)^2 U^2 x_t^2 - 2\frac{\sigma U}{\sqrt{n}}u_t \\
&\quad - 2\frac{\frac{\sigma}{\delta\sqrt{[L_\gamma](1)}}U}{n^{1/\gamma}}u_t x_t - 2\frac{\frac{\sigma^2}{\delta\sqrt{[L_\gamma](1)}}U^2}{n^{1/2+1/\gamma}}x_t\Bigg) \\
&=^d \frac{1}{n}\sum_{t=1}^{n}u_t^2 + \frac{\sigma^2 U^2}{n} + \frac{1}{n}\left(\frac{\sigma U}{\delta\sqrt{[L_\gamma](1)}}\right)^2 \frac{\delta^2}{n^{2/\gamma}\delta^2}\sum_{t=1}^{n}x_t^2 \\
&\quad - 2\frac{\sigma U}{n}\frac{\sigma}{\sqrt{n}\sigma}\sum_{t=1}^{n}u_t - 2\frac{\frac{\sigma}{\delta\sqrt{[L_\gamma](1)}}}{n^{1/\gamma+1}}U\sqrt{\frac{n^{2/\gamma}\delta^2}{n^{2/\gamma}\delta^2}\sum_{t=1}^{n}x_t^2\sigma U} \\
&\quad + 2\frac{\frac{\sigma^2}{\delta\sqrt{[L_\gamma](1)}}U^2}{n^{3/2}}\frac{\delta}{n^{1/\gamma}\delta}\sum_{t=1}^{n}x_t \\
&\overset{n\to\infty}{\longrightarrow} \sigma^2 + \frac{\sigma^2 U^2}{n} + \frac{1}{n}\left(\frac{\sigma U}{\delta^2\sqrt{[L_\gamma](1)}}\right)^2 [L_\gamma](1) - 2\frac{\sigma^2 U^2}{n} \\
&\quad - 2\frac{\frac{\sigma^2}{\sqrt{[L_\gamma](1)}}}{n}U^2\sqrt{[L_\gamma](1)} + 2\frac{\frac{\sigma^2}{\sqrt{[L_\gamma](1)}}}{n^{3/2}}U^2 X
\end{aligned}$$

Therefore σ^2 is the most dominant term. Then, as $n \to \infty$,

$$\hat{\sigma}_u^2 \overset{n\to\infty}{\longrightarrow} \sigma^2.$$

First we will focus on $t_{\hat{\mu}}$. It must be remembered that $t_{\hat{\mu}}$ becomes

$$
\begin{aligned}
t_{\hat{\mu}} &= \frac{N_\mu}{\sqrt{D}\sqrt{\sum_{t=1}^{n} x_t^2 \hat{\sigma}_u}} \\
&=^d \frac{N_\mu}{\sqrt{D}\sqrt{\frac{n^{2/\gamma}\delta^2}{n^{2/\gamma}\delta^2}\sum_{t=1}^{n} x_t^2 \hat{\sigma}_u}} \\
&\xrightarrow{d} \frac{n^{2/\gamma+1/2}\sigma\delta^2[L_\gamma](1)U}{n^{1/\gamma+1/2}\delta\sqrt{[L_\gamma](1)}n^{1/\gamma}\delta\sqrt{[L_\gamma](1)}\sigma} \\
&= U.
\end{aligned}
$$

Second, we must remember that $t_{\hat{\beta}}$ becomes

$$
\begin{aligned}
t_{\hat{\beta}} &= \frac{N_\beta}{\sqrt{D}\sqrt{n}\,\hat{\sigma}_u} \\
&= \frac{n^{1+1/\gamma}\delta\sigma\sqrt{[L_\gamma](1)}U}{n^{1/\gamma+1/2}\delta\sqrt{[L_\gamma](1)}\sqrt{n}\,\sigma} \\
&= U,
\end{aligned}
$$

which concludes the sketch of the proof.

Appendix D

Sketch of the proof It can easily be verified that

$$
(\hat{\mu}_n - \mu) = \frac{\sum_{t=1}^{n} x_t^2 \sum_{t=1}^{n} u_t - \sum_{t=1}^{n} x_t \sum_{t=1}^{n} u_t x_t}{n\sum_{t=1}^{n} x_t^2 - (\sum_{t=1}^{n} x_t)^2}
$$

and

$$
(\hat{\beta}_n - \beta) = \frac{n\sum_{t=1}^{n} u_t x_t - \sum_{t=1}^{n} x_t \sum_{t=1}^{n} u_t}{n\sum_{t=1}^{n} x_t^2 - (\sum_{t=1}^{n} x_t)^2}.
$$

Define

$$
N_\mu := \sum_{t=1}^{n} x_t^2 \sum_{t=1}^{n} u_t - \sum_{t=1}^{n} x_t \sum_{t=1}^{n} u_t x_t,
$$

$$
N_\beta := n\sum_{t=1}^{n} u_t x_t - \sum_{t=1}^{n} x_t \sum_{t=1}^{n} u_t
$$

and

$$
D := n\sum_{t=1}^{n} x_t^2 - \left(\sum_{t=1}^{n} x_t\right)^2.
$$

Let us first consider N_μ. We must determine the dominant term in order to see the behavior for $n \to \infty$. To this end, we condition on the known sequence x_t,

$$\begin{aligned} N_\mu &= \sum_{t=1}^n x_t^2 \sum_{t=1}^n u_t - \sum_{t=1}^n x_t \sum_{t=1}^n u_t x_t \\ &=^d \left(\frac{n^{2/\gamma}\delta^2}{n^{2/\gamma}\delta^2} \sum_{t=1}^n x_t^2\right) \left(\frac{n^{1/\alpha}\sigma}{n^{1/\alpha}\sigma} \sum_{t=1}^n u_t\right) \\ &\quad - \left(\frac{n^{1/\gamma}\delta}{n^{1/\gamma}\delta} \sum_{t=1}^n x_t\right) \left(\sum_{t=1}^n |x_t|^\alpha\right)^{1/\alpha} \sigma U, \end{aligned}$$

where $U \sim S\alpha S(1)$. Thus,

$$\begin{aligned} N_\mu &= \left(\frac{n^{2/\gamma}\delta^2}{n^{2/\gamma}\delta^2} \sum_{t=1}^n x_t^2\right) \left(\frac{n^{1/\alpha}\sigma}{n^{1/\alpha}\sigma} \sum_{t=1}^n u_t\right) \\ &\quad - \left(\frac{n^{1/\gamma}\delta}{n^{1/\gamma}\delta} \sum_{t=1}^n x_t\right) \left(\sum_{t=1}^n |x_t|^\alpha\right)^{1/\alpha} \sigma U \\ &\xrightarrow{d} n^{2/\gamma+1/\alpha}\sigma\delta^2[L_\gamma](1)U - n^{1/\gamma}\delta\sigma X \left(\sum_{t=1}^n |x_t|^\alpha\right)^{1/\alpha} U, \end{aligned}$$

where $X \sim S\gamma S(1)$. Set $S_t = x_t/\delta$ unconditionally.

$$N_\mu = n^{2/\gamma+1/\alpha}\sigma\delta^2[L_\gamma](1)U - n^{1/\gamma}\delta^2\sigma X \left(\sum_{t=1}^n |S_t|^\alpha\right)^{1/\alpha} U.$$

In order to find the right normalization of the corresponding sum, we must check the tail behavior of $\left(\sum_{t=1}^n |S_t|^\alpha\right)^{1/\alpha}$. Set $Z_t = |S_t|^\alpha$. Then,

$$\left(\sum_{t=1}^n |S_t|^\alpha\right)^{1/\alpha} = \left(\sum_{t=1}^n Z_t\right)^{1/\alpha}.$$

Therefore,

$$P(Z_t > \lambda) = P(|S_t| > \lambda^{1/\alpha}) \approx \frac{1}{2} C_\gamma (\lambda^{1/\alpha})^{-\gamma},$$

where

$$C_\gamma = \left(\int_0^\infty x^{-\gamma} \sin x \, dx\right)^{-1}$$

$$C_\gamma = \begin{cases} \frac{1-\gamma}{\Gamma(2-\gamma)\cos(\pi\gamma/2)} & \text{if } \gamma \neq 1, \\ \frac{2}{\pi} & \text{if } \gamma = 1, \end{cases}$$

is a constant. See Samorodnitsky and Taqqu (1994). But then, according to Feller (1966, p. 547), Z_t belongs to the normal domain of attraction of some stable random variable that has the index of stability of $\kappa = \gamma/\alpha$. Hence, the right normalization for $\sum_{t=1}^n |S_t|^\alpha = \sum_{t=1}^n Z_t$ is $n^{1/\kappa} = n^{\alpha/\gamma}$. In summary, N_μ becomes

$$\begin{aligned} N_\mu &= n^{2/\gamma+1/\alpha}\sigma\delta^2[L_\gamma])(1)U - n^{1/\gamma}\delta^2\sigma X\left(\frac{n^{1/\kappa}}{n^{1/\kappa}}\sum_{t=1}^n |S_t|^\alpha\right)^{1/\alpha} U \\ &= n^{2/\gamma+1/\alpha}\sigma\delta^2[L_\gamma](1)U - n^{1/\gamma}\delta^2\sigma X\left(\frac{n^{\alpha/\gamma}}{n^{\alpha/\gamma}}\sum_{t=1}^n |S_t|^\alpha\right)^{1/\alpha} U \\ &\xrightarrow{d} n^{2/\gamma+1/\alpha}\sigma\delta^2[L_\gamma](1)U - n^{2/\gamma}\delta^2\sigma X(V)^{1/\alpha}U, \end{aligned}$$

where V is a random variable that is in the normal domain of attraction of a κ-stable random variable. However, the dominant part is $n^{2/\gamma+1/\alpha}\sigma\delta^2[L_\gamma](1)U$. Then, as $n \to \infty$,

$$n^{-2/\gamma-1/\alpha}N_\mu \overset{n\to\infty}{\longrightarrow} \sigma\delta^2[L_\gamma](1)U.$$

Next, let us consider N_β.

$$\begin{aligned} N_\beta &= n\sum_{t=1}^n u_t x_t - \sum_{t=1}^n x_t \sum_{t=1}^n u_t \\ &= n\left(\sum_{t=1}^n |x_t|^\alpha\right)^{1/\alpha} \sigma U - \left(\frac{n^{1/\gamma}\delta}{n^{1/\gamma}\delta}\sum_{t=1}^n x_t\right)\left(\frac{n^{1/\alpha}\sigma}{n^{1/\alpha}\sigma}\sum_{t=1}^n u_t\right), \end{aligned}$$

conditioned on the known sequence x_t, where $U \sim S\alpha S(1)$.

$$\begin{aligned} N_\beta &= n\left(\sum_{t=1}^n |x_t|^\alpha\right)^{1/\alpha} \sigma U - \left(\frac{n^{1/\gamma}\delta}{n^{1/\gamma}\delta}\sum_{t=1}^n x_t\right)\left(\frac{n^{1/\alpha}\sigma}{n^{1/\alpha}\sigma}\sum_{t=1}^n u_t\right) \\ &= n\left(\frac{n^{\alpha/\gamma}\delta^\alpha}{n^{\alpha/\gamma}\delta^\alpha}\sum_{t=1}^n |x_t|^\alpha\right)^{1/\alpha} \sigma U - n^{1/\gamma+1/\alpha}\sigma\delta X U, \end{aligned}$$

where $X \sim S\gamma S(1)$. Recall the correct normalization of $\sum_{t=1}^{n}|x_t|^\alpha$ from above.

$$\begin{aligned} N_\beta &= n\left(\frac{n^{\alpha/\gamma}\delta^\alpha}{n^{\alpha/\gamma}\delta^\alpha}\sum_{t=1}^{n}|x_t|^\alpha\right)^{1/\alpha}\sigma U - n^{1/\gamma+1/\alpha}\sigma\delta XU \\ &\xrightarrow{d} n^{1/\gamma+1}\sigma\delta V^{1/\alpha}U - n^{1/\gamma+1/2}\delta\sigma XU. \end{aligned}$$

Therefore, the dominant term is $n^{1/\gamma+1}\sigma\delta V^{1/\alpha}U$. Then, as $n \to \infty$,

$$n^{-1/\gamma-1}N_\beta \overset{n\to\infty}{\longrightarrow} \sigma\delta V^{1/\alpha}U.$$

Next, let us consider D.

$$\begin{aligned} D &= n\sum_{t=1}^{n}x_t^2 - \left(\sum_{t=1}^{n}x_t\right)^2 \\ &= n\left(\frac{n^{2/\gamma}\delta^2}{n^{2/\gamma}\delta^2}\sum_{t=1}^{n}x_t^2\right) - \left(\frac{n^{1/\gamma}\delta}{n^{1/\gamma}\delta}\sum_{t=1}^{n}x_t\right)^2 \\ &\xrightarrow{d} n^{1+2/\gamma}\delta^2[L_\gamma](1) - n^{2/\gamma}\delta^2X^2, \end{aligned}$$

where $X \sim S\gamma S(1)$.

Therefore, the dominant part is $n^{1+2/\gamma}\delta^2[L_\gamma](1)$. Then, as $n \to \infty$,

$$n^{-1-2/\gamma}D \overset{n\to\infty}{\longrightarrow} \delta^2[L_\gamma](1).$$

So far, we have shown the convergence of the univariate marginal distributions of (N_μ, N_β, D). Using more advanced techniques of stochastic processes (see Paulauskas and Rachev (1998)) one can show that the joint distribution of (N_μ, N_β, D) converges as well. In summary we get

$$\frac{n^{-2/\gamma-1/\alpha}N_\beta}{n^{-2/\gamma-1}D} = n^{1-1/\alpha}(\hat{\mu}_n - \mu) \overset{n\to\infty}{\longrightarrow} \frac{\sigma\delta^2[L_\gamma](1)}{\delta^2[L_\gamma](1)}U = \sigma U,$$

and

$$\frac{n^{-1-1/\gamma}N_\beta}{n^{-2/\gamma-1}D} = n^{1/\gamma}(\hat{\beta}_n - \beta) \overset{n\to\infty}{\longrightarrow} \frac{\sigma\delta V^{1/\alpha}U}{\delta^2[L_\gamma](1)} = \frac{\sigma V}{\delta\sqrt{[L_\gamma](1)}}U.$$

From now on X, V and U are defined as above.

For the t-statistics, we must look at

$$\begin{aligned} t_{\hat{\mu}} &= \frac{\hat{\mu}-\mu}{s_{\hat{\mu}}} \quad \text{and} \quad t_{\hat{\beta}} = \frac{\hat{\beta}-\beta}{s_{\hat{\beta}}} \\ &= \frac{N_\mu}{Ds_{\hat{\mu}}} \quad \text{and} \quad \quad = \frac{N_\beta}{Ds_{\hat{\beta}}}, \end{aligned}$$

where

$$s_{\hat{\mu}}^2 = \frac{\hat{\sigma}_u^2 \sum_{t=1}^n x_t^2}{n\sum_{t=1}^n x_t^2 - (\sum_{t=1}^n x_t)^2}, \quad s_{\hat{\beta}}^2 = \frac{n\hat{\sigma}_u^2}{n\sum_{t=1}^n x_t^2 - (\sum_{t=1}^n x_t)^2}$$

$$s_{\hat{\mu}} = \frac{\hat{\sigma}_u \sqrt{\sum_{t=1}^n x_t^2}}{\sqrt{D}}, \qquad s_{\hat{\beta}} = \frac{\sqrt{n}\hat{\sigma}_u}{\sqrt{D}}.$$

Therefore,

$$\begin{aligned} t_{\hat{\mu}} &= \frac{N_\mu}{Ds_{\hat{\mu}}} & \text{and} \quad t_{\hat{\beta}} &= \frac{N_\beta}{Ds_{\hat{\beta}}} \\ &= \frac{N_\mu\sqrt{D}}{D\hat{\sigma}_u\sqrt{\sum_{t=1}^n x_t^2}} & \text{and} \quad &= \frac{N_\beta\sqrt{D}}{D\sqrt{n}\hat{\sigma}_u} \\ &= \frac{N_\mu}{\sqrt{D}\sqrt{\sum_{t=1}^n x_t^2}\hat{\sigma}_u} & \text{and} \quad &= \frac{N_\beta}{\sqrt{D}\sqrt{n}\hat{\sigma}_u}, \end{aligned}$$

where

$$\hat{\sigma}_u^2 := n^{-1}\sum_{t=1}^n \hat{u}_t^2 \quad \text{and} \quad \hat{u}_t^2 := y_t - \hat{\mu}_n - \hat{\beta}_n x_t.$$

For both $t_{\hat{\mu}}$ and $t_{\hat{\beta}}$ the term $\hat{\sigma}_u^2 := n^{-1}\sum_{t=1}^n \hat{u}_t^2$ is important.

$$\begin{aligned} \hat{\sigma}_u^2 &= n^{-1}\sum_{t=1}^n \hat{u}_t^2 = n^{-1}\sum_{t=1}^n (y_t - \hat{\mu}_n - \hat{\beta}_n x_t)^2 \\ &= n^{-1}\sum_{t=1}^n (\mu + \beta x_t + u_t - \hat{\mu}_n - \hat{\beta}_n x_t)^2 \\ &= n^{-1}\sum_{t=1}^n (u_t - (\hat{\mu}_n - \mu) - (\hat{\beta}_n - \beta)x_t)^2 \\ &\xrightarrow{d} n^{-1}\sum_{t=1}^n \left(u_t - \left(\frac{\sigma U}{n^{1-1/\alpha}}\right) - \left(\frac{1}{n^{1/\gamma}}\frac{\sigma V^{1/\alpha}U}{\delta\sqrt{[L_\gamma](1)}}\right)x_t\right)^2 \\ &= n^{-1}\sum_{t=1}^n \Bigg(u_t^2 + \frac{\sigma^2U^2}{n^{2-2/\alpha}} + \frac{1}{n^{2/\gamma}}\left(\frac{\sigma V^{1/\alpha}U}{\delta\sqrt{[L_\gamma](1)}}\right)^2 x_t^2 - 2\frac{\sigma U}{n^{1-1/\alpha}}u_t \\ &\quad - \frac{2}{n^{1/\gamma}}\left(\frac{\sigma V^{1/\alpha}U}{\delta\sqrt{[L_\gamma](1)}}\right)u_t x_t + \frac{2}{n^{1/2+1/\gamma}}\left(\frac{\sigma^2 V^{1/\alpha}U^2}{\delta\sqrt{[L_\gamma](1)}}\right)x_t\Bigg) \\ &= \frac{1}{n}\frac{n^{2/\alpha}\sigma^2}{n^{2/\alpha}\sigma^2}\sum_{t=1}^n u_t^2 + \frac{\sigma^2U^2}{n^{2-2/\alpha}} + \frac{1}{n}\left(\frac{\sigma V^{1/\alpha}U}{\delta\sqrt{[L_\gamma](1)}}\right)^2 \frac{\delta^2}{n^{2/\gamma}\delta^2}\sum_{t=1}^n x_t^2 \end{aligned}$$

$$
\begin{aligned}
& - 2\frac{\sigma U}{n^{2-1/\alpha}}\frac{n^{1/\alpha}\sigma}{n^{1/\alpha}\sigma}\sum_{t=1}^{n} u_t \\
& - \frac{2}{n^{1/\gamma+1}}\left(\frac{\sigma V^{1/\alpha}U}{\delta\sqrt{[L_\gamma](1)}}\right)\left(\frac{n^{\alpha/\gamma}\delta^\alpha}{n^{\alpha/\gamma}\delta^\alpha}\sum_{t=1}^{n}|x_t|^\alpha\right)^{1/\alpha}\sigma U \\
= & + \frac{2}{n^{3/2}}\left(\frac{\sigma^2 V^{1/\alpha}U^2}{\delta\sqrt{[L_\gamma](1)}}\right)\frac{\delta}{n^{1/\gamma}\delta}\sum_{t=1}^{n} x_t \\
\overset{n\to\infty}{\longrightarrow} & \frac{1}{n^{1-2/\alpha}}[L_\alpha](1)\sigma^2 + \frac{\sigma^2U^2}{n^{2-2/\alpha}} + \frac{1}{n}\left(\frac{\sigma V^{1/\alpha}U}{\sqrt{[L_\gamma](1)}}\right)^2[L_\gamma](1) - 2\frac{\sigma^2U^2}{n^{2-2/\alpha}} \\
& - \frac{2}{n}\left(\frac{\sigma V^{2/\alpha}U}{\sqrt{[L_\gamma](1)}}\right) + \frac{2}{n^{3/2}}\left(\frac{\sigma^2 V^{1/\alpha}U^2}{\sqrt{[L_\gamma](1)}}\right)X.
\end{aligned}
$$

Therefore $\frac{1}{n^{1-2/\alpha}}[L_\alpha](1)\sigma^2$ is the most dominant term. Then, as $n \to \infty$,

$$\hat{\sigma}_u^2 \overset{n\to\infty}{\longrightarrow} n^{2/\alpha-1}[L_\alpha](1)\sigma^2.$$

First we will focus on $t_{\hat{\mu}}$. It must be remembered that $t_{\hat{\mu}}$ becomes

$$
\begin{aligned}
t_{\hat{\mu}} &= \frac{N_\mu}{\sqrt{D}\sqrt{\sum_{t=1}^{n} x_t^2\hat{\sigma}_u}} \\
&= \frac{N_\mu}{\sqrt{D}\sqrt{\frac{n^{2/\gamma}\delta^2}{n^{2/\gamma}\delta^2}\sum_{t=1}^{n} x_t^2\hat{\sigma}_u}} \\
&\overset{d}{\longrightarrow} \frac{n^{2/\gamma+1/\alpha}\sigma\delta^2[L_\gamma](1)U}{n^{1/\gamma+1/2}\delta\sqrt{[L_\gamma](1)}\,n^{1/\gamma}\delta\sqrt{[L_\gamma](1)}\,n^{1/\alpha-1/2}\sqrt{[L_\alpha](1)}\sigma} \\
&= \frac{U}{\sqrt{[L_\alpha](1)}}.
\end{aligned}
$$

Second, we must remember that $t_{\hat{\beta}}$ becomes

$$
\begin{aligned}
t_{\hat{\beta}} &= \frac{N_\beta}{\sqrt{D}\sqrt{n}\,\hat{\sigma}_u} \\
&= \frac{n^{1+1/\gamma}\delta\sigma V^{1/\alpha}U}{n^{1/\gamma+1/2}\delta\sqrt{[L_\gamma](1)}\sqrt{n}\,n^{1/\alpha-1/2}\sqrt{[L_\alpha](1)}\sigma} \\
&= n^{1/2-1/\alpha}\frac{UV^{1/\alpha}}{\sqrt{[L_\gamma](1)}\sqrt{[L_\alpha](1)}}.
\end{aligned}
$$

Hence,

$$n^{1/\alpha-1/2}t_{\hat{\beta}} \xrightarrow{d} \frac{UV^{1/\alpha}}{\sqrt{[L_\gamma](1)}\,\sqrt{[L_\alpha](1)}},$$

which concludes the sketch of the proof.

References

1. Admati A.R., Bhattacharya, S., Pfleiderer P., and Ross, S. (1986). On Timing and Selectivity, *Journal of Finance*, 41.

2. Banz, R.W. (1981). The Relationship between Return and the Market Value of Common stocks, *Journal of Financial and Quantitative Analysis*, 14, 421–441.

3. Best, M. and Grauer, R. (1990). The Efficient Set Mathematics Subject When Mean-Variance Portfolio Problems are Subject to General Linear Constraints, *Journal of Economics and Business*, 42, 105–120.

4. Bhandari, L.C. (1988). Debt/Equity Ratios and Expected Common Stocks Returns, *Journal of Finance*, 43, 507–528.

5. Black, F., Jensen, M., and Scholes, M. (1972). The Capital Asset Pricing Model: Some Empirical Tests, in M. Jensen and Ed.: *Studies in the Theory of Capital Markets*, Praeger, New York.

6. Blattberg, R. and Sargent, T. (1971). Regression with Non-Gaussian Stable Disturbances: Some Sampling Results, *Econometrica*, 39 (3), 501–510.

7. Blume, M. (1968). The Assessment of Portfolio Performance: An Application of Portfolio Theory, Ph.D. dissertation, University of Chicago.

8. Brennan, M.J. (1971). Capital Market Equilibrium with Diverging Borrowing and Lending Rates, *Journal of Financial and Quantitative Analysis*, 6.

9. Carhart, M. (1997). On Persistence in Mutual Fund Performance, *Journal of Finance*, 52, 57–82.

10. Chamberlain, G. (1983). A Characterization of the Distributions that Imply Mean-Variance Utility Functions, *Journal of Economic Theory*, 29, 185–201.

11. Chen, C.R., Lee, C.F., Rahmann, S., and Chan, A. (1992). Cross-Sectional Analysis of Mutual Fund Market Timing and Security Selection Skill, *Journal of Business Finance*, 19.

12. Connor, G. (1984). A Unified Beta Pricing Theory, *Journal of Economic Theory*, 34, 13–31.

13. Connor, G. and Korajczyk, R.A. (1986). Performance Measurement with the Arbitrage Pricing Theory: A New Framework for Analysis, *Journal of Financial Economics*, 15, 373–394.

14. Connor, G. and Korajczyk, R.A. (1989). An Intertemporal Equilibrium Beta Pricing Model, *Review of Financial Studies*, 2, 373–392.

15. Connor, G. and Korajczyk, R.A. (1991), The Attributes, Behavior, and Performance of US Mutual Funds, *Review of Quantitative Finance and Accounting*, 1, 5–26.

16. Copeland, T.E. and Mayers, D. (July 1982). The Value Line Enigma (1965–1978): A Case Study of Performance Evaluation Issues, *Journal of Financial Economics*, 10.

17. Cornell, B. (December 1979). Asymetric Information and Portfolio Performance Measurement, *Journal of Financial Economics*, 7.

18. Daniel, K., Grinblatt, M., Titman, S., and Wermers, R. (1997). Measuring Mutual Fund Performance with Characteristic-Based Benchmarks, *Journal of Finance*, 52, 1035–1058.

19. Daniel, K. and Titman, S. (1997). Evidence on the Characteristics of Cross Sectional Variation in Stock Returns, *Journal of Finance*, 52, 1–11.

20. Dybvig, P. and Ross, S. (1985). The Analysis of Performance Measurement Using a Security Market Line, *Journal of Finance*, 40.

21. Elton, E., Gruber, M., Das, S., and Hlavka, M. (1993). Efficiency with costly information: A Reinterpretation of Evidence from Managed Portfolios, *Review of Financial Studies*, 1.

22. Elton, E., Gruber, M., Das, S., and Blake, C. (1995), The Persistence of Risk-Adjusted Mutual Fund Performance, *Journal of Business*, 69, 133–157.

23. Fama, E.F. (1972). Components of Investment Performance, *Journal of Finance*, 27, 551–576.

24. Fama, E.F. (1965). The Behavior of Stock Market Prices, *Journal of Business*, 38, 34–105.

25. Fama, E.F. (1976). *Foundations of Finance*, New York, Basic.

26. Fama, E.F. (1970). Multiperiod Consumption-Investment Decisions, *American Economic Review*, 60, 163–174.

27. Fama, E.F. and French, K.R. (1992). The Cross-Section of Expected Stock Returns, *Journal of Finance*, 47, 427–465.

28. Fama, E.F. and French, K.R. (1993). Common risk factors in the returns on stocks and bonds, *Journal of Financial Economics*, 33, 3–56.

29. Fama, E.F. and French, K.R. (1995). Size and Book-to-Market Factors in Earnings and Returns, *Journal of Finance*, 50, 131–167.

30. Fama, E.F. and French, K.R. (1996). Multifactor Explanations of Asset Pricing Anomalies, *Journal of Finance*, 51, 55–84.

31. Fama, E.F. and French, K.R. (1998). Value versus Growth: The International Evidence, *Journal of Finance*, 53, 1975–1993.

32. Fama, E.F. and MacBeth, J.D. (1973). Risk, Return and Equilibrium: Empirical Tests, *Journal of Political Economy*, 71, 607–636.

33. Feller, W. (1966). *An Introduction to Probability Theory and Its Applications*, Wiley, New York.

34. Ferson, W. and Harvey, C.R. (1991). The Variation of Economic Risk Premiums, *Journal of Political Economy*, 99, 385–415.

35. Ferson, W. and Harvey, C.R. (1993). Seasonality and Heteroskedasticity in Consumption-Based Asset Pricing: An Analysis of Linear Models, *Res. Finance*, 11, 1–35.

36. Ferson, W.E. and Harvey, C.R. (1999). Conditioning Variables and the Cross-Section of Stock Returns, *Journal of Finance*, to be published.

37. Ferson, W.E. and Schadt, R.W. (1996). Measuring Fund Strategy and Performance in Changing Economic Conditions, *Journal of Finance*, 51, 425–461.

38. Gamrowski, B. and Rachev, S.T. (1994). Stable Models in Testable Asset Pricing, in G. Anastassiou and S.T. Rachev, eds.: *Approximation, Probability and Related Fields*, Plenum, New York.

39. Ghysels, E. (1998). On Stable Factor Structures in the Pricing of Risk: Do Time-Varying Betas Help or Hurt? *Journal of Finance*, 53, 549–575.

40. Grauer, R.R. (1991) Further Ambiguity When Performance is Measured by the Security Market Line, *Financial Review*.

41. Green, R. (1986). Benchmark Portfolio Inefficiency and Deviations from the Security Market Line, *Journal of Finance*, 41.

42. Grinblatt, M., Titman, S., and Wermers, R. (1995) Momentum Investment Strategies, Portfolio Performance, and Herding: A Study of Mutual Fund Behavior, *American Economic Review*, 85, 1977–1984.

43. Grinblatt, M. and Titman, S. (1989). Mutual Funds Performance: An Analysis of Quarterly Portfolio Holdings, *Journal of Business*, 62.

44. Grinblatt, M. and Titman, S. (1992). The Persistence of Mutual Funds Performance, *Journal of Finance*, 47.

45. Grinblatt, M. and Titman, S. (1993). Performance Measurement without Benchmarks: An Examination of Mutual Funds, *Journal of Business*, 66.

46. Henriksson, R.D. and Merton, R.C. (1981) On Market Timing and Investment Performance. II. Statistical Procedures for Evaluating Forecasting Skills, *Journal of Business*, 54, 513–533.

47. Henriksson, R. (1984). Market Timing and Mutual Fund Performance: An Empirical Investigation, *Journal of Business*, 57, 73–96.

48. Ippolito, R. (1989). Efficiency with costly information: A Study of Mutual Fund Performance, *Quarterly Journal of Economics*, 104.

49. Jagannathan, R. and Korajczyk, R.A. (1986). Assessing the Market Timing Performance of Managed Portfolios, *Journal of Business*, 59.

50. Jagannathan, R. and Wang, Z. (1996). The Conditional CAPM and the Cross-Section of Expected Returns, *Journal of Finance*, 51, 3–55.

51. Jegadeesh, N. and Titman, S. (1993). Returns to Buying Winners and Selling Losers: Implications for Stock Market Efficiency, *Journal of Finance*, 48, 93–130.

52. Jensen, M.C. (1968). The Performance of Mutual Funds in the Period 1945–1964, *Journal of Finance*, 23.

53. Kim, D. (1997). A Reexamination of Firm Size, Book-to-Market, and Earnings Price in the Cross-Section of Expected Stock Returns, *Journal of Financial and Quantitative Analysis*, 32, 463–488.

54. Kothari, S.P., Shanken, J., and Sloan, R.G. (1995). Another Look at the Cross-Section of Expected Stock Returns, *Journal of Finance*, 50, 1605–1634.

55. Lakonishok, J., Shleifer, A., and Vishny, R. (1994). Contrarian Investment, Extrapolation, and Risk, *Journal of Finance*, 49, 1541–1578.

56. Lee, C. and Rahmann, S. (1991). Errors in Variables, Functional Form and Mutual Fund Returns, *Quarterly Review of Economics and Business*, 31.

57. Lee, C. and Rahmann, S. (1990). Market Timing, Selectivity and Mutual Fund Performance: An Empirical Investigation, *Journal of Business*, 63.

58. Lehmann, B.N. and Modest, D.M. (1987). Mutual Fund Performance Evaluation: A Comparison of Benchmarks and Benchmark Comparison, *Journal of Finance*, 42.

59. Lintner, J. (1965). Security Prices, Risk, and Maximal Gains from Diversification, *Journal of Finance*, 20.

60. Mandelbrot, B. (1963). The Variation of Certain Speculative Prices, *Journal of Business*, 36, 394–419.

61. Mandelbrot, B. (1967). The Variation of some other Speculative Prices, *Journal of Business*, 40, 393–413.

62. Markowitz, H. (1959). *Portfolio Selection: Efficient Diversification of Investments*, New York, NY: Wiley.

63. Mayers D. (1972). Nonmarketable Assets and Capital Market Equilibrium under Uncertainty, in M. Jensen, Ed.: *Studies in the Theory of Capital Markets*, Praeger, New York, 223–248.

64. Merton, R.C. (1973). An Intertemporal Capital Asset Pricing Model, *Econometrica*, 41, 867–887.

65. Merton, R.C. (1981). On Market Timing and Investment Performance: An Equilibrium Theory of Value for Market Forecasts, *Journal of Business*, 54.

66. Mittnik, S., Rachev, S., and Paollela, M. (1997). Stable Paretian Modeling in Finance: Some Empirical and Theoretical Aspects, in R. Adler et al, eds.: *A Practical Guide to Heavy Tails: Statistical Techniques for Analyzing Heavy Tailed Distributions*, Birkhauser, Boston.

67. Modilgiani F. and Miller, M. (1958). The Cost of Capital, Corporation Finance, and the Theory of Investment, *American Economic Review*, 53, 261–297.

68. Rachev, Z.T. (1999). Stable Paretian Models in Finance, forthcoming.

69. Reilly, F.K. and Akhtar, R.A. (1995). The Benchmark Error Problem with Global Capital Markets, *Journal of Portfolio Management.*

70. Richardson, M. and Smith, T. (1993). A Test for Multivariate Normality in Stock Returns, *Journal of Business*, 66, 295–321.

71. Roll, R. (1977). A Critique of the Asset Pricing Theory's Tests: Part I on Past and Potential Testability of the Theory, *Journal of Financial Economics.*

72. Roll, R. (1978). Ambiguity When Performance is Measured by the Security Market Line, *Journal of Finance*, 33.

73. Ross, S. and Roll, R. (1980). An Empirical Investigation of the Arbitrage Pricing Theory, *Journal of Finance*, 35, 1073–1103.

74. Ross, S. (1976). The Arbitrage Theory of Capital Asset Pricing, *Journal of Economic Theory*, 13, 341–360.

75. Ross, S. (1978). Mutual Fund Separation in Financial Theory — the Separating Distributions, *Journal of Economic Theory*, 17, 254–286.

76. Shanken J. (1982). The Arbitrage Pricing Theory: Is It Testable?, *Journal of Finance*, 37, 1129–1140.

77. Shanken J. (1990). Intertemporal Asset Pricing: An Empirical Investigation, *Journal of Econometrics*, 45, 99–120.

78. Sharpe, W. (1964). Capital Asset Prices: A Theory of Market Equilibrium under Conditions of Risk, *Journal of Finance*, 19, 425–442.

79. Sharpe, W. (1966). Mutual Funds Performance, *Journal of Business*, 39.

80. Treynor, J.L. and Mazuy, K.K. (1966). Can Mutual Funds Outguess the Market, *Harvard Business Review*, 44.

81. Vandall, R.F. and Stevens, J.L. (1989). Evidence on Superior Performance from Timing, *Journal of Portfolio Management*, 15.

82. Zolotarev, V.M. (1986). *One-dimensional Stable Distributions*, American Mathematical Society, 65. Translation from the original 1983 Russian edition.

EVALUATING STATISTICAL FUNCTIONALS BY MEANS OF PROJECTIONS ONTO CONVEX CONES IN HILBERT SPACES

Thomasz Rychlik
Institute of Mathematics, Polish Academy of Sciences, Chopina 12, 87100 Toruń, Poland. E-mail: trychlik@impan.gov.pl

We present a general method of determining sharp bounds on some statistical functionals, including quantiles, expectations of record and order statistics of independent and dependent samples, with sample distributions belonging to various nonparametric classes of importance in mathematical statistics and reliability theory: arbitrary distributions, symmetric and symmetric unimodal ones, distributions with monotone density and failure rate (also in average) etc. The method is based on representing classes of distributions as convex cones in properly chosen Hilbert spaces. The projections of the functionals onto the cones and their norms provide optimal bounds and describe distributions which attain the bounds. Specific projection problems are solved by use of various analytic, geometrical and numerical tools. The first part contains a general description of the method, and bounds on quantiles and expectations of order statistics of independent samples. The second part is devoted to order statistics of dependent samples and record statistics.

PART I

1 Introduction and Notation

1.1 Introduction

This work was intended as a first systematic attempt to present a method of use projections of functions onto convex cones in Hilbert spaces for determining sharp bounds on values of statistical functionals over general and restricted families of population distributions, expressed in terms of natural moment parameters of the distributions. The method is based on representing the statistical functionals and families of distributions as fixed elements and convex cones, respectively, in a common real Hilbert space. Then the norm of projection of the element onto the cone provides the optimal bound. The distribution for which the bound is attained is derived by a simple transformation of the projection. The idea of using projections for sharp evaluations of statistical functionals was formulated in Gajek and Rychlik [18]. Some earlier results discussed here were originally proved without reference to projections, but our unified approach provides simpler proofs and indicates mutual relations among various evaluations. In principle, this review provides an exposition of recent research: a significant number of papers we refer to has not been published yet, and some results were not presented elsewhere. Since the paper is entirely devoted to the projection method, we consistently omit discussing other approaches. Therefore references to results obtained by different methods are rare and laconic here.

The structure of the paper reflects the purpose of presenting bounds on various statistical functionals over various classes of distributions by means of the common method based on projections. Fundamental notions are introduced in Section 2. Some facts of the general Hilbert spaces theory which are applied in our method are collected in Subsection 2.1. Inner product formulae for some functionals with statistical interpretations are worked out in Subsection 2.2. In Subsection 2.3 special classes of distributions of practical importance are characterized by convex cones of respective quantile functions. Some partial orderings of distributions which are useful in defining the classes are introduced there. Consecutive Sections are devoted to specific functionals, and bounds on the functionals over various classes are discussed in respective Subsections. Titles of Subsections refer to families of distributions with intuitive interpretations. In fact, more general results are presented. We consider families of distributions related to a fixed general one with respect to partial orderings defined in Subsection 2.3. Some results are specified for specially chosen extreme elements of the families, e.g., for uniform and exponential distributions which generate some classes with natural intuitive properties.

The main results of the paper are presented in Sections 3–6 in a unified way: each sharp bound on a fixed functional over a fixed class of distributions is explicitly written, together with a formula for the distribution function for which the bound is attained. Different bounds for various classes of distributions are also compared numerically. Formal proofs are not presented here. These will be included in an extended book version of the work that is under preparation now. However, some informal arguments leading to solutions are outlined. The special emphasis is laid on constructions of projections that is the essence of our method. Since in the problems we study there are not known general construction methods for projecting functions, various tools are needed for solving specific problems. Usually we first describe the shape of projections up to several real parameters by means of geometric arguments, and then we determine the optimal parameters analytically. Some bounds are expressed by complicated formulae that should be evaluated by use of subtle tools of numerical analysis.

The results presented here are far from being complete. There are still many interesting open problems that can be formulated for statistical functionals and families of distributions described here. There are also other functionals and families for which the projection method would work, but we do not discuss them here.

It is important to become acquainted with the content of Section 2 before reading the following ones. These can be studied independently, with the exception of Section 5 that contains some references to auxiliary results of Section 3. For technical reasons the work was divided into 2 parts. We kept a consistent numeration of sections, theorems and formulae throughout both the parts. The first part contains common table of contents and introduction. The common list of references follows the second one.

We tried to set a unified notation for the whole work. A list of symbols is presented below. Only notions used locally are not included there. Generally, we preserved standard symbols for standard notions. Numerous convex cones of functions introduced in the paper are denoted by $\mathcal{C}$ with various superscripts and subscripts whose meaning is explained in the list (see also Subsection 2.3). The projection operator onto a given cone is written as P with the same upper and lower indices. Moreover, the indices appear in the notation of sharp bounds on functionals which are obtained by means of projections on specified convex cones. For instance, $B^0(j,n)$ denotes the sharp mean-variance bound on the expectation of the jth order statistic of independent identically distributed sample of size n with arbitrary marginal distribution that has a finite variance. The bound is obtained by means of projection P^0 of a properly chosen functional (see Subsection 2.2) onto the convex cone $\mathcal{C}^0$ of quantile functions of all distributions with finite variance, centered about the respective mean.

1.2 List of Symbols

$(\mathcal{H},(\cdot,\cdot))$	—	(real) Hilbert space with inner product $(\cdot,\cdot)$
$\|\|\cdot\|\|$	=	$(\cdot,\cdot)^{1/2}$ — norm in $(\mathcal{H},(\cdot,\cdot))$
$L^2([a,d),w(x)\,dx)$	—	Hilbert space of functions $g:[a,d)\mapsto\Re$ satisfying $\int_a^d g^2(x)w(x)\,dx<\infty$ with inner product $(g,h)=\int_a^d g(x)h(x)w(x)\,dx$ for a positive weight function $w(x)$
$\alpha,\beta,\dots$	—	scalars, parameters of functions
$\alpha_*,\beta_*,\dots$	—	optimal scalars, parameters of projections
$g,h,\dots$	—	functions, elements of Hilbert spaces
$G,H,\dots$	—	antiderivatives of $g,h,\dots$, respectively
$gH(x)$	=	$g(H(x))$ — composition of functions H and g
$\bar{H}(x)$	—	greatest convex minorant of $H(x)$
$\bar{h}(x)$	—	(right) derivative of $\bar{H}(x)$
$\mathbf{1}(x)$	—	constant function equal to 1
$\mathbf{1}_A(x)$	—	indicator of A (= 1 on A and 0 elsewhere)

x_+	$=$	$\max\{x,0\}$ — positive part of a number
$\lfloor x \rfloor$	$=$	$\max\{k \le x : k$ integer $\}$ — floor of a number
$T_h(\cdot)$	$=$	$(h,\cdot)$ — continuous linear functional on a Hilbert space
$T_h(\cdot)/\|\cdot\|$	—	normalized linear functional
$F(x)$	—	(marginal) distribution function of population
$f(x)$	—	density function of F
$\lambda_F(x)$	$=$	$f(x)/[1-F(x)]$ — failure rate of F
$F^{-1}(x)$	$=$	$\sup\{y : F(y) \le x\}$, $0 \le x < 1$, — quantile function of F
$F^{-1}(p)$	—	quantile of order p, $0 < p < 1$
$\mu = \mu_F$	$=$	$\int_0^1 F^{-1}(x)\,dx$ — mean of F
$m^2 = m_F^2$	$=$	$\int_0^1 [F^{-1}(x)]^2\,dx$ — second raw moment of F
$\sigma^2 = \sigma_F^2$	$=$	$m_F^2 - \mu_F^2$ — variance of F
$\mathcal{P}_n(F)$	—	family of all distributions on $\Re^n$ with common marginal F
$U(x)$	$=$	x, $0 \le x \le 1$, — standard uniform distribution function
$V(x)$	$=$	$1-\exp(-x)$, $x \ge 0$, — standard exponential distribution function
$W(x)$	—	a (distinguished) distribution function
$a = a_W$	—	(finite) left end-point of (interval) support of W
$d = d_W$	—	right end-point of (interval) support of W
$w(x)$	—	density function of W, weight function in $L^2([a_W,d_W),w(x)dx)$
$\mu_W(x)$	$=$	$\mathrm{E}_W(X\|X>x) = \int_x^d yw(y)\,dy/[1-W(x)]$
$m_W^2(x)$	$=$	$\mathrm{E}_W(X^2\|X>x) = \int_x^d y^2w(y)\,dy/[1-W(x)]$
$\sigma_W^2(x)$	$=$	$m_W^2(x) - \mu_W^2(x)$
$X_1,\ldots,X_n,\ldots$	—	independent identically F-distributed (i.i.d.) random variables
$Y_1,\ldots,Y_n,\ldots$	—	possibly dependent identically F-distributed random variables
$X_{j:n}$	—	jth order statistic of $X_1,\ldots,X_n$, $1 \le j \le n$
$Y_{j:n}$	—	jth order statistic of $Y_1,\ldots,Y_n$, $1 \le j \le n$
$\sum_{j=1}^n c_jX_{j:n}(Y_{j:n})$	—	L-statistic
$\mathbf{c} = (c_1,\ldots,c_n)$	—	vector of coefficients of L-statistic
L_n	—	nth occurrence time of (1st) record (increase in sequence of sample maxima $X_{j:j}$, $j \ge 1$)
R_n	$=$	X_{L_n} — nth value of (1st) record,

$L_n^{(k)}$	—	nth occurrence time of (kth) record (increase in sequence of kth greatest order statistics $X_{j+1-k:j}$, $j \ge k$), $k \ge 1$
$R_n^{(k)}$	$=$	$X_{L_n^{(k)}+1-k:L_n^{(k)}}$ — nth value of (kth) record
$f_{j:n}(x)$	$=$	$n\binom{n-1}{j-1}x^{j-1}(1-x)^{n-j}\mathbf{1}_{[0,1]}$ — density function of jth order statistic of i.i.d standard uniform sample of size n, $1 \le j \le n$, expectation functional for $X_{j:n}$
$F_{j:n}(x)$	$=$	$\sum_{k=j}^{n}\binom{n}{k}x^k(1-x)^{n-k}$, $0 \le x \le 1$, — distribution function of $f_{j:n}(x)$
$\tilde{G}_{j:n}(x)$	—	distribution function of jth order statistic of dependent identically distributed sample of size n
$G_{j:n}(x)$	$=$	$\frac{nx+1-j}{n+1-j}\mathbf{1}_{[\frac{j-1}{n},1]}$ — stochastically largest distribution function of jth order statistic of dependent U-distributed sample of size n
$g_{j:n}(x)$	$=$	$\frac{n}{n+1-j}\mathbf{1}_{[\frac{j-1}{n},1)}$ — density function of $G_{j:n}$, expectation functional for $Y_{j:n}$
$G_{\mathbf{c}}(x)$	—	the greatest convex function satisfying $G(j/n) \le \sum_{k=1}^{j} c_k$, $0 \le j \le n$, $0 \le x \le 1$,
$g_{\mathbf{c}}(x)$	$=$	$\sum_{j=1}^{n} d_j \mathbf{1}_{[\frac{j-1}{n},\frac{j}{n})}(x)$ — (right) derivative of $G_{\mathbf{c}}(x)$, expectation functional for $\sum_{j=1}^{n} c_j Y_{j:n}$
$f_n(x)$	$=$	$\frac{1}{n!}[-\ln(1-x)]^n\mathbf{1}_{[0,1)}$ — density function of nth value of (1st) record of i.i.d standard uniform sequence, expectation functional of R_n
$F_n(x)$	$=$	$(1-x)\sum_{j=0}^{n}\frac{1}{j!}[-\ln(1-x)]^j$, $0 \le x \le 1$, — distribution function of $f_n(x)$
$f_n^{(k)}(x)$	$=$	$\frac{k^{n+1}}{n!}[-\ln(1-x)]^n(1-x)^{k-1}\mathbf{1}_{[0,1)}$ — density function of nth value of (kth) record of i.i.d standard uniform sequence, expectation functional for $R_n^{(k)}$
$F_n^{(k)}(x)$	$=$	$(1-x)^k\sum_{j=0}^{n}\frac{k^j}{j!}[-\ln(1-x)]^j$, $0 \le x \le 1$, — distribution function of $f_n^{(k)}(x)$
$\preceq_c$	—	convex ordering of distribution functions: $F \succeq_c W$ if $F^{-1}W(x)$ is convex on $[a_W, d_W)$

$\preceq_*$	—	star ordering of distribution functions: $F \succeq_* W$ if $F^{-1}W(x)$ is starshaped, i.e. $F^{-1}W(x)/(x-a_W)$ is nondecreasing on $[a_W, d_W)$
$\preceq_s$	—	s-ordering of symmetric distribution functions: $F \succeq_s W$ if $F^{-1}W(x)$ is convex on $[\mu_W, d_W)$
$\mathcal{C}^{\nearrow}$	=	$\{g \in L^2([0,1), dx) : g$ is nondecreasing$\}$ — family of all quantile functions $F^{-1}(x)$ of distribution functions $F(x)$ with finite variance
$\mathcal{C}^0$	=	$\{g \in \mathcal{C}^{\nearrow} : \int_0^1 g(x)\,dx = 0\}$ — family of respective centered quantile functions $F^{-1}(x) - \mu_F$
$\mathcal{C}^+$	=	$\{g \in \mathcal{C}^{\nearrow} : g(0) = 0\}$ — family of quantile functions of life distributions with $a_F = 0$
$\mathcal{C}^s$	=	$\{g \in \mathcal{C}^{\nearrow} : g(x) = -g(1-x-)\}$ — family of centered quantile functions of symmetric distributions
$\mathcal{C}^\bullet$	—	any of $\mathcal{C}^{\nearrow}, \mathcal{C}^0, \mathcal{C}^+, \mathcal{C}^s$
$\mathcal{C}^\bullet_W$	=	$\{gW : g \in \mathcal{C}^\bullet\} \subset L^2([a_W, d_W), w(x)dx)$ — family of compositions of $W(x)$ with (centered) quantile functions of $\mathcal{C}^\bullet$ (with a_W replaced by μ_W in the last case)
$\mathcal{C}^\bullet_{\succeq_c(\preceq_c)W}$	=	$\{g \in \mathcal{C}^\bullet_W : g$ is convex (concave) on $[a_W, d_W)\}$ — family of compositions of $W(x)$ with (centered) quantile functions of $\mathcal{C}^\bullet$ for $F \succeq_c (\preceq_c) W$
$\mathcal{C}^\bullet_{\succeq_*(\preceq_*)W}$	=	$\{g \in \mathcal{C}^\bullet_W : g$ is (anti)starshaped on $[a_W, d_W)\}$ — family of compositions of $W(x)$ with (centered) quantile functions of $\mathcal{C}^\bullet$ for $F \succeq_* (\preceq_*) W$
$\mathcal{C}^s_{\succeq_s(\preceq_s)W}$	=	$\{g \in \mathcal{C}^s_W : g$ is convex (concave) on $[\mu_W, d_W)\}$ — family of compositions of symmetric $W(x)$ with centered quantile functions of $\mathcal{C}^s$ for symmetric $F \succeq_s (\preceq_s) W$

In particular:

$\mathcal{C}^\bullet_{\succeq_c(\preceq_c)U}$	—	(centered) quantile functions of distributions with decreasing (increasing) density
$\mathcal{C}^\bullet_{\succeq_*(\preceq_*)U}$	—	(centered) quantile functions of distributions with decreasing (increasing) density in average

$\mathcal{C}^s_{\succeq_s(\preceq_s)U}$	—	centered quantile functions of symmetric unimodal (U-shaped) distributions
$\mathcal{C}^\bullet_{\succeq_c(\preceq_c)V}$	—	compositions of $V(x) = 1 - \exp(-x)$ with (centered) quantile functions of distributions with decreasing (increasing) failure rate
$\mathcal{C}^\bullet_{\succeq_*(\preceq_*)V}$	—	compositions of $V(x) = 1 - \exp(-x)$ with (centered) quantile functions of distributions with decreasing (increasing) failure rate in average
$\mathcal{C}^\bullet_\bullet$	—	any of above defined convex cones
$P^\bullet_\bullet$	—	projection onto $\mathcal{C}^\bullet_\bullet$
$A = A^\bullet_\bullet(p)$	—	sharp bound on $F^{-1}(p)$ determined by projection onto $\mathcal{C}^\bullet_\bullet$
$B = B^\bullet_\bullet(j,n)$	—	sharp bound on expectation of $X_{j:n}$ (independent case) determined by projection onto $\mathcal{C}^\bullet_\bullet$
$C = C^\bullet_\bullet(j,n)$	—	sharp bound on expectation of $Y_{j:n}$ (dependent case) determined by projection onto $\mathcal{C}^\bullet_\bullet$
$D = D^\bullet_\bullet(k,n)$	—	sharp bound on expectation of $R^{(k)}_n$ determined by projection onto $\mathcal{C}^\bullet_\bullet$

2 Basic Notions

2.1 Elements of Hilbert Spaces Theory

We here recall some basic facts about the Hilbert spaces that will be used in the sequel. They can be found in textbooks on functional analysis (see, e.g., Balakrishnan [6]. A pair $(\mathcal{H}, (\cdot,\cdot))$ is called a real inner product space if $\mathcal{H}$ is a real linear space and function $(\cdot,\cdot) : \mathcal{H} \times \mathcal{H} \mapsto \Re$, referred further to as the inner product, is linear in each argument when the other is fixed, symmetric under rearrangement of arguments, and positive if the both arguments are identical and nonzero. The properties imply the Schwarz inequality

$$\forall g, h \in \mathcal{H} \quad (g,h) \le [(g,g)(h,h)]^{1/2}. \tag{2.1}$$

This is trivial when either of arguments is zero. Otherwise we conclude 2.1 from the relations

$$0 \le \left(g - \frac{(g,h)}{(h,h)}h, g - \frac{(g,h)}{(h,h)}h\right)(h,h) = (g,g)(h,h) - (g,h)^2.$$

This also shows that 2.1 becomes equality iff either $g = 0$ or $h = 0$ or $g = \alpha h$ for some $\alpha > 0$. We use 2.1 for verifying that function $h \mapsto ||h|| = (h,h)^{1/2}$ defines a norm in $\mathcal{H}$. If $(\mathcal{H}, ||\cdot||)$ is complete, then $(\mathcal{H}, (\cdot,\cdot))$ is called the Hilbert space. The Riesz representation theorem asserts that every linear continuous functional defined on a Hilbert space can be written as $T_h(g) = (g, h)$, $g \in \mathcal{H}$, for some $h \in \mathcal{H}$. By 2.1 again, $||T_h|| = ||h||$. One can see that the normalized nonzero functional $T_h(g)/||g||$, $0 \neq g \in \mathcal{H}$, attains its maximum $||h||$ at $g = \alpha h$ with $\alpha > 0$.

In numerous statistical problems, it is of importance to maximize a linear normalized functional over a convex cone in a Hilbert space. We say that $\mathcal{C} \subset \mathcal{H}$ is a convex cone if $f, g \in \mathcal{C}$ implies that $\alpha f + \beta g \in \mathcal{C}$ for arbitrary $\alpha, \beta \geq 0$. If $h \in \mathcal{C}$, then the solution of our restricted maximization problem coincides with that of the general one. Otherwise we show that h should be replaced by its projection Ph onto $\mathcal{C}$, i.e. the element of $\mathcal{C}$ that is least distant from h. This can be deduced from the following theorem (cf. Balakrishnan [6, Section 1.4]).

Theorem 1. *If h is an arbitrary element of a real Hilbert space $\mathcal{H}$ and $\mathcal{C}$ is a closed convex cone in $\mathcal{H}$, then there exists a unique projection Ph of h onto $\mathcal{C}$ that is characterized by two relations*

$$\forall\, g \in \mathcal{C} \qquad (g, h) \leq (g, Ph), \tag{2.2}$$

$$(Ph, h) = (Ph, Ph). \tag{2.3}$$

This is a refinement of the statement that there is a uniquely defined projection $\bar{P}h$ of arbitrary $h \in \mathcal{H}$ onto a closed convex set $\bar{\mathcal{C}} \subset \mathcal{H}$, and $\bar{P}h$ satisfies

$$\forall\, g \in \bar{\mathcal{C}} \quad (g, h - \bar{P}h) \leq (\bar{P}h, h - \bar{P}h). \tag{2.4}$$

(see Balakrishnan [6, Section 1.4]). The projection is the only point of $\bar{\mathcal{C}}$ that satisfies 2.4.

Observe that for $Ph \neq 0$ relation 2.2 combined with 2.1 gives

$$\forall\, 0 \neq g \in \mathcal{C} \quad T_h(g)/||g|| \leq ||Ph|| > 0. \tag{2.5}$$

Setting $g = \alpha Ph$ for some positive α and using 2.3, the equality holds in 2.5. If $Ph = 0$ then, due to 2.2, T_h is nonpositive on $\mathcal{C}$, and clearly $T_h(Ph) = 0$.

Here we concentrate on functionals on Hilbert space of a specific form $L^2([a, d), w(x)\, dx)$, for some $-\infty < a < d \leq +\infty$, $w : [a, d) \mapsto \Re_+$. The space consists of the square integrable functions on interval $[a, d)$ with a positive weight function w, and the inner product is defined by

$$(g, h) = \int_a^d g(x)h(x)w(x)\, dx. \tag{2.6}$$

Here the Schwarz inequality takes on the form

$$\int_a^d g(x)h(x)w(x)\,dx \le \left[\int_a^d g^2(x)w(x)\,dx \int_a^d h^2(x)w(x)\,dx\right]^{1/2}. \quad (2.7)$$

Below we present two examples of projections onto convex cones contained in $L^2([a,d), w(x)\,dx)$.

Example 1. Let $\mathcal{C}_+ = \{h \in L^2([a,d), w(x)\,dx) : h \ge 0\}$. Verifying 2.2–2.3 we deduce that $P_+h = h_+ = \max\{h, 0\}$ is the projection of h onto $\mathcal{C}_+$ for every $h \in \mathcal{H}$. Also, one can check directly that h_+ is actually the nonnegative function closest to h for arbitrary weight w.

Example 2. Consider the set $\mathcal{C}_W^{\nearrow}$ of all nondecreasing functions in $L^2([a,d), w(x)\,dx)$. We assume that the constant and linear functions belong to the space. Then

$$W(y) = \int_a^y w(x)\,dx \quad (2.8)$$

is strictly increasing absolutely continuous and finite function. The same holds for its well-defined inverse $W^{-1} : [0, W(d)) \mapsto \Re_+$. By finiteness of

$$\int_a^d h(x)w(x)\,dx = \int_0^{W(d)} hW^{-1}(x)\,dx, \quad (2.9)$$

we can define an absolutely continuous function

$$H_W(y) = \int_0^y hW^{-1}(x)\,dx, \qquad 0 \le y < W(d), \quad (2.10)$$

and its greatest convex minorant $\bar{H}_W$, with a nondecreasing derivative $\bar{h}_W$, say.

We prove that $h_W^{\nearrow} = \bar{h}_W W \in \mathcal{C}_W^{\nearrow}$ is the projection of h onto $\mathcal{C}_W^{\nearrow}$ by checking 2.2–2.3. For the former one, we need the following Lemma (cf Marshall and Olkin [32, Proposition A.2.(iii), p. 444]).

Lemma 1. *If $G \le H$ are functions of bounded variation on an interval $[A, D)$, which are equal at the end-points, then*

$$\int_A^D g(x)\,G(dx) \ge \int_A^D g(x)\,H(dx) \quad (2.11)$$

holds for every nondecreasing function g for which both the integrals exist.

Since $\bar{H}_W \le H_W$ satisfy the assumptions, for arbitrary $g \in \mathcal{C}_W^{\nearrow}$ we can write

$$
\begin{aligned}
(g,h) &= \int_0^{W(d)} gW^{-1}(y)hW^{-1}(y)\,dy = \int_0^{W(d)} gW^{-1}(y)\,H_W(dy) \\
&\le \int_0^{W(d)} gW^{-1}(y)\,\bar{H}_W(dy) \\
&= \int_a^d g(x)h_W^{\nearrow}(x)w(x)\,dx = (g, h_W^{\nearrow}).
\end{aligned} \tag{2.12}
$$

In order to derive 2.3, we should first deeper analyze relations in pairs $H_W, \bar{H}_W$, and $hW^{-1}, \bar{h}_W$, and $h, h_W^{\nearrow}$. Observe that the open set $\{\bar{H}_W < H_W\}$ is a (possibly empty) union of countably many (at most) disjoint open intervals, $\bigcup_i (W(b_i), W(c_i))$ say. Function $\bar{H}_W$ is linear on each interval, and coincides with H_W at the end-points. Therefore $\bar{h}_W$ is constant equal to $[H_W W(c_i) - H_W W(b_i)]/[W(c_i) - W(b_i)]$ on $(W(c_i), W(b_i))$, and we have

$$
\begin{aligned}
\int_{b_i}^{c_i} h(x)h_W^{\nearrow}(x)w(x)\,dx &= \int_{W(b_i)}^{W(c_i)} hW^{-1}(y)\bar{h}_W(y)\,dy \\
&= \frac{[H_W W(c_i) - H_W W(b_i)]^2}{W(c_i) - W(b_i)} \\
&= \int_{W(b_i)}^{W(c_i)} \bar{h}_W^2(y)\,dy = \int_{b_i}^{c_i} h_W^{\nearrow 2}(x)w(x)\,dx.
\end{aligned}
$$

Thus the equality holds for the integrals over the whole $W^{-1}(\{\bar{H}_W < H_W\})$. For the remaining part $W^{-1}(\{\bar{H}_W = H_W\})$ the conclusion holds, since $\bar{H}_W = H_W$ implies that $h_W^{\nearrow} = h$ there. Summing up, we have $(h, h_W^{\nearrow}) = (h_W^{\nearrow}, h_W^{\nearrow})$, which is the desired conclusion.

Construction of L^2-projection onto the family of monotone functions under uniform weighting was presented in Moriguti [34]. For the general case we refer, e.g., to Rychlik [50]. For simplicity, we treated elements of L^2-spaces as functions rather than equivalence classes up to almost sure equality, and we shall follow this convention later on. E.g., $h \in \mathcal{C}_+$ generally means almost sure nonnegativity, and monotonicity can be precisely defined by comparisons of integrals over the intervals of the same weight. In Examples 1, 2 we were able to determine projections for arbitrary h. Note that in the latter case the form of projection depends on weight function w. However, there are no general rules of constructing projections onto other convex cones. Then we apply arguments suitable for specific functionals h and cones. Usually, we first try to describe the shape of projection in a parametric way, and then precisely determine the parameters. We finally note that more often characterization 2.2–2.3 is used for determining projections. In the problems under study we try to find the projection by minimizing the distance of a functional to a cone. The solution satisfies 2.2–2.3 which are needed for estimating values of the functional over the cone.

2.2 *Statistical Linear Functionals*

Investigations of statistical procedures treated as functionals on distribution functions were initialized by von Mises [63]. The general theory is presented in Serfling [58] and Prakasa Rao [40]. Here we confine ourselves to some statistical functions which can be represented as linear functionals on Hilbert spaces. Assume that a random variable X has a distribution function F, finite mean $\mu = \mu_F$ and second raw moment $m^2 = m_F^2$. Changing variables, we write

$$\mu = \int_{-\infty}^{+\infty} y\, F(dy) = \int_0^1 F^{-1}(x)\, dx, \tag{2.13}$$

$$m^2 = \int_{-\infty}^{+\infty} y^2\, F(dy) = \int_0^1 [F^{-1}(x)]^2\, dx, \tag{2.14}$$

where

$$F^{-1}(x) = \sup\{y: \ F(y) \leq x\}, \qquad 0 \leq x < 1, \tag{2.15}$$

is the right-continuous quantile function of X. We can say that m_F is the norm of F^{-1} in $L^2([0,1), dx)$, and $\mu_F = (F^{-1}, \mathbf{1})$. The family of all possible quantile functions is identical with the convex cone of (right-continuous versions of) nondecreasing functions in $L^2([0,1), dx)$. For variance $\sigma^2 = \sigma_F^2$, we have

$$\sigma^2 = \int_{-\infty}^{+\infty} (y-\mu)^2\, F(dy) = \int_0^1 [F^{-1}(x) - \mu]^2\, dx = ||F^{-1} - \mu||^2. \tag{2.16}$$

Observe that $F^{-1} - \mu_F$ form the convex cone of nondecreasing functions integrating to 0. Our purpose is to determine sharp bounds for normalized statistical functionals represented as $T(F^{-1})/m_F$ and $T(F^{-1} - \mu_F)/\sigma_F$ for general and restricted classes of quantile functions. Now we present exemplary linear functionals of statistical importance acting on quantile functions in $L^2([0,1), dx)$. Considering bounds on narrower classes of distributions, it is convenient to transform quantile functions so that other L^2-spaces will be studied.

Quantiles $F^{-1}(p)$ *of order* $0 < p < 1$. They characterize distributions by describing levels that divide respective populations into subsets containing desired proportions of elements. Quantiles are often used for defining critical levels of tests and interval estimates. In order to evaluate $F^{-1}(p)$ in terms of moments, we represent it as a limit of L^2-functionals

$$F^{-1}(p) = \lim_{q \searrow p} \frac{1}{q-p} \int_p^q F^{-1}(x)\, dx = \lim_{q \searrow p} (F^{-1}, \frac{1}{q-p} \mathbf{1}_{[p,q)}), \tag{2.17}$$

where $\mathbf{1}_A$ denotes the indicator function of set A.

Expectations of order statistics of independent samples. Let $X_1, \ldots, X_n$ be independent identically distributed (i.i.d.) random variables with a common distribution function F. Then the jth order statistic $X_{j:n}$, $1 \leq j \leq n$, is the jth smallest value in the sequence $X_1, \ldots, X_n$. The respective distribution function is

$$\begin{aligned} \mathrm{P}(X_{j:n} \leq y) &= \mathrm{P}(\text{at least } j \text{ among } n \text{ independent } F - \text{distributed} \\ &\qquad \text{random variables is not greater than } y) \\ &= \sum_{k=j}^{n} \binom{n}{k} F^k(y)[1 - F(y)]^{n-k} = F_{j:n}F(y), \end{aligned} \tag{2.18}$$

say. Since

$$f_{j:n}(x) = F'_{j:n}(x) = n \binom{n-1}{j-1} x^{j-1}(1-x)^{n-j}, \tag{2.19}$$

we have

$$\mathrm{E}_F X_{j:n} = \int_{-\infty}^{+\infty} y\, F_{j:n}F(dy) = \int_0^1 F^{-1}(x) f_{j:n}(x)\, dx = (F^{-1}, f_{j:n}). \tag{2.20}$$

Note that $F_{j:n}$ and $f_{j:n}$ are the distribution and density functions, respectively, of the jth order statistic of the standard uniform i.i.d. sample of size n. Order statistics are directly used for estimating quantiles of order j/n and describing lifetime of j-out-of-n reliability systems which contain n independent identical elements and operate until at least $n+1-j$ of them are in working order. We can also study the expectations of linear combinations of order statistics (so called L-statistics)

$$\mathrm{E}_F \sum_{j=1}^{n} c_j X_{j:n} = \left(F^{-1}, \sum_{j=1}^{n} c_j f_{j:n} \right) \tag{2.21}$$

which have numerous applications in statistical inference. E.g., these with $\sum_{j=1}^n c_j = 1$ (including sample mean $\frac{1}{n}\sum_{j=1}^n X_{j:n}$, and median $X_{(n+1)/2:n}$ for odd n, and trimmed means $\frac{1}{n-2k}\sum_{j=k+1}^{n-k} X_{j:n}$ for $j \le n/2-1$) and ones satisfying $\sum_{j=1}^k c_j \le 0$, $1 \le k \le n-1$, and $\sum_{j=1}^n c_j = 0$ (including sample range $X_{n:n} - X_{1:n}$ and interquartile distance $X_{n+1-\lfloor n/4\rfloor:n} - X_{\lfloor n/4\rfloor:n}$) are used for estimating the location and dispersion of population, respectively. Moreover, projection method enables us to evaluate precisely uniform convergence rates of estimates for particular families of distributions. In exemplary case of quantile estimation, this is possible by analyzing $\mathrm{E}_F X_{j:n} - F^{-1}(j/n)$ and $\mathrm{E}_F(X_{j:n} - X_{kj:kn})$ for $k > 1$.

Order statistics of dependent samples. Suppose that $Y_1, \ldots, Y_n$ are possibly dependent identically F-distributed random variables, and $Y_{j:n}$, $1 \le j \le n$, are respective order statistics. As in the previous case, $Y_{j:n}$ may represent the lifetime of $(n+1-j)$-out-of-n system of elements with identical failure probability, but here each element affects the other ones somehow. It may happen in particular that $Y_1 = \ldots = Y_n$, if the damage of a single element causes the immediate damage of all remaining ones. It was shown in Rychlik [44] that for $\mathbf{c} = (c_1, \ldots, c_n) \in \Re^n$

$$\sup_{P \in \mathcal{P}_n(F)} \mathrm{E}_P \sum_{j=1}^n c_j Y_{j:n} = \int_0^1 F^{-1}(x) g_{\mathbf{c}}(x)\, dx = (F^{-1}, g_{\mathbf{c}}), \tag{2.22}$$

where $\mathcal{P}_n(F)$ denotes the family of all joint distributions P on $\Re^n$ with identical marginals F, and $g_{\mathbf{c}}$ is (the right) derivative of function $G_{\mathbf{c}}$, being the greatest convex one satisfying

$$G_{\mathbf{c}}(0) = 0, \qquad G_{\mathbf{c}}(j/n) \le \sum_{i=1}^j c_i, \qquad j = 1, \ldots, n. \tag{2.23}$$

Formula 2.22 is valid for arbitrary coefficients $c_1, \ldots, c_n$, and distribution function F with a finite expectation. The supremum is attained for some distributions in $\mathcal{P}_n(F)$. A detailed characterization of the distributions as well as arguments leading to 2.22 will be presented in Subsection 5.1. Maximizing 2.22 over a family of marginals F, we first determine the F providing the extreme value of $(F^{-1}, g_{\mathbf{c}})$, and then take the joint distributions in $\mathcal{P}_n(F)$, for which the expectation of the L-statistic is actually equal to $(F^{-1}, g_{\mathbf{c}})$ for the specified F. By definition, $g_{\mathbf{c}}$ is nondecreasing step function with $n-1$ jumps at most, located at some points of the form j/n, $j = 1, \ldots, n-1$. In particular, for arbitrary $1 \le j \le n$,

$$\sup_{P \in \mathcal{P}_n(F)} \mathrm{E}_P Y_{j:n} = \frac{n}{n+1-j} \int_{(j-1)/n}^{1} F^{-1}(x)\, dx$$
$$= \left(F^{-1}, \frac{n}{n+1-j} \mathbf{1}_{[(j-1)/n,1)} \right) = (F^{-1}, g_{j:n}), \quad (2.24)$$

say, (cf also Caraux and Gascuel [12], Rychlik [42]). One can easier determine projections of simple step functions presented in 2.22, 2.24 than of polynomials appearing in respective formulae 2.21, 2.20 for the i.i.d. case. This explains the paradox that we have more results and of simpler forms for arbitrarily dependent samples than for standard independent observations. Note finally that the projection method allows us to measure sensitivity of L-statistics upon dependence by evaluating

$$\sup_{P \in \mathcal{P}_n(F)} \mathrm{E}_P \sum_{j=1}^{n} c_j (Y_{j:n} - X_{j:n}) = \left(F^{-1}, g_{\mathbf{c}} - \sum_{j=1}^{n} c_j f_{j:n} \right) \quad (2.25)$$

in chosen classes of marginal distributions.

Record values. Record values in numerical sequences are ones that exceed all the preceding ones. For a random sequence X_j, $j \ge 1$, record values R_n, and respective occurrence times L_n, $n \ge 1$, are random increasing sequences. By convention, we assume

$$L_0 = 1, \qquad R_0 = X_1, \quad (2.26)$$

and further put

$$L_n = \min\{j > L_{n-1} : X_j > X_{L_{n-1}}\}, \quad (2.27)$$
$$R_n = X_{L_n}, \qquad n \ge 1. \quad (2.28)$$

Like extreme order statistics, records are applied in estimating strength of materials, predicting natural disasters, sport achievements etc. Formulae 2.26–2.28 are well defined for arbitrary original sequences, but further we assume that X_j, $j \ge 1$, are independent and have an identical continuous distribution function F, say. If F has not an atom at its right support end-point, then the sequence of records is infinite almost surely. It is obvious that if a current value of record is given, then the conditional distribution of the next one is identical with the distribution of the parent variable under condition that it exceeds the actual record value. In particular, for records R_n^V of a standard exponential sequence V_j, $j \geq 1$, with distribution function $V(x) = 1 - \exp(-x)$, $x \geq 0$, we have

$$\begin{aligned} \mathrm{P}(R_{n+1}^V - R_n^V > y | R_n^V = x) &= \mathrm{P}(V_1 > y + x | V_1 > x) \\ &= \exp(-y) = P(V_1 > y) \end{aligned} \tag{2.29}$$

for arbitrary $x, y \geq 0$. It follows that the record increments $R_0^V, R_1^V - R_0^V, \ldots, R_{n+1}^V - R_n^V, \ldots$ are independent V-distributed random variables, and R_n^V has the gamma distribution $\Gamma(n+1, 1)$. Transformation $X_j = F^{-1}V(V_j)$, $j \geq 1$, produces an independent F-distributed sequence. By strict monotonicity, it preserves the record times. Therefore

$$R_n = X_{L_n} = F^{-1}V(V_{L_n}) = F^{-1}V(R_n^V), \tag{2.30}$$

and

$$\begin{aligned} \mathrm{E}_F R_n = \mathrm{E}_V F^{-1}V(R_n^V) &= \int_0^\infty F^{-1}(1 - e^{-y}) \frac{y^n}{n!} e^{-y}\, dy \\ &= \int_0^1 F^{-1}(x) f_n(x)\, dx = (F^{-1}, f_n) \end{aligned} \tag{2.31}$$

with $f_n(x) = [-\ln(1-x)]^n / n!$, which is a desired inner product representation.

kth record values. An increasing sequence of record values arises from a nondecreasing sequence of sample maxima $(X_{n:n})$, $n \geq 1$, by crossing out all repetitions. For arbitrary fixed k, the sequence of kth greatest order statistics $(X_{n+1-k:n})$, $n \geq k$, is nondecreasing as well. By analogy, we can define occurrence times and values of kth records in the following way

$$L_0^{(k)} = k, \qquad R_0^{(k)} = X_{1:k}, \tag{2.32}$$

$$L_n^{(k)} = \min\{j > L_{n-1}^{(k)} :\ X_j > X_{L_{n-1}^{(k)}+1-k:L_{n-1}^{(k)}}\}, \tag{2.33}$$

$$R_n^{(k)} = X_{L_n^{(k)}+1-k:L_n^{(k)}}, \qquad n \geq 1. \tag{2.34}$$

In contrast with standard records, a random variable $X_{L_n^{(k)}}$ observed at kth record time $L_n^{(k)}$ for $k \geq 2$ does not necessarily becomes kth record value immediately. Generally, we have $X_{L_n^{(k)}} = X_{L_n^{(k)}+1-i:L_n^{(k)}} \geq X_{L_n^{(k)}+1-k:L_n^{(k)}}$ for some $1 \leq i \leq k$. Thus it becomes the ith record first and will become the kth record when $k - i$ greater values occur in the sequence. A key relation that allows to establish the distribution function of kth records is

$$P(R_{n+1}^{(k)} > y | R_n^{(k)} = x) = \left[\frac{1 - F(y)}{1 - F(x)}\right]^k, \qquad y > x, \tag{2.35}$$

(cf. 2.29), which means that the distribution of the next value of kth record, when the current one is known, is the same as the distribution of the minimum of k original variables X_j under condition that they are greater than the current record. Relation 2.35 shows that distributions of nth values of kth records from independent F-distributed sample and standard 1st records of the i.i.d. sample with distribution function $F_{1:k}F = 1 - (1-F)^k$ (i.e. that of $X_{1:k}$) coincide. By 2.31,

$$\begin{aligned} \mathrm{E}_F R_n^{(k)} &= \mathrm{E}_V (F_{1:k}F)^{-1} V(R_n^V) = \int_0^\infty F^{-1}(1-e^{-y/k}) \frac{y^n}{n!} e^{-y}\, dy \\ &= \int_0^1 F^{-1}(x) f_n^{(k)}(x)\, dx = (F^{-1}, f_n^{(k)}) \end{aligned} \tag{2.36}$$

where

$$f_n^{(k)}(x) = \frac{k^{n+1}}{n!}[-\ln(1-x)]^n (1-x)^{k-1}, \qquad k, n \geq 1, \tag{2.37}$$

is the density function of nth value of kth record of the i.i.d. standard uniform sequence, and $f_n^{(1)} = f_n$ defined in 2.31. Formula 2.35 is true for arbitrary F. However, for discontinuous F function $F^{-1}F_{1:k}^{-1}V$ is not strictly increasing and may transform a record in the exponential sequence into a number equalizing a previous score. Therefore 2.36 is true for continuous F merely. On the other hand, closed convex cones contain quantile functions of discontinuous distributions, and it is very likely that a normalized statistical functional is maximized at some F^{-1} which is not strictly increasing. In such a case, we can easily approximate the F^{-1} by quantiles of absolutely continuous distributions converging to F^{-1} in the L^2-norm.

We only mention a possibility of evaluating conditional expectations of order and record statistics by the projection method. Then we can exploit identity of the conditional distributions with unconditional distributions of some order and record statistics from smaller and censored samples. We shall not develop these ideas here. We confine ourselves to evaluating the upper bounds. To get the lower ones, it suffices to analyze the negatives of above defined functionals. Only that corresponding to the order statistics in the dependent case needs a more subtle transformation. Changing the signs of coefficients c_j, $1 \leq j \leq n$, in 2.23 results in constructing a functional $g_{-\mathbf{c}}$ different from $-g_{\mathbf{c}}$. Also, one should realize that generally projections of a functional and its negative cannot be derived one from the other in a simple way. Both the problems should be solved separately by use of specific tools.

We formally define the convex cone of quantile functions considered in the previous Subsection as

$$\mathcal{C}^{\nearrow} = \{g \in L^2([0,1), dx) : g \text{ -- nondecreasing, right continuous}\}. \quad (2.38)$$

Right continuity assumption is inessential for problems of L^2-projections. This is introduced here in order to preserve the consistency with definitions of quantile functions, and one can simply replace an arbitrary nondecreasing function by the right continuous version. We regard elements of convex cones defined here and statistical functionals as being right continuous. Observe that projecting functions onto 2.38 we maximize normalized statistical functionals in $||F^{-1}|| = m_F$ units. In order to get more subtle evaluations in terms of mean and variance, we should consider the class of $F^{-1} - \mu_F$ functions defined as

$$\mathcal{C}^0 = \{g \in \mathcal{C}^{\nearrow} : \int_0^1 g(x)\, dx = 0\}. \quad (2.39)$$

It is also of interest to study life distributions, which have the left support end-point at zero, and generate the following family of nonnegative quantile functions

$$\mathcal{C}^+ = \{g \in \mathcal{C}^{\nearrow} : g(0) = 0\}. \quad (2.40)$$

In this case we have another pair of location and scale parameters: the left end-point of support $a_F = 0$ and the square root of the second moment m_F, respectively.

The family of symmetric distributions (about the respective expectation $\mu = \mu_F$) is characterized by equivalent relations

$$F(x - \mu) = 1 - F(\mu - x-), \quad (2.41)$$
$$F^{-1}(x) - \mu = \mu - F^{-1}(1 - x-). \quad (2.42)$$

It is convenient to study the upper halves of $F^{-1} - \mu_F$, which form the cone

$$\mathcal{C}^s = \{g \in L^2([1/2,1), dx) : g \text{ -- nondecreasing, } g(1/2) = 0\}, \quad (2.43)$$

and extend the functions to the whole unit interval using 2.42. This implies the modification of functionals which consists in folding them about $\frac{1}{2}$. Indeed, by 2.42, yields

$$\begin{aligned} T_h(F^{-1} - \mu) &= \int_0^1 [F^{-1}(x) - \mu] h(x)\, dx \\ &= \int_{1/2}^1 [F^{-1}(x) - \mu] h^s(x)\, dx \quad (2.44) \end{aligned}$$

for

$$h^s(x) = h(x) - h(1 - x-). \tag{2.45}$$

Observe that the norm of $F^{-1} - \mu_F$ in $L^2([1/2, 1), dx)$ is $\sigma_F/\sqrt{2}$.

In various models, existence and monotonicity of the density function of the parent distribution F is assumed. If the density is nonincreasing (nondecreasing) then F is increasing concave (convex) on its support, and F^{-1} is increasing convex (concave, respectively). The closures of the families of the quantile functions contain elements that are constant at either of ends. Respective distribution functions have single jumps. We prefer studying closed convex cones of convex (concave) quantile functions which guarantees existence of projections. If a projection is a closure point of a cone (usually it is) then this can be approximated in the norm by a sequence of strictly increasing convex (concave) functions. Accordingly, the sequence attains the supremum of a linear functional in the limit. Motivated by the above considerations, we define

$$\mathcal{C}^{\nearrow}_{\succeq_c U} = \{g \in \mathcal{C}^{\nearrow} : g \ - \ \text{convex}\}, \tag{2.46}$$

$$\mathcal{C}^{\nearrow}_{\preceq_c U} = \{g \in \mathcal{C}^{\nearrow} : g \ - \ \text{concave}\}, \tag{2.47}$$

with $\mathcal{C}^0_{\succeq_c U}$ ($\mathcal{C}^0_{\preceq_c U}$) and $\mathcal{C}^+_{\succeq_c U}$ ($\mathcal{C}^+_{\preceq_c U}$) denoting intersections of 2.46 (2.47, respectively) with linear subspaces of functions integrating to 0 (vanishing at 0, respectively). The apparently awkward notation will be justified below. We say that $F(x)$ succeeds the uniform distribution function $U(x) = x$, $0 \le x \le 1$, in the convex ordering (written $F \succeq_c U$) iff $F^{-1}U = F^{-1}$ is convex on $[0, 1)$. The reversed relation $F \preceq_c U$ is defined by convexity of $U^{-1}F = F$ on the support of F, being equivalent with concavity of F^{-1}. The convex ordering , defined in van Zwet [62], is a partial ordering of (absolutely) continuous distribution functions, invariant under location and scale transformations (see Dharmadhikari and Joag-dev [16, Theorem 9.1, p. 217]). Therefore we can generalize 2.46, 2.47 as follows

$$\mathcal{C}^{\nearrow}_{\succeq_c W} = \{g \in L^2([a_W, d_W), w(x)dx) : g \ - \ \text{nondecreasing and convex}\}, \tag{2.48}$$

$$\mathcal{C}^{\nearrow}_{\preceq_c W} = \{g \in L^2([a_W, d_W), w(x)dx) : g \ - \ \text{nondecreasing and concave}\}, \tag{2.49}$$

by taking compositions $F^{-1}W$ for an arbitrarily fixed distribution function W with support $[a, d) = [a_W, d_W)$, and density w. We also introduce convex cones $\mathcal{C}^0_{\succeq_c W}$ ($\mathcal{C}^0_{\preceq_c W}$) and $\mathcal{C}^+_{\succeq_c W}$ ($\mathcal{C}^+_{\preceq_c W}$) by adding conditions $\int_a^d g(x)w(x)\,dx = 0$ and

$g(a) = 0$, respectively, to definition 2.48 (2.49, respectively). The definitions are justified by properties of compositions $F^{-1}W$. In particular, we have

$$\int_a^d [F^{-1}W(y)]^2 w(y)\, dy = \int_0^1 [F^{-1}(x)]^2\, dx < \infty, \tag{2.50}$$

$$\int_a^d [F^{-1}W(y) - \mu] w(y)\, dy = \int_0^1 [F^{-1}(x) - \mu]\, dx = 0, \tag{2.51}$$

$$F^{-1}W(a) = F^{-1}(0). \tag{2.52}$$

The families of life distributions being in the convex ordering with the exponential distribution are studied in the reliability theory. Observe that $F \preceq_c V$ implies convexity of $V^{-1}F(x) = -\ln[1 - F(x)]$, and nondecrease of $\lambda_F(x) = (V^{-1}F)'(x) = f(x)/[1 - F(x)]$. Function $\lambda_F(x)$, called the failure (hazard) rate, describes the infinitesimal probability of failure at a short period of time following x under the condition of surviving until x. Condition $F \preceq_c V$ defines the family of distributions with increasing failure rate (IFR, for short). Likewise, $F \succeq_c V$ coincides with nonincrease of λ_F, and defines the family of decreasing failure rate (DFR) distributions. Below we specify bounds on statistical functionals restricted to IFR and DFR distributions. Since $U \prec_c V$, every DFR distribution has a decreasing density, and the increase of density implies that of failure rate.

van Zwet [62] (see also Lawrence [29]) defined a counterpart of convex ordering for symmetric distributions, and called it s-ordering. We have $F \preceq_s W$ for F and W symmetric about μ_F and μ_W, respectively, if $F^{-1}W(x)$ is concave for $x \geq \mu_W$. This is equivalent with convexity of $F^{-1}W$ on the lower half of support of W, and concavity-convexity of $W^{-1}F$ on the support of F. Combining constructions leading to 2.43, 2.48 and 2.49, we define the convex cones

$$C^s_{\succeq_c W} = \{g \in L^2([\mu_W, d_W), w(x)\, dx) : \ g(\mu_W) = 0,\ g \ - \text{ nondecreasing, convex}\}, \tag{2.53}$$

and $C^s_{\preceq_c W}$ replacing convexity assumption by concavity in 2.53. It is worth pointing out here that $F \succeq_s (\preceq_s) U$ define the classes of symmetric unimodal (U-shaped) distributions.

The star ordering $\preceq_*$ of continuous life distributions, introduced in Barlow and Proschan [8], is more general partial ordering than the convex one (see, e.g., Dharmadhikari and Joag-dev [16, Theorem 9.1, p. 217]). By definition, $F \preceq_* W$ iff $W^{-1}F$ is starshaped, i.e. $W^{-1}F(x)/x$ is nondecreasing on the support of F. If $F \preceq_* V$ in particular, then

$$\Lambda_F(x) = \frac{-\ln[1 - F(x)]}{x} = \frac{1}{x}\int_0^x \lambda_F(y)\, dy \tag{2.54}$$

is nondecreasing, and we can say that F has increasing failure rate in average (IFRA, for brevity). DFRA life distributions F are defined by relation $F \succeq_* V$. Also, $F \preceq_* (\succeq_*) U$ means that $\frac{1}{x}\int_0^x f(y)\, dy$ is nondecreasing (nonincreasing, respectively). Accordingly, the relation defines the family of life distributions with increasing (decreasing) density in average: although f may be multimodal, the larger values in $[a_F, d_F)$ are more (less) probable than the smaller ones. The star ordering is scale invariant. In order to make it invariant with respect to translations as well, we generalize the definition as follows: $F \succeq_* (\preceq_*) W$ iff a_F, a_W are finite and $[F^{-1}W(x) - F^{-1}W(a_W)]/(x - a_W)$ is nondecreasing (nonincreasing) on $[a_W, d_W)$. The definition will enable us to establish mean-variance bounds on statistical functionals by projecting them onto convex cones of $F^{-1}W - \mu_F$ described by the generalized star relations:

$$\mathcal{C}^0_{\succeq_* W} = \{g \in L^2([a_W, d_W), w(x)dx) : \ g(x), \ \frac{g(x) - g(a_W)}{x - a_W} \ \text{are nondecreasing and} \ \int_{a_W}^{d_W} g(x)w(x)\, dx = 0\}, \tag{2.55}$$

$$\mathcal{C}^0_{\preceq_* W} = \{g \in L^2([a_W, d_W), w(x)dx) : \ g(x) \ - \ \text{nondecreasing}, \ \frac{g(x) - g(a_W)}{x - a_W} \ - \ \text{nonincreasing}, \ \int_{a_W}^{d_W} g(x)w(x)\, dx = 0\}. \tag{2.56}$$

The special emphasis will be laid on cases $W = U, V$. For a detailed treatment of stochastic orders and their applications, we refer the reader to monographs of Dharmadhikari and Joag-dev [16], and Shaked and Shantikumar [59].

We adopt the convention of denoting the projection onto a given convex cone by writing P with the same upper and lower indices that appear in the notation of the cone. E.g., $P^0_{\succeq_* W} h$ will denote the projection of h onto $\mathcal{C}^0_{\succeq_* W}$. Although the original quantile functions of restricted families of distributions discussed above form convex cones, we prefer considering compositions $F^{-1}W$. The reason is that the latter have natural analytic and geometric properties. Our process of determining the projection of a given functional consists in choosing an arbitrary starting point in the cone and constructing consecutive approximations improving the previous ones. We first try to describe the shape of projection function and then calculate optimal parameters. It is therefore essential for the first step that we are able to check immediately if a function proposed for an approximation actually belongs to the cone. The only assumption appearing in definitions of our cones that cannot be verified at the first glance is weighted integrability to 0 (see, e.g., 2.39, 2.53, 2.55, 2.56). One can overcome the problem using the following Lemma (cf Rychlik [51, Lemma 1]).

Lemma 2. *Suppose that $\mathcal{C}$ is a subset of a real Hilbert space $\mathcal{H}$ such that $g \in \mathcal{C}$ implies $g + cg_0 \in \mathcal{C}$ for some $g_0 \in \mathcal{H}$ and all real c. If $h_0 \in \mathcal{H}$ has a projection Ph_0 onto $\mathcal{C}$, then $(Ph_0, g_0) = (h_0, g_0)$.*

Proof. For arbitrary fixed $g \in \mathcal{C}$ and $c \in \Re$, $||g + cg_0 - h_0||^2$ is minimized by $c_0 = c_0(g) = (h_0 - g, g_0)/(g_0, g_0)$. The projection Ph_0 has necessarily the form $g + c_0 g_0$ for some $g \in \mathcal{C}$, and satisfies $(g + c_0 g_0, g_0) = (h_0, g_0)$. □

Let $\mathcal{C}^0_\bullet$ be any of above considered convex cones such that $\int_a^d g(x)w(x)\,dx = (g, \mathbf{1}) = 0$ for all $g \in \mathcal{C}^0_\bullet$ is assumed. Let $\mathcal{C}^\nearrow_\bullet$ denote the extension of $\mathcal{C}^0_\bullet$ by dropping the integral condition. Note that each $\mathcal{C}^\nearrow_\bullet$ is translation invariant, i.e. fulfills the assumption of Lemma 2 with $g_0 = \mathbf{1}$. Therefore $(h, \mathbf{1}) = 0$ implies that $P^\nearrow_\bullet h \in \mathcal{C}^0_\bullet$ and coincides with $P^0_\bullet h$. Otherwise we replace h by

$$h_0 = h - \frac{(h, \mathbf{1})}{(\mathbf{1}, \mathbf{1})}\mathbf{1}. \tag{2.57}$$

Then, by 2.2, 2.3,

$$\forall\, g \in \mathcal{C}^0_\bullet \qquad T_h(g) = (g, h) = \left(g, h - \frac{(h, \mathbf{1})}{(\mathbf{1}, \mathbf{1})}\mathbf{1}\right) = T_{h_0}(g) \le ||P^0_\bullet h_0||\,||g||,$$
$$T_h(P^0_\bullet h_0) = T_{h_0}(P^0_\bullet h_0) = ||P^0_\bullet h_0||^2.$$

Since $(h_0, \mathbf{1}) = 0$, we have $P^0_\bullet h_0 = P^\nearrow_\bullet h_0$. It follows that it suffices to replace functional h by 2.57 and project it onto $\mathcal{C}^\nearrow_\bullet$, without bothering about the integral condition. In fact, translation invariance of $\mathcal{C}^\nearrow_\bullet$ yields $P^\nearrow_\bullet(h_0) = P^\nearrow_\bullet(h) - \frac{(h,\mathbf{1})}{(\mathbf{1},\mathbf{1})}\mathbf{1}$ so that we are reduced to projecting the original h onto $\mathcal{C}^\nearrow_\bullet$, and subtracting the constant from the result.

3 Quantiles

The main results of this Section come from Rychlik [54]. The bounds for quantiles of general distributions were obtained by Moriguti [34]. These for symmetric and symmetric unimodal distributions may also be concluded from the Chebyshev and Gauss inequalities, respectively. Vysochanskii and Petunin [64] presented a refinement of the Gauss inequality for unimodal distributions. Further generalizations can be found in Dharmadhikari and Joag-dev [16, Section 1.5]. We also notice that the Markov inequality yields $F^{-1}(p) \le \mu_F/(1-p)$ for quantiles of nonnegative random variables. Another implication of the Markov inequality is the second moment bound $F^{-1}(p) \le m_F/(1-p)^{1/2}$.

3.1 General and Symmetric Distributions

For these two cases, there is no need to think of a quantile value as a limit of L^2-functionals (cf 2.17), because we can directly use results of Moriguti [34], summarized below. These enable us to analyze integrals of monotone functions with respect to arbitrary, not necessarily absolutely continuous distribution functions.

Lemma 3. *If $H : [a,d) \mapsto \Re$ is right continuous and of bounded variation, H^- denotes its left continuous version, and $\bar{h}$ is the right derivative of the greatest convex minorant $\bar{H}$ of H, then*

$$\int_a^d g(x)\, H(dx) \le \int_a^d g(x)\bar{h}(x)\, dx \le \left[\int_a^d g^2(x)\, dx \int_a^d \bar{h}^2(x)\, dx\right]^{1/2} \tag{3.1}$$

for every nondecreasing function g. The former relation in 3.1 becomes equality iff g is constant in every interval contained in $\{\bar{H} < \min\{H, H^-\}\}$, and right (left) continuous at every discontinuity point of H (if any) such that $H > H^-$ ($H < H^-$, respectively) there. The latter relation in 3.1 becomes equality iff either $\bar{h} \equiv 0$ or $g = \alpha\bar{h}$ for $\alpha \ge 0$. Also, under assumption $H(a) = H^-(d)$, function g can be replaced by arbitrary translation $g + c$ in the middle and last expressions of 3.1, and the conditions for equality.

We tacitly assume that all integrals in 3.1 are well-defined and finite. Observe that the conditions for the latter equality imply those for the former, because $\bar{H}$ is linear and $\bar{h}$ is constant on every interval of $\{\bar{H} < \min\{H, H^-\}\}$.

Theorem 2 (general distributions)) *For arbitrary distribution function F, with mean μ_F and variance σ_F^2, we have*

$$\frac{F^{-1}(p) - \mu_F}{\sigma_F} \le \left(\frac{p}{1-p}\right)^{1/2}, \qquad 0 < p < 1. \tag{3.2}$$

The equality in 3.2 holds iff F is the two point distribution valued at $\mu - [(1-p)/p]^{1/2}\sigma$ and $\mu + [p/(1-p)]^{1/2}\sigma$ with probabilities p and $1-p$, respectively.

The Theorem immediately follows from Lemma 3 by putting $H = \mathbf{1}_{[p,1)} - U$ with

$$\bar{h}(x) = \begin{cases} -1, & \text{if } 0 \le x < p, \\ p/(1-p), & \text{if } p \le x < 1, \end{cases}$$

and $||\bar{h}||^2 = p/(1-p)$, and $(F^{-1} - \mu)/\sigma = \bar{h}/||\bar{h}||$. Moriguti [34] also showed that

$$\frac{F^{-1}(q) - F^{-1}(p)}{\sigma_F} \le \left[\frac{1}{1-q} + \frac{1}{p}\right]^{1/2}, \qquad 0 < p < q < 1. \tag{3.3}$$

In special cases of 3.2 and 3.3 for the median and interquartile distance yields

$$F^{-1}(1/2) - \mu_F \le \sigma_F,$$
$$F^{-1}(3/4) - F^{-1}(1/4) \le 2^{3/2}\sigma_F.$$

Likewise for symmetric distributions and $1/2 \le p < 1$, we can write

$$\begin{aligned} F^{-1}(p) - \mu &\le \int_{1/2}^{1} [F^{-1}(x) - \mu]\, \mathbf{1}_{[p,1)}(dx) \le \int_{1/2}^{1} [F^{-1}(x) - \mu] \bar{\mathbf{1}}'_{[p,1)}(x)\, dx \\ &\le ||\bar{\mathbf{1}}'_{[p,1)}||\sigma/\sqrt{2} = [2(1-p)]^{-1/2}\sigma, \end{aligned} \tag{3.4}$$

because $\bar{\mathbf{1}}_{[p,1)}(x) = \frac{x-p}{1-p}\mathbf{1}_{[p,1)}(x)$ is the greatest convex minorant of $\mathbf{1}_{[p,1)}$. For $p < 1/2$ we trivially obtain $F^{-1}(p) - \mu_F \le F^{-1}(p) - F^{-1}(1/2) \le 0$.

Theorem 3. (symmetric distributions) *Let F be a distribution function of a symmetric random variable X with a finite variance. Then $F^{-1}(p) \le \mu_F$ for every quantile of order $p < 1/2$. For $p \ge 1/2$, we have*

$$[F^{-1}(p) - \mu_F]/\sigma_F \le [2(1-p)]^{-1/2}, \tag{3.5}$$

with the equality attained in 3.5 for

$$\mathrm{P}(X = \mu \pm [2(1-p)]^{-1/2}\sigma) = 1 - p,$$
$$\mathrm{P}(X = \mu) = 2p - 1.$$

Observe that for $p = 1/2$ bounds 3.2 and 3.5 coincide, and 3.5 is sharper for $p > 1/2$.

3.2 Distributions with Decreasing Density and Failure Rate

Our auxiliary projection problem is to find a function in $\mathcal{C}^{\nearrow}_{\preceq_c W}$ minimizing the distance to an indicator function of an interval.

Lemma 4. *Let* $h = M\mathbf{1}_{[b,c)} \in L^2([a_W, d_W), w(x)dx)$ *for* $a < b < c \leq d$ *and* $M > 0$. *Then for every* $g \in \mathcal{C}^{\nearrow}_{\succeq_c W}$ *there exists* $g_{\alpha\beta} \in \mathcal{C}^{\nearrow}_{\succeq_c W}$ *defined as* $g_{\alpha\beta}(x) = \alpha(x-\beta)_+$ *for some* $\alpha \geq 0$, $\beta \leq b$, $\alpha(b-\beta) < M$, *such that* $||g_{\alpha\beta} - h|| \leq ||g - h||$.

Lemma 5. *Let the assumptions of Lemma 4 hold and*

$$\int_a^d w(x)\,dx = \int_a^d h(x)w(x)\,dx = 1. \tag{3.6}$$

If

$$\int_a^d xw(x)\,dx \geq \int_b^c xw(x)\,dx \Big/ \int_b^c w(x)\,dx, \tag{3.7}$$

then $P^{\nearrow}_{\succeq_c W} h = \mathbf{1}$.

Otherwise there exists a unique $\beta_* < b$ *that solves*

$$\frac{\int_{\max\{\beta,a\}}^d (x-\beta)^2 w(x)\,dx}{\int_{\max\{\beta,a\}}^d (x-\beta)w(x)\,dx} = \frac{\int_b^c (x-\beta)w(x)\,dx}{\int_b^c w(x)\,dx} \tag{3.8}$$

and

$$\alpha_* = \alpha_*(\beta_*) = 1 \Big/ \left[\int_{\max\{\beta_*,a\}}^d (x-\beta_*)w(x)\,dx\right] > 0 \tag{3.9}$$

such that

$$P^{\nearrow}_{\succeq_c W} h(x) = \alpha_*(x-\beta_*)_+ = \frac{(x-\beta_*)_+}{\int_{\max\{\beta_*,a\}}^d (y-\beta_*)w(y)\,dy}. \tag{3.10}$$

Precisely, if relation " $\geq$ *" holds in 3.8 with* β *replaced by* a, *then* $\beta_* \leq a$, *and 3.10 is linear on* $[a, d)$. *In the opposite case,* $\beta_* > a$ *and the projection is actually a two-piece broken line.*

By 3.10,

$$||P^0_{\succeq_c W} h||^2 = ||P^{\nearrow}_{\succeq_c W} h||^2 - 1$$

$$= \frac{\int_{\max\{\beta_*,a\}}^d (x-\beta_*)^2 w(x)\,dx - [\int_{\max\{\beta_*,a\}}^d (x-\beta_*)w(x)\,dx]^2}{[\int_{\max\{\beta_*,a\}}^d (x-\beta_*)w(x)\,dx]^2}, \tag{3.11}$$

$$\frac{P^0_{\succeq_c W} h(x)}{||P^0_{\succeq_c W} h||}$$

$$= \frac{(x-\beta_*)_+ - \int_{\max\{\beta_*,a\}}^d (y-\beta_*)w(y)\,dy}{\left\{\int_{\max\{\beta_*,a\}}^d (y-\beta_*)^2 w(y)dy - [\int_{\max\{\beta_*,a\}}^d (y-\beta_*)w(y)dy]^2\right\}^{\frac{1}{2}}}. \tag{3.12}$$

Letting $c \searrow b$, we reduce the right-hand sides of 3.7, 3.8 to $b - a$, $b - \beta$, respectively. Analyzing the resulting formulae, we derive bounds for quantiles.

Theorem 4. *($F \succeq_c W$) If $W^{-1}(p) \le \mu_W$, then the same holds for all $F \succeq_c W$.*

If $\mu_W < W^{-1}(p) \le \mu_W + \sigma_W^2/(\mu_W - a_W)$, then

$$\forall F \succeq_c W \qquad \frac{F^{-1}(p) - \mu_F}{\sigma_F} \le \frac{W^{-1}(p) - \mu_W}{\sigma_W}, \tag{3.13}$$

and the equality in 3.13 holds for location-scale transformations of W, i.e. for

$$F(x) = W\left(\mu_W + \sigma_W \frac{x - \mu}{\sigma}\right). \tag{3.14}$$

Finally assume that $W^{-1}(p) > \mu_W + \sigma_W^2/(\mu_W - a_W)$. For $\beta \in (a_W, d_W)$ and W-distributed random variable X, write

$$\mu_W(\beta) = \mathrm{E}_W(X|X > \beta) = \int_\beta^d x w(x)\, dx/[1 - W(\beta)], \tag{3.15}$$

$$\begin{aligned}\sigma_W^2(\beta) &= \mathrm{Var}_W(X|X > \beta)\\ &= \int_\beta^d (x - \mu_W(\beta))^2 w(x)\, dx/[1 - W(\beta)],\end{aligned} \tag{3.16}$$

$$\tau_W(\beta) = [W^{-1}(p) - \mu_W(\beta)]/\sigma_W(\beta). \tag{3.17}$$

Then there exists a unique solution $\beta_* \in (a_W, W^{-1}(p))$ to equation

$$[W^{-1}(p) - \mu_W(\beta)][\mu_W(\beta) - \beta] = \sigma_W^2(\beta), \tag{3.18}$$

and for all $F \succeq_c W$ yields

$$\frac{F^{-1}(p) - \mu_F}{\sigma_F} \le A = A^0_{\succeq_c W}(p) = \left[\frac{\tau_*^2 + W(\beta_*)}{1 - W(\beta_*)}\right]^{1/2}. \tag{3.19}$$

The bound in 3.19 is attained by

$$F(x) = W\left(\beta_* + (\mu_* - \beta_*)[1 - W(\beta_*)]\left[1 + A\frac{x - \mu}{\sigma}\right]\right) \mathbf{1}_{[\mu - \sigma/A, \infty)}(x). \tag{3.20}$$

For brevity, symbols $\mu_*, \sigma_*^2, \tau_*$ in 3.19, 3.20 denote 3.15–3.17, respectively, valued at $\beta = \beta_*$.

Distribution function 3.20 has a jump of height $W(\beta_*) < p$ at the left end-point, and shares the shape of W on its support. This is a life distribution iff $\mu_F A^0_{\succeq_c W}(p) = \sigma_F$. Mean-variance bounds for $F \succeq_c U, V$ are specified in Propositions 1, 2, respectively.

Proposition 1. (decreasing density) *If $0 < p \leq 1/2$, then $F^{-1}(p) \leq \mu_F$.*
If $1/2 < p \leq 2/3$, then

$$[F^{-1}(p) - \mu_F]/\sigma_F \leq 2\sqrt{3}(p - 1/2), \tag{3.21}$$

which becomes equality for the uniform distribution on $[\mu - \sqrt{3}\sigma, \mu + \sqrt{3}\sigma]$.
If $2/3 < p < 1$, then

$$\frac{F^{-1}(p) - \mu_F}{\sigma_F} \leq \left[\frac{9p - 5}{9(1-p)}\right]^{1/2}. \tag{3.22}$$

This is an equality for the mixture of the Dirac distribution concentrated at $\mu - 3\sigma[(1-p)/(9p-5)]^{1/2}$ and the uniform one on $[\mu - 3\sigma[(1-p)/(9p-5)]^{1/2}, \mu + 3\sigma(3p-1)/[(1-p)(9p-5)]^{1/2}]$ with weights $3p - 2$ and $3(1-p)$, respectively.

Proposition 2. (decreasing failure rate) *If $0 < p \leq 1 - e^{-1} \approx 0.63212$, then $F^{-1}(p) \leq \mu_F$.*
If $1 - e^{-1} < p \leq 1 - e^{-2} \approx 0.86466$, then bound

$$[F^{-1}(p) - \mu_F]/\sigma_F \leq -\ln(1-p) - 1, \tag{3.23}$$

is attained by the exponential distribution function with location μ and scale σ.
If $1 - e^{-2} < p < 1$, then for $q = (1-p)e^2 < 1$, we get

$$[F^{-1}(p) - \mu_F]/\sigma_F \leq (2/q - 1)^{1/2}. \tag{3.24}$$

This is an equality if F is the combination of an atom at $\mu - \sigma[q/(2-q)]^{1/2}$, and the exponential distribution with location $[q(2-q)]^{1/2}\mu - q\sigma$ and scale σ, with respective probabilities $1 - q$ and q.

We see that bounds 3.22, 3.24 tend to infinity if $p \nearrow 1$, and the same holds generally for 3.19. It follows from the fact that for $b = W^{-1}(p)$ large enough, $\beta = \beta(b)$ satisfies

$$b = \frac{\int_\beta^d x(x-\beta)w(x)\,dx}{\int_\beta^d (x-\beta)w(x)\,dx} \tag{3.25}$$

(cf. 3.8 for $\beta > a$). Therefore $b(\beta) \nearrow d$, as $\beta \nearrow d$, and the same holds for the inverse. Furthermore, $1 - W(\beta) \searrow 0$, whereas the nominator of $A^0_{\succeq_c W}$ remains bounded below from zero, as $b \nearrow d$.

Lemma 6. *For h defined in Lemma 4, and for every $g \in \mathcal{C}^{\nearrow}_{\succeq_* W}$ there exists $g_{\alpha\beta} \in \mathcal{C}^{\nearrow}_{\succeq_* W}$ defined as $g_{\alpha\beta}(x) = \alpha(x-a)\mathbf{1}_{[b,d)}(x) + \beta$ for some $\alpha, \beta \geq 0$, $\alpha(b-a) + \beta < M$, such that $||g_{\alpha\beta} - h|| \leq ||g - h||$.*

Candidates for projection $P^{\nearrow}_{\succeq_* W} h$ have a constant value $\beta \in [0, h(b))$ in $[a, b)$, a jump at b to $g_{\alpha\beta}(b) < h(b)$, and a linear part in $[b, d)$ which can be extended to the left so to pass through $(a, \beta) = (a, g_{\alpha\beta}(a))$. A constant projection β is also possible.

Lemma 7. *Let assumptions of Lemma 6 and 3.6 hold. If*

$$\frac{\int_b^c (x-a) w(x)\, dx}{\int_b^c w(x)\, dx} \leq \int_b^d (x-a) w(x)\, dx, \tag{3.26}$$

then $P^{\nearrow}_{\succeq_* W} h = \mathbf{1}$. Otherwise

$$P^{\nearrow}_{\succeq_* W} h(x) = \alpha_*(x-a)\mathbf{1}_{[b,d)}(x) + \beta_* \tag{3.27}$$

for $\alpha_* > 0$, $0 \leq \beta_* < 1$ defined by

$$\alpha_* = \frac{\frac{\int_b^c (x-a) w(x)\, dx}{\int_b^c w(x)\, dx} - \int_b^d (x-a) w(x)\, dx}{\int_b^d (x-a)^2 w(x)\, dx - \left[\int_b^d (x-a) w(x)\, dx\right]^2}, \tag{3.28}$$

$$\beta_* = \frac{\int_b^d (x-a)^2 w(x)\, dx - \frac{\int_b^c (x-a) w(x)\, dx}{\int_b^c w(x)\, dx} \int_b^d (x-a) w(x)\, dx}{\int_b^d (x-a)^2 w(x)\, dx - \left[\int_b^d (x-a) w(x)\, dx\right]^2}. \tag{3.29}$$

By arguments of Subsection 2.3, $P^0_{\succeq_* W} h = P^{\nearrow}_{\succeq_* W} h - 1$. This is 0 if 3.26 holds. Otherwise, we have

$$||P^0_{\succeq_* W} h|| = \frac{\frac{\int_b^c (x-a) w(x)\, dx}{\int_b^c w(x)\, dx} - \int_b^d (x-a) w(x)\, dx}{\left\{\int_b^d (x-a)^2 w(x)\, dx - \left[\int_b^d (x-a) w(x)\, dx\right]^2\right\}^{1/2}}, \tag{3.30}$$

$$\frac{P^0_{\succeq_* W} h}{||P^0_{\succeq_* W} h||} = \frac{(x-a)\mathbf{1}_{[b,d)} - \int_b^d (x-a) w(x)\, dx}{\left\{\int_b^d (x-a)^2 w(x)\, dx - \left[\int_b^d (x-a) w(x)\, dx\right]^2\right\}^{1/2}}, \tag{3.31}$$

Letting $c \searrow b$, and using notation of Theorem 4, we state its analogue for $F \succeq_* W$.

Theorem 5. $F \succeq_* W$) *Using 3.15 and 3.16 with $\beta = b = W^{-1}(p)$, write*

$$\eta = \eta_W(a,b) = b - a - \int_b^d (x-a)w(x)\,dx$$
$$= b - pa - (1-p)\mu_W(b), \tag{3.32}$$
$$\vartheta^2 = \vartheta_W^2(a,b) = \int_b^d (x-a)^2 w(x)\,dx - \left[\int_b^d (x-a)w(x)\,dx\right]^2$$
$$= (1-p)\sigma_W^2(b) - p(1-p)a[2\mu_W(b) - a]. \tag{3.33}$$

Then

$$\frac{F^{-1}(p) - \mu_F}{\sigma_F} \le A^0_{\succeq_* W}(p) = \frac{[\eta_W(a,b)]_+}{\vartheta_W(a,b)}. \tag{3.34}$$

If $\eta_W(a,b) > 0$, then bound 3.33 is attained by

$$F(x) = \begin{cases} 0, & if\ \frac{x-\mu}{\sigma} < -\frac{(1-p)\mu_W(b)}{\vartheta}, \\ p, & if\ \frac{(1-p)\mu_W(b)}{\vartheta} \le \frac{x-\mu}{\sigma} \le \frac{\eta}{\vartheta}, \\ W\left(b - \eta + \vartheta\frac{x-\mu}{\sigma}\right), & if\ \frac{x-\mu}{\sigma} \ge \frac{\eta}{\vartheta}. \end{cases} \tag{3.35}$$

Obviously, $A^0_{\succeq_* W}(p) \ge A^0_{\succeq_c W}(p)$, because $F \succeq_c W$ implies $F \succeq_* W$. It follows that 3.34 tends to infinity as $p \nearrow 1$. The main difference between 3.20 and 3.35 is that the atom of the latter is distant from the smooth part. Special cases of distributions dominating the uniform and exponential ones in the star ordering are presented in Propositions 3 and 4.

Proposition 3. (decreasing density in average) *If $p \le \sqrt{2} - 1 \approx 0.41421$, then $F^{-1}(p) \le \mu_F$ for all F with nonincreasing density in average.*

If $p > \sqrt{2} - 1$, then

$$\frac{F^{-1}(p) - \mu_F}{\sigma_F} \le \frac{\sqrt{3}(p^2 + 2p - 1)}{\theta_U(p)} \tag{3.36}$$

with

$$\theta_U^2(p) = 12\vartheta_U^2(0,p) = 1 + 6p^2 - 4p^3 - 3p^4. \tag{3.37}$$

Relation 3.36 becomes equality for the mixture of an atom at $\mu - \sqrt{3}\sigma(1 - p^2)/\theta_U(p)$ with probability p, and the uniform distribution on $[\mu + \sqrt{3}\sigma(p^2 + 2p - 1)/\theta_U(p), \mu + \sqrt{3}\sigma(1 + p^2)/\theta_U(p)]$ with probability $1 - p$.

Proposition 4. (decreasing failure rate in average) *Let $p_0 \approx 0.553567$ be the unique zero of strictly increasing function*

$$\nu_V(p) = p[1 - \ln(1-p)] - 1, \qquad 0 < p < 1. \tag{3.38}$$

If $p \le p_0$, then $F^{-1}(p) \le \mu_F$ for all $F \succeq_* V$.

Otherwise

$$\frac{F^{-1}(p) - \mu_F}{\sigma_F} \le \frac{\nu_V(p)}{\theta_V(p)} \tag{3.39}$$

for

$$\theta_V^2(p) = \vartheta_V^2(0, -\ln(1-p)) = (1-p)p[1 - \ln(1-p)]^2 + 1 - p. \tag{3.40}$$

Bound 3.39 is sharp. This is attained by F being a combination of jump distribution at $\mu - \sigma(1-p)[1 - \ln(1-p)]/\theta_V(p)$ and the exponential distribution with location $\mu + \sigma\nu_V(p)/\theta_V(p)$ and scale $\sigma(1-p)/\theta_V(p)$ with probabilities p and $1-p$, respectively.

3.4 Symmetric Unimodal Distributions

Calculating mean-variance bounds for quantiles of symmetric unimodal distributions consists in projecting $h_{pq}(x) = \frac{1}{q-p}\mathbf{1}_{[p,q)}(x)$ onto $\mathcal{C}^s_{\succeq_c U}$, and determining $\lim_{q\searrow p} P^s_{\succeq_c U} h_{pq}$. We shall not decide on this approach here, but try to make use of results of Subsection 3.2 and obtain a partial solution. We can see that for $b = W^{-1}(p)$ large enough $P^{\nearrow}_{\succeq_c W} h_p = \lim_{q\searrow p} P^{\nearrow}_{\succeq_c W} h_{pq}$ vanishes at a_W, and so belongs to $\mathcal{C}^+_{\succeq_c W^s}$ for W^s being the symmetrized version of W. In order to apply Lemma 5, we consider $\frac{1}{2}h_{pq} \in L^2([\frac{1}{2}, 1), 2\,dx))$. From the last statement of Lemma 5, we conclude that $P^{\nearrow}_{\succeq_c 2U-1}\frac{1}{2}h_p = \frac{1}{2}P^s_{\succeq_c U}h_p$ if $p \ge 5/6$. It has the form $\alpha_*(x - \beta_*)_+$ for $\alpha_* = (3-3p)^{-2}$ and $\beta_* = 3p - 2 \ge 1/2$. Recalling arguments of Subsection 2.3, we determine the bound

$$[F^{-1}(p) - \mu_F]/\sigma_F \le ||P^s_{\succeq_c U}h_p||/\sqrt{2}$$

and the distribution function attaining it which is defined on $[1/2, 1)$ by

$$\frac{F^{-1}(x) - \mu_F}{\sigma_F} = \frac{P^s_{\succeq_c U}h_p(x)}{||P^s_{\succeq_c U}h_p||\sqrt{2}}.$$

Proposition 5 provides precise evaluations.

Proposition 5. (symmetric unimodal distributions) *For arbitrary symmetric unimodal distribution we have*

$$\frac{F^{-1}(p) - \mu_F}{\sigma_F} \le \frac{1}{3}\left(\frac{2}{1-p}\right)^{1/2}, \qquad \frac{5}{6} \le p < 1. \tag{3.41}$$

Table 1: Sharp uniform mean-variance bounds on quantiles for various families of distributions.

p	G	S	SUN	DD	DDA	DFR	DFRA
0.45	0.9045	0	0	0	0.1351	0	0
0.50	1	1	?	0	0.3216	0	0
0.55	1.1055	1.0541	?	0.1732	0.5091	0	0
0.60	1.2247	1.1180	?	0.3464	0.7024	0	0.1554
0.65	1.3628	1.1952	?	0.5196	0.9076	0.0498	0.3373
0.70	1.5275	1.2910	?	0.6939	1.1341	0.2040	0.5390
0.75	1.7321	1.4142	?	0.8819	1.3958	0.3863	0.7704
0.80	2	1.5811	?	1.1055	1.7178	0.6094	1.0506
0.85	2.3805	1.8257	1.2172	1.4011	2.1506	0.8971	1.4208
0.90	3	2.2361	1.4907	1.8559	2.8231	1.3064	1.9888
0.95	4.3589	3.1623	2.1082	2.8087	4.2408	2.1008	3.1853

The bound becomes equality for a mixture of atom at μ *with probability* $6p-5$, *and uniform distribution on interval* $[\mu - \sigma/\sqrt{2(1-p)}, \mu + \sigma/\sqrt{2(1-p)}]$ *with probability* $6(1-p)$.

Observe that 3.41 is 1.5 times less than respective bound 3.5 for symmetric distributions. In Table 1 we compare numerically bounds on standarized quantiles of orders $p = 0.45, (0.05), 0.95$, for seven families of distributions: general (G), symmetric (S), and symmetric unimodal (SUN), with decreasing density (DD) and failure rate (DFR), and those in average (DDA and DFRA, respectively).

4 Order Statistics of Independent Samples

Bounds for expectations of order statistics from general i.i.d. samples, due to Moriguti [34], are presented in Subsection 4.1. Bounds of Subsections 4.2 and 4.4 were obtained by Gajek and Rychlik [19]. A fuller discussion of results of Subsection 4.3 can be found in Rychlik [52]. We also mention another method of Blom [10] and van Zwet [62] for deriving quantile bounds for expectations of order statistics from restricted families of distributions defined by means of stochastic orderings which is based on the Jensen inequality. Bounds and approximations for moments of order statistics were reviewed in David [13, Chapter 4], Arnold and Balakrishnan [4, Chapters 4, 5], and Rychlik [49].

4.1 General and Symmetric Distributions

The problem of calculating mean-variance bounds on expectations of order statistics lies in determining projections of density functions 2.19 onto convex cone 2.39. Observe that

$$f_{n:n}(x) = nx^{n-1}, \qquad 0 < x < 1, \tag{4.1}$$

is actually nondecreasing, and so $P^{\nearrow} f_{n:n} = f_{n:n}$, and $P^0(f_{n:n} - 1) = f_{n:n} - 1$. Therefore

$$\frac{\mathrm{E}_F X_{n:n} - \mu_F}{\sigma_F} \leq ||f_{n:n} - 1|| = \frac{n-1}{(2n-1)^{1/2}} \sim (n/2)^{1/2} \tag{4.2}$$

with

$$F(x) = \left[\frac{1}{n}\left(1 + \frac{n-1}{(2n-1)^{1/2}} \frac{x-\mu}{\sigma}\right)\right]^{1/(n-1)},$$
$$-\frac{(2n-1)^{1/2}}{n-1} \leq \frac{x-\mu}{\sigma} \leq (2n-1)^{1/2}, \tag{4.3}$$

attaining the bound. Distribution 4.3 is a location-scale transformation of a power distribution, and this is uniform for $n = 2$. The result was published independently by Gumbel [22], and Hartley and David [24]. More generally, bound

$$\mathrm{E}_F \sum_{j=1}^{n} c_j (X_{j:n} - \mu_F)/\sigma_F \leq \left|\left| \sum_{j=1}^{n} c_j (f_{j:n} - 1) \right|\right| \tag{4.4}$$

is sharp if the argument of the norm is nondecreasing. By differentiation,

$$\sum_{j=1}^{n} c_j f'_{j:n}(x) = n \sum_{j=1}^{n-1} (c_{j+1} - c_j) f_{j:n-1}(x), \tag{4.5}$$

we conclude that nondecrease of the sequence of coefficients is a sufficient condition. This was implicitly exploited by Plackett [39] and Nagaraja [36] in calculating optimal bounds for the sample range $X_{n:n} - X_{1:n}$ and selection differentials $\frac{1}{k}\sum_{j=n+1-k}^{n} X_{j:n}$, respectively.

In the case of general bounds, the Moriguti projection method should be used (cf. Rychlik [49, Theorem 7, p. 121]). Here we confine ourselves to single order statistics. Observe first that $\mathrm{E}_F X_{1:n} \leq \mu_F$ is implied by relation $X_{1:n} \leq X_1$. Calculating bounds for nonextreme order statistics $X_{j:n}$, $2 \leq j \leq n-1$, we refer to Moriguti [34, Example 2, p. 111]. We first aim at determining $P^{\nearrow} f_{j:n}$. Note that $f_{j:n}(x) > 0$ for $0 < x < 1$, and increasing-decreasing with the maximum at $\frac{j-1}{n-1}$. Therefore $F_{j:n}$ is strictly increasing on $[0,1]$, and convex on $[0, \frac{j-1}{n-1}]$, and concave on $[\frac{j-1}{n-1}, 1]$. The slopes of tangent lines $l_\alpha(x) = f_{j:n}(\alpha)(x-\alpha) + F_{j:n}(\alpha)$ continuously increase for $\alpha \in [0, \frac{j-1}{n-1}]$ and so do $l_\alpha(1)$, ranging from 0 to $l_{\frac{j-1}{n-1}}(1) > F_{j:n}(1) = 1$. The line l_{α_*} for some $\alpha_* \in (0, \frac{j-1}{n-1})$ such that $l_{\alpha_*}(1) = 1$ becomes a part of the lower convex envelope of $F_{j:n}$ on $[\alpha_*, 1]$, and the remaining part coincides with $F_{j:n}$. Therefore, the greatest convex minorant $\bar{F}_{j:n}$ of $F_{j:n}$ has form

$$\bar{F}_{j:n}(x) = \begin{cases} F_{j:n}(x), & \text{if } 0 \le x \le \alpha_*, \\ f_{j:n}(\alpha_*)(x-1)+1, & \text{if } \alpha_* \le x < 1, \end{cases} \tag{4.6}$$

for a unique $\alpha_* \in (0, \frac{j-1}{n-1})$ satisfying

$$(1-\alpha_*) f_{j:n}(\alpha_*) = 1 - F_{j:n}(\alpha_*). \tag{4.7}$$

The derivative of 4.6 is

$$P^{\nearrow} f_{j:n}(x) = \bar{f}_{j:n}(x) = f_{j:n}(\min\{x, \alpha_*\}). \tag{4.8}$$

Using

$$f_{i:m}(x) f_{j:n}(x) = n \frac{\binom{i+j-2}{i-1}\binom{m+n-i-j}{m-i}}{\binom{m+n-1}{m}} f_{i+j-1:m+n-1}(x), \tag{4.9}$$

we calculate

$$\begin{aligned} ||\bar{f}_{j:n}||^2 &= \int_0^{\alpha_*} f_{j:n}^2(x)\,dx + (1-\alpha_*) f_{j:n}^2(\alpha_*) \\ &= n \frac{\binom{2j-2}{j-1}\binom{2n-2j}{n-j}}{\binom{2n-1}{n}} F_{2j-1:2n-1}(\alpha_*) + (1-\alpha_*) f_{j:n}^2(\alpha_*). \end{aligned} \tag{4.10}$$

We have thus proved

Theorem 6. (general distributions) *For arbitrary i.i.d. sample, we have* $\mathrm{E}_F X_{1:n} \le \mu_F$, *which is attained by the atom measure at* μ_F.

For $2 \le j \le n-1$, *inequality*

$$[\mathrm{E}_F X_{j:n} - \mu_F]/\sigma_F \le B = B^0(j,n) = (||\bar{f}_{j:n}||^2 - 1)^{1/2} \tag{4.11}$$

is sharp (see 4.7, 4.10), and this is attained by the distribution function

$$F(x) = \begin{cases} 0, & \text{if } \frac{x-\mu}{\sigma} \le -\frac{1}{B}, \\ f_{j:n}^{-1}\left(1 + B\frac{x-\mu}{\sigma}\right), & \text{if } -\frac{1}{B} \le \frac{x-\mu}{\sigma} < \frac{f_{j:n}(\alpha_*)-1}{B}, \\ 1, & \text{if } \frac{x-\mu}{\sigma} \ge \frac{f_{j:n}(\alpha_*)-1}{B}. \end{cases} \tag{4.12}$$

For $j = n$, 4.2 is the best bound, attainable by 4.3.

Distribution function 4.12 has an absolutely continuous component (inverse of a polynomial of degree $n-1$), and a jump of height $1-\alpha_*$ at the right end. With few exceptions, the distribution has not an explicit formula. Balakrishnan [7] used a relationship between binomial and negative binomial distributions for reducing polynomial equation 4.7 of degree n to one of degree $j-1$. This makes possible writing explicit values of solutions α_* and bounds $B^0(j,n)$ for $j = 2, 3, n-2, n-1$. For $j = 2$ in particular $\alpha_* = (n-1)^{-2}$, and 4.10 amounts to

$$||\bar{f}_{2:n}||^2 = \frac{n^2(n-1)}{(2n-3)(2n-1)}\left[1 - \frac{n^{2n-3}(n-2)^{2n-1}}{(n-1)^{4(n-1)}}\right] \sim \frac{1}{4n}. \tag{4.13}$$

The problem of determining bounds analogous to 4.11 for symmetric parent distributions can be solved by use of the trick of folding functional about 1/2. Case $j = n$ is the only one for which $s_{j:n}(x) = f_{j:n}(x) - f_{j:n}(1-x) = f_{j:n}(x) - f_{n+1-j:n}(x)$ is nondecreasing. Explicit bound

$$\begin{aligned}
\mathrm{E}_F X_{n:n} - \mu &= n \int_{1/2}^{1} [F^{-1}(x) - \mu][x^{n-1} - (1-x)^{n-1}]\, dx \\
&\leq n \left[\int_{1/2}^{1} [x^{n-1} - (1-x)^{n-1}]^2\, dx\right]^{1/2} \sigma/\sqrt{2} \\
&= \left\{\frac{n^2}{2(2n-1)}\left[1 - \binom{2n-2}{n-1}^{-1}\right]\right\}^{1/2} \sigma \\
&= B^s(n,n)\sigma \sim n^{1/2}/2
\end{aligned} \tag{4.14}$$

(cf. 4.2) was obtained by Moriguti [33] by use of the Schwarz inequality. The extreme distributions for which the equality holds have quantile functions equal to $x^{n-1} - (1-x)^{n-1}$ up to affine transformations.

Bounds for general order statistics of symmetric populations will be derived by projecting differences $s_{j:n}$ onto the family of nondecreasing functions in $L^2([1/2, 1), dx)$. For this purpose we first analyze variability of the differences, using the auxiliary results of Gajek and Rychlik [19, Lemma 3, p. 167].

Lemma 8. *Consider $s_{j:n}(x) = f_{j:n}(x) - f_{n+1-j:n}(x)$ for $x \in [1/2, 1)$ and $(n+1)/2 < j < n$. Then $s_{j:n}$ is nonnegative, and equal to 0 at 1/2 and 1, and increasing-decreasing. For $j \leq n-2$ function $s_{j:n}$ is concave-convex if $j \leq [n + (3n-5)^{1/2}]/2$, and convex-concave-convex otherwise. Also, $s_{n-1:n}$ is concave for $n \leq 7$, and convex-concave for $n \geq 8$.*

Now we merely apply the first statement of the Lemma. Further properties will be needed in Subsection 4.4. Observe first that, by symmetry, $s_{j:n} \leq 0$ for $j \leq (n+1)/2$, and so

$$\mathrm{E}_F X_{j:n} - \mu = \int_{1/2}^{1} [F^{-1}(x) - \mu] s_{j:n}(x)\, dx \leq 0, \tag{4.15}$$

since $F^{-1} - \mu_F \geq 0$ on $[1/2, 1)$. If $j = (n+1)/2$, then $s_{j:n} \equiv 0$, and 4.15 becomes equality for any symmetric F.

For $(n+1)/2 < j \leq n-1$, we repeat arguments leading to 4.8, and obtain

$$P^s f_{j:n}(x) = s_{j:n}(\min\{x, \alpha_*\}), \qquad 1/2 \leq x < 1, \tag{4.16}$$

for $\alpha_* \in [1/2, 1)$ defined by equation

$$(1-\alpha)[f_{j:n}(\alpha) - f_{n+1-j:n}(\alpha)] = F_{n+1-j:n}(\alpha) - F_{j:n}(\alpha) \tag{4.17}$$

(cf 4.7).Using 4.9, we calculate

$$\begin{aligned}||P^s f_{j:n}||^2 &= n\frac{\binom{2j-2}{j-1}\binom{2n-2j}{n-j}}{\binom{2n-1}{n}}[F_{2j-1:2n-1}(\alpha_*) + F_{2n-2j+1:2n-1}(\alpha_*) \\ &\quad - F_{2j-1:2n-1}(1/2) - F_{2n-2j+1:2n-1}(1/2)] \\ &\quad - 2n\frac{\binom{n-1}{j-1}^2}{\binom{2n-1}{n}}[F_{n:2n-1}(\alpha_*) - 1/2] + (1-\alpha_*)s_{j:n}^2(\alpha_*). \end{aligned} \tag{4.18}$$

Theorem 7. (symmetric distributions) *For $1 \leq j \leq (n+1)/2$, we have $\mathrm{E}_F X_{j:n} \leq \mu_F$ for all symmetric parent distributions of the sample. This becomes equality if either F is the Dirac measure at μ_F or $j = (n+1)/2$.*

For $(n+1)/2 < j \leq n-1$, we have

$$\frac{\mathrm{E}_F X_{j:n} - \mu_F}{\sigma_F} \leq B = B^s(j,n) = \frac{||P^s f_{j:n}||}{\sqrt{2}}, \tag{4.19}$$

defined in 4.17–4.18, with the equality for

$$F(x) = \begin{cases} 0, & if\ \frac{x-\mu}{\sigma} < -\frac{s_{j:n}(\alpha_*)}{2B}, \\ s_{j:n}^{-1}(2B\frac{x-\mu}{\sigma}), & if\ -\frac{s_{j:n}(\alpha_*)}{2B} \leq \frac{x-\mu}{\sigma} < \frac{s_{j:n}(\alpha_*)}{2B}, \\ 1, & if\ \frac{x-\mu}{\sigma} \geq \frac{s_{j:n}(\alpha_*)}{2B}, \end{cases} \tag{4.20}$$

(see 4.16, 4.18).

For $j = n$ we have 4.14 which becomes equality for 4.20 with $\alpha_* = 1$ and $s_{n:n}(1) = n$.

Distribution 4.20 has a smooth component with two atoms of measure $1 - \alpha_*$ ($= 0$ for $j = n$) at the end-points of support.

Unlike in the problems studied above, we present here bounds in terms of the square root of the second raw moment m_F. In fact, except of the scale unit we have also a location parameter in the model. This is the population minimal value which for the life distributions amounts to 0. The problem is to evaluate the expected lifetime of $(n+1-j)$-out-of-n systems of independent elements whose identical distribution functions satisfy $F \succeq_c W$ for some fixed W. The solution is based on determining the projection $P^+_{\succeq_c W} h$ of $h = f_{j:n}W$ onto convex cone

$$\mathcal{C}^+_{\succeq_c W} = \{g \in L^2([a_W, d_W), w(x)dx) : g \ - \ \text{nondecreasing, convex}, \quad g(a_W) = 0\}. \tag{4.21}$$

In the sequel we frequently abstract from the specific form of h and assume the following

$$h(x), w(x) > 0 \quad \text{for } a < x < d, \quad h(a) = 0, \tag{4.22}$$

$$\int_a^d h(x)w(x)\,dx = \int_a^d w(x)\,dx = 1, \tag{4.23}$$

$$\int_a^d x^2 w(x)\,dx < \infty, \tag{4.24}$$

$$h(x) \ - \ \text{bounded, } h'' \text{ exists,} \tag{4.25}$$

$$\exists\, a \le b < c \le d \quad \begin{array}{l} \forall\, a < x < b \ h'(x), h''(x) > 0, \\ \forall\, b < x < c \ h''(x) < 0 < h'(x), \\ \forall\, c < x < d \ h'(x) < 0. \end{array} \tag{4.26}$$

However, it is worth pointing out that for the representation $h = f_{j:n}W$, 4.22–4.26 are actually conditions on W, and some of them are naturally satisfied. For $j \ne 1$, we have 4.22, 4.23, and boundedness and monotonicity assumptions (with $c = d$ for $j = n$). An equivalent formulation of 4.24 is finiteness of $\mathrm{E}_W X_1^2$, and existence of h'' requires differentiability of density w on (a_W, d_W). Crucial assumptions are ones describing regions of convexity and concavity of h. These were chosen so to cover the important cases of uniform and exponential weights, without pretending to develop a general theory. One can easily check that the assumptions are satisfied for $2 \le j \le n-1$ and both $W = U, V$ (with $a = b$ in case $j = 2$), and for $j = n$ and $W = V$ with $c = d$. We remind that for $j = 1$ the trivial bound $\mathrm{E}_F X_{1:n} \le \mu_F \le m_F$ is valid for arbitrary F, being attainable for degenerate ones. In case $j = n$ and $W = U$, we have

$P^+_{\succeq_c U} f_{n:n} = f_{n:n} \in \mathcal{C}^+_{\succeq_c U}$, and so the bound for $F \succeq_c U$ is identical with that for general F.

We now present the solution to the projection problem. Lemma 9 describes geometrical properties of the projection.

Lemma 9. *For fixed h satisfying 4.22–4.26, set*

$$h_{\alpha\beta}(x) = \begin{cases} h(x), & \text{if } a \le x \le \beta, \\ h(\beta) + \alpha(x - \beta), & \text{if } \beta \le x < d. \end{cases} \tag{4.27}$$

For every $g \in \mathcal{C}^+_{\succeq_c W}$ there exists $h_{\alpha\beta} \in \mathcal{C}^+_{\succeq_c W}$ such that $||h_{\alpha\beta} - h|| \le ||g - h||$.

Observe that $h_{\alpha\beta} \in \mathcal{C}^+_{\succeq_c W}$ requires $a \le \beta \le b$, and either $\alpha \ge h'(\beta) > 0$ for $\beta > a$ or $\alpha \ge 0$ for $\beta = a$. If $a = b$ in particular, then the projection is a nondecreasing linear function $\alpha(x - a)$ whose slope can be easily determined. Note that for fixed $\beta \in [a, b]$

$$D(\alpha, \beta) = ||h_{\alpha\beta} - h||^2 = \int_\beta^d [\alpha(x - \beta) + h(\beta) - h(x)]^2 w(x)\, dx \tag{4.28}$$

is a convex quadratic function of α, and under restriction $h_{\alpha\beta} \in \mathcal{C}^+_{\succeq_c W}$ is minimized at

$$\hat{\alpha}(a) = \max\{0, \alpha_*(a)\}, \tag{4.29}$$

$$\hat{\alpha}(\beta) = \max\{h'(\beta), \alpha_*(\beta)\}, \qquad \beta > a, \tag{4.30}$$

with

$$\alpha_*(\beta) = \frac{\int_\beta^d (x - \beta)[h(x) - h(\beta)]w(x)\, dx}{\int_\beta^d (x - \beta)^2 w(x)\, dx}, \tag{4.31}$$

being the optimal slope without restrictions. Consecutive steps of reasoning consist in eliminating $\beta > a$ for which $h'(\beta) > \alpha_*(\beta)$, and analyzing $D(\alpha_*(\beta), \beta)$ with continuous derivative

$$\begin{aligned} \frac{dD(\alpha_*(\beta), \beta)}{d\beta} &= 2[\alpha_*(\beta) - h'(\beta)] \int_\beta^d [h(x) - h_{\alpha_*(\beta)\beta}(x)]w(x)\, dx \\ &= 2K(\beta)L(\beta), \end{aligned} \tag{4.32}$$

say. We minimize $D(\alpha_*(\beta), \beta)$ by determining the set $\mathcal{K} = \{\beta \in (a, b) : K(\beta) \ge 0\}$, and analyzing sign changes of L in $\mathcal{K}$. If $\mathcal{K} = (a, \kappa)$ for some $\kappa \in (a, b)$ (which actually holds in special cases $W = U, V$), then $L(\beta)$ is either positive or negative or negative-positive, with zero at some $\lambda \in \mathcal{K}$, and so optimal $\beta_* = a, \kappa$, and λ in the respective cases. Due to representation $h = f_{j:n}W = nB_{j-1,n-1}W$, the basic tool of analysis is the variation diminishing property of the Bernstein polynomials $B_{k,n}(x) = \binom{n}{k}x^k(1-x)^{n-k}$, $0 \le k \le n < \infty$, described in the Lemma below (cf. Schoenberg [56, p. 252]).

Lemma 10. *The number of zeros of a given combination of Bernstein polynomials* $\sum_{k=0}^{n} a_k B_{k,n}$ *in* $(0,1)$ *does not exceed the number of sign changes of the sequence* $a_0, \dots, a_n$. *The first and the last signs of* $\sum_{k=0}^{n} a_k B_{k,n}$ *are identical with the signs of the first and last nonzero elements of* $a_0, \dots, a_n$, *respectively.*

Theorem 8. (decreasing density) *If* $j \le \min\{\frac{2}{3}(n+1), n-1\}$, *then*

$$\frac{\mathrm{E}_F X_{j:n}}{m_F} \le \sqrt{3}\frac{j}{n+1}, \tag{4.33}$$

which is equality if F *is the uniform distribution on* $[0, \sqrt{3}m_F]$.

If $\frac{2}{3}(n+1) < j \le n-1$, *then*

$$\frac{\mathrm{E}_F X_{j:n}}{m_F} \le B = B^{+}_{\succeq_c U}(j,n) = ||(f_{j:n})_{\alpha_*\beta_*}||, \tag{4.34}$$

(cf. 4.27) for

$$\begin{aligned} ||(f_{j:n})_{\alpha_*\beta_*}||^2 &= n\frac{\binom{2j-2}{j-1}\binom{2n-2j}{n-j}}{\binom{2n-1}{n}}F_{2j-1:2n-1}(\beta_*) + (1-\beta_*)f_{j:n}^2(\beta_*) \\ &\quad + (1-\beta_*)^2\alpha_* f_{j:n}(\beta_*) + (1-\beta_*)^3\alpha_*^2/3, \qquad (4.35) \\ \alpha_* = \alpha(\beta_*) &= \frac{3}{(1-\beta_*)^3}\left\{\frac{j}{n+1}[1-F_{j+1:n+1}(\beta_*)] - \beta_*[1-F_{j:n}(\beta_*)]\right\} \\ &\quad - \frac{3}{2(1-\beta_*)}f_{j:n}(\beta_*), \qquad (4.36) \end{aligned}$$

and β_* *being the smaller of the smallest positive zeros of polynomials*

$$\begin{aligned} K_U &= 3\sum_{k=1}^{j}(j+1-k)f_{k:n+2} - \frac{(n-j+3)!}{(n-j)!}f_{j-1:n+2} \\ &\quad + (n-j-\frac{3}{2})\frac{(n-j+2)!}{(n-j)!}f_{j:n+2}, \qquad (4.37) \\ L_U &= \sum_{k=1}^{j-1}(n+1-\frac{3j-k}{2})f_{k:n+2} - \frac{(n-j)(n-j-1)}{4}f_{j:n+2}. \qquad (4.38) \end{aligned}$$

Bound 4.34 is attained by

$$F(x) = \begin{cases} 0, & if\ \frac{x}{m} \le 0, \\ f_{j:n}^{-1}\left(B\frac{x}{m}\right), & if\ 0 \le \frac{x}{m} \le \frac{f_{j:n}(\beta_*)}{B}, \\ \beta_* + \frac{B\frac{x}{m} - f_{j:n}(\beta_*)}{\alpha_*}, & if\ \frac{f_{j:n}(\beta_*)}{B} \le \frac{x}{m} \le \frac{f_{j:n}(\beta_*)+\alpha_*(1-\beta_*)}{B}, \\ 1, & if\ \frac{x}{m} \ge \frac{f_{j:n}(\beta_*)+\alpha_*(1-\beta_*)}{B}. \end{cases} \tag{4.39}$$

Finally,

$$\frac{\mathrm{E}_F X_{n:n}}{m_F} \le \frac{n}{(2n-1)^{1/2}}, \tag{4.40}$$

which becomes equality for the power distribution

$$F(x) = \left[(2n-1)^{-1/2}\frac{x}{m}\right]^{1/(n-1)}, \qquad 0 \le \frac{x}{m} \le (2n-1)^{1/2}. \tag{4.41}$$

(cf. 4.2–4.3).

Theorem 9. (decreasing failure rate) *Set*

$$\mu_V(j:n) = \mathrm{E}_V X_{j:n} = \sum_{k=1}^{j} \frac{1}{n+1-k}, \qquad 1 \le j \le n. \tag{4.42}$$

If $\mu_V(j:n) \le 2$, then

$$\frac{\mathrm{E}_F X_{i:n}}{m_F} \le \frac{\mu_V(j:n)}{\sqrt{2}}, \tag{4.43}$$

which becomes equality for the exponential distribution with scale parameter $m_F/\sqrt{2}$.

Otherwise

$$\frac{\mathrm{E}_F X_{j:n}}{m_F} \le B = B^{+}_{\preceq_c V}(j,n) = ||(f_{j:n}V)_{\alpha_* V^{-1}(\gamma_*)}||, \tag{4.44}$$

for

$$\begin{aligned} ||(f_{j:n}V)_{\alpha_* V^{-1}(\gamma_*)}||^2 &= n\frac{\binom{2j-2}{j-1}\binom{2n-2j}{n-j}}{\binom{2n-1}{n}} F_{2j-1:2n-1}(\gamma_*) \\ &\quad + (1-\gamma_*)[2\alpha_*^2 + 2\alpha_* f_{j:n}(\gamma_*) + f_{j:n}^2(\gamma_*)], \end{aligned} \tag{4.45}$$

$$\begin{aligned} \alpha_* = \alpha_*(V^{-1}(\gamma_*)) &= \frac{1}{2n(1-\gamma_*)} \sum_{k=1}^{j} \mu_V(j+1-k:n+1-k) f_{k:n+1}(\gamma_*) \\ &\quad - \frac{1}{2} f_{j:n+1}(\gamma_*) \end{aligned} \tag{4.46}$$

and γ_* being the minimum of the smallest positive zeros of

$$K_V = \sum_{k=1}^{j} \mu_V(j+1-k:n+1-k) f_{k:n+1} - 2\frac{(n-j+2)!}{(n-j)!} f_{j-1:n+1}$$
$$+ 2(n-j-\frac{1}{2})(n-j+1) f_{j:n+1}, \tag{4.47}$$

$$L_V = \sum_{k=1}^{j} [2-\mu_V(j+1-k:n+1-k)] f_{k:n+1}$$
$$- (n+1-j) f_{j:n+1}. \tag{4.48}$$

Bound 4.44 is attained by

$$F(x) = \begin{cases} 0, & \text{if } \frac{x}{m} \le 0, \\ f_{j:n}^{-1}\left(B\frac{x}{m}\right), & \text{if } 0 \le \frac{x}{m} \le \frac{f_{j:n}(\gamma_*)}{B}, \\ 1-(1-\gamma_*)\exp\left(-\frac{B\frac{x}{m}-f_{j:n}(\gamma_*)}{\alpha_*}\right), & \text{if } \frac{x}{m} \ge \frac{f_{j:n}(\gamma_*)}{B}. \end{cases} \tag{4.49}$$

Formulae 4.34–4.36, and 4.44–4.46 are concluded from 4.27 and 4.31. Moreover, 4.37-4.38 and 4.47–4.48 differ from K, L defined in 4.32 by positive functional factors common for all j, and original variables are replaced by $V(x)$ in the latter case. All bounds are achieved by absolutely continuous distributions. The left-hand parts of 4.39 and 4.49 are the inverses of polynomials equal to $f_{j:n}$ up to scale factors, and cannot be written explicitly. They have uniform and exponential right tails, respectively. Bound 4.40 was derived directly from the Schwarz inequality. Using the following integral approximations of harmonic series

$$\ln\frac{n+1}{n+1-j} < \mu_V(j:n) < \ln\frac{n+1/2}{n+1/2-j},$$

we deduce that condition $\mu_V(j:n) \le 2$ providing bound 4.43 is true for $j \le (1-e^{-2})(n+1/2)$ and false for $j \ge (1-e^{-2})(n+1)$. The difference between both the estimates is $(1-e^{-2})/2 \approx 0.43233$ which implies that for given n the condition has to be directly checked for one j at most. Bounds of Theorem 9 are tighter than those of Theorem 8, because a narrower class of distributions was treated there. This is confirmed by numerical comparisons presented in Gajek and Rychlik [19, Table I, p. 162].

4.3 Distributions with Monotone Density and Failure Rate in Average

The crucial points in calculating mean-variance bounds for $E_F X_{j:n}$ for F with decreasing density and failure rate in average lies in deriving projections $P^0_{\succeq_* U}(f_{j:n}-1)$ and $P^0_{\succeq_* V}(f_{j:n}V-1)$, respectively, onto convex cones of orthogonal to constants nondecreasing starshaped functions. In fact, we consider a more general problem of projecting $h = f_{j:n}W$ satisfying 4.22–4.26 onto $\mathcal{C}^{\nearrow}_{\succeq_* W}$ which gives $P^0_{\succeq_* W}(h-1) = P^{\nearrow}_{\succeq_* W}h - 1$.

Note first that $P^{\nearrow}_{\succeq_* W}h(a) \in [h(a), h(c))$. Functions starting from a higher level uniformly majorize h, and contradict the statement of Lemma 2. Ones with $g(a) < h(a)$ are eliminated by $\max\{g, h(a)\} \in \mathcal{C}^{\nearrow}_{\succeq_* W}$. Therefore in Lemma 11, describing the shape of projection functions, we confine ourselves to functions starting from a range point of h.

Lemma 11. *Define functions*

$$h_{\alpha\beta}(x) = \begin{cases} h(\beta), & if\ a \le x \le \beta, \\ h(x), & if\ \beta \le x \le \alpha, \\ \frac{h(\alpha)-h(\beta)}{\alpha - a}(x-a) + h(\beta), & if\ \alpha \le x < d, \end{cases} \tag{4.50}$$

for some $\alpha \neq a \le \beta \le \alpha < c$. For every $a \le \beta < c$ and $g \in \mathcal{C}^{\nearrow}_{\succeq_* W}$ satisfying $g(a) = h(\beta)$ there exists $a \neq \alpha \in [\beta, c)$ such that $h_{\beta\alpha} \in \mathcal{C}^{\nearrow}_{\succeq_* W}$ and $||h_{\beta\alpha} - h|| \le ||g-h||$.

Functions 4.50 are obviously nondecreasing. Rychlik [52, Lemma 2] showed that under assumptions 4.22–4.26 and for given β, the slope function $[h(\alpha) - h(\beta)]/(\alpha - a)$ is actually increasing for $\beta \le \alpha \le \bar{\alpha}(\beta)$ with some $\bar{\alpha}(\beta) \in [\max\{\beta, b\}, c)$. Under the condition, 4.50 is starshaped as well. Note that constant functions $h_{\beta\beta}(x) \equiv h(\beta)$, $a < \beta < c$, are also admissible here.

Below we determine the optimal parameters. For $a \le \beta \le c$ and $\beta \le \alpha \le c$, $\alpha \neq a$, set

$$K(\alpha, \beta) = \int_\alpha^d [h_{\alpha\beta}(x)) - h(x)]\,(x-a)w(x)dx, \tag{4.51}$$

$$L(\alpha, \beta) = \int_a^d [h_{\alpha\beta}(x) - h(x)]\,w(x)\,dx. \tag{4.52}$$

Then for $a < \beta \le c$ write

$$\hat{K}(\beta) = K(\beta, \beta) = \int_\beta^d [h(\beta) - h(x)]\,(x-a)w(x)\,dx, \tag{4.53}$$

$$\hat{L}(\beta) = L(\beta, \beta) = \int_a^d [h(\beta) - h(x)]\,w(x)\,dx = h(\beta) - 1. \tag{4.54}$$

Lemma 12. *Let $\hat{\beta}$ be the unique zero of 4.54 in (a,c). If*

$$\hat{K}(\hat{\beta}) = \int_{\hat{\beta}}^{d} [1-h(x)]\,(x-a)w(x)\,dx \geq 0, \tag{4.55}$$

then $P^{\nearrow}_{\succeq_* W} h = h_{\hat{\beta}\hat{\beta}} = \mathbf{1}$.

Otherwise there exists a unique pair (β_*, α_*), $a \leq \beta_* < \hat{\beta}$, $\max\{\beta_*, b\} < \alpha_* < c$, determined by equations

$$K(\alpha,\beta) = 0, \tag{4.56}$$

$$L(\alpha,\beta) = 0, \tag{4.57}$$

such that $P^{\nearrow}_{\succeq_* W} h = h_{\alpha_*\beta_*}$, defined in 4.50.

Since h is strictly increasing from $h(a) = 0$ to $h(c) = \sup h > 1$ (cf. 4.22 and 4.23), $\hat{\beta}$ is actually well defined. Taking $h = f_{j:n}W$, we deduce sharp bounds for the expectations of order statistics.

Theorem 10. ($F \succeq_* W$) *Suppose that the density w of W and $h = f_{j:n}W$ for some $2 \leq j \leq n$ satisfy 4.22–4.26.*

If for unique $\hat{\beta} = \hat{\beta}(j,n) \in (0, \frac{j-1}{n-1})$ satisfying

$$f_{j:n}(\hat{\beta}) = 1 \tag{4.58}$$

we have

$$\int_{W^{-1}(\hat{\beta})}^{d} [1 - f_{j:n}W(x)](x-a)w(x)\,dx \geq 0, \tag{4.59}$$

then $\mathrm{E}_F X_{j:n} \leq \mu_F$ for all $F \succeq_* W$.

Otherwise there exists a pair (α_*, β_*), $a_W \leq \beta_* < W^{-1}(\frac{j-1}{n-1})$, $a_W < \alpha_* \in [\beta_*, W^{-1}(\frac{j-1}{n-1}))$, determined by equations

$$\frac{f_{j:n}W(\alpha) - f_{j:n}W(\beta)}{\alpha - a} \int_{\alpha}^{d} (x-a)^2 w(x)\,dx + \int_{\alpha}^{d} [f_{j:n}W(\beta) - f_{j:n}W(x)](x-a)w(x)\,dx = 0, \tag{4.60}$$

$$f_{j:n}W(\beta)[W(\beta) + 1 - W(\alpha)] - F_{j:n}W(\beta) + F_{j:n}W(\alpha) + \frac{f_{j:n}W(\alpha) - f_{j:n}W(\beta)}{\alpha - a} \int_{\alpha}^{d} (x-a)w(x)\,dx - 1 = 0, \tag{4.61}$$

such that

$$[\mathrm{E}_F X_{j:n} - \mu_F]/\sigma_F \leq B = B^{0}_{\succeq_* W}(j,n) \tag{4.62}$$

for

$$
\begin{aligned}
B^2 &= [f_{j:n}W(\beta_*)]^2[W(\beta_*) + 1 - W(\alpha_*)] \\
&+ n\frac{\binom{2j-2}{j-1}\binom{2n-2j}{n-j}}{\binom{2n-1}{n}}[F_{2j-1:2n-1}W(\alpha_*) - F_{2j-1:2n-1}W(\beta_*)] \\
&+ 2f_{j:n}W(\beta_*)\frac{f_{j:n}W(\alpha_*) - f_{j:n}W(\beta_*)}{\alpha_* - a}\int_{\alpha_*}^{d}(x-a)w(x)\,dx \\
&+ \left[\frac{f_{j:n}W(\alpha_*) - f_{j:n}W(\beta_*)}{\alpha_* - a}\right]^2\int_{\alpha_*}^{d}(x-a)^2w(x)\,dx - 1. \qquad (4.63)
\end{aligned}
$$

Equality in 4.62 is attained by

$$
F(x) = \begin{cases} 0, & if\ \frac{x-\mu}{\sigma} < -\frac{1-f_{j:n}W(\beta_*)}{B}, \\ f_{j:n}^{-1}(B\frac{x-\mu}{\sigma}+1), & if\ -\frac{1-f_{j:n}W(\beta_*)}{B} \le \frac{x-\mu}{\sigma} \\ & < -\frac{1-f_{j:n}W(\alpha_*)}{B}, \\ W\left(a + \frac{(\alpha_*-a)[B\frac{x-\mu}{\sigma}+1-f_{j:n}W(\beta_*)]}{f_{j:n}W(\alpha_*)-f_{j:n}W(\beta_*)}\right), & if\ \frac{x-\mu}{\sigma} \ge -\frac{1-f_{j:n}W(\alpha_*)}{B} \end{cases} \qquad (4.64)
$$

Trivial bounds identical with general ones for the sample minimum are consequences of $P^0_{\succeq_* W}(h-1) = P^{\nearrow}_{\succeq_* W}h - 1 \equiv 0$ under 4.55, rewritten as 4.59. These apparently hold for small order statistics. Equations 4.60–4.61 follow from 4.56–4.57. Plugging $W = U, V$ we specify 4.59–4.64.

Proposition 6. (decreasing density in average) *Assume that $F(x)/(x - a_F)$ is nonincreasing for $x > a_F > -\infty$.*

If for given $2 \le j \le n-1$, and $\hat{\beta}$ defined in 4.58

$$
1 - \hat{\beta}^2 \ge \frac{2j}{n+1}[1 - F_{j+1:n+1}(\hat{\beta})], \qquad (4.65)
$$

holds, then $\mathrm{E}_F X_{j:n} \le \mu_F$.

Otherwise there are unique $0 < \beta_* < \hat{\beta}$, $\beta_* < \alpha_* < (j-1)/(n-1)$ that solve equations

$$
\frac{1-\alpha^3}{3\alpha}f_{j:n}(\alpha) - \frac{(1-\alpha)^2(2+\alpha)}{6\alpha}f_{j:n}(\beta) - \frac{j[1-F_{j+1:n+1}(\alpha)]}{n+1} = 0, \qquad (4.66)
$$

$$
\left[\beta - \frac{(1-\alpha)^2}{2\alpha}\right]f_{j:n}(\beta) + \frac{1-\alpha^2}{2\alpha}f_{j:n}(\alpha) - 1 - F_{j:n}(\beta) + F_{j:n}(\alpha) = 0, \qquad (4.67)
$$

and then

$$
[\mathrm{E}_F X_{j:n} - \mu_F]/\sigma_F \le B = B^0_{\succeq_* U}(j,n), \qquad (4.68)
$$

for

$$B^2 = \left[\beta_* + \frac{(1-\alpha_*)^3}{3\alpha_*^2}\right] f_{j:n}^2(\beta_*) + \frac{1-\alpha_*^3}{3\alpha_*^2} f_{j:n}^2(\alpha_*)$$
$$- \frac{(1-\alpha_*)^2(2+\alpha_*)}{3\alpha_*^2} f_{j:n}(\beta_*) f_{j:n}(\alpha_*) - 1$$
$$+ n\frac{\binom{2j-2}{j-1}\binom{2n-2j}{n-j}}{\binom{2n-1}{n}}[F_{2j-1:2n-1}(\alpha_*) - F_{2j-1:2n-1}(\beta_*)]. \qquad (4.69)$$

Equality in 4.68 holds for the location-scale family of distributions

$$F(x) = \begin{cases} 0, & \text{if } \frac{x-\mu}{\sigma} < -\frac{1-f_{j:n}(\beta_*)}{B}, \\ f_{j:n}^{-1}(B\frac{x-\mu}{\sigma}+1), & \text{if } -\frac{1-f_{j:n}(\beta_*)}{B} \le \frac{x-\mu}{\sigma} < -\frac{1-f_{j:n}(\alpha_*)}{B}, \\ \frac{\alpha_*[B\frac{x-\mu}{\sigma}+1-f_{j:n}(\beta_*)]}{f_{j:n}(\alpha_*)-f_{j:n}(\beta_*)}, & \text{if } -\frac{1-f_{j:n}(\alpha_*)}{B} \le \frac{x-\mu}{\sigma} \\ & \quad < \frac{\frac{1}{\alpha_*}f_{j:n}(\alpha_*)+(1-\frac{1}{\alpha_*})f_{j:n}(\beta_*)-1}{B}, \\ 1, & \text{if } \frac{x-\mu}{\sigma} \ge \frac{\frac{1}{\alpha_*}f_{j:n}(\alpha_*)+(1-\frac{1}{\alpha_*})f_{j:n}(\beta_*)-1}{B}. \end{cases} \qquad (4.70)$$

Proposition 7. (decreasing failure rate in average) *Assume that* $-\ln[1-F(x)]/(x-a_F)$ *is nondecreasing for* $x > a_F > -\infty$. *If for* $2 \le j \le n$ *and* $\hat{\beta}$ *defined in 4.58 we have*

$$(1-\hat{\beta})[1-\ln(1-\hat{\beta})] - \sum_{k=1}^{j} \frac{F_{n+1-k:n}(\hat{\beta})}{n+1-k} + \ln(1-\hat{\beta})F_{n+1-j:n}(\hat{\beta}) \ge 0, \quad (4.71)$$

then $\mathrm{E}_F X_{j:n} \le \mu_F$.

Otherwise there exist $0 < \beta_* < -\ln(1-\hat{\beta})$, $\beta_* < \alpha_* < \ln(n-1)/(n-j)$ *solving equations*

$$\left(\alpha + 2 + \frac{2}{\alpha}\right) e^{-\alpha} f_{j:n}V(\alpha) - \left(1 + \frac{2}{\alpha}\right) e^{-\alpha} f_{j:n}V(\beta)$$
$$- \sum_{k=1}^{j} \frac{1}{n+1-k} F_{k:n}V(\alpha) - \alpha F_{j:n}V(\alpha) = 0, \qquad (4.72)$$

$$\left(1 - e^{-\beta} - \frac{e^{-\alpha}}{\alpha}\right) f_{j:n}V(\beta) + \left(1 + \frac{1}{\alpha}\right) e^{-\alpha} f_{j:n}V(\alpha)$$
$$-1 + F_{j:n}V(\beta) - F_{j:n}V(\alpha) = 0, \qquad (4.73)$$

such that

$$[\mathrm{E}_F X_{j:n} - \mu_F]/\sigma_F \le B = B^0_{\succeq_* V}(j,n), \qquad (4.74)$$

where

$$
\begin{aligned}
B^2 = & \left(1 - e^{-\beta_*} + 2\alpha_*^{-2} e^{-\alpha_*}\right) [f_{j:n} V(\beta_*)]^2 \\
& - 2(2+\alpha_*)\alpha_*^{-2} e^{-\alpha_*} f_{j:n} V(\beta_*) f_{j:n} V(\alpha_*) \\
& + (2 + 2\alpha_* + \alpha_*^2)\alpha_*^{-2} e^{-\alpha_*} [f_{j:n} V(\alpha_*)]^2 - 1 \\
& + n \frac{\binom{2j-2}{j-1}\binom{2n-2j}{n-j}}{\binom{2n-1}{n}} [F_{2j-1:2n-1} V(\beta_*) - F_{2j-1:2n-1} V(\alpha_*)]. \qquad (4.75)
\end{aligned}
$$

Bound 4.74 becomes equality for

$$
F(x) = \begin{cases} 0, & if \ \frac{x-\mu}{\sigma} < -\frac{1-f_{j:n}V(\beta_*)}{B}, \\ f_{j:n}^{-1}(B\frac{x-\mu}{\sigma} + 1), & if \ -\frac{1-f_{j:n}V(\beta_*)}{B} \le \frac{x-\mu}{\sigma} \\ & < -\frac{1-f_{j:n}V(\alpha_*)}{B}, \\ 1 - \exp\left(-\frac{\alpha_*[B\frac{x-\mu}{\sigma}+1-f_{j:n}V(\beta_*)]}{f_{j:n}V(\alpha_*)-f_{j:n}V(\beta_*)}\right), & if \ \frac{x-\mu}{\sigma} \ge -\frac{1-f_{j:n}V(\alpha_*)}{B}. \end{cases} \tag{4.76}
$$

Forms of distribution functions 4.39, 4.49, and 4.70, 4.76 are similar. The essential difference is that the latter ones have jumps of height $\beta_*, V(\beta_*)$, respectively. If

$$\mu_F = [1 - f_{j:n}V(\beta_*)]\sigma_F / B^0_{\preceq_* V}(j,n), \tag{4.77}$$

then 4.76 is actually a DFRA life distribution starting at 0.

Under assumptions of Proposition 6, the sample maximum attains general bound 4.2, because the density of 4.3 is decreasing. Surprisingly, general bounds 4.11 and 4.4 are also attained for arbitrary order statistics from the samples with increasing density in average. The same holds for the distribution of the narrower class of IFRA distributions, and, more generally, for families of distributions satisfying $F \preceq_* W$, when conditions 4.22–4.26 hold. This is a consequence of the fact that nondecreasing functions $P^{\nearrow} f_{j:n}W$, $2 \le j \le n$ (see 4.1, 4.8) may be approximated in $L^2([a_W, d_W), w(x)dx)$ with any desired accuracy by sequences h_k, $k \ge 1$, of antistarshaped nondecreasing functions starting from sufficiently low levels $h_k(a)$. E.g., we can take

$$h_k(x) = \min\{\alpha_k(x - \beta_k), P^{\nearrow} f_{j:n}W(x)\},$$

where $\beta_k \searrow a$, and α_k are sufficiently large. Alternatively, we can also analyze the norm convergence of $h_k W^{-1}$ in $L^2([0,1), dx)$. For more details of reasoning we refer the reader to Rychlik [52]. Summarizing, we have

Theorem 11. *($F \preceq_* W$) For $2 \le j \le n-1$, $h = f_{j:n}W$ and $w = \frac{dW}{dx}$ satisfying 4.22–4.26, general bounds 4.11 are attained in limit by sequences of*

absolutely continuous distribution functions $F_k \preceq_* W$ whose quantile functions tend in $L^2([0,1), dx)$ to that of 4.12. In particular, the general bounds cannot be improved in the classes of distributions with increasing density and failure rate in average. Analogous conclusions hold for the sample extremes.

4.4 Symmetric Unimodal Distributions

For $j \le (n+1)/2$, we have $\mathrm{E}_F X_{j:n} \le \mu_F$ for all $F \succeq_s U$ which is guaranteed by the symmetry assumption only. Otherwise the bounds for standarized expectations of order statistics are nontrivial, because $\mathrm{E}_U(X_{j:n} - \mu_U)/\sigma_U = \sqrt{3}(\frac{2j}{n+1} - 1) > 0$. In the case $j = n$, equality in 4.14 holds for a symmetric unimodal distribution. In the remaining cases $(n+1)/2 < j < n$, we obtain the optimal bounds by projecting $s_{j:n}(x) = f_{j:n}(x) - f_{n+1-j:n}(x)$, $1/2 \le x < 1$, onto $\mathcal{C}^+_{\succeq_c 2U-1}$. Applying Lemma 8 for verification of 4.22–4.26, we are in a position to describe the shape of projections by means of Lemma 9. It immediately follows that for $(n+1)/2 < j \le \min\{n-2, [n+(3n-5)^{1/2}]/2\}$ and $j = n-1 \le 6$, the projection is linear, and the respective bounds are attained by uniform samples. In fact, the uniform distributions provide the optimal bounds for a wider range of order statistics. Necessary and sufficient conditions are presented in Theorem 12.

Theorem 12. (symmetric unimodal distributions) *For $(n+1)/2 < j \le n-1$, put*

$$K^s_U(x) = K_U(x) - K^-_U(x), \tag{4.78}$$

$$L^s_U(x) = L_U(x) - L^-_U(x), \tag{4.79}$$

for $1/2 \le x < 1$, where K_U, L_U are defined in 4.37, 4.38, respectively, and K^-_U, L^-_U are respective modifications of 4.37, 4.38 that consist in replacing j by $n+1-j$.

If L^s_U is positive on a right neighborhood of $1/2$, then

$$\mathrm{E}_F[X_{j:n} - \mu_F]/\sigma_F \le \sqrt{3}(\frac{2j}{n+1} - 1), \tag{4.80}$$

where equality holds for F uniformly distributed on $[\mu - \sqrt{3}\sigma, \mu + \sqrt{3}\sigma]$.

Otherwise, under notation 4.27 we obtain

$$\mathrm{E}_F(X_{j:n} - \mu(F))/\sigma(F) \le B = ||(s_{j:n})_{\alpha_*\beta_*}||/\sqrt{2}, \tag{4.81}$$

where

$$||(s_{j:n})_{\alpha_*\beta_*}||^2 = n\frac{\binom{2j-2}{j-1}\binom{2n-2j}{n-j}}{\binom{2n-1}{n}}[F_{2j-1:2n-1}(\beta_*) + F_{2n-2j+1:2n-1}(\beta_*)$$

$$
\begin{aligned}
&- F_{2j-1:2n-1}(1/2) - F_{2n-2j+1:2n-1}(1/2)] \\
&- 2n\frac{\binom{n-1}{j-1}^2}{\binom{2n-1}{n}}[F_{n:2n-1}(\beta_*) - 1/2] + (1-\beta_*)s_{j:n}^2(\beta_*) \\
&+ \alpha_*(1-\beta_*)^2 s_{j:n}(\beta_*) + \alpha_*^2(1-\beta_*)^3/3, \qquad (4.82)
\end{aligned}
$$

$$
\begin{aligned}
\alpha_* = \alpha_*(\beta_*) &= \frac{3j[F_{n+2-j:n+1}(\beta_*) - F_{j+1:n+1}(\beta_*)]}{(1-\beta_*)^3(n+1)} \\
&- \frac{3\beta_*[F_{n+1-j:n}(\beta_*) - F_{j:n}(\beta_*)]}{(1-\beta_*)^3} - \frac{3s_{j:n}(\beta_*)}{2(1-\beta_*)}, \qquad (4.83)
\end{aligned}
$$

$$
\beta_* = \min\{\beta > 1/2 : \; K_U^s(\beta)L_U^s(\beta) = 0\}. \qquad (4.84)
$$

The equality in 4.81 is achieved by

$$
F(x) = \begin{cases}
0, & if\ \frac{x-\mu}{\sigma} \le -\frac{s_{j:n}(\beta_*)+\alpha_*(1-\beta_*)}{\sqrt{2}B}, \\
1-\beta_* + \frac{s_{j:n}(\beta_*)}{\alpha_*} + \frac{\sqrt{2}B}{\alpha_*}\frac{x-\mu}{\sigma}, & if\ -\frac{s_{j:n}(\beta_*)+\alpha_*(1-\beta_*)}{\sqrt{2}B} \\
& \le \frac{x-\mu}{\sigma} \le -\frac{s_{j:n}(\beta_*)}{\sqrt{2}B}, \\
s_{j:n}^{-1}(\sqrt{2}B\frac{x-\mu}{\sigma}), & if\ -\frac{s_{j:n}(\beta_*)}{\sqrt{2}B} \le \frac{x-\mu}{\sigma} \le \frac{s_{j:n}(\beta_*)}{\sqrt{2}B}, \\
\beta_* - \frac{s_{j:n}(\beta_*)}{\alpha_*} + \frac{\sqrt{2}B}{\alpha_*}\frac{x-\mu}{\sigma}, & if\ \frac{s_{j:n}(\beta_*)}{\sqrt{2}B} \le \frac{x-\mu}{\sigma} \\
& \le \frac{s_{j:n}(\beta_*)+\alpha_*(1-\beta_*)}{\sqrt{2}B}, \\
1, & if\ \frac{x-\mu}{\sigma} \ge \frac{s_{j:n}(\beta_*)+\alpha_*(1-\beta_*)}{\sqrt{2}B}.
\end{cases}
\qquad (4.85)
$$

Analysis similar to that in the proof of Theorem 8 is applied here. Replacing $h = f_{j:n}$ by $h = s_{j:n} = f_{j:n} - f_{n+1-j:n}$ in 4.28 and 4.32 we write $D^s(\alpha, \beta)$, $K^s(\beta)$, $L^s(\beta)$, choosing optimal slopes 4.31 for various β, and try to determine $\beta \in [1/2, b)$ that minimizes $D^s(\alpha_*(\beta), \beta)$ under condition $K^s(\beta) \ge 0$. Since $K(\beta), L(\beta)$ of 4.32 are linear operators acting on h, we have

$$
\frac{dD^s(\alpha_*(\beta), \beta)}{d\beta} = 2K^s(\beta)L^s(\beta) = 2K_U^s(\beta)L_U^s(\beta)M(\beta)
$$

with a positive factor $M(\beta)$ independent of j (see 4.32 and comments following Theorem 9). A thorough analysis lead us to the conclusions that $K_U^s(\beta)$ is $+-$ on $(1/2, b)$, and $L_U^s(\beta)$ is either $+$ or $-+$ on $(1/2, c)$. If $L_U^s(1/2+) > 0$ then $\beta_* = 1/2$ is the solution which implies linearity of projection and resulting quantile function. Otherwise optimal $\beta_* > 1/2$ is the smaller of zeros of K_U^s and L_U^s. Then $P^+_{\succeq_c 2U-1} s_{j:n} = (s_{j:n})_{\alpha_*(\beta_*)\beta_*}$, and 4.82, 4.83, 4.85 are derived from general formulae by elementary calculations.

Table 2: Sharp uniform mean-variance bounds on expectations of order statistics from independent samples of size 20 for various families of distributions.

j	G	S	SUN	DDA	DFRA
10	0.5688	0	0	0.0253	0
11	0.6421	0.1777	0.0825	0.1944	0
12	0.7213	0.5015	0.2474	0.3602	0
13	0.8085	0.7469	0.4124	0.5245	0.0885
14	0.9071	0.9045	0.5773	0.6904	0.2419
15	1.0218	1.0015	0.7423	0.8631	0.4177
16	1.1605	1.0822	0.9073	1.0510	0.6253
17	1.3377	1.1810	1.0722	1.2691	0.8821
18	1.5845	1.3249	1.2464	1.5490	1.2256
19	1.9881	1.5736	1.5423	1.9773	1.7610
20	3.0424	2.2646	2.2646	3.0442	3.0301

Note that 4.85 has a density symmetric about μ_F, a finite support with uniform ends and a (principally infinite) peak at the center. The statement of Theorem 12 is weaker than those of Theorems 8 and 9 in that we were not able to determine explicitly the pairs (j, n) for which $L(a) > 0$, and resulting optimal bounds are determined by the minimal distributions W in the class with respect to the ordering.

In Table 2 numerical evaluations of mean-variance bounds are presented for the jth order statistics, $10 \leq j \leq 20$, of i.i.d. samples of size $n = 20$, coming from general (G), symmetric (S), symmetric unimodal (SUN) populations, and those with decreasing density and failure rate in average (DDA and DFRA, respectively). For $j \leq 9$, all bounds are trivial except of the first case. The values of the third column were presented in Gajek and Rychlik [19, Table III]. Numerical bounds for the DDA and DFRA samples of size 15 can be found in Rychlik [52].

PART II

5 Order Statistics of Dependent Observations

We assume that $Y_1, \ldots, Y_n$ are possibly dependent and identically distributed. Recalling arguments of Rychlik [44], in Subsection 5.1 we conclude sharp bounds 2.22 and 2.24 on expectations of general L-statistics and single order statistics, respectively, depending on the common marginal distribution of the observations. Next we apply the projections of functionals defined in 2.22

and 2.24 for establishing respective moment bounds over general and restricted families of marginals. General, symmetric and nonnegative observations are treated in Subsection 5.2. The results are cited from Rychlik [46], but some earlier partial solutions are also mentioned. In the reminder of Section 5 we confine ourselves to single order statistics, pointing out some extensions at the end of Subsection 5.5. In Subsection 5.3 we present mean-variance and second moment bounds on the expectations of order statistics for families of parent distributions related to a given one in the convex ordering. The mean-variance bounds for $F \succeq_c W$ were not published elsewhere, except of for DFR distributions, given in Rychlik [51]. The results for general W and $W = U$ (i.e. for the decreasing density distributions) are presented in Theorem 15 and Proposition 8, respectively. We further cite bounds for DFR distributions from Rychlik [51]. Rychlik [49] established second moment bounds for $F \succeq_c (\preceq_c) W$ with general W. The decreasing density and failure rate were studied in Gajek and Rychlik [18]. Analogous results for increasing ones come from Rychlik [49]. Subsection 5.4 deals with mean-variance bounds for order statistics based on samples with common marginal distributions being in a star relation with a fixed one W. For ones determined by the exponential $W = V$, which includes important classes of life distributions with monotone failure rate, respective bounds were established in Rychlik [51]. Especially, it was shown that general bounds 5.13 are attained by the IFRA distributions. In fact, the claim can be extended to families of distributions defined by $F \preceq_* W$ for general W. We also write explicitly the bounds for $F \succeq_* U$ which have decreasing densities in average. The mean-variance bounds for order statistics with symmetric unimodal and U-shaped distributions of parent variables, described in Subsection 5.5, were obtained in Gajek and Rychlik [18] and Rychlik [49], respectively. We call a symmetric distribution U-shaped if it has nonincreasing and nondecreasing density on the lower and upper halves of its support. The extreme deviation of expected order statistics under violating the independence assumption is evaluated in Subsection 5.6. For some quantile bounds on order statistics of dependent observations, we refer the reader to Rychlik [49, Section 5].

5.1 Dependent Observations with Given Marginal Distribution

Suppose that arbitrarily dependent $Y_1, \ldots, Y_n$ have a fixed common distribution function F with a finite mean μ_F. We aim in justifying bound 2.22 for the expectation of arbitrary combination of order statistics under all possible interdependencies of observations and specifying the conditions of its attainability. For the formal proof we refer the reader to Rychlik [44] (see also Rychlik [49]).

First we characterize vectors $(\tilde{G}_{1:n}, \ldots, \tilde{G}_{n:n})$ of all possible distribution functions of order statistics $Y_{1:n}, \ldots, Y_{n:n}$ by relations

$$\sum_{j=1}^{n} \tilde{G}_{j:n} = nF, \tag{5.1}$$

$$\tilde{G}_{1:n} \geq \tilde{G}_{2:n} \geq \ldots \geq \tilde{G}_{n:n}. \tag{5.2}$$

The former immediately follows from

$$\sum_{j=1}^{n} \mathbf{1}_{(-\infty,x]}(Y_{j:n}) = \sum_{j=1}^{n} \mathbf{1}_{(-\infty,x]}(Y_j), \qquad x \in \Re, \tag{5.3}$$

by taking expectations of both sides of 5.3. The latter is a consequence of relations $Y_{1:n} \leq Y_{2:n} \leq \ldots \leq Y_{n:n}$. There are many ways of constructing ordered variables $Y_{j:n}$, $1 \leq j \leq n$, with distributions satisfying 5.1 and 5.2 (the simplest one consists in taking $\tilde{G}_{j:n}^{-1}(X)$, $1 \leq j \leq n$, for X being a standard uniform variable). There are also many ways of constructing identically F-distributed random variables Y_j, $1 \leq j \leq n$, whose order statistics have given distribution functions satisfying 5.1–5.2 (the simplest one consists in random rearranging $Y_{j:n}$, $1 \leq j \leq n$).

Now we solve the problem of minimizing $\sum_{j=1}^{n} c_j \tilde{G}_{j:n}(x)$ for fixed coefficients $\mathbf{c} = (c_1, \ldots, c_n)$ of the L-statistic under study and distribution functions satisfying 5.1–5.2 valued at arbitrary point x. This is a linear programming problem that has the solution

$$\min \sum_{j=1}^{n} c_j \tilde{G}_{j:n}(x) = G_{\mathbf{c}} F(x) \tag{5.4}$$

for $G_{\mathbf{c}}$ being the greatest convex function on the unit interval satisfying 2.23. Recalling Lemma 1, we obtain

$$\begin{aligned} \mathrm{E}_F \sum_{j=1}^{n} c_j Y_{j:n} &= \int_{-\infty}^{+\infty} y \left(\sum_{j=1}^{n} c_j \tilde{G}_{j:n}(x) \right) \\ &\leq \int_{-\infty}^{+\infty} y G_{\mathbf{c}} F(dy) = \int_0^1 F^{-1}(x) G_{\mathbf{c}}(dx) \\ &= \int_0^1 F^{-1}(x) g_{\mathbf{c}}(x)\, dx = (F^{-1}, g_{\mathbf{c}}), \end{aligned} \tag{5.5}$$

which is the desired conclusion.

Analyzing values $\tilde{G}_{j:n}(x)$, $1 \le j \le n$, $x \in \Re$, for the solutions to 5.4, we are able to determine supports and some mutual relations of respective order statistics. They depend on properties of function $G_{\mathbf{c}}$, whose graph is a (broken) line with breaks (if any) at some multiplicities of $1/n$. Let $0 = k_0 < k_1 < \ldots < k_\kappa = n$, $1 \le \kappa \le n$, be the sequence of integers such that each k_i/n, $1 \le i \le \kappa - 1$, is a break point of $G_{\mathbf{c}}$. Let $0 = l_0 < l_1 < \ldots < l_\lambda = n$, $\kappa \le \lambda \le n$, be the sequence of integers for which $G_{\mathbf{c}}(j/n) = \sum_{k=1}^{n} c_k$ holds. We have $\{0, n\} \subset \mathcal{K} = \{k_i : 1 \le i \le \kappa\} \subset \mathcal{L} = \{l_i : 1 \le i \le \lambda\} \subset \{0, \ldots, n\}$. Then equality in 5.4 for all $x \in \Re$ and some $\tilde{G}_{j:n}$ satisfying 5.1–5.2 implies

$$\mathrm{P}(F^{-1}(k_{i-1}/n) \le Y_{l_{j-1}+1:n} = Y_{l_j:n} \le F^{-1}(k_i/n)) = 1 \tag{5.6}$$

for all $1 \le i \le \kappa$, and $1 \le j \le \lambda$ such that $k_{i-1} \le l_{j-1} < l_j \le k_i$. On one hand, relation $\mathcal{L} = \mathcal{K}$ uniquely determines the distributions of all order statistics that solve 5.4 and attain equality in 5.5. Then $\tilde{G}_{j:n}(x) = [nF(x) - k_{i-1}]/(k_i - k_{i-1})$, $F^{-1}(k_{i-1}/n) \le x < F^{-1}(k_i/n)$, for $k_{i-1} < j \le k_i$. If $\mathcal{L} \ne \mathcal{K}$, there are also other solutions. For instance, for the sample mean we have $G_{\mathbf{c}}(x) = x$, $\mathcal{K} = \{0, n\}$, $\mathcal{L} = \{0, \ldots, n\}$, and 5.6 simply means that each $Y_{j:n}$ should belong to the domain of Y_1. Indeed, $\mathrm{E}_F \frac{1}{n} \sum_{j=1}^{n} Y_{j:n} = \int_0^1 F^{-1}(x)\,dx = \mu_F$ for any type of dependence in variables.

In the special case of single order statistics $Y_{j:n}$, functions $G_{\mathbf{c}}$ and $g_{\mathbf{c}}$ have forms $G_{j:n}(x) = \frac{n}{n+1-j}(x - \frac{j-1}{n})_+$, $g_{j:n}(x) = \frac{n}{n+1-j}\mathbf{1}_{[\frac{j-1}{n},1)}(x)$, respectively, which implies 2.24. Since $\mathcal{K} = \{0, j-1, n\} \subset \mathcal{L} = \{0, 1, \ldots, j-1, n\}$,

$$\mathrm{P}(Y_{j-1:n} \le F^{-1}((j-1)/n) \le Y_{j:n} = Y_{n:n}) = 1 \tag{5.7}$$

combined with 5.1–5.2 are respective conditions for equality. The stochastically largest (i.e. uniformly smallest) distribution function $(nF + 1 - j)_+/(n + 1 - j)$ of jth order statistic is uniquely determined. There are various ways of constructing dependent F-distributed samples with the extreme distribution of $Y_{j:n}$. The simplest one is a random rearrangement of $Y_{1:n} = \ldots = Y_{j-1:n} = F^{-1}(\frac{j-1}{n}X) \le F^{-1}(\frac{j-1}{n}) \le F^{-1}(\frac{j-1}{n} + (1 - \frac{j-1}{n})X) = Y_{j:n} = \ldots = Y_{n:n}$ for some X uniformly distributed on $[0, 1]$. For the sample minimum, 2.24 and 5.7 imply the trivial claims $\mathrm{E}_F Y_{1:n} \le \mu_F$, becoming equality for identical observations $Y_{1:n} = Y_{n:n} = Y_1$. Condition 5.7 excludes the possibility of constructing an absolutely continuous joint distribution of the sample with stochastically maximal jth order statistic except of the sample maximum.

In the first paper in this field of research Mallows [30] constructed a density function of the sample with identical uniform marginals that provide the maximal expectation of the sample maximum. Lai and Robbins [27] extended the construction to arbitrary possibly nonidentical marginal distributions. Lai and

Robbins [28] and Tchen [60] constructed infinite sequences of variables with identical and arbitrary distributions, respectively, such that all sample maxima are stochastically maximal. Bounds 2.24 for general order statistics of identically distributed samples were proved independently in Caraux and Gascuel [12] and Rychlik [42]. In the former, some inequalities for nonidentically distributed observations were presented (with conditions for equality established in Rychlik [47]). In the latter, the problem of constructing sequences with stochastically extreme order statistics was also discussed. Asymptotic properties of sequences of stochastically extreme maxima and other order statistics were studied in Lai and Robbins [28] and Rychlik [43], respectively.

In the further parts of this Section we present tight bounds for expected order and L-statistics of dependent samples for various families of marginal distributions. We also indicate the marginals for which the bounds are attained. Formally, in each case one should say that these are attained by the joint distributions specified for arbitrary marginal and fixed L-statistic by 5.1–5.2 with 5.6 (replaced by 5.7 for single order statistics in particular), and the given marginal. However, we avoid the repeative reference to the construction of the joint probability and confine ourselves on describing the optimal marginal.

5.2 General and Symmetric Distributions

We first study the optimal bounds for general L-statistics based on dependent samples with arbitrary common marginal distribution F. For arbitrarily fixed sequence of coefficients $\mathbf{c} = (c_1, \ldots, c_n)$, we construct function $g_{\mathbf{c}}$ that defines the functional of expectation of respective L-statistic (see 2.22 and 5.5) by differentiating function $G_{\mathbf{c}}$ determined by 2.23. Using the notation of Subsection 5.1, we have

$$g_{\mathbf{c}} = \sum_{j=1}^{n} d_j \mathbf{1}_{[(j-1)/n, j/n)} = \sum_{i=1}^{\kappa} d_{k_i} \mathbf{1}_{[k_{i-1}/n, k_i/n)} \tag{5.8}$$

for a nondecreasing sequence d_j, $1 \le j \le n$, with the specified increasing subsequence d_{k_i}, $1 \le i \le \kappa \le n$, of distinct values. Precisely, we have

$$d_j = d_{k_i} = \frac{n\left[G_{\mathbf{c}}\left(\frac{k_i}{n}\right) - G_{\mathbf{c}}\left(\frac{k_{i-1}}{n}\right)\right]}{k_i - k_{i-1}} = \frac{n}{k_i - k_{i-1}} \sum_{k=k_{i-1}+1}^{k_i} c_k, \tag{5.9}$$

$k_{i-1} + 1 \le j \le k_i$, $1 \le i \le \kappa$. Also, put

$$\bar{d} = \frac{1}{n} \sum_{j=1}^{n} d_j = \sum_{j=1}^{n} c_j = G_{\mathbf{c}}(1) = \int_0^1 g_{\mathbf{c}}(x)\, dx. \tag{5.10}$$

We claim that the optimal mean-variance bound for the expectation of a given L-statistic amounts to the Euclidean norm of respective vector $\frac{1}{n}(d_j - \bar{d})$, $1 \leq j \leq n$.

Theorem 13. (general distributions) *Under notation 5.8–5.10 with $G_{\mathbf{c}}$ defined in 2.23, we have*

$$\mathrm{E}_F \sum_{j=1}^{n} c_j (Y_{j:n} - \mu_F)/\sigma_F \leq C = C^0(\mathbf{c}) = \left[\frac{1}{n} \sum_{j=1}^{n} d_j^2 - \bar{d}^2 \right]^{1/2}. \tag{5.11}$$

If $C = 0$, then 5.11 becomes equality for degenerate distributions. Otherwise equality holds for the κ-point marginal distribution

$$\mathrm{P}(Y_1 = \mu + \sigma(d_{k_i} - \bar{d})/C) = (k_i - k_{i-1})/n, \qquad 1 \leq i \leq \kappa. \tag{5.12}$$

Verification of 5.11

$$\begin{aligned}
\mathrm{E}_F \sum_{j=1}^{n} c_j (Y_{j:n} - \mu) &= \int_0^1 F^{-1}(x)[g_{\mathbf{c}}(x) - \bar{d}]\, dx \\
&= \int_0^1 [F^{-1}(x) - \mu][g_{\mathbf{c}}(x) - \bar{d}]\, dx \\
&\leq \left\{ \int_0^1 [F^{-1}(x) - \mu]^2 dx \int_0^1 [g_{\mathbf{c}}(x) - \bar{d}]^2 dx \right]^{1/2} = C\sigma
\end{aligned}$$

is based on 2.22, 5.10, 2.7 and 5.8. The Schwarz inequality provides the sharp bound here, because $g_{\mathbf{c}} - \bar{d}$ is nondecreasing, and so $F^{-1} - \mu = \alpha(g_{\mathbf{c}} - \bar{d})$ with $\alpha = \sigma/C > 0$ actually defines a quantile function with desired mean and variance. An elementary algebra enables us to derive respective distribution 5.12. Especially, for the single order statistics we have

Corollary 1. (general distributions) *Inequality*

$$\frac{\mathrm{E}_F Y_{j:n} - \mu_F}{\sigma_F} \leq C = C^0(j, n) = \left(\frac{j-1}{n+1-j} \right)^{1/2} \tag{5.13}$$

is sharp and becomes equality for the two-point marginal

$$\mathrm{P}(Y_1 = \mu - \sigma/C) = (j-1)/n = 1 - \mathrm{P}(Y_1 = \mu + \sigma C). \tag{5.14}$$

Bounds 5.13 for the sample maximum and general j were presented in Arnold [1] and Gascuel and Caraux [20], respectively. We proceed now to present analogous results for symmetrically distributed random variables.

Theorem 14. (symmetric distributions) *Inequality*

$$\mathrm{E}_F \sum_{j=1}^{n} c_j \frac{Y_{j:n} - \mu_F}{\sigma_F} \le C^s(\mathbf{c}) = \left[\frac{1}{2n} \sum_{j=\lfloor (n+3)/2 \rfloor}^{n} (d_j - d_{n+1-j})^2 \right]^{\frac{1}{2}}, \tag{5.15}$$

where $\lfloor \cdot \rfloor$ denotes the floor of a number, is tight. Equality is attained by a unique (up to location-scale transformations) marginal symmetric distribution supported on n points at most whose probabilities are multiplicities of $1/n$.

Inequality 5.15 follows from

$$\mathrm{E}_F \sum_{j=1}^{n} c_j (Y_{j:n} - \mu) = \int_{1/2}^{1} [F^{-1}(x) - \mu] g^s_{\mathbf{c}}(x)\, dx \tag{5.16}$$

with

$$g^s_{\mathbf{c}}(x) = g_{\mathbf{c}}(x) - g_{\mathbf{c}}(1 - x-) = \sum_{j=\lfloor (n+3)/2 \rfloor}^{n} (d_j - d_{n+1-j}) \mathbf{1}_{[\frac{j-1}{n}, \frac{j}{n})}(x), \tag{5.17}$$

$1/2 \le x < 1$, and the Schwarz inequality. Note that 5.17 is a nonnegative nondecreasing piecewise constant function with $\lfloor (n+1)/2 \rfloor$ values at most. Thus its antisymmetric extension onto $[0, 1/2)$ is also nondecreasing step function with n values at most and jumps at some points j/n, $1 \le j \le n-1$. The extension coincides with the quantile function of the extreme distribution providing equality in 5.15 up to an affine transformation. This justifies the latter claim of Theorem 14. We omitted formal presentation of the extreme distribution, because this needs introducing rather a complicated notation. Although the algorithm of determining $g_{\mathbf{c}}$ and $g^s_{\mathbf{c}}$ is simple, we are not able to write explicitly respective formulae for general combinations of order statistics. For single order statistics yields

$$g^s_{j:n}(x) = \frac{n}{n+1-j} \mathbf{1}_{[\max\{\frac{j-1}{n}, 1 - \frac{j-1}{n}\}, 1)}(x), \qquad 1/2 \le x < 1. \tag{5.18}$$

By an easy computation we conclude

Corollary 2. (symmetric distributions) *Inequality*

$$\frac{\mathrm{E}_F Y_{j:n} - \mu_F}{\sigma_F} \le C = \left[\frac{n}{2(n+1-j)} \min\left\{ \frac{j-1}{n+1-j}, 1 \right\} \right]^{1/2} \tag{5.19}$$

is sharp and becomes equality for the three-point marginal distribution

$$\mathrm{P}\,(Y_1 = \mu) = 2\left|\frac{j-1}{n} - \frac{1}{2}\right|, \tag{5.20}$$

$$\mathrm{P}\,(Y_1 = \mu \pm \sigma C) = \min\left\{\frac{j-1}{n}, 1 - \frac{j-1}{n}\right\}. \tag{5.21}$$

Inequality 5.19 and its special case for $j = n$ can be found in Gascuel and Caraux [20] and Arnold [1], respectively. Here we merely mention the second moment bounds for general L-statistics

$$\mathrm{E}_F \sum_{j=1}^{n} c_j Y_{j:n} \le \int_0^1 F^{-1}(x)[g_{\mathbf{c}}(x)]_+ \, dx$$

$$\le ||(g_{\mathbf{c}})_+|| m_F = \left[\frac{1}{n}\sum_{j=1}^{n}(d_j)_+^2\right]^{1/2} m_F \tag{5.22}$$

and for single order statistics

$$\mathrm{E}_F Y_{j:n} \le \left(\frac{n}{n+1-j}\right)^{1/2} m_F. \tag{5.23}$$

of nonnegative samples. Bounds 5.22 and 5.23 are attained by

$$\mathrm{P}(Y_1 = 0) = \frac{k_{i_0}}{n}, \tag{5.24}$$

$$\mathrm{P}\left(Y_1 = m_F \frac{d_{ki}}{||(g_{\mathbf{c}})_+||}\right) = \frac{k_i - k_{i-1}}{n}, \qquad i_0 < i \le \kappa, \tag{5.25}$$

with $i_0 = \max\{0 \le i \le \kappa : \ d_{k_i} \le 0\}$, and

$$\mathrm{P}(Y_1 = 0) = \frac{j-1}{n} = 1 - \mathrm{P}\left(Y_1 = \left[\frac{n}{n+1-j}\right]^{1/2} m_F\right), \tag{5.26}$$

respectively. If $d_n \le 0$ and so $||(g_{\mathbf{c}})_+|| = 0$, then 5.24–5.25 is clearly replaced by the Dirac measure at 0. In fact, 5.24–5.25 and 5.26 are not unique solutions. The second relation in 5.22 shows that F^{-1} may have various forms on $[0, k_{i_0}/n)$ provided that nonnegativity, nondecrease and moment conditions are not violated (see Rychlik [46] for more details).

Rychlik [46] (and Arnold [2] for the cases of sample maximum $Y_{n:n}$ and range $Y_{n:n} - Y_{1:n}$) presented more general sharp bounds in terms of central absolute moments of various orders based on the Hölder inequality instead of the Schwarz one. These are also attainable by discrete marginal distributions with probabilities k_i/n for some integer k_i. This form of solution allows us to deduce analogous inequalities for deterministic sequences. Randomly rearranging a sequence of (not necessarily distinct) numbers $y_1, \ldots, y_n$, we obtain a random sequence of dependent identically distributed random variables with expectation $\bar{y} = \frac{1}{n}\sum_{j=1}^{n} y_j$, second raw moment $m^2 = \frac{1}{n}\sum_{j=1}^{n} y_j^2$, variance $s^2 == \frac{1}{n}\sum_{j=1}^{n}(y_j - \bar{y})^2 = m^2 - \bar{y}^2$, and deterministic order statistics $y_{1:n} \leq \ldots \leq y_{n:n}$. Using 5.11, 5.15 and 5.22, we conclude optimal bounds

$$\sum_{j=1}^{n} c_j(y_{j:n} - \bar{y})/s \leq \left[\frac{1}{n}\sum_{j=1}^{n} d_j^2 - \bar{d}^2\right]^{1/2}, \tag{5.27}$$

$$\sum_{j=1}^{n} c_j(y_{j:n} - \bar{y})/s \leq \left[\frac{1}{2n}\sum_{j=\lfloor(n+3)/2\rfloor}^{n} (d_j - d_{n+1-j})^2\right]^{1/2}, \tag{5.28}$$

$$\sum_{j=1}^{n} c_j y_{j:n}/m \leq \left[\frac{1}{n}\sum_{j=1}^{n}(d_j)_+^2\right]^{1/2}, \tag{5.29}$$

for general, symmetric and nonnegative sequences of numbers, respectively. E.g., sharpness of 5.27 is verified by putting $y_j = d_j$, $1 \leq j \leq n$, and reference to 5.12.

The problem of best deterministic bounds for L-statistics in terms of various sample parameters has a long history. We confine ourselves to mention these derived by means of the Schwarz inequality. Samuelson [55] raised and solved the problem of how much can a single observation deviate from the sample mean in the standard deviation units. Samuelson's paper stimulated intensive investigations of the problem and its modifications: alternative proofs, rediscoveries of earlier results, and extensions. Six different proofs were reviewed by Arnold and Balakrishnan [4]. The earliest proofs found in literature were due to Thompson [61] and Scott [57]. We do not attempt to present a complete record of consecutive contributions, referring the reader to Arnold [3] for a comprehensive bibliography, and Olkin [38] for a recent review, with yet another proof. Scott [57] established the bound for deviations of $y_{n-1:n}$ from the mean in the standard deviation units. The bounds for arbitrary order statistics follow directly from Mallows and Richter [31], and were explicitly stated by Boyd [11] and Hawkins [25]. Mallows and Richter [31] established the inequalities for selection differentials $\frac{1}{j}\sum_{i=1}^{j} y_{i:n}$, and $\frac{1}{k}\sum_{i=n+1-k}^{n} y_{i:n}$, and their difference. The respective results for $y_{n:n} - y_{1:n}$, $y_{n-1:n} - y_{1:n}$, and the difference of arbitrary order statistics were derived by Nair [37], David *et al.* [15], and Fahmy and Proschan [17] (implicitly in Arnold and Groeneveld [5]), respectively, and for the L-estimates with nondecreasing coefficients by David [14]. Bounds 5.27–5.29 for general L-statistics come from Rychlik [46].

5.3 Distributions with Monotone Density and Failure Rate

Determining mean-variance bounds for $F \succeq_c W$ by the projection method, we look for the element of $\mathcal{C}^0_{\succeq_c W}$ least distant from function $h = \frac{n}{n+1-j}\mathbf{1}_{[(j-1)/n,1)}W = \frac{n}{n+1-j}\mathbf{1}_{[W^{-1}((j-1)/n),d_W)}$. Applying Lemmas 4 and 5 with $b = W^{-1}(\frac{j-1}{n})$, $c = d_W$ yields the solution. Note that there are no trivial zero projections and bounds for $b > a_W$, because

$$\mathrm{E}_W X < \mu_W(b) = \mathrm{E}_W(X|X > b)$$

contradicts 3.7. Hence the projection is linear increasing with a possible constant left part, and the condition for distinguishing the cases is given in the last part of Lemma 5.

Theorem 15. *($F \succeq_c W$) If for $b = W^{-1}(\frac{j-1}{n})$, we have*

$$\mu_W(b) \le \mu_W + \sigma_W^2/(\mu_W - a_W), \tag{5.30}$$

then inequality

$$(\mathrm{E}_F Y_{j:n} - \mu_F)/\sigma_F \le [\mu_W(b) - \mu_W]/\sigma_W \tag{5.31}$$

holds true and becomes equality for

$$F(x) = W(\mu_W + \sigma_W \frac{x-\mu}{\sigma}). \tag{5.32}$$

Otherwise there exists a unique $\beta \in (a_W, b)$ satisfying

$$\sigma_W^2(\beta) = [\mu_W(b) - \mu_W(\beta)][\mu_W(\beta) - \beta] \tag{5.33}$$

such that

$$\frac{\mathrm{E}_F Y_{j:n} - \mu_F}{\sigma_F} \le C^0_{\succeq_c W}(j,n) = \frac{\hat{\eta}_W(\beta)}{\hat{\vartheta}_W(\beta)} \tag{5.34}$$

for

$$\hat{\eta} = \hat{\eta}_W(\beta) = \int_\beta^d (x-\beta)w(x)\,dx = [\mu_W(\beta)-\beta][1-W(\beta)], \tag{5.35}$$

$$\begin{aligned}\hat{\vartheta}^2 = \hat{\vartheta}^2_W(\beta) &= \int_\beta^d (x-\beta)^2 w(x)\,dx - \left[\int_\beta^d (x-\beta)w(x)\,dx\right]^2 \\ &= [1-W(\beta)]\{\sigma^2_W(\beta) - W(\beta)[\mu_W(\beta)-\beta]\}.\end{aligned} \tag{5.36}$$

Bound 5.34 is attained by

$$F(x) = W(\beta + \hat{\eta} + \hat{\vartheta}\frac{x-\mu}{\sigma})\mathbf{1}_{[\mu-\sigma\hat{\eta}/\hat{\vartheta},\infty)}(x). \tag{5.37}$$

Note that $\sup \mathrm{E}_W Y_{j:n} = \mu_W(b)$ which confirms the first statement of Theorem 15. Condition 5.30 is certainly satisfied by small order statistics. Definitions 5.35 and 5.36 are analogues of the negative of 3.32 and 3.33, respectively, valued at (β,β). Distribution function 5.37 has a jump of height $W(\beta) < \frac{j-1}{n}$ at $\mu - \sigma\hat{\eta}/\hat{\vartheta}$ and density $\frac{\hat{\vartheta}}{\sigma}w(\beta+\hat{\eta}+\hat{\vartheta}\frac{x-\mu}{\sigma})$ right to the point.

Proposition 8. (decreasing density) *If* $\frac{j-1}{n} \le \frac{1}{3}$, *then*

$$(\mathrm{E}_F Y_{j:n} - \mu_F)/\sigma_F \le \sqrt{3}\frac{j-1}{n}, \tag{5.38}$$

which becomes equality for F *being the uniform distribution function on* $[\mu - \sqrt{3}\sigma, \mu + \sqrt{3}\sigma]$.

Otherwise

$$\frac{\mathrm{E}_F Y_{j:n} - \mu_F}{\sigma_F} \le \frac{1}{3}\left(\frac{9j-9-n}{n+1-j}\right)^{1/2}, \tag{5.39}$$

and the equality holds for the mixture of the atom at $\mu - 3\sigma\left(\frac{n+1-j}{9j-9-n}\right)^{1/2}$ *and the uniform distribution on* $[\mu - 3\sigma\left(\frac{n+1-j}{9j-9-n}\right)^{1/2}, \mu + \sigma\frac{3j-3+n}{[(n+1-j)(9j-9-n)]^{1/2}}]$ *with respective probabilities* $\frac{3}{2}\frac{j-1}{n} - \frac{1}{2}$ *and* $\frac{3}{2}(1-\frac{j-1}{n})$.

Proposition 9. (decreasing failure rate) *If* $\frac{j-1}{n} \le 1 - e^{-1} \approx 0.63212$, *then*

$$\frac{\mathrm{E}_F Y_{j:n} - \mu_F}{\sigma_F} \le \ln\frac{n}{n+1-j}, \tag{5.40}$$

and the equality holds in 5.40 for the exponential distribution with location $\mu - \sigma$ and scale σ.

Otherwise for

$$\gamma = \gamma_V(j,n) = (1 - \frac{j-1}{n})e,$$

we have

$$(\mathrm{E}_F Y_{j:n} - \mu_F)/\sigma_F \le C = C^0_{\succeq_c V}(j,n) = [2/\gamma_V(j,n) - 1]^{1/2}, \tag{5.41}$$

which becomes equality for

$$F(x) = 1 - \gamma \exp(-\gamma - \gamma C \frac{x-\mu}{\sigma}) \mathbf{1}_{[\mu - \sigma/C, \infty)}(x). \tag{5.42}$$

Note that 5.42 is the mixture of the exponential distribution with location $\mu - \sigma/C$ and scale $\sigma/(\gamma C)$, with probability γ, and the Dirac measure at $\mu - \sigma/C$ with probability $1-\gamma$. This is a DFR life distribution if $\mu = \sigma/C$. For small order statistics 5.40 is attained by a life exponential distribution with $\mu = \sigma$.

Rychlik [51, Lemmas 4 and 5] proved that the projection of $h = \frac{n}{n+1-j}\mathbf{1}_{[(j-1)/n,1)}V$ onto $\mathcal{C}^{\nearrow}_{\preceq_c V}$ has form

$$P^{\nearrow}_{\preceq_c V} h(x) = \frac{n}{n+1-j} + \min\{\alpha(x-\beta), 0\} \tag{5.43}$$

for unique $\beta > \ln \frac{n}{n+1-j}$ satisfying

$$\frac{1-(1+\beta)e^{-\beta}}{e^{-\beta} - 1 + \beta} = \frac{n+1-j}{j-1} \ln \frac{n}{n+1-j} \tag{5.44}$$

and

$$\alpha = \alpha(\beta) = \frac{j-1}{n+1-j} \frac{1}{e^{-\beta} - 1 + \beta}. \tag{5.45}$$

This enables us to estimate order statistics of dependent IFR samples.

Proposition 10. (increasing failure rate) *Set*

$$\theta^2 = \theta^2_V(\beta) = 1 - 2\beta e^{-\beta} - e^{-2\beta} \tag{5.46}$$

for β defined in 5.44. Then sharp bound

$$\frac{\mathrm{E}_F Y_{j:n} - \mu_F}{\sigma_F} \le \frac{j-1}{n+1-j} \frac{\theta_V(\beta)}{1-(1+\beta)e^{-\beta}}, \tag{5.47}$$

Table 3: Sharp uniform mean-variance bounds on quantiles for various families of distributions.

p	G	S	SUN	DD	DDA	DFR	DFRA
0.45	0.90453	0	0	0	0.135108	0	0
0.50	1	1	?	0	0.32163	0	0
0.55	1.10554	1.05409	?	0.17321	0.50913	0	0
0.60	1.22474	1.11803	?	0.34641	0.70235	0	0.15541
0.65	1.36277	1.19523	?	0.51962	0.90763	0.04982	0.33725
0.70	1.52753	1.29099	?	0.69389	1.13406	0.20397	0.53901
0.75	1.73205	1.41421	?	0.88192	1.39582	0.38629	0.77040
0.80	2	1.58114	?	1.10554	1.71781	0.60944	1.05060
0.85	2.38048	1.82574	1.21716	1.40106	2.15063	0.89712	1.42081
0.90	3	2.23607	1.49071	1.85592	2.82311	1.30641	1.98883
0.95	4.35890	3.16228	2.10819	2.80872	4.24076	2.10081	3.18532

holds, and is achieved by

$$F(x) = \begin{cases} 0, & if\ \frac{x-\mu}{\sigma} \le -\frac{1-e^{-\beta}}{\theta}, \\ 1-\exp(-1+e^{-\beta}-\theta\frac{x-\mu}{\sigma}), & if\ -\frac{1-e^{-\beta}}{\theta} \le \frac{x-\mu}{\sigma} < \frac{e^{-\beta}-1+\beta}{\theta}, \\ 1, & if\ \frac{x-\mu}{\sigma} \ge \frac{e^{-\beta}-1+\beta}{\theta}. \end{cases} \quad (5.48)$$

This is the right truncated at $\mu + \frac{\sigma}{\theta}(e^{-\beta} - 1 + \beta)$ exponential distribution with location and scale parameters $\mu - \frac{\sigma}{\theta}(1 - e^{-\beta})$ and $\frac{\sigma}{\theta}$,respectively. The jump probability is $e^{-\beta}$.

Likewise, for determining respective second moment bounds for life distributions $F \succeq_c W$ and $F \preceq_c W$ we need to find best approximations of $\frac{n}{n+1-j}\mathbf{1}_{[(j-1)/n,1)}W$ by nondecreasing convex and concave functions that additionally obey the condition of vanishing at a_W. We assume here that W is a life distribution and $a_W = 0$. The respective projections have similar parametric forms $\alpha(x-\beta)_+$ and $\frac{n}{n+1-j}\min\{\frac{x}{\alpha}, 1\}$. In the first case, under conditions of the latter statement of Theorem 15, the optimal parameter β defined in 5.33 is positive which implies that $P^{\nearrow}_{\succeq_c W}h \in \mathcal{C}^+_{\succeq_c W}$ provides the solution to our problem. Then we have

$$||P^+_{\succeq_c W}h||^2 = ||P^0_{\succeq_c W}h||^2 + 1 = \frac{\int_\beta^d (x-\beta)^2 w(x)\,dx}{\left[\int_\beta^d (x-\beta)w(x)\,dx\right]^2} \quad (5.49)$$

Table 4: Sharp uniform mean-variance bounds on expectations of order statistics from independent samples of size 20 for various families of distributions.

j	G	S	SUN	DDA	DFRA
10	0.56881	0	0	0.0252617	0
11	0.642107	0.17773	0.0825	0.19442156	0
12	0.721266	0.501546	0.2474	0.36023849	0
13	0.808549	0.746867	0.4124	0.52453361	0.0884843
14	0.907143	0.904474	0.5773	0.69044574	0.241894
15	1.02182	1.00151	0.7423	0.86309839	0.417728
16	1.16054	1.08224	0.9073	1.0509551	0.62534
17	1.33774	1.18095	1.0722	1.2691209	0.882055
18	1.58450	1.32488	1.2464	1.5490172	1.22562
19	1.98814	1.57364	1.5423	1.9773083	1.76097
20	3.04243	2.26455	2.2646	3.04423	3.03006

Table 5: Sharp uniform mean-variance bounds on expectations of order statistics from dependent samples of size 20 for various families of distributions.

j	G	S	SUN	SUS	DD	DFR	DDA	DFRA
2	0.22942	0.16644	0.15692	0.08660	0.08660	0.05129	0.09028	0.05250
3	0.33333	0.24845	0.23424	0.17321	0.17321	0.10536	0.18543	0.10999
4	0.42008	0.32219	0.30376	0.25981	0.25981	0.16252	0.28232	0.17247
5	0.50000	0.39529	0.37269	0.34641	0.34641	0.22314	0.37897	0.23995
6	0.57735	0.47141	0.44444	0.43341	0.43301	0.28768	0.47458	0.31254
7	0.65465	0.55328	0.52164	0.52398	0.51962	0.35668	0.56934	0.39046
8	0.73380	0.64359	0.60622	0.62188	0.60624	0.43078	0.66414	0.47409
9	0.81650	0.74536	0.69282	0.73080	0.69389	0.51083	0.76038	0.56408
10	0.90453	0.86244	0.77942	0.85499	0.78496	0.59784	0.85983	0.66139
11	1.00000	1.00000	0.86603	1.00000	0.88192	0.69318	0.96490	0.76749
12	1.10554	1.05409	0.95263	1.04499	0.98758	0.79851	1.07834	0.88455
13	1.22474	1.11803	1.03923	1.09620	1.10554	0.91629	1.20402	1.01574
14	1.36277	1.19523	1.12583	1.15493	1.24084	1.04984	1.34731	1.16593
15	1.52753	1.29099	1.21716	1.22261	1.40106	1.20521	1.51632	1.34278
16	1.73205	1.41421	1.33333	1.30023	1.59861	1.39393	1.72425	1.55911
17	2.00000	1.58114	1.49071	1.38564	1.85592	1.63670	1.99448	1.83837
18	2.38048	1.82574	1.72133	1.47224	2.21944	1.97612	2.37741	2.22936
19	3.00000	2.23607	2.10819	1.55885	2.80872	2.52143	2.99846	2.85796
20	4.35890	3.16228	2.98142	1.64545	4.09607	3.70340	4.35840	4.22194

(cf. 5.34–5.36). Otherwise $P^+_{\succeq_c W}h$ is the linear function crossing the origin, with the optimal slope and norm

$$\alpha_*(0) = \frac{\int_b^d xw(x)\,dx}{\int_b^d w(x)\,dx \int_0^d x^2 w(x)\,dx} = \frac{\mu_W(b)}{m_W^2}, \tag{5.50}$$

$$||P^+_{\succeq_c W}h|| = \frac{\int_b^d xw(x)\,dx}{\int_b^d w(x)\,dx \left[\int_0^d x^2 w(x)\,dx\right]^{1/2}} = \frac{\mu_W(b)}{m_W}, \tag{5.51}$$

respectively.

Theorem 16 ($F \succeq_c W$) *If for* $b = W^{-1}(\frac{j-1}{n})$

$$\mu_W(b) \le m_W^2/\mu_W, \tag{5.52}$$

(cf 5.30), then

$$\mathrm{E}_F Y_{j:n}/m_F \le \mu_W(b)/m_W, \tag{5.53}$$

(cf 5.51), which is attained for $F(x) = W(m_W \frac{x}{m_F})$.

Otherwise for $\beta \in (0, b)$ *determined by 5.33 with*

$$\nu^2 = \nu_W^2(\beta) = \int_\beta^d (x-\beta)^2 w(x)\,dx = [1 - W(\beta)]m_W^2(\beta) \tag{5.54}$$

we have

$$\frac{\mathrm{E}_F Y_{j:n}}{m_F} \le \frac{\nu_W(\beta)}{\mu_W(\beta) - \beta}, \tag{5.55}$$

which becomes equality for

$$F(x) = W(\beta + \nu\frac{x}{m})\mathbf{1}_{[0,\infty)}(x). \tag{5.56}$$

Again, we point out analogies between Theorems 15 and 16. Under equivalent conditions 5.30 and 5.52, we obtain analogous bounds 5.31 and 5.53, respectively. In the opposite case, the bounds are determined by the same parameter β and related by 5.49. Moreover, both the extreme distributions 5.37 and 5.56 are location-scale modifications of W with identical jump $W(\beta)$ at the left support end.

Proposition 11. (decreasing density) *If* $\frac{j-1}{n} \le \frac{1}{3}$, *then*

$$\frac{\mathrm{E}_F Y_{j:n}}{m_F} \le \frac{\sqrt{3}}{2}\left(1 + \frac{j-1}{n}\right), \tag{5.57}$$

where equality holds if F is the uniform distribution function on $[0, \sqrt{3}m_F]$.
Otherwise

$$\frac{\mathrm{E}_F Y_{j:n}}{m_F} \le \frac{2}{3}\left(\frac{2n}{n+1-j}\right)^{1/2}, \tag{5.58}$$

which becomes equality for the mixture of the atom at 0 with probability $\frac{3}{2}\frac{j-1}{n} - \frac{1}{2}$ and the uniform distribution on $[0, m_F\left(\frac{2n}{n+1-j}\right)^{1/2}]$ with probability $\frac{3}{2}(1-\frac{j-1}{n})$.

Proposition 12. (decreasing failure rate) *If $\frac{j-1}{n} \le 1 - e^{-1} \approx 0.63212$, then*

$$\frac{\mathrm{E}_F Y_{j:n}}{m_F} \le \left(\ln \frac{n}{n+1-j} + 1\right)/\sqrt{2}, \tag{5.59}$$

which is equality for the exponential life distribution with scale $m_F/\sqrt{2}$.
Otherwise

$$\frac{\mathrm{E}_F Y_{j:n}}{m_F} \le \left[\frac{2n}{(n+1-j)e}\right]^{1/2}. \tag{5.60}$$

The bound becomes equality for the mixture of exponential distribution with scale $m_F\left[\frac{n}{2e(n+1-j)}\right]^{1/2}$ and zero, with weights $(1-\frac{j-1}{n})e$ and $1-(1-\frac{j-1}{n})e$, respectively.

Projection of $h = M\mathbf{1}_{[b,d)}$ for $M = \frac{n}{n+1-j}$, $0 = a_W < b = W^{-1}(\frac{j-1}{n}) < d_W$ onto $\mathcal{C}^+_{\preceq_c W}$, desired for calculating second moment bounds for samples with life distributions satisfying $F \preceq_c W$, is described in Lemma 13 (cf. Rychlik [49, Theorem 11(b), p. 131]).

Lemma 13. *Projection of $h = M\mathbf{1}_{[b,d)}$ onto $\mathcal{C}^+_{\preceq_c W}$ in $L^2([0,d_W), w(x)\,dx)$ has form*

$$P^+_{\preceq_c W} h(x) = M \min\{x/\alpha_*, 1\}. \tag{5.61}$$

If

$$d \int_b^d x w(x)\,dx > \int_0^d x^2 w(x)\,dx, \tag{5.62}$$

then $\alpha_* \in (b,d)$ is the unique solution to

$$\alpha \int_b^\alpha x w(x)\,dx = \int_0^\alpha x^2 w(x)\,dx. \tag{5.63}$$

Otherwise

$$\alpha_* = \frac{\int_0^d x^2 w(x)\,dx}{\int_b^d x w(x)\,dx} \ge d. \tag{5.64}$$

Under 5.62 the projection is actually a broken line with break at α_* that belongs to the domain of h. Then

$$||P^+_{\preceq_c W}h||^2 = M^2\left[\frac{1}{\alpha_*^2}\int_0^{\alpha_*} x^2 w(x)\,dx + \int_{\alpha_*}^{d} w(x)\,dx\right]$$
$$= M^2\left[\frac{1}{\alpha_*}\int_b^{\alpha_*} xw(x)dx + \int_{\alpha_*}^{d} w(x)dx\right] = M^2\gamma^2, \quad (5.65)$$

say, and

$$\frac{P^+_{\preceq_c W}h(x)}{||P^+_{\preceq_c W}h||} = \frac{1}{\gamma}\min\left\{\frac{x}{\alpha_*}, 1\right\}. \quad (5.66)$$

If 5.62 is not true (which is possible for $d_W < \infty$ only), then the projection is actually linear and satisfies

$$||P^+_{\preceq_c W}h|| = M\frac{\int_b^d xw(x)\,dx}{\left[\int_0^d x^2 w(x)\,dx\right]^{1/2}}, \quad (5.67)$$
$$\frac{P^+_{\preceq_c W}h(x)}{||P^+_{\preceq_c W}h||} = \frac{x}{\left[\int_0^d x^2 w(x)\,dx\right]^{1/2}}, \quad (5.68)$$

Theorem 17. ($F \preceq_c W$) *If*

$$d\mu_W(b) > \frac{n}{n+1-j}m_W^2, \quad (5.69)$$

for $b = W^{-1}(\frac{j-1}{n})$ *then there exists unique* $b < \alpha_* < d_W$ *such that for*

$$\gamma^2 = \gamma_W^2(\alpha_*) = \frac{1}{\alpha_*^2}\int_0^{\alpha_*} x^2 w(x)\,dx + \int_{\alpha_*}^{d} w(x)\,dx$$
$$= W(\alpha_*)\mathrm{E}_W\left[\frac{X^2}{\alpha_*^2}\mathbf{1}_{[0,1)}\left(\frac{X}{\alpha_*}\right)\right] + 1 - W(\alpha_*) \quad (5.70)$$

we have

$$\frac{\mathrm{E}_F Y_{j:n}}{m_F} \le \frac{n}{n+1-j}\gamma_W(\alpha_*). \quad (5.71)$$

The equality in 5.71 holds for

$$F(x) = \begin{cases} W\left(\alpha_*\gamma\frac{x}{m_F}\right), & if\ \frac{x}{m_F} < \frac{1}{\gamma}, \\ 1, & if\ \frac{x}{m_F} \geq \frac{1}{\gamma}. \end{cases} \tag{5.72}$$

If 5.69 does not hold then bound

$$\frac{\mathrm{E}_F Y_{j:n}}{m_F} \leq \frac{\mu_W(b)}{m_W} \tag{5.73}$$

is attained by $F(x) = W(m_W \frac{x}{m_F})$.

Relation 5.69 is satisfied for small order statistics, and life distribution 5.72 has a finite support with probability mass $1 - W(\alpha_*)$ at the right end-point. Below we specify results for distributions with increasing density and failure rate.

Proposition 13. (increasing density) *If* $\frac{j-1}{n} < \frac{1}{\sqrt{3}} \approx 0.57735$, *then*

$$\frac{\mathrm{E}_F Y_{j:n}}{m_F} \leq \frac{n}{n+1-j}\left(1 - \frac{2}{\sqrt{3}}\frac{j-1}{n}\right)^{1/2}. \tag{5.74}$$

The equality holds for a mixture of the uniform distribution on interval $[0, m_F/(1 - \frac{2}{\sqrt{3}}\frac{j-1}{n})^{1/2}]$ *and the degenerate distribution concentrated at point* $m_F/(1 - \frac{2}{\sqrt{3}}\frac{j-1}{n})^{1/2}$ *with probabilities* $\sqrt{3}\frac{j-1}{n}$ *and* $1 - \sqrt{3}\frac{j-1}{n}$, *respectively.*

If $\frac{j-1}{n} > \frac{1}{\sqrt{3}}$, *then sharp bound*

$$\frac{\mathrm{E}_F Y_{j:n}}{m_F} \leq \frac{\sqrt{3}}{2}\left(1 + \frac{j-1}{n}\right) \tag{5.75}$$

is attained by the uniform distribution on the interval $[0, \sqrt{3}m_F]$.

Proposition 14. (increasing failure rate) *For* $\alpha_* > \ln\frac{n}{n+1-j}$ *uniquely defined by equation*

$$\frac{2}{\alpha}(1 - e^{-\alpha}) - e^{-\alpha} = \left(1 - \frac{j-1}{n}\right)\left(1 - \ln\frac{n}{n+1-j}\right) \tag{5.76}$$

and

$$\gamma^2 = \gamma_W^2(\alpha_*) = 2[1 - (1 - \alpha_*)e^{-\alpha_*}] \tag{5.77}$$

we have

$$\frac{\mathrm{E}_F Y_{j:n}}{m_F} \leq \frac{n}{n+1-j}\frac{\gamma}{\alpha_*} \tag{5.78}$$

The equality holds iff

$$F(x) = \begin{cases} 0, & if\ \frac{x}{m_F} \le 0, \\ 1 - \exp\left(-\gamma \frac{x}{m_F}\right) & if\ 0 \le \frac{x}{m_F} < \frac{\alpha_*}{\gamma}, \\ 1, & if\ \frac{x}{m_F} \ge \frac{\alpha_*}{\gamma}, \end{cases} \tag{5.79}$$

which is a combination of a right truncated exponential distribution with a pole at the truncation point $m_F \frac{\alpha_*}{\gamma}$.

5.4 Distributions with Monotone Failure Rate in Average

The results of this Subsection are based on assertions of Lemmas 6 and 7 for $b = W^{-1}(\frac{j-1}{n})$ and $c = d_W$. Therefore one can expect some analogies with inequalities for quantiles of order p derived for $b = W^{-1}(p)$ and $c \searrow b$. Observe that norm 3.30 does depend on c, and the normalized projection 3.31 does not. It follows that bounds for quantiles and expected order statistics substantially differ, but they are attained by the same distributions when $p = \frac{j-1}{n}$. Note that condition 3.26 is false for $c = d_W$, and there are no trivial bounds $\mathrm{E}_F Y_{j:n} \le \mu_F$ for $j \ge 2$. In Theorem 18 and Propositions 14 and 15 we omit explicit description of distributions for which bounds are attained, and refer the reader to respective assertions in Theorem 5 and Propositions 3 and 4 in which p and b should be replaced by $\frac{j-1}{n}$ and $W^{-1}(\frac{j-1}{n})$, respectively.

Theorem 18. ($F \succeq_* W$) *For arbitrary* $2 \le j \le n < \infty$, *with* $b = W^{-1}(\frac{j-1}{n})$ *and notation of 3.15 and 3.33, the following inequality is tight*

$$\frac{\mathrm{E}_F Y_{j:n} - \mu_F}{\sigma_F} \le \frac{j-1}{n} \frac{\mu_W(b)}{\vartheta_W(a,b)}. \tag{5.80}$$

Proposition 15. (decreasing density in average) *For arbitrary* $2 \le j \le n < \infty$, *bound*

$$\frac{\mathrm{E}_F Y_{j:n} - \mu_F}{\sigma_F} \le \frac{\sqrt{3}(j-1)(n+j-1)}{n^2 \theta_U(\frac{j-1}{n})}. \tag{5.81}$$

(cf 3.37) is the best possible.

Proposition 16. (decreasing failure rate in average) *For arbitrary* $2 \le j \le n < \infty$, *bound*

$$\frac{\mathrm{E}_F Y_{j:n} - \mu_F}{\sigma_F} \le \frac{(j-1)(1 + \ln \frac{n}{n+1-j})}{n \theta_V(\frac{j-1}{n})}. \tag{5.82}$$

(cf 3.40) is the best possible.

Theorem 19. ($F \preceq_* W$) *For arbitrary continuous distribution function W with a finite second moment there exists a sequence $F_k \preceq_* W$, $k = 1, 2, \ldots$, of distribution functions which have densities and finite second moments, and satisfy*

$$\lim_{k \to \infty} \sup_{P \in \mathcal{P}_n(F_k)} \frac{\mathrm{E}_{F_k} Y_{j:n} - \mu_{F_k}}{\sigma_{F_k}} = \left(\frac{j-1}{n+1-j} \right)^{1/2}. \tag{5.83}$$

In particular, bound 5.83 is sharp for distributions with increasing density and failure rate in average.

Theorem 19 asserts that general bound 5.13 is attained among $F \preceq_* W$ under mild assumptions on the maximal element W. The proof consists in constructing a sequence of antistarshaped superpositions $F_k^{-1} W - \mu_{F_k}$ (with $F_k^{-1} W(a_W) \searrow -\infty$) that integrate to 0, and tend to $\mathbf{1}_{[W^{-1}((j-1)/n), d_W)}$ in $L^2([a_W, d_W), w(x)dx)$. Actually, we can prove an analogous claim for arbitrary limiting function $\sum_{j=1}^{n} d_j \mathbf{1}_{[W^{-1}((j-1)/n), W^{-1}(j/n))}$ with nondecreasing coefficients d_j, $1 \le j \le n$, which means that general bound 5.11 for arbitrary L-statistic cannot be improved once we restrict the family of distributions to any of the form $\{F : F \preceq_* W\}$.

5.5 Symmetric Unimodal and U-Shaped Distributions

Theorem 20. (symmetric unimodal distributions) *If $0 < \frac{j-1}{n} \le \frac{1}{3}$, then*

$$\frac{\mathrm{E}_F Y_{j:n} - \mu_F}{\sigma_F} \le \frac{2[n(j-1)]^{1/2}}{3(n+1-j)}. \tag{5.84}$$

This becomes equality if $Y_1 = \mu$ with probability $1 - 3\frac{j-1}{n}$ and is uniformly distributed on $[\mu - \sigma(\frac{n}{j-1})^{1/2}, \mu + \sigma(\frac{n}{j-1})^{1/2}]$ with probability $3\frac{j-1}{n}$.

If $\frac{1}{3} \le \frac{j-1}{n} \le \frac{2}{3}$, then

$$\frac{\mathrm{E}_F Y_{j:n} - \mu_F}{\sigma_F} \le \sqrt{3}\frac{j-1}{n}, \tag{5.85}$$

which is equality iff Y_1 is uniformly distributed on $[\mu - \sqrt{3}\sigma, \mu + \sqrt{3}\sigma]$.

If $\frac{j-1}{n} > \frac{2}{3}$, then

$$\frac{\mathrm{E}_F Y_{j:n} - \mu_F}{\sigma_F} \le \frac{2}{3} \left(\frac{n}{n+1-j} \right)^{1/2}. \tag{5.86}$$

Here the equality is attained by the mixture of the atom at μ and the uniform distribution on $[\mu - \sigma(\frac{n}{n+1-j})^{1/2}, \mu + \sigma(\frac{n}{n+1-j})^{1/2}]$ with probabilities $3\frac{j-1}{n} - 2$ and $3(1 - \frac{j-1}{n})$, respectively.

Theorem 21. (symmetric U-shaped distributions) *If either* $\frac{j-1}{n} < \frac{1}{2} - \frac{1}{2\sqrt{3}} \approx 0.21132$ *or* $\frac{j-1}{n} > \frac{1}{2} + \frac{1}{2\sqrt{3}} \approx 0.78868$ *then inequality 5.85 holds true with the respective condition for equality.*

For $\frac{1}{2} - \frac{1}{2\sqrt{3}} < \frac{j-1}{n} < \frac{1}{2} + \frac{1}{2\sqrt{3}}$ *bound*

$$\frac{\mathrm{E}_F Y_{j:n} - \mu_F}{\sigma_F} \leq \frac{n}{n+1-j} \left(\frac{1}{4} - \frac{1}{\sqrt{3}} \left| \frac{j-1}{n} - \frac{1}{2} \right| \right)^{1/2} \tag{5.87}$$

is sharp, and this becomes equality for the combination of symmetric uniform distribution on $[\mu - \sigma/(1 - 4\sqrt{3}|\frac{j-1}{n} - \frac{1}{2}|)^{1/2}, \mu + \sigma/(1 - 4\sqrt{3}|\frac{j-1}{n} - \frac{1}{2}|)^{1/2}]$ *with probability* $2\sqrt{3}|\frac{j-1}{n} - \frac{1}{2}|$ *and two atoms at the ends of the interval, with identical probabilities* $\frac{1}{2} - \sqrt{3}|\frac{j-1}{n} - \frac{1}{2}|$.

Proofs of Theorems 20 and 21 are based on projecting folded functionals 5.18 onto convex cones $\mathcal{C}^+_{\succeq_c 2U-1}$ and $\mathcal{C}^+_{\preceq_c 2U-1}$. Observe that $g^s_{j:n} = \frac{j-1}{n+1-j} g^s_{n+1-j:n}$, and the same holds for the respective projections and their norms. This explains the relation between the bounds in both the Theorems and the fact that the distributions with extreme expectations of jth smallest and largest order statistics are identical. We can immediately conclude similar bounds for some L-statistics. E.g., we have

$$\begin{aligned} \sup_{P \in \mathcal{P}_n(F)} \mathrm{E}_F Y_{j:n} &= \int_0^1 F^{-1}(x) g_{j:n}(x)\, dx \\ &= \sup_{P \in \mathcal{P}_n(F)} \mathrm{E}_F \frac{1}{k+1-j} \sum_{l=j}^{k} Y_{l:n} \qquad (5.88) \\ &= \sup_{P \in \mathcal{P}_n(F)} \mathrm{E}_F \left[\frac{j}{n} Y_{j:n} + \frac{1}{n} \sum_{l=j+1}^{k-1} Y_{l:n} \right. \\ &\qquad \left. + \frac{n+1-k}{n} Y_{k:n} \right], \qquad (5.89) \end{aligned}$$

$1 \leq j \leq k \leq n$, (cf 2.22–2.23), and therefore all the bounds derived in this Section for a given jth order statistic coincide with the bounds for trimmed and Winsorized means (5.88 and 5.89, respectively) that reject $j-1$ smallest observations. Moreover,

$$\begin{aligned} \sup_{P \in \mathcal{P}_n(F)} \mathrm{E}_F (Y_{j:n} - Y_{j-1:n}) &= \frac{n}{j-1} \int_0^1 F^{-1}(x)[g_{j:n}(x) - 1]\, dx \\ &= \frac{n}{j-1} \sup_{P \in \mathcal{P}_n(F)} (\mathrm{E}_F Y_{j:n} - \mu_F), \qquad 2 \leq j \leq n. \end{aligned}$$

Table 6: Sharp uniform mean-variance bounds on expectations of order statistics from dependent samples of size 20 for various families of distributions.

j	G	S	SUN	SUS	DD	DFR	DDA	DFRA
2	0.2294	0.1664	0.1569	0.0866	0.0866	0.0513	0.0903	0.0525
3	0.3333	0.2485	0.2342	0.1732	0.1732	0.1054	0.1854	0.1100
4	0.4201	0.3222	0.3038	0.2598	0.2598	0.1625	0.2823	0.1725
5	0.5	0.3953	0.3727	0.3464	0.3464	0.2231	0.3790	0.2400
6	0.5774	0.4714	0.4444	0.4334	0.4330	0.2877	0.4746	0.3125
7	0.6547	0.5533	0.5216	0.5240	0.5196	0.3567	0.5693	0.3905
8	0.7338	0.6436	0.6062	0.6219	0.6062	0.4308	0.6641	0.4741
9	0.8165	0.7454	0.6928	0.7308	0.6939	0.5108	0.7604	0.5641
10	0.9045	0.8624	0.7794	0.8550	0.7850	0.5978	0.8598	0.6614
11	1	1	0.8660	1	0.8819	0.6932	0.9649	0.7675
12	1.1055	1.0541	0.9526	1.0450	0.9876	0.7985	1.0783	0.8846
13	1.2247	1.1180	1.0392	1.0962	1.1055	0.9163	1.2040	1.0157
14	1.3628	1.1952	1.1258	1.1549	1.2408	1.0498	1.3473	1.1659
15	1.5275	1.2910	1.2172	1.2226	1.4011	1.2052	1.5163	1.3428
16	1.7321	1.4142	1.3333	1.3002	1.5986	1.3939	1.7243	1.5591
17	2	1.5811	1.4907	1.3856	1.8559	1.6367	1.9945	1.8384
18	2.3805	1.8257	1.7213	1.4722	2.2194	1.9761	2.3774	2.2294
19	3	2.2361	2.1082	1.5589	2.8087	2.5214	2.9985	2.8580
20	4.3589	3.1623	2.9814	1.6455	4.0961	3.7034	4.3584	4.2219

Therefore we obtain bounds for spacings $Y_{j:n} - Y_{j-1:n}$ in standard deviation units by multiplying all respective mean-variance bounds for single jth order statistics of various families of sample distributions by factor $\frac{n}{j-1}$. In all these cases, the bounds are attained by the same joint distributions.

In contrast with the independent case, all the bounds are nontrivial for any order statistic except of the sample minimum. Table 3 contains numerical values of mean-variance bounds for order statistics from dependent samples of size $n = 20$ coming from the following families of distributions: general (G), symmetric (S), symmetric unimodal (SUN) and U-shaped (SUS), distributions with decreasing density and failure rate (DD and DFR, respectively), and with decreasing density and failure rate in average (DDA and DFRA, respectively). In fact, all the bounds depend on value $\frac{j-1}{n}$ only, because they were derived by means of projecting functions of $\frac{j-1}{n}$. Therefore all estimates for $Y_{j:n}$ hold also true for any $Y_{k:m}$ provided that $\frac{k-1}{m} = \frac{j-1}{n}$. In particular, the moment bounds for the sample medians $Y_{n:2n+1}$ do not change under increase of the sample size. Moreover, one can check that all the bounds for restricted families of distributions presented in this Section increase continuously in $\frac{j-1}{n}$. Consequently, Table 3 may provide fair approximations of bounds for $Y_{k:m}$ from various families of parent distributions with $\frac{k-1}{m} \approx \frac{j-1}{20}$ for some $2 \le j \le 20$. Some tables for second moment bounds were presented in Gajek and Rychlik [49]. Also, the third column (SUN) of Table 3 comes from the paper.

5.6 *Extreme Effect of Dependence*

The problem we consider here stems from the robust statistics whose domain is studying sensitivity of statistical procedures against violations of standard assumptions about statistical models. Robustness of L-statistics which are popular tools in robust and nonparametric inference was studied by many authors (we refer merely to monographs of Huber [26] and Hampel *et al.* [23] and references given there). For standard i.i.d. parametric models, many violations of marginals were studied. Dependence-robustness for location models was analyzed in Rychlik [45]. The projection method provides tools for determining sensitivity of expectation of arbitrary L-statistics against dependence of observations

$$\sup_{P \in \mathcal{P}_n(F)} \mathrm{E}_F \sum_{j=1}^{n} c_j Y_{j:n} - \mathrm{E}_F \sum_{j=1}^{n} c_j X_{j:n}$$
$$= \int_0^1 F^{-1}(x) \left[G_{\mathbf{c}}(x) - \sum_{j=1}^{n} c_j f_{j:n}(x) \right] dx, \tag{5.90}$$

for various families of parent marginals F.

We present some preliminary results of Rychlik [53], where robustness of single order statistics coming from general populations was studied. To evaluate robustness of the jth order statistic in standard deviation units, we first determine the projection of $h_{j:n} = g_{j:n} - f_{j:n}$ onto $\mathcal{C}^0$ in $L^2([0,1), dx)$. Since $h_{j:n}$ is the difference of density functions, $P^0 h_{j:n} = P^{\nearrow} h_{j:n}$. Thus we are reduced to establishing projections onto the cone of nondecreasing functions which can be determined by means of the greatest convex minorants. The solution is trivial for the sample minimum, since

$$h_{1:n}(x) = 1 - n(1-x)^{n-1} \tag{5.91}$$

is actually increasing and so $P^{\nearrow} h_{1:n} = h_{1:n}$. Another simple solution emerges for the sample maximum. Then

$$h_{n:n}(x) = \begin{cases} -nx^{n-1}, & \text{if } 0 \le x < 1 - 1/n, \\ n(1 - x^{n-1}), & \text{if } 1 - 1/n \le x \le 1, \end{cases} \tag{5.92}$$

has antiderivative

$$H_{n:n}(x) = \begin{cases} -x^n, & \text{if } 0 \le x \le 1 - 1/n, \\ n(x-1) + 1 - x^n, & \text{if } 1 - 1/n \le x \le 1, \end{cases} \tag{5.93}$$

which is decreasing on $[0, 1 - 1/n]$ and increasing on $[1 - 1/n, 1]$, and concave on both the intervals. Therefore

$$\bar{H}_{n:n}(x) = \begin{cases} -(1 - 1/n)^{n-1}x, & \text{if } 0 \le x \le 1 - 1/n, \\ n(1 - 1/n)^n(x - 1), & \text{if } 1 - 1/n \le x \le 1, \end{cases} \tag{5.94}$$

and

$$P^{\nearrow} h_{n:n}(x) = \begin{cases} -(1 - 1/n)^{n-1}, & \text{if } 0 \le x < 1 - 1/n, \\ n(1 - 1/n)^n, & \text{if } 1 - 1/n \le x \le 1. \end{cases} \tag{5.95}$$

A deeper analysis is needed for $2 \le j \le n-1$. Then $h_{j:n}$ decreases form 0 at 0 to $-f_{j:n}(\frac{j-1}{n})$, jumps up by $\frac{n}{n+1-j}$, runs down to $h_{j:n}(\frac{j-1}{n-1})$ where $f_{j:n}$ is maximized and eventually increases to $h_{j:n}(1) = \frac{n}{n+1-j}$. It is important to know the sign of $h_{j:n}(\frac{j-1}{n-1})$. It can be shown that this is negative for small and moderate j, and the proportion of j's for which this is true increases to 1, as n becomes large. In this case, a thorough analysis shows that either $H_{j:n}(\frac{j-1}{n})$ lies above the greatest convex minorant which consists the line tangent to $H_{j:n}$ at 0 and some $s > \frac{j-1}{n-1}$, and $H_{j:n}$ itself right to s, or $H_{j:n}(\frac{j-1}{n})$ spans the minorant which has two linear pieces joining at $\frac{j-1}{n}$ and coincides with $H_{j:n}$ in the right part. If $h_{j:n}(\frac{j-1}{n-1}) \ge 0$, then the convex minorant has the form analogous to the latter one, with the only difference that the slope of the second linear piece is positive. Analytic forms of assumptions and resulting projection are precisely described in Lemma 14.

Lemma 14. *Set*

$$S_{j:n}(x) = \frac{H_{j:n}(x)}{x}, \qquad 0 < x \le 1, \tag{5.96}$$

$$T_{j:n}(x) = \frac{H_{j:n}(x) - H_{j:n}(\frac{j-1}{n})}{x - \frac{j-1}{n}}, \qquad \frac{j-1}{n} < x \le 1. \tag{5.97}$$

If $h_{j:n}(\frac{j-1}{n-1}) < 0$, then there exist unique $r \in (\frac{j-1}{n-1}, 1)$ such that $h_{j:n}(r) = 0$ and $s \subset [\frac{j-1}{n-1}, r)$ such that either $s = \frac{j-1}{n-1}$ if $S_{j:n}(\frac{j-1}{n-1}) \le h_{j:n}(\frac{j-1}{n-1})$ or s is the solution to $S_{j:n}(x) = h_{j:n}(x)$ otherwise.

If

$$h_{j:n}(\tfrac{j-1}{n-1}) < 0 \text{ and } S_{j:n}(s) \le S_{j:n}(\tfrac{j-1}{n}), \tag{5.98}$$

then

$$P^{\nearrow} h_{j:n}(x) = \begin{cases} S_{j:n}(s), & \text{if } 0 \le x < s, \\ h_{j:n}(x), & \text{if } s \le x \le 1. \end{cases} \tag{5.99}$$

If

either $h_{j:n}(\frac{j-1}{n-1}) < 0$ *and* $S_{j:n}(s) > S_{j:n}(\frac{j-1}{n})$ *or* $h_{j:n}(\frac{j-1}{n-1}) \geq 0$ (5.100)

then there exists a unique $t \in (\frac{j-1}{n-1}, 1)$ *such that* $T_{j:n}(t) = h_{j:n}(t)$, *and*

$$P^{\nearrow} h_{j:n}(x) = \begin{cases} S_{j:n}(\frac{j-1}{n}), & \text{if } 0 \leq x < \frac{j-1}{n}, \\ h_{j:n}(t), & \text{if } \frac{j-1}{n} \leq x \leq t, \\ h_{j:n}(x), & \text{if } t \leq x \leq 1. \end{cases} \tag{5.101}$$

For the prevailing number of cases, we have 5.98 with $s > \frac{j-1}{n-1}$. Then 5.99 is continuous, and can be written as $P^{\nearrow} h_{j:n}(x) = h_{j:n}(\max\{s, x\})$. This form is easier to handle in numerical calculations. Function 5.101 has the only jump at $\frac{j-1}{n}$. Using projections 5.91, 5.95, 5.99 and 5.101, we are in a position to write optimal bounds for the extreme dependence effect.

Theorem 22. *Bounds*

$$\sup_{P \in \mathcal{P}_n(F)} (\mathrm{E}_F Y_{1:n} - \mathrm{E}_F X_{1:n})/\sigma_F = \Delta = \Delta(1,n) = (n-1)/(2n-1)^{1/2} \tag{5.102}$$

are attained by the marginal distribution functions

$$F(x) = 1 - \left[\frac{1}{n}\left(1 - \frac{n-1}{(2n-1)^{1/2}} \frac{x-\mu}{\sigma}\right)\right]^{1/(n-1)},$$
$$-(2n-1)^{1/2} \leq \frac{x-\mu}{\sigma} \leq (2n-1)^{1/2}/(n-1). \tag{5.103}$$

If 5.98 holds for $2 \leq j \leq n-1$, *then*

$$\sup_{P \in \mathcal{P}_n(F)} (\mathrm{E}_F Y_{j:n} - \mathrm{E}_F X_{j:n})/\sigma_F = \Delta = \Delta(j,n) \tag{5.104}$$

for

$$\Delta^2 = \frac{1}{s}\left[\frac{ns+1-j}{n+1-j} - F_{j:n}(s)\right]^2 + \frac{n^2(1-s)}{(n+1-j)^2} - \frac{2n}{n+1-j}[1 - F_{j:n}(s)]$$
$$+ n\frac{\binom{2j-2}{j-1}\binom{2n-2j}{n-j}}{\binom{2n-1}{n}}[1 - F_{2j-1:2n-1}(s)]. \tag{5.105}$$

Bound 5.104 is attained by

$$F(x) = \begin{cases} 0, & \textit{if } \frac{x-\mu}{\sigma} < \frac{S_{j:n}(s)}{\Delta}, \\ s, & \textit{if } \frac{S_{j:n}(s)}{\Delta} \leq \frac{x-\mu}{\sigma} \leq \frac{h_{j:n}(s)}{\Delta}, \\ h_{j:n}^{-1}\left(\Delta\frac{x-\mu}{\sigma}\right), & \textit{if } \frac{h_{j:n}(s)}{\Delta} \leq \frac{x-\mu}{\sigma} \leq \frac{n/(n+1-j)}{\Delta}, \\ 1, & \textit{if } \frac{x-\mu}{\sigma} \geq \frac{n/(n+1-j)}{\Delta}. \end{cases} \tag{5.106}$$

Under conditions 5.100 we have 5.106 with

$$\Delta^2 = \tfrac{n}{j-1}F_{j:n}^2(\tfrac{j-1}{n}) + (t - \tfrac{j-1}{n})\left[\tfrac{n}{n+1-j} - f_{j:n}(t)\right]^2 + \tfrac{n^2(1-t)}{(n+1-j)^2}$$
$$- \tfrac{2n}{n+1-j}[1 - F_{j:n}(t)] + n\frac{\binom{2j-2}{j-1}\binom{2n-2j}{n-j}}{\binom{2n-1}{n}}[1 - F_{2j-1:2n-1}(t)]. \quad (5.107)$$

The supremum is attained by 5.106 with s, $S_{j:n}(s)$ and $h_{j:n}(s)$ replaced by $\frac{j-1}{n}$, $S_{j:n}(\frac{j-1}{n})$ and $h_{j:n}(t)$, respectively.

Finally, bound

$$\sup_{P\in\mathcal{P}_n(F)} (\mathrm{E}_F Y_{n:n} - \mathrm{E}_F X_{n:n})/\sigma_F = (n-1)^{n-1/2}/n^{n-1}, \quad (5.108)$$

is attained by the two-point marginal distribution concentrated on points $\mu - \sigma/(n-1)^{1/2}$ and $\mu + \sigma(n-1)^{1/2}$ with respective probabilities $1 - \frac{1}{n}$ and $\frac{1}{n}$.

Note that 5.102 coincides with the mean-variance bound for the maximum in the i.i.d. sample (see 4.2), and these bounds are attained by distribution functions of variables mutually symmetric about μ (see 4.3 and 5.103). This is a consequence of relations $\sup_{P\in\mathcal{P}_n(F)} \mathrm{E}_F Y_{1:n} - \mathrm{E}_F X_{1:n} = \mu_F - \mathrm{E}_F X_{1:n} = \mathrm{E}_{F^-} X_{1:n}^- - \mu_{F^-}$, where $X_j^- = 2\mu - X_j$, $1 \le j \le n$, have distribution function defined by $F^-(x-\mu) = 1 - F(\mu - x-)$. Comparing 5.13–5.14 for $j = n$, and 5.108 we observe that

$$\frac{\sup \mathrm{E}_F Y_{n:n} - \mathrm{E}_F X_{n:n}}{\sup \mathrm{E}_F Y_{n:n} - \mu_F} = \left(1 - \frac{1}{n}\right)^{n-1} \searrow e^{-1} \approx 0.36788, \quad (5.109)$$

and both the suprema in 5.109, taken over all $P \in \mathcal{P}_n(F)$ are achieved by the same distribution function.

Numerical calculations based on Theorem 22 show that for general marginal distributions central order statistics are more robust under dependence than the extreme ones. We can also see that each jth smallest order statistic is more sensitive than the respective jth greatest one. In fact, Rychlik [53] measured dependence-robustness of order statistics in terms of scale parameters generated by central absolute moments of order $1 \le p \le \infty$. Numerical analysis shows similarity of conclusion for cases $p = 1$ and $p = 2$, which were presented above. For $p = +\infty$ (i.e. for bounded observations), the conclusions are diametrically opposite. An interesting fact to note here is that 5.109 holds true for the classes of marginal distributions with finite pth moment for arbitrary $1 \le p \le \infty$.

6 Records and kth Records

In comparison with evaluations of other statistical functionals discussed here, investigations for record values are still at a preliminary stage, and only few results will be presented now. Examples of Subsection 6.1 show that the range of record values can be arbitrarily large when all types of interdependence among the original variables are admitted. For the case of i.i.d. sequences with general and symmetric distributions, mean-variance bounds on standard and kth records, due to Nagaraja [35] and Raqab [41], respectively, are presented in Subsection 6.2. Finally, we discuss evaluations of Rychlik [48] for increments of 1st records coming from various families of parent distributions. Note that Nagaraja [35] used the Jensen inequality for deriving some quantile bounds on expectations of records (see also Arnold and Balakrishnan [4, Section 6.2]).

6.1 Dependent Identically Distributed Observations

The results of the previous Section prove that numerous nontrivial inequalities can be established for expectations of order statistics under the assumption that all observations have an identical distribution, but they are arbitrarily dependent. However, the assumption admits probabilistic models with peculiar properties of records. The examples presented below show that stating the problem of bounds for the expectation of record values of arbitrarily dependent observations makes no sense.

On one hand, in the sequence of identical variables $Y_1 = Y_2 = \dots$, the primary record value $R_0 = Y_1$ can never be improved, and kth records cannot be defined for $k \geq 2$. On the other hand, we can construct a sequence of identically distributed variables for which the first record value may be arbitrarily large. Precisely, the notion of being arbitrarily large depends here on properties of distribution F of single observation. If F has an atom at the right end-point of its support d_F, then $R_1 = d_F$. If d_F is finite and F is continuous at d_F, then R_1 can be arbitrarily close to d_F. If d_F is infinite, then R_1 may have an arbitrarily large value. The conclusions follow from the following Theorem.

Theorem 23. *For arbitrary positive integer n, there exists a sequence of standard uniform random variables Y_j, $j \geq 1$, for which $R_1 \geq 1 - \frac{1}{n}$.*

The proof is constructive. Combining the construction with transformation $F^{-1}(Y_j)$, $j \geq 1$, we obtain a sequence of F-distributed random variables with first record value $R_1 \geq F^{-1}(1 - \frac{1}{n})$, which is arbitrarily large in the sense described above.

Proof. Fix n, and take n random variables U_i, which are uniformly distributed on intervals $[\frac{i-1}{n}, \frac{i}{n}]$, $1 \le i \le n$. Here U_i may be independent or arbitrarily dependent, e.g., generated for a single uniform random variable U_0 by transforms $U_i = (i - 1 + U_0)/n$, $1 \le i \le n$. Create now n random vectors

$$\mathbf{U}_i = (U_i, U_{i-1}, \ldots, U_1, U_n, U_{n-1}, \ldots, U_{i+1}), \qquad 1 \le i \le n. \tag{6.1}$$

Let $(Y_1, \ldots, Y_n)$ coincide with a randomly chosen vector $\mathbf{U}_i$. We easily check that Y_j, $1 \le j \le n$, are uniformly distributed on $[0, 1]$. Now extend the sequence by adding i.i.d. random variables Y_j, $j > n$, with the same uniform distribution, and independent of $(Y_1, \ldots, Y_n)$. If $(Y_1, \ldots, Y_n) = \mathbf{U}_i$ for $1 \le i < n$, then $L_1 = i + 1$, and $R_1 = U_n \in [1 - \frac{1}{n}, 1]$. Otherwise $L_1 > n$ and $R_1 \in (U_n, 1] \subset [1 - \frac{1}{n}, 1]$. This completes the proof. □

Analogous assertions can be proved for general kth records. If suffices to replace single U_i by k-tuples of (independent, say) variables uniformly distributed on $[\frac{i-1}{n}, \frac{i}{n}]$, $1 \le i \le n$, generating vectors $\mathbf{U}_i$ (see 6.1) of length kn. There were some attempts of imposing stronger conditions on the dependence structure of observations (e.g., exchangeability in Arnold and Balakrishnan [4, Section 6.3], stationary Markov property in Biondini and Siddiqui [9]), but they do not lead to representations for which our projection method works. In the remainder of this Section we therefore confine ourselves to record values of independent sequences.

6.2 General and Symmetric Distributions

Bounds on expectations of kth records for general and symmetric distributions were analyzed in Raqab [41] by means of Moriguti projection of functions 2.37 based on greatest convex minorants. Special cases of 1st records for which 2.37 are convex functions varying from 0 to ∞, were solved in Nagaraja [35] by use of the Schwarz inequality. For general F, we therefore have

$$\begin{aligned} \frac{\mathrm{E}_F R_n - \mu_F}{\sigma_F} &\le \left\{ \frac{1}{(n!)^2} \int_0^1 [-\ln(1-x)]^{2n}\, dx - 1 \right\}^{1/2} \\ &= \left[\binom{2n}{n} \int_0^1 f_{2n}(x)\, dx - 1 \right]^{1/2} = \left[\binom{2n}{n} - 1 \right]^{1/2}, \end{aligned} \tag{6.2}$$

$n \ge 1$, which becomes equality for an affine transformation of the Weibull distribution

$$F(x) = 1 - \exp\left(- \left[n! \left(1 + D\frac{x-\mu}{\sigma} \right) \right]^{1/n} \right) \mathbf{1}_{[\mu - \sigma/D, \infty)}(x) \tag{6.3}$$

with shape parameter $1/n$ and $D = D^0(1,n) = \left[\binom{2n}{n} - 1\right]^{1/2}$. Observe that 6.3 has decreasing density and failure rate.

If $k \geq 2$, then 2.37 vanishes at 0 and 1, and increases on $[0, 1-\exp(-\frac{n}{k-1})]$, and ultimately decreases. Analysis similar to that in Subsection 4.1 (cf 4.6–4.8) gives the following result

Theorem 24. (general distributions) *For given $k \geq 2$ and $n \geq 1$, define $\alpha_* = \alpha_*(k,n) \in (0, 1-\exp(-\frac{n}{k-1}))$ as the unique solution to*

$$(1-x)f_n^{(k)}(x) = 1 - F_n^{(k)}(x) = (1-x)^k \sum_{j=0}^{n} \frac{k^j}{j!}[-\ln(1-x))]^j, \tag{6.4}$$

where $f_n^{(k)}$ and $F_n^{(k)}$ are the density function defined in 2.37 and the respective distribution function. Then

$$(\mathrm{E}_F R_n^{(k)} - \mu_F)/\sigma_F \leq D = D^0(k,n) \tag{6.5}$$

for

$$\begin{aligned} D^2 &= \int_0^{\alpha_*} [f_n^{(k)}(x)]^2\,dx + (1-\alpha_*)[f_n^{(k)}(\alpha_*)]^2 - 1 \\ &= k\binom{2n}{n}\left(\frac{k}{2k-1}\right)^{2n+1} F_{2n}^{(2k-1)}(\alpha_*) \\ &+ (1-\alpha_*)[f_n^{(k)}(\alpha_*)]^2 - 1. \end{aligned} \tag{6.6}$$

Equality in 6.6 holds for

$$F(x) = \begin{cases} 0, & if\ \frac{x-\mu}{\sigma} \leq -\frac{1}{D}, \\ (f_n^{(k)})^{-1}\left(1 + D\frac{x-\mu}{\sigma}\right), & if\ -\frac{1}{D} \leq \frac{x-\mu}{\sigma} < \frac{f_n^{(k)}(\alpha_*)-1}{D}, \\ 1, & if\ \frac{x-\mu}{\sigma} \geq \frac{f_n^{(k)}(\alpha_*)-1}{D}. \end{cases} \tag{6.7}$$

For $n = 2, 3$, equation 6.4 is solved by

$$\alpha_*(k,2) = 1 - \exp\left(-\frac{1}{k(k-1)}\right),$$

$$\alpha_*(k,3) = 1 - \exp\left(-\frac{1+(2k-1)^{1/2}}{k(k-1)}\right),$$

respectively, and then 6.6 has complicated explicit representations. Distribution function 6.7 has a finite support with a smooth density function, and a pole with probability $1-\alpha_*$ at the right end.

Since $s_n(x) = f_n(x) - f_n(1-x)$, $x \in [\frac{1}{2}, 1)$, is the difference of increasing and decreasing functions with identical values at $\frac{1}{2}$, this is strictly increasing from 0 to ∞. Evaluating expectations of 1st records in symmetrically distributed sequences we simply use the Schwarz inequality

$$\begin{aligned} \mathrm{E}_F R_n - \mu_F &\leq \int_{1/2}^{1} [F^{-1}(x) - \mu] s_n(x)\, dx \\ &\leq D^s(1,n)\sigma_F = D\sigma_F, \end{aligned} \tag{6.8}$$

where

$$2D^2 = \int_{1/2}^{1} s_n^2(x)\, dx = \binom{2n}{n} - \frac{1}{(n!)^2} \int_0^1 \ln^n x \ln^n(1-x)\, dx. \tag{6.9}$$

The bound in 6.8 is attained by

$$F(x) = s_n^{-1}(\sqrt{2}D\frac{x-\mu}{\sigma}), \tag{6.10}$$

which has a smooth symmetric density with the infinite support and symmetry center μ.

Raqab [41] also considered kth records in symmetrically distributed populations using the convex minorant approach. He proposed numerical evaluations that substantially improved bounds for general populations 6.5–6.6, and nonsharp ones derived in Grudzień and Szynal [21] based on direct application of the Schwarz inequality. However, Raqab's paper lacks a theoretical justification for use of specific constructions of the greatest convex minorant and precise description of the scope of applicability. Therefore we shall not present details here.

6.3 Improvements of Records

Here we present some standard deviation bounds on expectations of record increments

$$\mathrm{E}_F(R_n - R_{n-1}) = \int_0^1 [F^{-1}(x) - \mu][f_n(x) - f_{n-1}(x)]\, dx, \qquad n \geq 2. \tag{6.11}$$

We omit case $n = 1$, because estimates for $\mathrm{E}_F(R_1 - R_0) = \mathrm{E}_F R_1 - \mu_F$ were already presented. Properties of functions

$$\begin{aligned} \varphi_n(x) = f_n(x) - f_{n-1}(x) &= \frac{[-\ln(1-x)]^n}{n!} - \frac{[-\ln(1-x)]^{n-1}}{(n-1)!} \\ &= f_{n-1}(x)\left[\frac{-\ln(1-x)}{n} - 1\right] \quad n \geq 2, \end{aligned} \tag{6.12}$$

are crucial for determining its projection onto various convex cones. We easily check that each φ_n integrates to 0, and starts from the origin, decreases to $\varphi(1-e^{-n+1}) = -(n-1)^{n-1}/n!$, and increases to $+\infty$ at 1 passing through the horizontal axis at $1-e^{-n}$. Its antiderivative Φ_n that is needed for the Moriguti projection, is therefore concave decreasing, convex decreasing and convex increasing in $[0, 1-e^{-n+1}]$, $[1-e^{-n+1}, 1-e^{-n+1}]$, and $[1-e^{-n}, 1]$, respectively. This vanishes at 0 and 1, and is negative in the middle. Thus we deduce that its greatest convex minorant $\bar{\Phi}_n$ is linear in $[0, \alpha_*]$ for some $\alpha_* \in [1-e^{-n+1}, 1-e^{-n+1}]$, that is determined by equation

$$x\varphi_n(x) = \Phi_n(x) = F_n(x) - F_{n-1}(x) = f_n(x) \tag{6.13}$$

(cf. 6.4), and coincides with Φ_n elsewhere. Finally, $P^0\varphi_n(x) = \bar{\Phi}_n'(x) = \varphi_n(\max\{\alpha_*, x\})$.

Theorem 25. (general distributions) *For $n \geq 2$ we have*

$$\mathrm{E}_F(R_n - R_{n-1})/\sigma_F \leq \Delta = \Delta(n), \tag{6.14}$$

where

$$\Delta^2(n) = \binom{2n-2}{n-1}(1-\alpha_*)\left[\sum_{j=0}^{2n-1} \frac{(n\alpha_*)^j}{j!} + \left(1-\frac{1}{n}\right)\frac{(n\alpha_*)^{2n-1}}{(2n-1)!}\right] \tag{6.15}$$

for unique $\alpha_* \in (1-e^{-n+1}, 1-e^{-n+1})$ satisfying equation

$$-\ln(1-x) = nx. \tag{6.16}$$

Equality in 6.14 holds for

$$F(x) = \varphi_n^{-1}\left(\Delta\frac{x-\mu}{\sigma}\right)\mathbf{1}_{[\mu-\sigma\varphi_n(\alpha_*)/\Delta, \infty)}(x). \tag{6.17}$$

Formula 6.16 is a reduction of 6.13, and $||\bar{\Phi}_n'||^2$ is rewritten as 6.15. We can also show that $\Delta(n) \sim \binom{2n-2}{n-1}^{1/2} \sim 2^{n-1}/(n\pi)^{1/4}$ which is the rate of increase of the extreme expectation of $(n-1)$st record value (cf. 6.2). Distribution 6.17 has jump α_* and a density with infinite support right to the jump point. Observe that the contribution of the smooth component $1-\alpha_* < e^{-n+1}$ is very small.

We are able to establish similar bounds for populations with monotone density and failure rate functions, once we apply the following observation.

Lemma 15. *The decreasing-increasing functions $\varphi_n(x)$ and $\psi_n(x) = \varphi_n(1 - e^{-x})$, $n \geq 2$, are convex on their intervals of increase.*

Proof. For $\psi_n(x) = \frac{x^n}{n!} - \frac{x^{n-1}}{(n-1)!}$ that increases for $x \geq n-1$, we simply verify the claim by repeated differentiation. Since φ_n is its superposition with increasing convex function $V^{-1}(x) = -\ln(1-x)$, we have

$$\varphi_n''(x) = \psi_n'' V^{-1}(x)[V^{-1}(x)]^2 + \psi_n' V^{-1}(x)(V^{-1})''(x) > 0$$

for $x > 1 - e^{-n+1}$, which is the minimum point of φ_n. The relation is implied by positivity of all the above components of $\varphi_n''(x)$. □

Proposition 17. (decreasing density and failure rate) *For every $F \succeq_c U$ bound 6.14 is the best possible.*

The conclusion for $F \succeq_c V$ follows from $V \succ_c U$. The statement of the Proposition can be proved by checking that 6.17 has a decreasing density which is equivalent with convexity of $\varphi_n(\max\{\alpha_*, x\})$. Since $\alpha_* > 1 - e^{-n+1}$, this immediately follows from Lemma 15.

Proposition 18. (increasing density and failure rate) *For every $F \preceq_c U$*

$$\mathrm{E}_F(R_n - R_{n-1})/\sigma_F \leq \sqrt{3}/2^n, \tag{6.18}$$

with equality holding for the uniform distribution F on some interval of length $2\sqrt{3}\sigma$.

For every $F \preceq_c U$

$$\mathrm{E}_F(R_n - R_{n-1})/\sigma_F \leq 1, \tag{6.19}$$

which becomes equality for the exponential distribution with scale σ.

Both the assertions are deduced by means of analogous arguments. The crucial steps of proofs consist in showing that projections of φ_n and ψ_n onto the cones of nondecreasing concave functions are linear. By Lemma 2, neither of φ_n (ψ_n) and the projection can majorize the other. Since both φ_n and ψ_n are first decreasing and then strictly convex increasing, every nondecreasing concave function may cross either of them at two points at most. Taking linear functions that pass through the crossing points improves the approximations of φ_n and ψ_n. Standard arguments of the projection method lead us to the final conclusions.

Acknowledgements The work was done under the support of the State Committee for Scientific Research (KBN) Grant 2 P03A 014 13.

References

1. Arnold, B.C. (1980), Distribution-free bounds on the mean of the maximum of a dependent sample, *SIAM J. Appl. Math.* 38, 163–167.

2. Arnold, B.C. (1985), p-Norm bounds on the expectation of the maximum of possibly dependent sample, *J. Multivar. Anal.* 17, 316–332.

3. Arnold, B.C. (1988), Bounds on the expected maximum, *Commun. Statist. — Theor. Meth.* 17, 2135–2150.

4. Arnold, B. C. and N. Balakrishnan (1989), *Relations, Bounds and Approximations for Order Statistics*, Lecture Notes in Statistics, Vol. 53, Springer-Verlag, New York.

5. Arnold, B.C. and R.A. Groeneveld (1974), Bounds for deviations between sample population statistics, *Biometrika* 61, 387–389.

6. Balakrishnan, A. V. (1981), *Applied Functional Analysis*, 2nd ed., Springer-Verlag, New York.

7. Balakrishnan, N. (1993), A simple application of binomial-negative binomial relationship in the derivation of sharp bounds for moments of order statistics based on greatest convex minorants, *Statist. Probab. Lett.* 18, 301–305.

8. Barlow, R.E. and F. Proschan (1966), Inequalities for linear combinations of order statistics from restricted families, *Ann. Math. Statist.* 37, 1574–1591.

9. Biondini, R. and M.M. Siddiqui (1975), Record values in Markov sequences, in: *Statistical Inference and Related Topics*, Vol. 2, (M.L. Puri, ed.), Academic Press, New York, 291–352.

10. Blom, G. (1958), *Statistical Estimates and Transformed Beta-Variables*, Almqvist and Wiksells, Uppsala.

11. Boyd, A.V. (1971), Bound for order statistics, *Publ. of the Electrotechnical Faculty of Belgrade Univ., Math. Phys. Series* 365, 31–32.

12. Caraux, G. and O. Gascuel (1992), Bounds on distribution functions of order statistics for dependent variates, *Statist. Probab. Lett.* 14, 103–105.

13. David, H.A. (1981), *Order Statistics*, 2nd ed., Wiley, New York.

14. David, H.A. (1988), General bounds and inequalities in order statistics, *Commun. Statist. — Theor. Meth.* 17, 2119–2134.

15. David, H.A., H.O. Hartley and E.S. Pearson (1954), The distribution of the ratio, in a single normal sample, of range to standard deviation, *Biometrika* 41, 482–493.

16. Dharmadhikari, S. and K. Joag-dev (1988), *Unimodality, Convexity, and Applications*, Academic Press, New York.

17. Fahmy, S. and F. Proschan (1981), Bounds on differences of order statistics, *Amer. Statist.* 35, 46–47.

18. Gajek, L. and T. Rychlik (1996), Projection method for moment bounds on order statistics from restricted families. I. Dependent case, *J. Multivar. Anal.* 57, 156–174.

19. Gajek, L. and T. Rychlik (1998), Projection method for moment bounds on order statistics from restricted families. II. Independent case, *J. Multivar. Anal.* 64, 156–182.

20. Gascuel, O. and G. Caraux (1992), Bounds on expectations of order statistics via extremal dependences, *Statist. Probab. Lett.* 15, 143–148.

21. Grudzień, Z. and D. Szynal (1985), On the expected values of kth records and associated characterizations of distributions, in: *Proc. 4th Pannonian Symp. on Math. Statist., Bad Tatzmanndorf* 1983, (F. Konecny *et al.*, eds.), Reidel-Academiai Kiado, 119–127.

22. Gumbel, E.J. (1954), The maxima of the mean largest value and of the range, *Ann. Math. Statist.* 25, 76–84.

23. Hampel, F.R., E.M. Ronchetti, P.J. Rousseeuw and W.A. Stahel (1986), *Robust Statistics. The Approach Based on Influence Functions*, Wiley, New York.

24. Hartley, H.O. and H.A. David (1954), Universal bounds for mean range and extreme observation, *Ann. Math. Statist.* 25, 85–99.

25. Hawkins, D.M. (1971), On the bounds of the range of order statistics, *J. Amer. Statist. Assoc.* 66, 644–645.

26. Huber, P.J. (1981), *Robust Statistics*, Wiley, New York.

27. Lai, T.L. and H. Robbins (1976), Maximally dependent random variables, *Proc. Nat. Acad. Sci. U.S.A.* 73, 286–288.

28. Lai, T.L. and H. Robbins (1978), A class of dependent random variables and their maxima, *Z. Wahrsch. Verw. Gebiete* 42, 89–111.

29. Lawrence, M.J. (1975), Inequalities for s-ordered distributions, *Ann. Statist.* 3, 413–428.

30. Mallows, C.L. (1969), Extrema of expectations of uniform order statistics, *SIAM Rev.* 11, 410–411.

31. Mallows, C.L. and D. Richter (1969), Inequalities of Chebyshev type involving conditional expectations, *Ann. Math. Statist.* 40, 1922–1932.

32. Marshall, A.W. and I. Olkin (1979), *Inequalities: Theory of Majorization and Its Applications*, Academic Press, New York.

33. Moriguti, S. (1951), Extremal properties of extreme value distributions, *Ann. Math. Statist.* 22, 523–536.

34. Moriguti, S. (1953), A modification of Schwarz's inequality with applications to distributions, *Ann. Math. Statist.* 24, 107–113.

35. Nagaraja, H.N. (1978), On the expected values of records, *Austral. J. Statist.*, 20, 176-182.

36. Nagaraja, H.N. (1981), Some finite sample results for the selection differential, *Ann. Inst. Statist. Math.* 33, 437–448.

37. Nair, K.R. (1948), The distribution of the extreme deviate from the sample mean and its studentized form, *Biometrika* 35, 118–144.

38. Olkin, I. (1992), A matrix formulation on how deviant can an observation be, *Amer. Statist.* 46, 205–209.

39. Plackett, R.L. (1947), Limits of the ratio of mean range to standard deviation, *Biometrika* 34, 120–122.

40. Prakasa Rao, B.L.S. (1983), *Nonparametric Functional Estimation*, Academic Press, Orlando.

41. Raqab M.Z. (1997), Bounds based on greatest convex minorants for moments of record values, *Statist. Probab. Lett.* 36, 35–41.

42. Rychlik, T. (1992), Stochastically extremal distributions of order statistics for dependent samples, *Statist. Probab. Lett.* 13, 337–341.

43. Rychlik, T. (1992), Weak limit theorems for stochastically largest order statistics, in: *Order Statistics and Nonparametrics. Theory and Applications* (I.A. Salama and P.K. Sen, eds.), North-Holland, Amsterdam, 141–154.

44. Rychlik, T. (1993), Bounds for expectation of L-estimates for dependent samples, *Statistics* 24, 1–7.

45. Rychlik, T. (1993), Bias-robustness of L-estimates of location against dependence, *Statistics* 24, 9–15.

46. Rychlik, T. (1993), Sharp bounds on L-estimates and their expectations for dependent samples, *Commun. Statist. — Theory Meth.* 22, 1053–1068. Erratum in *Commun. Statist. — Theory Meth.* 23, 305–306.

47. Rychlik, T. (1995), Bounds for order statistics based on dependent variables with given nonidentical distributions, *Statist. Probab. Lett.* 23, 351–358.

48. Rychlik T. (1997), Evaluating improvements of records, *Appl. Math. (Warsaw)* 24, 315–324.

49. Rychlik T. (1998), Bounds on expectations of L-estimates, in: *Order Statistics: Theory & Methods*, (N. Balakrishnan and C.R. Rao, eds.), Handbook of Statistics, Vol. 16, North-Holland, Amsterdam, 105–145.

50. Rychlik, T. (1999), Error reduction in density estimation under shape restrictions, *Canad. J. Statist.*, to appear.

51. Rychlik T. (1999), Mean-variance bounds for order statistics from dependent DFR, IFR, DFRA and IFRA samples, submitted for publication.

52. Rychlik, T. (1999), Optimal mean-variance bounds on order statistics from families determined by star ordering, submitted for publication.

53. Rychlik, T. (1999), Stability of order statistics under dependence, submitted for publication.

54. Rychlik, T. (1999), Sharp mean-variance inequalities for quantiles of distributions determined by convex and star orderings, submitted for publication.

55. Samuelson, P.A. (1968), How deviant can you be? *J. Amer. Statist. Assoc.* 63, 1522–1525.

56. Schoenberg, I.J. (1959), On variation diminishing approximation methods, in: On Numerical Approximation: Proc. of Symp., Madison, 1958, (R.E. Langer, ed.), Univ. Wisconsin Press, Madison.

57. Scott, J.M.C. (1936), Appendix to paper by Pearson and Chandra Sekar, *Biometrika* 28, 319–320.

58. Serfling, R.J. (1980), *Approximation Theorems of Mathematical Statistics*, Wiley, New York.

59. Shaked, M. and J.G. Shantikumar (1994), *Stochastic Orders and Their Applications*, Academic Press, Boston.

60. Tchen, A. (1980), Inequalities for distributions with given marginals, *Ann. Probab.* 8, 814–827.

61. Thompson, W.R. (1935), On a criterion for the rejection of observations and the distribution of the ratio of deviation to sample standard deviation, *Ann. Math. Statist.* 6, 214–219.

62. van Zwet, W.R. (1964), *Convex Transformations of Random Variables*, Math. Centre Tracts, Vol. 7, Mathematisch Centrum, Amsterdam.

63. von Mises, M. (1947), On the asymptotic distribution of differentiable statistical functions, *Ann. Math. Statist* 18, 309–348.

64. Vysochanskii, D.F. and Yu. Petunin (1979), Justification of the three-sigma rule for unimodal distributions, *Theor. Probab. and Math. Statist.* 21, 25–36.

DYNAMICAL SYSTEMS AND DISCRETE METHODS FOR SOLVING NONLINEAR ILL-POSED PROBLEMS

Ruben G. Airapetyan and Alexander G. Ramm
Department of Mathematics, Kansas State University,
Manhattan, KS 66506-2602, E-mail: ramm@math.ksu.edu
Contact author: A. Ramm

1 Introduction

The theme of this chapter is solving of nonlinear operator equation

$$F(z) = 0, \quad F : H \to H, \tag{1}$$

by establishing a relation between the limiting behavior for large times of the trajectories of dynamical systems in H and the solutions to equation (1) in a real Hilbert space H. We consider a real Hilbert space for the sake of simplicity, equation (1) in a complex Hilbert space can be treated similarly, and a part of results can be generalized to a Banach space.

A standard approach to numerical solution of equation (1) consists of using one of the numerous iterative methods. These methods are also very useful for a theoretical investigation of problem (1). For some operators F they allow one to establish existence of a solution to problem (1), or existence and uniqueness theorems.

In the iterative methods one chooses some initial approximation z_0 and defines a sequence of points $\{z_n\}_{n=0,1,2,\ldots}$ having as a limit $\lim_{n\to\infty} z_n$ one of the solutions of equation (1). The choice of an initial point z_0 is very important especially for problems solution to which is not unique. Usually the choice of z_0 determines to what solution the iterative sequence converges.

The simplest iterative method is the method of simple iteration defined as follows:

$$z_n = z_{n-1} - \omega F(z_{n-1}), \quad n = 1, 2, \ldots \tag{2}$$

z_0 is given.

In the well-known Newton's method ([14, 18, 19]) one constructs a sequence $\{z_n\}$ by the following formula:

$$z_n = z_{n-1} - [F'(z_{n-1})]^{-1} F(z_{n-1}), \quad n = 1, 2, \ldots. \tag{3}$$

Here $F'(h)$ is the Fréchet derivative of the operator F, defined at a point $h \in H$ as a linear operator from H to H such that:

$$F(h + \xi) - F(h) = F'(h)\xi + o(||\xi||). \tag{4}$$

Thus, an important condition for applicability of Newton's method is invertibility of the Fréchet derivative operator F'. The important advantage of Newton's method is the quadratic convergence to the solution in a neighborhood of this solution:

$$||z_{n+1} - z_n|| \leq \text{const}||z_n - z_{n-1}||^2. \tag{5}$$

In many cases when Newton's method diverges one uses the damped Newton's method:

$$z_n = z_{n-1} - \omega_n[F'(z_{n-1})]^{-1}F(z_{n-1}), \quad n = 1, 2, \dots, \tag{6}$$

where ω_n is an appropriately chosen sequence of positive numbers.

In the gradient method one constructs an iterative sequence using the formula:

$$z_n = z_{n-1} - [F'(z_{n-1})]^* F(z_{n-1}), \quad n = 1, 2, \dots. \tag{7}$$

The iterative methods mentioned above are widely known representatives of a large family of iterative methods. A detailed description of many iterative methods one can find in [19]. See also an approach to construction of iterative process for solving nonlinear equation proposed in [20, 21].

The applicability of iterative methods is established by various convergence theorems (see [14, 18, 19]), which specify assumptions on the operator F which guarantee convergence of the iterative sequence to a solution of equation (1). The proofs of convergence theorems for iterative methods are usually based on the contraction mapping principle.

Another approach to solving problem (1) is based on a construction of a dynamical system with the trajectory starting from an initial approximation point z_0 and having a solution to problem (1) as a limiting point.

In [12] R. Courant proposed the dynamical system, which is a continuous analog of the gradient method, in the problem of minimization of some functionals. In [16] M.K. Gavurin proposed continuous Newton's method and established the corresponding convergence theorem. In continuous Newton's method one considers the following Cauchy problem for a nonlinear differential equation in a Hilbert space H:

$$\dot{z}(t) = -[F'(z(t))]^{-1}F(z(t)) \quad z(0) = z_0, \tag{8}$$

where z_0 is some initial approximation point. In [25] E.P. Zhidkov and I.V. Puzynin applied continuous Newton's method for solving nonlinear physical problems. Y. Alber used continuous methods for solving operator equations and variational inequalities ([6, 7, 8]). A modified continuous Newton's method is proposed in [1, 3]. In this method one avoids numerically difficult inversion of the Fréchet derivative F' by solving an expanded system of nonlinear differential equation in Hilbert space. In [2, 3] one can find the applications of this method to some physical problems.

The goal of this chapter is to develop a general approach to continuous analogs of discrete methods and to establish fairly general convergence theorems. This approach is based on an analysis of the solution to the Cauchy problem for a nonlinear differential equation in a Hilbert space. Such an analysis was done for well-posed and some ill-posed problems in [1, 4, 5], and was based on a usage of integral inequalities.

Let z_0 be an initial approximation for a solution to (1) and $z(t)$ be the trajectory of an autonomous dynamical system:

$$\dot{z}(t) = \Phi(z(t)), \quad 0 \leq t < \infty, \quad z(0) = z_0. \tag{9}$$

The main question investigated in this chapter is:

Under what assumptions on operators F in (1) and Φ in (9) one can guarantee that:

(i) Cauchy problem (9) is uniquely solvable for $t \in [0, +\infty)$,

(ii) the solution $z(t)$ tends to one of the solutions of (1) as $t \to \infty$,

(iii) there exists a step ω (or a sequence $\{\omega_n\}$) such that the corresponding discrete (iterative) method $z_{n+1} = z_n + \omega\Phi(z_n)$ (or $z_{n+1} = z_n + \omega_n\Phi(z_n)$) produces the sequence $\{z_n\}$, which converges to one of the solutions of (1).

The answers are given in Theorems 1, 2 and Corollaries 1, 2.

Thus an analysis of continuous processes is based on the investigation of the asymptotic behavior of nonlinear dynamical systems in Banach and Hilbert spaces. If a convergence theorem is proved for a continuous method, one can construct various discrete schemes generated by this continuous process. Thus construction of a discrete numerical scheme is divided into two parts: construction of the continuous process and numerical integration of the corresponding nonlinear differential equation in a Hilbert space.

The main assumption in Theorems 1, 2 and Corollaries 1, 2 is that the Fréchet derivative F' of the operator F has a trivial null-space at the solution to (1). If F' has a nontrivial null-space at the solution to (1) then, in general, one can use the classical Newton method for solving of (1) only under some strong special assumptions on the operator F (see [9, 13]). In order to relax these assumptions and to construct numerical method for solving (1) when F' has a nontrivial null-space at the solution to (1), one needs a regularized discrete Newton-like methods (see [6, 11, 15, 22, 24]).

In this chapter we consider the continuous Newton's method:

$$\dot{z}(t) = -[F'(z(t)) + \varepsilon(t)I]^{-1}[F(z(t)) + \varepsilon(t)(z(t) - \tilde{z}_0)], \quad z(0) = z_0, \tag{10}$$

where $\varepsilon(t)$ is a specially chosen positive function which tends to zero as $t \to +\infty$.

Thus, in the framework of our general approach, instead of the Cauchy problem (9) for autonomous equation the following Cauchy problem for nonautonomous equation has to be considered:

$$\dot{z}(t) = \Phi(z(t), t), \quad z(0) = z_0, \tag{11}$$

where Φ is a nonlinear operator, $\Phi : H \times [0, +\infty) \to H$.

An analysis of dynamical system (11) is more complicated. In this study we use new integral inequality (Theorem 3). Based on Theorem 3, a general convergence theorem for a regularized continuous process is proved (Theorem 4). Applying this theorem to the regularized Newton's and simple iteration methods (for monotone operators) we obtain convergence theorems under less restrictive conditions on the equation than the theorems known for the corresponding discrete methods. Statements of these theorems contain some useful recommendations for the choice of a regularizing operator and estimate the rate of convergence of the regularized process. Convergence theorems for regularized continuous Newton-like methods are established in [4, 7, 22]. The applications of these scheme to Gauss-Newton-type methods for nonmonotone operators can be found in [4, 5].

Throughout this chapter the Hilbert space is assumed real-valued.

2 Continuous Methods for Well Posed Problems

Let y be a solution to equation (1), z_0 be some point in H, considered as an initial approximation to a solution to equation (1), and $z(t)$ be some trajectory in H such that $z(0) = z_0$.

Definition 1. We say that a trajectory $z(t)$ converges to solution y of equation (1) exponentially if there exist positive constants c and c_1 such that

$$||z(t) - y|| \leq c_1 e^{-ct} ||F(z_0)||, \quad ||F(z(t))|| \leq ||F(z_0)|| e^{-ct}. \tag{12}$$

Consider Cauchy problem (8). Examples of the choice of an operator Φ are given in Remark 1.

Theorem 1. *Assume that there exist some positive real numbers r, c such that F, F', and Φ are Fréchet differentiable and bounded in $B_r(z_0)$ and the following conditions hold for every $h \in B_r(z_0)$:*

$$(F'(h)\Phi(h), F(h)) \leq -c||F(h)||^2, \tag{13}$$

and

$$||\Phi(h)|| \leq \frac{rc}{||F(z_0)||}||F(h)||. \tag{14}$$

Then

1) there exists a solution $z = z(t)$, $t \in [0, \infty)$, to problem (8) and $z(t) \in B_r(z_0)$ for all $t \in [0, +\infty)$;

2) there exists

$$\lim_{t \to +\infty} z(t) = y, \tag{15}$$

y is a solution of problem (1) in $B_r(z_0)$, and $z(t)$ converges to y exponentially.

Remark 1.

a) Choosing $\Phi(h) = -[F'(h)]^{-1}F(h)$ one gets Continuous Newton's method. In this case $c = 1$ and Theorem 1 yields the convergence theorem for Continuous Newton's method ([16]);

b) choosing $\Phi = -F$, one gets a simple iteration method for which condition (13) means strict monotonicity of F;

c) $\Phi(h) = -[F'(h)]^* F(h)$ corresponds to the gradient method.

Proof of Theorem 1. From the Fréchet differentiability of Φ in $B_r(z_0)$, we get the local existence of a solution of the problem (8). Then from (8) one gets:

$$F'(z(t))\dot{z}(t) = -F'(z(t))\Phi(z(t). \tag{16}$$

From (16) denoting $\lambda(t) = F(z(t))$, one gets:

$$\dot{\lambda}(t) = -F'(z(t))\Phi(z(t), \quad \lambda(0) = F(z_0). \tag{17}$$

Therefore, for sufficiently small t for which $z(t) \in B_r(z_0)$, one can use estimate (13) and get:

$$\frac{d}{dt}||\lambda(t)||^2 = 2(\dot{\lambda}(t), \lambda(t)) = -2(F'(z(t))\Phi(z(t), F(z(t)) \leq -2c||\lambda(t)||^2.$$

Thus, the following estimate holds at least for sufficiently small $t > 0$:

$$||\lambda(t)|| \leq ||F(z_0)||e^{-ct}. \tag{18}$$

For $0 \leq t_1 \leq t_2$ one has:

$$||z(t_2) - z(t_1)|| \leq || \int_{t_1}^{t_2} \dot{z}(s)ds|| \leq \int_{t_1}^{t_2} ||\Phi(z(s))||ds$$

$$\leq \frac{rc}{||F(z_0)||} \int_{t_1}^{t_2} ||\lambda(s)||ds \leq r(e^{-ct_1} - e^{-ct_2}) \leq re^{-ct_1}. \qquad (19)$$

Setting $t_1 = 0$ and $t_2 = t$, one concludes from (19) that $z(t) \in B_r(z_0)$ for all $t > 0$. Now let $t_1 = t$ and $t_2 \to +\infty$ in (19). Then one gets

$$||z(t) - y|| \leq re^{-ct}, \qquad (20)$$

where y is defined in (15) and the limit in (15) does exist due to (19). The exponential convergence of $z(t)$ to y follows now from (18) and (20). □

Remark 2. Note that in general the assumptions of Theorem 1 do not imply the uniqueness of a solution of equation (1) in $B_r(y)$. If (1) is not uniquely solvable then $z(t)$ converges to one of its solutions.

To establish convergence theorems for discrete methods we need a modified version of the statement of Theorem 1. We start with the following definition.

Definition 2. Let y be a unique solution to equation (1) in some set M. M is called Φ-attractive set for y if for any $z_0 \in M$ the trajectory $z(t)$ of the dynamical system (9) does not leave M and tends to y as $t \to +\infty$. If $z(t)$ converges to y exponentially we call M an exponential Φ-attractive set for y. If constants c and c_1 in Definition 1 do not depend on a point $z_0 \in M$ we call M a uniformly-exponential Φ-attractive set for y.

Let us formulate a corollary to Theorem 1.

Corollary 1. *Assume that there exist some positive numbers r, c such that y is the unique solution to problem (1) in the ball $B_r(y)$, $z_0 \in B_r(y)$, and the assumptions of Theorem 1 are satisfied in $B_r(y)$. Then $B_r(y)$ is a uniformly-exponential Φ-attractive set for y with c defined in (13) and*

$$c_1 = \frac{r}{||F(z_0)||}. \qquad (21)$$

3 Discretization Theorems for Well-posed Problems

Consider the following Cauchy problem:

$$\dot{z}(t) = \Phi(z(t)), \quad z(0) = z_0, \tag{22}$$

and the corresponding discrete process:

$$z_{n+1} = z_n + \omega\Phi(z_n). \tag{23}$$

Theorem 2. *Let $B_r(y)$ be a uniform-exponential Φ-attractive set for y for some $r > 0$ such that $r \geq c_1||F(z_0)||$, where c_1 is the same as in Definition 1, $z_0 \in B_r(y)$. Assume that*

1. *F and Φ are Fréchet differentiable in $B_r(y)$, and*

$$||\Phi(h)|| \leq N_0||F(h)||, \quad ||F'(h)|| \leq N_1,$$
$$||\Phi'(h)|| \leq N_2 \quad \textit{for} \quad h \in B_r(y), \tag{24}$$

2. *c_2 and ω are some positive constants satisfying the following inequalities:*

$$(c_1 e^{-c\omega} + N_0 N_2 \omega^2)||F(z_0)|| \leq r e^{-c_2\omega}, \tag{25}$$

$$e^{-c\omega} + N_0 N_1 N_2 \omega^2 \leq e^{-c_2\omega}. \tag{26}$$

Then all $\{z_n\}$, $n = 1, 2, \ldots$, defined by formula (23) belong to the ball $B_r(y)$ and

$$||z_n - y|| \leq r e^{-c_2 n\omega}, \quad ||F(z_n)|| \leq ||F(z_0)||e^{-c_2 n\omega}, \quad n = 0, 1, 2, \ldots. \tag{27}$$

Proof of Theorem 2. The idea of the proof is illustrated in Figure 1. Let z_0 be an initial approximation point. Since $B_r(y)$ is an exponential attractive set for y integral curve $\psi_1(t)$ of equation (22) connects z_0 with y. Let ω be a step. Consider points $\psi_1(\omega)$ and z_1 defined by (22) and (23) correspondingly. The point $\psi_1(\omega)$ is located on an integral curve and the point z_1 is located on a tangent line to this integral curve passing through z_0. Since $\psi_1(t)$ converges exponentially to y and distance between $\psi_1(\omega)$ and z_1 decreases as ω^2 for $\omega \to 0$, one shows that for a sufficiently small step ω, z_1 is closer to y than z_0. Therefore z_1 also belongs to $B_r(y)$. Moreover using the triangle inequality one can estimate the distance between z_1 and y. Then take the integral curve $\psi_2(t)$, which connects z_1 with y, points $\psi_2(\omega)$ and z_2, show that z_2 belongs to $B_r(y)$, and estimate the distance between z_2 and y. Repeating this process

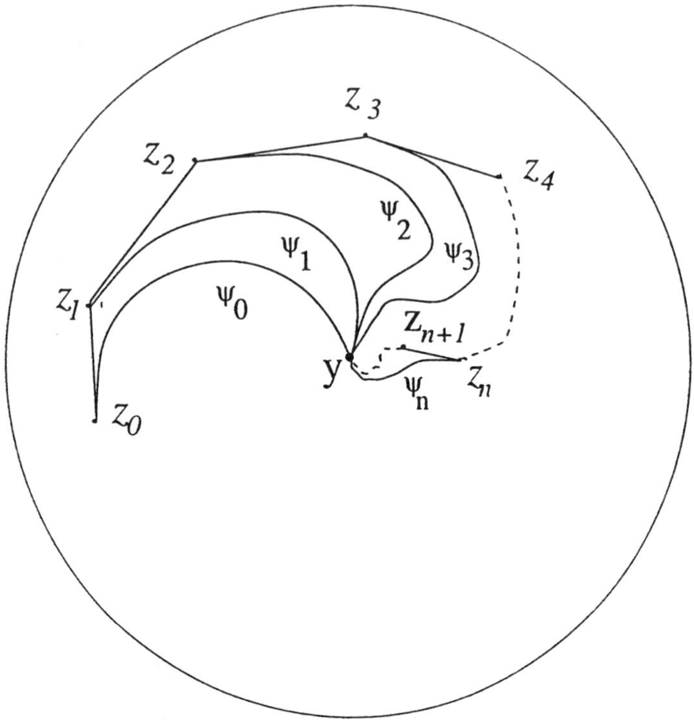

Figure 1: Discretization scheme of a continuous process.

one estimates the distance between $\{z_n\}$ and y and shows that this distance exponentially tends to zero.

We prove (27) using mathematical induction. For $n = 0$ conditions (27) are satisfied. Assume that (27) are satisfied for $n = m - 1$.

Denote by $\psi_m(t)$ the solution to the following Cauchy problem:

$$\dot{z}(t) = \Phi(z(t)), \quad 0 < t \leq \omega, \quad z(0) = z_{m-1}. \tag{28}$$

Then one has:

$$||z_m - y|| \leq ||\psi_m(\omega) - y|| + ||\psi_m(\omega) - z_m||. \tag{29}$$

Since $B_r(y)$ is a uniform-exponential Φ-attractive set for y one gets:

$$||F(\psi_m(t))|| \leq ||F(z_{m-1})||e^{-ct}, \quad t \in [0, \omega], \tag{30}$$

and

$$||\psi_m(\omega) - y|| \leq c_1 e^{-c\omega} ||F(z_{m-1})|| \leq c_1 e^{-c\omega} e^{-c_2\omega(m-1)} ||F(z_0)||. \tag{31}$$

From (22) and (23) one obtains:

$$\psi_m(\omega) - z_m = \int_0^\omega [\Phi(z(\tau)) - \Phi(z_{m-1})] d\tau = \int_0^\omega d\tau \int_0^1 \frac{d}{ds} \Phi(z(s\tau)) ds =$$

$$= \int_0^\omega \tau d\tau \int_0^1 \Phi'(z(s\tau)) \dot{z}(s\tau) ds \tag{32}$$

Therefore

$$||\psi_m(\omega) - z_m|| \leq \omega \int_0^\omega ||\Phi'(z(s))|| \, ||\Phi(z(s))|| \, ds. \tag{33}$$

Using (30) one gets:

$$||\psi_m(\omega) - z_m|| \leq \omega N_0 N_2 \int_0^\omega ||F(z(s))|| ds$$

$$\leq \omega^2 N_0 N_1 ||F(z_{m-1})|| \leq \omega^2 N_0 N_2 e^{-c_2\omega(m-1)} ||F(z_0)||. \tag{34}$$

From (29), (31), and (34) one gets:

$$\|z_m - y\| \leq (c_1 e^{-c\omega} e^{-c_2(m-1)\omega} + \omega^2 N_0 N_2 e^{-c_2\omega(m-1)}) \|F(z_0)\|. \tag{35}$$

Using condition (21) one obtains:

$$||z_m - y|| \leq r e^{-c_2 m\omega}. \tag{36}$$

Also

$$||F(z_m)|| \leq ||F(\psi_m(\omega))|| + N_1 ||\psi_m(\omega) - z_m|| \leq e^{-c\omega} ||F(z_{m-1})||$$

$$+\omega^2 N_0 N_1 N_2 e^{-c_2\omega(m-1)} ||F(z_0)|| \leq e^{-c_2\omega(m-1)} (e^{-c\omega} + \omega^2 N_0 N_1 N_2) ||F(z_0)||$$

$$\leq (e^{-c\omega} + \omega^2 N_0 N_1 N_2) e^{-c_2\omega(m-1)} ||F(z_0)||. \tag{37}$$

From (37) and (26) one gets:

$$||F(z_m)|| \leq ||F(z_0)|| e^{-c_2 m\omega}. \tag{38}$$

Thus the estimate (21) and (26) hold also for $n = m$. Theorem 2 is proved. □

Corollary 2. *Assume that there exist some positive numbers r, c such that:*

1. *y is the unique solution to problem (1) in the ball $B_r(y)$, and an initial approximation point $z_0 \in B_r(y)$,*

2. *the assumptions of Theorem 1 are satisfied in $B_r(y)$,*

3. $$||F'(h)|| \leq N_1, \; ||\Phi'(h)|| \leq N_2 \quad \textit{for} \quad h \in B_r(y), \tag{39}$$

4. *c_2 and ω are some positive constants satisfying the following inequality:*

$$e^{-c\omega} + cN_2\omega^2 \max\left\{1, \frac{rN_1}{||F(z_0)||}\right\} \leq e^{-c_2\omega}, \tag{40}$$

5. *z_0 is an initial approximation point and the sequence $\{z_n\}$, $n = 1, 2, \ldots$, is defined recursively:*

$$z_n = z_{n-1} + \omega\Phi(z_{n-1}), \tag{41}$$

Then all $\{z_n\}$, $n = 1, 2, \ldots$, belong to the ball $B_r(y)$ and

$$||z_n - y|| \leq re^{-c_2 n\omega}, \quad ||F(z_n)|| \leq ||F(z_0)||e^{-c_2 n\omega}, \quad n = 0, 1, 2, \ldots. \tag{42}$$

Proof of Corollary 2. Since the assumptions of Theorem 1 are satisfied in $B_r(y)$, choosing

$$N_0 = \frac{rc}{||F(z_0)||} \tag{43}$$

one gets the first inequality in (24) from (14). The second and the third inequalities in (24) follow from (39).

From Corollary 1 one gets that $B_r(y)$ is a uniformly-exponential Φ-attractive set for y with c defined in condition (13) and

$$c_1 = \frac{r}{||F(z_0)||}. \tag{44}$$

For such N_0 and c_1 condition (40) is equivalent to conditions (21) and (26). To finish the proof one refers to Theorem 2. □

4 An Non-linear Inequality

The main result of this section is Theorem 3 which is used throughout the paper.

The following lemma is a version of some known results concerning integral inequalities (see e.g. Theorem 22.1 in [23]). For convenience of the reader and to make the presentation essentially self-contained we include a proof.

Lemma 1. *Let $f(t,w)$, $g(t,u)$ be continuous on region $[0,T) \times D$ ($D \subset R$, $T \leq \infty$) and $f(t,w) \leq g(t,u)$ if $w \leq u$, $t \in (0,T)$, $w,u \in D$. Assume that $g(t,u)$ is such that the Cauchy problem*

$$\dot{u} = g(t,u), \quad u(0) = u_0, \quad u_0 \in D \tag{45}$$

has a unique solution. If

$$\dot{w} \leq f(t,w), \quad w(0) = w_0 \leq u_0, \quad w_0 \in D, \tag{46}$$

then $u(t) \geq w(t)$ for all t for which $u(t)$ and $w(t)$ are defined.

Proof of Lemma 1. Step 1. Suppose first $f(t,w) < g(t,u)$, if $w \leq u$. Since $w_0 \leq u_0$ and $\dot{w}(0) \leq f(t,w_0) < g(t,u_0) = \dot{u}(0)$, there exists $\delta > 0$ such that $u(t) > w(t)$ on $(0,\delta]$. Assume that for some $t_1 > \delta$ one has $u(t_1) < w(t_1)$. Then for some $t_2 < t_1$ one has

$$u(t_2) = w(t_2) \quad \text{and} \quad u(t) < w(t) \quad \text{for} \quad t \in (t_2, t_1].$$

One gets

$$\dot{w}(t_2) \geq \dot{u}(t_2) = g(t,u(t_2)) > f(t,w(t_2)) \geq \dot{w}(t_2).$$

This contradiction proves that there is no point t_2 such that $u(t_2) = w(t_2)$.

Step 2. Now consider the case $f(t,w) \leq g(t,u)$, if $w \leq u$. Define

$$\dot{u}_n = g(t,u_n) + \varepsilon_n, \quad u_n(0) = u_0, \quad \varepsilon_n > 0, \quad n = 0,1,...,$$

where ε_n tends monotonically to zero. Then

$$\dot{w} \leq f(t,w) \leq g(t,u) < g(t,u) + \varepsilon_n, \quad w \leq u.$$

By Step 1 $u_n(t) \geq w(t)$, $n = 0,1,\ldots$. Fix an arbitrary compact set $[0,T_1]$, $0 < T_1 < T$.

$$u_n(t) = u_0 + \int_0^t g(\tau, u_n(\tau))d\tau + \varepsilon_n t. \tag{47}$$

Since $g(t, u)$ is continuous, the sequence $\{u_n\}$ is uniformly bounded and equicontinuous on $[0, T_1]$. Therefore there exists a subsequence $\{u_{n_k}\}$ which converges uniformly to a continuous function $u(t)$. By continuity of $g(t, u)$ we can pass to the limit in (47) and get

$$u(t) = u_0 + \int_0^t g(\tau, u(\tau))d\tau, \quad t \in [0, T_1]. \tag{48}$$

Since T_1 is arbitrary (48) is equivalent to the initial Cauchy problem that has a unique solution. The inequality $u_{n_k}(t) \geq w(t)$, $k = 0, 1, ...$ implies $u(t) \geq w(t)$. If the solution to the Cauchy problem (45) is not unique, the inequality $w(t) \leq u(t)$ holds for the maximal solution to (45). □

Our second lemma is a key to the basic result, namely to Theorem 4.

Theorem 3. *Let $\gamma(t), \sigma(t), \beta(t) \in C[t_0, \infty)$ for some real number t_0. If there exists a positive function $\mu(t) \in C^1[t_0, \infty)$ such that*

$$0 \leq \sigma(t) \leq \frac{\mu(t)}{2}\left(\gamma(t) - \frac{\dot{\mu}(t)}{\mu(t)}\right), \quad \beta(t) \leq \frac{1}{2\mu(t)}\left(\gamma(t) - \frac{\dot{\mu}(t)}{\mu(t)}\right), \tag{49}$$

then a nonnegative solution to the following inequalities:

$$\dot{v}(t) \leq -\gamma(t)v(t) + \sigma(t)v^2(t) + \beta(t), \quad v(t_0) < \frac{1}{\mu(t_0)}, \tag{50}$$

satisfies the estimate:

$$v(t) \leq \frac{1 - \nu(t)}{\mu(t)} < \frac{1}{\mu(t)}, \tag{51}$$

for all $t \in [t_0, \infty)$, where

$$\nu(t) = \left(\frac{1}{1 - \mu(t_0)v(t_0)} + \frac{1}{2}\int_{t_0}^t \left(\gamma(s) - \frac{\dot{\mu}(s)}{\mu(s)}\right) ds\right)^{-1}. \tag{52}$$

Remark 3. Without loss of generality one can assume $\beta(t) \geq 0$.

In [8] a differential inequality $\dot{v} \leq -A(t)\psi(v(t)) + \beta(t)$ was studied under some assumptions which include, among others, the positivity of $\psi(v)$ for $v > 0$. In Theorem 3 the term $-\gamma(t)v(t) + \sigma(t)v^2(t)$ (which is analogous to some extent to the term $-A(t)\psi(v(t))$) can change sign. Our Theorem 3 is not covered by the result in [8]. In particular, in Theorem 3 an analog of $\psi(v)$, for the case

$\gamma(t) = \sigma(t) = A(t)$, is the function $\psi(v) := v - v^2$. This function goes to $-\infty$ as v goes to $+\infty$, so it does not satisfy the positivity condition imposed in [8].

Unlike in the case of Bihari integral inequality ([10]) one cannot separate variables in the right hand side of the first inequality (50) and estimate $v(t)$ by a solution of the Cauchy problem for a differential equation with separating variables. The proof below is based on a special choice of the solution to the Riccati equation majorizing a solution of inequality (50).

Proof of Theorem 3. Denote:

$$w(t) := v(t)e^{\int_{t_0}^{t} \gamma(s)ds}, \tag{53}$$

then (50) implies:

$$\dot{w}(t) \le a(t)w^2(t) + b(t), \quad w(t_0) = v(t_0), \tag{54}$$

where

$$a(t) = \sigma(t)e^{-\int_{t_0}^{t} \gamma(s)ds}, \quad b(t) = \beta(t)e^{\int_{t_0}^{t} \gamma(s)ds}.$$

Consider Riccati's equation:

$$\dot{u}(t) = \frac{\dot{f}(t)}{g(t)}u^2(t) - \frac{\dot{g}(t)}{f(t)}. \tag{55}$$

One can check by a direct calculation that the the solution to problem (55) is given by the following formula [17, eq. 1.33]:

$$u(t) = -\frac{g(t)}{f(t)} + \left[f^2(t)\left(C - \int_{t_0}^{t} \frac{\dot{f}(s)}{g(s)f^2(s)}ds\right)\right]^{-1}. \tag{56}$$

Define f and g as follows:

$$f(t) := \mu^{\frac{1}{2}}(t)e^{-\frac{1}{2}\int_{t_0}^{t} \gamma(s)ds}, \quad g(t) := -\mu^{-\frac{1}{2}}(t)e^{\frac{1}{2}\int_{t_0}^{t} \gamma(s)ds}, \tag{57}$$

and consider the Cauchy problem for equation (55) with the initial condition $u(t_0) = v(t_0)$. Then C in (56) takes the form:

$$C = \frac{1}{\mu(t_0)v(t_0) - 1}.$$

From (49) one gets

$$a(t) \le \frac{\dot{f}(t)}{g(t)}, \quad b(t) \le -\frac{\dot{g}(t)}{f(t)}.$$

Since $fg = -1$ one has:

$$\int_{t_0}^{t} \frac{\dot{f}(s)}{g(s)f^2(s)} ds = -\int_{t_0}^{t} \frac{\dot{f}(s)}{f(s)} ds = \frac{1}{2}\int_{t_0}^{t} \left(\gamma(s) - \frac{\dot{\mu}(s)}{\mu(s)}\right) ds.$$

Thus

$$u(t) = \frac{e^{\int_{t_0}^{t} \gamma(s)ds}}{\mu(t)} \left[1 - \left(\frac{1}{1-\mu(t_0)v(t_0)} + \frac{1}{2}\int_{t_0}^{t} \left(\gamma(s) - \frac{\dot{\mu}(s)}{\mu(s)}\right) ds\right)^{-1}\right]. \quad (58)$$

It follows from conditions (49) and from the second inequality in (50) that the solution to problem (55) exists for all $t \in [0, \infty)$ and the following inequality holds with $\nu(t)$ defined by (52):

$$1 > 1 - \nu(t) \geq \mu(t_0)v(t_0). \quad (59)$$

From Lemma 1 and from formula (58) one gets:

$$v(t)e^{\int_{t_0}^{t} \gamma(s)ds} := w(t) \leq u(t) = \frac{1-\nu(t)}{\mu(t)} e^{\int_{t_0}^{t} \gamma(s)ds} < \frac{1}{\mu(t)} e^{\int_{t_0}^{t} \gamma(s)ds}, \quad (60)$$

and thus estimate (51) is proved. □

To illustrate conditions of Theorem 3 consider the following examples of functions γ, σ, β, satisfying (49) for $t_0 = 0$.

Example 1. Let

$$\gamma(t) = c_1(1+t)^{\nu_1}, \quad \sigma(t) = c_2(1+t)^{\nu_2}, \quad \beta(t) = c_3(1+t)^{\nu_3}, \quad (61)$$

where $c_2 > 0$, $c_3 > 0$. Choose $\mu(t) := c(1+t)^{\nu}$, $c > 0$. From (49), (50) one gets the following conditions

$$c_2 \leq \frac{cc_1}{2}(1+t)^{\nu+\nu_1-\nu_2} - \frac{c\nu}{2}(1+t)^{\nu-1-\nu_2},$$

$$c_3 \leq \frac{c_1}{2c}(1+t)^{\nu_1-\nu-\nu_3} - \frac{\nu}{2c}(1+t)^{-\nu-1-\nu_3}, \quad cv(0) < 1. \quad (62)$$

Thus one obtains the following conditions:

$$\nu_1 \geq -1, \quad \nu_2 - \nu_1 \leq \nu \leq \nu_1 - \nu_3, \quad (63)$$

and

$$c_1 > \nu, \quad \frac{2c_2}{c_1 - \nu} \leq c \leq \frac{c_1 - \nu}{2c_3}, \quad cv(0) < 1. \quad (64)$$

Therefore for such γ, σ, β a function μ with the desired properties exists if

$$\nu_1 \geq -1, \quad \nu_2 + \nu_3 \leq 2\nu_1, \tag{65}$$

and

$$c_1 > \nu_2 - \nu_1, \quad 2\sqrt{c_2 c_3} \leq c_1 + \nu_1 - \nu_2, \quad 2c_2 v(0) < c_1 + \nu_1 - \nu_2. \tag{66}$$

In this case one can choose $\nu = \nu_2 - \nu_1$, $c = \frac{2c_2}{c_1+\nu_1-\nu_2}$. However in order to have $v(t) \to 0$ as $t \to +\infty$ (the case of interest in Theorem 4) one needs the following conditions:

$$\nu_1 \geq -1, \quad \nu_2 + \nu_3 \leq 2\nu_1, \quad \nu_1 > \nu_3, \tag{67}$$

and

$$c_1 > \nu_2 - \nu_1, \quad 2\sqrt{c_2 c_3} \leq c_1, \quad 2c_2 v(0) < c_1. \tag{68}$$

Example 2. If

$$\gamma(t) = \gamma_0, \quad \sigma(t) = \sigma_0 e^{\nu t}, \quad \beta(t) = \beta_0 e^{-\nu t}, \quad \mu(t) = \mu_0 e^{\nu t},$$

then conditions (49), (50) are satisfied if

$$0 \leq \sigma_0 \leq \frac{\mu_0}{2}(\gamma_0 - \nu), \quad \beta_0 \leq \frac{1}{2\mu_0}(\gamma_0 - \nu), \quad \mu_0 v(0) < 1.$$

Example 3. Here and throughout the paper log stands for the natural logarithm. For some $t_1 > 0$

$$\gamma(t) = \frac{1}{\sqrt{\log(t+t_1)}}, \quad \mu(t) = c\log(t+t_1),$$

conditions (49), (50) are satisfied if

$$0 \leq \sigma(t) \leq \frac{c}{2}\left(\sqrt{\log(t+t_1)} - \frac{1}{t+t_1}\right),$$

$$\beta(t) \leq \frac{1}{2c\log^2(t+t_1)}\left(\sqrt{\log(t+t_1)} - \frac{1}{t+t_1}\right), \quad v(0)c\log t_1 < 1.$$

In all considered examples $\mu(t)$ can tend to infinity as $t \to +\infty$ and provide a decay of a nonnegative solution to integral inequality (50) even if $\sigma(t)$ tends to infinity. Moreover in the first and the third examples $v(t)$ tends to zero as $t \to +\infty$ when $\gamma(t) \to 0$ and $\sigma(t) \to +\infty$.

5 Regularization Procedure for Ill-posed Problems

In the well-posed case (when the Fréchet derivative F' of the operator F is a bijection in a neighborhood of the solution of equation (1)) in order to solve equation (1) one can use the following continuous processes:

- simple iteration method:

$$\dot{z}(t) = -F(z(t)), \quad z(0) = z_0 \in H, \tag{69}$$

- Newton's method:

$$\dot{z}(t) = -[F'(z(t))]^{-1}F(z(t)), \quad z(0) = z_0 \in H. \tag{70}$$

However if F' is not continuously invertible (ill-posed case) one has to replace the Cauchy problems (69) and (70) by the corresponding regularized Cauchy problems (see [4, 5, 6]):

- regularized simple iteration method:

$$\dot{z}(t) = -[F(z(t)) + \varepsilon(t)(z(t) - \tilde{z}_0)], \quad z(0) = z_0 \in H, \tag{71}$$

- regularized Newton's method:

$$\dot{z}(t) = -[F'(z(t)) + \varepsilon(t)I]^{-1}[F(z(t)) + \varepsilon(t)(z(t) - \tilde{z}_0)], \quad z(0) = z_0 \in H. \tag{72}$$

Here $\tilde{z}_0 \in H$ is some element, $\varepsilon(t) > 0$ is a suitable function, with the properties specified in Theorems 7 and 8, and I is the identity operator. The equations in (71) and (72) are no longer autonomous.

Our goal is to develop a uniform approach to such regularized methods. Let us consider the Cauchy problem:

$$\dot{z}(t) = \Phi(z(t), t), \quad z(0) = z_0 \in H, \tag{73}$$

with an operator $\Phi : H \times [0, \infty) \to H$.

Let y be a solution to equation (1) (as it was mentioned in Introduction, we assume that this equation is solvable). Denote: $B_R(y) := \{h : h \in H, ||h - y|| < R\}$.

Theorem 4. *Assume that there exists $r > 0$ such that $\Phi(h, t)$ is Fréchet differentiable with respect to h in the ball $B_{2r}(y)$ for any $t \in [0, \infty)$ and satisfy the following condition:*

there exists a differentiable function $x(t)$, $x : [0, +\infty) \to B_r(y)$, such that for any $h \in B_{2r}(y)$, $t \in [0, +\infty)$

$$(\Phi(h,t), h - x(t)) \le \alpha(t)||h - x(t)|| - \gamma(t)||h - x(t)||^2 + \sigma(t)||h - x(t)||^3, \quad (74)$$

where $\alpha(t)$ is a continuous function, $\alpha(t) \ge 0$, $\gamma(t)$ and $\sigma(t)$ satisfy conditions (49) of Theorem 3 with

$$\beta(t) := ||\dot{x}(t)|| + \alpha(t), \quad (75)$$

$\mu(t)$ in (49) tends to $+\infty$ as $t \to +\infty$, and

$$||z_0 - x(0)|| < \frac{1}{\mu(0)}, \qquad \inf_{t \in [0,+\infty)} \mu(t) \ge \frac{1}{r}. \quad (76)$$

Then problem (73) has a unique solution $z(t) \in B_{2r}(y)$ for all $t \in [0, \infty)$, and

$$||z(t) - x(t)|| < \frac{1 - \nu(t)}{\mu(t)} < \frac{1}{\mu(t)}, \quad (77)$$

where $\nu(t)$ is defined by (52), and

$$\lim_{t \to +\infty} ||z(t) - x(t)|| = 0. \quad (78)$$

Remark 4. One can choose the regularizing operator $\Phi(h, t)$ in (73) such that condition (74) holds in the case when $F'^*(h)F'(h)$ is not boundedly invertible (see Section 7, Lemmas 6, and 7).

Proof of Theorem 4. Since $\Phi(h, t)$ is Fréchet differentiable with respect to h problem (73) is locally uniquely solvable. Denote by $[0, T)$ the maximal interval on which the solution $z(t)$ to problem (73) exists and $z(t) \in B_{2r}(y)$. One has to show that $T = +\infty$. Assume $T < +\infty$, then the trajectory $z(t)$ hits the boundary of $B_{2r}(y)$ at $t = T$: $||z(T) - y|| = 2r$. Since H is a real Hilbert space one has:

$$\begin{aligned}\frac{1}{2}\frac{d}{dt}||z(t) - x(t)||^2 &= (\dot{z} - \dot{x}, z(t) - x(t)) \\ &= (\Phi(z(t), t), z(t) - x(t)) - (\dot{x}, z(t) - x(t)). \quad (79)\end{aligned}$$

Therefore from (74) and (75) for $t \in [0, T)$ one obtains

$$\begin{aligned}\frac{1}{2}\frac{d}{dt}||z(t) - x(t)||^2 \le -\gamma||z(t) - x(t)||^2 &+ \sigma(t)||z(t) - x(t)||^3 \\ &+ \beta(t)||z(t) - x(t)||. \quad (80)\end{aligned}$$

Denote

$$v(t) := ||z(t) - x(t)||.$$

From (80) one has:

$$v(t)\dot{v}(t) \leq -\gamma(t)v^2(t) + \sigma(t)v^3(t) + \beta(t)v(t).$$

If $v > 0$, one gets:

$$\dot{v}(t) \leq -\gamma(t)v(t) + \sigma(t)v^2(t) + \beta(t). \tag{81}$$

If $v = 0$ on some interval, then inequality (81) is satisfied trivially because $\beta(t) \geq 0$. Thus (81) holds for all $t > 0$.

By Theorem 3 using (76) one obtains

$$||z(t) - x(t)|| \leq \frac{1}{\mu(t)} \leq r, \qquad \text{for} \qquad t \in [0, T). \tag{82}$$

Thus, since $x(t) \in B_r(y)$, one gets

$$||z(t) - y|| \leq ||z(t) - x(t)|| + ||x(t) - y|| < 2r, \qquad \text{for} \qquad t \in [0, T). \tag{83}$$

Therefore there exists a sequence $\{t_n\} \to T$ such that $\{z(t_n)\}$ converges weakly to some z^*. From equation (73) one derives the uniform boundedness of the norm $||\dot{z}(t)||$ on $[0, T)$ since $||\Phi(z(t), t)|| \leq \text{const} < \infty$ for $||z(t)|| \leq \text{const}$ and $0 \leq t \leq \text{const}$. Thus there exists $\lim_{t\to T} ||z(t) - z^*|| = 0$. Since

$$||z^* - y|| \leq ||z^* - x(T)|| + ||x(T) - y|| < 2r \qquad \text{for} \qquad t \in [0, T), \tag{84}$$

the conditions for the unique local solvability of the Cauchy problem for equation (73) with initial condition $z(T) = z^*$ are satisfied. Therefore one can continue the solution to (73) through the point T. This contradicts the assumption of maximality of T, thus $T = +\infty$. Moreover, from (51) one gets:

$$\lim_{t\to+\infty} ||z(t) - x(t)|| \leq \lim_{t\to+\infty} \frac{1}{\mu(t)} = 0. \tag{85}$$

□

In order to establish the discretization theorem in the next section we need to estimate $||\Phi||$ along the trajectory $z(t)$. The following theorem gives such an estimate.

Theorem 5. *Let the assumptions of Theorem 4 hold and $z(t)$ solve problem (73). Assume that the following two conditions hold:*

1. *for any h belonging to the trajectory $z(t)$, $\Phi(h,t)$ is differentiable with respect to t and the following two inequalities hold:*

$$(\Phi'(h,t)\xi,\xi) \le \left(-a_1(t) + a_2(t)||\Phi(h,t)||\right)||\xi||^2 \text{ for any } \xi \in H, \quad (86)$$

$$\left|\left|\frac{\partial\Phi}{\partial t}(h,t)\right|\right| \le \beta_1(t) + b_1(t)||\Phi(h,t)|| + b_2(t)||\Phi(h,t)||^2, \quad (87)$$

where $\Phi'(h,t)$ is the Fréchet derivative with respect to h, $\partial\Phi/\partial t$ is the derivative with respect to t, $a_1(t)$, $a_2(t)$, $\beta_1(t)$, $b_1(t)$, and $b_2(t)$ are continuous functions;

2. *the functions $\gamma_1(t) := a_1(t) - b_1(t)$, $\sigma_1(t) := a_2(t) + b_2(t)$ and $\beta_1(t)$ are continuous and satisfy conditions (49) of Theorem 3 with $v_1(0) := ||\Phi(z_0,0)|| < \frac{1}{\mu_1(0)}$ and with $\mu_1(t) > 0$, which tends to $+\infty$ as $t \to +\infty$.*

Then the following estimate holds:

$$||\Phi(z(t),t)|| \le \frac{1}{\mu_1(t)}. \quad (88)$$

Proof of Theorem 5. Denote $v_1(t) := ||\Phi(z(t),t)||$. Recall that H is a real Hilbert space. From (73) one gets:

$$v_1\frac{dv_1}{dt} = \left(\frac{d}{dt}\Phi(z(t),t), \Phi(z(t),t)\right) = \left(\frac{\partial\Phi}{\partial t} + \Phi'(z(t),t)\dot{z}(t), \Phi(z(t),t)\right). \quad (89)$$

From (89) and (73) one gets:

$$v_1\frac{dv_1}{dt} = \left(\Phi'(z(t),t)\Phi(z(t),t), \Phi(z(t),t)\right) + \left(\frac{\partial\Phi}{\partial t}, \Phi(z(t),t)\right). \quad (90)$$

Using (86) and (87) one obtains from (90) the following inequality:

$$v_1\dot{v}_1 \le (-a_1(t) + a_2(t)v_1)v_1^2 + (\beta_1(t) + b_1(t)v_1 + b_2(t)v_1^2)v_1,$$

or

$$\dot{v}_1 \le \beta_1 + (b_1 - a_1)v_1 + (a_2 + b_2)v_1^2 = \beta_1 - \gamma_1 v_1 + \sigma_1 v_1^2, \quad (91)$$

where $\gamma_1(t) = a_1(t) - b_1(t)$ and $\sigma_1(t) = a_2(t) + b_2(t)$.

It follows from the assumptions of Theorem 5 that $v_1(t)$ satisfies the integral inequality:

$$\frac{dv_1}{dt} \le -\gamma_1(t)v_1(t) + \sigma_1(t)v_1^2(t) + \beta_1(t). \quad (92)$$

To finish the proof of Theorem 5 one uses Theorem 3. □

6 Discretization Theorem for Ill-posed Problems

The theorem of this section gives an answer to the following question: under what assumptions on Φ and $\{\omega_n\}$ the convergence of a continuous process

$$\dot{z}(t) = \Phi(z(t), t), \quad z(0) = z_0, \tag{93}$$

implies the convergence of the corresponding discrete process

$$z_n = z_{n-1} + \omega_n \Phi(z_{n-1}, t_{n-1}), \tag{94}$$

$$t_n = t_{n-1} + \omega_n, \quad t_0 = 0, \quad n = 1, 2, \dots, \tag{95}$$

with z_0 is the same as in (93).

Theorem 6. *Let Φ satisfy conditions of Theorems 4 and 5 with a function $x(t)$, which tends to y as $t \to +\infty$.*

Assume that

1.
$$||\Phi'(h,t)|| \le a_3(t) + a_2(t)||\Phi(h,t)||, \tag{96}$$
where $a_3(t)$ is a nonnegative continuous function and $a_2(t)$ is the same as in Theorem 5;

2. *the sequence $\{z_n\}$ is defined by formulas (94) and (95);*

3.
$$A = \sup_{t \in [0,+\infty)} (a_1(t) + a_3(t)) < +\infty, \tag{97}$$
where $a_1(t)$ is defined in (86);

4.
$$0 < \omega_n \le \frac{1}{A}, \quad \sum_{n=1}^{\infty} \omega_n = \infty; \tag{98}$$

5.
$$\frac{\omega_n}{\nu_n(\omega_n)} < \frac{\mu_1(t_n)}{\mu(t_n)}, \tag{99}$$
where
$$\frac{1}{\nu_n(t)} = \frac{1}{1 - \mu(t_{n-1})||z_{n-1} - x(t_{n-1})||} + \frac{1}{2}\int_{t_{n-1}}^{t} \left(\gamma(s) - \frac{\dot{\mu}(s)}{\mu(s)}\right) ds, \tag{100}$$
the continuous positive functions $\mu(t)$, $\mu_1(t)$ are defined in Theorems 4 and 5;

6. the function $\mu_1(t)$ is monotonically increasing.

Then the following conclusions hold:

i) all z_n, $n = 1, 2, \ldots$, belong to the ball $B_{2r}(y)$;

ii)

$$||z_n - x(t_n)|| < \frac{1}{\mu(t_n)}; \tag{101}$$

iii)

$$t_n \to +\infty, \quad \frac{1}{\mu(t_n)} \to 0 \quad as \quad n \to \infty; \tag{102}$$

iv)

$$\lim_{n\to\infty} ||z_n - y|| = 0. \tag{103}$$

Proof of Theorem 6. Statements (102) follow from (98) and from the assumption that $\mu_1(t) \to \infty$ as $t \to \infty$. We prove (101) by induction. It follows from (76) that $||z_0 - x(0)|| < \frac{1}{\mu(0)}$. Suppose that

$$||z_{n-1} - x(t_{n-1})|| < \frac{1}{\mu(t_{n-1})}. \tag{104}$$

We want to prove that (104) with n replacing $n-1$ is true. Denote by $\psi_n(t)$ the solution to the following Cauchy problem:

$$\dot{z}(t) = \Phi(z(t), t), \quad t_{n-1} < t \le t_n, \quad z(t_{n-1}) = z_{n-1}. \tag{105}$$

For problem (105) the conditions of Theorem 4 are satisfied by assumption. Therefore from (77) one gets:

$$||\psi_n(t) - x(t)|| \le \frac{1 - \nu_n(t)}{\mu(t)}, \tag{106}$$

with $\nu_n(t)$ defined in (100). Using (93) and (94) one gets:

$$\psi_n(t_n) - z_n = \int_{t_{n-1}}^{t_n} [\Phi(z(\tau), \tau) - \Phi(z_{n-1}, t_{n-1})] d\tau =$$

$$= \int_{t_{n-1}}^{t_n} d\tau \int_0^1 \frac{d}{ds} \Phi(z(t_{n-1} + s(\tau - t_{n-1})), t_{n-1} + s(\tau - t_{n-1})) ds =$$

$$= \int_{t_{n-1}}^{t_n} d\tau \int_0^1 [[\Phi'(z(t_{n-1} + s(\tau - t_{n-1})))\dot{z}(t_{n-1} + s(\tau - t_{n-1}))(\tau - t_{n-1}) +$$

$$+ \frac{\partial \Phi}{\partial t}(z(t_{n-1} + s(\tau - t_{n-1})))(\tau - t_{n-1})\Big] d\tau$$

$$= \int_0^{\omega_n} \theta d\theta \int_0^1 [\Phi'(z(t_{n-1}+s\theta), t_{n-1}+s\theta)\dot{z}(t_{n-1}+s\theta) + \frac{\partial \Phi}{\partial t}(z(t_{n-1}+s\theta), t_{n-1}+s\theta)]ds. \tag{107}$$

Replacing $t_{n-1} + \theta$ by τ and using (93), one gets:

$$||\psi_n(t_n) - z_n|| \leq \int_0^{\omega_n} d\tau \int_{t_{n-1}}^{t_n} \left(||\Phi'(z(s), s)\Phi(z(s), s) + \frac{\partial \Phi}{\partial t}(z(s), s))|| \right) ds \leq$$

$$\leq \omega_n \int_{t_{n-1}}^{t_n} \left(||\Phi'(z(s), s)||\, ||\Phi(z(s), s)|| + \left|\left| \frac{\partial \Phi}{\partial t}(z(s), s) \right|\right| \right) ds. \tag{108}$$

From (87) and (96) one obtains:

$$||\Phi'(z(s), s)||\, ||\Phi(z(s), s)|| + \left|\left| \frac{\partial \Phi}{\partial t}(z(s), s) \right|\right|$$
$$\leq (a_3(s) + a_2(s)\, ||\Phi(z(s), s)||)\, ||\Phi(z(s), s)||$$
$$+ \beta_1(s) + b_1(s)\, ||\Phi(z(s), s)|| + b_2(s)\, ||\Phi(z(s), s)||^2 . \tag{109}$$

Using estimate (88) and the notation $\sigma_1 = a_2 + b_2$ from Theorem 5, one gets:

$$||\psi_n(t_n) - z_n|| \leq \omega_n \int_{t_{n-1}}^{t_n} \left(\frac{a_3(s)}{\mu_1(s)} + \frac{a_2(s)}{\mu_1^2(s)} + \beta_1(s) + \frac{b_1(s)}{\mu_1(s)} + \frac{b_2(s)}{\mu_1^2(s)} \right) ds. \tag{110}$$

According to the assumptions of Theorem 5 conditions (49) of Theorem 3 are satisfied with $\gamma_1(t) = a_1(t) - b_1(t)$, $\sigma_1(t)$, $\beta_1(t)$ and $\mu_1(t)$ in place of γ, σ, β and μ respectively. Thus one gets:

$$||\psi_n(t_n) - z_n|| \leq \omega_n \int_{t_{n-1}}^{t_n} \left(\beta_1(s) + \frac{a_3(s) + b_1(s)}{\mu_1(s)} + \frac{a_2(s) + b_2(s)}{\mu_1^2(s)} \right) ds$$

$$\leq \omega_n \int_{t_{n-1}}^{t_n} \left(\beta_1(s) + \frac{a_1(s) + a_3(s) - \gamma_1(s)}{\mu_1(s)} + \frac{\sigma_1(s)}{\mu_1^2(s)} \right) ds. \tag{111}$$

Using inequalities (49), assumption (97), first assumption (98), the positivity and the monotonicity of $\mu_1(t)$, one gets the estimate:

$$||\psi_n(t_n) - z_n|| \leq \omega_n \int_{t_{n-1}}^{t_n} \left(\frac{a_1(s) + a_3(s)}{\mu_1(s)} - \frac{\dot{\mu}_1}{\mu_1^2} \right) ds$$

$$\leq \omega_n \left(\frac{A\omega_n - 1}{\mu_1(t_{n-1})} + \frac{1}{\mu_1(t_n)} \right) \leq \omega_n \frac{1}{\mu_1(t_n)}. \tag{112}$$

From (112) and (99) one gets:

$$||\psi_n(t_n) - z_n|| < \frac{\nu_n(\omega_n)}{\mu(t_n)}. \tag{113}$$

From this inequality and (106) one obtains:

$$||z_n - x(t_n)|| \leq ||z_n - \psi_n(t_n)|| + ||\psi_n(t_n) - x(t_n)||$$

$$< \frac{\nu_n(\omega_n)}{\mu(t_n)} + \frac{1 - \nu_n(\omega_n)}{\mu(t_n)} = \frac{1}{\mu(t_n)}. \tag{114}$$

Thus (101) is proved.

Finally, (103) follows follows from (101) and (102). Theorem 6 is proved. □

Remark 5. We prove now that a sequence $\{\omega_n\}$ which satisfies (98) and (99) does always exist. Choose an arbitrary $c \in (0, 1)$ and consider a continuous function:

$$\chi(s) := s - c\frac{\mu_1(t_{n-1} + s)}{\mu(t_{n-1} + s)}\nu_n(s) \tag{115}$$

on the interval $[0, 1/A]$ with ν_n defined in (100). Clearly $\chi(0) < 0$. Choose $\omega_n = 1/A$ if $\chi(s) < 0$ on $[0, 1/A]$ and $\omega_n = s_0$ otherwise, where s_0 is the smallest zero of the function $\chi(s)$ on $[0, 1/A]$. Therefore

$$\omega_n = \frac{1}{A} \quad \text{or} \quad \omega_n = c\frac{\mu_1(t_{n-1} + \omega_n)}{\mu(t_{n-1} + \omega_n)}\nu_n(\omega_n). \tag{116}$$

By (95) $t_n = t_{n-1} + \omega_n$. Since $0 < c < 1$, the constructed sequence $\{\omega_n\}$ satisfies the first condition in (98) and condition (99). To show that the second condition in (98) is also satisfied assume that

$$\sum_{n=1}^{\infty} \omega_n = C < \infty. \tag{117}$$

Then (94) implies $t_n \leq C$ for all n. Therefore it follows from (116) that for every n either $\omega_n = 1/A$ or

$$\omega_n = c \frac{\mu_1(t_n)}{\mu(t_n)(\mu(t_n)} \nu_n(\omega_n) \geq \tilde{c} > 0.$$

Thus one has inequality $\omega_n \geq \min\{\tilde{c}, 1/A\} > 0$ for all n. This is a contradiction to (117).

7 Regularized Continuous Methods for Monotone Operators

In this section we apply the regularization procedure described in Section 3 to solve nonlinear operator equation (1). Assume that $F(y) = 0$, F is Fréchet differentiable in a ball $B_{2r}(y)$ and

$$(F'(h)\xi, \xi) \geq 0 \quad \text{for all} \quad h \in B_{2r}(y), \quad \xi \in H. \tag{118}$$

Under this assumption the operator $F'(h) + \varepsilon(t)I$ is boundedly invertible for $h \in B_{2r}(y)$ and for all positive ε. Define Φ as follows:

$$\Phi(h,t) := -[F'(h) + \varepsilon(t)I]^{-1}[F(h) + \varepsilon(h - \tilde{z}_0)], \tag{119}$$

where $\tilde{z}_0$ is a point belonging to $B_r(y)$, and $\varepsilon(t)$ is some positive function on the interval $[0, \infty)$. Some restrictions on $\varepsilon(t)$ will be stated in Theorem 7.

An outline of the convergence proof is the following. One considers an auxiliary well-posed problem:

$$F_\varepsilon(x) := F(x) + \varepsilon(t)(x - \tilde{z}_0) = 0, \quad \varepsilon > 0, \tag{120}$$

and shows that the difference between its solution $x(t)$ and the solution $z(t)$ to problem (73) tends to zero as $t \to +\infty$. On the other hand one shows that $x(t)$ converges to the exact solution y of equation (1). Thus one proves the convergence of $z(t)$ to y as $t \to +\infty$.

We recall first some definitions from nonlinear functional analysis which are used below. The most essential restrictions on the operator F imposed in this section are (118) and Fréchet differentiable in the ball $B_{2r}(y)$. In particular they imply w-closeness of the operator F.

Definition 3. A mapping φ is monotone in a ball $B_r(y) \subset H$ if

$$(\varphi(h_1) - \varphi(h_2), h_1 - h_2) \geq 0, \quad \forall h_1, h_2 \in B_r(y).$$

Definition 4. A mapping φ is hemicontinuous at $h_0 \in H$ if the map $t \to (\varphi(h_0 + th_1), h_2)$ is continuous in a neighborhood of $t = 0$ for any $h_1, h_2 \in H$.

Definition 5. A mapping φ is strongly monotone in a ball $B_r(y)$ if there exists a constant $k > 0$ such that

$$(\varphi(h_1) - \varphi(h_2), h_1 - h_2) \geq k||h_1 - h_2||^2, \quad \forall h_1, h_2 \in B_r(y).$$

Let $\rightharpoonup$ denote weak convergence and $\to$ denote strong convergence in H.

Definition 6. φ is w-closed in a ball $B_r(y)$ if $\{ \{x_n\}_{n=1,2,\ldots} \subset B_r(y),\ x_n \rightharpoonup \xi$ and $\varphi(x_n) \to \eta \}$ implies $\varphi(\xi) = \eta$.

Let $\phi : [0, +\infty) \to [0, +\infty)$ be a continuous monotone increasing function, $\phi(0) = 0$, $\phi(t) > 0$ for $t \in (0, +\infty)$.

Definition 7. φ is ϕ-monotone in a ball $B_r(y)$, if

$$(\varphi(h_1) - \varphi(h_2), h_1 - h_2) \geq \phi(||h_1 - h_2||) \quad \text{for all} \quad h_1, h_2 \in B_r(y). \tag{121}$$

Lemma 2. *If φ is ϕ-monotone and continuous in a ball $B_r(y)$, then φ is w-closed in $B_r(y)$.*

Proof of Lemma 2. Consider a sequence $\{x_n\}_{n=1,2,\ldots} \subset B_r(y)$ such that $x_n \rightharpoonup \xi$ and $\varphi(x_n) \to \eta$. φ is ϕ-monotone in a ball $B_r(y)$, therefore one has:

$$\phi(||x_n - \xi||) \leq (\varphi(x_n) - \varphi(\xi), x_n - \xi). \tag{122}$$

Since $\varphi(x_n) - \varphi(\xi)$ strongly converges and $x_n - \xi \rightharpoonup 0$, one concludes that $(\varphi(x_n) - \varphi(\xi), x_n - \xi) \to 0$, and equation (122) implies $\phi(||x_n - \xi||) \to 0$. By continuity of ϕ and positivity of ϕ on $(0, +\infty)$ this implies $||x_n - \xi|| \to 0$. To finish the proof one uses the continuity of φ. □

Lemma 3. *If F is monotone and hemicontinuous in a ball $B_{2r}(y)$ then problem (120) is uniquely solvable in $B_{2r}(y)$.*

Proof of Lemma 3. Since F_ε is strongly monotone and hemicontinuous in a ball $B_r(y)$ the unique solvability of equation (120) follows from the proof of Theorem 11.2 in [14]. □

Lemma 4. *Suppose that F is w-closed in a ball $B_{2r}(y)$, all the assumptions of Lemma 3 are satisfied, and y is the unique solution to (1) in $B_{2r}(y)$. Let $x(t)$ solve (120) for $\varepsilon = \varepsilon(t)$, and $\varepsilon(t)$ tend to zero as $t \to +\infty$. Then $x(t) \in B_r(y)$ for all $t \in [0, +\infty)$ and*

$$\lim_{t\to+\infty} ||x(t) - y|| = 0. \tag{123}$$

Proof of Lemma 4. First let us show that $x(t)$ is bounded. Indeed, it follows from (120) that

$$F(x(t)) - F(y) + \varepsilon(t)[x(t) - y] = \varepsilon(t)(\tilde{z}_0 - y).$$

Therefore

$$(F(x(t)) - F(y), x(t) - y) + \varepsilon(t)||x(t) - y||^2 = \varepsilon(t)(\tilde{z}_0 - y, x(t) - y). \tag{124}$$

This and (118) imply

$$||x(t) - y|| \le ||\tilde{z}_0 - y|| < r, \tag{125}$$

and therefore $F(x(t)) \to 0 = F(y)$ as $t \to +\infty$.

Also, it follows from (125) that there exists a sequence $\{x(t_n)\}$, $t_n \to \infty$ as $n \to \infty$, which converges weakly to some element $\tilde{y} \in H$. Since F is w-closed one gets that $F(\tilde{y}) = 0$, and, by the uniqueness of the solution to (1) in $B_{2r}(y)$, it follows that $\tilde{y} = y$. Let us show that the sequence $\{x(t_n)\}$ converges strongly to y. Indeed, from (124), (118) and the relation $x(t_n) \rightharpoonup y$, one gets:

$$||x(t_n) - y||^2 \le (\tilde{z}_0 - y, x(t_n) - y) \to 0 \text{ as } n \to \infty. \tag{126}$$

Thus

$$\lim_{n\to\infty} ||x(t_n) - y|| = 0. \tag{127}$$

From (127) it follows by the standard argument that $x(t) \to y$ as $t \to \infty$. Lemma 4 is proved. □

Lemma 5. *Assume that F is continuously Fréchet differentiable in a ball $B_{2r}(y)$,*
$\sup_{x\in B_{2r}(y)} ||F'(x)|| \le N_1$, and condition (118) holds. If $\varepsilon(t)$ is continuously differentiable, then the solution $x(t)$ to problem (120) with $\varepsilon = \varepsilon(t)$ is continuously differentiable in the strong sense and one has

$$||\dot{x}(t)|| \le \frac{|\dot{\varepsilon}(t)|}{\varepsilon(t)}||y - \tilde{z}_0|| \le r\frac{|\dot{\varepsilon}(t)|}{\varepsilon(t)}, \quad t \in [0, +\infty). \tag{128}$$

Proof of Lemma 5. Fréchet differentiability of F implies hemicontinuity of F. Therefore problem (120) with $\varepsilon = \varepsilon(t)$ is uniquely solvable. The differentiability of $x(t)$ with respect to t follows from the implicit function theorem [2]. To derive (128) one differentiates equation (120) and uses the estimate $\left\|[F'(x(t)) + \varepsilon(t)I]^{-1}\right\| \leq \frac{1}{\varepsilon(t)}$. The result is:

$$||\dot{x}(t)|| = |\dot{\varepsilon}(t)| \cdot ||[F'(x(t)) + \varepsilon(t)I]^{-1}(x(t) - \tilde{z}_0)|| \leq \frac{|\dot{\varepsilon}(t)|}{\varepsilon(t)}||x(t) - \tilde{z}_0||. \quad (129)$$

Here we have used the estimate

$$||x(t) - \tilde{z}_0|| \leq ||y - \tilde{z}_0||, \quad (130)$$

which can be derived from (120) similarly to the derivation of (125). Indeed,

$$F(x(t)) - F(y) + \varepsilon(x(t) - \tilde{z}_0) = 0.$$

It follows from the monotonicity of F that $(x(t) - \tilde{z}_0, x(t) - y) \leq 0$. Therefore

$$(x(t) - \tilde{z}_0, x(t) - \tilde{z}_0) \leq (x(t) - \tilde{z}_0, y - \tilde{z}_0) \leq ||x(t) - \tilde{z}_0|| \cdot ||y - \tilde{z}_0||.$$

From this estimate and (125) one gets (130).

Finally, estimate (128) follows from (129) and (130). □

Remark 6. Lemma 4 and formula (127) do not give a rate of convergence to y. This rate, in general, can be arbitrary slow.

To estimate $||x(t) - y||$ one may try to use the following inequality:

$$||x(t) - y|| \leq \int_t^{+\infty} ||\dot{x}(\tau)||d\tau \leq r \int_t^{+\infty} \frac{|\dot{\varepsilon}(\tau)|}{\varepsilon(\tau)}d\tau. \quad (131)$$

Since $\varepsilon(t) \to 0$ as $t \to +\infty$ monotonically, then $\int_t^{+\infty} \frac{|\dot{\varepsilon}(\tau)|}{\varepsilon(\tau)}d\tau = -\log \varepsilon(t)|_t^{\infty} = +\infty$, so in fact one can not use the above estimate in order to estimate the rate of convergence of $||x(t) - y||$.

This illustrates the strength of conclusion (123).

Lemma 6. *Assume that* $\varepsilon = \varepsilon(t) > 0$, *$F$ is twice Fréchet differentiable in* $B_{2r}(y)$, *condition (118) holds, and*

$$||F'(x)|| \leq N_1, \quad ||F''(x)|| \leq N_2 \quad \forall x \in B_{2r}(y). \quad (132)$$

Then for the operator Φ defined by (119) and $x(t)$, the solution to (120) with $\varepsilon = \varepsilon(t)$, estimate (74) holds with

$$\alpha(t) \equiv 0, \quad \gamma(t) \equiv 1, \text{ and } \sigma(t) := \frac{N_2}{2\varepsilon(t)}. \tag{133}$$

Proof of Lemma 6. Since $x(t)$ is the solution to (120) applying Taylor's formula one gets:

$$(\Phi(h,t), h - x(t)) = -([F'(h) + \varepsilon(t)I]^{-1}[F(h) - F(x(t)) + \varepsilon(t)(h - x(t))], h - x(t))$$

$$\leq -\left([F'(h) + \varepsilon(t)I]^{-1}[F'(h)(h - x(t)) + \varepsilon(t)(h - x(t))], h - x(t)\right)$$

$$+ \frac{N_2||h - x(t)||^3}{2\varepsilon(t)} = -||h - x(t)||^2 + \frac{N_2||h - x(t)||^3}{2\varepsilon(t)}. \tag{134}$$

From (134) and (74) the conclusion of Lemma 6 follows. □

Let us state the main result of this section.

Theorem 7. *Assume:*

1. *problem (1) has a unique solution y in $B_{2r}(y)$;*

2. *F is twice Fréchet differentiable in $B_{2r}(y)$ and inequalities (118), (132) hold;*

3. $$z_0, \tilde{z}_0 \in B_r(y); \tag{135}$$

4. *$\varepsilon(t) > 0$ is continuously differentiable, monotonically decreases to 0 as $t \to +\infty$, and*
$$C_\varepsilon := \max_{t \in [0,\infty)} \frac{\varepsilon(0)|\dot{\varepsilon}(t)|}{\varepsilon^2(t)} < 1; \tag{136}$$

5. $$\varepsilon(0) > \frac{N_2}{1 - C_\varepsilon}||z_0 - x(0)||; \tag{137}$$

6. $$\varepsilon(0) \geq \frac{2N_2 C_\varepsilon}{(1 - C_\varepsilon)^2}||\tilde{z}_0 - y||; \tag{138}$$

7.

$$r \geq \frac{\varepsilon(0)(1 - C_\varepsilon)}{N_2}; \tag{139}$$

8. Φ *is defined by (119).*

Then the following conclusions hold:

i) Cauchy problem (73) has a unique solution $z(t) \subset B_{2r}(y)$ *for* $t \in [0, +\infty)$,

ii)

$$||z(t) - x(t)|| \leq \frac{1 - C_\varepsilon}{N_2}\varepsilon(t), \quad \lim_{t \to +\infty} ||z(t) - y|| = 0, \tag{140}$$

iii)

$$\begin{aligned} ||F(z(t))|| &\leq e^{-t}\left(||F(z_0)|| + \varepsilon(0)||\tilde{z}_0 - z_0||\right) + \frac{3r}{1 - C_\varepsilon}\varepsilon(t) \\ &\leq \left(\frac{||F(z_0)||}{\varepsilon(0)} + ||\tilde{z}_0 - z_0|| + \frac{3r}{1 - C_\varepsilon}\right)\varepsilon(t). \end{aligned} \tag{141}$$

Remark 7. First notice that Theorem 7 establishes convergence for any initial approximation point z_0 if r is big enough and $\varepsilon(t)$ is appropriately chosen. To make an appropriate choice of $\varepsilon(t)$ one has to choose some function $\varepsilon(t)$ satisfying condition (136). Examples of such functions $\varepsilon(t)$ are given below. One can observe that condition (136) is invariant with respect to a multiplication $\varepsilon(t)$ by a constant. Therefore one can choose $\varepsilon(t)$ satisfying conditions (137) and (138) by a multiplication of the original $\varepsilon(t)$ by a sufficiently large constant. If $\frac{|\dot{\varepsilon}(t)|}{\varepsilon^2(t)}$ is not increasing, then in conditions (137) and (138) $C_\varepsilon := \max_{t \in [0,\infty)} \frac{\varepsilon(0)|\dot{\varepsilon}(t)|}{\varepsilon^2(t)}$ can be replaced by $C_\varepsilon := \frac{|\dot{\varepsilon}(0)|}{\varepsilon(0)}$.

Remark 8. Inequalities (141) give the estimates for $||F(z(t))||$, the discrepancy of the continuous process. The first estimate in (141) shows that decay of $||F(z(t))||$ is estimated by the some of two terms. The first term depends on $||F(z(0))||$ and on the distance between points z_0 and $\tilde{z}_0$. This term decreases with the rate e^{-t} as in well posed cases. The second term decreases slower because of the ill-posedness of the problem.

Remark 9. In order to get an estimate of the convergence rate for $||x(t) - y||$ one has to make some additional assumptions either on $F(x)$ or on the choice of the initial approximation z_0. Without such assumptions one cannot give an estimate of the convergence rate. Indeed, as a simple example consider the scalar equation $F(x) := x^m = 0$. Then one gets the following algebraic equation for $x(\varepsilon)$:

$$F_\varepsilon(x) := x^m + \varepsilon(x - z_0) = 0. \tag{142}$$

Assume m is a positive integer and $z_0 > 0$. It is known that the solution to this equation is an algebraic function which can be represented by the Puiseux series: $x = \sum_{j=1}^{\infty} c_j \varepsilon^{\frac{j}{p}}$ in some neighborhood of zero. Thus $x = c_1\varepsilon^{\frac{1}{p}}(1+O(\varepsilon))$ as $\varepsilon \to 0$. Now from (142) one gets:

$$c_1^m \varepsilon^{\frac{m}{p}}(1 + O(\varepsilon)) + c_1\varepsilon^{1+\frac{1}{p}}(1 + O(\varepsilon)) = z_0\varepsilon.$$

Thus $p = m$, $c_1 = z_0^{\frac{1}{m}}$ and $x(\varepsilon) = z_0^{\frac{1}{m}}\varepsilon^{\frac{1}{m}}(1 + O(\varepsilon))$. For $\varepsilon = 0$ one gets the solution $y = 0$. Therefore

$$|x(\varepsilon) - y| \sim \varepsilon^{\frac{1}{m}}, \quad \varepsilon \to 0, \tag{143}$$

where m can be chosen arbitrary large.

Below in Propositions 1 and 2 some sufficient conditions are given that allow one to obtain the estimates for $||x(t) - y||$.

Proof of Theorem 7. For $\alpha(t)$, $\gamma(t)$, $\sigma(t)$ defined in (133), $\beta(t)$ defined in (75), and $v(t) := ||z(t) - x(t)||$ we are looking for a function $\mu(t)$ satisfying inequalities (49) and the second inequality in (50). Choose $\mu(t) = \frac{\lambda}{\varepsilon(t)}$, where λ is a constant. Then rewrite first inequality (49) as

$$N_2 \le \lambda\left(1 - \frac{|\dot\varepsilon(t)|}{\varepsilon(t)}\right). \tag{144}$$

Since $\alpha(t) = 0$, using estimate (128) one gets:

$$\beta(t) = ||\dot x(t)|| \le \frac{|\dot\varepsilon(t)|}{\varepsilon(t)}||y - \tilde z_0||. \tag{145}$$

Therefore the second inequality in (49) follows from the following inequality:

$$2||\tilde z_0 - y||\frac{|\dot\varepsilon(t)|}{\varepsilon^2(t)} \le \frac{1}{\lambda}\left(1 - \frac{|\dot\varepsilon(t)|}{\varepsilon(t)}\right). \tag{146}$$

Also, the the second inequality in (50) can be rewritten as:

$$\lambda||z_0 - x(0)|| < \varepsilon(0). \tag{147}$$

Choose

$$\lambda := \frac{N_2}{1 - C_\varepsilon}. \tag{148}$$

Then inequality (147) is the same as (137). It follows from (136) and the monotone decay of $\varepsilon(t)$ that

$$\frac{|\dot{\varepsilon}(t)|}{\varepsilon(t)} \le C_\varepsilon \frac{\varepsilon(t)}{\varepsilon(0)} \le C_\varepsilon. \tag{149}$$

From (148) and (149) one gets:

$$\frac{N_2}{\lambda} = 1 - C_\varepsilon \le 1 - \frac{|\dot{\varepsilon}(t)|}{\varepsilon(t)}, \tag{150}$$

and (144) holds.

From (136), (138), and (148) one gets:

$$\lambda = \frac{N_2}{1 - C_\varepsilon} \le \frac{(1 - C_\varepsilon)\varepsilon(0)}{2C_\varepsilon||\tilde{z}_0 - y||} \le \frac{1 - \frac{\varepsilon(0)|\dot{\varepsilon}(t)|}{\varepsilon^2(t)}}{2||\tilde{z}_0 - y||\frac{|\dot{\varepsilon}(t)|}{\varepsilon^2(t)}}. \tag{151}$$

Hence one obtains:

$$2||\tilde{z}_0 - y||\frac{|\dot{\varepsilon}(t)|}{\varepsilon^2(t)} \le \frac{1}{\lambda}\left(1 - \frac{\varepsilon(0)|\dot{\varepsilon}(t)|}{\varepsilon^2(t)}\right) \le \frac{1}{\lambda}\left(1 - \frac{|\dot{\varepsilon}(t)|}{\varepsilon(t)}\right). \tag{152}$$

Therefore the first inequality in (146) also follows from the conditions of Theorem 7.

It follows from the monotonicity of $\varepsilon(t)$ that for the chosen function $\mu(t)$ the assumption (139) of Theorem 7 is the same as the second inequality in (76). Thus all assumptions of Theorem 4 are satisfied. Applying Theorem 4 one concludes that $z(t) \in B_{2r}(y)$ and inequality (138) holds. The second relation (140) follows from (123), inequality (140) and the triangle inequality:

$$||z(t) - y|| \le ||z(t) - x(t)|| + ||x(t) - y||.$$

To prove (141), one uses (119) and gets:

$$[F'(z(t)) + \varepsilon(t)I]\dot{z}(t) = -[F((z(t)) + \varepsilon(t)(z(t) - \tilde{z}_0)]. \tag{153}$$

Denote:

$$\rho(t) := ||F((z(t)) + \varepsilon(t)(z(t) - \tilde{z}_0)||. \tag{154}$$

Then it follows from (153) and (154) that

$$\frac{d}{dt}\rho^2(t) = 2\Bigg([F'(z(t)) + \varepsilon(t)]\dot{z}(t) + \varepsilon(t)(z(t) - \tilde{z}_0), \rho(t)\Bigg)$$

$$\leq -2\rho^2(t) + 2\rho(t)|\dot{\varepsilon}(t)|\,||z(t) - \tilde{z}_0||. \tag{155}$$

Since $\rho(t)$ is a positive function, one obtains the following integral inequality:

$$\frac{d}{dt}\rho(t) \leq -\rho(t) + |\dot{\varepsilon}(t)|\,||z(t) - \tilde{z}_0||. \tag{156}$$

It was already proved that $z(t) \in B_{2r}(y)$. Therefore from assumption (135) one gets:

$$||z(t) - \tilde{z}_0|| \leq ||z(t) - y|| + ||\tilde{z}_0 - y|| \leq 3r. \tag{157}$$

Using Lemma 1 one gets:

$$\rho(t) \leq e^{-t}\rho(0) + 3re^{-t}\int_0^t e^s|\dot{\varepsilon}(s)|ds. \tag{158}$$

Denote:

$$g(t) = \int_0^t e^s|\dot{\varepsilon}(s)|ds, \quad f(t) := \frac{C_\varepsilon}{1 - C_\varepsilon}e^t\varepsilon(t). \tag{159}$$

It follows from (149) that for $t \in [0, \infty)$ the following inequality holds:

$$f'(t) = \frac{C_\varepsilon}{1 - C_\varepsilon}e^t[\varepsilon(t) + \dot{\varepsilon}(t)] \geq \frac{C_\varepsilon}{1 - C_\varepsilon}e^t\left(\frac{|\dot{\varepsilon}(t)|}{C_\varepsilon} - |\dot{\varepsilon}(t)|\right) = e^t|\dot{\varepsilon}(t)| = g'(t). \tag{160}$$

Since $g(0) = 0$, and by (159) $f(0) > 0$, it follows that $f(0) > g(0)$. Thus $f(t) > g(t)$ for all $t \in [0, \infty)$ and one obtains the following inequality:

$$e^{-t}\int_0^t e^s|\dot{\varepsilon}(s)|ds \leq \frac{C_\varepsilon}{1 - C_\varepsilon}\varepsilon(t). \tag{161}$$

From (158) and (161) one obtains:

$$\rho(t) \leq e^{-t}\rho(0) + 3r\frac{C_\varepsilon}{1 - C_\varepsilon}\varepsilon(t). \tag{162}$$

It follows from (154) and the monotonicity of $\varepsilon(t)$ that

$$\rho(0) \leq ||F(z_0)|| + \varepsilon(0)||z_0 - \tilde{z}_0||, \tag{163}$$

and

$$||F(z(t))|| \leq \rho(t) + \varepsilon(t)||z(t) - \tilde{z}_0||. \tag{164}$$

Thus from (162) and (164) one gets

$$||F(z(t))|| \le e^{-t}\left(||F(z_0)|| + \varepsilon(0)||z_0 - \tilde{z}_0||\right) + 3r\frac{C_\varepsilon}{1 - C_\varepsilon}\varepsilon(t), \tag{165}$$

and the first inequality in (141) is proved.

It follows from (149) that

$$[\log(\varepsilon(t))]' \ge \left[\log(\varepsilon(0)e^{-C_\varepsilon t})\right]'. \tag{166}$$

Therefore:

$$\log(\varepsilon(t)) \ge \log(\varepsilon(0)e^{-C_\varepsilon t}), \tag{167}$$

and, since $C_\varepsilon < 1$, one obtains:

$$\varepsilon(t) \ge \varepsilon(0)e^{-t}. \tag{168}$$

The second inequality in (141) follows from (168) and the first inequality (141). Theorem 7 is proved. □

Example 4.

1. Let $\varepsilon(t) = \varepsilon_0(t_0 + t)^{-\nu}$, ε_0, t_0 and ν are positive constants. Then $C_\varepsilon = \frac{\nu}{t_0}$ and condition (136) is satisfied if $\nu \in (0, 1]$ and $t_0 > \nu$.

2. If $\varepsilon(t) = \frac{\varepsilon_0}{\log(t_0+t)}$, then $C_\varepsilon = \frac{1}{t_0 \log t_0}$ and condition (136) is satisfied if $t_0 \log t_0 > 1$.

Note that if $\varepsilon(t) = \varepsilon_0 e^{-\nu t}$ then condition (136) is not satisfied.

Proposition 1. *Let all the assumptions of Theorem 7 hold. Suppose also that the following inequality holds:*

$$(F(h), h - y) \ge c||h - y||^{1+a}, \quad a > 0. \tag{169}$$

Then for the solution $z(t)$ to problem (73) with Φ defined in (119), the following estimate holds:

$$||z(t) - y|| = O\left(\varepsilon^{\min\{\frac{1}{a},1\}}(t)\right). \tag{170}$$

Proof of Proposition 1. Denote $||x(t) - y|| := \varrho(t)$. Since $F(y) = 0$, inequality (124) implies

$$c\varrho^{1+a}(t) + \varepsilon(t)\varrho^2(t) \le \varepsilon(t)||\tilde{z}_0 - y||\varrho(t) \tag{171}$$

and $\varrho(t) \to 0$ as $t \to +\infty$. This inequality can be reduced to

$$c\varrho^a(t) + \varepsilon(t)\varrho(t) \le \varepsilon(t)||\tilde{z}_0 - y||. \tag{172}$$

Thus $\varrho^a(t) \le \frac{||\tilde{z}_0 - y||}{c}\varepsilon(t)$, and

$$\varrho(t) = ||x(t) - y|| \le \left(\frac{||\tilde{z}_0 - y||}{c}\right)^{\frac{1}{a}} \varepsilon^{\frac{1}{a}}(t). \tag{173}$$

Combining this estimate with estimate (140) for $||z(t) - x(t)||$ and using the triangle inequality one gets:

$$||z(t) - y|| \le ||z(t) - x(t)|| + ||x(t) - y|| \le$$

$$\le \frac{1 - C_\varepsilon}{N_2}\varepsilon(t) + \left(\frac{||\tilde{z}_0 - y||}{c}\right)^{\frac{1}{a}} \varepsilon^{\frac{1}{a}}(t). \tag{174}$$

Proposition 1 is proved. □

Example 5. In the case of a scalar function $f(h)$ and even integer $a > 0$ the estimate $f(h)(h - y) \ge c|h - y|^{1+a}$ means that $f(h) = (h - y)^a g(h)$, where $g(h) \ge c > 0$, and hence y is a zero of multiplicity a for f.

Proposition 2. *Let all the assumptions of Theorem 7 hold and there exists $v \in H$ such that*

$$z_0 - y = F'(y)v, \quad ||v|| < \frac{2}{N_2}. \tag{175}$$

Then the solution $z(t)$ to problem (73) satisfies the following convergence rate estimate:

$$||z(t) - y|| \le \left[\frac{1 - C_\varepsilon}{N_2} + \frac{4||v||}{2 - N_2||v||}\right] \varepsilon(t). \tag{176}$$

Proof of Proposition 2. Note that $F(y) = 0$. Therefore from (120) one gets:

$$F(x(t)) - F(y) + \varepsilon(t)(x - y) = \varepsilon(t)(z_0 - y).$$

By the Lagrange formula one has:

$$\left\{\int_0^1 (F'(y + s(x(t) - y))ds + \varepsilon(t)I\right\} (x(t) - y) = \varepsilon(t)(z_0 - y). \tag{177}$$

Denote $Q_\varepsilon(x) := \int_0^1 (F'(y + s(x - y))ds + \varepsilon I$. From (177) it follows that

$$||x - y|| = \varepsilon||Q_\varepsilon^{-1}(x)Q_0(y)v|| \le \varepsilon||Q_\varepsilon^{-1}(x)(Q_0(y) - Q_\varepsilon(x))v|| + \varepsilon||Q_\varepsilon^{-1}(x)Q_\varepsilon(x)v||.$$

Since $Q_\varepsilon(x) = Q_0(x) + \varepsilon I$, one obtains

$$||x - y|| \leq \varepsilon ||Q_\varepsilon^{-1}(x)(Q_0(y) - Q_0(x))v|| + \varepsilon ||Q_\varepsilon^{-1}(x)\varepsilon v|| + \varepsilon ||v||$$

$$\leq \frac{N_2}{2}||x - y|| \, ||v|| + 2\varepsilon ||v||. \tag{178}$$

Here we have used assumption (118) which implies the inequality $||Q_\varepsilon^{-1}(x)|| \leq \frac{1}{\varepsilon}$.

From (178) one has:

$$||x - y|| \leq \frac{4\varepsilon ||v||}{2 - N_2 ||v||}. \tag{179}$$

Using the triangle inequality from the first inequality (140), the second inequality (175), and (179), one gets:

$$||z(t) - y|| \leq ||z(t) - x(t)|| + ||x(t) - y|| \leq \frac{1 - C_\varepsilon}{N_2}\varepsilon(t) + \frac{4||v||}{2 - N_2||v||}\varepsilon(t). \tag{180}$$

For $\varepsilon = \varepsilon(t)$ and $x = x(t)$ satisfying the assumptions of Theorem 7 one concludes that estimate (176) holds. □

Now we describe the simple iteration scheme for solving nonlinear equation (1). Define:

$$\Phi(h, t) := -[F(h) + \varepsilon(t)(h - \tilde{z}_0)], \tag{181}$$

where $\tilde{z}_0 \in B_r(y)$ and $\varepsilon(t) > 0$ is defined on $[0, +\infty)$. Some restrictions on $\varepsilon(t)$ will be stated in Theorem 8.

Lemma 7. *Assume that F is monotone, Φ is defined by (181), and $x(t)$ is a solution to problem (120) with $\varepsilon = \varepsilon(t) > 0$, $t \in [0, +\infty)$. Then for $\gamma(t) := \varepsilon(t) > 0$ and for $\sigma(t) = \alpha(t) \equiv 0$ inequality (74) holds.*

Proof of Lemma 7. Since $x(t)$ solves (120), by the monotonicity of F one has:

$$(\Phi(h, t), h - x(t)) = -(F(h) - F(x(t)), h - x(t)) - \varepsilon(t)(h - x(t), h - x(t))$$

$$\leq -\varepsilon(t)||h - x(t)||^2. \tag{182}$$

Lemma 7 is proved. □

Lemma 7 together with Lemma 8 presented below allow one to formulate the convergence result concerning the simple iteration procedure (see Theorem 8).

Lemma 8. *Let $\nu(t)$ be integrable on $[0, +\infty)$. Suppose that there exists $T \geq 0$ such that $\nu(t) \in C^1[T, +\infty)$ and*

$$\nu(t) > 0, \quad -\frac{\dot{\nu}(t)}{\nu^2(t)} \leq C, \quad \textit{for} \quad t \in [T, +\infty). \tag{183}$$

Then

$$\lim_{t \to +\infty} \int_0^t \nu(\tau) d\tau = +\infty. \tag{184}$$

Proof of Lemma 8. One can integrate (183)

$$-\int_T^t \frac{\dot{\nu}(\tau)}{\nu^2(\tau)} d\tau \leq \int_T^t C d\tau, \quad t \in [T, +\infty)$$

and get

$$\frac{1}{\nu(t)} \leq C(t - T) + \frac{1}{\nu(T)}.$$

Without loss of generality we can assume that $C > 0$, and then

$$\nu(t) \geq \frac{1}{C(t - T) + \frac{1}{\nu(T)}}.$$

Integrating this inequality one gets (184) and completes the proof. □

Lemmas 3–5 and 7–8 imply the following result.

Theorem 8. *Assume that:*

1. *problem (1) has a unique solution y in a ball $B_{2r}(y)$;*
2. *F is monotone;*
3. *F is continuously Fréchet differentiable and*

$$||F'(h)|| \leq N_1, \quad \textit{for all} \quad h \in B_{2r}(y); \tag{185}$$

4. *$\varepsilon(t) > 0$ is continuously differentiable, tends to zero monotonically as $t \to +\infty$, and $\lim_{t \to +\infty} \frac{\dot{\varepsilon}(t)}{\varepsilon^2(t)} = 0$.*

Then, for Φ defined by (181), Cauchy problem (73) has a unique solution $z(t)$ for all $t \in [0, +\infty)$ and

$$\lim_{t\to\infty} ||z(t) - y|| = 0.$$

Proof of Theorem 8. In order to verify the assumptions of Theorem 4 we use estimate (182) to conclude that $\alpha(t) = \sigma(t) = 0$ and $\gamma(t) = \varepsilon(t)$ in formula (74). By (75) $\beta(t) = ||\dot{x}(t)||$ because $\alpha(t) = 0$. By (128)

$$\beta(t) = ||\dot{x}(t)|| \leq \frac{|\dot{\varepsilon}(t)|}{\varepsilon(t)} ||y - \tilde{z}_0||.$$

To apply Theorem 4 one has to find a function $\mu(t) \in C^1[0, +\infty)$ satisfying (49) and the second inequality in (50). This will be so if

$$\frac{|\dot{\varepsilon}(t)|}{\varepsilon(t)} ||y - \tilde{z}_0|| \leq \frac{1}{2\mu(t)} \left(\varepsilon(t) - \frac{\dot{\mu}(t)}{\mu(t)} \right), \quad \mu(0) ||x(0) - z_0|| < 1. \tag{186}$$

The function $\mu(t)$ can be chosen as the solution to the differential equation

$$-\frac{\dot{\mu}(t)}{\mu^2(t)} + \frac{\varepsilon(t)}{\mu(t)} = \frac{A|\dot{\varepsilon}(t)|}{\varepsilon(t)}, \tag{187}$$

where $A := 2||y - \tilde{z}_0||$. Denote $\rho(t) := \frac{1}{\mu(t)}$. Then

$$\dot{\rho}(t) + \varepsilon(t)\rho(t) = \frac{A|\dot{\varepsilon}(t)|}{\varepsilon(t)}.$$

Solving this equation, one gets:

$$\rho(t) = \left[A \int_0^t \frac{|\dot{\varepsilon}(s)|}{\varepsilon(s)} e^{\int_0^s \varepsilon(\tau) d\tau} ds + \frac{1}{\mu(0)} \right] e^{-\int_0^t \varepsilon(\tau) d\tau}. \tag{188}$$

By Lemma 8 $e^{\int_0^t \varepsilon(\tau) d\tau} \to \infty$ as $t \to \infty$. Applying L'Hôspital's rule to (188) and using condition 4 of Theorem 8 one gets:

$$\lim_{t\to+\infty} \rho(t) = \lim_{t\to+\infty} \frac{|\dot{\varepsilon}(t)|}{\varepsilon^2(t)} = 0.$$

Therefore $\mu(t) = 1/\rho(t)$ tends to $+\infty$ as $t \to +\infty$. To complete the proof one can take $\mu(0)$ sufficiently small for the second inequality in (186) to hold. By

Theorem 4 one concludes that $||z(t) - x(t)|| \to 0$ as $t \to +\infty$ and by Lemma 4 that $||x(t) - y|| \to 0$ as $t \to +\infty$. Therefore it follows from the estimate:

$$||z(t) - y|| \leq ||z(t) - x(t)|| + ||x(t) - y||,$$

that $||z(t) - y|| \to 0$ as $t \to +\infty$. □

Remark 10. One has the estimate $||z(t) - x(t)|| \leq \frac{1}{\mu(t)} \to 0$ as $t \to +\infty$. For the term $||x(t) - y||$ one can get the rate of convergence if some additional assumptions are made on F or on z_0 (see Propositions 1, 2, and also Remark 9).

Remark 11. An interesting result similar to our Theorem 8 was established in [6, Theorem 8] for accretive operators in Banach space (in the case of Hilbert space accretive means monotone). The dynamical system considered in [6] is different from the one we study. In contrast to Theorem 8 in [6], where the existence of the global solution to the Cauchy problem for the corresponding nonlinear differential equation is one of the assumptions, we prove the existence and uniqueness of the solution to the corresponding Cauchy problem. The method of investigation in [6] is based on a linear differential inequality which is a particular case of (50) with $\sigma(t) \equiv 0$. This linear differential inequality has been used in the literature by many authors.

Example 6.

1. Let $\varepsilon(t) = \varepsilon_0(1+t)^{-\nu}$, ε_0 and ν are positive constants. Then the assumptions of Theorem 8 are satisfied if $\nu \in (0,1)$.

2. If $\varepsilon(t) = \frac{\varepsilon_0}{\log(1+t)}$, then the assumptions of Theorem 8 are satisfied.

If $\varepsilon(t) = \varepsilon_0 e^{-\nu t}$ then condition 4 of Theorem 8 is not satisfied.

8 Regularized Discrete Methods for Monotone Operators

In this section we apply the results of Sections 6 and 7 to derive convergence theorems for regularized discrete methods.

First we consider the regularized Newton's method:

$$z_{n+1} = z_n - \omega_{n+1}[F'(z_n) + \varepsilon(t_n)I]^{-1}[F(z_n) + \varepsilon(t_n)(z_n - \tilde{z}_0)], \quad n = 0,1,2,\dots, \tag{189}$$

where $z_0, \tilde{z}_0 \in B_r(y)$, $t_{n+1} = t_n + \omega_{n+1}$, $t_0 = 0$.

Applying Theorem 6 to the regularized Newton's method one gets the following theorem.

Theorem 9. *Assume:*

1. *problem (1) has a unique solution y in $B_{2r}(y)$;*

2. *F is twice Fréchet differentiable in $B_{2r}(y)$, and inequalities (118), (132) hold;*

3.

$$z_0, \tilde{z}_0 \in B_r(y); \tag{190}$$

4. *$\varepsilon(t) > 0$ is continuously differentiable and monotonically tends to 0,*

$$0 < C_\varepsilon := \max_{t\in[0,\infty)} \frac{\varepsilon(0)|\dot{\varepsilon}(t)|}{\varepsilon^2(t)} \le \frac{1}{4}\left(2 + \frac{N_2||\tilde{z}_0 - y||}{\varepsilon(0)}\right)^{-1}; \tag{191}$$

5.

$$\varepsilon(0) > \frac{N_2}{1 - C_\varepsilon}||z_0 - x(0)||; \tag{192}$$

$$\varepsilon(0) > \frac{2N_2}{1 - 2C_\varepsilon}||z_0 - \tilde{z}_0|| + \sqrt{\frac{2N_2}{1 - 2C_\varepsilon}||F(z_0)||}; \tag{193}$$

6.

$$r \ge \frac{\varepsilon(0)(1 - C_\varepsilon)}{N_2}; \tag{194}$$

7. *Φ is defined in (119),*

8.

$$\sum_{n=1}^{\infty} \omega_n = \infty, \tag{195}$$

9.

$$\omega_n \le \frac{1}{2}, \quad \frac{\omega_n}{\nu_n(\omega_n)} < \frac{2(1 - C_\varepsilon)}{1 - 2C_\varepsilon}, \quad n = 1, 2, \ldots, \tag{196}$$

where the continuous positive functions $\mu(t)$, $\mu_1(t)$ are defined in Theorems 4 and 5, $\mu_1(t)$ is monotone, $\mu_1(t) \to \infty$ as $t \to +\infty$, and $\nu_n(t)$ is defined in (100).

Then the following conclusions hold:

i) all z_n, $n = 1, 2, \ldots$, defined in (189) belong to the ball $B_{2r}(y)$,

ii)

$$||z_n - x(t_n)|| \le \frac{1 - C_\varepsilon}{N_2}\varepsilon(t_n), \tag{197}$$

iii)

$$t_n \to +\infty, \quad \varepsilon(t_n) \to 0 \quad as \quad n \to \infty, \tag{198}$$

iv)

$$\lim_{n\to\infty} ||z_n - y|| = 0. \tag{199}$$

Proof of Theorem 9. Clearly the assumptions of Theorem 9 imply the assumptions of Theorem 7. Therefore the assumptions of Theorem 4 with $\mu(t) = \frac{\lambda}{\varepsilon(t)}$, $\lambda = \frac{N_2}{1-C_\varepsilon}$ are also satisfied (see the proof of Theorem 7).

To check the assumptions of Theorem 5, consider Φ defined by formula (119). Since

$$\Phi'(h,t) = -[F'(h) + \varepsilon(t)I]^{-1}F''(h)[F'(h) + \varepsilon(t)I]^{-1}[F(h) + \varepsilon(t)(h - \tilde{z}_0)] -$$

$$- [F'(h) + \varepsilon(t)I]^{-1}[F'(h) + \varepsilon(t)I], \tag{200}$$

for $h \in B_{2r}(y)$ one gets:

$$(\Phi'(h,t)\xi, \xi) \le -||\xi||^2 + \frac{||F''(h)||}{\varepsilon(t)}||\Phi(h,t)|| \cdot ||\xi||^2$$

$$\le -||\xi||^2 + \frac{N_2}{\varepsilon(t)}||\Phi(h,t)|| \cdot ||\xi||^2. \tag{201}$$

Since $z(t) \in B_{2r}(y)$ for any $t \in [0,\infty)$ Φ' satisfies condition (86) on $[0,+\infty)$ with

$$a_1(t) = 1, \quad a_2(t) = \frac{N_2}{\varepsilon(t)}. \tag{202}$$

From (119) one has:

$$[F'(h) + \varepsilon(t)I]\Phi(h,t) = -F(h) - \varepsilon(t)(h - \tilde{z}_0). \tag{203}$$

Differentiating (203) with respect to t one gets:

$$[F'(h) + \varepsilon(t)I]\frac{\partial\Phi}{\partial t}(h,t) = -\dot{\varepsilon}(t)[\Phi(h,t) + h - \tilde{z}_0]. \tag{204}$$

Thus, with $h = z(t)$ in (204), using the estimate $||F'(h) + \varepsilon I|| \le \frac{1}{\varepsilon}$ and the triangle inequality, one gets:

$$\left|\left|\frac{\partial\Phi}{\partial t}(z(t),t)\right|\right| \le \frac{|\dot{\varepsilon}(t)|}{\varepsilon(t)}(||\Phi(z(t),t)|| + ||z(t) - x(t)|| + ||x(t) - \tilde{z}_0||). \tag{205}$$

Using (130) and (140) one obtains:

$$\left\|\frac{\partial \Phi}{\partial t}(z(t),t)\right\| \leq \frac{|\dot{\varepsilon}(t)|}{\varepsilon(t)}\left(||\Phi(z(t),t)|| + \frac{1-C_\varepsilon}{N_2}\varepsilon(t) + ||\tilde{z}_0 - y||\right). \tag{206}$$

Since

$$\frac{|\dot{\varepsilon}(t)|}{\varepsilon(t)} \leq C_\varepsilon \frac{\varepsilon(t)}{\varepsilon(0)} \quad \text{and} \quad 0 < \frac{\varepsilon(t)}{\varepsilon(0)} \leq 1, \tag{207}$$

condition (87) holds with

$$\beta_1(t) = \left(\frac{C_\varepsilon(1-C_\varepsilon)}{N_2} + \frac{C_\varepsilon||\tilde{z}_0 - y||}{\varepsilon(0)}\right)\varepsilon(t),$$

$$b_1(t) = C_\varepsilon \frac{\varepsilon(t)}{\varepsilon(0)}, \quad \text{and} \quad b_2(t) = 0. \tag{208}$$

Thus the functions $\gamma_1(t)$ and $\sigma_1(t)$ in Theorem 5 are:

$$\gamma_1(t) := a_1(t) - b_1(t) = 1 - C_\varepsilon \frac{\varepsilon(t)}{\varepsilon(0)} > \frac{1}{2},$$

$$\sigma_1(t) := a_2(t) + b_2(t) = a_2(t) = \frac{N_2}{\varepsilon(t)}, \tag{209}$$

Choosing $\mu_1(t) = \frac{\lambda_1}{\varepsilon(t)}$, where λ_1 is a constant, one rewrites the assumptions of Theorem 5, that is, conditions (49) and the second inequality (50) with $v(0) := v_1(0) := ||\Phi(z_0, 0)||$, Φ defined in (191), and $t_0 = 0$, as:

$$N_2 \leq \frac{\lambda_1}{2}\left(1 - C_\varepsilon \frac{\varepsilon(t)}{\varepsilon(0)} - \frac{|\dot{\varepsilon}(t)|}{\varepsilon(t)}\right), \tag{210}$$

$$\frac{C_\varepsilon(1-C_\varepsilon)}{N_2} + \frac{C_\varepsilon||\tilde{z}_0 - y||}{\varepsilon(0)} \leq \frac{1}{2\lambda_1}\left(1 - C_\varepsilon \frac{\varepsilon(t)}{\varepsilon(0)} - \frac{|\dot{\varepsilon}(t)|}{\varepsilon(t)}\right), \tag{211}$$

$$\lambda_1||[F'(z_0) + \varepsilon(0)]^{-1}||\,||F(z_0) + \varepsilon(0)(z_0 - \tilde{z}_0)|| < \varepsilon(0). \tag{212}$$

Choose

$$\lambda_1 := \frac{2N_2}{1 - 2C_\varepsilon}. \tag{213}$$

From (191) one gets:

$$\frac{|\dot{\varepsilon}(t)|}{\varepsilon(t)} \leq C_\varepsilon \frac{\varepsilon(t)}{\varepsilon(0)} \leq C_\varepsilon. \tag{214}$$

Inequality (214) and definition (213) imply (210). From (214) and (191) one gets the following inequality:

$$4C_\varepsilon(1-C_\varepsilon)+4\frac{C_\varepsilon N_2}{\varepsilon(0)}||\tilde{z}_0-y||\leq(1-2C_\varepsilon)^2. \tag{215}$$

Formula (211) is a consequence of (215).

Because of the monotonicity of F one has:

$$||[F'(z_0)+\varepsilon(0)I]^{-1}||\leq\frac{1}{\varepsilon(0)}. \tag{216}$$

Therefore inequality (212), with λ_1 as in (213), follows from the inequality:

$$\varepsilon(0)^2-\frac{2N_2||\tilde{z}_0-z_0||}{1-2C_\varepsilon}\varepsilon(0)-\frac{2N_2}{1-2C_\varepsilon}||F(z_0)||>0. \tag{217}$$

Inequality (217) and therefore (212) is satisfied if

$$\varepsilon(0)>\frac{N_2||\tilde{z}_0-z_0||}{1-2C_\varepsilon}+\sqrt{\frac{N_2^2||\tilde{z}_0-z_0||^2}{(1-2C_\varepsilon)^2}+\frac{2N_2}{1-2C_\varepsilon}||F(z_0)||}. \tag{218}$$

This inequality is satisfied if (193) holds. Thus (193) implies (212). We have shown that the assumptions of Theorem 5 are satisfied.

Formula (200) implies assumption (96) of Theorem 6 with $a_3(t)=1$ and $a_2(t)$ defined in (202). Thus $A=2$ in (97). Therefore for $\mu(t)=\lambda/\varepsilon(t)$ and $\mu_1(t)=\lambda_1/\varepsilon(t)$ with λ and λ_1 defined in (148) and (213), conditions (195) and (196) of Theorem 9 are the same as conditions (98) and (99) of Theorem 6. Hence one finishes the proof of Theorem 9 by applying Theorem 6. □

Now consider a simple iteration method:

$$z_{n+1}=z_n-\omega_{n+1}[F(z_n)+\varepsilon(t_n)(z_n-\tilde{z}_0)],\quad n=0,1,2,\ldots, \tag{219}$$

where $z_0,\tilde{z}_0\in B_r(y)$, $t_{n+1}=t_n+\omega_{n+1}$, $t_0=0$.

Theorem 10. *Assume that:*

1. *problem (1) has a unique solution y in a ball $B_{2r}(y)$;*
2. *F satisfies condition (118);*
3. *F is continuously Fréchet differentiable and*

$$||F'(h)||\leq N_1,\quad\forall h\in B_{2r}(y); \tag{220}$$

4. *$\varepsilon(t) > 0$ is continuously differentiable, tends to zero monotonically as $t \to +\infty$, and $\lim_{t\to+\infty} \frac{\dot\varepsilon(t)}{\varepsilon^2(t)} = 0$;*

5.
$$\varepsilon(0) > C_\varepsilon, \tag{221}$$
where C_ε is defined in (191);

6.
$$\frac{||F(z_0)||}{\varepsilon(0)} + ||z_0 - \tilde z_0|| < \frac{6rC_\varepsilon}{\varepsilon(0) - C_\varepsilon}; \tag{222}$$

7. *the sequence $\{z_n\}$ is defined in (219), $z_0, \tilde z_0 \in B_r(y)$, and*
$$\sum_{n=1}^{\infty} \omega_n = \infty, \tag{223}$$

8.
$$0 < \omega_n \le \frac{1}{N_1 + 2\varepsilon(0)}, \quad \frac{\omega_n}{\nu_n(\omega_n)} < \frac{2(1 - C_\varepsilon)}{1 - 2C_\varepsilon}, \quad n = 1, 2, \ldots, \tag{224}$$
where the continuous positive functions $\mu(t)$, $\mu_1(t)$ are defined in Theorems 4 and 5, $\mu_1(t)$ is monotone, and $\nu_n(t)$ is defined in (100).

Then the following conclusions hold:

i) *all z_n, $n = 1, 2, \ldots$, belong to the ball $B_{2r}(y)$,*

ii)
$$||z_n - x(t_n)|| \le \frac{1 - C_\varepsilon}{N_2} \varepsilon(t_n), \tag{225}$$

iii)
$$t_n \to +\infty, \quad \varepsilon(t_n) \to 0 \quad as \quad n \to \infty, \tag{226}$$

iv)
$$\lim_{n\to\infty} ||z_n - y|| = 0. \tag{227}$$

Proof of Theorem 10. The assumptions of Theorem 10 imply the assumptions of Theorem 8. Therefore the assumptions of Theorem 4 with $\mu(t)$ defined in (187) are also satisfied (see the proof of Theorem 8).

To check the assumptions of Theorem 5 consider Φ defined by formula (181). Since

$$\Phi'(h,t) = -F'(h) - \varepsilon(t)I, \tag{228}$$

condition (118) implies

$$(\Phi'(h,t)\xi,\xi) = -(F'(h)\xi,\xi) - \varepsilon(t)||\xi||^2 \leq -\varepsilon(t)||\xi||^2, \forall \xi \in H. \tag{229}$$

Therefore Φ' satisfies condition (86) with

$$a_1(t) = \varepsilon(t), \quad \text{and} \quad a_2(t) = 0. \tag{230}$$

For $h \in B_{2r}(y)$ and $\tilde{z}_0 \in B_r(y)$ one has:

$$\left|\left|\frac{\partial \Phi}{\partial t}(z(t),t)\right|\right| = ||\dot{\varepsilon}(t)(h-\tilde{z}_0)|| \leq |\dot{\varepsilon}(t)|(||h-y||+||\tilde{z}_0-y||) \leq 3r|\dot{\varepsilon}(t)|. \tag{231}$$

Thus condition (87) also holds with

$$\beta_1(t) = 3r|\dot{\varepsilon}(t)|, \quad b_1(t) = b_2(t) = 0. \tag{232}$$

Since $\gamma_1(t) = \varepsilon(t)$, $\sigma_1(t) \equiv 0$, and in Theorem 5 $v(0) = ||\Phi(z_0,0)|| = ||F(z_0) + \varepsilon(0)(z_0 - \tilde{z}_0)||$ in order to satisfy the assumptions of Theorem 5 the function $\mu_1(t)$ should satisfy the following conditions:

$$3r|\dot{\varepsilon}(t)| \leq \frac{1}{2\mu_1(t)}\left(\varepsilon(t) - \frac{\dot{\mu}_1(t)}{\mu_1(t)}\right), \tag{233}$$

$$\mu(0)||F(z_0) + \varepsilon(0)(z_0 - \tilde{z}_0)|| < 1. \tag{234}$$

Choose

$$\mu_1(t) = \frac{\lambda_1}{\varepsilon(t)}, \quad \lambda_1 = \frac{\varepsilon(0) - C_\varepsilon}{6rC_\varepsilon}. \tag{235}$$

Hence assumptions (221) and (222) of Theorem 10 imply conditions (233) and (234). Therefore the assumptions of Theorem 5 are also satisfied.

From (228), (220) and the monotonicity of $\varepsilon(t)$ one concludes that assumption (96) of Theorem 6 holds with $a_3(t) = N_1 + \varepsilon(t)$ and $a_2(t) \equiv 0$, and the assumption (97) with $A = N_1 + 2\varepsilon(0)$.

To finish the proof of Theorem 10 one refers to Theorem 6. □

References

1. Airapetyan, R.G. [2000] *Continuous Newton method and its modification*, Applicable Analysis, **73**, 3–4, pp. 463–484.

2. Airapetyan, R.G. [2000] *On new statement of inverse problem of Quantum Scattering Theory,* Operator theory and its applications, Amer. Math. Soc., Providence RI, 2000, Fields Inst. Comm., 25.

3. Airapetyan, R.G. and Puzynin, I.V. [1997] *Newtonian iterative scheme with simultaneous iterations of inverse derivative,* Comp. Phys. Comm., **102**, pp. 97–108.

4. Airapetyan, R.G., Ramm, A.G. and Smirnova, A.B. [1999] *Continuous analog of Gauss-Newton method*, Math. Models and Meth. in Appl. Sci., **9**, N3, pp. 463–474.

5. Airapetyan, R.G., Ramm, A.G. and Smirnova, A.B. [2000] *Continuous methods for solving nonlinear ill-posed problems*, Operator theory and its applications, Amer. Math. Soc., Providence RI, 2000, Fields Inst. Comm., **25**, pp. 111–138.

6. Alber, Ya.I. [1975] *On a solution of operator equations of the first kind with accretive operators in Banach spaces,* Diffferen. Uravneniya, **11**, N12, 2242–2248.

7. Alber, Ya.I. [1993] *The regularization method for variational inequalities with nonsmooth unbounded operators in Banach space,* Appl. Math. Lett., **6**, N4, 63–68.

8. Alber, Ya.I. [1994] *A new approach to the investigation of evolution differential equations in Banach spaces,* Nonlin. Anal., Theory, Methods & Appl., **23**, N9, 1115–1134.

9. Argyros, I.K. [1998] *Polynomial operator equations in abstract spaces and applications,* CRC Press, Boca Raton.

10. Beckenbach, E. and Bellman R. [1961] *Inequalities,* Springer-Verlag, Berlin.

11. Blaschke, B., Neubauer, A. and Scherzer O. [1997] *On convergence rates for the iteratively regularized Gauss-Newton method,* IMA J. Num. Anal., **17**, 421–436.

12. Courant, R [1943] *Variational methods for the solution of problems of equilibrium and vibrations,* Bull. Amer. Math. Soc., **49**, 1–23.

13. Decker, D.W., Keller, H.B. and Kelley, C.T. [1983] *Convergence rates for Newton's method at singular points,* SIAM J. Numer. Anal., **20**, N2, 296–314.

14. Deimling, K. [1985] *Nonlinear functional analysis,* Springer-Verlag, New York.

15. Engl, H.W., Hanke, M. and Neubauer, A. [1996] *Regularization of inverse problems,* Kluwer Acad. Publ. Group, Dordrecht.

16. Gavurin, M.K. [1958] *Nonlinear functional equations and continuous analogies of iterative methods,* Izv. Vuzov. Ser. Matematika. **5**, pp. 18–31.

17. Kamke, E. [1974] *Differentialgleichungen. Lösungmethoden und Lösungen,* Chelsea, New York.

18. Kantorovich, L.V. and Akilov, G.P. [1982] *Functional Analysis,* Pergamon Press.

19. Ortega, J.M. and Rheinboldt, W.C. [1970] *Iterative Solution of Nonlinear Equations in Several Variables,* Academic Press.

20. Ramm, A.G. [1999] *A numerical method for some nonlinear problems*, Math. Models and Meth. in Appl.Sci., 9, N2, pp. 325-335.

21. Ramm, A.G. and Smirnova, A.B. [1999] *A numerical method for solving nonlinear ill-posed problems*, Nonlinear Funct. Anal. and Optimiz., 20, N3, pp. 317-332.

22. Ryazantseva, I.P. [1994] *On some continuous regularization methods for monotone equations,* Comput. Math. Math. Phys., **34**, N1, 1–7.

23. Szarski, J. [1967] *Differential inequalities,* PWN, Warszawa.

24. Vasin, V.V. and Ageev, A.L., [1995] *Ill-posed problems with a priori information*, VNU, Utrecht.

25. Zhidkov, E.P. and Puzynin, I.V. [1967] *Solving of the boundary problems for second order nonlinear differential equations by means of the stabilization method,* Soviet Math. Dokl. **8**, pp. 614-616.

EXTRAPOLATION: FROM CALCULATION OF π TO FINITE ELEMENT METHOD OF PARTIAL DIFFERENTIAL EQUATIONS

Xiaoping Shen
Department of Mathematics & Computer Science, Eastern Connecticut State University, Willimantic, CT 06226
E-mail: shenx@ecsu.ctstateu.edu

Dedicated to Professor Qun Lin for his pioneering work in the analysis of extrapolation method for finite elements.

The research on extrapolation for the finite element method (FEM) in solving partial differential equations (PDE) began in the earlier 1980s. The pioneering work in this area was done mainly by the Chinese mathematician Qun Lin and his collaborators ([25], [28], [30], and [34]). In the last 20 years, their results have been developed further in [5], [32], [33], [44]. The results during this period of time were on elliptic and parabolic equations. An excellent survey paper by R. Rannacher [45] provides a concise summary of the most of these results.
The latest development at this time (1999) are due to Q. Lin *et al*, who extended these resuts to hyperbolic equations in [20], [21], [36].
The purpose of this paper is twofold, to give a brief review of the history of the method and an introduction to the latest developments. The presented materials are based on the above references and some unpublished articles by Q. Lin *et al.*

1 Introduction

Historical Remark

The idea of extrapolating a numerical sequence could be traced back to the era of Archimedes (250 BC). To approximate π, we use the perimeter of a regular polygon inscribed in a unit circle. For example, the perimeter of a inscribed regular hexagon gives a rough approximation of π, which is 3. In general, if we denote the perimeter of n-sided regular polygons by π_n, then

$$\pi_6 = 3,$$
$$\pi_{96} = 3.1410319.$$

Though $\pi_{192} = 3.1414524$, it is still not a very good approximation. The sequence π_n converges to π but very slowly. However, if we use the two consecutive approximations π_{24} and π_{48} in the extrapolation formula:

$$\pi_n^{(1)} = \frac{4\pi_{2n} - \pi_n}{3},$$

we have $\pi_{96}^{(1)} = 3.1415926$. Without extrapolating, we have to calculate the perimeter of a 12288-sided regular polygon to get the same precision (see [24] for more details)!

As in this example, the extrapolation is a numerical process which accelerates the convergence of a given sequence of numerical approximations. It was the British scientist L. F. Richardson who systematically developed the method as a powerful tool in numerical analysis (see [48], [49]). Now this method is called by Richardson extrapolation to the limit.

Algorithms based on the extrapolation principle are used almost every where in mathematical computation, from numerical quadrature (see [10]) to the finite difference method (FDM) for differential equations (see [41]).

As is well known, the finite element method (FEM) is a powerful numerical method commonly adopted in scientific and engineering computing. Naturally, the idea of extrapolation has attracted the interests of many mathematicians in the area of FEM. However, for several decades, people believed that to extrapolate the finite element was impossible (except the right triangular elements (Courant elements), which is essentially a 5 point difference scheme).

The research on the extrapolation for the finite element method (FEM) began about 20 years ago. Pioneering work was done by Chinese mathematician Qun Lin and his collaborators in [25], [28], [30], and [34]. Their results have been developed futher [5], [32], [33], [44].

In higher dimensional cases, the method, referred to as the splitting extrapolation method (SEM), is more useful. It is one way to overcome the so call "dimension effect", the phenomenon that the computational time and computer storage requirement increase exponentially with respect to the dimension of the problem (see [24], [19])

The SEM has developed rapidly and has been combined with other methods, such as multigrid method and other multilevel methods. Numerous recent research papers have been published. Many mathematicians, such as H. Blum, H. C. Huang, T. Lü, R. Rannacher, U. Rüde, V. Shaidurov and C. Zenger, have made remarkable contributions to the method ([2], [4], [5], [14], [15], [16], [49], [54] and [58]).

There are not many reviews and books that document the research results in this area. [45] and [46] are excellent survey articles. So far, to the knowledge of the author, [19], [36] and [54] are the only references in book format. [19] and [54] devote one chapter to this topic. The recent book [36] (in Chinese) is an excellent reference that covers the materials in both theoretical and practical aspects. It should be remarked that one reason for this lack of references is that many papers are published in Chinese, and have not drawn enough attention in the world.

We conclude this section with some mathematical preliminaries, treated as briefly as possible. The curious reader can easily find the detailed background related to the finite element method in [1], [9] and [55]. In Section 2, we start with the simplest case, one dimensional case, to illustrate the idea. We then discuss the two dimensional case in the Section 3. Section 4 is devoted to the work on rectangular partition which related to the most recent developments in this area.

Preliminary

We shall need the following:

Notations. Define the following norm:

$||u||_{p,\Omega} = (\int_\Omega |u|^p dx))^{1/p}$,

$||u||_{m,p,\Omega} = \begin{cases} \left\{\sum_{0\le|\alpha|\le m} ||D^\alpha u||_p^p\right\}^{1/p}, & 1 \le p < \infty \\ \max_{0\le|\alpha|\le m} ||D^\alpha u||_\infty, & p = \infty \end{cases}$, where $D^\alpha u$ is the weak (or distributional partial derivative).

The following Sobolev spaces are defined based on the above norm:

$L^p(\Omega) := \{u \mid \text{if } ||u||_{p,\Omega} < \infty\}$.
$C^m(\Omega) := \{u \mid \text{if } ||u||_{m,p,\Omega} < \infty\}$
$H^{m,p}(\Omega) :=$ the completion of $C^m(\Omega)$.
$H_0^{m,p}(\Omega) :=$ the completion of $C_0^\infty(\Omega) := \{u \in C^\infty, \text{with compact support}\}$

Ω will be dropped if there is no confusion on the domain from the context.

Finite element method. The finite element method (for details, see [9]) is a special case of the Ritz-Galerkin method. Instead of using smooth functions, the piecewise polynomials are used in approximation. Beginning with a partition of the domain, denote $\mathfrak{T}_h = \{e\}$, where e is a partition element, h is mesh parameter. In the two dimensional case, the most important elements are the triangular and rectangular. We will restrict ourselves to those two cases in this paper. To illustrate the idea of extrapolation, we will use the linear finite element space, which is defined by

$$S_h := \left\{u_h \in H_0^1(\Omega) : u_{hT} \in P_1(T), \quad \forall T \in \mathfrak{T}_h\right\}$$

where $P_1(T)$ is the set of piecewise linear polynomials.

2 The One-dimensional Case

Before going to the latest developments on the extrapolation for the first order problem, let us review the proof from the early paper by Q. Lin and J. Liu [22] for the following second order problem

$$\begin{aligned} -u'' + u &= f \quad \text{in } (0,1), \\ u(0) &= u(1) = 0. \end{aligned}$$

In order to motive the idea, as with all FEMs, we begin with the linear finite element space S_h and the variational formulation of the model problem:

$$\begin{aligned} (u, v)_1 &= \int_0^1 u'v' + uv \\ &= \int_0^1 fv, \quad \forall v \in S_h, \end{aligned}$$

where $(\cdot, \cdot)$ is the inner product of the Hilbert space H^1. Let $u^h \in S_h$ be the linear finite element approximation to u:

$$(u^h, v)_1 = \int_0^1 fv, \quad \forall v \in S_h.$$

The theoretical basis for the extrapolation method for linear FEM is the following expansion:

$$u^h - u^I = h^2 w + o(h^2),$$

where u^I is the linear interpolation of u and u^h is linear finite element approximation of u.

Based on past history, one expected the expansion

$$u^h - u = h^2 w + o(h^2),$$

should be true on every point in the domain. It would lead, however, to the fact that the linear element

$$\frac{4u^{h/2} - u^h}{3}$$

would approximate u up to the order of $o(h^2)$, which is impossible.

It took some time for mathematicians to recongnize the absurdity. The idea to consider the pair u^h and u^I rather than u^h and u, is the milestone pointing toward the correct direction to get the extrapolation formula.

Now if we compare u^h and u^I we get:

$$\begin{aligned}(u^h - u^I, v)_1 &= (u - u^I, v)_1 \\ &= \int_0^1 (u - u^I)'v' + \int_0^1 (u - u^I)v, \quad \forall v \in S_h.\end{aligned}$$

The first integral disappears and the last integral can be expanded; consequently, we have

$$(u^h - u^I, v)_1 = \frac{h^2}{12} \int_0^h u'v' + \frac{h^4}{12} \int_0^1 u'''v' + \cdots, \quad \forall v \in S_h.$$

Now we rewrite the first integral on the right into a form as follows:

$$\frac{1}{12} \int_0^h u'v' = (w, v)_1, \quad \forall v \in H_0^1;$$

where the existence of $w \in H_0^1$ is guaranteed by Riesz's representation theorem. We proceed to consider the finite element approximation $w^h \in S_h$ to w:

$$(w^h, v)_1 = \frac{1}{12} \int_0^h u'v', \quad \forall v \in S_h.$$

We then have

$$(u^h - u^I - h^2 w^h, v)_1 = O(h^4)\|_3 \|v\|_1, \quad \forall v \in S_h.$$

Let $v = u^h - u^I - h^2 w^h$, we obtain the H^1-estimate

$$\|u^h - u^I - h^2 w^h\|_1 \leq c h^4 \|u\|_3$$

and hence, the L^∞-estimate

$$\max |u^h - u^I - h^2 w^h| \leq c h^4 \|u\|_3.$$

Since

$$\max |w^h - w| \leq c h^2 \|w\|_{2,\infty} \leq c h^2 \|u\|_3$$

we get

$$\max |u^h - u^I - h^2 w| \leq c h^4 \|u\|_3;$$

particularly, at the nodes p,

$$u^h(p) - u(p) = h^2 w(p) + O(h^4)\|u\|_3,$$

which leads, by Richardson extrapolation, to a higher order estimate:

$$\left|\frac{4u^{h/2}-u^h}{3}(p)-u(p)\right|\le ch^4\|u\|_3.$$

We see from the above proof that, compared to the finite difference methods, the FEM approach greatly reduce the smoothness demand and, more important, the procedure is easy to extend; e.g., Chen Chuanmiao and Huang Yunying (see [7]) have given, for the more general second order two point boundary value problems and high order finite elements, a complete asymptotic expansion formula. Furthermore, through the FEM approach, Helfrich (see [11]) worked out the asymptotic formula for 1-d parabolic equations.

More interesting insight is given by a non self adjoint case: for the first order equation

$$u'+u=f,\quad \text{in } (0,x_n),\quad u(0)=0. \tag{1}$$

Lin Qun, *et al*, have derived the expansion formula in a similar way.

Before getting into the derivation, we would like to make a few comments here. Even though the linear continuous finite elements could be applied to the equation (2.1), the method has the following shortcomings:

1. Lessening the convergence rate (See [56]). That is, it can not reach the general convergence rate $o(h^2)$ for the hyperbolic equation except for the uniform partition (see [55]).

2. Losing the explicitly of the solution. Even for the simplest model problem, the expression is only implicit.

3. Losing the possibility of using extrapolaton to improve the convergency rate.

To overcome the shortcomingsis discontinous finite elements have been introduced. Lesiant and Raviart have proved, that for discontinous finite elements, the convergence rate is second order, that is one order higher than for continous finite elements(See [17]). Furthermore, using the superconvergence and posterior treatments, Lin JiaFu and Qun Lin have proved that the convergence rate can be increased by $\frac{1}{2}$ for the case of uniform partition (See [38]).

To illustrate this approach, we begin with the piecewise constant finite element space:

$$W_h=\{v: v\in L^2(0,x_n), v(x)=v_i,\quad \forall x\in[x_i,x_{i+1}],\ 0\le i\le n-1\}$$

and the variational formulation of (1)

$$B(u,v):=\sum\left(\int_{x_i}^{x_{i+1}}u'v+\int_{x_i}^{x_{i+1}}uv\right)=\int_0^{x_n}fv,\quad \forall v\in W_h.$$

Then define a finite element solution $u^h \in W_h$:

$$B(u^h, v) = \int_0^{x_n} fv \ \ \forall v \in W_h, \ u^h(0) = 0;$$

or

$$[u^h(x_{i+1}^-) - u^h(x_i^-)]v(x_{i+1}^-) + \int_{x_i}^{x_{i+1}} u^h v$$
$$= \int_{x_i}^{x_{i+1}} fv, \quad 0 \le i \le n-1, \ \forall v \in W_h,$$

with $u^h(0) = 0$. Here we have extended the definition of $B(w, v)$ to $w \in W_h$:

$$B(w, v) = \sum_{i=0}^{n-1} \left\{ [w(x_{i+1}^-) - w(x_i^-)]v(x_{i+1}^-) + \int_{x_i}^{x_{i+1}} wv \right\}, \quad \forall v \in W_h. \tag{2}$$

Such B is still positive definite if $v(0) = 0$:

$$B(v, v) = \sum_{i=0}^{n-1} \left[v(x_{i+1}^-) - v(x_i^-)\right] v(x_{i+1}^-) + \int_0^{x_n} v^2$$
$$= \frac{1}{2}\sum_{i=0}^{n-1} \left[v(x_{i+1}^-) - v(x_i^-)\right]^2 + \frac{1}{2}v^2(x_n^-) - \frac{1}{2}v^2(0) + \int_0^{x_n} v^2. \tag{3}$$

Now take a special constant interpolation $u^I \in W_h$ of u:

$$u^I(x) = u(x_{i+1}), \quad \forall x \in [x_i, x_{i+1}], \ 0 \le i \le n-1 \tag{4}$$

and $u^I(0) = 0$. Compare it with $u^h \in W_h$,

$$B(u^h - u^I, v) = B(u - u^I, v)$$
$$= \sum \left[(u - u^I)(x_{i+1}^-) - (u - u^I)(x_i^-)\right] v(x_i^+) + \int_0^{x_n} (u - u^I)v$$
$$= \int_0^{x_n} (u - u^I)v.$$

The last integral can be expanded:

$$\int_0^{x_n} (u - u^I)v = \sum v_i \int_{x_i}^{x_{i+1}} (u - u^I)$$
$$= -\frac{h}{2}\sum v_i \int_{x_i}^{x_{i+1}} u' - \frac{h^2}{12}\sum v_i \int_{x_i}^{x_{i+1}} u'' + \cdots$$
$$= -\frac{h}{2}\int_0^{x_n} u'v - \frac{h^2}{12}\int_0^{x_n} u''v + \cdots.$$

Now we rewrite the first integral on the right as

$$-\frac{1}{2}\int_0^{x_n} u'v = B(w,v), \quad w(0)=0, \tag{5}$$

where w is an element of H^1 which exists according to the Riesz representation theorem; and the corresponding finite element approximation of w is denoted $w^h \in W_h$. We have

$$B(u^h - u^I - hw^h, v) = O(h^2)\|u\|_2\|v\|_0, \tag{6}$$

which leads to

$$\|u^h - u^I - hw^h\|_0 \le ch^2\|u\|_2. \tag{7}$$

Furthermore, if we take some kind of extrapolation, we can arrive at a second order accuracy:

$$\|2I_h u^{h/2} - I_{2h}u^h - u\|_0 \le ch^2\|u\|_2$$

where $I_{2h}u$ is the linear interpolation of u defined on the big element of size $2h$ consisted of two elements (of size h) and determined by the value of u at the two vertices of the right element.

We now consider the most interesting case where

$$W_h = \{v : v \in L^2(0,1), \quad v|_{(x_i,x_{i+1})} \in P_1\} \tag{8}$$

is the discontinous linear element space and the bilinear form is

$$B(w,v) = \sum_{i=0}^{n-1}\left\{[w(x_i^+) - w(x_i^-)]v(x_i^+) + \int_{x_i}^{x_{i+1}} (w_x + w)v\right\}, \quad \forall w,v \in W_h. \tag{9}$$

We take a special linear interpolation $u^I \in W_h$ of u: $u^I(0) = 0$,and for $[x_i, x_{i+1}]$

$$u^I(x_{i+1}) = u(x_{i+1}), \tag{10}$$

$$u^I(\frac{2}{3}x_i + \frac{1}{3}x_{i+1}) = u(\frac{2}{3}x_i + \frac{1}{3}x_{i+1}).$$

If

$$x_{i+1} - x_i = h, \quad 0 \le i \le n-1.$$

we have the following superclose result:

$$||u^h - u^I||_0 \le ch^{2.5}||u||_4. \tag{11}$$

Proof. By definition, for $w = u - u^I$,

$$\begin{aligned}B(u^h - u^I, v) &= B(w, v)\\ &= \sum \left[w(x_i^+) - w(x_i^-)\right] v(x_i^+) + \sum \int_{x_i}^{x_{i+1}} (w_x + w)v\\ &= \sum \left[w(x_i^+) - w(x_i^-)\right] v(x_i^+) + \sum [\int_{x_i}^{x_{i+1}} (wv - wv_x)\\ &\quad +(wv)(x_{i+1}^-) - (wv)(x_i^+)]\\ &= \sum [w(x_{i+1}^-)v(x_{i+1}^-) - w(x_i^-)v(x_i^+)] + \sum [\int_{x_i}^{x_{n+1}} (wv - wv_x).\end{aligned}$$

Since

$$w(x_{i+1}^-) = 0 = w(x_i^-),$$

we have

$$B(u^h - u^I, v) = \sum \int_{x_i}^{x_{i+1}} (wv - wv_x).$$

The integrals in right hand side can be expended as

$$\int_{x_i}^{x_{i+1}} wv = -\frac{h^3}{72} \int_{x_i}^{x_{i+1}} u_{xx}v_x + O(h^3)||u||_{3,[x_i,x_{i+1}]}||v||_{0,[x_i,x_{i+1}]}$$

and

$$-\int_{x_i}^{x_{i+1}} wv_x = \frac{h^3}{216} \int_{x_i}^{x_{i+1}} u_{xxx}v_x + O(h^3)||u||_{4,[x_i,x_{i+1}]}||v||_{0,[x_i,x_{i+1}]}.$$

We have then

$$\begin{aligned}B(u^h - u^I, v) &= \frac{h^3}{216} \sum \int_{x_i}^{x_{i+1}} (u_{xxx} - 3u_{xx})v_x + O(h^3)||u||_4||v||_0\\ &= \frac{h^3}{216} \sum \int_{x_i}^{x_{i+1}} (3u_{xxx} - u_{xxxx})v\\ &\quad +\frac{h^3}{216} \sum [(u_{xxx} - 3u_{xx})(x_{i+1})v(x_{i+1}^-)\\ &\quad -(u_{xxx} - 3u_{xx})(x_i)v(x_i^+)] + O(h^3)||u||_4||v||_0\\ &= \frac{h^3}{216} \sum [(u_{xxx} - 3u_{xx})(x_i)[v(x_i^-) - v(x_i^+)]\\ &\quad +\frac{h^3}{216}(u_{xxx} - 3u_{xx})(x_n)v(x_n^-) + O(h^3)||u||_4||v||_0.\end{aligned}$$

By the trace theorem

$$|(u_{xxx} - 3u_{xx})(x_i)| \leq \frac{c}{\sqrt{h}}||u||_{4,[x_i,x_{i+1}]},$$

and

$$|(u_{xxx} - 3u_{xx})(x_n)| \leq \frac{c}{\sqrt{h}}||u||_{4,[x_{n-1},x_n]}.$$

We then have

$$\begin{aligned} B(u^h - u^I, v) &\leq ch^{2.5}(\sum ||u||_{4,[x_i,x_{i+1}]}|v(x_i^-) - v(x_i^+)| \\ &\quad +||u||_{4,[x_{n-1},x_n]}|v(x_n^-)|) + O(h^3)||u||_4||v||_0 \\ &\leq ch^{2.5}||u||_4\{\sum [v(x_i^-) - v(x_i^+)]^2 + v^2(x_n^-)\}^{1/2} \\ &\quad +O(h^3)||u||_4||v||_0 \\ &\leq ch^{2.5}||u||_4\sqrt{B(v,v)}. \end{aligned}$$

Taking $v = u^h - u^I$, we obtain (1.13).

After a certain postprocessing for u^h, we can get $\overline{u}^h$, which satisfies

$$||\overline{u}^h - u||_0 \leq ch^{2.5}||u||_4.$$

This gives us the suprising result that a linear element can have the accuracy of order 2.5.

From the above we see that, in the framework of FEM, the proofs of expansion formulae for problems (for the 1-d case so far) of different kinds can be unified to certain extent.

3 Two-dimensional Triangulation

For two dimensional problems, initially we confine ourselves to triangulations (and L^∞-estimates). Did the asymptotic formula still hold in this case? The first obstacle we meet is: for linear triangular elements there arises an error loss in L^∞ norm:

$$\max|u^h - u| \leq ch^2|\log h| \cdot \|u\|_{2,\infty},$$

where the factor $|\log h|$ can not be improved in general. So, sadly for the general linear triangular elements there does not seems to exist an asymptotic formula with a first term of second order. But, Lü Tao noticed that, Zhu Qiding had been able to improve $|\log h|$ to be $|\log h|^{\frac{1}{2}}$, for $u \in C^2 \cap H^3$, under the restriction of uniform triangulation [61]. Thus, if the u's were smoother, e.g., $u \in C^3$ or $u \in C^{2+\varepsilon}$, could the factor be removed? By 1983, this problem had acquired an affirmative answer, and a group of theses (by Lin, Lü, Shen, Wang, Xu, Blum, Rannacher, Shaidurov, Schatz, etc., see [35], [36], [45], [53], and [54]) appeared subsequently to give asymptotic expansions similar to those in the 1-d case; under the uniform triangulation, at the nodes,

$$u^h(p) - u(p) = h^2 w(p) + O(h^4 |\log h|) \|u\|_{4+\varepsilon,\infty}$$

(see [3]). So, for the linear triangular elements, under the restriction of uniform triangulation, the asymptotic formula has been confirmed. Also we know such expansion formulae hold true for piecewise uniform triangulations if the nodes p under consideration are bounded away from the vertices of the triangular subdomain. For a domain with smooth boundary, if making a uniform triangulation in the interior and a general regular partition close to the boundary region, an inner estimate can be obtained: for arbitrary $p \in \Omega_0 \subset \Omega$,

$$u^h(p) - u(p) = h^2 w(p) + O(h^3 |\log h|) \|u\|_{3,\infty}$$

(See [3]). However, through an "6 pc" triangulation, Chen Chuanmao and Huang Yunqing obtained a global estimate, thereby overcoming the handicap of the inner estimates.

The prime defect of extrapolation theory has been its heavy dependence on the regularity of triangulation; how to relax the regularity limits will be a major subject of research in the future.

Also the global smoothness demand on the exact solution would be too restrictive; thus, some local expansions (see [37]) were obtained for locally smooth solutions.

For derivatives, there exist similar expansions (see [35], [59]):

$$\overline{D}(u^h - u)(p) = h^2 w(p) + o(h^2),$$

where $\overline{D}$ is an average of derivatives. As a corollary, by preserving the dominant term, we can derive some superconvergence results about the derivatives. The relation between the superconvergence and error expansion is just like one between the mean value formula and Taylor formula. In contrast to the 1-d case, the proofs of asymptotic formulae for linear triangular elements were, though simplified several times (see [35] and the recent book [36, p. 204–206], too complicated to be given here. For quadratic triangular elements the asymptotic formula has just been announced (Zhu Junming, 1999), the proof of which was formidable. As for higher order triangular elements, the corresponding results can hardly be imaged. The situation on asymptotic expansions of eigenvalues is, however, different; even for quadratic elements, the proof of the following expansion is not complicated (see [35]):

$$\lambda_h - \lambda = ch^4 + o(h^4).$$

For the first order hyperbolic problem, we have not seen any work about the expansion formulas for triangular case.

4 The Case of Rectangular Partition

It was realized in 1984 (see [8], [31], [61]) that the situation was greatly simplified by using rectangular elements (and L^2 estimates). First of all, the proof framework need not suffer any essential change in the extension from one dimension to two dimensional rectangular elements; then, the occurrence of the building of "integral identities" in the early 1990s (see [36]) largely eased the understanding of the entire proofs.

Consider, for the Poisson equation on a rectangular domain Ω:

$$-\Delta u = f \text{ in } \Omega, \quad u = 0 \text{ on } \partial\Omega$$

and the corresponding bilinear finite element solution $u^h \in S_h$:

$$(u^h, v)_1 = (u, v)_1, \quad \forall v \in S_h$$

where $(u, v)_1 = (\nabla u, \nabla v)$; in combination with "integral identities", we have

$$\begin{aligned}(u^h - u^I, v)_1 &= (u - u^I, v) \\ &= \frac{h^2}{3} \sum_e \frac{h_e^2 + k_e^2}{h^2} \int_e u_{xxyy} v + \begin{cases} O(h^4)\|u\|_5 \|v\|_5 \\ O(h^3)\|u\|_4 \|v\|_1, \end{cases}\end{aligned}$$

where $e = [x_e - h_e, x_e + h_e] \times [y_e - k_e, y_e + k_e]$, $h = \max(h_e, k_e)$, u^I is the bilinear interpolation of u. Let $w \in H_0^1$ and $w^h \in S_h$ be such that

$$\begin{aligned}\frac{1}{3} \sum_e \frac{h_e^2 + k_e^2}{h^2} \int_e u_{xxyy} v = (w, v)_1, \quad \forall v \in H_0^1; \quad (w^h, v)_1 \\ = \frac{1}{3} \sum_e \frac{h_e^2 + k_e^2}{h^2} \int_e u_{xxyy} v, \quad \forall v \in S_h.\end{aligned}$$

It follows that

$$(u^h - u^I - h^2 w^h, v)_1 = \begin{cases} O(h^4)\|u\|_5 \|v\|_1 \\ O(h^3)\|u\|_4 \|v\|_1, \end{cases}$$

$$\|u^h - u^I - h^2 w^h\|_1 = \begin{cases} O(h^4)\|u\|_5 \\ O(h^3)\|u\|_4. \end{cases}$$

We now follow a procedure similar to the above and replace w^h by w. Consider the maximum norm estimates

$$|w^h - w| \le ch^2 |\log h|^2 \|u\|_{4,\infty};$$
$$\max |v| \le c|\log h| \cdot \|u\|_1, \quad \forall v \in S_h.$$

It follows that

$$\max |u^h - u^I - h^2 w| \le ch^4 |\log h|^2 (\|u\|_5 + \|u\|_{4,\infty});$$

particularly at the nodes p,

$$u^h(p) - u(p) = h^2 w(p) + O(h^4 |\log h|^2)(\|u\|_5 + \|u\|_{4,\infty}).$$

If one adopts the finite difference methods, however, $u \in C^6$ would be required *a priori*. Still, in the FEM setting, the proofs were unified.

Only recently, for general elliptic equations, specially those with mixed terms, the expansion formulae for bilinear rectangular elements has been given by Zhu Qiding (see [59]). For higher order rectangular elements there also exist asymptotic formulae; Chen Chuanmiao and Jin Jicheng, for example, have proved for the quadratic rectangular elements

$$u^h(p) - u(p) = h^4 w(p) + O(h^5 |\log h|).$$

Thanks to the simplicity of extrapolation theory for rectangular elements and particularly with the coming of "integral identities", work has been done to extend the above results to more general partitions, such as the piecewise deformed rectangular partition, the rectangular partition with local refinements, and the non-coupled partition between distinct subdomains, i.e., mortar elements (for details see [36]). Hence, for high accuracy, a rectangular partition appears to be more advisable than triangulation.

For various types of equations other than elliptic, the rectangular elements can also be applied. In fact, among the many different kinds of finite elements, they are superior for deriving the asymptotic expansion formulae. Besides elliptic and parabolic equations, the rectangular element trick is even available for first order hyperbolic equations defined on rectangular domains:

$$\begin{aligned} u_x + u_y + u = f, \quad &\text{in } \Omega, \\ u(x,y) = 0 \quad &\text{on } \Gamma_- \end{aligned} \tag{12}$$

where Γ_- is the left hand side boundary of Ω. Lin Jiafu and Lin Qun ([22]) have proved that, if the piecewise constant finite element solutions of (12) and the solution of the auxiliary equation

$$w_x + w_y + w = -(2u_{xy} + u_x + u_y), \quad \text{in } \Omega,$$
$$w(0) = 0 \quad \text{on } \Gamma_-$$

are u^h and w^h, respectively, then

$$\|u^h - u^I - hw^h\|_0 \le ch^2\|u\|_3$$

where u^I is the piecewise constant interpolation of u and assumes constant value on each element equal to the value of u at the upper right corner point. Furthermore, if we take some kind of extrapolation, we can arrive at a second order accuracy.

The proof for 2-d equation (12) is the generalization of the proof for 1-d equation (1). The formulas (2)-(7) are generalized to be

$$\begin{aligned}
B(w,v) &= \sum_e \Big\{ \int_{y_e-k_e}^{y_e+k_e} [w(x_e+h_e-0,y) - w(x_e-h_e-0,y)]v\,dy \\
&\quad + \int_{x_e-h_e}^{x_e+h_e} [w(x,y_e+k_e-0) - w(x,y_e-k_e-0)v\,dx \Big\} \\
&\quad + \int_\Omega wv\,dx\,dy, \\
B(v,v) &= 2\sum_e \{k_e[v(x_e+h_e,y_e+k_e) \\
&\quad - v(x_e-h_e,y_e+k_e)]v(x_e+h_e,y_e+k_e) \\
&\quad + h_e[v(x_e+h_e,y_e+k_e) - v(x_e+h_e,y_e-k_e)]v(x_e+h_e,y_e+k_e)\} \\
&\quad + \int_\Omega v^2 \\
&= \sum_e \{k_e[v(x_e+h_e,y_e+k_e) - v(x_e-h_e,y_e+k_e)]^2 \\
&\quad + h_e[v(x_e+h_e,y_e+k_e) - v(x_e+h_e,y_e-k_e)]^2\} \\
&\quad + \frac{1}{2}\int_{\Gamma_+} v_-^2\,dy + \int_\Omega v^2, \quad \text{if } v_-|_{\Gamma_-} = 0, \\
u^I|_e &= u(x_e+h_e,y_e+k_e), \ u^I|_{\Gamma_-} = 0,
\end{aligned}$$

$$\begin{aligned}
B(u^h - u^I, v) &= B(u - u^I, v) \\
&= \sum_e \{ \int_{y_e-k_e}^{y_e+k_e} ([u(x_e+h_e,y) - u(x_e+h_e,y_e+k_e)]
\end{aligned}$$

$$
\begin{aligned}
&- [u(x_e - h_e, y) - u(x_e - h_e, y_e + k_e)])vdy \\
&+ \int_{x_e-h_e}^{x_e+h_e} ([u(x, y_e + k_e) - u(x_e + h_e, y_e + k_e)] \\
&- [u(x, y_e - k_e) - u(x_e + h_e, y_e - k_e)])vdx\} \\
&+ \sum_e \int_e (u - u^I)v, \qquad (13)
\end{aligned}
$$

where the last integral can be expanded as:

$$
\begin{aligned}
\int_e (u - u^I)v &= \int_e [u(x, y) - u(x_e + h_e, y_e + k_e)]v \\
&= \int_e [u_x(x, y)(x - x_e - h_e) + u_y(x, y)(y - y_e - k_e)]v \\
&+ O(h^2)\|u\|_{2,e}\|v\|_{0,e}.
\end{aligned}
$$

Again the first term on the right can be simplified as

$$
\begin{aligned}
&\int_e u_x(x, y)(x - x_e - h_e)v \\
&= -\int_e h_e u_x v + \int_e u_x E' v \\
&= -\int_e h_e u_x v - \int_e u_{xx} E v = -\int_e h_e u_x v + O(h^2)\|u\|_{2,e}\|v\|_{0,e},
\end{aligned}
$$

etc. and hence we have a simple expansion:

$$
\int_e (u - u^I)v = -\int_e (h_e u_x + k_e u_y)v + O(h^2)\|u\|_{2,e}\|v\|_{0,e}.
$$

The integrals in the first sum in (13) can be simplified as

$$
\begin{aligned}
&\int_{y_e-k_e}^{y_e+k_e} \int_{x_e-h_e}^{x_e+h_e} [u_x(x, y) - u_x(x, y_e + k_e)]v \\
&\quad + \int_{x_e-h_e}^{x_e+h_e} \int_{y_e-k_e}^{y_e+k_e} [u_y(x, y) - u_y(x_e + h_e, y)]v \\
&= \int_e u_{xy}(x, y)(y - y_e - k_e)v + \int_e u_{yx}(x, y)(x - x_e - h_e)v \\
&\quad + O(h^2)\|u\|_{3,e}\|v\|_{0,e} \\
&= -k_e \int_e u_{xy}(x, y)v - h_e \int_e u_{yx}(x, y)v + \int_e u_{xy}(x, y)F'(y)v
\end{aligned}
$$

$$+ \int_e u_{yx}(x,y)E'(x)v + O(h^2)\|u\|_{3,e}\|v\|_{0,e}, \tag{14}$$

$$\begin{aligned} F(y) &= \frac{1}{2}[(y-y_e)^2 - k_e^2], \\ E(x) &= \frac{1}{2}[(x-x_e)^2 - h_e^2]. \end{aligned} \tag{15}$$

The last two integrals in (14), by integration by parts, are of higher order:

$$-\int_e u_{xyy}Fv - \int_e u_{yxx}Ev = O(h^2)\|u\|_{3,e}\|v\|_{0,e}.$$

Finally we get a simple expansion:

$$\begin{aligned} B(u^h - u^I, v) = -\sum_e [(h_e + k_e)\int_e u_{xy}v + \int_e (h_e u_x + k_e u_y)v] \\ + O(h^2)\|u\|_3\|v\|_0. \end{aligned}$$

We assume below that the partition is uniform: $h_e = k_e = h$. We then define an auxiliary equation:

$$\begin{aligned} w_x + w_y + w &= -(2u_{xy} + u_x + u_y), \quad \text{in } \Omega, \\ w &= 0 \quad \text{on } \Gamma_- \end{aligned}$$

with a piecewise constant finite element solution $w^h \in W_h$:

$$\begin{aligned} B(w^h, v) &= -\int_\Omega (2u_{xy} + u_x + u_y)v, \quad \forall v \in W_h, \\ w_-^h|_{\Gamma_-} &= 0. \end{aligned}$$

We then have

$$B(u^h - u^I, v) = hB(w^h, v) + O(h^2)\|u\|_3\|v\|_0,$$

and

$$\|u^h - u^I - hw^h\|_0 \le ch^2\|u\|_3.$$

Because

$$\begin{aligned} \|w^h - w^I\|_0 &\le ch\|w\|_2 \\ &\le ch\|2u_{xy} + u_x + u_y\|_1 \le ch\|u\|_3, \end{aligned}$$

we have

$$\|u^h - u^I - hw^I\|_0 \le ch^2 \|u\|_3.$$

Thus we see that if we take some kind of extrapolation, we can arrive at a second order accuracy

$$\|u^h - u^I - hw^h\|_0 \le ch^2 \|u\|_3.$$

Finally, we turn to the latest unpublished result:

$$||u^h - u^I||_0 \le ch^{2.5}||u||_4.$$

for the discontinous linear finite element methods for the hyperbolic equation in the two dimensional case. Its proof is a generalization of the one dimensional case in the second section. Unfortunately, it is much more complicated and we have to omit it in this review paper. The interested readers can find the details in [36].

Concluding Remark

We conclude the review by pointing out the following major differences between the extrapolation for FDM and FEM.

First, the difference frame relies on discrete maximum principles (see, [41]) which limits its applicability. On the other hand, the latter is based on the principle of variation, which is suitable for a wide class of problems, from second order to higher order elliptic equations, from a single equation to a system of equations, from differential equations to integral equations.

Second, the extrapolating for FDM starts with the exact solution of the equation, while the FEM starts with the weak solution of the equation; this reduces the requirements to certain extent.

However, there is still ample room for improvement in both theory and practice for the FEM extrapolation. Although it is too difficult to predict the future research directions in this area, we expect to see more attention being paid to hyperbolic equations and multidimensional problems. It is also challenging to apply the principle to other multiscale (such as wavelet) methods.

Acknowledgment. The author is indebted to Professor Qun Lin for his encouragement and help in preparing this article. She is also grateful to Professor Rolf Rannacher, who has provided several valuable references and to Professor Gilbert Walter, who has made suggestions to the improvement of this paper.

References

1. Adams, R. A., *Sobolev Spaces*, Academic Press, 1975.

2. Blum, H., On Richardson extrapolation of linear finite elements on domains with reentrant corners. *Z. Angew. Math. Mech.* **67** (1987), T351–T353.

3. Blum, H., Lin, Q., and Rannacher, R., Asymptotic error expansion and Richardson extrapolation for linear finite elements, *Numer. Math.*, **49** (1986), 11–37.

4. Blum, H. and Rannacher, R., Extrapolation techniques for reducing the pollution effect of reentrant corners in the finite element method, *Numer. Math.*, **52** (1988), 539–564.

5. Blum, H. and Rannacher, R., Finite element eigenvalue computation on domains with reentrant corners suing Richardson's extrapolation, *J. Comp. Math.*, **8** (1990), 321–332.

6. Bramble, J. H. and Hilbert, S. R., Estimation of linear functional on Sobolev spaces with application to Fourier transforms and spline interpolation, *SIAM J. Numer. Anal.* **7** (1970), 112–124.

7. Chen, C. and Huang, Y., *High Accuracy Theory for Finite Element Methods*, Hunan Scientific and Technology Publishing House, 1995.

8. Chen, C. and Q. Lin, Extrapolation of finite element approximations in rectangular domain, *J. of Comp. Math.*, Vol. 7, 1989.

9. Ciarlet, P. G., *The Finite Element Method for Elliptic Problems*, North-Holland, 1978.

10. Davis, P. J. and Rabinowitz, P., Methods of numerical integration, *Computer Science and Applied Mathematics*, Academic Press [A subsidiary of Harcourt Brace Jovanovich, Publishers] New York, London, 1975.

11. Helfrich, H., Asymptotic expansion for the finite element approximation of parabolic problems, *Bonn. Math. Schrift.* **158** (1984), 11–30.

12. Hackbusch, W., *Multigrid Methods and Applications*, Springer-Verlag, Berlin, 1985.

13. Huang, H. and Liu, G., Stepwise refining methods of solving elliptic boundary value problems, *Numer. Math Sinica.*, Part. 1 (1978), 41–52. Part. 2,3 (1978), 28–35.

14. Huang, H., Er, W., and Mu, M., Extrapolation combined with MG method for solving finite element equations, *J. Comp. Math.*, **4** (1986).

15. Jung, M. and Rüde, U., Implicit extrapolation methods for variable coefficient problems, *SIAM J. Sci. Comput.* **19**, No. 4 (1998), 1109–1124 (electronic).

16. Jung, M. and Rüde, U., Multigrid algorithms with implicit extrapolation for solving finite element equations, Proceeding of the International Conference on the Optimization of the Finite Element Approximations (St. Petersburg, 1995). *Mat. Model*, **8**, No. 9 (1996), 53–62.

17. P. Lesaint and P. A. Raviart [in Mathematical aspects of finite elements in partial differential equations (Madison, WI, 1974),89-123, Academic Press, New York, 1974; MR 58#31918]

18. Liao, X. and Zhou, A., A multi-parameter splitting extrapolation and a parallel algorithm for elliptic eigenvalue problems, *J. Comput. Math.* **16**, No. 3 (1998), 213–220.

19. Liem, C. B., Lü, T., and Shih, T. M., *The Splitting Extrapolation Method*, World Scientific Publishing Co. Pte. Ltd, Singapore, 1995.

20. Lin, J. and Lin, Q., High superconvergence for discrete bilinear finite element methods for fist order hyperbolic equations, to submit.

21. Lin, Q., The extrapolation of discrete finite elements for the neutron transport equation, *Control Theory and its Application*, Vol. 16, Suppl., 1999.

22. Lin, Q. and Liu, J., A discussion for Extrapolation Method for Finite Elements, Techn. Rep. Inst. Math. Academia Sinica, Peking, 1980.

23. Lin, Q. and Liu, Jia Quan, An introduction to finite element extrapolation techniques starting from calculations of ratios of circumferences, in Chinese, *Math. Practice Theory* No. 4 (1989), 62–72.

24. Lin, Q. and Lü, T., Splitting extrapolations for multidimensional problems, *J. Comp. Math.*, **1** (1983) 45–51.

25. Lin, Q., Lü, T., and Shen, S. M., Maximum norm estimate extrapolation and optimal point of stresses for finite element methods on strongly regular triangulation, *J. Comp. Math.*, **1** (1963), 376–383.

26. Lin, Q., Lü, T., and Shen, S. M., Asymptotic expansions for finite element approximations, in Research Report IMS-11, Academia Sinica, Chengdu, 1983.

27. Lin, Q. and Lü, T., Asymptotic expansions for finite element approximations of elliptic problem on polygonal domains, *Comp. Math. Appl. Sci. Eng.* (Proc. Sixth Int. Conf. Versailles, 1983), LN in Comp. Sci., North-Holland, INRIA, 1984, 317–321.

28. Lin, Q. and Lü, T., Asymptotic expansions for finite element eigenvalues and finite element solution, in Proc. Int. Conf., Bonn, 1983, *Bonner Math. Scherift.*, 158 (1984), 1–10.

29. Lin, Q. and Liu, J., An introduction to finite element extrapolation techniques starting from calculations of ratios of circumferences, in Chinese, *Math. Practice Theory*, No. 4 (1989), 62–72.

30. Lin, Q. and Wang, J., Some expansions of the finite element approximation. Shuli Kexue [Mathematical Sciences. Research Reports IMS], 15. Academia Sinica, Institute of Mathematical Sciences, Chengdu, 1984. 11 pp.

31. Lin, Q. and Xu, J., Linear finite elements with high accuracy, *J. Comput. Math.*, **3** (1985), 115–133.

32. Lin, Q. and Xie, R., Extrapolation of bilinear finite element solution with local refinement mesh, in Chinese, *Ann. of Math.*, **11B**:2 (1990), 179–190.

33. Lin, Q. and Xie, R., Some advances in the study of error expansion for finite elements. I. Eigenvalue error expansion, *J. Comput. Math.*, **4**, No. 4 (1986), 368–382.

34. Lin, Q. and Xie, R., How to recover the convergent rate for Richardson extrapolation on bounded domains, *J. Comput. Math.*, **6**, No. 1 (1988), 68–79.

35. Lin, Q. and Xie, R., Error expansions for FEM and superconvergence under natural assumption, *J. Comput. Math.*, **7** (1989) 402–411.

36. Lin, Q. and Yan, N., *Construction and Analysis for High Effective Finite Element Methods*, Hebie. Univ. Publishing house, 1996.

37. Lin, Q. and Zhu, Q. D., Local asymptotic expansion and extrapolation for finite elements, *J. Comput. Math.*, **4**, No. 3 (1986), 263–265.

38. Lin, Q. and Zhu, Q. D., *The Preprocessing and Postprocessing for the Finite Element Method*, Shanghai Scientific & Technical Publishers, 1994 (in Chinese).

39. Lü, T., Correction and splitting extrapolation methods for the collocation solutions of two-point boundary value problems, (in Chinese), *Av. in Math.* (Beijing) **16**, No. 4 (1987), 391–396.

40. Lü, T., Asymptotic expansion and extrapolation for the finite element approximation of non linear elliptic partial differential equations, (in Chinese), *Math. Numer. Sinica*, **9**, No. 2 (1987), 194–199.

41. Marchuk, G. and Shaidurov, V., *Difference Methods and Their Extrapolation*, Springer, New York, 1983.

42. Neittanmäki, P. and Lin, Q., Acceleration of the convergenc ein finite difference method by predictor-corrector and splitting extrapolation methods, *J. Comput. Math.*, **5**, No. 2 (1987), 181–190.

43. Rabinowitz, P., Extrapolation methods in numerical integration. Extrapolation and rational approximation (Puerto de la Cruz, 1992). *Numer. Algorithms*, **3**, No. 1–4 (1992), 17–28.

44. Rannacher, R., Richardson extrapolation with finite elements, in *Notes in Numerical Fluid Mechanics*, Hackbusch. W. and Witsch, C., eds., Vieweg, Braunschweig, Wiesbaden, 1987.

45. Rannacher, R., Extrapolation techniques in the finite element method (A survey), Summer School on Numerical Analysis, Helsinki, Univ. of Tech., MATC7, (1988), pp. 80–113.

46. Rannacher, R., Defect correction techniques in the finite element method. *Progress in Partial Differential Equations: The Metz Surveys*, pp. 184–200, Pitman Res. Notes Math. Ser., 249, Longman Sci. Tech., Harlow, 1991.

47. Rannacher, R., Richardson extrapolation for a mixed finite element approximation of a plate bending problem, *Z. Angew. Math. Mech.*, **67**, No. 5 (1987), T381–T383

48. Richardson, L. F., The approximation of physical problems involving differential equations, with an application to the stress in a masonry dam. *Philos. Trans. Roy. Soc. London Ser. A*, **210** (1927), 299–349.

49. Richardson, L. F., The deferred approach to the limit. 1: The single lattice. *Philos. Trans. Roy. Soc. London Ser. A*, **226** (1910), 307–357.

50. Rüde, U., Extrapolation and related techniques for solving elliptic equations, in Bericht I-9135, Institute fur Informatic, TU Muncher, January 1991.

51. Rüde, U., The hierarchical basis extrapolation method, *SIAM J. Sic. Statist. Comput.*, **13**, No. 1 (1992), 307–318.

52. Rüde, U. and Zhou, Aihui, Multi-parameter extrapolation methods for boundary integral equations, Numerical treatment of boundary integral equations, *Adv. Comput. Math.*, **9**, No. 1–2 (1998), 173–190.

53. Schatz, A., Pointwise error estimates and asymptotic error expansion inequalities for the finite element method on irregular grids, *Math. of Comp.*, **67**, No. 233 (1998), 877–899.

54. Shaidurov, V., Multigrid methods for finite elements (Translated from the 1989 Russian original by N. B. Urusova), *Mathematics and its Applications*, **318**, Kluwer Academic Publishers Group, Dordrecht.

55. Strang, G; Fix; Fix, G. J., An analysis of the finite element method, Prentice-Hall Series in Automatic computation, Prentice-Hall, Inc., Englewood Cliffs, N. J., 1973.

56. Wang, J. and Lin, Q., Asymptotic expansions and extrapolation for the finite element method, (in Chinese), *J. Systems Sci. Math. Sci.*, **5** (1985).

57. Xie, R., Pointwise estimates for finite-element approximations to Green functions on a concave polygonal domain, and finite-element extrapolation, (in Chinese), *Math. Numer. Sinica*, **10**, No. 3 (1988), 232–241.

58. Zerger, C., Sparse grids, in *Parallel Algorithms for Partial Differential Equations*, Proceedings of the Sixth GAMM-Seminar, Kiel January 19–21, 1990, (Hackbusch, ed.), Vieweg Verlag, 1991.

59. Zhu, Q. etc., Some notes for Lin's integral identity, submitted to *J. Comput. Math.*, 1999.

60. Zhu, Q., Point by point estimation and maximum norm inner estimation in the element method, *Math. Numer. Sinica*, **3** (1981).

61. Zhu, Q. and Lin, Q., Unidirectional extrapolations of finite difference and finite elements, *J. of Engin. Math.*, **1**, No. 2, (1984), 1–12.

A SURVEY ON SCALING FUNCTION INTERPOLATION AND APPROXIMATION

En-Bing Lin
Department of Mathematics, The University of Toledo,
Toledo, Ohio 43606. E-mail: elin@math.utoledo.edu

This survey aims at providing a self-contained introduction to notions and results connected with approximation order, accuracy and scaling function. We will outline scalar scaling function, multi-scaling function and nonseparable scaling function interpolation schemes and present the corresponding approximation theory. The interpolation methods will be illustrated by some numerical experiments. Relationship with approximation order from shift-invariant spaces will be mentioned. Several applications on solving partial differential equations and image processing will also be presented.

1 Introduction

A well known approximation theorem can be dated back to 1886 published by Weierstrass, namely, any continuous function can be approximated by a polynomial. Since that time, there are many approximation theorems one can find in the literature. Accuracy and efficiency are key to improve the approximation theory. On the other hand, wavelets have shown their ability to analyze different parts of a function at different scales and the fact that they can represent polynomials up to a certain order exactly. As a consequence, functions with fast oscillations, or even discontinuities, in localized regions may be approximated well by a linear combination of relatively few wavelets or scaling functions. Approximation methods based on wavelets have become a powerful tool in the applications. There are many possible ways to get faster or better approximation. Namely, one is to choose some special wavelet systems in the approximation procedure; or one can take certain weighted sampling values. We are mainly to combine both approaches to approximate a given function. With weighted sampling, the multiwavelet interpolation greatly simplifies the computation and preserves the approximation order. The sampling theorem obtained by using scaling functions was first introduced by Wells and Zhou [28]. It was later generalized or improved by using multi-scaling functions and nonseparable scaling functions [18], [19]. In [28], Wells and Zhou showed that a wavelet approximation theorem valid for degree 1 wavelet systems in which one obtains second-order approximation by sampling on a precisely defined perturbation of the standard lattice. For Coifman wavelet systems, Tian and Wells obtained exponential approximation in the degree by sampling on the standard lattice [27]. Lin and Zhou generalized the above results for scalar scaling function interpolation [21]. Lin and Xiao further obtained more generalized results for multiwavelets [19]. Nonseparable cases are discussed by Lin and Ling [18].

Wavelet-based methods for solving partial differential equations(PDEs) have become more powerful and useful for the numerical solution of PDEs. The use of multiresolution techniques and wavelets has become increasingly popular in the development of numerical schemes for the solution of partial differential equations [20], [21]. The multilevel and approximation properties of the scaling function of the multiresolution analysis provide computational efficiency and accuracy of numerical solutions of PDEs. Numerical techniques for the solution of differential equations usually fall into the following classes: finite-difference, finite element and spectral methods. Sometimes the latter two methods are considered as subsets of the method of weighted residuals. Galerkin method is one of the method of weighted residual. Based on Galerkin method, we present two different approaches in solving the Dirichlet problem. We conclude the survey by showing another application on image processing.

The organization of this paper is as follows. In Section 2, an introduction of the general theory of wavelet is given. Approximation order and accuracy are defined and their equivalent conditions are stated. The scaling function interpolation and approximation theorems are summarized in Section 3. The interpolation formulas are outlined for scalar scaling, multi-scaling, nonseparable scaling as well as nonseparable multi-scaling functions. Section 4 presents the application of wavelet approximation on the numerical solutions of the Dirichlet problem. Using wavelet-Galerkin approximation and the penalty method, the unkown solution of the Dirichlet problem is represented by the expansion of scaling functions. Numerical examples show that this approach providese more accurate approximate solutions than scalar scaling function method. In Section 5, we apply the multiscaling function interpolation to the image denoising. It has been very successful to apply the discret wavelet transform on image and signal processing. We omit most of the proofs which can be found in [18], [19], [20], [21].

2 Wavelet Analysis

In this section we briefly review wavelet transform, the framework of multiresolution analysis, and the basic ideas of the orthonormal bases of compactly supported wavelets. The typical examples are Daubechies wavelets and coiflets [6], [7], [8]. At the end of this section, we define approximation order and accuracy and present their relationships.

2.1 Wavelet Transforms

The continuous wavelet transform of a function $f(x) \in L^2(\mathbf{R})$ with respect to $\psi(x) \in L^2(\mathbf{R})$ is defined by

$$(Wf)(a,b) = \int_{-\infty}^{\infty} f(x)\psi^{a,b}(x)dx,$$

where $a, b \in \mathbf{R}$, $a \neq 0$ and $\psi^{a,b}(x)$ is from the function $\psi(x)$ by translating and dilating:

$$\psi^{a,b}(x) = |a|^{1/2}\psi(\frac{x-b}{a})$$

This transform is useful when one wants to recognize or extracts features of the function $f(x)$ from the transform domain.

If a and b are restricted to a special sequence of discrete numbers, such as $a = 2^j$ and $b = k$, $j, k \in \mathbf{Z}$, we have the following commonly used **Discrete Wavelet Transform**:

$$(DWTf)(j,k) = \int_{-\infty}^{\infty} f(x)\psi_k^j(x)dx,$$

where

$$\psi_k^j(x) = 2^{j/2}\psi(2^j x - k).$$

If $\{\psi_k^j;\ k, j \in \mathbf{Z}\}$ constitutes an othornormal (or Riesz) basis of $L^2(\mathbf{R})$, the function $\psi(x)$ is called a **Wavelet** or **Wavelet Function**.

The oldest example of wavelets is the Haar function [15],

$$\psi^{Haar}(x) = \begin{cases} 1, & \text{for } 0 \leq x < \frac{1}{2}, \\ -1, & \text{for } \frac{1}{2} \leq x < 1, \\ 0, & \text{otherwise.} \end{cases}$$

The family of functions

$$\psi_k^{Haar,j}(x) = 2^{j/2}\psi^{Haar}(2^j x - k) = \begin{cases} 2^{j/2}, & \text{for } \frac{k}{2^j} \leq x < \frac{k+1/2}{2^j}, \\ -2^{j/2}, & \text{for } \frac{k+1/2}{2^j} \leq x < \frac{k+1}{2^j}, \\ 0, & \text{otherwise,} \end{cases}$$

generated from $\psi^{Haar}(x)$ by the operation of dilation and translations, constitutes an orthonormal basis of $L^2(\mathbf{R})$. The Haar basis is a compactly supported unconditional basis, but it is not a suitable basis for smoother function spaces, such as the Sobolev spaces. However it provides a good example to introduce the so-called Multiresolution Analysis, which will lead to the construction of smoother wavelet bases for much wider range of functional spaces.

Let W_j be the linear subspaces spanned by $\{\psi_k^{Haar,j},\ k \in \mathbf{Z}\}$ and $V_j = \cup_{i<j} W_i$. Then $V_{j+1} = V_j \oplus W_j$. Associated with the Harr function $\psi^{Haar}(x)$ is another piecewise constant function, called Haar scaling function, defined by

$$\phi^{Haar}(x) = \begin{cases} 1, & \text{for } 0 \le x < 1, \\ 0, & \text{otherwise.} \end{cases}$$

The linear subspaces spanned by $\{\phi_k^{Haar,j},\ k \in \mathbf{Z}\}$ is exactly the subspace V_j. The Haar wavelet function $\psi^{Haar}(x)$ can be represented by the Haar scaling function

$$\psi^{Haar}(x) = \phi^{Haar}(2x - 0) - \phi^{Haar}(2x - 1)$$

This suggests that we construct the wavelet from the subspaces V_j.

2.2 Multiresolution Analysis and Wavelets

A multiresolution analysis (MRA) consists of a sequence of successive approximation spaces $\{V_j\}_{j \in Z}$ of $L^2(R)$ with the following properties:

(i) $V_j \subset V_{j+1}$,

(ii) $\lim\limits_{j \to \infty} V_j = \bigcup\limits_{j \in Z} V_j$ is dense in $L^2(R)$,

(iii) $\bigcap\limits_{j \in Z} V_j = \{0\}$,

(iv) $f(x) \in V_j \Longleftrightarrow f(2x) \in V_{j+1}$,

(v) $f(x) \in V_j \Longleftrightarrow f(x + 2^{-j}k) \in V_j,\ \ \forall k \in Z$,

(vi) There exists a function $\phi \in V_0$ so that $\{\phi(x - j)\}_{j \in Z}$ is an orthonormal basis of V_0.

ϕ is called a scaling function that generates a MRA with the above properties.

Through translation and dilation of ϕ, a Riesz basis $\{\phi_{j,k}(x)\}_{k \in Z}$ is obtained for the subspace $V_j \subset L^2(R)$ by the properties (iv) and (v), where

$$\phi_k^j(x) = 2^{\frac{j}{2}} \phi(2^j x - k), \quad j, k \in Z. \tag{2.1}$$

Since $V_0 \subset V_1$, there is a set of coefficients $\{a_k\}_{k\in Z}$, so that ϕ satisfies the two–scale equation or refinement equation

$$\phi(x) = \sum_k a_k \phi(2x-k). \tag{2.2}$$

For every $j \in Z$, we define W_j to be the orthonormal complement of V_j in V_{j+1}, we then have

$$V_{j+1} = V_j \bigoplus W_j \tag{2.3}$$

and

$$W_j \perp W_{j'} \quad if \quad j \neq j'. \tag{2.4}$$

It follows that, for $j > J$

$$V_j = V_J \bigoplus (\bigoplus_{k=0}^{J-j+1} W_{J-k}). \tag{2.5}$$

By virtue of (ii) and (iii) above, this implies

$$L^2(R) = \bigoplus_{j\in Z} W_j \tag{2.6}$$

which is a decomposition of $L^2(R)$ into mutually orthogonal subspaces. It turns out that a basis for W_0 can be obtained by dilating and translating a single function $\psi(x)$ called basic (mother) wavelet which is defined by

$$\psi(x) = \sum_k b_k \phi(2x-k) \tag{2.7}$$

where $b_k = (-1)^k a_{-k+1}$. In fact, $\{\psi_k^j(x) = 2^{\frac{j}{2}}\psi(2^j x - k)\}_{k\in Z}$ forms an orthonormal basis for W_j.

Let P_j, Q_j denote the orthogonal projection $L^2 \to V_j$, $L^2 \to W_j$, respectively. Then

$$P_j f(x) = \sum_k \alpha_{j,k} \phi_k^j(x), \tag{2.8}$$

$$Q_j f(x) = \sum_k \beta_{j,k} \psi_k^j(x), \tag{2.9}$$

where the coefficients $\alpha_{j,k}$, $\beta_{j,k}$ are given by the inner product:

$$\alpha_{j,k} = < f, \phi_k^j > = \int_{-\infty}^{\infty} f(x)\phi_k^j(x)dx, \tag{2.10}$$

$$\beta_{j,k} =< f,\ \psi_k^j >= \int_{-\infty}^{\infty} f(x)\psi_k^j(x)dx. \tag{2.11}$$

$P_j f$ converges to f in the L^2 norm which is the best approximation of f in V_j.

By $V_j = V_{j-1} \bigoplus W_{j-1}$ we have

$$\begin{aligned} P_j f(x) &= \sum_k \alpha_{j,k}\phi_k^j(x) \\ &= \sum_k \alpha_{j-1,k}\phi_k^{j-1}(x) + \sum_k \beta_{j-1,k}\psi_k^{j-1}(x). \end{aligned} \tag{2.12}$$

By multiplying ϕ_k^{j-1} to both sides of (2.12) and integrating, and applying two-scale relation and orthogonality of the scaling and wavelet functions we find that, for a fixed j and k,

$$\begin{aligned} \alpha_{j-1,k} &= \sum_\ell \alpha_{j,\ell} \int_{-\infty}^{\infty} \phi_k^{j-1}(x)\phi_\ell^j(x)dx \\ &= \frac{1}{\sqrt{2}} \sum_{\ell,n} \alpha_{j,\ell} a_n \int_{-\infty}^{\infty} \phi_\ell^j(x)\phi_{2k+n}^j(x)dx \\ &= \frac{1}{\sqrt{2}} \sum_\ell a_{\ell-2k}\alpha_{j,\ell} \end{aligned} \tag{2.13}$$

and similarly

$$\beta_{j-1,k} = \frac{1}{\sqrt{2}} \sum_\ell b_{\ell-2k}\alpha_{j,\ell}. \tag{2.14}$$

By applying (2.12) to the right hand side of (2.10), we obtain the following reconstruction algorithm:

$$\alpha_{j,k} = \frac{1}{\sqrt{2}} \sum_\ell a_{k-2\ell}\alpha_{j-1,\ell} + b_{k-2\ell}\beta_{j-1,k}. \tag{2.15}$$

The relations (2.12)–(2.15) are called Mallat's transform [22].

2.3 Orthonormal Bases of Compactly Supported Wavelets

If there are only finite non-zero terms $a_{N_1}, a_{N_1+1}, \cdots, a_{N_2}$ ($N_1 < N_2$) in the sequence of the coefficients $\{a_k\}_{k\in Z}$ for the two-scale relation (2.2), then the scaling function ϕ and wavelet function ψ will be of compact support $[N_1, N_2]$, and $\{a_{N_1}, \cdots, a_{N_2}\}$ satisfies

$$\sum_{k=N_1}^{N_2} a_k = 2, \tag{2.16}$$

$$\sum_{k=N_1}^{N_2} a_k a_{k+2\ell} = 2\delta_{0\ell}, \quad \ell \in Z, \tag{2.17}$$

$$\sum_{k=N_1}^{N_2} (-1)^k k^m a_k = 0, \; m = 0, 1, \cdots, N-1, \tag{2.18}$$

where (2.17) is the condition for orthogonality, N is an integer ($N \geq 1$) called the order of the wavelet, (2.18) is called the sum rules.

Daubechies wavelets of order N ($N_1 = 0, N_2 = 2N-1$) and coiflets of order N ($N = 2K$, $N_1 = -2K$, $N_2 = 3K-1$) are important examples of the compactly supported wavelets.

For j, k $\in Z, x \in R$, we recall

$$\phi_k^j(x) = 2^{\frac{j}{2}} \phi(2^j x - k), \; x \in R, \tag{2.19}$$

$$\psi_k^j(x) = 2^{\frac{j}{2}} \psi(2^j x - k), \; x \in R. \tag{2.20}$$

Then $\{\psi_k^j, j, k \in Z\}$ is an orthonormal bases of compactly supported wavelets for $L^2(R)$. We have

$$supp(\phi_k^j) = supp(\psi_k^j) = [\frac{k+N_1}{2^j}, \frac{k+N_2}{2^j}], \tag{2.21}$$

$$\|\phi_k^j\|_{L^2(R)} = 1. \tag{2.22}$$

The conditions (2.18) has several consequences as stated in [26]:

(i) The polynomials $1, x, \cdots, x^{N-1}$ are linear combinations of $\{\phi(x-k)\}_{k \in Z}$.

(ii) Smooth functions can be approximated with error $O(h^N)$ by combinations at every scale $h = 2^{-j}$:

$$\|f - \sum c_k \phi(2^j x - k)\| \leq C 2^{-jN} \|f^{(N)}\| \tag{2.23}$$

where

$$c_k = 2^j \int f(x) \phi(2^j x - k) dx. \tag{2.24}$$

(iii) The first N moments of the wavelet $\psi(x)$ are zero:

$$\int x^l \psi(x) dx = 0, \quad l = 0, 1, \cdots, N-1. \tag{2.25}$$

(iv) The wavelet coefficients of a smooth function f decay like

$$\int f(x)\psi(2^j x)dx \le C2^{-jN}. \tag{2.26}$$

(v) The recursion matrix L_N that determines ϕ at the integers has the eigenvalues $1, \frac{1}{2}, \cdots, (\frac{1}{2})^{N-1}$, where the N x N matrix

$$L_N = \begin{pmatrix} \cdots & a_0 & 0 & 0 & 0 & \cdots \\ \cdots & a_2 & a_1 & a_0 & 0 & \cdots \\ \cdots & \cdots & a_3 & a_2 & a_1 & \cdots \\ \cdots & \cdots & \cdots & \cdots & a_3 & \cdots \end{pmatrix}$$

By (2.2), (2.16), (2.23) we have

$$\sum_k \phi_k^j(x) = 2^{\frac{j}{2}}, \tag{2.27}$$

$$\int \phi(x)dx = 1, \tag{2.28}$$

$$\int x\phi(x)dx = \frac{1}{2}\sum_{k=N_1}^{N_2} k\, a_k. \tag{2.29}$$

2.4 Approximation Order and Accuracy

Applying the Fourier Transform on both sides of (2.2), we can rewrite the dilation equation as

$$\hat{\phi}(\xi) = \frac{1}{\sqrt{2}}\sum a_k e^{-ik\xi/2}\hat{\phi}(\xi/2),$$

or

$$\hat{\phi}(\xi) = m_0(\xi/2)\hat{\phi}(\xi/2) \quad in\ L^2,$$

where

$$m_0(\xi) = \frac{1}{2}\sum a_k e^{-ik\xi} \quad a.e..$$

The order of zero of the function $m_0(\xi)$ at $\xi = \pi$ is related to many important properties of the scaling function and the wavelet function. It has several equivalent conditions as shown in the next theorem. This provides the relationship between the accuracy of the scaling function and the order of the approximation.

Definition 2.4.1. Let $V_j, j \in \mathbf{Z}$, be the multiresolution analysis associated with the a scaling function $\phi(x)$ and P_j is the orthogonal projection onto V_j. $\phi(x)$ is said to have approximation order N if

$$\|f - P_j f\|_{L^2} \leq C2^{-jN}\|f\|_{H^n}$$

for every $f \in H^n(\mathbf{R})$, where C is independent of j and H^n is the Sobolev space.

Definition 2.4.2. A scaling function $\phi(x)$ is said to have approximation accuracy N if each polynomial up to degree $N-1$ can be a locally finite series of the integer translation $\phi(x+k)$, $k \in \mathbf{Z}$, i.e., there are some constants $y_k^{(p)}$, such that

$$x^p = \sum_k y_k^{(p)} \phi(x+k) \ \ a.e.,\ p = 0, \cdots, N-1.$$

Theorem 2.1. *Suppose a scaling function $\phi(x)$ is compactly supported and its translations $\phi(x-k)$, $k \in \mathbf{Z}$ are linearly independent. Then the following statements are equivalent:*

(i) *$\phi(x)$ has approximation accuracy N.*

(ii) *$\phi(x)$ has approximation order N.*

(iii)

$$\hat{\phi}(0) = 1,$$

$$\frac{d^p}{d\xi^p}\hat{\psi}|_{\xi=0} = 0, \quad p = 0, 1, \cdots, N-1.$$

(iv)

$$\sum_k c_k = 2,$$

$$\sum_k (-1)^k k^p c_k = 0, \quad p = 0, 1, \cdots, N-1.$$

(v)

$$\int_{\mathbf{R}} \phi(x)dx = 1,$$

$$\int_{\mathbf{R}} x^p \psi(x)dx = 0, \quad p = 0, 1, \cdots, N-1.$$

The condition (v) is called the vanishing moment condition.

The wavelet system with vanishing moments on both the scaling function ϕ and the wavelet function ψ was first suggested by R. Coifman and hence it is called the Coiflets. In a generalized convergence theorem (Theorem 3.2), E. B. Lin and X. Zhou [21] introduced generalized Coiflets (in 3.32) and generalized the result of J. Tian and R. O. Wells [27].

Compactness, smoothness and multiresolution properties of wavelets allow us to give a good microlocal analysis for a given function. One of the main approximation questions here is how to decompose a function into wavelet coefficients, and how to reconstruct a function from the coefficients. The efficiency and accuracy of calculating coefficients are also our main concerns. In the following section, we introduce wavelet approximation methods which will serve these purposes.

3 Wavelet Interpolation Formulas

3.1 Scalar Scaling Function Interpolation

In this section we present a generalized convergence theorem in R^2 which is also true in R^n. We then show some examples. The interpolation theorem in [28] is an immediate consequence of Theorem 3.2. A similar interpolation formula using multiwavelets is presented in Section 3.2.

With the additional vanishing moment condition imposed on the scaling function $\phi(x)$, Jun Tian and R. O. Wells Jr. [27] established the following wavelet interpolation (sampling approximation) theorem:

Theorem 3.1. *Suppose a scaling function $\phi(x)$ is compactly supported and has approxi mation accuracy N. If*

$$\int_{\mathbf{R}} x^p \phi(x) dx = 0, \quad p = 1, 2, \cdots, N-1,$$

then for every $f(x) \in C_0^N(\mathbf{R})$,

$$\|f(x) - 2^{-j/2} \sum_k f(\frac{k}{2^j}) \phi_k^j(x)\|_{L^2} \leq C 2^{-jN}.$$

If in addition, $\phi(x) \in C^n(\mathbf{R})$, where $n \in \mathbf{Z}$, $0 \leq n \leq N-1$, then

$$\|f(x) - 2^{-j/2} \sum_k f(\frac{k}{2^j}) \phi_k^j(x)\|_{H^n} \leq C 2^{-j(N-n)}$$

where Cis independent of j.

A generalization of Theorem 3.1 is stated in the following theorm.

3.1.1. Convergence of the Scalar Wavelet Interpolation

Let $\phi, \psi \in C^1$ be the orthonormal multiresolution analysis with compact support as described in Section 2. Let c and $\{M_\ell\}$ denote the moments of ϕ as follows.

$$c = \int x\phi(x)dx \quad \ell = 1,\ 2, \cdots, L-1,$$

$$M_\ell = \int x^\ell \phi(x)dx, \quad \ell = 1,\ 2, \cdots, L-1.$$

Theorem 3.2. [21] *Assume the function $f \in C^L(\overline{\Omega})$, where Ω is a bounded open set in R^2, $L \geq 2$. Let, for $j \in Z$,*

$$f^j(x,y) := \frac{1}{2^j} \sum_{(p,q)\in\Lambda} f(\frac{p+c}{2^j}, \frac{q+c}{2^j})\phi_p^j(x)\phi_q^j(y), \ (x,y) \in \Omega, \tag{3.1}$$

where the index set

$$\Lambda = \{(p,q)|(supp(\phi_p^j) \bigotimes supp(\phi_q^j)) \cap \Omega \neq \emptyset\}. \tag{3.2}$$

In addition the moments of ϕ and ψ satisfy

$$M_\ell = (c)^\ell, \quad \ell = 1,\ 2, \cdots, L-1, \tag{3.3}$$

i.e.,

$$\int x^\ell \phi(x)dx = (\int x\phi(x)dx)^\ell, \quad \ell = 1,\ 2, \cdots, L-1. \tag{3.4}$$

Then

$$\|f - f^j\|_{L^2(\Omega)} \leq C\|f^{(L)}\|_\infty(\frac{1}{2^j})^L, \tag{3.5}$$

$$\|f - f^j\|_{H^1(\Omega)} \leq C\|f^{(L)}\|_\infty(\frac{1}{2^j})^{L-1}, \tag{3.6}$$

where C is a constant depending only on L and diameter of Ω.

$$\|f^{(L)}\|_\infty := max_{(x,y)\in\Omega,\ m=0,1,\cdots,L}|\frac{\partial^L f}{\partial x^m \partial y^{L-m}}(x,y)|. \tag{3.7}$$

Remark 3.1. The condition (3.4) is equivalent to the following condition which first appeared in G. Beylkin, R. Coifman, and V. Rokhlin [4]: there is a real constant τ_L such that

$$\int \phi(x+\tau_L)x^m dx = 0, \ m = 1, 2, \cdots, L-1 \tag{3.8}$$

Remark 3.2. Under the additional assumption $\phi,\ \psi \in C^L$, one can derive the following estimates:

$$\|f - f^j\|_{H^m(\Omega)} \leq C\|f^{(L)}\|_\infty(\frac{1}{2^j})^{L-m},\ m = 0, 1, \cdots, L-1. \tag{3.9}$$

3.1.2. Some Examples of Scalar Scaling Function Interpolation

First example of ϕ, ψ that satisfy the conditions in Theorem 3.2 is the Daubechies wavelets of order N ($N > 2$). This family of wavelets ϕ, ψ satisfy (3.3) for L = 3.

The second example of ϕ, ψ that satisfy the conditions in Theorem 3.2 is the coiflets. This is a family of wavelets ϕ, ψ of order N so that

$$\int x^\ell \psi(x)dx = 0, \quad \ell = 0, \cdots, N-1, \tag{3.10}$$

$$\int \phi(x)dx = 1, \quad \int x^\ell \phi(x)dx = 0, \quad \ell = 1, \cdots, N-1. \tag{3.11}$$

So (3.3) is valid for L = N.

Numerical examples of wavelet interpolation. We apply the wavelet interpolation (3.1) to Daubechies wavelet of order 3 (Figure 1)and coiflets of order 4 (Figure 2 and Figure 3) to approximate the following functions:

$$\begin{aligned} f_1(x, y) &= (x+y)^4, \\ f_2(x, y) &= \cos(\pi x y), \\ f_3(x, y) &= \cosh(x^2 y + 1), \end{aligned}$$

on the rectangle [0,1]x[0,1], the results are listed on the Table 1 and Table 2, where the relative error of the interpolation formula f^j for a function f is defined by

$$E = \frac{\|f - f^j\|_{L^2(\Omega)}}{\|f\|_{L^j(\Omega)}}. \tag{3.12}$$

3.2 Multiwavelet Approximation and Interpolation

In this section, we generalize the theory of (scalar) wavelet approximation and interpolation to the Multiwavlet case.

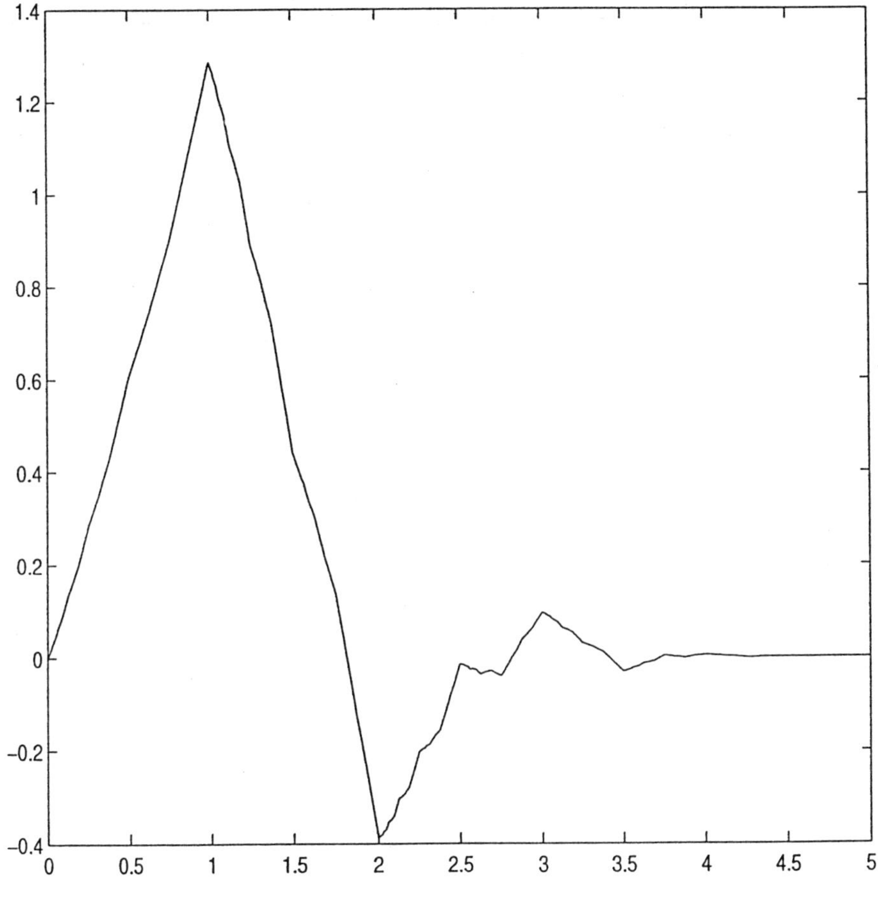

Figure 1: The scaling function of Daubechies wavelet of order 3.

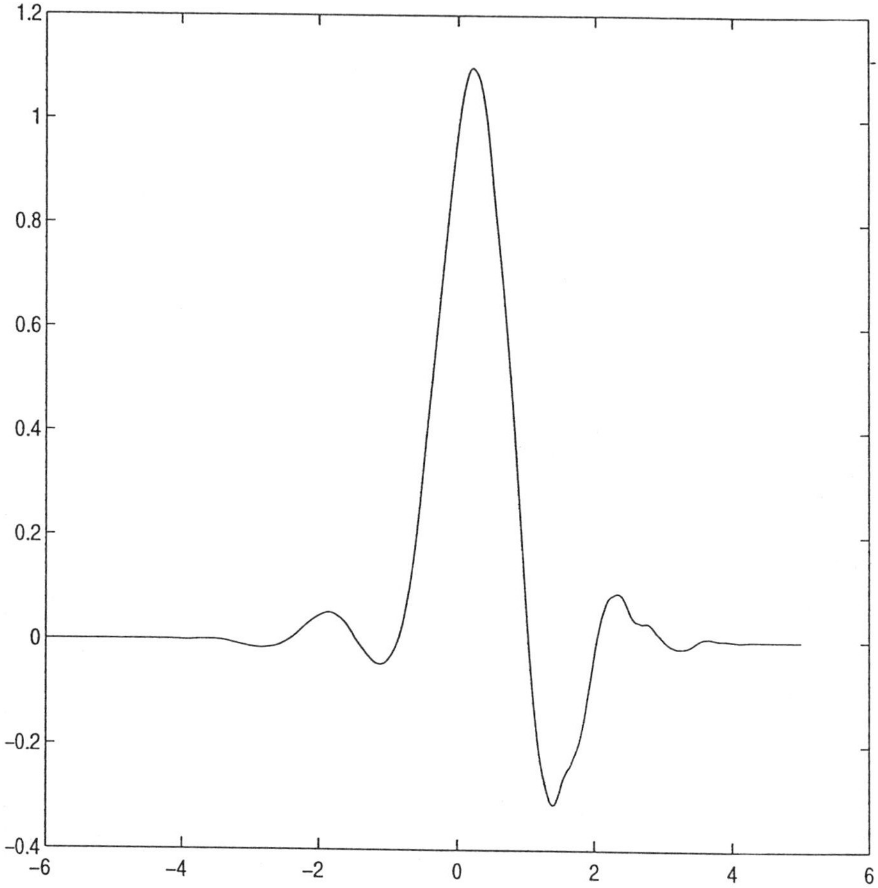

Figure 2: The scaling function of Coiflet of order 4.

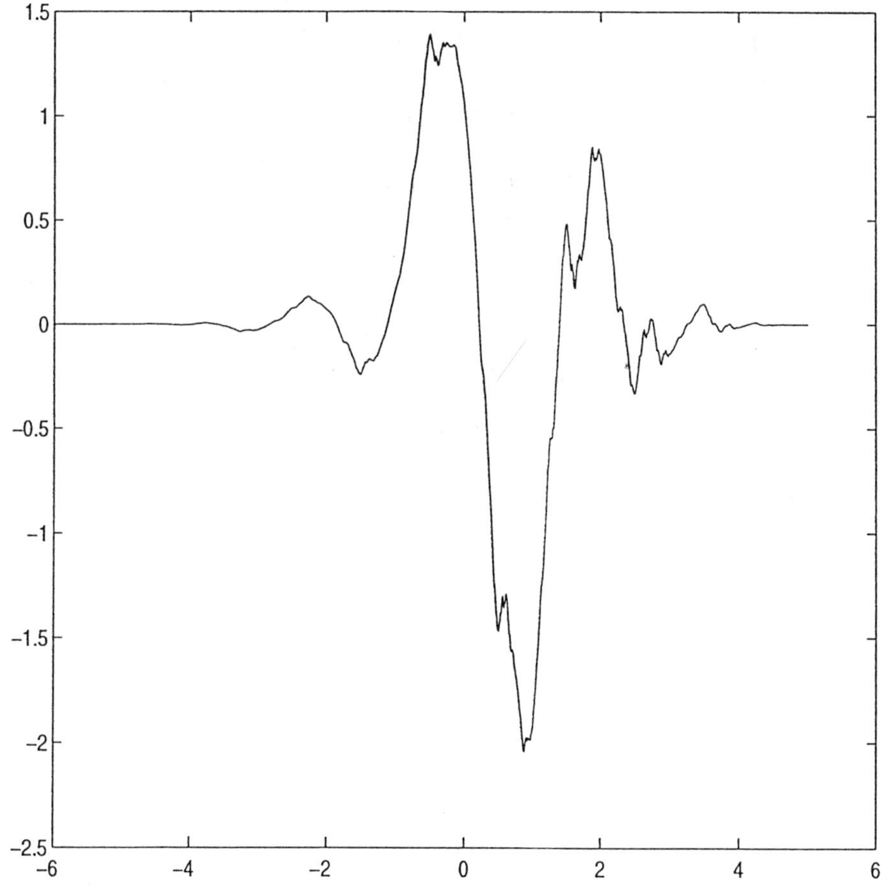

Figure 3: The derivative of the scaling function of Coiflet of order 4.

Table 1: Relative Error of Approximation by Using Daubechies Wavelet of Order 3

	$f_1(x,y)$	$f_2(x,y)$	$f_3(x,y)$
j = 1	0.136848	0.081605	0.017593
j = 3	0.003624	0.001563	0.000302
j = 5	0.000062	0.000026	5.994E-6
j = 7	9.964E-7	4.112E-7	9.961E-8

Table 2: Relative Error of Approximation by Using Coiflet of Order 4

	$f_1(x,y)$	$f_2(x,y)$	$f_3(x,y)$
j = 1	0.033627	0.035825	40688.3
j = 3	0.000197	0.000113	0.000067
j = 5	7.706E-7	4.644E-7	2.092E-7
j = 7	3.010E-9	1.835E-9	7.785E-9

3.2.1. Multiscaling Function and Multiwavelet

In the definition of multiresolution analysis, the function $\phi(x)$ is replaced by two or more functions

$$\phi_0(x), \phi_1(x), \cdots, \phi_{r-1}(x)$$

which form a **Multi-scaling Function**

$$\Phi(x) = [\phi_0(x), \phi_1(x), \cdots, \phi_{r-1}(x)]^T,$$

and the dilation equation becomes

$$\Phi(x) = \sum_{k \in \mathbf{Z}} C_k \Phi(2x - k) \tag{3.13}$$

where C_k are r by r constant matrices.

Correspondingly, the **Multiwavelet** associated with $\Phi(x)$ is a vector valued function

$$\Psi(x) = [\psi_0(x), \psi_1(x), \cdots, \psi_{r-1}(x)]^T$$

satisfying the othorgonal condition and the **Multiwavelet Equation**

$$\Psi(x) = \sum_{k \in \mathbf{Z}} D_k \Phi(2x - k) \tag{3.14}$$

where $D_k{}'s$ are r by r constant matrices. The special case of $r = 1$ is the (scalar) wavelet.

One of the first multiwavelet constructions is due to Alpert and Roklin [2]. The simplest Alpert multi-scaling function Φ^{Alpert} can be described as follows. The first component of Φ^{Alpert} is a box function and the second has a piece of linear function. The corresponding dilation equation is

$$\begin{bmatrix} \phi_0(x) \\ \phi_1(x) \end{bmatrix} = \begin{bmatrix} 1 & 0 \\ \sqrt{3}/2 & 1/2 \end{bmatrix} \begin{bmatrix} \phi_0(2x-0) \\ \phi_1(2x-0) \end{bmatrix} + \begin{bmatrix} 1 & 0 \\ -\sqrt{3}/2 & 1/2 \end{bmatrix} \begin{bmatrix} \phi_0(2x-1) \\ \phi_1(2x-1) \end{bmatrix}.$$

Using fractal interpolation, Geronimo, Hardin and Massopust [11] constructed a continuous multi-scaling function with short support, symmetry and second approximation accuracy. Plonka and Strela [25] developed an algorithm to construct multi-scaling functions with higher approximation accuracy and symmetry. In the following section, we will characterize the relationship between accuracy of multi-scaling function and approximation order of a given function which is approximated by the multi-scaling function interpolation. We now give the definition of accuracy of a multi-scaling function.

Definition 3.2.1. A multi-scaling function $\Phi(x) = [\phi_0(x), \phi_1(x), \cdots, \phi_{r-1}(x)]^T$ is said to have approximation accuracy N if each polynomial x^p, $p = 0, \cdots, N-1$, is a locally finite series of the integer translates $\phi_\alpha(x+k)$, $k \in \mathbf{Z}$, $\alpha = 0, \cdots, r-1$, i.e.

$$x^p = \sum_k y_k^{(p)} \ \Phi(x+k) \ \ a.e., \quad p = 0, \cdots, N-1 \tag{3.15}$$

where $y_k^{(p)}$ are constant row vectors and $y_0^{(p)}$ are called *starting vectors.*

3.2.2. Multi-scaling Function Interpolation

In this section, we generalize the scalar interpolation to the following statement, namely, the necessary and sufficient condition for the asymptotic behavior of MSF interpolation with error 2^{-jN} is that Φ has approximation accuracy N. The idea here is to choose the sampling values by using suitable weighted average of the values of given function at the sampling points.

Using some standard arguments one can extend scalar interpolation theorem in R^1 to higher dimensional cases. More work is needed in higher dimensional cases for multi-scaling functions. The theorem is stated in $\mathbf{R}^2$ and can be generalized to $\mathbf{R}^d$, $d > 2$.

Let $\lambda_{m\alpha}$, $u_{m\alpha}$, $\alpha = 0, 1, \cdots, r-1$, $m = 0, 1, \cdots, M-1$, be constants. We choose the sampling values as follows.

For $j \in N$, let

$$f^j_{k,\alpha,l,\beta} = \sum_{n,m=0}^{M-1} u_{n\alpha}\, u_{m\beta}\, f(\frac{k+\lambda_{n\alpha}}{2^j}, \frac{l+\lambda_{m\beta}}{2^j}) \tag{3.16}$$

and

$$f^j(x,y) = \frac{1}{2^j}\sum_{k,l}\sum_{\alpha,\beta=0}^{r-1} f^j_{k,\alpha,l,\beta},\phi^j_{\alpha k}(x)\phi^j_{\beta l}(y). \tag{3.17}$$

In fact, the weighted coefficients $u_{m\alpha}$ are related to the starting vectors as follows:

$$y_0^{(p)} = (\sum_{m=0}^{M-1} u_{m0}\lambda^p_{m0}, \sum_{m=0}^{M-1} u_{m1}\lambda^p_{m1}, \cdots, \sum_{m=0}^{M-1} u_{m,r-1}\lambda^p_{m,r-1}), \tag{3.18}$$

$$p = 0,\ \cdots,\ N-1, \quad \lambda^0_{m\alpha} = 1.$$

The $y_k^{(p)}$ relates to the starting vector $y_0^{(p)}$ as follows.

$$y_k^{(p)} = \sum_{m=0}^{p} \binom{p}{m} (-k)^{p-m} y_0^{(m)} \tag{3.19}$$

Theorem 3.3. [19] *Suppose $\Phi(x)$ is compactly supported. Given $\lambda_{m\alpha}$, $u_{m\alpha}$ and define f^j as (3.17). Then, for every $f \in C_0^N(\mathbf{R}^2)$,*

$$||f - f^j||_{L^2} \le C2^{-jN}, \tag{3.20}$$

if and only if Φ has approximation accuracy N and satisfies (3.18), (3.19). The constant C is independent of j.

Theorem 3.4. [19] *Suppose a multi-scaling function $\Phi(x)$ is compactly supported and $\Phi(x) \in H^n(\mathbf{R})$, $n \ge 0$. If $\Phi(x)$ has approximation accuracy N ($N \ge n$) and satisfies (3.18) and (3.19) then, for every $f \in C_o^N(\mathbf{R}^2)$,*

$$||f - f^j||_{H^n} \le C2^{-j(N-n)} \tag{3.21}$$

where C is independent of j. Conversely, if (3.21) holds, then Φ has approximation accuracy N.

3.2.3. Special Cases

The constants $\lambda_{m\alpha}, u_{m\alpha}$ in Theorem 3.3 give us some flexibility in choosing the sampling points and the weighted coefficients. The following four cases are special choices which give rise to several immediate consequences.

(1) $M = N$

In this case, we can choose λ's so that, for each $\alpha, \lambda_{m\alpha} \neq \lambda_{n\alpha}, m \neq n$, then one can always solve (3.18) for $u_{m\alpha}$. For example, we set

$$\lambda_{m\alpha} = m, \lambda_{m\alpha} = -[\frac{M}{2}] + m, \quad \lambda_{m\alpha} = M - m \quad or \quad \lambda_{m\alpha} = m - M.$$

In these settings, the sampling points can be kept at lever j. In particular, if we choose $\lambda_{m\alpha} = m$, then we have the following result.

Corollary 3.1. *Suppose a multi-scaling function $\Phi(x)$ is compactly supported and has approximation accuracy N with starting vectors $y_0^{(0)}, \cdots, y_0^{(N-1)}$. Let vectors $Z_0, \cdots, Z_{N-1}$ satisfy*

$$\sum_{n=0}^{N-1} Z_n n^p = y_0^{(p)}, \quad p = 0, \cdots, N-1.$$

Then for ever $f \in C_0^N(\mathbf{R}^2)$,

$$||f(x,y) - \frac{1}{2^j} \sum_{k,l} \sum_{n,m=0}^{N-1} f(\frac{k}{2^j}, \frac{l}{2^j}) Z_n \Phi_{k-n}^j(x) Z_m \Phi_{l-m}^j(y)||_{L^2} \leq C 2^{-jN}$$

and

$$||f(x,y) - \frac{1}{2^j} \sum_{k,l} \sum_{n,m=0}^{N-1} f(\frac{k}{2^j}, \frac{l}{2^j}) Z_n \Phi_{k-n}^j(x) Z_m \Phi_{l-m}^j(y)||_{H^n} \leq C 2^{-j(N-n)}$$

provided $\Phi(x) \in H^n$ and $N \geq n$ where C is independent of j.

(2) $M = 1$. $\lambda_{0\alpha} = \lambda, \ \alpha = 0, \cdots, r-1$.

In this case, (3.18) becomes $y_0^{(p)} = \lambda^p y_0^{(0)}$.

Corollary 3.2. *Suppose a multi-scaling function $\Phi(x)$ is compactly supported. Then for every $f \in C_0^N(\mathbf{R}^2)$,*

$$||f(x,y) - \frac{1}{2^j} \sum_{k,l} f(\frac{k+\lambda}{2^j}, \frac{l+\lambda}{2^j}) y_0^{(0)} \Phi_k^j(x) y_0^{(0)} \Phi_l^j(y)||_{L^2} \leq C 2^{-jN}$$

where C is independent of j, if and only if Φ has approximation accuracy N and satisfies (3.19) such that $y_0^{(p)} = \lambda^p y_0^{(0)}$.

In addition, if $\Phi(x) \in H^n$, $n \leq N$, then

$$||f(x,y) - \frac{1}{2^j}\sum_{k,l} f(\frac{k+\lambda}{2^j}, \frac{l+\lambda}{2^j}) y_0^{(0)} \Phi_k^j(x) y_0^{(0)} \Phi_l^j(y)||_{H^n} \leq C2^{-j(N-n)}$$

This generalizes the results in [21].

(3) $M = 1, \ \ \lambda_{m\alpha} = 0$.

In this case, (3.18) gives rise to $y_o^{(p)} = 0, \ \ p = 1, \cdots, N-1$. As an immediate consequence with respect to Coiflets, we have the following theorem:

Theorem 3.5. *Suppose a multi-scaling function Φ has compact support and $\{\phi_\alpha(x+k);\ \alpha = 0, \cdots, r-1,\ k \in Z\}$ are linearly independent. Then the following statements are equivalent.*

(i) $\Phi(x)$ has approximation accuracy N and satisfies

$$y_k^{(p)} = (-k)^p y_0^{(0)}, \quad k \in Z, p = 0, \cdots, N-1.$$

(ii) For every $f \in C_0^N(\mathbf{R}^2)$,

$$||f(x,y) - \frac{1}{2^j}\sum_{k,l} f(\frac{k}{2^j}, \frac{l}{2^j}) y_0^{(0)} \Phi_k^j(x) y_0^{(0)} \Phi_l^j(y)||_{L^2} \leq C2^{-jN}$$

where C is independent of j.

(iii)

$$\sum_k (-1)^k k^p y_0^{(0)} C_k = 0$$

$$\sum_k k^p y_0^{(0)} C_k = 2\delta_{p0}, \quad p = 0, \cdots, N-1$$

(iv)

$$\sum_k (x+k)^p y_0^{(0)} \Phi(x+k) = \delta_{p0} \quad a.e., \ p = 0, \cdots, N-1.$$

When $r = 1$, it is shown in [27] that (ii) is the sum rule condition for Coifman wavelet system of degree N. Hence, we not only generalize Theorem 3.3. and its converse in the previous section to the case of multi-scaling functions, but also obtain four equivalent conditions for scalar Coifman wavelet system of degree N. Therefore, one can define multi-Coiflets by using one of the above conditions (i)–(iv).

(4) M=1, N= 2. In this case, (3.18) is reduced to

$$y_0^{(0)} = (u_{00}, \cdots, u_{0,r-1})$$

and

$$y_0^{(1)} = (\lambda_{00}u_{00}, \cdots, \lambda_{0,r-1}u_{0,r-1}).$$

Corollary 3.3. *Let $\Phi(x)$ be compactly supported and have approximation accuracy 2 with starting vectors*

$$y_0^{(0)} = (u_0, \cdots, u_{r-1})$$

and

$$y_0^{(1)} = (v_0, \cdots, v_{r-1})$$

If $u_\alpha \neq 0, \quad \alpha = 0, \cdots, r-1$, then for every $f \in C_0^N(\mathbf{R}^2)$

$$\|f(x,y) - \frac{1}{2^j}\sum_{k,l}\sum_{\alpha,\beta=0}^{r-1} u_\alpha u_\beta f(\frac{k+\frac{v_\alpha}{u_\alpha}}{2^j}, \frac{l+\frac{v_\beta}{u_\beta}}{2^j})\phi^j_{\alpha k}(x)\phi^j_{\beta l}(y)\|_{L^2} \leq C2^{-2j}. \tag{3.22}$$

Moreover, if in addition $\Phi \in H^1$, then

$$\|f(x,y) - \frac{1}{2^j}\sum_{k,l}\sum_{\alpha,\beta=0}^{r-1} u_\alpha u_\beta f(\frac{k+\frac{v_\alpha}{u_\alpha}}{2^j}, \frac{l+\frac{v_\beta}{u_\beta}}{2^j})\phi^j_{\alpha k}(x)\phi^j_{\beta l}(y)\|_{H^1} \leq C2^{-j},$$

where C is independent of j.

As an immediate consequence, the case $r = 1$ is reduced to Theorem 3.1. in [28].

Remark 3.3. The hypothesis on nonzero components of the first starting vector can be relaxed. In fact, for any nondegenerate $r \times r$ matrix C, $\tilde{\Phi}(x) = C\Phi(x)$ has the same approximation accuracy and its starting vectors are

$$y_0^{(0)}C^{-1}, \cdots, y_0^{(N-1)}C^{-1}.$$

Therefore, one can choose C so that all the components of $y_0^{(0)}C$ will differ from 0 and apply the corollary to $\tilde{\Phi}(x)$.

3.2.4. Numerical Examples of MSF Interpolation

Example 1. Let $f(x,y) \in C_0^4(\mathbf{R}^2)$ be a smooth extension of

$$g(x,y) = (x+y)^4$$

which is defined on the unit square $(x,y)||x| \leq 1,\ |y| \leq 1$. The MSF approximation of f can be obtained as follows.

We use Corollary 3.1 and multi-scaling function $\Phi(t) = (\phi_0(t), \phi_1(t))$ which is defined by [25]

$$\phi_0(t) = \begin{cases} -2t^3 + 3t^2, & t \in [0,1), \\ (t-2)^2(2t-1), & t \in [1,2], \\ 0, & otherwise. \end{cases}$$

$$\phi_1(t) = \begin{cases} -3t^3 + 3t^2, & t \in [0,1), \\ (t-2)^2(-3t+3), & t \in [1,2], \\ 0, & otherwise. \end{cases}$$

This Φ has accuracy 4, and its starting vectors are

$$y_0^{(0)} = (1,0), \quad y_0^{(1)} = (1, -\frac{1}{3}), \quad y_0^{(2)} = (1, -\frac{2}{3}), \quad y_0^{(3)} = (1,-1).$$

Follow Corollary 3.1, we have

$$Z_0 = (0, \frac{1}{9}), \quad Z_1 = (1, \frac{1}{6}), \quad Z_2 = (0, -\frac{1}{3}), \quad Z_3 = (0, \frac{1}{18}).$$

Let

$$f^j(x,y) = \frac{1}{2^j} \sum_{k,l} \sum_{n,m=0}^{3} f(\frac{k}{2^j}, \frac{l}{2^j}) Z_n \Phi_{k-n}^j(x) Z_m \Phi_{l-m}^j(y) \tag{3.23}$$

$$\triangle f_j = ||f(x,y) - f^j(x,y)||_{L^2([0,1]\times[0,1])}$$

The numerical results are in Table 3.

Example 2. The one dimensional version of (3.22) is

$$||f(x) - \frac{1}{2^j} \sum_k \sum_{\alpha=0}^{r-1} u_\alpha f(\frac{k + \frac{v_\alpha}{u_\alpha}}{2^j}) \phi_{\alpha k}^j(x)||_{L^2} \leq C 2^{-2j}.$$

Use this interpolation formula, we approximate $F(x) \in C_0^2(R^1)$ which is defined as a smooth extension of

$$f(t) = e^{t^3+t+2.12} \sin t, |t| \leq 2.$$

Table 3: Coiflet and MSF interpolation results for $f(x, y) = (x + y)^4$

Level j	Δf_j with Coiflet	Δf_j with MSF
1	0.0165214	0.0130342
2	7.18854×10^{-4}	8.14619×10^{-4}
4	4.77539×10^{-6}	3.18205×10^{-6}
4	1.86539×10^{-8}	1.24301×10^{-8}

Consider the multi-scaling function $\Phi(t) = (\phi_0(t), \phi_1(t))$ given in Example 1.

Since its first starting vector is $(1, 0)$, we need to create another multi-scaling function $\Phi(t) = (\phi_0(t), \phi_1(t))$ which is obtained by using a nondegenerate transform applied to $\Phi(t)$ in Example 1.

$$\phi_0(t) = \begin{cases} -5t^3 + 6t^2, & t \in [0,1), \\ (2-t)^2(-t+2), & t \in [1,2], \\ 0, & otherwise. \end{cases}$$

$$\phi_1(t) = \begin{cases} -t^3, & t \in [0,1), \\ (2-t)^2(-5t+4), & t \in [1,2], \\ 0, & otherwise. \end{cases}$$

The first two new starting vectors of this new $\Phi(t)$ are $(\frac{1}{2}, -\frac{1}{2})$ and $(\frac{1}{3}, -\frac{2}{3})$. Let

$$\triangle F_j = ||F - F^j||_{L^2[0,1]}.$$

where

$$F^j(x) = \sum_{k=-2}^{2^j} \{\frac{1}{2}F(\frac{k+\frac{2}{3}}{2^j})\phi_0(2^jx - k) - \frac{1}{2}F(\frac{k+\frac{4}{3}}{2^j})\phi_1(2^jx - k)\}.$$

Using Coiflet interpolation (3.1) and MSF interpolation (3.17), we obtain numerical results which are in the Table 4.

3.3 2-D Nonseparable Scaling Function Interpolation

In this section, we present similar results of Sections 3.1 and 3.2 by usin nonseparable scaling functions interpolation. Separable wavelets in $L^2(\mathbf{R}^2$ which are constructed by tensor products of wavelets in $L^2(\mathbf{R})$, have g horizontal and vertical resolution properties but they do not perform s in other directions. Much effort has been spent on constructing nonse bivariate wavelets. For example, in [16], He and Lai gave some ex

Table 4: Coiflet and MSF interpolation results for $f(t) = e^{t^3+t+2.12}\sin t$

Level j	ΔF_j with Coiflet	ΔF_j with MSF
2	45.6215	0.616548
3	0.399638	9.77608×10^{-3}
4	1.26958×10^{-2}	7.75455×10^{-4}
5	6.35856×10^{-4}	8.62305×10^{-5}
6	3.96620×10^{-5}	4.39215×10^{-6}
7	2.36074×10^{-6}	2.29386×10^{-7}

bivariate nonseparable compactly supported orthonormal continuous wavelets. In [3], Belogay and Wang constructed a family of bivariate orthogonal wavelets with compact support that are nonseparable and have vanishing moments of order p or less. We will use He and Lai's nonseparable scaling function to illustrate our numerical experiments.

3.3.1. 2-D MRA

Given a dilation matrix $\mathbf{M} \in \mathbf{Z}^{2\times 2}$ with integer entries and its singular values σ_1, σ_2 satisfy $\sigma_1 \geq \sigma_2 > 1$. The **2-D MRA** (multiresolution analysis) with such a dilation matrix is a ladder of closed subspaces $\{V_j\}_{j\in\mathbf{Z}}$ which approximates $L^2(\mathbf{R}^2)$ and satisfies

$$\{0\} \rightarrow \cdots V_{-1} \subset V_0 \subset V_1 \cdots \rightarrow L^2(\mathbf{R}^2) \tag{3.24}$$

$$f(\mathbf{x}) \subset V_{j-1} \Leftrightarrow f(\mathbf{M}\mathbf{x}) \subset V_j \tag{3.25}$$

$$\exists\ \phi \in V_0\ \ s.t.\ \ the\ \ set\ \ \{\phi(\mathbf{x}-\mathbf{k})\}_{\mathbf{k}\in\mathbf{Z}^2} \tag{3.26}$$

is an orthonormal basis for V_0. The function $\phi(\mathbf{x})$ is called *scaling function* and since $V_0 \subset V_1$, $(\mathbf{x})$ has to be the solution of a dilation equation of the form

$$\phi(\mathbf{x}) = |det\,\mathbf{M}| \sum_{\mathbf{k}\in\mathbf{Z}^2} c_{\mathbf{k}}\phi(\mathbf{M}\mathbf{x}-\mathbf{k}) \tag{3.27}$$

where

$$\sum_{\mathbf{k}\in\mathbf{Z}^2} c_{\mathbf{k}} = 1. \tag{3.28}$$

The associted wavelet is then derived from the scaling function by the formula

$$\psi(\mathbf{x}) = |det\,\mathbf{M}| \sum_{\mathbf{k}\in\mathbf{Z}^2} d_{\mathbf{K}}\phi(\mathbf{M}\mathbf{x}-\mathbf{k}). \tag{3.29}$$

Obviously, the wavelet has the same smoothness as the scaling function. If the *wavelet coefficients* $d_{\mathbf{K}}$ are give by

$$d_{\mathbf{K}} = (-1)^{k_1} c_{\mathbf{e}-\mathbf{k}},$$

where $\mathbf{k} = (k_1, k_2)$ and $\mathbf{e} = (1, 0)$, then the wavelet $\psi(\mathbf{x})$ is orthogonal to $\{\phi(\mathbf{x} - \mathbf{k})\}_{\mathbf{k}\in\mathbf{Z}^2}$.

In what follows, we shall assume ϕ, ψ are sufficiently smooth and $c_{\mathbf{k}}$ are real, $c_{\mathbf{k}} \neq 0$ for only finitely many $\mathbf{k} \in \mathbf{Z}^2$ (ensuring that $\phi(\mathbf{x})$ has compact support). Also, we adopt the standard notation for multi-integers , writing $\mathbf{j}, \mathbf{k}, \cdots$ for elements of $\mathbf{Z}^2$ and $\alpha, \beta, \cdots$ for elements of $\mathbf{Z}^2_+$. Thus, each component j_v is an integer, each α_v is a non-negative integer, and

$$|\alpha| = \alpha_1 + \alpha_2 \tag{3.30}$$

$$\mathbf{x}^\alpha = x_1^{\alpha_1} x_2^{\alpha_2} \tag{3.31}$$

$$\alpha! = \alpha_1!\alpha_2! \tag{3.32}$$

$$\alpha \geq \beta \Leftrightarrow \alpha_1 \geq \beta_1, \quad \alpha_2 \geq \beta_2 \tag{3.33}$$

$$\binom{\alpha}{\beta} = \frac{\alpha!}{\beta!(\alpha-\beta)!} \quad (\alpha \geq \beta) \tag{3.34}$$

$$D^\alpha f = \frac{\partial^{|\alpha|} f}{\partial x_1^{\alpha_1} \partial x_2^{\alpha_2}}. \tag{3.35}$$

3.3.2. 2-D Accuracy

We first define accuracy of scaling function. Then we provide equivalent conditions of accuracy of a scaling function. There is a close relationship between accuracy of a scaling function and its interpolation which will be given in the following section.

Definition 3.3.1. A scaling function $\phi(\mathbf{x})$ is said to have **accuracy N** if each polynomial $\mathbf{x}^\alpha$, $|\alpha| = 0, 1, \cdots, N-1$, is a locally finite series of the integer translates $\phi(\mathbf{x} + \mathbf{k})$, $\mathbf{k} \in \mathbf{Z}^2$, i,e.

$$\mathbf{x}^\alpha = \sum_{\mathbf{k}\in\mathbf{Z}^2} b_{\mathbf{k}} \phi(\mathbf{x} + \mathbf{k}) \quad \text{a.e.} \quad |\alpha| = 0, 1, \cdots, N-1. \tag{3.36}$$

Definition 3.3.2. A scaling function $\phi(\mathbf{x}) \in L^2(\mathbf{R}^2)$ is called orthogonal if $\{\phi(\mathbf{x} - \mathbf{k})\}_{\mathbf{k}\in\mathbf{Z}^2}$ is orthogonal.

Let $M(\alpha)$ denote the moments of the scaling function ϕ,

$$M(\alpha) := \int_{\mathbf{R}^2} \mathbf{x}^\alpha \phi(\mathbf{x}) d\mathbf{x}, \qquad |\alpha| = 1, 2, \cdots.$$

In particular,

$$c_1 = M(1,0) = \int_{\mathbf{R}^2} x_1 \phi(\mathbf{x}) d\mathbf{x}. \tag{3.37}$$

$$c_2 = M(0,1) = \int_{\mathbf{R}^2} x_2 \phi(\mathbf{x}) d\mathbf{x}. \tag{3.38}$$

Thus, we have the following property of dilation matrix.

Let $\mathbf{M} = \begin{pmatrix} a & b \\ c & d \end{pmatrix}$, and $\phi(\mathbf{x})$ satisfies (3.27) and $\int_{\mathbf{R}^2} \phi(\mathbf{x}) d\mathbf{x} = 1$, then

$$c_1 = \frac{1}{Q}[(|det\,\mathbf{M}|d - det\mathbf{M}) \sum k_1 c_{\mathbf{k}} - det\,\mathbf{M}\ \ b \sum k_2 c_{\mathbf{k}}],$$

$$c_2 = \frac{1}{Q}[-c\ \ det\,\mathbf{M} \sum k_1 c_{\mathbf{k}} + (|det\,\mathbf{M}|a - det\,\mathbf{M}) \sum k_2 c_{\mathbf{k}}],$$

where $Q = |det\,\mathbf{M}|^2 - |det\,\mathbf{M}|(c+d) + det\,\mathbf{M} \neq 0$ and $\mathbf{k} = (k_1, k_2)$.

Theorem 3.6. *Suppose $\phi(\mathbf{x})$ satisfies the orthonormal* **MRA** *with $\int_{\mathbf{R}^2} \phi(\mathbf{x}) d\mathbf{x} = 1$. Then the following conditions are equivalent:*

(i) The $\phi(\mathbf{x})$ has accuracy N.

(ii) $$\mathbf{x}^\alpha = \sum_{\mathbf{l} \in \mathbf{Z}^2} \sum_{\beta \leq \alpha} \binom{\alpha}{\beta} M(\alpha - \beta) \mathbf{l}^\beta \phi(\mathbf{x} - \mathbf{l}). \qquad |\alpha| \leq N-1.$$

(iii) $\hat{\phi}(0) = 1$, *and* $D^\alpha \hat{\phi}(2\pi \mathbf{j}) = 0$, *if* $0 \neq \mathbf{j} \in \mathbf{Z}^2$, $|\alpha| \leq N-1$.

(iv) $$\sum_{\mathbf{l} \in \mathbf{Z}^2} \mathbf{l}^\alpha \phi(\mathbf{t} - \mathbf{l}) = \int_{\mathbf{R}^2} (\mathbf{t} - \mathbf{x})^\alpha \phi(\mathbf{x}) d\mathbf{x} \quad |\alpha| \leq N-1.$$

3.3.3. Nonseparable Interpolation

We define the *nonseparable scaling function interpolation* of an $L^2(\mathbf{R}^2)$ function $f(\mathbf{x})$ at the j-th level by

$$f^j(\mathbf{x}) = |det\,\mathbf{M}|^{-j/2} \sum_{\mathbf{q} \in \Lambda} f[\mathbf{M}^{-j}(\mathbf{q} + \mathbf{c})] \phi^j_{\mathbf{q}}(\mathbf{x}) \quad j \in \mathbf{Z} \tag{3.39}$$

where $\mathbf{c} = (c_1, c_2)$ which are given in (3.37) and (3.38) and

$$\phi^j_{\mathbf{q}}(\mathbf{x}) = |det\,\mathbf{M}|^{j/2} \phi(\mathbf{M}^j \mathbf{x} - \mathbf{q}) \tag{3.40}$$

$$\bigwedge = \{\mathbf{q} = (q_1, q_2) \in \mathbf{Z}^2 | supp\,\phi^j_{\mathbf{q}}(\mathbf{x}) \cap supp\, f \neq \phi\}. \tag{3.41}$$

We have the following approximation theorem for Nonseparable Scaling Function Interpolation.

Theorem 3.7. [18] *Assume ϕ satisfies the orthonormal* **MRA** *with $\int_{\mathbf{R}^2} \phi(\mathbf{x})d\mathbf{x} = 1$. And let Ω is a bounded open set in $\mathbf{R}^2$ with $supp\,\phi \subseteq \Omega$. Then, for all $f(\mathbf{x}) \in C^N(\overline{\Omega})$,*

$$\| f - f^j \|_{L^2(\Omega)} \leq C \| f^{(N)} \|_\infty \sigma_2^{-jN} \tag{3.42}$$

if and only if $\phi(\mathbf{x})$ has accuracy N and the moments satisfy

$$M(\alpha) = \mathbf{c}^\alpha \qquad 1 \leq |\alpha| \leq N-1,$$

where C is a constant depending only on N, diameter of Ω and $supp\,\phi$,and

$$\| f^{(N)} \|_\infty = max_{\mathbf{x}\in\Omega,\ |\alpha|=N} |D^\alpha f(\mathbf{x})|.$$

Theorem 3.8. [18] *Assume ϕ satisfies the orthonormal* **MRA** *with $\int_{\mathbf{R}^2} \phi(\mathbf{x})d\mathbf{x} = 1$. And let Ω is a bounded open set in $\mathbf{R}^2$ with $supp\,\phi \subseteq \Omega$. If $\phi(\mathbf{x})$ has accuracy N and the moments satisfy*

$$M(\alpha) = \mathbf{c}^\alpha \qquad 1 \leq |\alpha| \leq N-1,$$

Then, for all $f(\mathbf{x}) \in C^N(\overline{\Omega})$,

$$\| f - f^j \|_{H^1(\Omega)} \leq C \| f^{(N)} \|_\infty \sigma_1^j \sigma_2^{-jN} \quad , \tag{3.43}$$

where C is a constant depending only on N, diameter of Ω and $supp\phi$,and

$$\| f^{(N)} \|_\infty = max_{\mathbf{x}\in\Omega,\ |\alpha|=N} |D^\alpha f(\mathbf{x})|.$$

Conversely, if (3.43) holds, then ϕ has accuracy N.

3.3.4. Examples of 2-D NSFI

To illustrate the above approximation theorem, we verify the result by using the following examples:

Here we use a nonseparable scaling function in [16]. Let $\mathbf{M} = \begin{pmatrix} 2 & 0 \\ 0 & 2 \end{pmatrix}$, then (3.27) becomes

$$\phi(\mathbf{x}) = 4 \sum_{\mathbf{k}\in\mathbf{Z}^2} c_{\mathbf{k}} \phi[(2x_1, 2x_2) - \mathbf{k}]. \tag{3.44}$$

Table 5: $\|e^j\|_{L^2(\Omega)}$ of test functions

Function	NONSEPARABLE	SEPARABLE
$f(x,y) = cosh(x^2 - 4y + 1)$	0.312265	0.326993
$f(x,y) = (x - 4y)^4$	6.12595	6.36174
$f(x,y) = sin(x^2 - 12y)$	0.0880386	0.0855612
$f(x,y) = (x + y)^4$	0.0814454	0.0378736

Also, let $N = 1$ and suppose $\Omega = [0,1] \times [0,1], f \in C^4(\overline{\Omega})$, then the estimate (3.42) can be written as

$$\| f - f^j \|_{L^2(\Omega)} \leq C \| f^{(1)} \|_\infty 2^{-j} \ . \tag{3.45}$$

Consider the test functions: $f(x,y) = \cosh(x^2 - 4y + 1)$, $f(x,y) = (x - 4y)^4$, $f(x,y) = \sin(x^2 - 12y)$, $f(x,y) = (x + y)^4$.
Let

$$f^j(x,y) = \frac{1}{2^j} \sum_{\mathbf{q} \in \bigwedge} f(\frac{q_1 + c_1}{2^j}, \frac{q_2 + c_2}{2^j})\phi^j_{\mathbf{q}}(\mathbf{x}).$$

and

$$e^j = f - f^j.$$

We calculate $\|e^j\|_{L^2(\Omega)} = \|f - f^j\|_{L^2(\Omega)}$. The error estimates are in the following table.

Remark 3.3. It would be interesting to use other nonseparable scaling functions to implement the interpolation and approximation. In some cases, the nonseparable scaling function interpolation has better estimates than scalar scaling function interpolation. It is also interesting to apply nonseparable scaling function interpolation in solving PDEs and image denoising.

3.4 Nonseparable Multi-scaling Function

Let Γ be a *lattice* in $\mathbf{R}^2$, i.e., $\Gamma = \{m_1 u_1 + m_2 u_2 : m_i \in \mathbf{Z}\}$ is the collection of integer linear combinations of 2 independent vectors $u_1, u_2 \in \mathbf{R}^2$. Equivalently, Γ is the image of $\mathbf{Z}^2$ under some nonsingular linear transformation.

A *dilation matrix* associated with Γ is a 2×2 matrix $\mathbf{A}$ such that

(a) $\mathbf{A}(\Gamma) \subset \Gamma$, and

(b) $\mathbf{A}$ is expansive; i.e., all eigenvalues satisfy $|\lambda_k(\mathbf{A})| > 1$.

Since $\Gamma = \mathbf{W}(\mathbf{Z}^2)$, where $\mathbf{W}$ is the invertible matrix with u_1, u_2 as columns, the matrix $\mathbf{W}^{-1}\mathbf{A}\mathbf{W}$ maps $\mathbf{Z}^2$ into itself. Therefore $\mathbf{W}^{-1}\mathbf{A}\mathbf{W}$ has integer entries and integer determinant. Hence $\mathbf{A}$ has integer determinant as well. By applying the similarity transform $\mathbf{W}^{-1}\mathbf{A}\mathbf{W}$, it is always possible to take $\Gamma = \mathbf{Z}^2$ if desired. We shall only consider the case where $\Gamma = \mathbf{Z}^2$.

The vector-valued function $\mathbf{\Phi} = (\phi_1, \cdots, \phi_r)^T$ with $\phi_1, \cdots, \phi_r \in \mathbf{L}_2(\mathbf{R}^2)$ is *refinable* with respect to $\mathbf{A}$ if it satisfies a *refinement equation, dilation equation*, or *two-scale difference equation* of the form

$$\mathbf{\Phi}(\mathbf{x}) = \sum_{\mathbf{k}\in\mathbf{Z}^2} \mathbf{C}_{\mathbf{k}}\mathbf{\Phi}(\mathbf{A}\mathbf{x} - \mathbf{k}), \quad \mathbf{x} \in \mathbf{R}^2, \tag{3.46}$$

for some $r \times r$ matrices $\mathbf{C}_{\mathbf{k}}$ with constant elements. $\mathbf{\Phi}$ is the *multiscaling function* in $\mathbf{R}^2$. If the integer translates of $\mathbf{\Phi}$ form an orthonormal set, i.e.,

$$\int_{\mathbf{R}^2} \mathbf{\Phi}(\mathbf{x}+\alpha)\mathbf{\Phi}^T(\mathbf{x}+\beta)d\mathbf{x} = \delta_{\alpha,\beta}\mathbf{I}_r, \qquad \forall \alpha, \beta \in \mathbf{Z}^2, \tag{3.47}$$

where $\delta_{\alpha,\beta} = \{ \begin{array}{ll} 1, & if \;\; \alpha = \beta \\ 0, & otherwise \end{array}$ and $\mathbf{I}_r$ is the identity matrix of size $r \times r$, then $\mathbf{\Phi}$ is an *orthonormal refinable function.*

In what follows, we adopt the standard notation for multi-integers , writing $\mathbf{j}, \mathbf{k}, \cdots$ for elements of $\mathbf{Z}^2$ and $\alpha, \beta, \cdots$ for elements of $\mathbf{Z}^2_+$. Thus , each component j_v is an integer, each α_v is a non-negative integer, and

$$|\alpha| = \alpha_1 + \alpha_2 \tag{3.48}$$

$$\mathbf{x}^\alpha = x_1^{\alpha_1} x_2^{\alpha_2} \tag{3.49}$$

$$\alpha! = \alpha_1!\alpha_2! \tag{3.50}$$

$$\alpha \geq \beta \Leftrightarrow \alpha_1 \geq \beta_1, \quad \alpha_2 \geq \beta_2 \tag{3.51}$$

$$\begin{pmatrix} \alpha \\ \beta \end{pmatrix} = \frac{\alpha!}{\beta!(\alpha-\beta)!} \qquad (\alpha \geq \beta) \tag{3.52}$$

$$D^\alpha f = \frac{\partial^{|\alpha|} f}{\partial x_1^{\alpha_1} \partial x_2^{\alpha_2}}. \tag{3.53}$$

3.4.1. Approximation Accuracy

A multi-scaling function $\mathbf{\Phi} = (\phi_1, \cdots, \phi_r)^T$ is said to have *approximation accuracy* N if each polynomial $\mathbf{x}^\alpha$, $|\alpha| = 0, 1, \cdots, N-1$, is a locally finite series of the integer translates $\mathbf{\Phi}(\mathbf{x}+\mathbf{k})$, $\mathbf{k} \in \mathbf{Z}^2$, i.e.

$$\mathbf{x}^\alpha = \sum_{\mathbf{k}\in\mathbf{Z}^2} Y_{\mathbf{k}}^\alpha \mathbf{\Phi}(\mathbf{x}+\mathbf{k}) \quad a.e. \quad |\alpha| = 0, 1, \cdots, N-1 \tag{3.54}$$

where $Y_{\mathbf{k}}^{\alpha}$ are constant row vectors (each of length r) and $Y_{\mathbf{0}}^{\alpha}$ are called *staring vectors*.

Further, if translates of $\mathbf{\Phi}$ are *independent*, i.e.,

$$\sum_{\mathbf{k}\in\mathbf{Z}^2} w_{\mathbf{k}}\mathbf{\Phi}(\mathbf{x}+\mathbf{k}) \equiv 0 \Longleftrightarrow w_{\mathbf{k}} \textit{ is the zero } 1\times r \textit{ vector},$$

then

$$Y_{\mathbf{k}}^{\alpha} = \sum_{\beta\le\alpha} \binom{\alpha}{\beta} (-\mathbf{k})^{\alpha-\beta} Y_{\mathbf{0}}^{\beta}, \quad \mathbf{k}\neq\mathbf{0}, \quad |\alpha| = 0,1,\cdots,N-1. \tag{3.55}$$

In fact, by (3.54) we have

$$\begin{aligned}
&\sum_{\mathbf{l}\in\mathbf{Z}^2} Y_{\mathbf{l}+\mathbf{k}}^{\alpha}\mathbf{\Phi}(\mathbf{x}+\mathbf{l})\\
&= \sum_{\mathbf{l}\in\mathbf{Z}^2} Y_{\mathbf{l}}^{\alpha}\mathbf{\Phi}(\mathbf{x}-\mathbf{k}+\mathbf{l})\\
&= (\mathbf{x}-\mathbf{k})^{\alpha}\\
&= \sum_{\beta\le\alpha}\binom{\alpha}{\beta}\mathbf{x}^{\beta}(-\mathbf{k})^{\alpha-\beta}\\
&= \sum_{\beta\le\alpha}\binom{\alpha}{\beta}(-\mathbf{k})^{\alpha-\beta}\sum_{\mathbf{l}\in\mathbf{Z}^2} Y_{\mathbf{l}}^{\beta}\mathbf{\Phi}(\mathbf{x}+\mathbf{l})
\end{aligned}$$

Thus, the independence assumption implies that

$$Y_{\mathbf{l}+\mathbf{k}}^{\alpha} = \sum_{\beta\le\alpha} \binom{\alpha}{\beta} (-\mathbf{k})^{\alpha-\beta} Y_{\mathbf{l}}^{\beta}, \qquad |\alpha| = 0,1,\cdots,N-1. \tag{3.56}$$

In particular, let $\mathbf{l}=\mathbf{0}$, we get (3.55).

Clearly, (3.55) may be slightly weaker than the independency assumption.

Theorem 3.9. *Suppose that a 2-D nonseparable multiscaling function $\mathbf{\Phi}(\mathbf{x})$ is compactly supported, then $\mathbf{\Phi}(\mathbf{x})$ has approximation accuracy N and satisfies (3.55) if and only if*

$$\sum_{\mathbf{k}}\sum_{\beta\le\alpha}\binom{\alpha}{\beta}(-1)^{|\beta|}(\mathbf{x}+\mathbf{k})^{\alpha-\beta}Y_{\mathbf{0}}^{\beta}\mathbf{\Phi}(\mathbf{x}+\mathbf{k}) = \delta_{\alpha\mathbf{0}}, \tag{3.57}$$

a.e. $|\alpha| = 0,1,\cdots,N-1$ *where* $\alpha,\beta\in\mathbf{Z}_+^2$.

3.4.2. 2-D Nonseparable Multiscaling Function Interpolation

Let $u(t)$, $\mathbf{v}(t)$, $t = 1, 2, 3, \cdots, r$, be constants and $u(t) \in \mathbf{R}$, $\mathbf{v}(t) \in \mathbf{R}^2$, we define the $2-D$ *nonseparable multiscaling function interpolation* with respect to dilation matrix $\mathbf{A}$ and the vector-valued multiscaling function $\mathbf{\Phi} = (\phi_1, \cdots, \phi_r)^T$ of an $L^2(\mathbf{R}^2)$ function $f(\mathbf{x})$ at the j-th level by

$$f^j(\mathbf{x}) = \sum_{\mathbf{q}\in\bigwedge} \sum_{t=1}^{r} u(t) f[\mathbf{A}^{-j}(\mathbf{q} + \mathbf{v}(t))]\phi_t(\mathbf{A}^j\mathbf{x} - \mathbf{q}) \quad j \in \mathbf{Z} \tag{3.58}$$

where

$$\bigwedge = \{\mathbf{q} = (q_1, q_2) \in \mathbf{Z}^2 | supp\, \phi_t(\mathbf{x}) \cap supp\, f \neq \phi, t = 1, \cdots, r\}. \tag{3.59}$$

and the weighted coefficients $u(t), \mathbf{v}(t)$ are related to the starting vectors as follows:

$$Y_0^\alpha = (u(1)\mathbf{v}(1)^\alpha, \cdots, u(r)\mathbf{v}(r)^\alpha), \quad |\alpha| = 0, 1, \cdots, N-1. \tag{3.60}$$

Theorem 3.10. *Assume $\mathbf{\Phi}$ is compactedly supported. Given $u(t), v(t)$ and define f^j as (3.58). And let Ω is a bounded open set in $\mathbf{R}^2$ with supp$(\phi_t) \subseteq \Omega$. Then, for all $f(\mathbf{x}) \in C^N(\overline{\Omega})$,*

$$\| f - f^j \|_{L^2(\Omega)} \leq C \| f^{(N)} \|_\infty \sigma_2^{-jN} \tag{3.61}$$

if and only if $\mathbf{\Phi}(\mathbf{x})$ has approximation accuracy N and satisfies (3.55), (3.60). where C is a constant depending only on N, diameter of Ω and supp(ϕ_t) ,and

$$\| f^{(N)} \|_\infty = \max_{\mathbf{x}\in\Omega,\ |\alpha|=N} |D^\alpha f(\mathbf{x})|.$$

The following statements will use the fact that $\mathbf{\Phi}$ is orthonormal implies that $\mathbf{\Phi}$ is independent.

Theorem 3.11. *Assume $\mathbf{\Phi}$ is compactedly supported and orthonormal. If $\mathbf{\Phi}$ has approximation accuracy N, then*

$$Y_{\mathbf{k}}^\alpha = \int_{\mathbf{R}^2} \mathbf{x}^\alpha \mathbf{\Phi}(\mathbf{x} + \mathbf{k}) dx \tag{3.62}$$

where $\mathbf{k} \in \mathbf{Z}^2$ and $|\alpha| = 0, 1, \cdots, N-1$.

Theorem 3.12. *Assume $\mathbf{\Phi}$ is compactedly supported and orthonormal. Given $u(t)$, $v(t)$ and define f^j as (3.58). And let Ω is a bounded open set in $\mathbf{R}^2$ with supp$(\phi_t) \subseteq \Omega$. Then, for all $f(\mathbf{x}) \in C^N(\overline{\Omega})$,*

$$\| f - f^j \|_{L^2(\Omega)} \leq C \| f^{(N)} \|_\infty \sigma_2^{-jN} \tag{3.63}$$

if and only if $\mathbf{\Phi}(\mathbf{x})$ has approximation accuracy N and satisfies

$$\begin{aligned}(u(1), \cdots, u(r)) &= \int_{\mathbf{R}^2} \mathbf{\Phi}(\mathbf{x}) d\mathbf{x} \\ (u(1)v_1(1), \cdots, u(r)v_1(r)) &= \int_{\mathbf{R}^2} x_1 \mathbf{\Phi}(\mathbf{x}) d\mathbf{x} \\ (u(1)v_2(1), \cdots, u(r)v_2(r)) &= \int_{\mathbf{R}^2} x_2 \mathbf{\Phi}(\mathbf{x}) d\mathbf{x} \\ \mathbf{v}(t) &= (v_1(t), v_2(t)), \qquad t = 1, \cdots, r\end{aligned}$$

where C is a constant depending only on N, diameter of Ω and supp(ϕ_t) ,and

$$\| f^{(N)} \|_\infty = \max_{\mathbf{x}\in\Omega,\ |\alpha|=N} |D^\alpha f(\mathbf{x})|.$$

Remark 3.4. In the principal shift-invariant spaces S_ϕ, generated by the multi-integer translates of one single function $\phi \in L_2(R^d)$, or more generally finitely generated shift-invariant spaces S_Φ [17], under some conditions, one has L_2 approximation order for functions in S_Φ. In this case, one uses projection operator $P_{\Phi,h}$ as a scale version of a given function. In practice, it is more efficient to calculate scaling fuction interpolation than using projection as approximation.

4 Wavelet Solutions for The Dirichlet Problems

4.1 The Dirichlet Problem

Let Ω be an bounded domain in R^2 with Lipschitz boundary $\partial\Omega$. Given functions f $\in L^2(\Omega)$, g $\in H^{\frac{1}{2}}(\partial\Omega)$, a well-known Dirichlet problem is to find a function u $\in H^1(\Omega)$ so that

$$-\Delta u + u = f, \quad in \ \Omega \tag{4.1}$$

$$u = g, \quad on \ \partial\Omega. \tag{4.2}$$

Here $H^p(\Omega)$ denotes the Sobolev space of L^2 functions in Ω whose p-th distribution derivative is also in L^2 (see [1]).

There are a number of different ways of treating such a problem numerically. In [29] a penalty method (see [13, Appendix I, Section 4]) combined with wavelet-Galerkin methods (see [14]) was formulated and a convergernce theorem was proved. In this section we use the Theorem 3.2 to derive a better convergent error estimate.

To this end, we suppose that f $\in H^{L+1}(\Omega)$ (L was given by (3.3)) and g can be extended to be a function in $H^{L+1}(\Omega)$, i.e., there exists a $\tilde{g} \in H^{L+1}(\Omega)$, so that $g = \gamma_0 \tilde{g}$, where $\gamma_0 : H^1(\Omega) \to H^{\frac{1}{2}}(\partial\Omega)$ is the continuous trace operator.

Let D be a square containing Ω. Extend f, g to $H^{L+1}(D)$ by letting f $=$ 0, g $= 0$ on $D - \overline{\Omega}$. Then for any positive number ϵ, consider the following penalized variational problem

$$\int_D \nabla u_\epsilon \cdot \nabla v \, dx\, dy + \int_D u_\epsilon v \, dx\, dy + \frac{1}{\epsilon} \oint_{\partial\Omega} u_\epsilon v \, ds$$
$$= \int_D f\, v \, dx\, dy + \frac{1}{\epsilon} \oint_{\partial\Omega} g\, v\, ds, \tag{4.3}$$

where u_ϵ is the desired solution and (4.3) is satisfied for all $v \in H^1(D)$.

For each $\epsilon > 0$, there is a unique solution $u_\epsilon \in H^1(D)$ satisfying (4.3). This follows from the "direct method" of calculus of variations for its equivalent problem [13].

$$\inf_{v \in H^1(D)} \{ \frac{1}{2} \int_D |\nabla v|^2 \, dx\, dy + \frac{1}{2} \int_D v^2 \, dx\, dy$$
$$- \int_D f v \, dx\, dy + \frac{1}{2\epsilon} \oint_{\partial\Omega} |v - g|^2 \, ds \}, \tag{4.4}$$

We have the following theorem on the relation between the solutions of the above problems.

Theorem 4.1. *Let u_ϵ be the solution to (4.3), and $u \in H^2(\Omega)$ be the solution to (4.1)–(4.2). Extend u to $H^2(D)$ by letting $u = 0$ in $D - \overline{\Omega}$. Then*

$$\|u_\epsilon - u\|_{H^1(D)} = O(\epsilon). \tag{4.5}$$

Remark 4.1. Since $\partial\Omega$ is Lipschitz, we have

$$\|w_\epsilon - u\|_{L^2(\partial\Omega)} \le C \|w_\epsilon - u\|_{H^1(\Omega)} \le C\epsilon \|u^1\|_{H^1(D)}. \tag{4.6}$$

So we also have the estimate:

$$\|u - u_\epsilon\|_{L^2(\partial\Omega)} \le C\epsilon \|u^1\|_{H^1(D)}. \tag{4.7}$$

4.1.1. Convergence of the Wavelet-Galerkin Approximation

Next we want to derive an error estimate for the Wavelet-Galerkin approximation to the solution of the penalized problem (4.3). To this end we will consider a wavelet basis consisting of the special scaling function ϕ with the condition (3.3) and consider the space $V_j(R^2)$ defined by the multiresolution given in Section 2.2. Let f^j, g^j be the wavelet interpolations for f and g as in the Theorem 3.2:

$$f^j(x,y) := \sum_{p,q\in\Lambda_D^j} f_{p,q}^j\, \phi_p^j(x)\phi_q^j(y), \tag{4.8}$$

$$g^j(x,y) := \sum_{p,q\in\Lambda_D^j} g_{p,q}^j\, \phi_p^j(x)\phi_q^j(y), \tag{4.9}$$

where Λ_D^j is the index set defined as

$$\Lambda_D^j = \{(p,q) \in Z \bigotimes Z : supp\,(\phi_p^j(x)\phi_q^j(y)) \bigcap D \neq \phi\}.$$

Replace f, g in (4.3) by f^j, g^j and let

$$u_\epsilon^j(x,y) := \sum_{p,q\in\Lambda_D^j} u_{p,q}^j\, \phi_p^j(x)\phi_q^j(y) \tag{4.10}$$

be the Wavelet-Galerkin solution to

$$\begin{aligned}\int_D \nabla u_\epsilon^j \cdot \nabla v\, dx\, dy + \int_D u_\epsilon^j v\, dx\, dy + \frac{1}{\epsilon}\oint_{\partial\Omega} u_\epsilon^j v\, ds \\ = \int_D f^j\, v\, dx\, dy + \frac{1}{\epsilon}\oint_{\partial\Omega} g^j\, v\, ds, \quad \forall\, v \in H^1(D)\end{aligned} \tag{4.11}$$

where $u_{p,q}^j$ is to be determined by a system of linear equations. We then have

Theorem 4.2. *If $u_\epsilon \in H^{L+1}(D)$ is the solution to (4.3), and u_ϵ^j is the Wavelet-Galerkin solution to (4.11), then*

$$\|u_\epsilon - u_\epsilon^j\|_{H^1(D)} \leq C(\frac{1}{2^j})^N, \tag{4.12}$$

where C depends on the diameter of D, L, $\|f\|_{H^{L+1}(D)}$ and $\|g\|_{H^{L+1}(D)}$.

Remark 4.2. Without loss of generality, we assume that

$$\|w^1\|_{H^1(\Omega)} \leq C(\|f - f^j\|_{H^1(\Omega)} + \|g - g^j\|_{H^1(\Omega)}), \tag{4.13}$$

i.e., the solution of the Dirichlet problem depends continuously on the data $f - f^j$ and $g - g^j$ as $\|f - f^j\|_{H^1(\Omega)} \to 0$, $\|g - g^j\|_{H^1(\Omega)} \to 0$.

Remark 4.3. We have proved that the Wavelet-Galerkin approximation u_ϵ^j differs from the solution u of the original differential equation by an estimate of the following type

$$\|u_\epsilon - u_\epsilon^j\|_{H^1(D)} \le O(\epsilon) + O(2^{-Nj}). \tag{4.14}$$

4.1.2. Wavelet Geometric Measure Theory

In this section, we will briefly review some basic geometric measure theory which we will need in the following sections. Readers are refered to [10], [12] for more details.

We will say that a Lebesgue integrable function u is of bounded variation over a domain $U \in R^2$ if

$$\sup_{g \in C^1(U,R^2), |g| \le 1} \int_U u \, div \, g \, dxdy \le +\infty. \tag{4.15}$$

Suppose that u is of bounded variation in $U \subset R^2$. Then, by Riesz representation theorem, there exists a Radon measure μ and a μ-measurable function ν with $|\nu(x, y)| = 1$, for μ-almost every $(x, y) \in U$, so that, for any $g \in C_0^1(U, R)$,

$$\int_U u \, divg \, dxdy = \int_U g \cdot \nu d\mu. \tag{4.16}$$

We will denote by $|\nabla u|$ the Radon measure μ and by $|\nabla \mu|$ the vector-valued measure ν. Note that if μ is continuously differentiable, then μ is the 2-dimensional Lebesgue measure weighted by the norm of the gradient of u and $\nu = \dfrac{\nabla u}{|\nabla u|}$. Assume Ω is a bounded domain in R^2 whose boundary $\partial\Omega$ admits a representation of the form

$$\partial\Omega = \{(x, y) \in R^2 : F(x, y) = c\} \tag{4.17}$$

for some Lipschitz function F and constant c. Then the unit normal vector along the boundary $\partial\Omega, \vec{n}$, can be written as

$$\vec{n} = \frac{\nabla F}{|\nabla F|}. \tag{4.18}$$

Extend $\vec{n}$ to R^2 smoothly, we have arc length formula of $\partial\Omega$

$$L(\partial\Omega) = \int_{\partial\Omega} ds = -\int_{R^2} \nabla\chi_\Omega \cdot \vec{n} \, dxdy, \tag{4.19}$$

where χ_Ω is the characteristic function of Ω. In general, one has for any integrable function f defined on $\partial\Omega$, after extending f to R^2,

$$\int_\Omega f\, ds = -\int_{R^2} f\, \|\partial\Omega\|, \tag{4.20}$$

where we use $\|\partial\Omega\|$ to denote this numerical boundary measure $-\nabla\chi_\Omega \cdot \vec{n}$.

Interpolate the characteristic function χ_Ω at a certain level j by using the interpolation formula given in (3.1), we have

$$\chi^j_{p,q} = \frac{1}{2^j}\chi_\Omega\left(\frac{p+c}{2^j}, \frac{q+c}{2^j}\right). \tag{4.21}$$

Using the connection coefficients

$$\Gamma^k_{i,j} =: \int_R \phi_k\phi_i\phi_j\, dx \tag{4.22}$$

and (4.19), one obtains that

$$(\|\partial\Omega\|)^j_{p,q} = \sum_{k,l,s,t} \frac{\left((\frac{\partial\chi}{\partial x})_{k,l}\,(\frac{\partial F}{\partial x})_{s,t} + (\frac{\partial\chi}{\partial y})_{k,l}\,(\frac{\partial F}{\partial y})_{s,t}\right)\Gamma^p_{k,s}\Gamma^q_{l,t}}{\sqrt{(\frac{\partial F}{\partial x})^2_{s,t} + (\frac{\partial F}{\partial y})^2_{s,t}}}, \tag{4.23}$$

where F is the defining function in (4.17).

4.1.3. Wavelet-Galerkin Approximation of the Dirichlet Problem

Consider a wavelet basis in $L^2(R^2)$ and consider the space $V_j(R^2)$ defined by the multiresolution given in Section 2.2. Let us consider an approximate solution to the actual solution u_ϵ given by

$$u^j_\epsilon(x,y) := \sum_{p,q\in\Lambda^j_D} u^j_{p,q}\phi^j_p(x)\phi^j_q(y), \tag{4.24}$$

where Λ^j_D is the index set defined as

$$\Lambda^j_D = \{(p,q) \in Z\times Z : supp(\phi^j_p(x)\phi^j_q(y)) \cap D \neq \emptyset\}. \tag{4.25}$$

Replacing u_ϵ, v by u^j_ϵ and a fixed test function $\phi^j_m(x)\phi^j_n(y)$, $(m,n) \in \Lambda^j_D$ repectively, and using the connection coefficients

$$\Gamma^l_m =: \int_R \phi_l'(x)\phi_m(x)dx, \tag{4.26}$$

we can derive out a linear system for the unknown coefficients $u^j_{p,q}$'s from (4.3). Suppose that f, g admit the expansion

$$f(x,y) = \sum_{p,q} f^j_{p,q} \phi^j_p(x) \phi^j_q. \tag{4.27}$$

$$g(x,y) = \sum_{p,q} g^j_{p,q} \phi^j_p(x) \phi^j_q. \tag{4.28}$$

Then we have

$$\int_D \nabla u^j(x,y) \cdot \nabla(\phi^j_m(x)\phi^j_n(y))dxdy$$
$$= \sum_{p,q} u^j_{p,q}(2^{2j} \sum_k \Gamma^m_k \Gamma^p_k \delta^q_n + \Gamma^n_k \Gamma^q_k \delta^p_m), \tag{4.29}$$

$$\int_D f(x,y)\phi^j_m(x)\phi^j_n(y)dxdy = \sum_{p,q} u^j_{p,q} \delta^p_m \delta^q_n, \tag{4.30}$$

$$\int_D u^j(x,y)\phi^j_m(x)\phi^j_n(y)dxdy = f^j_{m,n}, \tag{4.31}$$

$$\int_{\partial\Omega} u^j \phi^j_m \phi^j_n ds = \sum_{p,q} u^j_{p,q}(2^j \sum_{k,l} (\|\partial\Omega\|)^j_{k,l} \Gamma^p_{k,m} \Gamma^q_{l,n}), \tag{4.32}$$

$$\int_{\partial\Omega} g\phi^j_m \phi^j_n ds = 2^j \sum_{p,q,k,l} g^j_{p,q} (\|\partial\Omega\|)^j_{k,l} \Gamma^p_{k,m} \Gamma^q_{l,n}. \tag{4.33}$$

Combinig all these together and letting (m, n) go over Λ^j_D, we obtain a linear system for unknown $u^j_{p,q}$'s.

4.1.4. Numerical Experiments

Let us consider two Dirichlet problems $-\Delta u + u = f_i, i = 1,2$, with the corresponding exact solutions as follows.

$$u_1 = x + y, \tag{4.34}$$

$$u_2 = x^2 + y^2, \tag{4.35}$$

on the solution domain

$$\Omega = \{(x,y) : x^2 + y^2 < 1\}, \tag{4.36}$$

Table 6: The Relative Error in the L^2 Norm of the Wavelet-Galerkin Penalty Solution for the Disk

Solution	Epsilon	Relative Error
u_1	$\epsilon = 1$	$E = 4.88628 \times 10^{-1}$
	$\epsilon = 10^{-3}$	$E = 2.60786 \times 10^{-1}$
	$\epsilon = 10^{-6}$	$E = 7.28893 \times 10^{-2}$
	$\epsilon = 10^{-9}$	$E = 1.03427 \times 10^{-3}$
	$\epsilon = 10^{-12}$	$E = 8.80106 \times 10^{-7}$
u_2	$\epsilon = 1$	E = 2.70987
	$\epsilon = 10^{-3}$	E = 1.65577
	$\epsilon = 10^{-6}$	$E = 4.90068 \times 10^{-1}$
	$\epsilon = 10^{-9}$	$E = 6.10831 \times 10^{-3}$
	$\epsilon = 10^{-12}$	$E = 5.28888 \times 10^{-6}$

with a fictitious domain

$$D = \{(x, y) : |x|, |y| < 2\}. \tag{4.37}$$

Let ϕ be the Daubechies scaling function of order 3. We solve for the penalty-wavelet-Galerkin solution u_ϵ^j in each case at level 0, and let ϵ vary. The results are tabulated in the the following table. Here we define the relative error as

$$E = \frac{\|u - u_\epsilon^j\|_{L^2(\Omega)}}{\|u\|_{L^2(\Omega)}}. \tag{4.38}$$

4.2 Multiwavelet Solutions for the Dirichlet Problem

In the following section, we obtain some multiwavelet characterizations of domains in $\mathbf{R}^2$ which give rise to some simplifications in calculating integrals. We then discuss the Dirichlet problem and present some numerical experiments. It is shown that our multi-scaling function interpolation method provides more accurate approximate solutions than scalar scaling function interpolation method [29].

4.2.1. Multiwavelet Characterization of Domains in $\mathbf{R}^2$

In what follows, we will extend the multiwavelet interpolation theorems to more general cases including continuous and Holder continuous functions which will be used to simplify the calculation of boundary integrals PDEs.

Given a domain $\Omega \subset \mathbf{R}^2$, positive integer N, $0 \le \gamma \le 1$ and $h > 0$, define

$$D^N_{f,\Omega}(h) = \sup\{|\frac{\partial^N f}{\partial x^\mu \partial y^{N-\mu}}(x,y) - \frac{\partial^N f}{\partial x^\mu \partial y^{N-\mu}}(x',y')|$$

$$|\ (x,y) \in \Omega, (x',y') \in \Omega, |x-x'| \le h, |y-y'| \le h,\ 0 \le \mu \le N\}.$$

Remark 4.4. $f \in H^\gamma_\Omega, 0 \le \gamma \le 1$ if and only if $D^0_{f,\Omega}(h) \le C_{f,\Omega} h^\gamma$.

Theorem 4.3. *Suppose a multi-scaling function $\Phi(x) \in H_0^{(N-1)}(R)$ and $\Phi(x)$ has approximation accuracy N with starting vector*

$$y_0^{(p)} = (\sum_{m=0}^{M-1} u_{m0}\lambda^p_{m0}, \cdots, \sum_{m=0}^{M-1} u_{m,r-1}\lambda^p_{m,r-1}),\ p = 0, \cdots, N-1.$$

Let $\Omega \subset \mathbf{R}^2$ be a bounded domain,

$$\Lambda = \max\{|\lambda_{n\alpha}||\alpha = 0, \cdots, r-1, n = 0, \cdots, M-1\}$$

and

$$\Omega_j = \{(x,y) \in \mathbf{R}^2 \mid dist((x,y),\Omega) \le 2^{-j}\Lambda\}$$

then for every $f(x,y)$ having (N-1)th measurable partial derivatives on Ω_j,

$$||f - f^j||_{H^n(\Omega)} \le C D^{N-1}_{f,\Omega_j}(\eta 2^{-j}) 2^{-j(N-n-1)}, \quad 0 \le n \le N-1,$$

where C and η depend on $f, \Omega, N, \lambda_{n\alpha}, n$ and Φ but independent of j.

Theorem 4.4. *In addition to the assumption of Theorem 4.3, we assume that Ω has finite perimeter in R^2. Then*

$$||\chi_\Omega - \chi^j_\Omega||_{L^2(\mathbf{R}^2)} \le C 2^{-\frac{j}{2}}.$$

Corollary 4.1. *Under the same assumption of Theorem 4.4, If $f \in L^2(\Omega_{j_0})$ for some integer j_0. Then*

$$\int\int_\Omega f dxdy = \lim_{j\to+\infty} \int\int_{R^2} \chi^j_\Omega f(x,y) dxdy.$$

Corollary 4.2. *Under the same assumption of Theorem 4.4, we have*

$$\int\int_\Omega f dxdy = \lim_{j\to+\infty} \int\int_{R^2} \chi^j_\Omega f^j(x,y) dxdy,$$

provided $f \in C^0(\Omega_{j_0})$ for some integer j_0.

This corollary shows that the integral of f over the domain Ω can be approximated by the integral of $\chi_\Omega^j f^j$ over $\mathbf{R}^2$. However, $\chi_\Omega^j f^j$ involves the product of the two expansion series. In what follows, we will further simplify the approximation of the integral.

Theorem 4.5. *Under the same assumption of Theorem 4.4, we have*

$$||(\chi_\Omega f)^j - \chi_\Omega f^j||_{L^2} \le C||f||_{L^\infty(\Omega_{j_0})}2^{-\frac{j}{2}}.$$

Corollary 4.3. *Under the same assumptions of Theorem 4.4, we have*

$$\int\int_\Omega f\nu dxdy = \lim_{j\to+\infty}\int\int_{R^2}(\chi_\Omega f)^j \nu dxdy,$$

provided $f \in C(\bar{\Omega}_{j_0})$ and $\nu \in L^2(\Omega_{j_0})$ for some integer j_0.

Remark 4.5. Corollary 4.3 can be used to simplify the calculation of the boundary integral:

$$\begin{aligned}\int_{\partial\Omega} f\nu ds &= \int_{\partial\Omega} f\nu\vec{n}\cdot\vec{n}ds \\ &= \int\int_\Omega div\ (f\nu\vec{n})dxdy \\ &= \int\int_\Omega div(f\vec{n})\nu dxdy + \int\int_\Omega f\vec{n}\bigtriangledown \nu dxdy \\ &= \lim_{j\to\infty}[\int\int_{R^2}(\chi_\Omega div(f\vec{n}))^j\nu dxdy + \int\int_{R^2}(\chi_\Omega f\vec{n})^j\cdot\bigtriangledown\nu dxdy].\end{aligned}$$

The convergence theory allows us using wavelet-Galerkin approximation, namely, following Corollary 3.1, we let

$$u_\epsilon^j(x,y) = \frac{1}{2^j}\sum_k\sum_l\sum_{n=0}^{N-1}\sum_{m=0}^{N-1} u_{\epsilon,k,l}^j Z_n\Phi_{k-n}^j(x)Z_m\Phi_{l-m}^j\ (y)$$

and similarly we interpolate f and g as follows.

$$f^j(x,y) = \frac{1}{2^j}\sum_k\sum_l\sum_{n=0}^{N-1}\sum_{m=0}^{N-1} f_{k,l}^j Z_n\Phi_{k-n}^j(x)Z_m\Phi_{l-m}^j\ (y)$$

$$g^j(x,y) = \frac{1}{2^j}\sum_k\sum_l\sum_{n=0}^{N-1}\sum_{m=0}^{N-1} g^j_{k,l} Z_n \Phi^j_{k-n}(x) Z_m \Phi^j_{l-m}(y).$$

Apply the Galerkin method to (4.3) and find an approximate solution. In other words, find the wavelet-Galerkin solution to

$$\int_D \nabla u^j_\epsilon \cdot \nabla v dxdy + \int_D u^j_\epsilon v dxdy + \frac{1}{\epsilon}\oint_{\partial\Omega} u^j_\epsilon v ds$$

$$= \int_D f^j v dxdy + \frac{1}{\epsilon}\oint_{\partial\Omega} g^j v ds, \quad \forall v \in H^1(D), \tag{4.39}$$

where $u^j_{\epsilon,k,l}$ are to be determined by a system of linear equations and the boundary integrals are calculated as mentioned in the previous section. In a similar fashion to Theorem 6.1 of [20], we have the following convergence result.

Theorem 4.6. *If $u_\epsilon \in H^{N+1}(D)$ is the solution to (4.3), and u^j_ϵ is the wavelet-Galerkin solution to (4.44), then*

$$||u_\epsilon - u^j_\epsilon||_{H^1(D)} \leq C(\frac{1}{2^j})^N,$$

where C depends on the diameter of D, on N and on $||f||_{H^{N+1}(D)}$, $||g||_{H^{N+1}(D)}$.

4.2.2. Numerical Experiments

Let us consider two Dirichlet problems $-\Delta u + u = f_i, i = 1, 2$, with the corresponding exact solutions as follows.

$$u_1 = x + y, \tag{4.40}$$

$$u_2 = x^2 + y^2, \tag{4.41}$$

on the solution domain

$$\Omega = \{(x,y) : x^2 + y^2 < 25\}, \tag{4.42}$$

with a fictitious domain

$$D = \{(x,y) : |x|, |y| < 15\}. \tag{4.43}$$

We solve for the penalty-wavelet-Galerkin solution u^j_ϵ in each case at level j = 0, and let ϵ vary. Here we use Corollary 3.1 and multi-scaling function in the Example 1 of Section 3.2.4. The results are tabulated in the Table 7.

Table 7: The Relative Error in the L^2 Norm of the Wavelet-Galerkin Penalty Solution for the Disk

Solution	Epsilon	Relative Error(scalar)	Relative Error(multi-scaling)
u_1	$\epsilon = 1$	$E = 2.762 \times 10^{-4}$	$E = 3.7419 \times 10^{-7}$
	$\epsilon = 10^{-3}$	$E = 9.435 \times 10^{-5}$	$E = 4.85868 \times 10^{-8}$
	$\epsilon = 10^{-6}$	$E = 2.826 \times 10^{-6}$	$E = 4.81125 \times 10^{-8}$
	$\epsilon = 10^{-9}$	$E = 4.025 \times 10^{-8}$	$E = 1.05609 \times 10^{-7}$
u_2	$\epsilon = 1$	E = 0.001001	$E = 5.44192 \times 10^{-7}$
	$\epsilon = 10^{-3}$	$E = 3.219 \times 10^{-4}$	$E = 7.44594 \times 10^{-8}$
	$\epsilon = 10^{-6}$	$E = 1.052 \times 10^{-5}$	$E = 7.70807 \times 10^{-8}$
	$\epsilon = 10^{-9}$	$E = 1.739 \times 10^{-7}$	$E = 8.90417 \times 10^{-8}$

A comparison with scalar case (the Daubechies scaling function of order 3) is also given in the Table. Here we define the relative error as

$$E = \frac{\|u - u_\epsilon^j\|_{L^2(\Omega)}}{\|u\|_{L^2(\Omega)}}. \tag{4.44}$$

In this example, it shows that our multi-scaling function interpolation method provides more accurate approximate solutions than scalar scaling function interpolation method [29].

5 Application of Multiwavelet Approximation on Image Denoising

Image denoising is one of the important topics of the image analysis. In many applications, such as biomedical images and SAR (Synthetic Aperture Radar) images, the observed images are often degraded by the noise. Hence one needs to remove the noise to enhance the images. In this chapter, we use the multi-scaling function interpolation to approximate the observed image. The resulting image is a smooth version of the observed image with less noise.

5.1 Preliminary

A gray-scale image with size $M \times N$ can be represented by a function of two variables $f(x, y)$ with

$$f(i,j) = \text{the pixel gray value at pixel position } (i,j), \tag{5.1}$$

$$i = 0, 1, \cdots, M-1, j = 0, 1, \cdots, N-1.$$

If the image is degraded by some noise, the observed image, represented by function $g(x, y)$, is different from the function $f(x, y)$ by the error of noise, namely,

$$e(x, y) = g(x, y) - f(x, y).$$

The difference or distortion is measured by SNR (Signal to Noise Ration) as defined in the following:

$$\text{SNR(dB)} = 10 \log_{10} \frac{\sigma_f}{\sigma_e} \tag{5.2}$$

where

$$\sigma_f = \frac{1}{MN} \sum_{i=0}^{M-1} \sum_{j=0}^{N-1} (f(i,j) - \bar{f})^2,$$

$$\bar{f} = \frac{1}{MN} \sum_{i=0}^{M-1} \sum_{j=0}^{N-1} f(i,j),$$

$$\sigma_e = \frac{1}{MN} \sum_{i=0}^{M-1} \sum_{j=0}^{N-1} (e(i,j))^2 \ .$$

Image denoising largely depends on statistical analysis of the image data, which can be considered as the sampling values of a random variable. Let ξ for the original (without noise) image and η be the random variable for the noise and ζ for the observed (noisy) image. Then

$$\zeta = \xi + \eta. \tag{5.3}$$

In most cases, η is a Gaussian random variable with zero mean and independent of ξ. Let

$$\eta \sim \frac{1}{\sqrt{2\pi}\sigma} \exp(-\frac{x^2}{2\sigma^2}).$$

Applying the DWT on (5.3) and let $W\xi$, $W\eta$, $W\zeta$ to denote the corresponding random variables in the transformed domain. Then

$$W\zeta = W\xi + W\eta \tag{5.4}$$

and $W\eta$ still satisfies the same distribution as η, i.e.

$$W\eta \sim \frac{1}{\sqrt{2\pi}\sigma} \exp(-\frac{x^2}{2\sigma^2}). \tag{5.5}$$

It is known that, for a large set of test images, the wavelet coefficients in each subband satisfies a generalized Laplacian distribution. For simplicity, we assume the Laplacian pdf for $W\xi$:

$$W\xi \sim \frac{1}{2}\sqrt{2\alpha}\exp(-\sqrt{2\alpha}|x|). \tag{5.6}$$

5.2 *Wavelet Thresholding*

Image denoising in wavelet transform domain using thresholding is one of the most effective techniques of image enhancement. A series of work have been contributed to the image denoising based on wavelet transform. Donoho and Johnston [9] have proposed several wavelet thresholds and demonstrated their asymptotic optimality conditions. They showed that wavelets, which are unconditional bases for a large number of function spaces are optimal bases for compression, estimation (noise reduction) and recovery. In all its simplicity the discovery was to observe that signals corrupted with additive noise satisfy the following heuristics [23], [24]: (i) Signals are represented with a few large coefficients.

(ii) Noise is evenly distributed across wavelet coefficients and is generally small.

Therefore, for the reduction of noise, one can 'kill or shrink' the 'small' wavelet coefficients and 'keep' the 'large' wavelet coefficients. This is a process called **Wavelet Thresholding**. More specifically, let Y be signal of the observed (noisy). In the wavelet transformed domain of Y, apply a **Soft Thresholding** T_t to each wavelet coefficient c:

$$T_t(c) = \begin{cases} c - sgn(c)t, & \text{for } |c| > t, \\ 0, & \text{for } |c| \leq t. \end{cases} \tag{5.7}$$

or a **Hard Thresholding** T_t :

$$T_t(c) = \begin{cases} c, & \text{for } |c| > t, \\ 0, & \text{for } |c| \leq t. \end{cases} \tag{5.8}$$

Then we reconstruct the image from the thresholded wavelet coefficient by the Inverse Wavelet Transform.

There are also many works of thresholding/shrinkage based on standard techniques, such as Bayesian, cross-validation, but most of them are not suitable for images. In [5], Chang, Yu and Vetterli presented an excellent scheme based on wavelet thresholding. For each subband of the wavelet coefficients, a soft-threshold

$$t = \sigma^2\sqrt{\alpha}$$

is used, where σ and α are given as in (5.5) and (5.6). It is also listed the oracle thresholds for some test images [5]. The oracle thresholds were obtained by minimizing the mean square errors, assuming the original images without noise were given. Therefore those oracle thresholds can be thought as the best results while applying the usual wavelet thresholding. It is shown that the SNRs resulting from are very closed to those of oracle thresholds.

5.3 Denoising by Multiscaling Function Interpolation

Even though the thresholding on the wavelet transform domain generally works well, one can still visually note the remaining noise in the resulting image, especially in the non-edge areas. It is well known that mean filter is best for removing the Gaussian noise if the background of the original image is relatively smooth. In this case, denoising is basically a smoothing process. Since Multiscaling Function Interpolation (MSF Interpolation) is also a smooth process to approximate a function by the scaling function, the MSF Interpolation can be used to smooth the noisy image. The following explains the procedures.

Step 1: Construct a smooth function $f(x,y)$ in the subspace V_0 in the multiresolution analysis to represent the image data as in (5.1). For example, if we use the multiscaling function in the Example 1 of Section 3.2.4, the function can be set as follows:

$$f(x,y) = \sum_{k=0}^{M-1}\sum_{l=0}^{N-1} f(k,l) \sum_{n,m=0}^{3} Z_n \Phi(x+n) Z_m \Phi(y+m), \tag{5.9}$$

Step 2: Approximate the function in a coarser subspace V_J. We choose $J = 4$. In the formula (3.23), we use the mean value of window $(2^J + 1) \times (2^J + 1)$ at pixel position $((k+1)2^J, (l+1)2^J)$, $k = 0, \cdots, M2^{-J} - 1$, $l = 0, 1, \cdots, N2^{-J} - 1$, as the sampling values. More specifically, replace the sampling values $f(\frac{k}{2^J}, \frac{l}{2^J})$ by

$$\bar{f^J}(k,l) = \frac{1}{(2^J+1)^2} \sum_{i,j=-2^J}^{2^J} f((k+1)2^J + i, (l+1)2^J + j) \tag{5.10}$$

Then, by using (3.23), we reconstruct an image with the same size as the input image, i.e. the reconstructed image is represented by the function

$$f^J(x,y) = \sum_{k=0}^{M2^{-J}-1} \sum_{l=0}^{N2^{-J}-1} \bar{f^J}(k,l)$$

$$\sum_{n,m=0}^{3} Z_n \Phi(2^{-J}x - k + n) Z_m \Phi(2^{-J}y - l + m)$$

Of course, this reconstructed image over-smooths the input noisy image.More details will be adaptively added in the following steps.

Step 3: Subtract the first reconstructed image from the input image, we get the first difference image, which mainly contains the noise and edges. Therefore we can estimate the noise level from this difference image. We use the minimum local variance of window 9×9 as the estimation of the input noise level.

Step 4: Now we use the first difference image as the new input image and apply (5.10) at level J-1 = 3. At this step, we have to remove the noise from the sampling values. For Gaussian noise with variance σ^2, the mean value of the noise with window size $(2^{J-1} + 1) \times (2^{J-1} + 1)$ is a Gaussian random variable with variance $\sigma^2/(2^{J-1} + 1)^2$.

From the standard statistics, the sample value out of 3 times the standard deviation can be ignored, Therefore we apply the Hard-thresholding with

$$t = 3\sigma/(2^{J-1} + 1)$$

over the mean values from the window. Then we use the thresholded mean values as the sampling values to form a similar function $f^{J-1}(x, y)$ and reconstruct the second image from this function.

Step 5: Repeat step 4 at level J-2, and J-3.

Step 6: Add all the reconstructed images, we get the denoised image with reduction of noise.

5.4 Simulation results

We use the Lena 512×512 and Goldhill 576×720 as the test images. The following two tables show the SNRs when the noise variance is 10^2, 15^2 and 20^2. The noisy column is the SNRs of the observed image before denoising. The MSFI column is the SNRs of the denoised images with the multiscaling function interpolation. The last column WT is the SNRs of denoised images by the wavelet thresholding in [5].

The results in Table 8 and 9 show that the SNRs from the MSFI are close to those from WT when the noise is strong ($\sigma = 20$). But when the noise strength is light ($\sigma = 10$), the SNRs of Lena by MSFI are significantly inferior to those by WT. However, since the lost of SNRs are mainly from the over-smoothness in the busy (edge) areas, the resulting images by MSFI still have good visual appearances. In the non-edge areas, the MFSI removes more noise

Table 8: SNRs for the denoised Lena 512 × 512 image

Noise Level	Noisy	MSFI	WT
$\sigma = 10$	13.60	17.45	18.89
$\sigma = 15$	10.08	16.31	16.99
$\sigma = 20$	7.58	15.11	15.71

Table 9: SNRs for the denoised Goldhill 576 × 720 image

Noise Level	Noisy	MSFI	WT
$\sigma = 10$	13.84	17.38	17.61
$\sigma = 15$	10.32	16.44	15.63
$\sigma = 20$	7.82	15.47	14.35

than the WT does. This is confirmed by the increase of SNRs if we combine the two resulting images based on their local variance. For example, when $\sigma = 10$, at each pixel, we use the local deviation within a window of size 9 by 9 to calculate the weight for the images denoised by MSFI and WT. (In this experiment, if the deviation ≥ 15, set the weight $= 1$ for WT, else, set the weight $=$ deviation/15 for WT). Combining the two images with the weight produces an image with much higher SNR $= 19.56$ and better visual quality.

It would be interesting to apply nonseparable scaling function interpolation on image denoising and compare with the results given in this section.

References

1. Adams, K. A. (1975), Sobolev Spaces, Academic Press, Orlando.

2. Alpert, B. K. and Rokhlin, V. (1991), A Fast Algorithm for the Evaluation of Legendre Expansions, SIAM J. Sci. Stat. Comput. 12, 158–179.

3. Belogay, E. and Wang, Y. (1999), *Arbitrarily Smooth Orthogonal Nonseparable Wavelets in* $\mathbf{R}^2$, SIAM J. Math. Anal., 30, 678–697.

4. Beylkin, G., Coifman, R., and Rokhlin, V. (1991), Fast wavelet transforms and numerical algorithms, Comm. Pure. Appl. Math., 44, pp. 141–183.

5. Chang, S. G., Yu, B., and Vetterli, M. (1997) Image denoising via Lossy Comp ression and Wavelet Thresholding, IEEE. 7, pp. 604–607.

6. Daubechies, I. (1988), Orthonormal Bases of Compactly Supported Wavelets, Comm. Pure. Appl. Math., 41, pp. 909–996.

7. Daubechies, I. (1992), Ten Lectures on Wavelets, SIAM, Philadelphia, PA.

8. Daubechies, I. (1993), Orthonormal Bases of Compactly Supported Wavelets II. Variations on a Theme, SIAM J. Math. Anal., 24, pp. 499–519.

9. Donoho, D. L. and Johnston, I. M. (1994), Ideal Spatial Adaptation via Wavelet Shrinkage, Biometrika, vol 81, pp. 425-455.

10. Federer, Herbert (1969), Geometric Measure Theory. Springer-Verlag, New York.

11. Geronimo, J. S., Hardin, D. P., and Massopust, P. R. (1994) Fractal Functions and Wavelet Expansion Based on Several Scaling Functions, J. Approx. Theory 78, 373–401.

12. Ginsti, Enrico (1984), Minimal Surfaces and Functions of Bounded Variation. Birkhauser, Boston.

13. Glowinski, Roland (1984), Numerical Methods for Nonlinear Variational Problems. Springer Series in Computational Physics. Springer Verlag, New York.

14. Glowinski, R., Lawton, W., Ravachol, M., and Tenenbaum, E. (1990), Wavelet Solutions of Linear and Nonlinear Elliptic, Parabolic and Hyperbolic Problems in One Space Dimension, Proceedings of International Conference on Computing Methods in Applied Science and Engineering, Philadephia,

15. Haar, A. (1910), Zur Theorie der Orthogonalen Funktionen-systeme, Math. Ann., 69:331–371.

16. He, W. and Lai, M. J. (1997), *Examples of Bivariate Nonseparable Compactedly Supported Orthonormal Continuous Wavelets*, Wavelet Applications in Signal and Image Processing IV, proceedings of SPIE, vol. 3169, 303–314.

17. Jia, R. Q., Stability of the shifts of a finite number of a functions, J of Approximation Theory, to appear.

18. Lin, E. B. and Lin, Y. (1999), Nonseparable Scaling Function Interpolation and Approximation, preprint.

19. Lin, E. B. and Xiao, Z. (1998), Multi-scaling function interpolation and approximation, AMS contemporary Mathematics, vol 216, pp. 129–148.

20. Lin, E. B. and Zhou, X. (1997), Wavelet Geometric Analysis of the Dirichlet Problem, Geometry from the Pacific Rim., Walter de Gruyter & Co., Berlin, 219–236.

21. Lin, E. B. and Zhou, X. (1997), Coiflet Interpolation and Approximate Soluti on of Elliptic Partial Differential Equations Numerical Methods for Partial Differential Equations, 13, 303–320.

22. Mallat, S. (1989), Multiresolution approximations and wavelet orthonormal bases of $L^2(R)$, Trans. Amer. Math. Soc., 315, pp. 69–87.

23. Mallat, S. and Hwang, W. L. (1992), Singularity Detection and Processing with Wavelets. IEEE Trans. Inform. Theory, 28, pp. 617–643.

24. Odegard, J. E. (1995), Image Enhancement by Nonlinear Wavelet Processing, Rice University, TX., (preprint).

25. Plonka, G. and Strela, V. (1998), Construction of Multi-scaling Functions with Approximation and Symmetry, SIAM J Math. Anal. 29, 481–510.

26. G. Strang, G. (1989), Wavelets and Dilation Equations: A Brief Introduction. SIAM Review, Vol. 31, No. 4, pp. 614–627.

27. Tian, Jun and Wells, R. O., Jr.(1998), *Vanishing Moments and Biorthogonal Systems*, . Tian, and R. O. Wells, Jr., in Mathematics in Signal Processing IV, J. G. McWhirter, I. K. Proudler, Editors, Oxford Univer sity Press, pp. 301–314.

28. Wells, R. O., Jr. and Zhou, Xiaodong (1994), Wavelet Interpolation and Approximate Solutions of Elliptic Partial Differential Equations. in Noncompact Lie Groups and Some of their Applications (R. Wilson and E. A. Tanner, editors), Kluwer Acad. Press, pp. 349–366.

29. Wells, R. O., Jr. and Zhou, Xiaodong (1995), Wavelet Solutions for the Dirichlet Problem. Numer. Mathematik, 70, 379–396.

List of Contributors

Ruben G. Airapetyan
Department of Mathematics, Kansas State University
Manhattan, KS 66506-2602

George A. Anastassiou
Department of Mathematical Sciences, The University of Memphis,
Memphis, TN 38152
E-Mail: anastasg@msci.memphis.edu; ganastss@memphis.edu

Ioannis K. Argyros
Department of Mathematics, Cameron University,
Lawton, OK 73505
E-mail: ioannisa@cameron.edu

Carlo Bardaro
Department of Mathematics, University of Perugia,
Via Vanvitelli 1, 06123 Perugia, Italy
E-mail: bardaro@dipmat.unipg.it

B. L. Chalmers
Department of Mathematics, University of California at Riverside,
Riverside, CA 92521
E-mail: blc@ucrmath.ucr.edu

Augustine O. Esogbue
Intelligent Systems and Controls Lab, School of Industrial and
Systems Engineering, Georgia Institute of Technology,
Atlanta, GA 30332
E-mail: augustine.esogbue@isye.gatech.edu

S. Gal
Department of Mathematics, University of Oradea, Str. Armatei
Romane 5, 3700 Oradea, Romania
E-mail: galso@math.uoradea.ro

G. Götzenberger
Institute of Statistics and Mathematical Economics,
School of Economics, University of Karlsruhe,
Postfach 6980, D-76128, Karlsruhe, Germany
E-mail: Gero.Goetzenberger@daimlerchrysler.com

Yingkang Hu
Department of Mathematics and Computer Science, Georgia Southern University, Statesboro, GA 30460-8093
E-mail: yhu@gasou.edu

Theodore Kilgore
Department of Mathematics, Auburn University, Auburn University, AL 36849
kilgota@banach.math.auburn.edu

En-Bing Lin
Department of Mathematics, The University of Toledo, Toledo, Ohio 43606
E-mail: elin@math.utoledo.edu

Ilaria Mantellini
Department of Mathematics, University of Perugia, Via Vanvitelli 1, 06123 Perugia, Italy

Carlo Marinelli
Department of Economics, University of California at Santa Barbara, Santa Barbara, CA 93106
E-mail: carlo@pstat.ucsb.edu

F. T. Metcalf
Department of Mathematics, University of California at Riverside, Riverside, CA 92521

Svetlozar T. Rachev
Institute of Statistics and Mathematical Economics, School of Economics, University of Karlsruhe, Kollegium and Schloss, Bau II, 20.12, R210, Postfach 6980, D-76128, Karlsruhe, Germany
E-mail: rachev@lsoe.uni-karlsruhe.de

Alexander G. Ramm
Department of Mathematics, Kansas State University Manhattan, KS 66506-2602
E-mail: ramm@math.ksu.edu

Thomasz Rychlik
Institute of Mathematics, Polish Academy of Sciences, Chopina 12, 87100 Torun, Poland, E-mail: trychlik@impan.gov.pl

Hans-Jürgen Schmeisser
Mathematisches Institut, F.-Schiller-Universität, Jena, D-07743 Jena, Germany
E-mail: mhj@minet.uni-jena.de

E. Schwarz
Anderson School of Management, University of California at Los Angeles, Los Angeles, CA 90095

Winfried Sickel
Mathematisches Institut, F.-Schiller-Universität, Jena, D-07743 Jena, Germany
E-mail: sickel@minet.uni-jena.de

B. Shekhtman
Department of Mathematics, University of South Florida, Tampa, FL 33620
E-mail: boris@tarski.math.usf.edu

Xiaoping Shen
Department of Mathematics & Computer Science, Eastern Connecticut State University, Willimantic, CT 06226
E-mail: shenx@ecsu.ctstateu.edu

Xiang Ming Yu
Department of Mathematics, Southwest Missouri State University Springfield, MO 65804
E-mail: xmy944f@mail.smsu.edu